A COMPANION TO THE ARCHAEOLOGY OF THE ROMAN REPUBLIC

BLACKWELL COMPANIONS TO THE ANCIENT WORLD

This series provides sophisticated and authoritative overviews of periods of ancient history, genres of classical literature, and the most important themes in ancient culture. Each volume comprises approximately twenty-five and forty concise essays written by individual scholars within their area of specialization. The essays are written in a clear, provocative, and lively manner, designed for an international audience of scholars, students, and general readers.

Ancient History

Published

A Companion to the Roman Army
Edited by Paul Erdkamp

A Companion to the Roman Republic
Edited by Nathan Rosenstein and Robert Morstein-Marx

A Companion to the Roman Empire
Edited by David S. Potter

A Companion to the Classical Greek World
Edited by Konrad H. Kinzl

A Companion to the Ancient Near East
Edited by Daniel C. Snell

A Companion to the Hellenistic World
Edited by Andrew Erskine

A Companion to Late Antiquity
Edited by Philip Rousseau

A Companion to Ancient History
Edited by Andrew Erskine

A Companion to Archaic Greece
Edited by Kurt A. Raaflaub and Hans van Wees

A Companion to Julius Caesar
Edited by Miriam Griffin

A Companion to Byzantium
Edited by Liz James

A Companion to Ancient Egypt
Edited by Alan B. Lloyd

A Companion to Ancient Macedonia
Edited by Joseph Roisman and Ian Worthington

A Companion to the Punic Wars
Edited by Dexter Hoyos

A Companion to Augustine
Edited by Mark Vessey

A Companion to Marcus Aurelius
Edited by Marcel van Ackeren

A Companion to Ancient Greek Government
Edited by Hans Beck

A Companion to the Neronian Age
Edited by Emma Buckley and Martin T. Dinter

Literature and Culture

Published

A Companion to Classical Receptions
Edited by Lorna Hardwick and Christopher Stray

A Companion to Greek and Roman Historiography
Edited by John Marincola

A Companion to Catullus
Edited by Marilyn B. Skinner

A Companion to Roman Religion
Edited by Jörg Rüpke

A Companion to Greek Religion
Edited by Daniel Ogden

A Companion to the Classical Tradition
Edited by Craig W. Kallendorf

A Companion to Roman Rhetoric
Edited by William Dominik and Jon Hall

A Companion to Greek Rhetoric
Edited by Ian Worthington

A Companion to Ancient Epic
Edited by John Miles Foley

A Companion to Greek Tragedy
Edited by Justina Gregory

A Companion to Latin Literature
Edited by Stephen Harrison

A Companion to Greek and Roman Political Thought
Edited by Ryan K. Balot

A Companion to Ovid
Edited by Peter E. Knox

A Companion to the Ancient Greek Language
Edited by Egbert Bakker

A Companion to Hellenistic Literature
Edited by Martine Cuypers and James J. Clauss

A Companion to Vergil's *Aeneid* and its Tradition
Edited by Joseph Farrell and Michael C. J. Putnam

A Companion to Horace
Edited by Gregson Davis

A Companion to Families in the Greek and Roman Worlds
Edited by Beryl Rawson

A Companion to Greek Mythology
Edited by Ken Dowden and Niall Livingstone

A Companion to the Latin Language
Edited by James Clackson

A Companion to Tacitus
Edited by Victoria Emma Pagán

A Companion to Women in the Ancient World
Edited by Sharon L. James and Sheila Dillon

A Companion to Sophocles
Edited by Kirk Ormand

A Companion to the Archaeology of the Ancient Near East
Edited by Daniel Potts

A Companion to Roman Love Elegy
Edited by Barbara K. Gold

A Companion to Greek Art
Edited by Tyler Jo Smith and Dimitris Plantzos

A Companion to Persius and Juvenal
Edited by Susanna Braund and Josiah Osgood

A Companion to the Archaeology of the Roman Republic
Jane DeRose Evans

A COMPANION TO THE ARCHAEOLOGY OF THE ROMAN REPUBLIC

Edited by

Jane DeRose Evans

WILEY-BLACKWELL

A John Wiley & Sons, Ltd., Publication

This edition first published 2013

Blackwell Publishing was acquired by John Wiley & Sons in February 2007. Blackwell's publishing program has been merged with Wiley's global Scientific, Technical, and Medical business to form Wiley-Blackwell.

Registered Office
John Wiley & Sons Ltd, The Atrium, Southern Gate, Chichester, West Sussex, PO19 8SQ, UK

Editorial Offices
350 Main Street, Malden, MA 02148-5020, USA
9600 Garsington Road, Oxford, OX4 2DQ, UK
The Atrium, Southern Gate, Chichester, West Sussex, PO19 8SQ, UK

For details of our global editorial offices, for customer services, and for information about how to apply for permission to reuse the copyright material in this book please see our website at www.wiley.com/wiley-blackwell.

Library of Congress Cataloging-in-Publication Data

A companion to the archaeology of the Roman Republic / edited by Jane DeRose Evans.
pages cm. – (Blackwell companions to the ancient world)
Includes bibliographical references and index.
"A John Wiley & Sons, Ltd., publication."
ISBN 978-1-4051-9966-7 (hardback : alkaline paper) – ISBN 978-1-118-55712-9 (Wiley Online Libary) – ISBN 978-1-118-55713-6 (Bro) – ISBN 978-1-118-55714-3 (eMobi) – ISBN 978-1-118-55715-0 (ePDF) – ISBN 978-1-118-55716-7 (ePub) 1. Rome–History–Republic, 510–265 B.C. 2. Rome–History–Republic, 265–30 B.C. 3. Rome–Antiquities. 4. Social archaeology–Rome. 5. Material culture–Rome. 6. Landscape archaeology–Rome. 7. Technology–Social aspects–Rome. 8. Group identity–Rome. 9. Imperialism–Social aspects–Rome. I. Evans, Jane DeRose, 1956–
DG77.C583 2013
937'.02–dc23

2012042774

A catalogue record for this book is available from the British Library.

Cover image: Largo di Torre Argentina. Rome, Italy © Marka / SuperStock
Cover design by Workhaus

Set in 10/12.5pt Galliard by SPi Publisher Services, Pondicherry, India

1 2013

Contents

List of Illustrations

Maps

Figures

Tables

Notes on Contributors

Albert J. Ammerman is a Research Professor at Colgate University. He took his PhD in European Archaeology at the Institute of Archaeology, University of London (under the supervision of Prof. John D. Evans). Between 1985 and 2004, he carried out fieldwork at more than 15 different sites in the heart of ancient Rome, in close collaboration with both the Superintendency of Archaeology in Rome and the City of Rome. He has taught at Stanford University, SUNY Binghamton and the Universities of Parma and Trento. His publications include the results of his work on the origins of the Forum, the early Comitium, the Capitoline Hill, the Palatine Hill, the Velabrum and the clay beds there, and on the characterization of the earliest roof tiles and architectural terracottas at early sites in Rome, Latium and southern Etruria. He has also directed projects in Calabria (Neolithic), Venice (origins of the city) and Cyprus (pre-Neolithic sites and the origins of seafaring in the eastern Mediterranean).

Michael Anderson is an archaeologist and Assistant Professor in Classics at San Francisco State University who has studied ancient Pompeii since 1996. He directs the Via Consolare Project in Pompeii, which conducts excavations in Insula VII 6 and the area of the Villa delle Colonne a Mosaico. His research and publications relate to the use of GIS and computer technologies in archaeological research.

Jeffrey A. Becker is a Mediterranean archaeologist. He earned both an MA and a PhD at the University of North Carolina at Chapel Hill. Since 2007 he has been one of the principal investigators of the archaeological fieldwork at Gabii in central Italy undertaken by the University of Michigan and the Kelsey Museum of Archaeology. He is the co-editor of *Roman Republican Villas: Architecture, Context, and Ideology* (2011) and has held teaching appointments at the College of William & Mary, Boston University, McMaster University, and the Joukowsky Institute for Archaeology and the Ancient World at Brown University. Becker is presently Acting Director of the Ancient World Mapping Center at the University of North Carolina at Chapel Hill.

Penelope J.E. Davies is Associate Professor in Roman art and architecture at the University of Texas at Austin. Author of *Death and the Emperor: The Funerary Monuments of the Roman Emperors from Augustus to Marcus Aurelius* (2000, 2004), and co-author of *Janson's History of Art* (2007, 2010), she currently focuses her research on public art and politics in Republican Rome.

Sylvia Diebner, a classical archaeologist, is the director of the photography archive at the German Archaeological Institute in Rome. Her publications include *Aesernia – Venafrum: Untersuchungen zu den römischen Steindenkmälern zweier römischer Landstädte*, 2 vols. (1979); "Frühkaiserzeitliche Urnen aus Picenum" (1982); *Reperti funerari in Umbria a sinistra del Tevere. I sec. A.C –I sec. D.C.* (1986); "Landstädtische Sepulkraldenkmäler aus Picenum" (2007); and "Aschenkisten aus Amelia: Handwerkstraditionen im südlichen Umbrien" (2008).

Michael Dobson studied archaeology at the University of Exeter and worked as an archaeologist in Exeter before returning to the University to teach Archaeological Computing. He is now Director of Flexible Combined Honours there. Interest in the Roman army and particularly Republican and early Imperial military installations began with his doctoral thesis. This continued research has included a number of publications and collaborations with archaeologists in Spain and Germany, including *The Army of the Roman Republic: The Second Century BC, Polybius and the Camps at Numantia, Spain* (2008).

Stephen L. Dyson is the Park Professor of Classics at the University of Buffalo. He has published extensively on the history of American archaeology in the Mediterranean Basin, (e.g. *In Pursuit of Ancient Pasts: A History of Classical Archaeology in the Nineteenth and Twentieth Centuries*, 2006). He has also written on the archaeology of the city of Rome (*The Roman Countryside*, 2003 and *Rome: A Living Portrait of an Ancient City*, 2010) and is preparing the final reports for his excavation in Capalbiaccio, near Cosa. Dyson has served as director for a National Endowment for the Humanities Summer Seminar for College and University Teachers at the American Academy in Rome, is a Fellow of the Society of Antiquaries of London, and is a Corresponding Member of the German Archaeological Institute.

Ingrid Edlund-Berry is Professor Emerita in the Department of Classics at the University of Texas at Austin. Her research interests are the history and archaeology of ancient Italy, with special emphasis on Etruscan archaeology. Her excavation experience includes Poggio Civitate (Murlo) and Poggio Colla (Vicchio), S. Angelo Vecchio, Metaponto, and Morgantina. Among her publications are *The Gods and the Place: Location and Function of Sanctuaries in the Countryside of Etruria and Magna Graecia (700–400 B.C.)* (1987), *The Seated and Standing Statue Akroteria from Poggio Civitate (Murlo)* (1992), with Lucy Shoe Meritt, *Etruscan and Republican Roman Mouldings* (2000), and with G. Greco and J. Kenfield (eds), *Deliciae Fictiles III: Architectural Terracottas in Ancient Italy: New Discoveries and Interpretations* (2006).

Jane DeRose Evans is a numismatist and archaeologist; she is a Professor in the Art History Department, and affiliated with the Classics Department, of Temple University. Evans has worked on Roman sites in Turkey, France and Israel and is currently the numismatist for the Harvard Sardis Expeditions. Her publications include *The Art of Persuasion: Political Propaganda from Aeneas to Brutus* (1992), and *The Joint Expedition to*

Caesarea Maritima: Excavation Reports v.6, The Coins and the Hellenistic, Roman and Byzantine Economy (2006).

Helena Fracchia is Professor of Classics in the Department of History and Classics, University of Alberta Edmonton, Canada. She is also the Director of the University of Alberta School in Italy (Cortona) and the Archaeological Field school at Ossaia. Her recent publications and fieldwork address various aspects of cultural and social history in pre-Roman southern Italy, the ceramic and survey evidence from the Upper Bradano Valley and Roccagloriosa as well as the Etruscan site and successive Roman villa at Ossaia in the Val di Chiana, Tuscany.

Andrew L. Goldman is an Associate Professor in the History Department of Gonzaga University, Spokane, Washington, and contributor to the Classical Civilizations Department. His areas of scholarly interest include the Roman military and the material culture of the Roman provinces, in particular those of Anatolia. Since 1992, his archaeological fieldwork has concentrated on the investigation of the Roman settlement at Gordion, where he has identified and excavated the first early Imperial auxiliary fort in Turkey.

Helen Goodchild is a landscape archaeologist and GIS specialist working in the Department of Archaeology at the University of York. She has worked as part of a number of large-scale landscape projects, including the *Wroxeter Hinterland Project* (United Kingdom), the *Tiber Valley Project* (Italy) and the *Cyrene Archaeological Project* (Libya). Her PhD in 2007 investigated Roman agricultural production using computer modeling techniques, from which she has published a number of papers.

Alison B. Griffith is a senior lecturer in Classics at the University of Canterbury in Christchurch, New Zealand. Her research and publications focus primarily on archaeological evidence for Roman religion and cults, especially the Roman cult of Mithras. She has also published on aspects of the topography of ancient Rome and the application of cognitive science to the study of ancient religious practice.

Maurizio Gualtieri, PhD, University of Pennsylvania and Professor Emeritus in the Department of History and Classics, University of Alberta, Edmonton (Canada), is presently teaching Classical Archaeology at the University of Perugia (Italy). His fieldwork includes the Etruscan site of Artimino (Florence), as coordinator of the University of Pennsylvania University Museum team in Tuscany and then Roccagloriosa (Salerno) as scientific collaborator of the Department of Antiquities of Salerno, first, and then as co-director of the University of Alberta team. More recently he has conducted excavations at the Roman site of Oppido Lucano and co-directed the joint University of Perugia University of Alberta excavations at the late Republican villa of Ossaia (Cortona, Arezzo). His recent publications include *La Lucania Romana. Cultura e Società nella Documentazione Archeologica* (Quaderni di Ostraka, vol. 8) (2004) and a chapter on "Roman villas in southern Italy" for the forthcoming volume edited by G. Métraux and A. Marzano on *The Roman Villa in the Mediterranean Basin.*

Shelley Hales is a Senior Lecturer in Art and Visual Culture in the Department of Classics and Ancient History at the University of Bristol. She works on Roman domestic and personal art and its relation to identity-making in the ancient world as well as its reception (particularly in terms of the excavations at Pompeii) in the nineteenth century. She is the author of *Roman Houses and Social Identity* (2003) and has co-edited several volumes exploring these issues, including *Material Culture and Social Identities in the Ancient World* (2010) and *Pompeii in the Public Imagination* (2011).

A. Trevor Hodge (1930–2012) was a classical archaeologist (BA, MA, PhD, Dipl. Cl. Arch. [Cantab]), who from 1960 to 1997 was a Professor at Carleton University, Ottawa, Canada, from which (2010) he held the title of Distinguished Research Professor. His publications include *The Woodwork of Greek Roofs* (1960, repr. 2011), *Roman Aqueducts and Water Supply* (1992), *Ancient Greek France* (1998) and, as joint editor, *Frontinus' Legacy* (2001). After retirement he often broadcasted on Canadian national radio and lectured on cruise ships.

Michael C. Hoff is Professor of Art History at the University of Nebraska. As a field archaeologist in Greece, Hoff has participated in excavations at the Athenian Agora, Corinth, Kavousi and Nemea. From 1997 to 2004, Hoff co-directed the architectural survey of the Rough Cilicia Archaeological Research Project, and since 2005 he has been director of excavations of the Antiochia ad Cragum Archaeological Research Project in Turkey. His publications include *The Romanization of Athens* (co-edited with Susan Rotroff, 1997); and *Rough Cilicia. New Historical and Archaeological Approaches* (co-edited with Rhys Townsend, 2011).

Marie D. Jackson is a Researcher in the Department of Civil and Environmental Engineering at the University of California at Berkeley. She received a Doctorat d'Université from Université de Nantes, France, in Structural Geology, and a PhD from Johns Hopkins University in Earth Sciences. She works on ancient Roman concretes from monumental buildings and seawater harbors, and their volcanic ash aggregates and pozzolanic cements, to describe the expertise of Roman builders and apply their methods to modern sustainable concretes. She has numerous publications in journals such as *Archaeometry*, *American Journal of Archaeology*, *Journal of Archaeological Science* and *Geoarchaeology*.

Cynthia K. Kosso is a Professor in the Department of History at Northern Arizona University. She received her PhD in History and Classics from the University of Illinois at Chicago, and her Master's degrees in Ancient History and Classical Archaeology from the University of Minnesota. Her professional interests are interdisciplinary in nature – from the translation of ancient languages and the study of innovation in ancient societies, to public and economic history as reflected in material evidence. She has co-edited several recent collections of essays, including *The Nature of Water, Baths, Bathing, and Hygiene from Antiquity through the Renaissance* (2009).

Fanette Laubenheimer is the Research Director (emeritus) at the Centre National de la Recherche Scientifique (CNRS), where her research projects centered on the economic history of the Roman world and especially amphoras. For 25 years she directed the excavation of a potters' complex which produced amphoras in southern France, at Sallèles d'Aude. Part of the work was to ensure that the excavations be protected and presented to the public; the museum above the site evokes a bird with outstretched wings protecting the archaeological structures. She is the author of several books and a number of articles.

Ray Laurence is Professor of Roman History and Archaeology at the University of Kent (United Kingdom). He is the author of a number of books, including

The Roads of Roman Italy: Mobility and Cultural Change (1999) and *Roman Pompeii: Space and Society* (2nd edn, 2007), co-author of *The City in the Roman West* (2011) and co-editor of *Rome, Pompeii and Ostia: Movement and Space* (2011).

Elio Lo Cascio is Professor of Roman History at Sapienza Università di Roma. He has published extensively on the demography and economy of the ancient Roman world, including *Il princeps e il suo impero: Studi di storia amministrativa e finanziaria romana* (2000) and *Crescita e declino: Studi di storia dell'economia romana* (2009). His many articles and edited volumes show his main areas of interest: the history of the administration of the Roman Empire; the institutional history of the Roman Republic; the economic and social history of Rome, from the Republic to the late Empire, with particular emphasis on monetary history; and Roman population history and the impact of demographic change on the economy and society of the Roman Empire. He has also written on late Republican municipal legislation; on the city of Rome and on Roman cities and their elites, especially on Pompeii; and on the dynamics of Romanization.

Margaret M. Miles is the Andrew W. Mellon Professor of Classical Studies at the American School of Classical Studies in Athens, Greece, and Professor of Art History and Classics at the University of California, Irvine. Her previous publications include a study of the Temple of Nemesis at Rhamnous (1989), a volume on the City Eleusinion in the Athenian Agora (1998), and *Art as Plunder* (2008), a study of Cicero's *Verrines* and their impact on ideas about cultural property.

J. Andrew Overman is the Harry M. Drake Distinguished Professor in the Humanities and Fine Arts in the Department of Classics at Macalester College, St. Paul, MN. He has directed archaeological excavations in northern Israel, as well as in the Black Sea region. He has published widely on the archaeology of Galilee, the Roman East, and the beginnings of Judaism and Christianity in the eastern Mediterranean.

Isabel Rodá is the Director of the ICAC (Catalan Institute of Classical Archaeology) and Professor of Archaeology of the UAB (Autonomous University of Barcelona). She has written on the ceramics of Hispania (*Ceramicas exoticas y de imitacion en el poblado iberico de Sant Julia de Ramis*, 1974) and authored and co-authored several corpora concerning Roman inscriptions on the peninsula.

David Gilman Romano, PhD, is the Karabots Professor of Greek Archaeology in the School of Anthropology at the University of Arizona. He began the Corinth Computer Project, (http://corinthcomputerproject.org) in 1987 as a long-term study of the urban and rural landscape of the Roman colony of Corinth. Since 2004 he has been Co-Director and Field Director of the Mt. Lykaion Excavation and Survey Project in Arcadia, Greece (http://lykaionexcavation.org). His publications include *Mapping Augustan Rome* (2002) in collaboration with Lothar Haselberger, *The Catalogue of the Classical Collection of the Glencairn Museum* (1999) with Irene Bald Romano and *Athletics and Mathematics in Archaic Corinth: The Origins of the Greek Stadion* (1993), as well as a series of publications on the city and landscape planning of the Roman colony of Corinth. Romano directs the Archaeological Mapping Lab (http://archaeologicalmappinglab.org).

Roman Roth teaches Classics at the University of Cape Town. His primary research interests concern the cultural history of ancient Italy, with a particular focus on the material and visual culture of the Republican period. He is the author of *Styling Romanisation: Pottery and Society in Central Italy* (2007), as well as several articles and book chapters, and also directs the current excavations on the site of ancient Capena (Provincial di Roma).

Rafael Scopacasa (PhD, University of Exeter) is a research fellow at the British School at Rome. His work focuses on combining written and material evidence to understand key historical developments in Italy from the Iron Age to the Roman conquest, most recently in his dissertation, "Identity as social practice in the funerary sites of central Apennine Italy ('Samnium'), 6th–3rd century BC."

Susan Kirkpatrick Smith is a biological anthropologist in the Department of Geography and Anthropology at Kennesaw State University in Kennesaw, Georgia. She has conducted skeletal analysis of human remains from the Bronze Age through the Roman period in Greece. Her areas of interest include warfare and social change and their effects on the body.

Tesse D. Stek is Golding Junior Research Fellow at Brasenose College, Oxford University. He is the author of *Cult Places and Cultural Change in Republican Italy* (2009) and conducts archaeological fieldwork in the south Italian region of Molise, ancient Samnium. This work is funded by the Netherlands Organization of Scientific Research (NWO).

David L. Stone is Assistant Professor of Classics at Florida State University. He is the co-editor of *Mortuary Landscapes of North Africa* (2007) and *Leptiminus (Lamta). Report no. 3, the Field Survey, JRA* Suppl. 87 (2011). He has also published several articles on epigraphy and landscape archaeology in North Africa.

Marlene Suano (PhD, University of London) is a lecturer in Archaeology and Ancient History at the University of São Paulo, Brazil. She is interested in using archaeological evidence to answer questions about cultural identity and cultural change in ancient Italy, especially that of the Samnite area. Her book, *I vivi fra i loro morti: frequentazione rituale nella necropoli di Tornareccio* (in press), explores cultural change in the Central Apennines.

Steven L. Tuck is Associate Professor in Classics and the History of Art at Miami University. His scholarship includes a dissertation on Roman Imperial harbor monuments and articles on the Tiber and its installations and the triumphal monuments in ports around the Roman world. He is also the author of a history of Roman art (forthcoming from Wiley-Blackwell).

Miguel John Versluys is Associate Professor of Classical Archaeology at Leiden University. His research focuses on cultural interaction in the Hellenistic-Roman Mediterranean and Near East. He is the author of *Aegyptiaca Romana. Nilotic Scenes and the Roman Views of Egypt* (2002) and the co-editor of two volumes of Isis studies: *Nile into Tiber. Egypt in the Roman World* (2007) and *Isis on the Nile. Egyptian Gods in Hellenistic and Roman Egypt* (2010). A monograph on Nemrud Dağ and late Hellenistic Commagene is forthcoming.

P. Gregory Warden, President of Franklin College (Lugano), formerly University Distinguished Professor of Art History and Associate Dean for Research and Academic Affairs at Southern Methodist University, has authored/co-authored five books as well as over 70 articles and reviews on areas ranging from Greek archaeology to Etruscan art, archaeology and ritual, and Roman architecture. Warden is the founder, Principal Investigator, and co-Director of the Mugello Valley Archaeological Project and excavations at Poggio Colla, an Etruscan settlement north-east of Florence, a joint mission of SMU, Franklin and Marshall College, and the University of Pennsylvania Museum of Archaeology. Warden is also the former editor of *Etruscan Studies* and a Trustee of the Etruscan Foundation, and has been elected to the Istituto di Studi Etruschi e Italici. In 2011 he was awarded the Stella della Solidarietà Italiana and the title of Cavaliere by the Republic of Italy.

R.J.A. Wilson is Professor of the Archaeology of the Roman Empire at the University of British Columbia and Director of the Centre for the Study of Ancient Sicily there. He has been Humboldt Fellow at the University of Bonn (1987–9), Visiting Professor at McMaster University (1998), Balsdon Fellow at the British School at Rome (2003) and Guest Scholar in Residence at the Getty Villa, Malibu (2012). His books include *Guide to the Roman Remains in Britain* (1975, 1980, 1988, 2002; 5th edn in preparation), *Piazza Armerina* (1982) and *Sicily under the Roman Empire* (1990).

Fikret K. Yegül is an architect and a Professor of the History of Art and Architecture at the University of California, Santa Barbara. A scholar of Roman architecture, he has been a member of the Harvard Sardis Excavations in Turkey and the Ohio State University Isthmia Excavations in Greece. Specializing in the baths and bathing culture of antiquity, Yegül is the author of articles and books on Roman architecture, notably *Baths and Bathing in Classical Antiquity* (1992, 1995) which received the Alice D. Hitchcock Award from the Society of Architectural Historians in 1994. His most recent book is *Bathing in the Roman World* (2010). He is working on a book on Roman architecture and urbanism.

Mantha Zarmakoupi received her MSt and DPhil in Classical Archaeology from Oxford University. Prior to this she studied Architecture at the National Technical University of Athens and gained a Master of Design Studies in History and Theory from Harvard University. Her research in archaeology is informed by her background in architectural practice, history and theory of architecture as well as digital visualizations. Her dissertation on the architecture of Roman luxury villas around the Bay of Naples is in press for Oxford University Press. She has also edited a volume on the Villa of the Papyri (2010), and is presently studying the urban growth of Late Hellenistic Delos, the "Quartier du Stade."

Abbreviations

Abbreviations of the names and works of ancient authors, as well as for collections of inscriptions, are as found in the *Oxford Classical Dictionary* (3rd edn, 1996). Abbreviations of journal titles and modern works abbreviated by initial titles of books (e.g. *NTDAR*) are as found in the online abbreviations for the *American Journal of Archaeology* (www.ajaonline.org); if the journals are not listed there, the titles are spelled out in full. All dates are BCE, unless otherwise specified.

Preface

Roman Republican archaeology, while overshadowed by its larger sibling the archaeology of the Roman Empire, has a growing body of evidence that is changing the way that archaeologists, art historians, historians, classicists and anthropologists think about a culture that is nominally recorded by ancient historians. This book, which came about through the suggestion of Haze Humbert at Wiley-Blackwell, is intended to speak to those archaeologists, art historians, historians, classicists and anthropologists who are interested in the sixth through first centuries on the Italian Peninsula and in the empire of the Republic. The work of the archaeologist has changed dramatically in the last 30 years, due to the introduction of new technologies, the explosion of information available (from satellite photos to DNA analysis), and the declining interest of organizations in funding "big digs." Archaeologists have thus expanded the types of questions they ask, the manner in which they can answer old and ask new questions, and the fora in which they publish. One example of how archaeological research has changed is the interpretation of a site through interdisciplinary teams. Thus, the time for such a *Companion* is now, as scholars in differing disciplines publishing in a range of journals and presses develop the picture of the Republic. I have asked a wide variety of scholars, from anthropologists to ancient historians to field archaeologists, to help illuminate broad swaths of this field, allowing readers to see what particular disciplines are contemplating. By giving both a guide to further reading and a detailed bibliography, the reader can move into specialized studies on a particular topic of interest. I have also asked both more established authors and younger authors to contribute, to benefit from their combined wisdom and new thinking. The Roman world, by the end of the Republican period, was an intensely pan-Mediterranean one and in order to bring a fuller discussion of these cultures which comprised the Republic, I have turned to scholars in North America, the United Kingdom, Europe, Africa and New Zealand, and have had contributions from French, German, Italian and Spanish translated for the English-speaking audience of the *Companion* – though the reference list will quickly alert the reader to the global nature of this study.

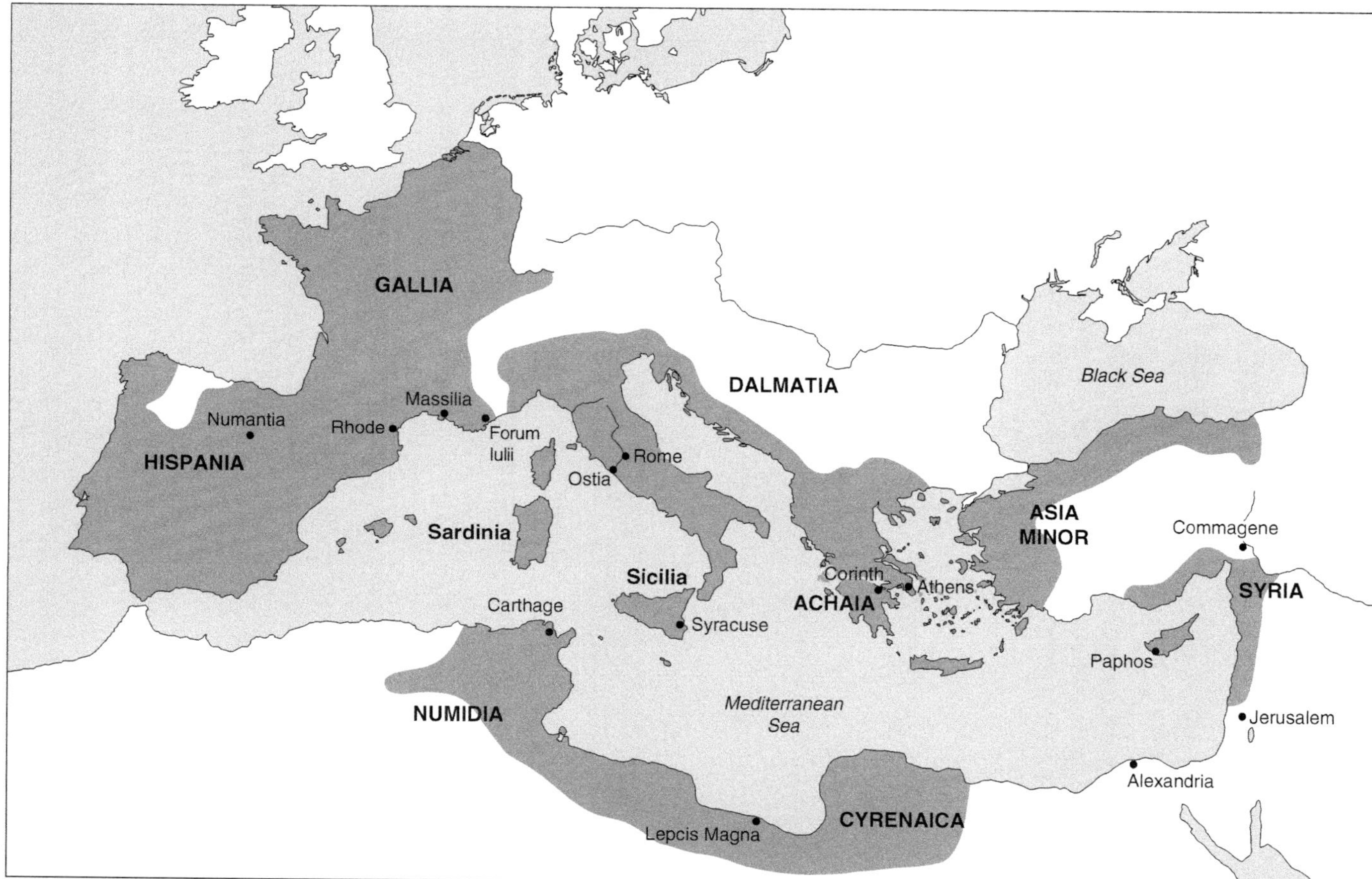

Figure A Map of the extent of the Republican empire at the time of the death of Caesar. Source: Drawing by J.D. Evans.

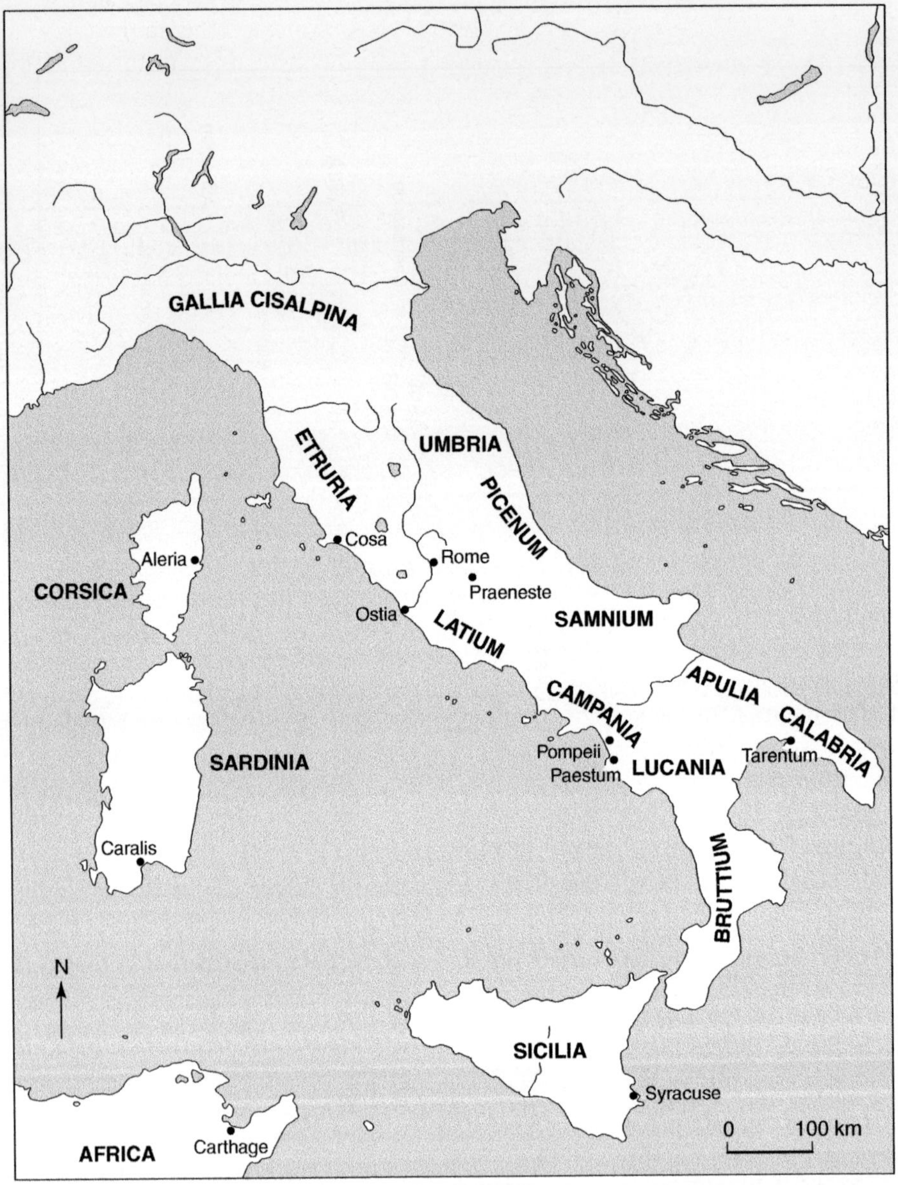

Figure B Map of the major cultural groups in Republican Italy. Source: Drawing by J.D. Evans.

Introduction

Jane DeRose Evans

In a traditional sense, the Roman Republic begins in 509 and ends with the death of Caesar in 44, or the Battle of Actium in 31, or the reforms of Octavian in 27. Yet what would be the Republic begins to take shape under the rule of its kings, and indeed, even earlier, with the development of the landscape that was to become Rome. Thus, Ammerman and Jackson and Kosso return us to the formation of the very earth itself, to help us understand what the Romans were undertaking as they built their city beside the Tiber. Edlund-Berry moves us into the proto-historical period, technically before the Republic is inaugurated; we cannot understand the later development of the city without this work. In the provinces, Stone (for Africa), Rodá (for Spain), Overman (for Israel) and Hoff (for Greece) explore reforms initiated in the Augustan period, in order to help us see what was begun in the Republican period that would come to full maturity in Imperial Rome.

The contributions of the book are organized into six parts. Those grouped under "Material Culture and Its Impact on Social Configuration" are intended to show how archaeologists are rethinking the physical objects that have been, or are being, excavated. Instead of grouping objects into material categories ("architecture," "frescoes," or "silver"), these scholars have undertaken the task of showing how objects can illuminate the society's formation, laws, behavior and habits. Thus, the most Roman of all habits – the daily bath – is shown by Yegül (Chapter 1) to be a complex mixture of input from Greek customs of education and indigenous interests which may have been connected to preventative medicinal procedures (the latter is also explored by Griffith in her essay on religious ritual). The custom develops its own internal stresses which the Romans express in unease over "luxuria" and assaults on female modesty. Yet baths were considered necessary parts of Roman colonies, as Dyson notes for Cosa; and by the end of the Republic, as Hales and Dyson show us, private bath suites were a normal part of every

A Companion to the Archaeology of the Roman Republic, First Edition. Edited by Jane DeRose Evans.

aristocratic house (though our evidence for bathing suites in villas comes in the first century CE at Settefinestre and Francolise; see Carandini, 1985a; Cotton, 1979).

Circuses, said to be a vital part of the Roman Imperial world, are explored in their infancy by Zarmakoupi (Chapter 2), along with other structures used for entertainment in the Roman world. The flexibility of the place of entertainment in the Republican world gives way to distinctive buildings, which become one more way for a Roman politician to emblazon his name across Rome and use manubial booty or proceeds from the developing economy. As a result, entertaining Romans became an increasingly important duty of ambitious politicians, whether they lived in Rome, Pompeii or Puteoli.

The development of the aristocratic atrium house is explored by Hales (Chapter 3), who shows how the neighborhood in which it was built, the carefully controlled view offered to the visitor, and the decoration of the rooms all contribute to the deeply held identity of the aristocrat and his public image. Indeed, she argues, the house becomes the stage upon which the elite male plays out his life, and by the end of the Republic, aristocrats increasingly depend on more luxurious surroundings to maintain their political status. We return to the houses of the rich and famous in Becker's essay on villas, where he focuses on the development of the villa form and the role of the villa in the agricultural landscape.

Yet another material manifestation of the making of a Roman's identity is in the building – or marking – of his tomb. Diebner (Chapter 4) takes us on a tour of Republican Italy to explore the various forms of these tombs, highlighting local tastes and trends to show a bewildering variety of form and decoration. But common to all was an inscription naming the individual dead. Tombs of aristocrats are also noted by Edlund-Berry and Davies as an important part of the landscape of Early and Mid-Republican Rome, respectively. Earlier tombs of the Greeks and Lucanians around Paestum allow Gualtieri to explore the changing society of the elite in the fifth and fourth century, as these cultures came in contact with the Romans. And to the north of Rome, Warden uses tombs to discuss the changing nature of elite identity in Etruria, as they, too, come in contact with the Romans. Farther afield, Stone remarks that while the study of tombs is an important component of the study of the indigenous people in North Africa, there is much still to be learned about the chronologies of the tomb types and the people who built them.

Black-gloss pottery, often used as the primary indication of chronology within the strata of a site, is shown by Roth (Chapter 5) to have less stability for the archaeologist than once thought. Archaeologists have long used the groundbreaking work of Lamboglia (1952), which was given an important and systematic overhaul by Morel (1981a). But as new excavations have isolated fabrics and production sites, our knowledge of what was produced where – and the variations that are possible on basic forms outlined by Lamboglia – means that we can have a fuller understanding of the trade patterns, the influences on various local productions, and the "Romanization" of pottery production in the peninsula (for a brief discussion, see also Chapter 22). Roth's work has had a wide impact as archaeologists begin to reassess the pottery evidence in their surveys and excavations, especially when it is used as a chronological determinant, as can be seen in the essay by Fracchia. But his greater point is that the pottery should not only be used as a chronological indicator; it is better studied as a cultural artifact, one that can contribute a great deal to our discussion of the character of "Roman Italy."

This discussion about pottery and what it means in its Roman cultural context is also taken up by Laubenheimer (Chapter 6), for a different class of pottery. Focusing on the wine trade between Italy and Gaul, Laubenheimer notes that these amphoras cannot be thought of as being used in the same way in both places. On the surface, the storage vessels look to be straightforward indications of the adoption of Roman culture, in an area that will soon be labeled "Togate Gaul" for its embrace of Roman ways of life. Yet, as Laubenheimer shows, the culture of wine drinking, and even the use and reuse of amphoras, looks very different in Gaul than in Italy, sounding a warning to archaeologists not to interpret the material culture by itself as a simple indication of "Romanization" (see also Dietler, 2010, who discusses the same issue in relation to Greek culture and southern Gaul).

In Chapter 7, I discuss another chronological indicator in strata, coins. Here, again, a warning is sounded to archaeologists against relying too heavily on coins as a sole indicator of chronology. The case for the impact of the Republican monetary system is assessed through Roman Hispania, where arguments still swirl around the date of the introduction of the denarius, and thus the military and economic impact of Rome on the peninsula – which is noted as well by Rodá.

The weapons of the legionary soldiers – one of the reasons often cited for the successful military expansion of Republican Rome – are surveyed by Goldman (Chapter 8). This period is briefly treated in handbooks on Roman weaponry, but the story of Rome's adaptations of neighboring cultures' forms and technologies and her own innovations show again the complex interaction between Rome and her competitors-turned-allies. The physical forms of the weapons depict changing political forces as well, as the army becomes a professional fighting force by the end of the first century, a formidable tool in the hands of strong leaders, and a contributing factor in the collapse of the Republic. The work of Dobson on Roman camps can be usefully consulted in conjunction with Goldman's essay, and the changing shape of weapons as the Lucanians are confronted with Rome is briefly addressed by Gualtieri.

For the last two chapters in this section, we turn to the bodies of the Romans themselves. Osteological studies are included in only a few final publications of archaeological excavations, but their potential for unlocking many mysteries about lifestyle in the time of the Roman Republic is great. Since so few studies have concentrated on this period, Smith (Chapter 9) needs to range farther than the other authors in the *Companion* to find interpretive parallels to make sense of the Republican bodies that have been studied. In doing so, she can assess physical stress (as shown by diet or trauma) to begin to decide if the coming of the Romans meant any major lifestyle changes to individuals in the east (for a parallel study of Imperial era bodies and malaria, see Gowland and Garnsey, 2010). Smith also appropriately pleads for more attention to the study of bones on the part of archaeologists; Warden notes that preliminary studies of the DNA of Etruscan individuals has already reminded scholars that the arguments concerning the origin of the Etruscans are still open in many minds. The origin of Imperial inhabitants of Rome is currently being studied by strontium isotope analysis (Killgrove, 2010), a method which may be able to be used on populations of an earlier age as well.

Lo Cascio's entry on population and demographic study brings together aspects of several of the chapters on material culture, and points to problems that are addressed in specific regions of Italy (Chapter 10). His analysis of the "high" and "low" estimates of the population in Italy is crucial to understanding the development of cities, towns and villas, the role that agriculture plays in the economic growth of the peninsula, and the expansion of the number of men who could be called "cives Romani." These numbers have an impact, as can be seen in the chapters by Fracchia, Gualtieri, Suano and Scopacasa, on the idea of identity, Romanization, and the problem of slave labor on latifundia. But they also have an impact on the growth of the city of Rome (as in Davies) and the needs that prompted colonization (as in Dyson).

We move from the social to the landscape in Part II, "Archaeology and the Landscape," where the authors use the physical features of the Roman world to explore how the Romans were formed by, and thereby formed, the landscape. Ammerman (Chapter 11) explores how the inhabitants of Rome largely reshaped the valleys (especially) and the hills to build their city. This view of earliest Rome can only be dimly seen, due to the deep, deep deposits that now overlie this era, and Ammerman cautions us about hewing too closely to the ancient authors' description – or our very modern ideas – about the ancient topography of Rome. His work is complemented and illuminated by the chapters by Jackson and Kosso on architectural building materials; Edlund-Berry and Davies on Early and Middle Republican Rome, respectively; Hodge on aqueducts and water supply; and Laurence on roads and bridges.

The landscape is used in a very different way by archaeologists who work on surveys, as Fracchia (Chapter 12) demonstrates. This form of "excavation," which has become popular especially since the 1980s, is still plagued by problems of interpreting the data obtained by fieldwalking. As a consequence, as Fracchia notes, survey results are not always taken into consideration when it is time to write the history of the Italian Peninsula. The results of surveys do help the archaeologist and ancient historian discuss the impact of Roman colonization and agricultural practices and legislation. However, to be fully understood the survey results must be used in conjunction with excavation of sites. Gualtieri and Suano and Scopacasa explicitly argue that survey results are evidence for comprehending the cultures of the Italian Peninsula.

Chapter 13 on agriculture and the environment by Goodchild is directly connected to the evidence found in surveys. Goodchild casts her net wider than surveys, using ancient authors, annual rainfall, man-made irrigation and drainage systems, botanical data, evidence of processing centers and animal bones to make her argument. By so doing, she can reflect on Late Republican agrarian "crises" and the development of large-scale agriculture worked primarily by slaves or tenants. These problems are touched on by Becker, in his discussion on the development of villas.

Dobson's Chapter 14 on the development of the legionary camp is an admirable example of how an archaeologist must wrestle with the interpretation of both the physical evidence and the ancient authors to understand the political and societal changes occurring in the Republic. Few Republican camps have been identified and fewer still excavated. The best-known camps, circling the city of Numantia, were built and rebuilt in the same spots during successive military campaigns. But Dobson can show the development from the manipular double-consular camp to the cohort-organized single-consular camp, a development that, as noted above, should be read in conjunction with the changes to the weaponry of the army that were outlined by Goldman.

The Romans felt strongly that the landscape held the sacred, while they were also bent on reforming the landscape to make it sacred. Griffith (Chapter 15) first explores how we can reconstruct religious ritual and then discusses how this ritual was informed by or changed by the great sanctuaries on hills or beside lakes, built in the Mid- to Late Republic. In doing so, she explores the role of votives and of ritual dramas that are incorporated into the architecture of these sanctuaries. Religious ritual permeated Roman life, as it is mentioned as a necessary part of the orientation of towns or colonies (see Romano and Dyson), especially Rome (Edlund-Berry), and as forming the landscape of Rome itself (see Ammerman and Davies).

Part III concentrates on "Archaeology and Ancient Technology"; the broad subject has been addressed by Oleson (2008), but the authors in this section home in on the Republican period to provide focused chapters on how this technology changed life in the fourth through first centuries. Romano (Chapter 16) explores the processes behind the division of land, and the idea of urbanization in colonies; he pays special attention to the two centuriation systems that can still be traced in the area around ancient Corinth. The interest in the Roman philosophy of urban design – and how this is married to the layout of actual cities – continues to be a fruitful topic of research, as can be seen in Sewell (2010), and in this volume, in the chapters by Dyson, Laurence, Gualtieri, Edlund-Berry, Hoff, Anderson and Dobson.

Jackson and Kosso (Chapter 17) remind us that building materials were a major factor in the appearance, and hence use, of ancient buildings. Their careful analysis shows how the area quarried by the Romans expanded as they gained control over more of the peninsula, then the area surrounding the Mediterranean Sea. They couple this with an analysis of the knowledge of the materials, as handed down to us in Vitruvius, who shows an uneasy balance between his empirical knowledge and his adherence to natural philosophy. Their work, then, is read profitably in conjunction with Edlund-Berry, Davies, and the chapter by Evans on the Late Republican city, as these authors explore the city where Vitruvius forms his ideas.

Vitruvius remains an important source of our information about water systems and aqueducts; thankfully, his confusion about the technology of those systems can be offset by the work of the engineer Frontinus, as Hodge (Chapter 18) makes clear. Hodge explains the process of providing water to the Roman household which, judging by the remains we have of the Republican water systems, must have been an essential part of civic life. It certainly allowed the construction of bath buildings (see Yegül) and allowed for the development of luxury in private bath-suites and well-watered gardens in the homes of aristocrats (see Hales). The control of water was an important part of the development of the city of Rome (see Ammerman, Edlund-Berry and Davies).

The building of roads and bridges is closely linked to this development of the infrastructure of Rome, and, like aqueducts, extended the physical presence of the city throughout the peninsula. Laurence (Chapter 19) points to the long-term impact of roads and bridges on the development of the Republican peninsula – from the movement of agricultural products and building supplies and consequent opening of new economic opportunities, to the development of pilgrimage routes to sanctuaries, to the tying of colonies and established cities in the newly conquered provinces to the city. The impact of the road can especially be seen in the settlements along the Via Domitia, which stretched from Spain through southern Gaul to Italy, linking these areas to Rome. Along the Via Domitia, cities in Gaul, such as Nemausus (Nîmes), do show some impact of Roman culture, but that impact does not extend to towns far away from the road (Monteil, 1999; unfortunately, the excavations of Arelate (Arles) remain unpublished).

Indeed, Strabo (IV.5) notes that Roman control of southern Gaul was tenuous. The impact of the Roman road system also figures in the contributions of Fracchia, Goodchild, Gualtieri and Suano and Scopacasa.

Becker's Chapter 20 on villas wrestles with the problem of defining a villa and thus finding the origins of the structures we are so familiar with, in remains around the Bay of Naples, wall paintings, and Pliny's description of his own villa. It is not until the Late Republican period that we have evidence of sprawling luxury homes cum farms, but the problem of defining the villa is intricately tied to the problem of agrarian crises in the Late Republic, as Goodchild notes. A parallel study, which focuses on the choice of terrain and the resulting agricultural activities, can be found in Fracchia and Dyson. The study of the villa is also tied to the lifestyles of the wealthy urban elite, as Hales noted for urban houses and Diebner noted for tombs.

The last chapter in Part III is Tuck's exploration of what we know about the ports of the Republic (Chapter 21). The Roman historians overplayed the Roman dismissal of overseas commerce and the need for a navy, Tuck argues, for we can find evidence for port facilities that were surely necessary for the large urban population of Rome in the fourth century. And by the first century the Romans built ports that relied not only on their own inventions in material structures (see Jackson and Kosso), but also on the decorative and functional models presented by large harbors in the eastern Mediterranean – another instance of changes in material structure and decorative models wrought in both the west and east, which Versluys brings out in such an interesting manner.

Now that we have discussed Roman innovations in technology, the use of landscape, and material culture, we can tackle the fraught issue of "The Archaeology of Identity" (Part IV). Stek (Chapter 22) reminds us the very posing of the question in terms of "Roman expansion" and "Italic identities/peoples" has already begun to answer the question in a particular way. He proposes to "de-center Rome" in order to understand more fully the processes underway on the peninsula in the Early and Middle Republic. In doing so, we begin to recognize how the Italic identities were changed, manipulated and absorbed by the end of the Republic and we are able to explicate the complex nature of the word "Roman" as it was applied by its own citizens. Work on identity of the various peoples living on the Italian Peninsula enters into almost every chapter in the *Companion*, as various authors have sought to distinguish the "Roman" nature of baths and other entertainment buildings, houses, tombs, temples, pottery and legionary equipment. Or they have sought to explain the problem of changes in identity during the Republic by the various people groups, such as the Etruscans, the Oscans, the Samnites, the Lucanians, and even the early Romans.

The Etruscans are explored by Warden (Chapter 23), who concentrates (as he must) on the identity of the Etruscan elite, showing how that identity morphed during the period of most intense contact, and conquest, by the Romans – if the "Etruscans" ever thought of themselves as such, rather than simply as members of a polis. In his exploration of architecture, metalwork, sculpture and tomb painting (all markers of elite identity), Warden is careful to distinguish between local (tomb painting in Tarquinia) and pan-Italic forms (the Etrusco-Italian temple type), giving us a nuanced portrait of a people who may never have self-identified as "a people." Read together with Edlund-Berry, we can begin to appreciate more fully the very local nature of the Etruscan phase at Rome.

The complex character of identity is explored in a different way for the city of Poseidonia, as it was known to the Greeks, or Paestum, as it was known to the Romans. The Greek colony was founded in the sixth century by colonists from nearby Sybaris and was planted in an area inhabited by the Lucanians, who later controlled the city. We do not know if the Lucanians called the city by a third name. Gualtieri (Chapter 24) begins with the Greek city, moves to the Lucanian phase and explores the Latin colony installed on the site in 273. He finds a city, in Ennius's phrase, that beats with three hearts, a point of view that dovetails with that of Stek.

The cities of Samnium are much smaller than Paestum, but the people who were identified by the Romans as "Samnites" contributed much to the identity of the Roman in the Late Republic. Suano and Scopacasa (Chapter 25) bring together evidence from ancient authors, surveys, excavations of settlements and cemeteries, landscape archaeology, metalwork, pottery and sculpture to explore the identity of the people who described themselves as "*safin-*." Yet, they caution, this identity is not unchanging, and may have looked very different depending on the Samnite settlement in which one resided. In many ways, scholars working in the area north of Rome and those working in the Central Apennines find themselves on common ground, for not only are "all politics local" but, indeed, all cultural definers are, too.

With Edlund-Berry's contribution (Chapter 26), we move into the sensitive problem of the identity and form of Early Rome – which, as explained above, is not technically part of the Republic, but which needs to be explored in order to understand the city of the Republic. Edlund-Berry has to work at the borders of history and myth as she discusses the formation of the identity of Rome from the time of its supposed founding through the proto-historical end of the reign of the kings. But she uses the archaeology that we do have – contentious as the chronology might be – to give us a tapestry of the culture of Early Rome. In doing so, she has admirably brought together material that Ammerman, Jackson and Kosso, and Laurence also explore, but she concentrates on what this material can tell us about what the Romans thought about Early Rome.

As we move into Part V, "The Archaeology of Empire during the Republic," Versluys (Chapter 27) asks us to place Rome within the Hellenistic *koine* of empires. The parallels between Nemrud Dağ and Praeneste are instructive here, as the forms of the buildings reflect very similar intentions on the part of the patrons. Yet it is only the Roman site that is called a "hybrid." By acknowledging that influences from outside the city of Rome were essential components of Roman culture, we can come to a better understanding of Roman culture and hence "Romanization." Stek's chapter should be read in conjunction with Versluys's, as one concentrates on the earlier period, in the Italian Peninsula, and Versluys on the later Republic, in a pan-Mediterranean setting. Stek's essay is also fundamental to the understanding of the archaeology in the provinces in the last two centuries BCE.

In order to understand the archaeology of the Republican empire, we begin with the city of Rome itself. In order to understand the changes that were wrought in Rome it is first necessary to comprehend the changes wrought in the city due to her burgeoning empire (see also the chapter by Miles). Davies (Chapter 28) uses the evidence provided by ancient authors, excavations, inscriptions and building materials to help us understand the transformations occurring in Roman infrastructure, temples, porticoes, civic buildings, and arches and honorary columns in the fourth through second centuries. She argues that the Middle Republic was when Rome took its place on the world stage, a move that could only have happened because of the military victories which allowed new revenue and new ideas to pour into the city. Jackson and Kosso, Hales, Hodge and Tuck also contribute to the picture of mid-Republican Rome outlined by Davies, in their respective essays on building materials and specific building types.

The study of first-century Rome has focused on the large personalities that increasingly took center stage and used their fortunes and political prestige to build and rebuild Rome. In so doing they intensified the competition for wealth, honors and offices while providing Rome with spectacular examples of entertainment buildings, temples, basilicas and luxuriously appointed aristocratic houses. In Chapter 29 I have concentrated on the last decade of research and publication of sites in Rome in order to give the reader a broad lens with which to consider the problems of the builders' intents and the beginnings of Imperial building programs – problems which are unlikely to be answered only by new excavations, but which can be discussed in more depth with the evidence from archaeologists.

As we turn our gaze away from the city of Rome, we look to the colony that the excavators believed contained "our purest example of a republican forum" (Brown, Richardson and Richardson, 1960). Dyson (Chapter 30) walks us through the history of the excavations of Cosa, excavations which have had, as Dyson (2005) himself has said, an outsized impact on our idea of what a Republican town looks like. Reassessing the evidence of what was present and when these buildings were built is an essential part of understanding the archaeology of the Republican period. Dyson also brings in the excavations at Settefinestre and Le Colonne to help us understand the environs of Cosa, and to see how the city had an impact on the agricultural development of the area. The colony is briefly explored by Romano, Tuck, Hales, Laubenheimer and Evans ("Coins"), testifying to the impact that the site has had on our interpretation of the Republican period.

We then turn from mainland Italy to the first provinces of Rome, as explored by Wilson (Chapter 31), who brings to bear his long experience in the field in both Sicily and Sardinia. Here we find that the coming of Rome seems to have little impact on the material culture, at least initially. Although Sardinia remained more isolated than Sicily, Wilson reminds us of the economic opportunities for the latter island that surely had an impact on the way the land was organized and farmed. Eventually the Romans – or at least Cicero – would call it "the first ornament of empire" and laud it for its agricultural riches. The Sardinians were given no such place in the Roman mind, and thus the impact of Rome is less distinctly felt in this former Punic colony.

While Carthage played an enormous role in the formation of the Roman Republic, finding the Roman colonies and cities in North Africa is hugely complicated by the overlying Imperial remains which archaeologists do not wish to disturb. Stone (Chapter 32) argues that the period, although it is often seen as static or one of slow development (especially in contrast to the Augustan period), is rightly seen as an important foundation for developments in the Imperial period. For it is in the second and first centuries that we see expanded urban settlements, the erection of new public buildings, and new exploitation of economic opportunities, especially agricultural ones. Although the inhabitants of cities in North Africa used Rome as a model for innovation, Rome was certainly not the only model; North Africa, like Rome, was part of the larger Hellenistic *koine*. Like Rome itself, the North African cities had other cultural references as part of their material and social context: Libyan, Punic, Egyptian, Greek or Roman. The cities of North Africa are clearly in need of more study, and in return, the study promises rich rewards.

From North Africa we turn to the European province closely connected to the continent, as Rodá (Chapter 33) guides us through the complex issues concerning the archaeology of the Republican period. She begins with the responses made by the people living in the north-east of the peninsula to the Roman campaigns of the First Punic War. By the second century, Roman presence is felt over much of the eastern and southern coasts. We can see from inscriptions, coins, sculpture and public buildings of the first century how the material culture of Hispania changed, partly in response to the intense military presence due to the revolt of Sertorius and the fighting during two civil wars. In order to clarify this picture, Rodá introduces us to some of the significant transformations that occur in the archaeology of the peninsula during the time of Augustus, although the seeds of these changes were certainly sown in the previous centuries.

In Overman's Chapter 34, we begin with the crucial transformation of Judea under the Seleucid kings, moving to the nationalistic Hasmonean government, which turned to Pompey to settle the debate for the throne between Hyrcanus and Aristobulus; as in Athens, the Romans were first encountered as necessary allies, not as invaders. Thereafter, the Hasmonean monarchy was essentially controlled by the Romans, until they replaced Hyrcanus with Herod. The archaeology of this tumultuous period is only recently coming into focus. Overman shows how a series of building projects, culminating with those of the last Republican-period general, Herod, shows us a province that is assimilating larger Mediterranean models, while incorporating structures such as *miqva'ot* (Jewish ritual baths) into the Hellenistic/Republican architectural vocabulary. By the death of Herod the Great in 4, all of Judea had been transformed, as can be seen, for example, in the Early Imperial temple at Omrit (for the date, see Berlin, in press).

Before we close the loop across the Mediterranean, it seems appropriate to explain why three areas are only tangentially approached in this sweep across the empire of the Republic: Gaul, Illyria and Asia Minor. It appears that Roman interest in southern Gaul was confined to the construction of a road to facilitate travel between Spain and Italy – until Julius Caesar decided to make its conquest an integral part of his political ambitions. Thus, the material culture of the settlements along the coast is impacted more by the (originally Phocaean) city and colonies of Massalia (Marseilles) than by Rome. There is little evidence of the importation of Roman goods except for wine; minimal adoption of Roman building materials such as roof tiles; and (even far into the Empire) few examples of the use of that very Roman custom of erecting inscriptions (see the now-standard text, Woolf, 1998). Indeed, an essay on Republican Gaul would have been very short and not very illuminating (see e.g. Long and Picard, 2009, which is titled *César: Le Rhône pour mémoire*, but very few Republican-era objects are in the catalog). It is only in the Augustan period that archaeology begins to show the changes wrought by the imposition of a new political order.

The archaeological record of Illyricum is a good parallel to that of Gaul. The region was ruled by client-kings controlled by Rome in the second century; it appears that Rome was more interested in keeping open the Via Egnatia, the land route between Italy and Byzantium, than making the province part of their empire (even Strabo VII.7.3 said this; see Chapter 19). Thus, some imported materials only reached cities and settlements along the route. The area became a province of Rome in Octavian's settlement in 27.

The archaeology of the region shows a major shift in the Augustan period, but not in the Republican phase (see e.g. Hansen and Hodges, 2007; Royal, 2012).

The archaeology of Asia Minor during the period of Roman control of the provinces cannot be differentiated from the Late Hellenistic phase – as Versluys has pointed out in this volume. In fact, he might say, the distinction is an artificial one, bounded by the change in rulers. The "Roman" aspects in the area can only then be noted with the coming of the imperial cult, the use of Latin in (some) inscriptions, and the face of the emperor on local coinages (which often retain their Greek legends) – that is, the material culture is only impacted in the Imperial period. For instance, research in the archaeology of Delos, where we have the best information about the presence of Italians in the last two centuries BCE, gives a nod to the Romans in the discussion of housing (e.g. Trümper, 2003). The Romans are considered to be a minor part of the cultural landscape of Hellenistic Delos.

It is Hoff's job to explore the changes wrought in the province of Achaia, which did change during the Republican period to reflect Roman culture (Chapter 35). We are able to see some of those changes due to the new colony imposed on the old city of Corinth (see also Romano). Surveys also help to illuminate the growth of large estates, and the loss of population in the rural areas of Greece. And from the excavations in Athens, we can begin to understand the complex narrative of the shift from the rule of the Hellenistic kings to the new rulers, as public spaces begin to reflect Roman political and religious realities.

For the final section, Part VI, we bridge the gap between the Republican period and the twenty-first century by looking at the problem of looting and its effects, and how technology is changing our ability to visualize the ancient world and helping us store and share the data from our excavations. Anderson (Chapter 36) explores the use of technology as archaeologists try to define the changes that occurred in Republican Pompeii as successive waves of Etruscan, Greek, Oscan and Roman inhabitants came to the city. Archaeologists warn that Pompeii – like Cosa – has an overwhelming influence on our thinking about the archaeology of the Republic and Early Empire, but Anderson shows that this attention allows – and now almost necessitates – experimentation in data retrieval, data storage and publication. Anderson's chapter carries an important warning about the use of new technologies, which can sometimes overload the archaeologist's budget or be used in a manner that is not entirely helpful. The need for standardization of platforms and the central location of data is a major concern and one that needs the attention of scholars. But archaeology is still a young discipline and it is not entirely surprising that it is still undergoing growing pains. As archaeologists begin to adjust their budgets to incorporate new realities of technology, I suspect that we will begin to see an agreement on how to store and share electronic data.

Miles (Chapter 37) looks at another way that the field is changing – in a manner that is rooted in past practices. She outlines for us how the ancient Roman society was altered by the fact of looting, in this case, mostly of the Greek cities it conquered. And she reminds us that the Romans themselves discussed the advantages and disadvantages that looting brought to their society. We move into the beginnings of archaeology in Italy, which had little to differentiate it from looting except perhaps in scale to the twenty-first century, where the problem of looting has been taken up by the United Nations. Since all major importing European countries and the United States have signed the UNESCO 1970 Convention on the Means of Prohibiting and Preventing the Illicit Import, Export and Transfer of Ownership of Cultural Property, the fight against looting has intensified and museums collecting antiquities have felt a major impact on new import controls, especially in the United States. World opinion has moved far away from the norm of "to the victor belong the spoils" (thus Senator William L. Macy in 1832 after a sweeping Democratic victory in the polls, though more than a few Roman senators would have agreed with the sentiment). But archaeologists must continue to explain why their work is important, and why archaeological strata need to be intact in order for the work to go forward; this is a global issue, sure to be debated far into the twenty-first century.

The authors of this *Companion* have combined their careful reading of specialist literature with their ability to explain specialist issues to a broader audience. We hope you use these chapters, and the guides to further reading, as springboards to thinking about old problems in a new way or considering in a new way the varied and complicated world of Republican Rome.

Acknowledgments

My thanks to my graduate assistant, Amy Yandek, whose invaluable help on many levels kept the project moving forward. Ab imo pectore, my grateful thanks to my husband Kevin who endured all with a sympathetic ear.

PART I

Material Culture and Its Impact on Social Configuration

PART I

Material Culture and Its Impact on Social Configuration

CHAPTER 1

Development of Baths and Public Bathing during the Roman Republic

Fikret K. Yegül

1 Introduction

Material remains of baths – basins, pools, terracotta, stone and metal tubs – and well-developed hydraulic systems found in Crete, Cyprus and Anatolia, such as the second millennium palaces at Knossos, Boğazköy and Zincirli, attest to the remarkable degree of technical and aesthetic sophistication attained in bathing, at least by the privileged classes of Bronze Age civilizations (Naumann, 1971: 195, 442–3; Lloyd, 1967: 78–9). Looking back to the legendary past, in Homer's world bathing in warm water was a luxurious and refreshing experience, a special reward reserved for heroes at the end of a trip or battle. In the *Iliad* men who returned from reconnaissance "washed away in the sea the thick sweat from their skins and necks" (*Il.* 10.572), and as retold by Athenaeus centuries later, after having this way refreshed themselves "they went to the polished tubs and bathed, smearing themselves with oil, they sat down to their meal" (*Deipnosophistae* 1.24 d, trans. C. Gulick, 1927, Loeb). In these aristocratic contexts hot bathing was a fairly simple matter of washing from a cauldron of hot water mixed with cold to comfortable warmth. However, already in the fifth century the conservative view in Greece criticized hot, pleasure bathing as a sign of decadence and aspired towards the frugal lifestyle of the Spartans, who supposedly only bathed in cold water ("Laconian style").

Among the ancient Mediterranean civilizations, Romans deserve to be singled out in their extraordinary devotion to bathing as a social and recreational activity deeply rooted in daily life. Bathing in public was a central event in Roman culture and a very important part of the Roman day. Demonstrating little interest in cold water bathing (unless it was a medical recommendation), Romans developed and perfected the heating and water supply technologies of baths, which changed bathing from a simple act of hygiene to one

A Companion to the Archaeology of the Roman Republic, First Edition. Edited by Jane DeRose Evans.

of pleasure. For the average Roman, who spent a considerable part of the day in the public baths, bathing was an activity which incorporated sports, games, massage, body culture and relaxation much like a social club or community center. Some of the larger baths even included quasi-intellectual facilities such as libraries and lecture halls as well as colonnades, galleries and exedrae for the exhibition of works of art. Roman baths embodied the ideal of Roman urban living. By the Late Republic, Rome had nearly 200 small baths as recorded in urban census documents (*Notitia Romae (Notitia regionem urbis)*, ca. 334–57 CE and *Curiosum urbis Romae*, ca. 357–403 CE). By the end of the fourth century CE their numbers had swelled to a staggering 856, plus ten or eleven Imperial thermae, which were exceptionally large and luxurious bathing complexes ordinarily owned and operated by the state.

What accounts for the obvious delight Romans had in their baths and the intense popularity of public bathing? There is no simple and definite answer to this question. One overarching observation is that bathing was popular simply because it had become a habit. It had become a comforting part of urban life and national identity – and the more Romans liked bathing the more likeable it became. Not to bathe would have been un-Roman. Still, why did bathing become a daily habit in the first place?

2 Bathing as Pleasure

We believe that among the most important reasons for the popularity of bathing is the pleasure factor: bathing is a physically and psychologically pleasing activity that relaxes the body and soothes the mind. The bathing experience – warm, clean water, shining smooth marble surfaces, steamy atmosphere, the aroma of perfumed ointments, the nudity and intimacy of massage – awakened the senses and created a state of comfort and well-being the Romans called "voluptas." A freshly bathed person felt light and optimistic. These sensory pleasures were admired aspects of baths and bathing. A fifth-century CE dedicatory inscription from a bath in Serdjilla, in Syria, is one of many that announced that the bath could bring "comfort and happiness" to the entire village (Prentice, 1901). The renovated Baths of Faustina in Miletus, in Asia Minor, were described as an "ornament" to the city where the citizens enjoying the delights provided by the refreshing bath waters "all are relieved of toils and labor" (Gerkan and Krischen, 1928: no. 339). Often these waters and their delights were identified with nymphs and *charites* (Graces), minor goddesses that protected the waters and personified beauty and charm.

3 Bathing as Luxury

Enhancing the sense of pleasure and delight was the sumptuous material world created by public baths. As observed by Cicero, in the first half of the first century, Romans disliked private luxuries but "delighted in public magnificence" (*Mur.* 76). The interiors of baths, decorated in fine polychromatic marbles, intricate mosaics, stucco ornament, gleaming bronze hardware and marble statuary, all under high and well-lit vaults and domes, were perhaps the most appropriate candidates for such public display of wealth and luxury. And this was a luxury that could be enjoyed by all, or almost all. Wealth in private life was for the enjoyment of the privileged few; public baths, dubbed as "people's palaces," brought this bounty to the masses. As a rule even baths run as lucrative enterprises which charged a fee were not outside middle-class affordability. There were, however, a few like the Baths of Etruscus in Rome which must have been a small but exclusive establishment with a select clientele. Martial recommended it to a friend admiringly: "If you do not bathe in the *thermulae* (small bath) of Etruscus, you will die unbathed!" (*Epigrams* 6.42, trans. W. Ker, 1968, Loeb; see also Stat. *Silv*. 1.5). The world of material magnificence offered by public baths, especially the great thermae of the Imperial period, afforded urban populations the opportunity to escape their overcrowded and cramped living conditions and the dusty streets for a few hours a day and bathe in style. For a few hours a day the privilege of public bathing took the individual out of his shell and gave him a place among others. But most importantly, it gave him an opportunity to share the empire's wealth and, perhaps, to believe in its ideologies.

4 Bathing and Ancient Medicine

An important and practical explanation for the popularity of baths is medical: bathing was believed to be good for health and particular forms of bathing were recommended as cures for particular ailments. In a world that lacked the modern means of diagnosis and cure, where almost every surgical intervention risked the patient's life, it is logical that preventive medicine, as represented by thermal and curative bathing and which aimed to maintain a healthy body, was far preferred to drastic and intrusive methods of treatment (Yegül, 1992: 352–5; Fagan, 1999: 85–103).

Bathing as a therapeutic measure received its initial authority from the recommendations of the Hippocratic School of Greek medicine in the fifth century. These recommendations, often promulgated by itinerant doctors and popular medical lecturers, were largely based on empirical and sometimes inconsistent assumptions. According to the *Regimen in Health* and *Regimen in Acute Diseases*, the two fundamental treatises of the Hippocratic corpus, one should bathe more frequently in the summer than in the winter; fresh water bathing moistens and cools the body while salt (sea) bathing dries and warms it up. Bathing was not recommended in cases of strong fevers or for those suffering from diarrhea or constipation. If not taken the right way bathing could actually harm the person, but "if preparations are satisfactory and the patient welcomes the idea, a bath should be taken every day" (*Acute Diseases*, trans. in Lloyd *et al.*, 1978: 205; cf. 186–206, 272–7; see also Vallance, 1996). It is interesting to note that the Roman habit of daily bathing seems to have its roots in the recommendations of Greek medicine as do the customary exercises as a prelude to baths.

According to Diocles of Carystus, a famed physician and dietician of the fourth century, an ideal daily routine included a relentless series of morning and afternoon exercises, massage and bathing (Jaeger, 1938). The views and practices on exercise and bathing for health were introduced to the Roman world through the works of various Hellenistic writers. Of particular significance is Asclepiades of Prusa (Bursa, Turkey), a Greek physician who practiced in Rome at the end of the second and the beginning of the first century (Green, 1955). This was just the period when bathing in public was making significant gains in Rome and Italy with concomitant development of baths and heating technology. Asclepiades employed a strict program of diet, exercise and baths and included cold water bathing in his regimen. His most influential followers, who established exercise and bathing as a quasi-scientific method, were Celsus, a medical writer of the first half of the first century CE, and Galen, a doctor from Pergamon and a physician who achieved great success in Rome during the second half of the second century CE. Both of these influential sources emphasized the benefits of hydrotherapy by specifying the nature and extent of bathing in an easy-to-understand style and in considerable detail. In treating a wide variety of maladies, they recommended bathing in close association with physical exercise, massage and sweating – activities that constitute the core of the Roman bathing routine.

It is also interesting to note that the explicit programs of bathing outlined by Galen involved a progression from cold to lukewarm to hot areas; he also recommended sweating and massage and a final plunge into the cold pool (to close the pores of the skin, to brace and energize the body). This progression roughly coincides with the tepidarium, caldarium, laconicum and frigidarium of the Roman baths already reflected in the architectural order and plan of many Republican establishments. Although we do not know to what extent the patients or the general public followed such a strict therapeutic regimen, it appears logical to assume that both thermo-mineral and regular city baths were used for curative purposes. Public baths were better equipped for treatment since even the better houses did not have elaborate bathing facilities. Nor could one get at home the professional attendants and health technicians necessary to administer the bath correctly. These points had been emphasized already in the Hippocratic corpus: "The basin should afford easy getting in and out. The patient should remain quiet and passive during the process of curative bathing; he should not attempt to wash himself, but let others do it for him" (trans. in G. Lloyd *et al.*, 1978: 204–5).

5 Bathing Ritual and Activities

Bathing as a social ritual occupied the larger part of the afternoon. Martial recommends the "eighth hour" (ca. 2–3 p.m.) as the best time for bathing when the heat of the baths was tempered (*Epigrams* 10.48). Since the Roman day was divided into 12 hours from sunset to sunrise (midday was the sixth hour), the length of the hours and the time to bathe varied from season to season. Patrons spent several hours in the baths, which in the shorter winter days might have necessitated artificial illumination (and explains the large number of lamps found in some bath excavations), but as a rule night bathing was rare and considered unsafe. Bathing was also enjoyed more in brightly lit interiors.

The essentials and basic routine of bathing as stated by Pliny the Younger appears simple: "I am oiled, I take my exercise, I have my bath" (*Ep.* 9.36). A wealthy person was accompanied to his bath by his servants and slaves carrying his towels, bathing garments and toilet kit; a poor person carried his own bundles. Upon arriving at the baths, one undressed usually in a special room reserved for this purpose called the apodyterium, which in some of the Republican era baths was characterized by rows of small cubby-holes to hold clothes (Figure 1.1). Putting on a short tunic, patrons normally engaged in some kind of physical exercise in an open courtyard or palaestra. It would be misleading, however, to imagine that the average Roman undertook a strenuous workout like the young

Figure 1.1 Reconstruction of a typical apodyterium in use in the Late Republican period. Source: Pasquinucci (1988: fig. 24). Drawing by A. Fremura. Used by permission of Franco Cosimo Panini Editore.

athletes of the Greek gymnasium. Roman gymnastics in the context of bathing were often light, such as the numerous ball games intended to loosen the muscles and build up sweat – activities that appealed to all age groups and body shapes. Martial mentions no less than five different ball games for the palaestra. One, called "triple ball" (pila trigonalis) was, as the name implies, a game for three where the opponents stood roughly at the three points of a triangle and tried to confuse each other by passing several balls back and forth (*Epigrams* 12.82, 14.46, 4.19, 7.32). Among other palaestral sports, running, boxing and wrestling are mentioned but their practice was not restricted to the open-air courtyard; many of the larger baths (especially of the Imperial period) had special halls (basilica thermarum; see below) that could be used for indoor athletics. Women did participate in sports and games, though they probably preferred the lighter variety. Swimming was also a popular bath exercise but it is unlikely that any serious or competitive swimming or diving was done. With few exceptions, such as the deep lap-pool of the military baths at Isca (Caerleon, Wales; see below), even the large swimming pools, the natationes, of the thermae only reached a depth over 1.0 to 1.20 m. Most of the bathers must have enjoyed a few easy strokes, leisurely wading and splashing. The easy and inclusive nature of sports welcomed even the elderly and the less athletically inclined to participate and made the public baths a pleasure for the masses.

Time spent in the palaestra was pleasant but few were so enthused with their exercise as not to drop everything and hurry inside when the tintinnabulum, the bell that announced the opening of the hot baths, sounded. Inside, the general order of bathing required a movement from warm to hot through a number of intercommunication rooms at varying degrees of warmth. Some bathers also spent time in one of the special rooms for sweating – the laconicum for dry heat and the sudatorium for steamy, wet heat – but identification of these spaces in actual baths is difficult. Bathing terminated in a cold plunge in one of the pools of the frigidarium. Some of the larger baths even had special sun rooms (heliocamini) with large, unglazed windows oriented west or south-west and unpaved, sand-covered floors. It must have been a pleasant experience to pretend to be at the beach and enjoy broad views across land and sea while sunbathing or immersed in a pool.

Among the most engaging of the plethora of social activities in the baths must have been the pleasure of being with friends, conversation and gossip, while perhaps enjoying a relaxing massage. Poetry reading, music and singing – not always welcome – were common. There were professional performers, too – jugglers, animal trainers, conjurors, jesters, mimes and gymnasts. Also pleasurable, at least for some, was eating and drinking, which would have added merriment to any group. A price list scribbled on a wall in the Suburban Baths in Herculaneum includes for sale "nuts, drinks, hog's feet, bread, meat and sausage" (*CIL* IV.10674). Vendors of food and drink could be found in any bath, but in Magnesia-on-the-Meander, there was a restaurant annexed to the bath which offered cheese, barley, olives, wine, fish, vegetables and a kind of "lyre-shaped" pretzel (Cousin and Deschamps, 1888: 203). The pastime engaged in the baths was supposed to be a light snack, a prelude to dinner proper, an important social event culminating the day, often taken in the company of others. Martial was a cordial host who strove to arrange a fine dinner for his bathing friends: "You will dine nicely, Julius Cerialis, at my house. You will be able to observe the eighth hour [time for bathing]; we will first bathe together at Stephanus' Baths … [then at dinner] there will be lettuce for relaxing the bowels …" followed by a long list of delectable treats (*Epigrams*, 11.52, trans. W. Ker, 1968, Loeb). A good bath called for a good dinner and public baths were the best places to invite somebody, or to get an invitation at a well-appointed household with a cook famed for his culinary skills. As mockingly narrated by Martial (*Epigrams* 2.14), some luckless characters, in order to avoid the social disgrace of dining alone, moved from one bath to another until they secured an invitation.

6 Ethical and Moral Concerns and Criticism of Roman Baths

Since ethical and moral concerns persisted as a topos of Roman literature, it is not surprising that baths with their luxurious and hedonistic world provided a likely subject for criticism. Their permissive and populist atmosphere could easily tolerate, or even encourage, coarse, irritating and occasionally violent behavior. The tramp, the vagabond and the ordinary lout found excellent opportunities for uncouth and boorish ways. The newly rich showed off their wealth; the half-educated subjected their victims to their grandiose literary ambitions; thieves and tricksters circled the crowd.

Seneca, a conservative writer and philosopher who lived in the first half of the first century CE, compared the well-lighted baths of his day with the dark and austere interiors of old ones, such as the one reputedly used by Scipio Africanus, the hero of the Punic Wars:

> [Today] we think ourselves poor and mean if the walls [of our baths] are not resplendent with large and costly mirrors [made of white polished marble]; if our marbles from faraway Alexandria are not set off by rich mosaics of rich, yellow stone from Numidia [*giallo antico*, an expensive marble] ... if our vaulted ceilings are not buried in glass [mosaics]; if our swimming pools are not lined with marble from Thasos, once a rare and wonderful sight in any temple ... What a vast number of statues, of columns that support nothing but are built for mere decoration ... Our ancestors did not think that one could have a hot bath except in darkness.
>
> (*Ep*. 86, trans. R. Gummer, 1962, Loeb)

Taken at face value Seneca's criticism is unjust. Excessive luxury for the sake of luxury may be condoned but good light and quality materials in architecture are a virtue; so is the habit of daily bathing. But Seneca was giving voice to a well-known moralizing position that disapproved of the increasing luxuries of the Empire contrasted with the frugal and modest virtues of the Republic. One should realize that writers often used the subject of bathing and baths as a metaphor in the service of other moralizing and polemical issues. Alternatively, baths could be used as literary trope to provide comical relief in describing the vibrant and unpredictable social mix of a growing metropolis such as Rome. In Plautus's plays (ca. 200–180), public baths of Rome often provide a colorful and familiar setting for courtesans, mischievous slaves and pickpockets. The popular New Comedy playwright thus provides us with some of the earliest literary evidence confirming that public bathing was widely accepted in Rome and Campania by the early second century.

The subject that attracted the greatest public criticism was the alleged immoral conduct associated with baths. The sexually permissive atmosphere tolerated in baths and spas made them the ideal playgrounds for sensual delights and questionable behavior. Sites with thermal baths and curative natural springs, such as the world-famous Baiae in the Bay of Naples, offered the greatest opportunities for misdemeanor (Yegül, 1992: 92–4, 1996: 137–62). Martial was delighted in the unrestricted opportunities for otium offered by the bewitching Baiae and described the place as a gift of nature "on the golden shore of blessed Venus" (*Epigrams* 11.80, trans. H.G. Bohn, 1897, Bohn's Classical Library). But, Seneca, the moralist, allowed himself to visit the famed spa once and denounced it as "a resort of vice" (*deversorium vitiorum*) (*Ep*. 51). Even at ordinary bathing establishments, there is little doubt that the presence of women (if mixed bathing was allowed) and nudity encouraged sexuality. Drawing on rich but often ambiguous evidence, the issues of mixed bathing, nudity and women in Roman baths elicit much scholarly debate.

The wet, hot and steamy atmosphere of baths probably encouraged a far greater degree of nudity than the palaestra where exercise garments, as specified by Martial, were clearly the norm (*Epigrams* 7.67, 4.19). A skimpy bathing costume, such as a cotton kilt wrapped around the waist or the chest (like the *pestamal* worn in modern Turkish baths), would have covered as well as exposed much in the course of washing and massage. Thus, whether men or women bathed in the nude may be a moot question; even if they had some cover, it revealed enough. A number of well preserved and remarkable paintings from a small neighborhood bath in the Esquiline region in Rome show nude or semi-nude women bathing or attending their toilettes in the baths. One particularly graceful figure simply holds in front of her nude body a loose, flimsy bathing wrap like a veil (Barbera and Paris, 1996: 16–37, figs. 16–17, 20–3).

Despite the Romans' famed sense of modesty and propriety, mixed bathing appears to have been acceptable during some periods of the Republic, at least in some baths. Quintilian, the famous rhetorician who was active in Rome in mid-first century CE, made the point that the habit of women bathing with men (lavari cum viris) was not necessarily an indication of adultery (*Institutio Oratoria* V.9.14). But his contemporary, Pliny the Elder, contrasted the alleged virtues of the long-gone Republic with the vulgar tastes of his day (*HN* 33.153, 36.121). Martial, on the other hand, referred to the practice with a degree of approval if not apparent delight, as an unremarkable social custom (e.g. *Epigrams* 11.47).

The architecture of public baths could give us some clues. Varro, writing around mid-first century, informs us that the first public baths introduced into Rome were designed as two connected buildings, one for men and the other for women (unum ubi viri, alterum ubi mulieres lavarentur; *Ling.* IX.41.68). Vitruvius, writing a few decades after Varro, lays down the architectural rule that hot baths for men and women should be separate, but, with admirable practicality, they should share the same furnaces and service yard (*De arch.* V.10.1). We do not know which baths in Rome Varro was referring to or where they were introduced from, but the rich evidence from Campania and now from Latium largely supports his observations. Three baths from Pompeii, the Republican Baths, Stabian Baths and Forum Baths, and the Forum Baths from Herculaneum, were all double baths for men and women with shared services (Figure 1.2) (Yegül, 1992: 57–66). These baths date from the mid-second to late first century. Displaying perhaps the earliest use of the same arrangement the Republican Baths in Fregellae in Latium are dated ca. 200–175 (see below). It is interesting to note that with the advancing decades of the Empire, the double bath was almost entirely replaced by establishments with a single set of bathing facilities, such as the Suburban Baths in Pompeii (early first century CE) or the Central Baths from the same city (still unfinished in 79 CE).

Implying that mixed bathing was *de rigueur* before his time, Hadrian prohibited mixed bathing with his famed decree "lavacra pro sexibus separavit," which was ratified under the Antonine emperors but discarded under Elagabalus (*SHA Hadrian* 18.10; *Marcus Aurelius* 23.8; *Heliogabalus* 31.7). Numerous inscriptions mention "baths of men" (balneum virile) and "baths of women" (balneum muliebre) but what is meant by these designations is not clear (Yegül, 1992: 429, n. 21). If mixed bathing was banned and the architecture of baths did not allow for the separation of sexes, men and women presumably bathed at different times of the day, normally women before noon, and men after. This arrangement is confirmed by a bath inscription of Hadrianic date from Vipascum, a small mining community in Portugal: women bathed in the morning and men in the afternoon (*CIL* II.5.181). For a habit that crossed continents and centuries, and involved vast numbers and types of baths, there could have been no single mode or practice. However, these rules and regulations imposed on bathing must have varied by time and place, by the type of clientele to which a bath catered, and by one's own sense of modesty and morality.

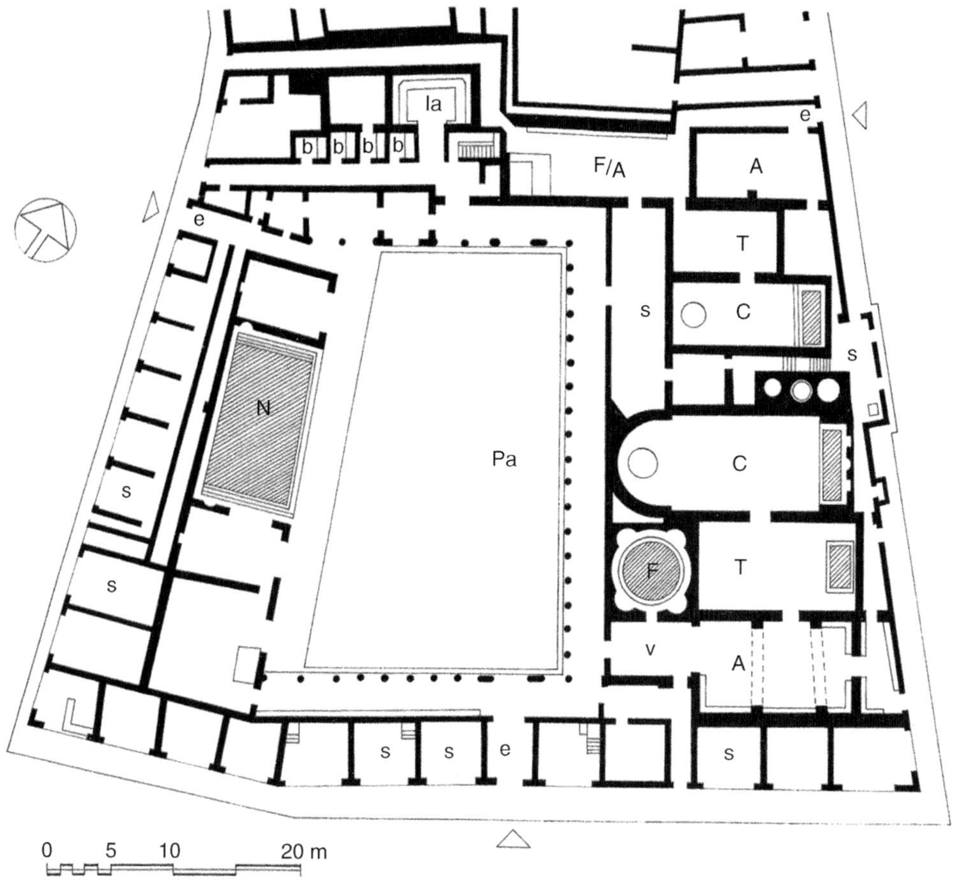

Figure 1.2 Plan of the Stabian Baths, Pompeii, *c.* 80. A=apodyterium; b=hipbaths; C=caldarium; e=entry; F=frigidarium (formerly a laconicum); la=lavatrina; N=natatio; Pa=palaestra; S=service yard; s=shop; T=tepidarium; v=vestibule. The women's bath-suite is found in the north-east corner of the baths (hence the replication of rooms). Source: Drawing by F. Yegül, adapted from Eschebach (1979).

7 Baths of the Greeks

In his study of Roman bathing Fagan underlines the influence of Greek and ethnic populations of the eastern Mediterranean upon the Romans: "the arrival at Rome of larger numbers of people [during the Late Republic] from regions where public bathing had long been practiced is surely significant and may well have contributed to the growing appreciation of baths among the lower orders" (Fagan, 1999: 82) – and, of course, also in the "higher orders," the urban elite. Indeed, Romans must have first learned about public baths and appreciated bathing as a social custom through their contacts with the Greeks of Sicily and south Italy in the early days of the Republic. There were Greek baths on Italian soil dating from the late fourth and early third centuries. Their plans, such as the early third-century baths at Gela, Sicily, typically display clusters of rooms, often circular in shape with individual hip baths with covered sides (for privacy and protection from splashing) arranged around the perimeter of the wall. Outside of Italy Greek baths of the Classical and Early Hellenistic period are characterized by the tholos type. Some of these, such as the baths at Piraeus, or the baths in the Sanctuary of Apollo at Cyrene, carved wholly or partially in soft rock, could have been known to Italian builders. It is important to note, however, that Roman baths in the subsequent centuries followed the proper Greek baths neither in plan typology nor in simple heating technology. With the exception of the second stage of the baths at Gortys in Arcadia (first half of the third century), which utilized one of the earliest applications of heating canals under the floor, Greek baths were heated by simple means, such as the use of braziers, unlike the more advanced central floor heating systems of Roman baths (see below). More significantly, they also reveal little concern for plans based on a sequencing of controlled and graded heating of spaces, the hallmark principle of Roman bath planning.

8 Bathing in the Context of the Gymnasium

Another important line of influence on Roman baths comes from the bathing facilities that were normally a part of the Greek gymnasium or palaestra. According to Vitruvius, our best written authority on the subject, the only element of the traditional Greek gymnasium connected with bathing was the *loutron*, either an open-air area or a special room of the palaestra equipped with elevated basins for cold water washing (*De arch.* V.11.1–2). Such cold water washing facilities have been preserved in many Hellenistic era palaestrae, such as those at Delphi in Greece and Priene in Asia Minor, where a series of marble basins are supplied with running cold water. In Delphi there is also a very large circular open-air pool in which the athletes could refresh themselves after their exercises, and perhaps even swim. Greek vase paintings provide many examples of palaestra scenes showing athletes washing around a raised, round basin (labrum) or showering under panther-head spouts (Yegül, 1992: figs. 15, 20, 21, 22). There is also a red-figure vase from the Berlin Staatliche Museum depicting women washing in cold water in what appears to be a public fountain (Yegül, 1992: fig. 19). Since athletes normally covered their bodies with oil before taking exercise, the advantages of hot water over cold in removing the muddy and greasy mixture from the skin must have encouraged hot water bathing in a gymnastic context. Vitruvius's description of the palaestra in fact includes an elaborate facility for hot water bathing. This was a semi-independent bath-suite located in a corner of the building which included a frigidarium, a warm-water room (calda lavatio) and a laconicum for sweating. A furnace servicing the suite was reached by a corridor (*De arch.* V.11.2). Writing at the end of the first century, Vitruvius was undoubtedly describing a palaestra of his own time, probably from Italy. This hypothesis is supported by epigraphic record which indicates that by the end of the Republic many of the old-fashioned gymnasia were renovated to include hot water bathing facilities; this was also in keeping with the fusion of the balnea with gymnasia and palaestrae in Campania and Latium. The gradual disappearance of traditional gymnasia and the transformation of others by the addition of fairly elaborate hot baths gave rise to confusion in the conception and terminology between a *gymnasion* and a *balaneion*; sometimes the same establishment was called by different names in two different inscriptions (Delorme, 1960: 244–5, n. 4; Yegül, 1992: 21–4). These changes, especially the growing interest in hot bathing as a pleasurable and sensuous activity in itself, can be seen as a part of the larger trend towards individualism and love of personal comforts affecting the entire post-Alexandrian world of the eastern Mediterranean.

9 Italian Farm Bathing

As demonstrated by literary and archaeological evidence the Greek component occupies a solid position in the development of the baths and bathing habit in Italy. Yet this appears to be only a part of the story: there is another important local source of influence enriching the many-sided debate on the origins of Roman baths – this is the contribution of the rural or farm traditions of bathing in Italy. This pro-Italian view (not surprisingly developed by Italian scholars) traces the origins of hot bathing to folk medicine of central Italian villages (Di Capua, 1939/40; Fabbricotti, 1976: 29–11; Yegül, 1992: 50–5). Farm houses in Early Republican Latium and Campania commonly had a special room with a stove or stove-boiler combination, often shared with the farm kitchen, to induce a "good sweat" effective in curing seasonal colds and alleviating rheumatic pains. To protect the patient from the cold this "sweat chamber" was joined by another moderately heated room, which also served as a changing room, and connected to a cold-water unit, or lavatrina. Thus, such rustic bath-suites established the three fundamental elements and sequences of the proper Roman public bath: the apodyterium/tepidarium, frigidarium/lavatrina and caldarium/laconicum. The early examples dating to the second and first centuries are simple and directly heated by kitchen stove-boilers; only the bath-suites of the later farm houses and villas, such as those in the House of the Faun and the House of the Labyrinth, both in Pompeii and both ca. 70–9 CE, utilize sophisticated floor heating systems. In the latter, even the walls were heated by employing special terracotta tiles (tegula mammatae; see below) which allowed the circulation of hot gasses between the wall and its revetment. Although the practice of therapeutic bathing in an agrarian society makes sense, it is obvious from the elaborate and luxurious layouts of some villa baths – which often encompass three or four interconnected rooms in a row – that domestic bathing could not have been solely and simply for curative purposes. Rather, bathing in Republican Italy was becoming a hygienic and recreational activity catering to the rising bourgeoisie and its ideals of the good life, its cherished otium.

10 Heating and Water Supply Systems

The development of effective heating and water supply systems played a major, perhaps decisive role in the creation of the Roman bath. The effectiveness of these systems appeals to popular imagination even today as one of the outstanding achievements of ancient technology, nothing short of a mysterious invention whose details are somewhat exaggeratedly believed lost to modern scholarship. Of course, this is not completely true – we know quite well how the heating and water systems of Roman baths worked, and admire them for their excellence and ingenuity.

The heating of a Roman bath was achieved mainly through its floors and its walls, and rarely its vaulted ceilings; it is categorically a radiant, panel heating system. The floor heating known as the hypocaust (literally "a furnace that heats from below") was achieved by raising

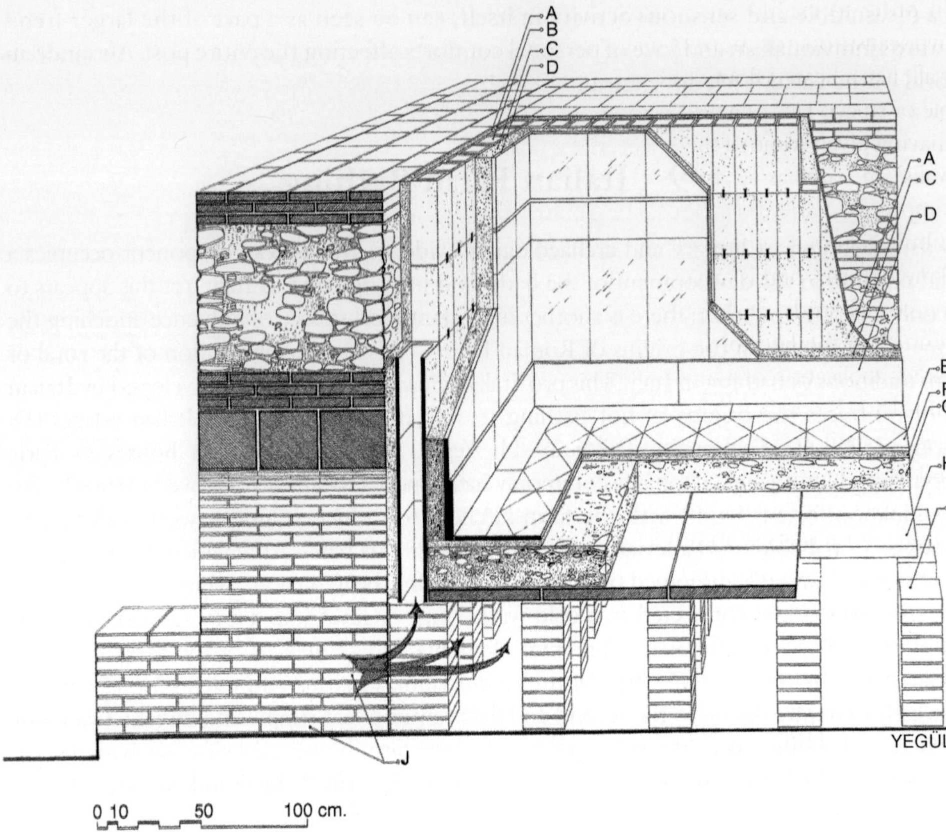

Figure 1.3 Schematic hypocaust system of a Roman bath. A=lime mortar; B=tubuli; C=crushed brick mortar; D=marble revetment; E=floor surface; F=mortar; G=subfloor; H=bipedales tiles; I=pilae; J=opening to furnace. Source: Drawing by F. Yegül.

the floor some 70 to 90 cm on small brick or stone pillars (pilae) and circulating hot gasses generated by a furnace (praefurnium) directly connected to the hypocaust space and ordinarily stoked from the outside. The system could deliver controlled and finely graded levels of temperature to different rooms depending on the extent of floor and wall space to be heated. Although it is impossible to determine exactly when this ingenious system came to use, starting from the fifth century, the development of various primitive forms of subfloor heating can be documented throughout the entire eastern Mediterranean region, such as the Greek baths at Gortys (Ginouvès, 1959: 331–4, 1962: 185, 206–7). The true hypocaust emerged in central Italy by the beginning of the second century in the baths at Fregellae and later in the same century in the fourth phase of the Stabian Baths in Pompeii. A century or so later than its first applications, Vitruvius described the system thus:

> The hanging floors of the hot baths (*suspensurae caldariorum*) are to be made as follows: first the ground is to be paved with eighteen-inch tiles sloping towards the furnace … On this pavement, piers of eight-inch bricks are to be built at such intervals that two-foot tiles (*bipedales*) can be placed above. The piers (*pilae*) are to be two feet high … and upon them two-foot tiles are to be placed to carry the pavement.
>
> (*De arch.* V.10, trans. F. Granger, 1931, Harvard University Press; see Figure 1.3)

In order to extend the effectiveness of the surface heating or radiation generated by the hypocaust, one or more of the walls of a space were also heated. The principle of a heated wall, like a heated floor, is based on the creation of a capillary space behind the finished marble or stucco face of a wall through which the hot gasses from the hypocaust, though cooled (having given some of their heat to the floor), could circulate. Seneca talks about the "hollow walls" in baths as a fairly new discovery (*Ep.* 90.25). The method described by Seneca as "impressos parietibus tubos," or tubes placed inside the walls, recognizable by us as the tubuli system, is the most practical and advanced of wall heating systems (*Ep.* 90.25). Simpler forms of this method by using tegulae mammatae, or flat terracotta tiles with projecting corner studs, or "nipples" affixed on the wall by metal clamps, came to use by the first century in the later phases of the Stabian Baths and Forum Baths in Pompeii. There is an earlier application of a heated, hollow wall using real tubes in the baths at Fregellae dated ca. 200–180 (Battaglini, 2009). Tubuli proper, known from innumerable archaeological examples, are hollow, box-shaped tiles that could be stacked on top of each other against a wall, and form, in effect, tall, vertical tubes. Their open bottoms that are connected to the hypocaust space and holes in their sides allowed the free circulation of hot air and gasses, creating a heated wall surface. Only a few of the vertical rows were open on the top and connected to a flue of exhaust, resulting in a low draft, slow burning, economical furnace.

The furnaces of Roman baths ordinarily served the dual purpose of heating the bath rooms and heating the water used in these rooms. This was achieved simply by placing a bronze boiler over the wood-burning furnace which supplied, through a complicated arrangement of pipes and valves, hot water to various rooms, basins and pools. A separate system connected the boiler to a large roof-top tank of cold water which could replenish the boiler as it was depleted. In some baths there were series of furnaces with interconnected boilers. Vitruvius describes a basic system for a double bath (much like the Stabian Baths of Pompeii) where the same set of furnaces and boilers should effectively serve both the women's and men's adjoining baths: "Three bronze cauldrons are to be set over the furnace; one for hot, another for tepid, and the third for cold water, placed in such positions that the amount of water which flows out of the hot water cauldron may be replaced from that for tepid water, and the tepid water one would be supplied from the cold water" (*De arch.* V.10.1, trans. M. Morgan, 1914, Loeb). In the same passage he also describes a device called "testudines alveolorum" to heat and maintain the water in hot pools – normally a difficult task due to the extreme thickness of the masonry at the bottom and the sides of a pool. The testudo ("turtle") was a semi-cylindrical metal container open at one end and closed at the other. The flat metal bottom of the container came in direct contact with the fire and the open end was open to the pool itself so that a constant water circulation was maintained.

11 Sergius Orata and the Invention of the Hypocaust

Consideration of the early development of heating systems, especially the hypocaust, sheds light upon the nature and origins of Roman baths. The subject has particular appeal because ancient writers, especially Pliny the Elder in *HN* 9.168, credit Sergius Orata, a Campanian businessman who was active around 90–80, as the "inventor" of the balnae pensiles (baths with hanging floors), hence the hypocaust system (cf. Val. Max. IX.1.1). Orata used this system not for heating baths but for commercial oyster beds and for upper-end country villas which he promoted. However, archaeology provides us with several instances of the use of the hypocaust system at least a century before this date. Among the early examples of the hypocaust are the previously mentioned late second-century Stabian Baths in Pompeii. Another possible contender for the privilege is Phase IV of the Greek Baths at the Sanctuary of Zeus at Olympia, dated ca. 100 by the excavators, although the archaeological evidence for the hypocaust is slim and elusive (Kunze and Schleif, 1944: 32–56, 51–6, 70–96; Mallwitz, 1972: 270–2; Yegül, 1992: 377–9). The dating of the Phase IV Olympia Baths, the last in a series of sanctuary baths used by athletes, is critical because if true, one could argue that this important technological breakthrough is Greek rather than Roman – an emotional issue. The controversy, however, seems to have settled in favor of the Italian side by the recent discovery of early second-century baths replete with a proper hypocaust and wall heating system in the Latin colony of Fregellae in Latium (Figure 1.4). Perhaps no specific example or moment should be identified as the origin or discovery of a broadly definable technology such as the hypocaust. Crude but effective versions of floor heating existed across the eastern Mediterranean as early as the first century, but it was the Romans who first exploited the system in heating their baths and shaping their bathing culture. One should also consider the input from thermo-mineral sites and spas, such as Baiae in the Gulf of Naples, which at an early stage offered technically viable models for heating baths by utilizing and controlling the naturally hot steam and gasses abundant in volcanic Campania. It is thus immaterial whether the hypocaust was "invented" or perfected by Sergius Orata as Pliny claimed, or by another practical and shrewd Campanian or Latian like him who lived several generations earlier.

12 Physical Evidence

The Stabian Baths of Pompeii provide us with the best preserved and most distinctive example of a Roman bath which illustrates the unbroken development from its fourth-century beginnings as a Greek *balaneion*, when the city was under Greek control, to its well-developed Late Republican phase ca. 80 (Figure 1.2) (Eschebach, 1979). The nucleus of the establishment started as a row of unheated chambers housing hip-baths supplied with water drawn from a well and heated by simple braziers. The fourth phase, late in the second century, represents a double bath for men and women arranged along a colonnaded palaestra. The bath block displays six barrel-vaulted rooms identified as the apodyterium, tepidarium and caldarium separated by a shared service yard. The baths were enlarged by adding a circular, domed laconicum in 80 which was later converted into a frigidarium by building a circular pool fully occupying the middle of the room. In the first century CE, the baths received direct running water from the aqueduct that supplied the town. The significance of the palaestra and exercise as a part of the bathing agenda was underlined by the addition of a very large swimming pool or natatio along its west side. The design characterized by a row of parallel, rectangular rooms represents an orderly movement from the unheated to heated rooms and back and defines the single-axis row-type plan loosely known as the Campanian or Pompeian type. Its usage, and technology based on back-to-back barrel-vaulted rooms, reflects structural and functional efficiency. This basic arrangement and its numerous variants remained the most logical and popular plan type in the west and the east well into the Byzantine era.

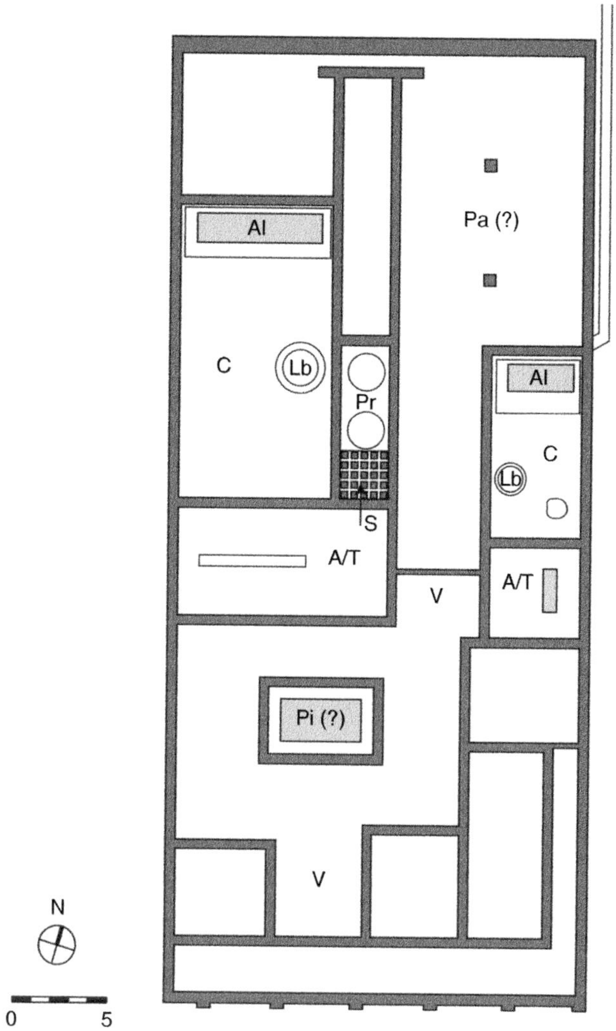

Figure 1.4 The Republican Baths in Fregellae, restored plan. A=apodyterium; Al=alveus (tub or basin); C=caldarium; Lb=labrum (basin); Pa=palaestra; Pi=piscine (pool); Pr=praefurnium (furnace); S=service yard; T=tepidarium; V=vestibule. Source: Drawing by F. Yegül, adapted from Tsiolis (2001: figs. 3 and 4).

Besides the Stabian Baths, Pompeii and its neighbor Herculaneum provide us with no less than four or five examples of the single-axis row-type. The Republican Baths (ca. 100) and Forum Baths (ca. 80) in Pompeii and the contemporary Forum Baths at Herculaneum are all double baths with palaestrae and shared services between men's and women's sections. The Suburban Baths at Pompeii, located outside the Porta Marina and dated to the beginning of the first century CE, are one of the first examples of a new concept of design: a single row of bathing rooms (without the somewhat reduced and cramped women's section) next to the palaestra; and a caldarium with a wide projecting apse and large windows confirming an appreciation of well lighted interiors and perhaps views of the shoreline. Another clear statement of the new taste in bath design is the Central Baths in Pompeii which were still unfinished at the time of the eruption in 79 CE. They are single baths characterized by an orderly arrangement of a row of vaulted rooms presenting a handsome window façade onto the palaestra.

Recent excavations at Fregellae, a Republican Latin colony in Latium (ca. 95 km south-east of Rome), uncovered one of the earliest and most technically advanced baths known to Roman archaeology (Figure 1.4) (Coarelli, Batocchioni and Romagnoli, 1995: 133–4; Tsiolis, 2001: 85–114; Coarelli, 2004: 73–6). Dating to the beginning of the second century, the Fregellae baths occupy a single large, rectangular city block on decumanus 1. The block is divided into men's and women's sections each composed of a vaulted apodyterium/tepidarium and a caldarium with a large, rectangular, heated pool and a free-standing, round labrum. Between the caldaria is a shared service yard with furnace-boiler combinations. A small, square room, probably a laconicum, was supported on hypocaust pillars and directly heated by a furnace. The barrel vaults of the apodyteria and caldaria appear to have been fashioned in terracotta tiles hanging from a series of transverse ribs, similar to the system described by Vitruvius (*De arch.* V.10.3). Below the cornice of the vault was a row of decorative terracotta telamones between cubby-holes for storing bathers' clothes as in the apodyterium/tepidarium of the Forum Baths in Pompeii (Tsiolis, 2006: 246–9, 2008a: 134–5; Battaglini, 2009: 354–6). Directly entered from the street, the men's and women's baths also shared a large vestibular hall, or atrium, with a shallow square pool (impluvium). The north end of the bath insula facing decumanus 2 is occupied by what appears to have been a service yard and a separate, larger, porticoed courtyard possibly for outdoor recreation. Shaped into a hard rectangular urban plot, the plan appears looser and more elementary than the tight spatial rows of the Pompeian type. However, the fully developed hypocaust system with additional wall heating by the use of simple terracotta tubes illustrates that the advanced heating technology that shaped Roman baths developed in Latium as much as it had in Campania. There is even an earlier and smaller bathing edifice on the southern end of the plot composed of several intercommunicating rooms under the existing foundations. One of these, with a large, heated pool (alveus), much like the arrangement of Phase 2, was probably the caldarium. Also like Phase 2, the main rooms appear to have been covered by terracotta vaults and floors decorated in opus signinum and beautiful opus sectile panels or polychrome terracotta with lozenges and shields. Datable to the middle of the third century, and with a simpler version of floor heating served directly by a furnace, Fregellae Phase 1 baths substantiate the leadership position of the region around Rome in the early development of Italian public baths (Tsiolis, 2006: 250–2, 2008a: 137–9; Vincenti, 2008: 408–16; Battaglini, 2009: 357–61).

13 Dissemination of the Row-Type Baths

Once established in central Italy the new single-axis row-type plan was introduced into northern Italy and the western provinces during the last decades of the Republic and the first decades of the Early Empire (Yegül, 1992: 66–90,

2009: 58–65). Successful adaptations of the plan emerged almost overnight in three Gallic sites: Glanum (Saint-Rémy-de-Provence), Derventum (Drevant) and Lugdunum Convenarum (Saint-Bertrand-de-Comminges), where the familiar double-zone Pompeian scheme – bath block next to the palaestra – is clearly recognized. Models for these baths must have come from intermediaries in northern Italian cities, such as the Late Republican baths at Velleia, where a small colonnaded courtyard with a central outdoor pool serves the combined functions of a frigidarium and palaestra. In Cemenelum (Nice), an important garrison town in the Maritime Alps, a cluster of three baths on either side of the town's decumanus all employ a remarkably uniform row design. Dating from the Severan age, these baths illustrate the astonishing longevity and popularity of this practical design type. Further evidence of the adaptability and change of the basic plan type comes from a group of military baths in the north-west provinces. In these examples, such as the Legionary Baths at Vindonissa (mid-first century CE) and the Legionary Baths at Isca (Caerleon, Wales, ca. 70–80 CE), the bathing halls, even the palaestrae, are arranged as a row on a strong, single axis presaging the symmetrical-axial bath planning culminating in the Imperial thermae of Rome. At Isca, there is a large, three-aisled, wooden-truss-roofed, basilical hall at the end of the main axis. This large space, and others like it in baths, is recognized as a basilica thermarum, a term that also appears in bath inscriptions. This probably served multiple functions including indoor athletic activities like a kind of roofed palaestra, convenient in cold climates.

Rome's position in the early development of public baths has been compromised by its success as a continuously occupied urban center. The city has yielded little evidence in physical remains of its many small neighborhood baths reported in the census records of the late first century. The paucity in archaeological remains, however, is partly offset by the identification of the inscribed names of at least three small baths represented on the Marble Plan (Forma Urbis), namely *[Ba]lneum Ampelidis* (fr. 47); *[Ba]lneum Caesaris* (fr. 43); and *Balne[um] Cotini* (fr. 48). A more informative example is the fully preserved *Bal Surae* (fr. 21), a bath where the row of parallel rooms arranged in a row alongside a colonnaded palaestra, with shops on one side, is a clear demonstration of the Pompeian plan type (Staccioli, 1961: 92–102; Yegül, 1979: 108–31). Although impossible to date (the Marble Plan dates from ca. 200 CE, but it is probably based on a late first-century CE version), small baths located in densely built business, commercial and residential neighborhoods, and catering to the tastes and economies of all classes, had become among the most important urban features of the growing metropolis. Whether Rome played an active role in the creation and development of public baths, or remained largely a recipient of ideas from outside, mainly from Campania and its own backyard Latium, there is little doubt that it provided a broad and fertile ground for the reception and assimilation of ideas. Starting with Augustus and continuing through the Imperial era, Rome made its incomparable and innovative mark on bath design and technology – as it did on much else on matters of art and architecture – through the creation of a remarkable series of Imperial thermae that raised the experience of bathing in public to a new dimension.

FURTHER READING

The primary and general studies on baths and bathing in the ancient world are Brödner (1992), Yegül (1992), Nielsen (1993), DeLaine and Johnston (1999) and Yegül (2010). Fuller bibliographies on baths and bathing in the ancient world can be found in DeLaine (1988), Manderscheid (2004) and Yegül (2010). Special studies on the heating and water supply systems of baths include Rook (1978), Nielsen (1985), Manderscheid (1988), Fagan (1996) and Yegül and Couch (2003). For Greek baths and gymnasia, see Ginouvès (1959), Delorme (1960) and Ginouvès (1962). For other special subjects on baths and bathing, see Di Capua (1939/40), Yegül (1979), DeLaine (1989), Dunbabin (1989), Ward (1992), Yegül (1996), Battaglini (2009) and Yegül (forthcoming).

CHAPTER 2

Public Entertainment Structures

Mantha Zarmakoupi

1 Introduction

In the Republic period there were two types of public entertainment, the religious public games (ludi publici), in the theater or the circus, and the secular private spectacles (munera). The public games accompanied the religious festivals that took place regularly every year in order to celebrate officially certain divinities. They consisted of three types: theatrical performances (ludi scaenici), animal hunts (venationes) and chariot racing (ludi circenses). Gladiatorial combats were first associated with funerals and were held irregularly, when a private benefactor took the initiative in his own name. Temple dedications, military triumphs and aristocratic funerals also provided opportunities for public entertainment. The buildings associated with these public entertainments were at first temporary wooden structures. The wealth achieved by some Italians in the Middle and Late Republic facilitated the construction of permanent structures for public entertainment as elites identified benefaction for them as a means to enhance their self-advancement (Veyne, 1990: 201–91). During the second century generals celebrated games for which they paid ex manubiae and by the middle of the second century supervising magistrates were appointed, who oversaw the production of the regular ludi (Livy XL.44.12; Veyne, 1990: 210). In this chapter we will examine the permanent structures (theaters, amphitheaters and circuses) that were developed during the Republican period in order to house the variety of public entertainment events. The definition of the form and the monumentalization of the structures used for public entertainment during this period point to the increasing importance that entertainment, and in particular funding for it, acquired in Roman society at this time.

A Companion to the Archaeology of the Roman Republic, First Edition. Edited by Jane DeRose Evans.

2 Theaters

Roman theater design, although Greek in origin as was Roman drama, differs in a number of ways from Greek and Hellenistic theater design (Gros, 1996–2001: 272–4; Sear, 2006: 1–10). The cavea (Plaut. *Amph.* 66; Cic. *Amic.* 24; *koilon* in Greek), the place where the audience sat, was semicircular in shape and was arranged around the orchestra, as in the Greek theater, but the stage, or pulpitum (Vitr. *De arch.* V.6.1, V.7.2; *logeion* in Greek), and the stage building, or scaenae frons (Vitr. *De arch.* V.6.1, V.7.1, VII.5.2), were joined to the cavea forming an enclosed space. As a result, the shape of the orchestra was semicircular as well. The action did not take place on the orchestra, or dancing floor, that was now reduced to half but on the raised stage in front of the elaborately ornamental façade of the scaenae frons. The orchestra was now part of the auditorium as movable seating was placed there for distinguished members of the audience. The lateral entrances to the theater, the itinera (Vitr. *De arch.* V.6.5; *parodoi* in Greek), were not open-air but vaulted passages under either end of the cavea, where the cavea met the scene building. Furthermore, the construction of the theater no longer depended on the choice of a good hillside as vaulted substructures were employed, where needed, to raise its cavea.

The distinctive design of the Roman theater as a unified building with a free-standing cavea and a prominent scaenae frons developed in the Republican period when wooden structures were constructed in the city of Rome due to a senatorial ban against permanent theater structures. There were permanent theaters in the Greek colonies in the south and in Sicily with which Romans would have been familiar, and these must have informed the development of Roman theater design. The theater at Syracuse is the earliest attested theater in Italy. Its architect was Damocopus of Myrilla (Eust. *Od.* 3.68). Its cavea probably consisted of the seventeen rectilinear rock-cut steps situated west of the later Roman theater (G. Gentili, 1952; Coarelli and Torelli, 1984: 247–55). The oldest surviving theater seems to be the one of Morgantina (*c.* 325) with a trapezoidal cavea, which was transformed in a semicircular shape in the later third century, laid against a slope that was contained by rectilinear retaining walls (Sjöquist, 1962: 137–8; Stillwell, 1964, 1967: 245–6; Allen, 1970: 363–4; Bell, 1988; Mitens, 1988: 105–8; Sear, 2006: 189).

The form of the stage building and less so of the orchestra and the cavea was important in the development of Roman theater design. Nearly all the late fourth-century theaters of Sicily and Magna Graecia featured a raised stage, which had a façade behind it, flanked by two projecting wings, the *paraskenia*. This form appeared in stone-built stage houses of theaters dating to the end of the fourth century in mainland Greece, such as the Lykourgan Theater in Athens, and they might have been based on earlier experiments in wood (F. Winter, 2006: 102). The projecting wings were eventually eliminated in the theaters of mainland Greece, which came to feature a straight façade with openings into which decorative panels could be placed (the *thyromata* type stage). Yet projecting wings continued to be used in southern Italy, possibly due to their use as the scenic conventions related to specifically western-type theatrical practice, such as the farces of the Greek *phlyakes* (Beacham, 2007: 213–15).

All but one of the theaters of Magna Graecia and Sicily were built against a natural hillside. The theater of Metapontum (end of fourth century; Mertens and De Siena,

1982; Mitens, 1988: 140–4; Sear, 2006: 147–8) is an early example of a semicircular theater whose cavea (diameter 77.4 m) was built on a flat site and rests on an earth fill contained in polygonal walls that form a 10-sided polygon. The outer façade of the polygonal walls featured an order of Doric columns carrying a triglyph frieze, cornice and low attic, which may be seen as a precedent of the Roman type of outer theater façade (Sear, 2006: 48). The design of the theater may have been informed by Greek models, just as the tombs of the region were subject to Macedonian influence; however, few theaters in mainland Greece stand on a flat site (e.g. the theater at Elis) and Metapontum is near Tarentum, which Romans captured in the early third century.

The design of the stage buildings of the Sicilian theaters at Segesta (Bieber, 1961: 168–70; Mitens, 1988: 109–12; Sear, 2006: 190) and Tyndaris (Bieber, 1961: 168–70; Mitens, 1988: 125–7; Sear, 2006: 194) provides a link for the transition from the Hellenistic to the Roman type of theater in Italy. These theaters have oblique *paraskenia*; the back wall of their stage featured three openings; and it was decorated with two stories of half-columns (Sear, 2006: 48–9). These stylistic features of the stage buildings bear strong resemblance to the stage buildings of several stone theaters, some as part of sanctuaries, that were built in Campania and Samnium during the second century (Lauter, 1976), for example at Sarno (D'Andria, 1989; Mitens, 1988: 170–1; Sear, 2006: 138) and Bovianum Vetus (Pietrabbondante; Strazzulla, 1973; Mitens, 1988: 162–5; Sear, 2006: 153), and the Large Theater at Pompeii (Figure 2.1; Mau, 1906; Maiuri, 1951; de Vos and de Vos, 1982: 64–7; L. Richardson, 1988: 75–80, 216–18; Pesando and Guidobaldi, 2006: 62–3; Sear, 2006: 130–2).

The late second-century theater at Sarno had a raised stage, rectilinear scaenae frons with five doorways and oblique *paraskenia* framing the proscaenium, as in the theater at Segesta (Lauter, 1976: 416–17). Its orchestra was surrounded by three rows of seats made by grey Nucera tufa, each row terminating in figured arm-rests, the lower two of which are in the shape of a griffin's or lion's foot and the last one in the shape of a winged sphinx. This decoration recalls the theater at Iaitas in western Sicily (end of fourth century; Isler, 1981), where a griffin's foot adorned the end of the second and third rows of *prohedria*, the seats of honor directly in front of and around the orchestra. It is possible that the theater of Iaitas was modeled after the theater of Syracuse (R. Wilson, 1990: 69–71).

The Large Theater at Pompeii was laid out on a natural slope and in its initial second-century phase featured a semicircular cavea (diameter 49 m) that consisted first of limestone and then of wooden seating. It had a horseshoe-shaped orchestra, a high stage that was detached from the cavea and a rectilinear scaenae frons that was flanked by oblique *paraskenia*, similar to the ones of the theater at Segesta. The upper tiers of the seats stood at the same level as the adjacent Temple of Athena in the Triangular Forum, whose colonnade probably operated as a monumental entrance to the theater, especially during ceremonial processions (pompae) that were associated with religious festivals. At a later stage, after the establishment of the Sullan colony, its stage was lowered and pairs of columns flanked the doors of the scaenae frons while the lateral entrances to the orchestra were vaulted over and the seating extended over them to form a cavea that had a horseshoe shape.

The theater at Pietrabbondante (*c.* 100) is part of a temple-theater complex, a building type that developed in the Late Republic in central Italy, Campania and Samnium and

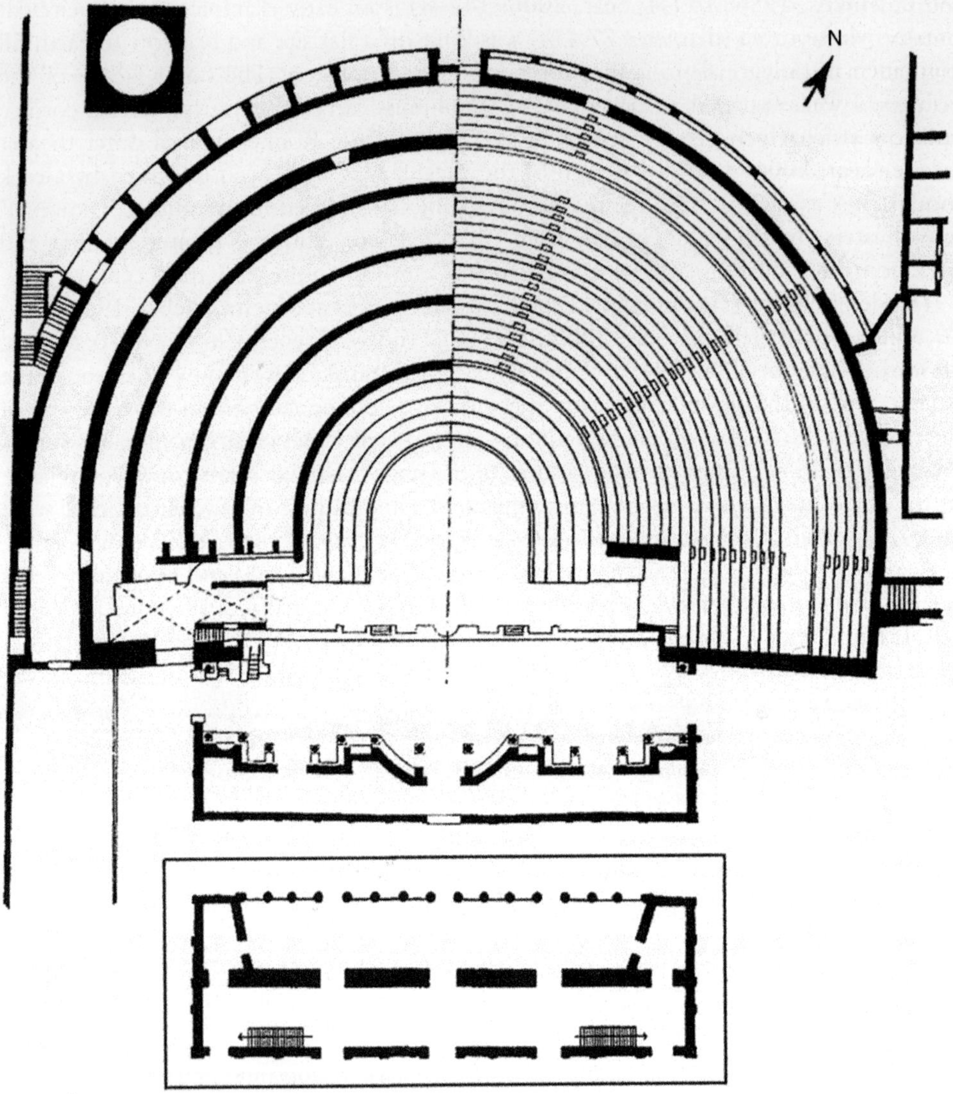

Figure 2.1 Plan of the Large Theater at Pompeii. Source: Sear (2006: plan 22). Used by permission of the author.

had no Greek antecedents. The influx of wealth from overseas conquests in combination with the availability of cheaper and faster building techniques facilitated the construction of sanctuaries built on stepped terraces in the Hellenistic manner. Although the association between religion and drama was eminent in the Greek world, this type of building was new, and its significance as a type will be addressed in the chapter on theater-temple sanctuaries (see Chapters 15 and 22). For our discussion the theaters of these complexes provide unique examples for the early permanent theaters in Italy; other notable examples include the theaters in the sanctuaries at Gabii, Tivoli and Praeneste (Palestrina).

The sanctuary at Pietrabbondante features a theater (dating to the end of the second century), whose cavea (diameter 54 m) is built in front of the large temple. Although originally a Samnite sanctuary, there are Hellenistic influences in its architecture. The rectangular *parodoi*, which recall the ones of the third-century theater at Syracuse, were covered with a flat wooden roof and were supported by a central column. The roofed *parodos* was an essential step in the development of a unified design of cavea and scaenae frons in Roman theater design. The entire embankment of the theater at Pietrabbondante is supported by polygonal retaining walls (*analemmata*), built of irregular limestone blocks, that on both sides of the orchestra end with kneeling telamones carved in stone.

The Small Theatre of Pompeii features similar kneeling telamones at the end of its retaining walls. A praecinctio and low wall, which terminate in winged lions' feet, divide the *prohedria* from the main part of the auditorium (Johannowsky, 1976: 272; de Vos and de Vos, 1982: 69–71; L. Richardson, 1988: 131–4; Zanker, 1999: 65–8; Pesando and Guidobaldi, 2006: 64–6; Sear, 2006: 132; Parslow, 2007: 213–14). It was built in the early years of the Sullan colony by the duumviri C. Quinctius Valgus and M. Porcius and was identified as a roofed theater ("theatrum tectum") in its dedicatory inscription (*CIL* X.844 = *ILLRP* 646). Its design was an architectural innovation, modeled after the *bouleuteria* of Hellenistic cities, in that the semicircular rows of seats were inserted in a rectangular form (27.75–28.60 by 34.80 m). The thicker east and west walls supported trusses that carried a hipped roof that probably extended as far as the scene wall (Murolo, 1959; Meinel, 1979: 36–44; Izenour, 1992: 91–101). It is an early example of a free-standing theater, with vaulted *parodoi* and special box seats (tribunalia) for organizers of the events arranged above them.

The areas in and around Campania were in close contact with the Greek colonies and it is understandable that only a few Late Republican theaters in Latium seem to conform to the model that Vitruvius describes for the stage of the theater of the Roman type, which has puzzled researchers to the point of being disregarded (Frézouls, 1982: 368; D. Small, 1983). Vitruvius uses four triangles circumscribed in the circle of the orchestra to define the line of the scaenae frons and the front of the stage (the bases of two of them), the staircases of the cavea (seven points of the triangles) and the layout of the scaenae frons (the other five points) (*De arch.* V.6.1; Sear, 1990: 250). The stage building of the theaters at Tusculum (first half of first century; Canina, 1841: 118, pl. XI; Ashby, 1910: 353–4; Sear, 2006: 141–2), and Alba Fucens (early first century; Mertens and De Ruyt, 1969, vol. 1: 76–80; Sear, 2006: 149–50) have features that are consistent with this method, and it is possible that Vitruvius's description reflects Late Republican building practices in central Italy (Sear, 1990, 2006: 27–9).

In addition to the evidence provided by the architectural remains of the second-century south Italian theaters, scenes of the Greek *phlyakes*, or farces, on a large number of fourth-century vases from southern Italy and Sicily provide invaluable information for the stage building (Trendall, 1967, 1976; Bieber, 1961: 129–46). These so-called *phlyax*-type vases show actors on a raised platform wooden stage, varying in height, supported by Doric or Ionic columns and with a decorated rear wall made of panels of wood or canvas, and present an exchange of actions taking place between the actual stage and open ground in front of it. They have been associated with performances of Greek Middle or Old Comedy (Taplin, 1993; Hughes, 1996). It has been argued that the stages and the exchange of actions shown on them agree with what the texts of early

Roman comedy suggest the earliest Roman stages may have looked like (Beacham, 1991: 59–65, 2007: 213–17; see also Beare, 1964: 176–83). It is possible that the farces of the Greek *phlyakes* together with the Oscan Atellanae (the stages for the Atellanae are thought to be identical with those for the *phlyakes*), which were popular in south Italy and Rome, informed stage design in this period (Bieber, 1961: 167–89; Frézouls, 1982: 349–53; see also Rawson, 1985a). The representation of scaenae frontes in Early Imperial wall paintings has been considered as providing a missing link between the early temporary stages and permanent theater structures (Beacham, 1991: 69–84). However, it is unlikely that first-century wall paintings would still reflect the temporary stages and not contemporary permanent theaters in southern Italy (Marshall, 2006: 32–3).

Nearly all of the second-century theaters had an auditorium that was formed from the natural topography and in most cases an artificial embankment was used especially at the ends of the cavea, for instance in the theater at Pietrabbondante. Towards the end of the second century designers attempted to erect free-standing theaters on flat sites, as seen in the theater at Cales where radial barrel-vaulted substructures were used to support its cavea (diameter 75.4 m; late second century; Johannowsky, 1963; Lauter, 1976: 272; Sear, 2006: 121). In the theater at Teanum Sidicinum (part of a temple-complex) two superimposed tiers of barrel-vaults supported the cavea (diameter 78 m, late second century; Johannowsky, 1963: 152–63, 1976: 267–88; Sear, 2006: 139). By the middle of the first century it became the norm to use two superimposed sets of barrel-vaults in order to support the seating (Sear, 2006: 50–2).

In Rome temporary wooden theaters were constructed every year for the ludi scaenici as theater spectacles were alleged to be valueless and morally corrupting; hence there was a prohibition against permanent structures for them until 55 (Val. Max. 2.4.2; Tac. *Ann.* 14.20; Vell. Pat. 1.15.3; App. *B Civ.* 1.28; Gruen, 1992: 183–222, esp. 205–10; Manuwald, 2011: 55–63). Wooden theaters were built as early as 179, when M. Aemilius Lepidus built a wooden cavea and stage near the Temple of Apollo (Livy XL.51.3), approximately where the later Theater of Marcellus was planned by Caesar and completed by Augustus. Vitruvius informs us that many wooden theaters (theatra lignea) were put up in Rome every year (Vitr. *De arch.* V.5.7–8) and the planning and construction of these structures was surely one of the most significant activities in the city, an event in its own right. These non-permanent structures were surely opportunities for design and technical experiments, which unfortunately we are not able to evaluate.

A few examples from the literary record are worth mentioning here for the information they provide about the opulence of these temporary theaters and the mastery of craftsmanship that the Romans had achieved. In 99, C. Claudius Pulcher created the first scaenae frons decorated with painted panels, which used illusionistic techniques to depict realistic architecture so that even "crows were seen trying to alight on the roof tiles represented on the scenery" (Plin. *HN* 35.23, trans. H. Rackham, 1952, Loeb; cf. Val. Max. 2.4.6). During the next decades the shows grew more ostentatious when, according to Valerius Maximus, C. Antonius decorated the whole stage in silver, Petreius one in gold and Q. Lutatius Catulus one in ivory (Val. Max. 2.4.6). In addition Catulus is credited with introducing Rome to the luxury of an awning or linen roof that was placed over the auditorium "in imitation of Campanian luxury" (Val. Max. 2.4.6; Rawson, 1985a: 104, n. 40). In 58, M. Aemilius Scaurus built a three-storied scaenae frons that was supported by about 360 columns, the lower tier of which were *c.* 12 m high; the scaenae frons featured a great number of bronze statues. The lower level of the façade was made of Lucullean marble, the middle one was faced with glass mosaics, and the uppermost with gilded boards (Plin. *HN* 36.114–15; *LTUR* 5: 38–9 [Pollard]). In 52, C. Scribonius Curio built two large wooden theaters back-to-back that could revolve on pivots and, when conjoined, create a single double theater, an "amphi"-theater for the presentation of gladiatorial events (Plin. *HN* 36.117; Golvin, 1988: 31–2, pl. IV; on the word "amphitheater," see Gros, 1996–2001: 317).

Rome's first permanent theater, the Theater of Pompey in the Campus Martius, was constructed in 55. Pompey circumvented possible senatorial criticism by building a temple to Venus Victrix, to whom Pompey credited his victories in the east, at the top of the cavea, thus inserting it in the tradition of Republican theater complexes. It was the largest Roman theater ever built and was sometimes referred to as "theatrum" by contemporary sources (Cic. *Att.* 4.1.6; Hor. *Carm.* 1.20.3, App. *B Civ.* 5.15). According to Plutarch, Pompey conceived the idea of building a theater in 63 while at Mytilene, where he was impressed by the theater and he ordered plans to be made of it in order to "build one like it in Rome, only larger and more splendid" (Plut. *Pomp.* 42.4). We only know that the orchestra of the theater of Mytilene had a diameter of 25 m, which suggests that it was large although not larger than Pompey's, and it was built against a hillside. It is certain that it could not have been a unified-type theater of the Roman manner and even if it was used as the formal model, Pompey's theater established a new type of civic building with its use of concrete vaulting to support its free-standing cavea (Rakob, 1976: 366–76).

Pompey's theater was subsequently built over and some of its substructures are preserved in the cellars and walls of later buildings. We know its plan from the Severan Marble Plan and recent excavations have adjusted our knowledge of the site (see Chapter 29; Galliardo and Packer, 2006; Beacham, 2007: 218–23; 3D reconstructions at www.pompey.cch.kcl.ac.uk; http://3dvisa.cch.kcl.ac.uk/project81.html). The diameter of the cavea was approximately 158 m and the diameter of the orchestra measured about 44 m. In addition to the theater itself, a quadriporticus 135 m wide and 180 m long (featuring rare works of art, including paintings of Polygnotos, and an enclosed grove) formed its stage backdrop. The long colonnaded enclosure featured a central avenue, which was defined by two lines of thickly planted plane trees (Prop. II.32.20) which led from the Curia, where Caesar was assassinated in 44, to the theater complex. This was the first public park of Rome (Gleason, 1994; Kuttner, 1999: 343–73; *LTUR* 4: 148–9 [Gros]). Pompey's theater complex was a unique structure and the theater itself was probably the largest ever built, at least in the city of Rome.

3 Amphitheaters

Gladiatorial games (munera) seem to have developed in connection with aristocratic funeral games, although there is some confusion in regards to their origin, even among Romans themselves. There are two hypotheses as to their origin. On the one hand, accounts of Livy (IX.40.17), Strabo (V.4.13) and Silius Italicus (*Pun.* 11.51–4) about the celebration of a victory of the Romans and their Campanian allies over the Samnites in 308 suggest that they originated in south Italy. This hypothesis is corroborated by representations of single combats in Osco-Samnite tombs and on south Italian vase paintings (*c.* 350–300; Nicolet, 1962). On the other hand, some Roman sources advance the theory of an Etruscan origin for such games (Nicolaus of Damascus, cited by Athenaeus *Deipnosophistae* 4.153f–154a; and in a fragment attributed to Suetonius: Reifferscheid, 1860: 320), and wall paintings from Etruscan tombs of the sixth century have been taken as evidence for this (e.g. Poulsen, 1922: 12–14). The literary and archaeological evidence is not conclusive and the question of origins remains open (Welch, 2007: 11–18).

The first recorded gladiatorial show in Rome was in 264, given by Decimus Junius Brutus in the Forum Boarium in honor of his father, in which three pairs of gladiators took part (Livy *Per.* 16; Val. Max. II.4.7). The combatants were war captives (Serv. III.67.10–14), a fact which points to the early association of the games not only with funerals but also with warfare (Ville, 1969). Over the second century the scale and frequency of gladiatorial shows gradually increased (cf. Livy XL.28). Gladiatorial combats were linked to the Roman army and the rise of gladiatorial games as a regular feature of Roman life coincided with the greatest period of Roman military expansion (Hopkins, 1983: 27–31; Welch, 2007: 22–9). Limitations were placed upon games with the reign of Augustus and the imposition of the Pax Romana, when funding them became a means of self-promotion in the competitive social arena of the Imperial period.

The Forum Romanum was the prime setting of gladiatorial combat in Rome up to the building of an amphitheater by Statilius Taurus in 29 (Ville, 1981: 42–6; Golvin, 1988: 298–313; Welch, 1991: 274–7). Most, if not all, of the gladiatorial displays before the Late Republic were funerary; for instance, the first attested gladiatorial game at the Forum Romanum was associated with the funeral of M. Aemilius Lepidus in 216 (Livy XXIII.30.15). For the setting of the gladiatorial games in the Forum, Golvin proposed straight runs of seats against the façade of the Basilica Sempronia (yet only after 170) on one side and the Basilica Aemila et Fulvia (179) on the other, forming a splayed layout that could perhaps be joined with circular ends (Golvin, 1988: 18–21; Golvin and Landes, 1990: 58–63). His reconstruction, however, is based on the plan of the Forum at the time of Caesar and Augustus (for which, see Figure 29.2). Thus, the neat shape that Golvin proposes makes little sense in the wider and more open irregular area of the Forum in its pre-Caesarian phase (Gros, 1996–2001: 318; Welch, 2007: 47–8). Following Golvin's suggestion that the oval shape of the amphitheater structure may be related to the oblong space of the Forum Romanum, Welch proposed that the arrangement was elliptical/oval formed by straight segments, which constituted the individual wooden seating structures. Her reconstruction takes into consideration the technical aspects of the scaffolding supporting the wooden seats and makes use of all the available area of the Forum at this time. The oval arrangement may be seen as a design solution to the formal conflict of the irregular shape of the Forum area (Welch, 1994, 2007: 43–55).

Welch's oval reconstruction of the wooden seating arrangement for the mid-first century embraces neatly the hypogea system that was constructed sometime in the first century in the Forum Romanum (Welch, 2007: 51–2, fig. 27 at 56). The network of subterranean galleries, or hypogea, consisting of one central corridor at right angle with four smaller ones, functioned probably as a "backstage" for the combat spectacles taking place in the Forum (Carettoni, 1956). The 12 vertical "elevator" shafts must have been used to bring up animals that participated in the venatio and the damnatio ad bestias, as similar shafts were used later in the Colosseum and the Early Imperial amphitheater at Puteoli. The remains of the wooden armature for a winch system were visible at the bottom of the shafts at the time of the excavation in 1904. The underground galleries and shafts might have been created by Caesar in order to accommodate the hunting and gladiatorial displays that he organized in relation to his triumph of 46 in the Forum, which is the first attested instance of animals being used in spectacles in the Forum (Suet. *Iul.* 37; Cass. Dio 43.19–24; *LTUR* 2: 325–36 [Purcell]; Welch, 2007: 38–42; for a Sullan date, see Wiseman, 1990).

Republican permanent amphitheaters had the essential features of what later became the established design but they were simpler structures than their Imperial counterparts (Golvin, 1988: 42–4; see also Wilson Jones, 1993). Their form was defined by the seating arrangement that surrounded and shaped an approximately elliptical arena (from the Latin word harena, "sand"). As a structure the form of the amphitheater was much more demanding than that of the theater, because support was needed for the all-around seating. Designers employed the natural topography and also crafted it so as to accommodate the construction of the amphitheater. Republican amphitheaters were partially excavated out of the existing terrain combined with extensive vaulting substructures at the upper zone of the cavea (the summa cavea). They did not have basement facilities and had a relatively plain façade featuring retaining walls and vaults that served the purpose of containing and supporting the cavea.

The form of the amphitheater had no Greek architectural precursors. It was – until recently – commonly believed that it was a Campanian invention, as the earliest securely datable amphitheater is that of Pompeii (e.g. Böethius and Ward-Perkins, 1970: 170–1; Rakob, 1976: 370; Golvin, 1988: 23–4, 42–4) and the fourth-century depictions of gladiatorial combat in south Italian tomb paintings were taken as supporting evidence for this view. This, however, does not explain how the oval architectural form suddenly appeared in Campania at the end of the second century whereas the gladiatorial games were held as early as the third century (see Figures 2.2 and 2.3). There must have been a first step in the creation of the form that the archaeological record has not preserved.

It has been recently suggested that the architectural form of the amphitheater was developed in Rome during the Middle Republican period (Golvin, 1988: 56–8; Welch, 1994: 274–6, 2007: 30–71, 91–4). The term "maenianum," which in its plural form (maeniana) became the standard word for the grandstand of an amphitheater by the Late Republic, originally designated a kind of cantilevered wooden balcony projecting from buildings around the Forum Romanum for the purpose of viewing spectacula (Vitr. *De arch.* V.1.1–2; for the origin of the term, see Golvin, 1988: 19; Welch, 2007: 32–5). The Agora of the Italians on Delos (*c.* 120; Lapalus, 1939; Trümper, 2008), which could have been the setting for the gladiatorial shows, featured an upper floor that may have served such a purpose (Rauh, 1992: esp. 317, 1993: 81–3, 289–338; see also Hatzfeld and Roussel, 1910: 403–5, no. 54; contra, Bruneau, 1995: 45–54; see also Mastino, 2008: 237–8; Trümper, 2008: 7–8, 96–8).

Another architectural term possibly originating in Rome is "spectacula," referring in general to the temporary wooden seating arrangement in the Forum (Vitr. *De arch.* V.1.1–2; Festus 22). This term also appears in the dedicatory inscription of the

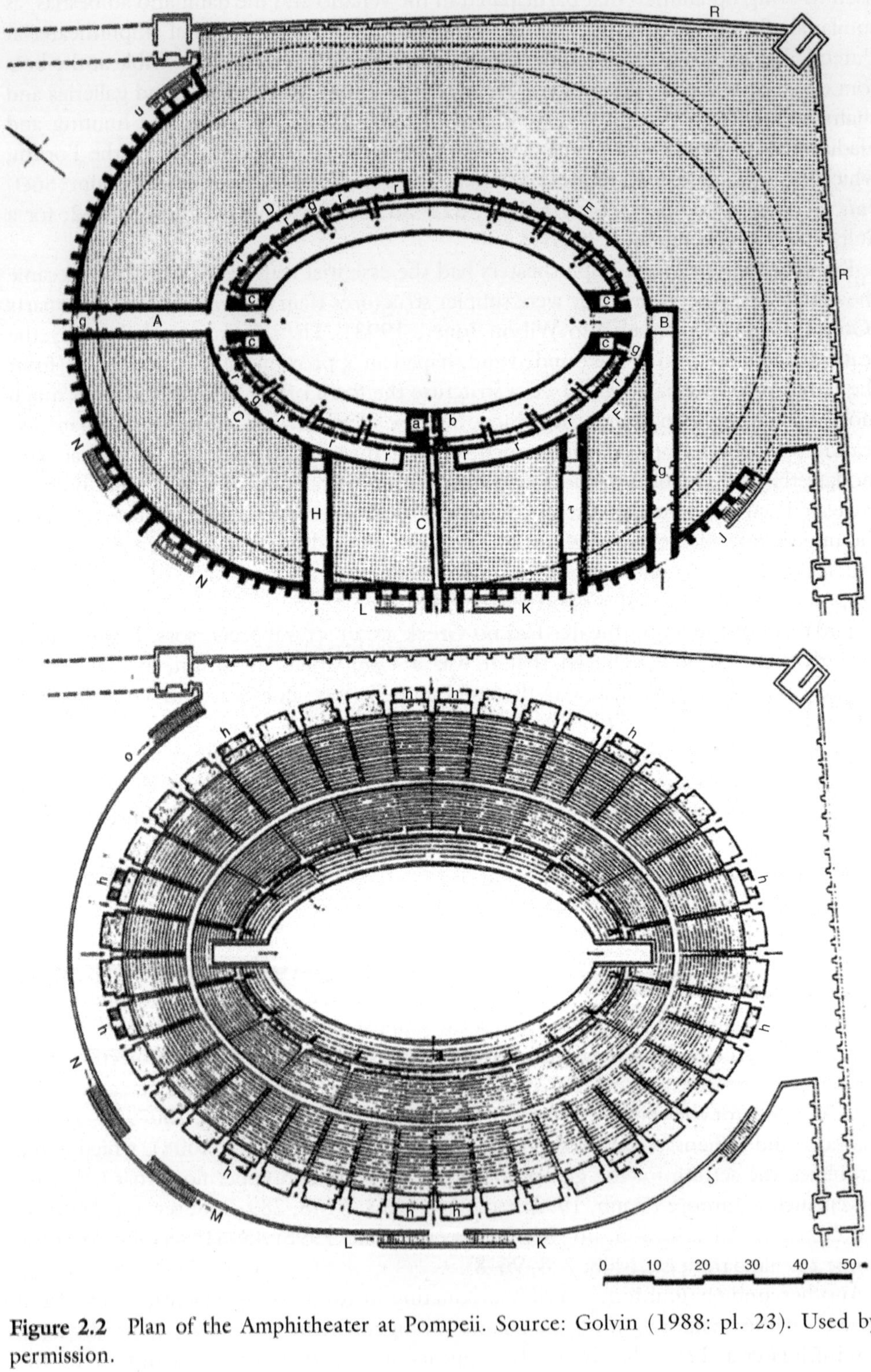

Figure 2.2 Plan of the Amphitheater at Pompeii. Source: Golvin (1988: pl. 23). Used by permission.

Figure 2.3 View of the arena and seating of the Amphitheater at Pompeii. Source: Photo by M. Zarmakoupi.

amphitheater in Pompeii. Welch has therefore suggested that the formal prototypes of the amphitheater would have been the wooden grandstands that were used in the Forum, on the basis of her reconstruction of the seating arrangement discussed above (Welch, 1994, 2007: 43–55, 91–4). This is an attractive solution to the question about the origin and the form of the amphitheater that takes into account the importance of the temporary wooden structures in Rome during the Republican period (Gros, 1996–2001: 318–23; Anderson, 1997: 279–80).

The temporary wooden structures may well have provided the model for the form of the amphitheater, but it was certainly the structural know-how of regional theaters as well as comitia that underpinned the realization of this form as a permanent structure. The comitia and theater structures often included a conical depression and took advantage of the surrounding embankment, as did the comitium at Paestum (Greco and Théodorescu, 1987: 37–9, figs. 37–42) and the theater at Cales (Johannowsky, 1961; Mitens, 1988: 154–5; Sear, 2006: 121, plan 6, fig. 10, pls. 1–2), and the amphitheater in Pompeii, as well as other early amphitheaters (Taylor, 2008: 444). In addition, the architecture of some Republican theaters provides some formal parallels, such as the outer façade of the theater at Bononia (Figure 2.4, end of second/beginning of first century; Sear, 2006: 171–2, plan 81 and fig. 7; Ortalli, 1986: figs. 8, 10, 12) that featured blind arches in opus incertum like the amphitheater at Pompeii.

Figure 2.4 Theater at Bononia, reconstructed outer theater façade. Source: Ortalli (1986: fig. 12).

There are at least 20 known Republican stone amphitheaters (Golvin, 1988: 24–5, 32–44; Welch, 2007: 72–101, 189–263). Most of them were built in Campania (e.g. Cumae, Liternum, Cales, Abellinum), where the technical advancements during the second and first centuries enabled the construction of this structurally demanding edifice. However, even in Campania architects used largely the natural terrain and only in a few examples, at Puteoli and Telesia, were radial vaults even partially used. There are also examples from Etruria (Sutrium, Ferentium) and outside Italy in the colony of Caesar at Corinth, at the probable municipium at Carmo (Spain) as well as at the capital of the province of Syria, Antioch. Only the amphitheater at Pompeii is securely datable (*c.* 75–70). The amphitheaters at Capua and Puteoli seem to be the earliest ones, dating to the end of second century. They are, however, badly preserved and their date has been recently questioned on the basis of the dating of the opus incertum and opus quasi-reticulatum facing (the latter a modern term signifying a facing technique that seems a transition phase between the opus incertum and opus reticulatum: Adam, 1984: 142–7; Welch, 2007: 84–8).

The Republican amphitheater of Capua was located just outside the city walls, like most amphitheaters (Golvin, 1988: 408–12; Spina, Arciero and Sampaolo, 1997; Welch, 2007: n. 3 at 287), at the north-west part of Capua with a north–south orientation. It was excavated out of the existing terrain (arena, *ima* and *media cavea* (lower and middle part of the cavea)). The upper part of the summa cavea was supported by vaults, whose foundations have partially survived, and its walls were faced in opus incertum (contra, Welch, 2007: 200). It was about 100 m long on its longitudinal axis. In the Flavian period a new amphitheater was built just to the north-east of the Republican structure, with the same orientation to it, but it is unclear whether they were used simultaneously (Johannowsky, 1961: 66–7; De Caro and Greco, 1981: 215–16; Golvin, 1988: 25, pl. 7.1; Welch, 2007: 198–202). Gladiatorial training schools (ludi) are attested since the late second century at Capua, where elite Romans drew gladiator trainers for Rome: for instance, Rutilius Rufus hired from the gladiatorial school of C. Aurelius Scaurus in 105 (Val. Max. II.3.2). These ludi were owned by Roman rather than Capuan aristocratic families (Frederiksen, 1984: 307–11; Welch, 2007: 91). For instance, the ludus where Spartacus's slave revolt originated in 74 was owned by a certain Cn. Lentulus Vatia (Livy *Per.* 95; Plut. *Crass.* 8.1; Oros. V.24.1; App. *B Civ.* 1.14), probably of noble status (Bailey, 1960: 258–9), and Julius Caesar was the owner of a ludus in Capua in 49 (Cic. *Att.* 7.14, 8.21).

The Republican amphitheater of Puteoli was located about 300 m north of the forum of Puteoli, roughly on a north–south axis. Its arena was dug out of the ground and partially built up against a hillside (north-east part of the cavea), while its summa cavea was supported by vaults and featured an annular vaulted walkway (7.5 m wide). The radial walls supporting the vaults were placed 3.60 m apart and faced in opus quasi-reticulatum. Its overall size is about 130 by 95 m (arena 69 by 35 m) (Beloch, 1890: 90–1; Spinazzola, 1915; Sommella, 1978: 54–7; Golvin, 1988: 38, pl. VII.2, XXII.2; Welch, 2007: 221–5). Restoration work took place probably in the Augustan period (due to the opus reticulatum facing; see Sommella, 1978: 54). A second amphitheater was built in the Flavian period similar to Capua's, south-west of the Republican structures and perpendicular to the first amphitheater in Puteoli (Somella, 1978: 58–9). The representation of both amphitheaters on glass flasks from late Antiquity is thought to suggest that they were both in use simultaneously (Maiuri, 1934: 34; Ostrow, 1981).

The amphitheater at Pompeii is the best-preserved Republican amphitheater (Figures 2.2 and 2.3; Overbeck and Mau, 1884: 176–86; Mau, 1908: 212–26; Girosi, 1936; de Vos and de Vos, 1982: 150–4; Richardson, 1988: 134–8; Zanker, 1999: 68–72; Golvin, 1988: 33–7, pl. 22–3; Pesando and Guidobaldi, 2006: 72–4; Welch, 2007: 74–9, 192–8). It was built soon after the establishment of the Sullan colony at Pompeii in 80 by Quinctius Valgus and Marcus Porcius. The two sponsors were the same duumviri who built the Small Theater (see above). The dedicatory inscription indicates that they built the amphitheater, which they refer to as spectacula, with their own funds ("de sua pecunia"; *CIL* X.852 = *ILLRP* 645 = *ILS* 5627). The amphitheater was built at the east corner of the city, where previously middle-class houses were located, against the north-east and south-east city walls (Nappo, 1997). The arena, *ima* and *media cavea* were dug into the ground, while the summa cavea rested on the mound of the excavated earth. This mound was retained by the surrounding wall-façade of the building (*c.* 6 m tall) that was reinforced by blind arcades and which merged with the city walls at the north-east and south-east.

Two series of double-ramped staircases located at the west part of the building further supported the exterior wall and the stability of the structure from this side and provided access to the seating of the *media* and summa cavea through an annular walkway at the top of the latter. Two single staircases at the point where the semicircular walkway joined the city walls gave access at the north-east and the south-east part of the *media* and summa cavea. The *ima cavea*, which was physically separated from the rest of the cavea by a parapet (a balteus 0.80 m high), was accessed by two vaulted corridors (2.98 m wide) located at the west side of the amphitheater and perpendicular to its longitudinal axis. These gave access to two annular cryptae, from which the *ima cavea* was reached through stairs.

The seats of the *ima cavea* were larger by 0.30 m (total 0.88 m wide) from the seats of the *media* and summa cavea in order to accommodate the bisellia (portable wooden seats), used by the most important spectators, such as the magistrates and the sponsors of the games. The arena (66.8 by 34.50 m), which was delimited by a podium (2.18 m high), was accessible by two vaulted galleries (4.26 m wide) from the north-west and south-east extremities of its longitudinal axis (the so-called "porta triumphalis" and "porta libitinensis" respectively). The north-west gallery was aligned with the longitudinal axis whereas the south-east one bent at a right angle and was accessible from the south side of the amphitheater near the city wall. Both sloped down 4.50 m into the arena and were used for the entry and the exit of the combatants. Pairs of rooms (erroneously called carceres) accessible from the galleries and located just before entering the arena were used as service rooms. A smaller gallery, which was perpendicular to the middle of the longitudinal axis, gave access to the arena through a small opening from the south-west and also had a service room opening onto it. The overall dimensions of the structure are 134.80 by 102.50 m and it is estimated that it had a seating capacity of about 22,000 spectators.

The construction of the amphitheater of Pompeii soon after the foundation of the Sullan colony points to the importance of gladiatorial combats for the colonists, to whom in fact the building was donated ("coloneis ... in perpetuom") and in particular with the army veterans among them (Zanker, 1999: 70–2; contra, Zevi, 1996: 131–2). It should be noted that the earliest amphitheaters appeared in towns that had close ties with Rome, in particular coloniae settled by army veterans as well as old Latin and maritime colonies and municipia. Examples include the amphitheaters of Cales and Sutrium (Latin colonies of the late fourth century), Liternum and Puteoli (maritime colonies founded in 194), and Cumae (awarded the status of municipium sine suffragio in 338). While army veterans seem to have brought the military habits of the gladiatorial combat in the colonies that they settled, the non-veteran inhabitants of the colonies and municipia found a way to express their allegiance to Rome by appropriating its habits (Welch, 2007: 88–91).

4 Circuses

The circus was the structure that was created during the Republican period in order to accommodate athletic and equestrian events, primarily chariot races with four-horse (quadrigae) or two-horse (bigae) chariots. Although the structure was principally used for chariot races in the Imperial period, it also served athletic and equestrian events as well as venationes in the Regal and Early Republican period on the basis of the information that we have for the Circus Maximus (Thuillier, 1982; Crowther, 1983; Humphrey, 1986: 71–2). Romans drew on the racing practices of the Etruscans, which in turn were informed by the Greek practice of chariot racing (Bronson, 1965). At the beginning the organization and financing of the events followed the Greek pattern, where aristocrats entered the competition in their own name and sometimes drove their own teams. This model was abandoned after the middle of the fifth century (Rawson, 1981). Subsequently, the State financially supported professional horse-breeders who provided horses for the races as attested by Livy during the period of the Second Punic War (Livy XXIX.18.10); this is when professional charioteers entered the competition. This change may be associated with the introduction of the colors or factiones under which the chariots competed. Tertullian tells us that there were originally only Red and White teams; it is possible that the Blues and Greens were only added right at the end of the Republic (Tert. *De spect.* 9.5; Rawson, 1981: 6; Cameron, 1976: 59).

The circus was different from its Greek equivalent, the hippodrome, in that it became a permanent structure whose organization was conceived with an emphasis on the viewing of the spectacle (Thuillier, 1985; Gros, 1996–2001: 346). This emphasis is related to Etruscan practices, for which our evidence is primarily based on representations in tomb paintings, reliefs, terracotta plaques and vases. In these images, the spectators seated on wooden stands are the principal subjects, for example in the Tomba delle Bighe at Tarquinia (*c.* 500), while the charioteers are of secondary importance. Etruscan arenas for equestrian events consisted of minimal features, as did the Greek hippodrome, and we have no archaeological evidence for them (Humphrey, 1986: 6–17). It is therefore difficult to trace Etruscan or Greek architectural influences on the Roman circus.

The earliest known circus is the one in Rome, the Circus Maximus, which was located in the vallis Murcia, the valley between the Aventine and the Palatine. It was developed, according to Livy, in the sixth century by the Etruscan kings Tarquin and Tarquinius Superbus, and Tarquinius Priscus was said to have brought boxers and horses from Etruria to Rome (Livy I.35.8–10, I.56.2; Dion. Hal. *Ant. Rom.* IV.44.1; Thuillier, 1975; Humphrey, 1986: 64–7). The execution of the first drainage project by the Etruscan kings mentioned by Livy in the same context may be related to the construction of the Circus Maximus. The floor of the valley between the Aventine and the Palatine featured a stream that collected run-off water from the nearby hills, which was converted to a drain that flows into the Tiber some 200 m south of the outlet of the Cloaca Maxima (see Chapters 11 and 26; Holland, 1961: 34–5, 348; Lanciani, 1967: 54–6). This literary tradition is supported by the chronology of the representations of horse races in Etruscan friezes (as on the plaques from Poggio Civitate, from the early sixth century; Root, 1973) and the representations of chariot races on vases (e.g. a black-figure amphora from the second half of the sixth century in Bronson, 1965: 93, no. 19, 95; see also Thuillier, 1985: 82–7). Furthermore, Herodotus mentions equestrian competitions that took place in the Etruscan city of Caere in about 540 (*Hist.* 1.167). Only Tacitus claims that horse races were introduced to Rome from the Greek colony of Thurii in south Italy (*Ann.* 14.21.1). Contact with the Greek cities of Italy, which participated regularly in the Olympic games, must have stimulated the Etruscan interest in the contest, but Thurii was founded in the second half of the fifth century and thus could not have inspired the Etruscan interest in horse races (Thuillier, 1975: 366, n. 3; Humphrey, 1986: 64). Another literary tradition holds that the first games in the Circus were held at the beginning of Roman history, as Romulus presided over the games at the festival of the Consualia, at the altar of the god Consus; these games ended with the rape of the Sabine women (Plut. *Rom.* 14.3; Tert. *De spect.* 5.5; Varro *Ling.* 6.20). It is doubtful whether this account should be taken as evidence for racing games in this period (Humphrey, 1986: 61–2; Coarelli, 2007b: 323).

During the Republican period the Circus Maximus must have been a very simple structure, resembling the Greek hippodromes rather than the much more defined design of the circuses of the Empire (Humphrey, 1986: 67–77; on the design of the circus: Humphrey, 1986: 18–24). At first wooden stands, called fori publici, would have been arranged around the track for the spectators. The construction of starting gates for chariots or carceres in 329 (Livy XXIX.37.2), probably wooden, suggests that the races were becoming more sophisticated and there was a need for the formalization of the arena's design. It is possible that towards the end of the same century the spina, the continuous barrier around which the races were run, was installed (Coarelli, 2007b: 323; Gros, 1996–2001: 347; contra, Humphrey, 1986: 69, 175–294). During the second century the number of games increased with the increasing number of triumphal processions and more attention was paid to their accommodation as magistrates handled their organization. In 196 L. Stertinus added an arch (fornix) to the entrance for his triumphs in Spain (Livy XXXIII.27.3–4) and in 174 M. Fulvius Flaccus and A. Postumius Albinus, the censors of the year, rebuilt the masonry of the carceres, and placed turning posts (metae) at the end of the spina as well as egg-shaped markers that counted the laps that the competitors had completed (Livy XLI.27.6).

But the monumentalization of the structure was undertaken at the time of Pompey and Caesar (Plin. *HN* 36.102; Suet. *Iul.* 39.2). Caesar directed a significant expansion of the Circus, possibly in order to celebrate his triumph in 46, and Augustus continued work on the Circus after the fire of 31 (Cass. Dio 50.10.3). As the excavated remains of the Circus Maximus generally date to the Trajanic phase of the monument (for a summary, see Humphrey, 1986: 106–15), Dionysius of Halicarnassus is the primary source for our understanding of the Circus in this period (Dion. Hal. *Ant. Rom.* 3.68.1–4; Thuillier, 1975). For him it was one of the most beautiful monuments of Rome. The arena was bordered by continuous seating around the three sides of the track, stone seats next to the arena for the most important spectators and wooden stands placed behind. The seating was surmounted by a one-story portico, containing shops and houses above, running along the exterior side of the seating. In front of the seating on all three sides of the arena stood a canal (euripus) that protected the spectators from the animals (Plin. *HN* 8.21; Suet. *Iul.* 39.2; Dion. Hal. *Ant. Rom.* 3.68.2). According to Dionysius, the Circus was 650 m long and 125 m wide including the seating. It had a seating capacity of about 150,000 spectators.

The Circus Flaminius was constructed in 220 by C. Flaminius Nepos (Livy *Per.* 20; Festus 79 L), who was also responsible for the construction of the Via Flaminia. Since it was built in the Campus Martius, it seems to have had a different use than the Circus Maximus (Wiseman, 1974, 1976; Humphrey, 1986: 540–5). The uses attested for it are similar to those found in the Forum: markets, financial transactions, funeral orations and assemblies (Wiseman, 1974: 4). Varro indicated that only the ludi Taurii took place in the Circus Flaminius (Varro *Ling.* 5.153–4). He also states that the circus took its name from the Flaminian fields, which existed before C. Flaminius. The Circus Flaminius was located in the area just north-west to the Theater of Marcellus, where many generals chose to build monuments, such as porticoes and temples, commemorating or giving thanks for their victories (Coarelli, 2007b: 266–7). For this reason it has been suggested that it served as the assembly point of the triumphal processions.

5 Conclusions

The refinement and monumentalization of the public entertainment structures from the Middle to the Late Republican period coincided with Rome's military expansion and the victorious wars abroad. The influx of wealth from overseas conquests and the wish to compete in the political arena led to the increasing number of public games as well as the establishment of state organization for them. The importance acquired by public entertainment promoted the advancement of permanent structures for the various types of events. Elites, and in particular victorious generals, recognized euergetism as a way to establish their political persona in the turbulent period of the Middle and the Late Republic and funded the construction of permanent structures for entertainment. The senatorial ban against permanent theater structures within the city of Rome suggests that the Senate recognized a potential threat in the visual potency of these structures.

Gradually permanent structures replaced the wooden ones but the archaeological record has not preserved information for this transition. On the basis of the literary sources we may hypothesize on the reconstruction of the wooden structures and their relation to the permanent ones, but such proposals remain speculative. In the first permanent entertainment buildings designers took advantage of the natural topography and they gradually started to experiment in partly free-standing structures. Important for this development was the availability of cheaper materials and faster building methods that facilitated the construction of sophisticated structures on a monumental scale. The new techniques enabled designers to refine Greek theater design in order to satisfy the specifics of the Roman spectacle and audience, as well as to create the novel architectural forms of the amphitheater and the circus to accommodate the needs of the games that were taking place in them. In doing so, they created monumental structures that dominated the landscape and defined the urban charsacter of Republican settlements.

FURTHER READING

For an overview of Republican theaters, Sears (2006, esp. ch. 5) is the most recent study in English, while Scaparro *et al.* (1994) is a complete catalogue of both Greek and Italian theaters that addresses the question of the development of architectural form. Lauter (1976) is a key article on the transition from Hellenistic to Roman theater design in Italy. For a discussion of the architectural history of Roman theaters, including their early development, see Frézouls (1982). Bieber (1961) analyzes the archaeological evidence on Greek and Roman theaters in relation to extant plays, while the recent treatments of Beacham (1991, 2007) present up-to-date approaches and new interpretations on the subject. See also Manuwald (2011) for a discussion of Republican dramatic performance. For Republican amphitheaters, Golvin (1988) is a comprehensive catalogue in French with discussion on questions of use and architectural design. Welch's recent articles (1991, 1994) as well as her book (2007) provide a lucid treatment with a catalogue in English and discuss insightfully the questions of origin and development of Republican amphitheater design. There is very little archaeological evidence for Republican circuses. Humphrey (1986, esp. ch. 3) discusses literary and visual representations, especially on the basis of the Circus Maximus. Articles by Thuillier (1975, 1982, 1985) address the literary and visual evidence on Republican equestrian events and their relation to Etruscan practices. Gros (1996–2001) (vol. 1 in his two-volume publication on Roman architecture) has chapters dedicated to each of the structures that address their early development. See also Tosi and Lucia (2003) on Roman entertainment structures, with thematic essays on these structures.

CHAPTER 3

Republican Houses

Shelley Hales

1 Introduction

In Livy's account of the sack of Rome in 390, the rampaging Gauls are temporarily overawed by the majesty of the elderly senators enthroned in their family atria, where they are awaiting death (V.41.1–8). The awe is momentary and the massacre is quickly underway but the moment is enough to show the importance of the house in Republican Rome. The domus was where the senator should be, where his roles as private paterfamilias and public magistrate were reconciled. It created his and his family's place in society. The degree to which the leading individuals of the Republic were associated with their homes is evident from their high profile: the house that Pompey built "like a tug boat" behind his Theater on the Campus Martius, the gable on Julius Caesar's house which proclaimed his authority – and foretold his downfall. Nowhere is the importance of the domus more painstakingly detailed than in the saga of Cicero's house, bought as a mark of his arrival at the heart of political life, lost to his arch-enemy Clodius at the moment of his exile and laboriously fought for on his return to Rome (Plut. *Pomp.* 4; *Caes.* 68.6; Cic. *Dom.*; Wiseman, 1987; Hales, 2003: 11–60). Elite houses were reflections of the success of the current, and past, homeowner and their families. Reminders of their success were draped around the façade and atria: Pompey displayed captured ship-prows in his atrium; Gaius Gracchus was struck down by spoils from the atrium of Fulvius (Cic. *Phil.* 2.28.68; Plut. *C. Gracch.* 15.1). And overlooking everything were the ancestors themselves, preserved in the form of the masks that paraded family history to the present generation (Flower, 1996: 93–7).

Houses also acted as the venue for two of the most important social traditions of Roman life, the salutatio and the convivium. Every morning the great atria of the elite

A Companion to the Archaeology of the Roman Republic, First Edition. Edited by Jane DeRose Evans.

were open for the salutatio, the ritual essential to the patronage network, when clients presented themselves to their patrons (Wallace-Hadrill, 1989: 63–88; Clarke, 1991: 1–29). In the evening, the good and the great, or those with pretensions to being so, entertained at home, sharing a meal with a few lucky clients, family members and political allies. In playing host to these daily rituals, the Roman domus became the medium through which the familia communicated with the wider community and expressed and justified their place in society.

In playing host to these rituals, houses should be understood as being just as important to understanding social and political discourse in the Republic as public monuments. Writing under Augustus, the architect Vitruvius must be drawing on Republican practice when he describes the ideal Roman house. In prescribing its layout and proportions, he conjures a home that is made Roman by being specifically the opposite of the Greek *oikos* and emphasizes its social role; the house should reflect the dignity and status of its owner and it should be engineered with an eye to his public display (Vitr. *De arch.* VI.5). The Republican elite home is not a domestic retreat but a building constantly penetrated by outsiders and it is for managing this penetration that the atrium domus was designed.

2 Where to Find the Republican Domus

Given the emphasis on the house as a socio-political tool, its immediate context was crucial. The quiet suburb was no place for the leaders of a Roman city to live. Politically advantageous house building entailed the acquisition of a highly conspicuous plot. The location of the house of P. Scipio Africanus, hero of the Battle of Zama in 202, behind the Tabernae Veteres next to the Roman Forum, is a reflection of how near the political action the elite preferred to be (Livy XLIV.16.10–11; *NTDAR* 134). The location of the house could also reflect ideological stances: both Marius and Gaius Gracchus lived around the Forum, a reflection of their popularis politics, while the young Julius Caesar lived in the great slum, the Subura, accruing "man of the people" status in the process (Plut. *Mar.* 32, *C. Gracch.* 12.1; Suet. *Iul.* 46).

The Palatine Hill, meanwhile, was the preferred location of the more aristocratic senator (Figure 3.1). Its associations with Rome's earliest settlement and heroes (the Hut of Romulus was a preserved heritage site) and proximity to the Forum made it very attractive to Rome's leading families (Dion. Hal. 1.79.11). Carandini's investigations have confirmed its dense occupation of substantial homes, dating from the time of the kings during the sixth century, at which point the lower, northern slopes of the hill were terraced. By the Late Republic, competition for space up here was intense and Royo's analysis of the names associated with the residences suggests the extent to which a few families had the area sewn up (Cass. Dio 53.27.5; Cic. *Cael.* 18 and 59; Plut. *Crass.* 2.4–5; Suet. *Aug.* 72; Carandini, 1986/8; Royo, 1999: 9–117; *LTUR* 4: 22–8 [Papi]). After his eventful consulship in 63, Cicero marked his elevated political profile by buying a prestigious house here (Cic. *Fam.* 5.6.2; Gell. *NA* 12.12). The house had once belonged to M. Livius Drusus, who had brought his political career right into the domus when he was stabbed there by an unknown assassin who concealed himself in the crowd of Drusus's supporters (Vell. Pat. 2.14). Among Cicero's contemporaries, Mark Antony, Caecilius Metellus Celer (cos. 60), Lutatius Catulus (cos. 78), the triumvir Crassus and

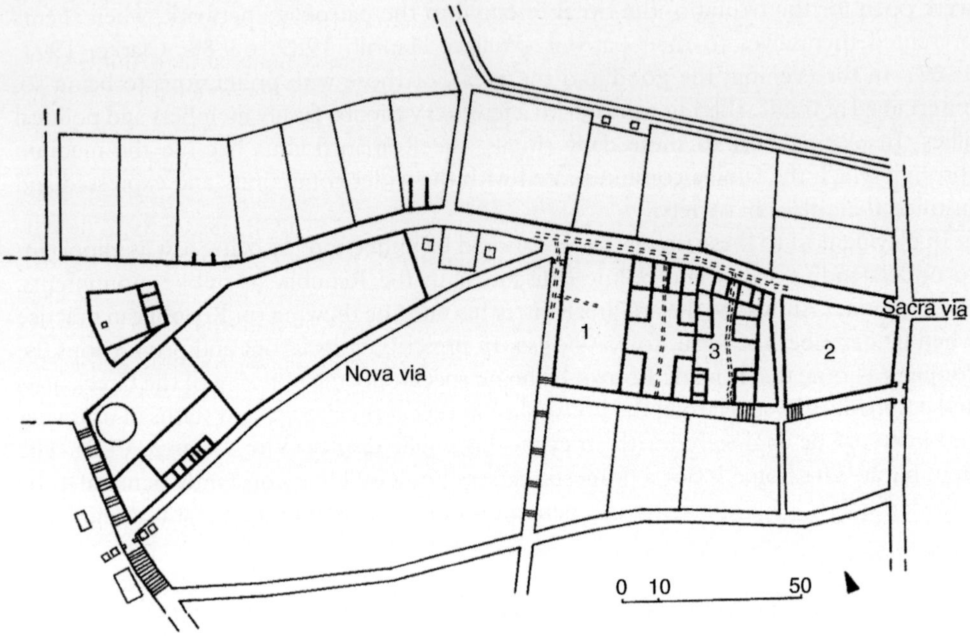

Figure 3.1 Plan of the house blocks on the lower slopes of the Palatine, Rome. (1) proposed house of Clodius Pulcher; (2) proposed house of Aemilius Scaurus; (3) third- to second-century atrium house. Source: Drawing by S.J. Hales, after Carandini (1986/8).

the orator Hortensius all had Palatine addresses, and Cicero's bitterest enemy, P. Clodius Pulcher, lived right next door (tentatively identified by Carandini as 1 on Figure 3.1).

Slightly further down the northern slopes of the Palatine lie the remains of another huge house (2 on Figure 3.1), which has become associated with M. Aemilius Scaurus (pr. 56). By securing a spot along the Sacra Via, the main ceremonial route through the Forum, these homeowners ensured their houses were visible and in view of the public monuments with which their family's exploits were associated – in Scaurus's case, the Senate House, and the Temples of the Castors and of Vesta (Cic. *Scaur.* 46–8). Across the road, too, the area now covered by Hadrian's Temple of Venus and Rome supported senatorial housing, the remains of one of which, in the tempting rush to match remains to personalities, have been assigned to Cn. Domitius Ahenobarbus, censor in 92 (Carandini, 1988: 370–1; see also Chapter 29).

The evidence of the Roman colonies reveals that this association between forum and aristocratic housing was not simply a Late Republican affectation but a strong ideological tradition, which could be carefully inscribed into the fabric of the city in new territory. At Cosa, the forum is surrounded by large, equal-sized atrium houses, which must have accommodated the city's leaders (Figure 30.1), while at Fregellae, atrium houses line the main streets entering the forum, including the Via Latina, which joined the colony to its mother city. In both colonies, the dimensions of these houses are similar, leading Fentress to expect that they are built of a standard pattern and size. Yet these central houses are almost double the size of the smaller houses in the insulae away from the forum at Cosa, which are presumably for ordinary colonists and are oriented on quieter streets (Fentress, 2003).

Figure 3.2 Aerial view of the insulae of Regio VI, Pompeii. Source: GoogleEarth.

At Pompeii, too, the owners of the bigger houses seem to prefer locations on busy thoroughfares. The façades of the Houses of the Faun (VI.12.2) and of Sallust (VI.2.4) are typical, both opening directly onto busy streets. Regio VI contains a number of old, grand houses (including the House of Pansa three insulae to the left of the House of the Faun) that enjoy close proximity to the Forum. Pompeii, however, is also a more realistic portrayal of the mix of housing in the urban landscape. A glance at the aerial view (Figure 3.2) shows the mix of housing stock even in this region of the city, with atrium houses of various orientations and sizes filling the insulae and smaller properties tucked around them. The colonies and the Palatine might be better understood as exceptions to this general rule and the rest of Rome better imagined as being a similar mélange of properties.

An inevitable result of hogging prime real estate in the city center is that, in Rome, much Republican housing is long gone. The disappearance of many senatorial homes under the palace structures of successive emperors is sign enough of the political reorientation of the city. Nevertheless, archaeologically, the Imperial buildings are also the mode of preservation. Substantial remains of the House of the Griffins, conventionally dated to the second century, and the later Aula Isiaca both survive because they became part of the fill for Domitian's palace. Scaurus's house too was a victim of Imperial overbuilding – this time, rather insultingly, nothing more glamorous than a Flavian horreum.

The best remains of domestic architecture are inevitably in places where life stopped violently and suddenly. When Pompeii was destroyed in 79 CE, the Samnite town had been a Roman colony for just over 150 years and still had a good deal of Republican-era housing stock. The sheer scale and integrity of the remains provide an unparalleled look into Roman domestic life but, more recently, the excavation of purpose-built Roman colonies has made it possible to view the ideologies of Roman domestic space in sealed Republican contexts. Cosa, about 140 km from Rome, was founded in 273 and destroyed in the second quarter of the first century (see Chapter 30). Fregellae, in Latium, established and refounded in the course of the fourth century, was sacked after its revolt from Rome in 125.

3 The Layout of the Atrium House

Cosa and Fregellae have excellent examples of the ideal atrium house, arranged symmetrically around a strong central axis (Coarelli and Monti, 1998: 62–5). The House of Diana at Cosa (Figure 3.3), located on the south-western side of the forum and dated to the early second century, is entered via a narrow, steep corridor (A) flanked by rooms (C and D), recognized as having been independent tabernae, presumably let by the homeowner (Fentress, 2003: 14–26). The corridor space itself was articulated by a threshold, indicating that it was divided into a fauces and vestibule, a feature also seen at Fregellae. Marks on the opus signinum floor of the outer vestibule suggest that it once had benches. Benches are found outside houses in Pompeii and are usually associated with the arrival of clients, to wait for access to their patron. Given that the shops dominate the façade of the house, the entrance is crucial as the point at which the owner can project himself to the outside world. The two-meter-long limestone threshold of the later House of the Skeleton in Cosa shows the emphasis put on grand entrances (Bruno and Scott, 1993: 114). In Pompeii, where doorways are preserved to a much greater height, it is possible to observe impressive portals, like that of the House of the Faun, flanked by half-columns or pilasters. That house further lays claim to the public space in front of it through the Latin welcome, *Have*, set in the pavement.

The corridor leads into the main circulatory space of the house, the atrium (B), associated, in the texts, with the salutatio and display of family trophies. Its central feature is the impluvium, the pool which collects water from the hole in the roof, the compluvium, which admits both rainwater and light. In the House of Diana, the impluvium was clearly used for its original purpose since it was connected to a large cistern. Elsewhere, in Pompeii and in Fregellae, where water was available from aqueducts, the presence of impluvia is more a testament to the importance of tradition than functionality. The two recesses on either side of the atrium are conventionally known as alae (H and I) – in the House of Diana, ala I was perhaps used for storage which Fentress (2003) suggests may have been for holding ancestor masks. The suggestion is perhaps hopeful. Despite the prominence of ancestor masks in texts no evidence to suggest their presence in atria has ever come to light. At the rear of the atrium, and opening full onto it, is the tablinum (J), the most symbolic room of the house. It is thought that the tablinum was originally the master bedroom, placing the union of the paterfamilias and his wife literally and symbolically at the heart of the house. As the domus developed, it retained its connection with family history, becoming a depository of family records.

Around the atrium there are a number of smaller rooms, termed *cubicula* (E, F, G). Often translated as bedrooms, it is likely that the rooms served a variety of uses though, in this case, G showed the scars of where the bed couch might once have been placed. In

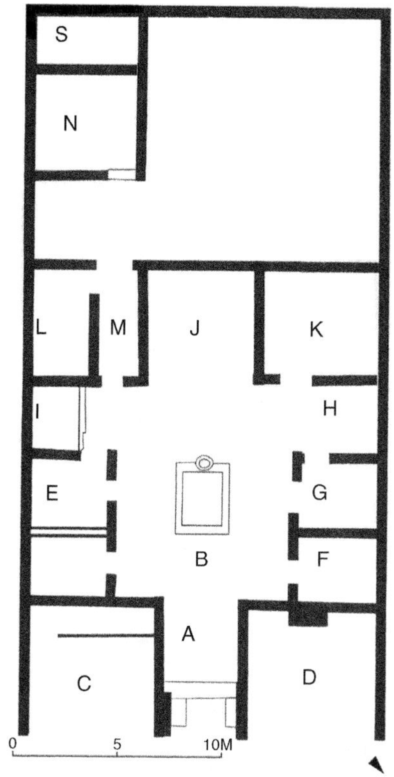

Figure 3.3 Plan of the House of Diana at Cosa. Source: Drawing by S.J. Hales, after Fentress (2003).

the rear corner of the atrium, the larger room K is identified as the triclinium, the room for the convivium.

Corridor M joins the atrium to the hortus, the garden, at the rear of the house. Room L is identified as a kitchen and bathroom space, since it had a drain leading to a soakaway pit. Combined kitchen and bathrooms, often indicated by hearth platforms, drains and even the remains of bath tubs, are well attested in Cosa (see also Bruno and Scott, 1993: 33–4, 67), though this house is unusual in that rooms S and N in the far corner of the hortus may have been a bath house. The garden itself was clearly tended as it had planting pits and a manure heap. Work on the garden of the early first-century House of the Skeleton also tells us something of the nature of the Republican hortus. It had planting pits and olive trees, and was embellished with pathways and a possible pergola (Bruno and Scott, 1993: 148–52).

4 Development of the Atrium House

Just when the atrium house began is unclear. Certainly the type was already well established in third-century Pompeii and the evidence from Fregellae may push that date further back. Though the majority of houses excavated there seem to be of second-century date, the smaller house underneath Domus 7 clearly has an impluviate atrium along with opus signinum floors and First Style wall painting. Given that it seems to have been the earliest building on the plot it may well date from the first stages of the colony, at the end of the fourth century.

The excavations on the northern slopes of the Palatine may potentially push back the type still further (Carandini, 1990). The first houses here were constructed around 530/520 and developed until a fire destroyed them in 210, after which they were rebuilt. One of these houses (3 on Figure 3.1) has a typical long entrance between rooms that do not connect to the interior; these appear to be shops. The corridor leads to a large, roughly cross-shaped space, which could be imagined as a forerunner of the atrium with alae and a tablinum at the rear and which had steps to an upper floor. There is no trace of an impluvium but Carandini imagines it to have had one since there is a large cistern to receive water. The smaller side rooms to the front of the atrium may well be classed cubicula. To the rear of the atrium at either side are two distinct suites, one which he thinks is a dining room (imagined as a Greek-style *andron*) and the other the quarters of the materfamilias. There is a rough axial symmetry here (which Carandini interprets in terms of a gender division of the house) but nothing comparable to the vistas we meet in the developed atrium-peristyle house. Like the later House of Diana, the house has a hortus, this time arranged along the side of the plot.

The house plots on the Palatine, as laid out in the sixth century, seem to have been roughly equal. At some time in the third century, though, Carandini suspects that the entrance to the house we have been describing was deliberately widened, absorbing one of the tabernae to create a much grander vestibule and, by extension, to have a greater impact on the street outside. But that is nothing compared to the thorough rebuilding of the block after the third-century fire, when these huge houses (round about 700 or 800 square meters) received concrete walls, opus signinum floors and First Style wall paintings (*LTUR* 4: 22–8 [Papi]).

This kind of showy embellishment is seen beyond Rome, too. During the second century, two cubicula in the House of Diana were merged to create a more impressive room (E), perhaps for reception purposes. Alterations in Domus 7 at Fregellae achieved a similar effect. The wall between two cubicula is knocked down to make one larger room, the function of which as a triclinium is suggested by the design of the mosaic floor. At the same time, the alae were given smarter floors, including, in one, a mosaic with a representation of a city wall.

These changes are small but reflect a much greater change happening in domestic space, particularly evident at Pompeii: the arrival of the peristyle garden. Houses, like the Houses of the Faun or of the Labyrinth (see Figure 3.2), extended the axial arrangement of the house through the addition of a peristyle. Their appearance as an integral display feature is accompanied by an increased range of room types, such as oeci and exedrae (room types developed by the Greeks), which, like the new rooms in the houses at Cosa and Fregellae, seem destined for reception. Their exotic forms are usually understood as affectations appearing from the Hellenistic east, from other domestic buildings such as those on Delos, and ultimately perhaps from the palaces of Hellenistic kings (see e.g. Nielsen, 1994: 164–70 on palaces; Trümper, 1998 on Delos). They are taken to reflect the increasing influence of Hellenistic culture in Italy, the increased wealth to pay for it and the increased social competition to stimulate the need for it. But, if the source of the fashion was Greek, it is clear that the ends to which it was put, supporting social rituals within the home and bolstering political careers, was entirely Roman (Wallace-Hadrill, 2008: 190–6). No second-century Roman house would have an ***andron*** or the type of gender division Carandini suggests for his sixth-century homes (contra Wallace-Hadrill, 1996).

At Cosa, the strict ideological arrangement of the colony begins to break down as market forces take over. The rear gardens of houses in the smaller insulae start to get gobbled up to provide new plots while other houses expand over neighboring space. This trend picks up in the early first century, when two anonymous houses are united to become the House of the Treasure and several back gardens are built over to create the House of the Skeleton, with the first impluviate atrium in these blocks (Bruno and Scott, 1993: 77–152). Changes are not always for the better: despite its apparent upward trajectory, Domus 7 at Fregellae had become a fullery by the time of the colony's destruction.

In Pompeii, the second century sees an explosion of elaboration and the emergence of super-sized houses (Dickmann, 1997; Nappo, 1997). The House of the Faun had already absorbed its entire insula by 180 or so (Pesando, 1997: 80, 267–74). Towards the other end of the century, the owners of the House of the Labyrinth, in the insula behind, bought up space from at least two neighboring houses in order to build an enormous peristyle, flanked by fashionable reception room types, such as the columned Corinthian oecus (Pesando, 1997: 78–80; Strocka, 1991). At Cosa, there is nothing so ambitious or early but, in the first-century phase of the House of the Skeleton, the tablinum-type space and the triclinium were opened onto a portico of three columns which looked onto the garden, in a manner that would be followed by the House of Diana in the Augustan era (see Figure 3.3). This trend is observable at around the same time too, in Pompeii, where smaller houses like the House of Sallust (VI.2.4) and of the Ceii (I.6.15) make an attempt to embellish their garden spaces with porticoes (Bruno and Scott, 1993: 137–42; Pesando, 1997: 267–74).

In Rome, more money and higher stakes might lead to much grander experiments. The grand, several-story houses built on terraces into the side of the Cispian Hill are testament to the possible scale of domestic architecture. The Esquiline Odyssey paintings belong to one of these houses and Coarelli suggests that the paintings were set as high as 4 m above floor level under a vault, which was supported by a colonnade on three sides (Coarelli, 1998a). Even (or perhaps especially) on the packed Palatine, homeowners were clearly making provision for peristyles and elaborate niched fountains and muscling into neighboring plots in order to make room for them (Papi, 1998). The most notorious example of such activity is doubtlessly that of Clodius in the first century, whose house became steadily bigger as he extended over other people's properties, perhaps occupying as many as eight units, most notably his neighbor Cicero's, whom he had forced into exile. Cicero's house was razed and a Temple of Libertas erected over it as part of Clodius's house (Cic. *Dom.* 62, 116; Carandini, 1986/8). The story receives most interest in terms of the political symbolism of the temple but its presence is also a great reflection of the tendency of domestic space to swallow public architecture, sacralize space and incorporate external landscapes (Royo, 1999: 21).

As houses grew, so did the household that served them. The basement of the House of Scaurus, preserved under its atrium, consists of about 50 tiny cubicles, presumably slave quarters. That basements were a popular place to house a growing domestic slave population is shown by another house across the road, by the Temple of Antoninus and Faustina, whose basement likewise consists of a series of small rooms with herringbone floors. The inclusion in the basement of Scaurus's house of a small bath-suite is further evidence of growing status differentiation. A far cry from the combined kitchens and bathrooms of the colonies, this house's suite imitated the sequence of the public baths (Claridge, 1998: 112; *LTUR* 2: 26 [Papi]; George, 1997).

5 Other Types of Housing

The atrium house is a privileged house type and its further elaboration increased the distance between it and other types of housing. At Cosa, the second-century houses in the smaller insulae, themselves of a divergent recurrent type, are clearly differentiated from the atrium houses in the forum, though they share some features (Bruno and Scott, 1993). Long entrance corridors, flanked by a room on either side, led to a large central area with a room in each rear corner, creating between them an open recess that could pass as a tablinum. Behind lies a transverse room with access to a lower garden. They have no impluvia and Bruno and Scott suggest they are roofed over to create a testudinate atrium, with water being collected not from a compluvium but from the eaves. It has been suggested that similar houses found at Pompeii should be ascribed to the presence of veterans settled at the time of the creation of the Sullan colony in 80 but Pesando thinks they are older and that the type is simply standard for lower-status housing (Pesando, 1997: 211–15; Zanker, 1999).

Pesando (1997) has attempted to articulate the Republican housing stock at Pompeii by creating a number of categories. At the top are the atrium-peristyle Houses of the Faun and the Labyrinth and slightly smaller houses, like the House of Sallust, which operate as the great houses on a strict symmetrical axis, with the atrium-tablinum complex, but retain a hortus. Between these atrium houses and the non-atrium houses, he places homes like the House of the Ceii, which in its earliest form was a small house built around an atrium, with just two cubicula in the front corners and in the rear corners, two rooms on either side of a corridor through to the hortus, which could have served as a tablinum and triclinium. The effect of this arrangement was that the axial view had little importance. As Bruno and Scott (1993) observed of their houses in Cosa, one of the most marked differences between these houses and the large atrium houses is the lack of attention given to the outside world. In this model, privacy is preserved, but the views are generally uninteresting and uninfluential. Of course, even these houses are not the smallest – Pompeii and Herculaneum also reveal multiple family dwellings. The smallest living spaces among our archaeological remains are the mezzanine levels in the shop units, witnessed by the remains of staircases in the shops in the front of the House of Diana and those at Pompeii (Fentress, 2003: 32). It is worth noting that the slave quarters of the Scaurus basement, apparently with signs of fabric nailed to walls and provision of drains, may well have provided a more comfortable existence than the life lived by many of the free poor (Carandini, 1994).

Any type of single property, whether small or large, was something of a luxury in a rapidly expanding city like Rome itself. There, the term "insula" takes on quite another meaning of the rickety tenements in which thousands of the city's citizens lived. The multiple, high-rise dwelling is clearly a practical response to lack of space as it is not found in the colonies, and Claridge suggests that its appearance in Rome was a "natural" response to immigration during the second century, an indication of how all parts of Roman society were affected by Roman expansion (Claridge, 1998: 56). Architecturally, these blocks were a great feat and probably home to a people of a wide variety of income and status, like their Imperial counterparts (Vitr. *De arch.* II.8.17). Nevertheless, the poor condition of many insulae is a recurrent theme in the texts. Built close together, probably made mostly of wood and having perhaps as many as 12 stories (Augustus was the first to impose legal height restrictions), these apartment houses were often short-lived. The fact is that all remaining insulae date from the second century CE, testament enough to the inadequacy of their forebears. Cicero complains that even the mice have fled from his own rotten insulae; his neighbor Crassus, the wealthiest man in Rome, turned a profit on buying up burned properties (Cic. *Att.* 14.9; Plut. *Crass.* 2.4). Livy's story of an ox that escaped in the Forum Boarium, made its way into a neighboring insula and consequently fell out of an upper-story window, is a reminder of how we should slip these blocks into our imaginative cityscape of Republican Rome (Livy XXI.62). Insulae probably fitted wherever they could in the city. They jutted onto public spaces just as much as the elite domus; it was just that nobody was looking at them.

The idea that levels of privacy and public availability are tied to the acquisition and maintenance of status is clearly demonstrated in the look at these other house types. In many ways, the privilege of the richest house owners was to win more privacy. The poorer members of Pompeian society were doomed to live their whole lives in public, eating at taverns, bathing in the town baths and using the public latrines. The owners of a rich domus dined at home presiding over their own convivia and even bathed in private suites. But the privacy won by the elite was put to work to controlling a public view of themselves. The poor remained invisible and "private" even when living in public.

6 Spatial Syntax

By the Late Republic, many atrium houses were extensive, complex spaces that far exceeded the necessities of shelter and, as a result, challenge attempts to understand them. The names given to the new rooms both parade their exoticism and reflect their form rather than function. In asking what they were used for, we are reminded to interrogate more closely the functions of the atrium house's more traditional rooms – what exactly might an ala be for? (Leach, 1997: 50–72). The prompt to investigate the functionality of the Roman house, to consider how it marries Greek fashion and Roman ideology, formal social ritual and everyday life, has provided a major challenge that has been met by a number of approaches over the last 20 years.

In terms of understanding the functions of individual rooms, Allison's consideration of how the artifacts found in them tell us about the ways in which spaces were used has been particularly enlightening (Allison, 1993: 4ff; cf. Berry, 1997: 183–95). Very quickly, her technique underlined the multiple uses of an atrium in Pompeii. While smart decor and marble artifacts stress the display capacity of this important space, strongboxes and cupboards demonstrate its role as storage space and the many whorls and loomweights attest to its use as workroom. As well as being the social and spiritual center of the house, the atrium was, clearly, at other times of the day, a center for the women and even slaves to work. The evidence suggests what Laurence has called the "temporal logic of space," by which the dynamics of the house changed during the day, with functions and status of spaces allocated by activities or people rather than by architectural type (Laurence, 1994: 154–66). Allison's work is hard to apply to Republican contexts since few artifacts survive, but her findings should inform the way we interpret the plans of these earlier houses.

Wider-scale spatial analysis has also been conducted in order to map movements of people around the house. Traditionally, the concept of the syntax of the domus has relied on readings of Vitruvius, who tells us the crucial, "public" rooms (the atrium, ala, tablinum, peristyle and triclinium, the rooms associated with male, social activity, in particular the morning salutatio and the evening convivium), leaving the rest to be assumed to be clearly unimportant and domestic. Several attempts have been made at a more sophisticated analysis by applying anthropological and sociological models of domestic space. Laurence, for example, with mixed results, applied the system devised by Hillier and Hanson by which the privacy or exclusivity of a room is judged by how many doorways have to be crossed to reach the room (Laurence, 1994: 126–9).

7 Decoration

Interior décor has proved to be a major focus in the investigation of the spatial syntax of the house. Art historians and archaeologists have considered the relationship between form and function, moving from analyzing subject (for instance, does an erotic painting signal a bedroom?) to consideration of how different styles of painting may suit different areas of the house (the difference, say, between the stern First Style in atria and the fantastic Second Style in reception rooms) or how the different schemes in wall painting and even patterns in threshold mosaics might act as signposts to differentiate space and guide people around the house (Clarke, 1991; Wallace-Hadrill, 1994; Swift, 2009: 28–104). These investigations have helped us understand the complexities of the house and have refined further the dogmatic view of Vitruvian public and private space, but they do have some shortcomings, not least because they can lead to circular argumentation: the use of space is determined by the mosaics, which are thus determined by the nature of the space.

Investigations into decoration have also informed our understanding of the types of second-century architectural embellishments we have mapped above. Textual evidence makes clear that interior décor became increasingly lavish as more exciting materials and objects become available during the first century. The houses of the stars of the Late Republic were luxuriously kitted out and Pliny blames a sequence of celebrities for turning up the heat. It is no surprise that Palatine residents lead the way, as the orator Lucius Crassus and Scaurus are both picked out for the installation of marble columns in their atria (Plin. *HN* 36.2.4–7; see Chapter 17). Such high-end decoration is intimated in some of the grander houses of Pompeii, particularly in the Nilotic mosaics and Alexander mosaic in the House of the Faun. As with the architectural embellishments, the ultimate source of these innovations is associated with the Hellenistic east. In Rome itself, the popularity of Hellenistic culture is reflected in the famous Esquiline Odyssey paintings that deliver a continuous narrative of the Odyssey set in an extensive landscape and the extravagant Graeco-Egyptian motifs of the Aula Isiaca, perhaps painted around 30 (Iacopi, 1997; Claridge, 1998: 135).

But if the texts, reflecting the anxieties of the time, focus on the Late Republic as source of innovation, it is clear that the trends exhibited in them go further back. Unsurprisingly, the evidence of the second-century colonial houses is simpler but its study gives some intimation of the particularly Roman role to which such decoration was put. Floors are more likely to be of opus signinum than mosaic, usually red or sometimes black, and often packed with limestone tesserae laid to make regular geometric patterns. At Pompeii, there are many more mosaic floors – the simplest effects might be monochrome mosaic interspersed by irregular-shaped off-cuts of colored marbles or limestone, a technique also found in a Late Republican house under the Ludus Magnus in Rome (Dunbabin, 1999: 38–59; Welch, 2006a: 527–8). Black and white mosaics with pattern or vegetal borders were also popular.

All the wall paintings at Cosa and Fregellae seem to have been of the First Style, even in the later built houses. This style was common across the Mediterranean and imitates ashlar masonry. Walls are covered with stucco which is either molded or incised to give the illusion of being composed of individual blocks, which are painted to add variety and to give the illusion of different types of stone, particularly marbles. The houses at Cosa show color ranges of purple, yellow and marbling of green and yellow blotches on a yellow ground. Further architectural details might be intimated, particularly friezes – the House of the Skeleton's First Style triclinium included a green meander band and a painted garland, while at Fregellae it seems that terracotta friezes were inserted into the tablinum wall (Bruno and Scott, 1993: 133; Coarelli and Monti, 1998: 62–5). The sheer monumentality of the First Style is most easily shown by its coverage of atrium, tablinum and alae in the Houses of the Faun and of Sallust in Pompeii. Its application across Italy and across house types shows that painting was an accessible medium through which to promote the illusion of grandeur (Laidlaw, 1985; Richardson, 1988: 108–11).

The persistence of the First Style in Cosa is noteworthy because, by the time of its destruction, the so-called Second Style was well underway in Pompeii and Rome. Also heavily reliant on making an architectural impression, the Second Style does so entirely with paint and gives the illusion that the wall has been opened. The House of the Griffins on the Palatine has an early example, currently dated 110–100. The walls are painted to imitate a colonnade in front of a wall decked out in typical First Style fashion complete with marbling. From this rather somber beginning, Pompeian walls show the style's imaginative possibilities: golden, jewel-studded columns form colonnades behind which lavish sanctuaries open up with *tholoi* and tripods. The overwhelming effects of these paintings are exaggerated by the sophisticated ways in which such painting merges with the real architectural fancies of houses. The Odyssey Landscapes, for instance, are interspersed with painted pillars as if viewed through a colonnade, a colonnade that was probably duplicated by the real colonnade in front of them. The hefty columns of the Corinthian oecus in the House of the Labyrinth at Pompeii likewise mirror the columns of the painted colonnade that frames the temple precinct on the walls behind (Figure 3.4) (Strocka, 1991: 44–8). As with the architectural elements of these houses, real sources for these fantastical, painted scenes have long been sought: the Roman theater and

Figure 3.4 Second Style wall in the Corinthian oecus of the House of the Labyrinth, Pompeii. Source: Photo by S.J.Hales.

Hellenistic royal palaces are favorite candidates (Royo, 1999: 26–7; Leach, 2004: 93–122). But, like those architectural elements, they are probably better appreciated as fantastic subversions than faithful imitations of real elements. If there is a connection of Second Style painting to dramatic performance, it is perhaps best understood not as a direct reference to the theater itself but to the theatricality of Roman social interaction performed in the home. As such, the increasingly popular theme of theatricality could well be applied to the architectural substance of the domus as well as to its decoration (Bryson, 1990; Elsner, 1995: 49–87; Hales, 2003: 135–66; Swift, 2009: 81–5 extends these ideas to mosaic).

The decoration of colonial houses also shows a very different kind of Roman interaction with the Hellenistic east in showcasing the aggressive conquests which brought back its luxuries to Italy in the first place. The self-referential display of booty and trophies in urban atria and façades was a familiar sight in Rome and Sulla famously had his martial exploits painted in his Tusculum villa (Plin. *HN* 26.2.12). The houses at Fregellae show that such display was not confined to generals (Coarelli and Monti, 1998: 62–5). Several have revealed terracotta fragments which appear to allude to land and sea battles; among the most extensive are the remains of a frieze 18 cm high showing Victories, trophies and prisoners that was probably inserted into the First Style wall of the tablinum in Domus 2. It is a distinct possibility that some second-century settlers, veterans of the wars against

Antiochus III of Syria, were celebrating their part in the conquest of the east. At first sight, these decorations seem to offer a very different kind of aesthetic from the fantasies of the Second Style or the extravagances of domestic sculptural collections, but Welch's investigation into the Republican household reminds us of their interrelation and suggests that the aesthetics of the former are directly inspired by the display of captured Greek booty, another reminder of the very Roman nature of the display of Hellenistic luxury in the domestic sphere (Welch, 2006b; see Chapter 27).

8 The View

The syntax of the atrium house, however, is not dependent on architecture and decoration alone. Its axial layout and the eagerness of homeowners to seek out prominent locations remind us that the view was a major part of the domestic experience. The emphasis on claiming and becoming a view must offer further motivation for the acquisition of homes perched on the Palatine. The biggest house was a waste of investment if it was tucked out of sight. Velleius Paterculus (2.14) describes Drusus commanding his architect to maximize the visibility of his home and hence his own public profile. As with decoration, the power of the view is seen by the suspicion that it could engender. Public opinion forced Publius Valerius to demolish his home on the highest part of the Velian Hill in 509 and he was careful to rebuild his domus on a more equalizing level (Livy II.7.5–12). His story is a telling reflection of the perceived power of domestic display and the danger of overstepping the mark.

Drerup was the first to endow the view through the atrium domus with a life of its own and to explore its effects on the visitor (Drerup, 1959, developed in Jung, 1984). When the front doors of these houses were open, the viewer saw the sequential arrangement of fauces, atrium, tablinum. The view takes a highly symbolic, privileged path through the house, passing straight over the impluvium and onto the tablinum, the apparent center of the home and place of the paterfamilias, before continuing through the windows of the tablinum to the peristyle and even reception rooms beyond. The importance of the central axis along which this view travelled was such that key features, such as the exedra in the House of the Faun that contained the Alexander mosaic, are aligned along them. The strong, carefully planned axial view allowed sprawling houses at Pompeii, like the Houses of the Faun and the Labyrinth, to intimate the complete symmetrical arrangement of the houses at Cosa and Fregellae.

This penetrating view thwarts physical realities, gaining for the outside viewer symbolic access that cannot physically be enjoyed by the inhabitants of the houses who are forced to walk around the impluvium, around the colonnades of the peristyle and through the corridor(s) to the side of the tablinum. The central axis gives the impression to this viewer standing on the threshold that the whole house has been arranged around his standpoint. The house is apparently laid bare and made public. But of course, the house is anything but. The view is carefully contrived and privacy carefully regulated. That area of movement in the atrium, and the wall decoration which frames it, is largely lost to the viewer because the long vestibule walls act as blinkers, channeling the sightlines along the central axis and denying visual access to the areas of the house that flank it (Jung, 1984: 76, 78ff; Grahame, 1997: 163). And entry was not so easy to achieve in reality. Even without chained guard dogs or scowling porters, the steep, narrow fauces create a highly restrictive, easily policed entrance. The interior looks as distanced and as superior as possible, making the paterfamilias in his tablinum who seemed so near both physically and socially remote.

This remoteness is confirmed by the priorities of sights inside the house. The central axis often loses its central importance as the wider context around that axis gains dominance. From within the house, particularly from the reception rooms around the peristyle, new concerns affect the viewer (Bek, 1983; Jung, 1984: 98). These privileged views reward that inside viewer with a spectacular picture to admire. The entrance from the street through the atrium to the peristyle extension of the House of the Labyrinth offers an axial view straight across the garden, through the colonnade and then between the columns in the Corinthian oecus. Inside that oecus, however, the axial view out of the room is only one of a series of views opening up on every side as the painted walls draw the eye into lavish temple precincts, which promise so much more than the mundane view of the street.

These views might explain further the articulation of the atrium house, which appears not to be divided tidily into public and private spaces, but between aspects of spaces available to outsiders and accessible to those inside, whether family or visitors. Inside the house, rooms of ambiguous function appear to have had the capacity to become public at any moment. Those members of the public who were admitted through the front door of the domus were faced with an impression of grandeur as they were brought into the owner's private world. The importance of these views is such that it is one of the major ways Bruno and Scott distinguish the gulf between the House of the Treasure and the House of the Skeleton at Cosa. While the latter does not have the axial symmetry of the House of Diana, the provision of a view from the triclinium across the portico to a garden which appears to have been landscaped deliberately to slope upwards, thus emphasizing its visibility from the house, shows a homeowner with an eye for the visual effects of his home (Bruno and Scott, 1993: 148).

9 Horti

From the second century, then, the atrium domus provided a powerful spectacle, pulling into its orbit exotic, civic, and even sacred elements and exploiting the view into and out of it. But urban houses remained restricted by space and by tradition. Other types of property were needed in order to exploit the principles of domestic display to a greater extent. The elite of Rome had always owned multiple properties in the country since the majority of their income was tied up in land ownership. Generally bigger properties, free of the restrictions of the city, country villas became favorite places for extravagant display but they were disappointing in one respect – they were invisible to the voting audience back in Rome (for the excavation of a villa very close to Rome, see Carandini *et al.*, 1997; see also Chapter 20). By far the most visible, and thereby effective, residences were the horti, the parkland residences that combined the landscape of the villa with the visibility of the domus. Neither in the city nor in the country, they sat in the liminal space of the *sub urbe*, sometimes literally straddling the city walls (Purcell, 1987). As with the domus, their elaboration probably dates back to the second century, since the Scipios seem to have owned them as early as the mid-century (Cic. *Nat. D.* 2.4.11).

Cicero himself tried to buy the Horti Drusi just as he had taken over the house of Livius Drusus on the Palatine (*Att.* 12.21.2). Caesar had two horti, one at the Colline Gate and one across the Tiber, so his name might be present at every stage of the journey to and from Rome (Obsequens 71; Suet. *Iul.* 83.2; cf. *LTUR* 3: 55 [Papi]). Though the horti are associated with otium and some were built by famous retirees, many others were certainly exploited by the politically ambitious as settings for public euergetism, particularly banqueting, on a scale that far exceeded the domestic convivium (App. *B Civ.* 3.14; Plut. *Pomp.* 44.4; Val. Max. 9.15.1; D'Arms, 1998; Wallace-Hadrill, 1998). By the Late Republic, these horti ringed the city. Lucullus had built his on the Pincian Hill and Sallust's, inherited from Caesar, were north of the Quirinal (Lucullus: *LTUR* 3: 67–70 [Broise and Jolivet]; Sallust: *LTUR* 3: 79–81 [Innocenti and Leotta]; Hartswick, 2004; Talamo, 1998: 113). The Horti of Maecenas were south of the Esquiline gate on the Oppian Hill while, over to the west, the Horti Scipionis overlooked the Campus Martius. Raised up on the hills and ramparts, many of these horti enjoyed a grand view, and could see and be seen across the whole of Rome and beyond to the surrounding countryside. The texts suggest they deliberately exploited the view, since they mention towers both in the Horti of Maecenas (from which Nero watched Rome burn) and of Caesar (struck by lightning in 17; Suet. *Ner.* 38.2).

Today, very little of the Republican horti remain. The majority of the remains date from alterations made when many of them had passed into the hands of the Imperial family. Plutarch, though, reckons that the Horti of Lucullus continued to hold their own (Plut. *Luc.* 39.2). The last remaining Republican vestiges of his horti are the remains of a terrace with a huge hemicycle, from where one could survey, as Welch suggests, the sunset, but perhaps more importantly the rest of Rome beneath it (Welch, 2006a: 517). The most well-known remnant of the Republican horti is the so-called Auditorium of Maecenas, perhaps better labeled a viridarium. It is a walled enclosure with an apse formed of steps which recall theater seats but were more likely steps for pots of plants or even a cascade. In its earliest form it also had an opening out of the flat side, which may have taken advantage of the view (Häuber, 1990: 62–4; *LTUR* 3: 74–5 [de Vos]). The horti became the setting for terraces, porticoes, pavilions, temples, water features and extensive sculptural arrangements. Of the many marble sculptures from the Horti of Sallust, the Gauls are now thought to have been arranged by Caesar, their new conqueror, as he rearranged the mighty displays of the Attalids (Talamo, 1998).

Various familiar sources of inspiration have been suggested for these first-century residences, from the Hellenistic palaces to the huge Italic sanctuaries. It seems better to suggest that, like the wall paintings and architectural details of the urban domus, they deliberately combined different elements to create fantastic settings for the performance of Republican life. The development of domus and horti alike reflect the extent to which Hellenistic influences infused Roman culture during the second century but the tasks to which these influences were put were decidedly Roman. In the last 20 years, our understanding of domestic space has increased dramatically with both the discovery of fresh material, particularly at Cosa, the application of new approaches to the atrium domus and renewed interest in other house types. Investigation of the houses has proven so rewarding precisely because the ultimate ambition of their owners was not to make them private retreats but public statements of power.

FURTHER READING

In terms of examining literary references to competitive, Republican domestic building, Wiseman's (1987) influential article demonstrates the importance of the domus in providing a public face for its owners. Edwards (1993) also has an extremely useful chapter on the morals of building. Wallace-Hadrill (1994) has done the most to pursue the sociological importance of domestic space and decoration, swapping emphasis from the literature of Rome to the archaeological remains of Pompeii and Herculaneum (see also Hales, 2003). Alongside Wallace-Hadrill, Laurence (1994) and Laurence and Wallace-Hadrill (1997) have analyzed the spatial syntax of Pompeian houses.

The houses of Pompeii are well examined and another important treatment is Zanker (1999). Pesando (1997) is rare in concentrating solely on the Republican buildings of Pompeii. The excavations of Cosa and Fregellae have added greatly to our knowledge of atrium houses. Bruno and Scott (1993), Fentress (2003) and Coarelli and Monti (1998: 62–5) publish the results. Carandini's excavations on the Palatine are published in Italian (1986/8, 1988, 1990) but a very brief English overview can be found in Carandini (1994). The other full account of the domestic houses on the Palatine is in French (Royo, 1999). Wallace-Hadrill (2008) and Welch (2006b) consider the ways in which new Greek influences emerging during the Republic come to serve a very Roman end in Republican houses.

The bibliography on domestic art is large and here I concentrate only on works that consider its relation to architectural space. For an overview of mosaic and wall painting, the principal media, see Dunbabin (1999) and Leach (2004), whose major study of wall painting is the first to break away from traditional, formal categorization and includes a whole chapter on the theatrical style. Bryson (1990) and Elsner (1995: 49–87) provide exciting ideas about wall painting's tendency to fantasy and subversion. Tybout (2001) considers the social role of wall painting. Clarke (1991) focuses on the ways in which decoration might react to movement around the Roman house and Swift (2009: 28–104) joins him in looking at mosaic patterns. Allison (2004) continues her approach of working with artifacts. Drerup's (1959) early work on the view was updated in English by Jung (1984) and Bek (1983).

The bibliography on horti is not too extensive. The remains of the Horti Lamiani published in Cima and La Rocca (1986) are highly evocative of the wealth of the horti though the remains are Imperial. Cima and La Rocca (1988) collect ruminations on the horti phenomenon in various languages. Purcell's (1987) essay on the nature of the suburbium is a very useful insight into the liminal nature of the space which the horti occupied.

CHAPTER 4

Tombs and Funerary Monuments

Sylvia Diebner

1 Introduction

Archaeologists have long subjected the burial customs of the Roman Republic to detailed investigations, but the great majority of their work has focused on individual sites. This flood of research has somewhat threatened to obscure the overarching trends that have come to define the field, assuming that such trends can be recognized at all. This is not to discourage such efforts, of course, for many sites are still in need of comprehensive analysis. The possible types of analysis are remarkably diverse, moreover, for tombs and funerary monuments exist in many different forms and display a great degree of regional variety. They are encountered throughout all of Italy and they number among the most significant antiquities that exist, as they often do, in plain sight. Expansive burial grounds and simpler funerary practices, involving inhumation or entombment, have existed since the earliest times, and it is only a matter of historical chance that more graves and monuments have survived from the Roman Empire than from the Republican period. Only a few surviving burial sites provide us with an impression of how the earliest Republican necropolises or collective graves might have appeared. That having been said, numerous grave goods – urns, glass vessels, dishes, for example – have been collected, cataloged according to their different materials, and been placed in chronological brackets. As is the case with the better-known tombs and funerary monuments of the Roman Empire, those of the Republic have likewise been interpreted as reflections of social and cultural structures. Such interpretations, however, should be approached with caution in light of the fragmentary state of the tradition and the often inadequate information at our disposal.

A Companion to the Archaeology of the Roman Republic, First Edition. Edited by Jane DeRose Evans.

The burial sites of antiquity were typically grouped together to form larger complexes; isolated funerary monuments are far less common. The large-scale necropolises on the outskirts of Roman cities, which often extend for miles along exurban streets, testify to the desire to establish permanent burial grounds. It must not be forgotten, however, that the majority of the population in the ancient Mediterranean world was either cremated or buried without any permanent sign of commemoration. If not placed directly into the earth, corpses were cremated and their ashes were collected in terracotta or glass vessels; graves were normally marked with a stone, with the top of an amphora, or with a large fragment of the same.

The term "funerary monument" typically designates a commemorative structure that was constructed above ground. However, underground structures also exist, and they are characterized by large chambers with walls carved out of bedrock. Beyond being mere places of interment, these tombs also served to aggrandize and prolong the memory of the deceased and the stature of their respective families. The transition that was made to simpler structures – such as steles or monolithic tombstones – has yet to be examined in detail.

In general, the state of the field is both quantitatively and qualitatively uneven. Whereas a great deal is known about the tombs of central and northern Italy, Rome, Ostia and Pompeii, comprehensive archaeological investigations are still lacking for strips of the Adriatic coast and for large portions of southern Italy. As regards the Roman Republic and the Early Empire, in any case, it is important to distinguish between the localized burial customs that varied from one region to the next. It was often the case that the earlier necropolises, which date as far back as the second century, were later abandoned and used as foundations for subsequent sites, and thus our knowledge of things from the early and middle years of the Republic is often incidental and fragmentary at best. Such is the starting point of the present survey, and it is important to keep this in mind. Like that of the latest research, the presentation below will be made according to rather fluid districts that do not quite correspond with either the Republican *regiones* or with the modern Italian states (Hesberg, 1992; Valenti, 2010: 13–22).

2 The Situation before the Second Century

Regarding the early years of Rome, it is known that a series of cemeteries was established outside of the city gates some time after the construction of the so-called Servian Wall. One of these early necropolises, which existed in the vicinity of the Roman Forum, contained the graves of those who had lived on the Palatine and Capitoline Hills (see Chapters 11 and 26). This swampy area ceased to be used as a necropolis once it had been transformed into an urban center. Between the late sixth and the second centuries, the structure of burial grounds remained essentially the same, and grave sites were strewn rather haphazardly across the countryside. It was predominantly the case that corpses were simply buried directly into the earth or placed into underground chambers. This was followed by a stage characterized by the use of unadorned tumuli or other grave markers, both of which could take several forms: small mounds, enclosures, or wooden steles, among other simple grave constructions that may or may not have borne the name of the deceased. Monuments of a larger sort do not seem to have existed at the time.

Our knowledge of Republican burial customs is based almost exclusively on the archaeological evidence from cemeteries and single graves. Because written accounts of such customs are lacking, the implications of these sites for our understanding of the religion or general beliefs of the people is rather limited. Nothing precise is known, for instance, about the complex rituals that would accompany a burial, though it is possible that certain ritual behaviors – such as the adornment and display of the corpse, the lamentation of the dead, the construction of a pyre, and the performance of sacrifices – may have been of more importance to a funeral than the burial itself. Recent advances in excavation methods – along with the collaboration of anthropologists, ethnologists, zoologists and botanists – have yielded an impressive number of detailed analyses with relevance to burial sites, but even these fine studies have to be read with a degree of caution. Whereas a copious grave inventory undoubtedly reflects the wealth of the deceased, for instance, it would be rash to assume that an unadorned grave is necessarily an indication of low social status. That having been said, a comprehensive study of the grave goods of a given region or settlement can certainly lead to significant insights into social, economic and religious matters. Such studies can also reveal valuable information about the process of acculturation that a given population may have been undergoing.

In addition to simple inhumations, a necropolis discovered on the Esquiline Hill has provided evidence for the widespread existence of small square burial chambers (Albertoni, 1983). The most famous of the Roman chamber tombs is that of the Cornelii Scipiones, which lies on the Via Appia and dates to the beginning of the third century (Coarelli, 1996: 178–238; *LTUR* 4: 281–5 [Coarelli]). Such underground burial chambers were necessarily dependent on the existing tufa formations and therefore followed the lay of the land. The family members of the Cornelii Scipiones were entombed in sarcophagi, and only that of Scipio Barbatus, who was elected consul in 298, has a decorative form. Nine epitaphs have also been discovered within this burial complex, each of which was carved between the first half of the third century and the first third of the second century, though there is a long gap between the years 230 and 180. These early chambers were simply carved out of the bedrock, and there is no indication that they were ever painted. This is also true of the southern Etruscan chamber tombs near Norchia (Colonna Di Paolo and Colonna, 1978; Quilici Gigli, 1976; Ceci, 2005), but things seem to have been somewhat different in the other Roman tombs from the period. The latter sites, which have been found throughout the countryside, have yet to be investigated in detail, and so there is no reliable information about their appearance. Nevertheless, several painted chambers have been discovered; although fragmentary and poorly preserved, it is clear that the paintings depict specific scenes from the lives of Roman magistrates. Whereas the latter chambers were constructed by the leading families of Rome, the Esquiline necropolis, which did not fall into disuse until the Augustan period, also contains less well-adorned burial sites, mostly earthen graves. Many areas of the necropolis, in fact, consist of mass graves for the bodies of the poorer population of the city. It is unlikely that the cemeteries of the fourth and third centuries featured any tombs or monuments above ground, given that nothing of that sort has survived. The location of an underground chamber was probably marked with a modest enclosure or stele.

During the funerals of illustrious men, according to Polybius (6.53), the virtues and exploits of the deceased and their ancestors were expounded upon in speeches by their surviving male relatives. Although large monuments do not appear until the third century, it is not unlikely that the grave sites of Roman aristocrats might have been distinguished by larger tumuli or mausoleums. Our picture of the fourth and third centuries remains incomplete. Detailed archaeological studies have yet to be devoted, for instance, to the simple stone monument marking the cemetery of S. Cesareo in Rome, to the grave steles in tarentum and elsewhere in southern Italy, or to the many tombstones (cippi) found in Latium and the Volscian cities (Franchi de Bellis, 1997; *ILLRP* 83–872; Palmieri, 2009: 387–96). Despite this lack of data, it can of course be assumed that simple burials far outnumbered the few elaborate tombs that have come to light.

3 Roman and Italian Necropolises of the Second Century

The structure and appearance of necropolises changed fundamentally in the second century, and the likely reason for this was the period of economic growth that Italy enjoyed in the wake of the Second Punic War. Merchants and military leaders had become wealthy, and it was this stratum of society that furnished the cities with durable and extravagant buildings. The tendency to build expensive and architecturally elaborate tombs emerged toward the end of the second century. This can be illustrated by the development of the Tomb of the Scipios mentioned above, which was furnished with a series of hewn chambers as early as the third century. Originally, the tomb contained only a simple vault, but after the Second Punic War – and probably before the middle of the second century – it was equipped with a monumental façade, consisting of a high podium adorned with Ionic or Corinthian semi-columns. The lateral surfaces of the podium were covered with frescoes depicting scenes of the family's military exploits, and statues of the deceased were placed in the niches between the columns. Such a grave was probably not unique, though it is difficult to know for sure. An approximate conception of the Roman funerary monuments of the second century can only be gained from the few remains of architectural structures and from the extant fragments of figurative decorations. Written documentation of such tombs is scarce, and a comparative examination of other Italian cities is lacking. Because the Republican necropolises have seldom been fully excavated, our knowledge of their inner organization is rather deficient; at any rate, their long period of use can leave the impression that their arrangement was somewhat haphazard.

From what is currently known, it seems as though a large number of Roman necropolises were first furnished with architectural grave constructions only toward the end of the second century. In the city of Rome, at least, this new historical context resulted in the construction of different types of tombs, and these monumental graves, which began to appear with greater frequency, represented something entirely novel compared to what had come before. It was at this time, too, that tomb façades were more frequently designed to accommodate statues of the entombed. This development was followed by the trend of tombs being built in closer proximity to the thoroughfares of the city, a trend that itself enhanced the importance of a tomb's architectural adornment. Of great significance is the fact that none of the tomb façades in Rome, Etruria or Apulia had any local precedent; as largely original creations, they can be said to resemble only the Lagrasta tombs in Apulia (Lippolis , 1987: 151–2), the Tomb of the Scipios, or that of Hildebrand of Sovana (Steingräber, 1981: 209–17). A close inspection of the second-century façades reveals innovative combinations of various architectural styles. They were built for the leading political families of the time, and the shift toward greater splendor was motivated by a number of different factors, including an increase in wealth, the emergence of more prominent families, and the growing influence of Greek culture. In this regard it should be pointed out that the more elaborate grave monuments were designed according to Hellenistic models that had begun to appear at nearly the same time, most notably the "middle class" mausoleums found in Rhodes, Cyrene and Miletus, among other places (Hesberg, 1992: 25). Certain decorative elements of these designs were also borrowed from the familiar public monuments in Rome – triumphal paintings, sacred statues, and so on. In terms of their numbers, however, elaborate tombs were rather rare. It can rightly be presupposed that the great majority of the nobility continued to be interred in the traditional underground chambers, hypogea, or burial mounds.

4 Streetside Tombs in Late Republican and Early Augustan Times

The second century ended in civil war. Whereas the culture of the Roman aristocracy had previously exercised a unifying effect over society, it was now the case that more and more individuals sought to make a mark in politics. This mentality was accompanied by the desire to create more conspicuous forms of self-representation, which served to enhance the public image of political figures. Not only were more members of the aristocracy eager to build lavish tombs, but many representatives of the middle classes were as well. This great demand led to outright competition both among equals and between different groups of society. Funerary monuments began to be built in more public and prominent locations, and new types were devised: tumuli that became more structured; pyramids; and multi-level aediculae, among other examples. These novel constructions were in turn looked upon as models by broader swaths of the population, and thus it comes as no surprise that far more diverse tomb types have survived from these years than from prior eras. The great variety of types is indeed rather surprising. Although many of these are conscious imitations of precedents from Greece, it is impossible to identify a linear development of styles from these years to the beginning of the Roman Empire. The many political upheavals between the third and the first centuries prevented the continuous importation of new styles and models into Italy, and those models that did arrive were treated differently in different localities, where they were combined with native forms to produce entirely new results. Toward the end of the second century, at any rate, tombs were being built directly along the streets in order to leave a more forceful public impression. It was not long before such streetside monuments were joined with others to form what are now referred to as "streets of tombs." Admittedly, such streets were interspersed with suburban villas, sanctuaries, businesses, fields and gardens. There were also a number of streetside tombs that stood without any others nearby, such as that of the consul Servius Sulpicius Galba on the Via Ostiensis (*ILLRP* 339; *LTUR* 4: 299 [Coarelli]). This tomb was located in the immediate vicinity of his family's estate and its vast storehouses in southern Rome, the so-called Horrea Galbana. Other examples of streetside funerary monuments include the tomb of

C. Publicius Bibulus, which lies at the base of the Capitoline Hill (*LTUR* 4: 295 [Gallitto]; Tomassetti, 2000); that of the Sempronii at the base of the Quirinal Hill (*LTUR* 4: 297 [Lega]); and various others along the Via Flaminia and the Via Appia.

Though its beginnings can be traced to the end of the second century, a high point of this development occurred in the Late Republican and Early Augustan periods. The architectural variety of the tombs built in these years can be illustrated with four well-known examples, each of which exists in Rome: the tomb of C. Publicius Bibulus, mentioned above, which dates to the first half of the first century; that of Cornelia, the daughter of Scipio (*LTUR* 4: 281 [Chioffi]); the pyramid of Gaius Cestius (*LTUR* 4: 278–9 [Krause]; Neudecker, 2005); and the tomb of Marcus Virgilius Eurysaces (Ciancio Rossetto, 1973). Noteworthy, too, is the tomb of the Horatii and Curatii near Albano, the design of which was modeled after the mausoleum of the Etruscan king Lars Porsenna in Chiusi (Ghini, 2010: 79–88).

The competitive nature of the construction projects during this period was only able to maintain any meaning as long as the political and societal claims to leadership were clearly defined. At the moment when the principate established itself as the secure form of government, such competitive behavior was necessarily put into a new perspective, and it was indeed the case that the appearance of necropolises fundamentally changed during the Augustan age.

Based on the evidence that has already been discovered and analyzed, the tombs and other funerary monuments outside of Rome present numerous deviations from the picture that had emerged from the Roman evidence itself. Most of Italy, in fact, exhibits regional and local developments both in the general architectural forms and the decorative elements of tombs and other funerary monuments, even though the burial rites and the cults of the dead had remained essentially unchanged throughout the centuries. As recent investigations in central Italy have shown, more and more significance came to be attached to local craftsmanship and to the formal and stylistic repertoire that such craftsmanship had to offer. The remainder of this survey will focus on such regional variations.

5 Ostia

The early necropolises of Ostia, which lie along the Via Ostiensis near the Porta Romana, can be dated to the second century (Squarciapino Floriani, 1958; Boschung, 1987). They consist of simple interred urns containing – in addition to cremated remains – fragments of wood, nails and small pieces of bone carvings. These remnants were the remains of the *klinai* of the deceased that were burned along with the corpse. It is clear in this case that pomp and splendor, if displayed at all, were displayed during the funerals themselves and not by the adornment of the graves. The important markers of the individual graves, upon which the names of the dead were inscribed, were most likely made of perishable material, but not a single example has survived. Although the excavation of these necropolises began as early as the nineteenth century, most of the findings have yet to be subjected to a thorough analysis. Their early, non-architectural stages can be identified with the simple cremation burials of the second and first centuries, and it was not until the Sullan period that an architectural phase emerged. Although the remains of enclosed grave constructions have survived from the first century, earlier sites were destroyed and built upon by later generations. In all, four stages of development can be identified, the second of which included a square monument that was visible from afar and was purely decorative. The necropolis outside of the Porta Laurentina, the graves of which date only to the Caesarian or Early Augustan age, lacks an early stage altogether. Two memorial graves exist outside of the Porta Marina that, though exhibiting many familiar elements of earlier necropolises, are distinguished by their unusually elaborate adornment and their arrestingly large size (Panciera, 1966).

6 Northern and Eastern Italy

Cisalpine Gaul, which represents an important link between central Italy and the north-west provinces, was not made part of Rome until the year 42 (Martin-Kilcher, 1998). In Aquileia, the burial sites from the second-century colony until the middle of the first century were nearly devoid of grave goods, which was customary in the Republican period. Their chronology is extremely difficult if not impossible to establish, which can also be said of the sites in Pompeii (Kockel, 1987: 187; Verzár-Bass, 1998).

Cremation had been the predominant burial practice in Republican Altino, which did not become subject to Roman burial laws until the imperiatorial period. Numerous grave goods have survived from the end of the second century and the beginning of the first century. Whereas their design indicates that they were made in Roman workshops, their arrangement is unique to the Venetian area. It was also the case, incidentally, that vessels of local design would occasionally be commissioned to be made in Rome. Greek Hellenistic and other eastern influences, which greatly expanded the stylistic repertoire of tombs, began to make their mark here in the second half of the first century. In fact, the wealth of discoveries and the combinations of styles in Altino far exceed those of other regions (Tirelli, 1993, 1997, 2003; Cresci Marrone and Tirelli, 2005). There are many similarities with the burial sites along the central Italian coastline, but a detailed comparative study is unfortunately lacking.

7 Etruria

Because several Italian universities offer programs in Etruscan Studies and Italic Archaeology, a vast number of publications on these fields has accumulated over the decades. In terms of archaeology, Etruria thus represents one of the best-known areas of Italy.

Funerary monuments were a significant aspect of the Etruscan cult of the dead. Cremated remains were collected in biconical or house-shaped urns (so-called "pozzo graves") as early as the ninth century, and in the middle of the next century it became common for corpses to be inhumed in rectangular graves beneath the ground ("fossa graves"). The practices of inhumation and cremation, however, continued to exist side by side, and the selection of one rite over the other was likely determined by family tradition or individual preferences. Any type of jar or jug could serve as a cinerary receptacle. Canopic jars were used from the late eighth to the early sixth century; hut-shaped bronze urns are attested from the early fifth century; and terracotta and alabaster urns are also in evidence. Both cremations and inhumations were covered with small burial mounds, and different social statuses – at least at first – were primarily indicated by grave goods. The tholos tombs of Populonia represent a special case of their own. During the so-called Orientalizing Period, which extended from the end of the eighth through the whole of the seventh century, a transition took place between singular graves to collective burials in multi-room chamber tombs, the number of which increased dramatically. In northern Etruria, these were constructed of squared blocks, whereas in the south they were carved into the soft, volcanic bedrock. An outstanding and famous example of the latter is the necropolis of Cerveteri, which consists of a clear succession of tombs that continued to be developed into the second century. Beginning in the seventh century, graves appear that are half embedded into the earth and half constructed above ground, such as the Regolini-Galassi tomb, but soon exceptionally large tumuli were introduced that frequently contained multiple burial chambers (Steingräber, 1981: 419–55). The tumulus grave, however, was not the only sort. The façades of tombs were designed in entirely novel ways in southern Etruria, for instance, and painted burial chambers were preferred in Tarquinia and Vulci (Steingräber, 1981: 181–203, 2006b: 369–401). The many distinct chamber tombs throughout the area, which are cut from canyon-like rock formations, are variously adorned with decorative columns, figural capitals, vine scroll patterns and features of the Doric order. Behind such decorations, the burial chambers themselves are embedded into the rock, a phenomenon on full display in Blera, Sovana and Norchia, among other places (Giuliani, 1966). The pediments of other necropolises, especially those in Vulci, exhibit relief ornamentation.

The transition from traditional Etruscan burial practices to those of Rome is a period of great interest. Although it is clear from the urns discovered in late Etruscan tombs, such as that of the Volumnii near Perugia, that family traditions involving burial chambers and underground graves had extended into the Early Augustan era, there is also evidence of the newfound popularity of marble urns, which were then fashionable in Rome (Körte, 1909; Gerkan, Messerschmidt and Ronczewski, 1930; Sinn, 1987).

8 Campania

A surprising number of funerary monuments have survived in southern Italy, especially from the third to first centuries. These frequently take the form of aediculae, rotundas or columnal structures with pyramidal tops. In the third-century necropolises of Capua there are burial chambers in the form of a tholos, and similar hypogea are also found in Caivano, Nola and Cuma (Valenza Mele, 1996; Caputo, 2000; Benassai, 2000, 2001; Libertini, 2005). The steles in Capua generally derive from the greater Greek tradition (Eckert, 1988). In Paestum, the sites of which have been well documented, there are richly painted burial chambers that date back to the fifth century (see Chapter 24; Pontrandolfo Greco, 1994).

In the rural cities of Rome, such as Pompeii, the image is much the same as in the capital, and this can also be said of the smaller towns (Kockel, 1983: 9). Pompeii numbers among the most highly researched places in Italy, and its necropolises have accordingly been the subject of multiple investigations (D'Ambrosio and De Caro, 1987; Kockel, 1987: 183–98).

Figure 4.1 Tomb of the Flavii by the Porta Nocera, Pompeii. Source: Photo by Sichterman, D-DAI-Rom 1958.1899, used with permission.

The old necropolises with inhumations, such as the deep graves outside of the Porta Ercolana, were in use until the second century, at which point they were abandoned and houses were built upon them. No large funerary monuments have come down to us from the so-called Tufa Period, and there seems to have been no continuity with earlier times. The only evidence of the appearance of pre-Sullan graves is the unassuming Fondo Azzolino, which is south of the Porta Stabiana. The aniconic steles in Pompeii, which take the form of herms and which mark the location of urns or graves, derive from pre-colonial times and remained in use until the destruction of the city. Steles of this sort are found exclusively in southern Campania, though a somewhat similar phenomenon can be observed in Tarantum (Taranto) (Pensabene, 1975; Kilmer, 1977).

It has not yet been possible to identify with certainty any funerary monuments in Pompeii that predate the year 80. After this year, the most popular architectural element of tombs was a narrow rectangular pedestal crowned with an aedicula. An early example of this from the Sullan period is the so-called Tomba delle Ghirlande (Kockel, 1983: 126–51). Here the statues of the dead once towered impressively over the adjacent road (e.g. Figure 4.1).

9 Apulia

In the Hellenistic city of Tarantum there are individual graves that date back to the fourth century and were probably part of a larger system within the city's necropolis. Elsewhere this phenomenon appeared much later on, typically in the first century.

In contrast to the great number of known graves, however, relatively few funerary monuments have been discovered, and these seem to have been commissioned by elite members of society. Burial chambers are well attested on the immediate outskirts of the city itself, but these were not marked by any visible steles; in the more distant countryside, however, such markers are rather common but burial chambers are extremely rare. Both types of grave are found within the city itself, but the situation is in need of further study (Graepler, 1990, 2002). Apulian red-figure vases from the fourth century depict a diverse array of tomb types (Lohmann, 1979), and yet a lack of additional evidence from the pre-Sullan age makes it difficult to identify any direct or indirect influences from the eastern Mediterranean. Features unique to Tarantum began to appear in the third century, including Corinthian capitals and unprecedented bases. The typology of the grave shrines (*naiskoi*) also became more diverse. They typically contained a figural statue, the nude elements of which were made of marble and the clothed elements of painted stucco. To some degree, such architectural and typological innovations continued to be made in Tarantum until the second century.

In Canosa it is clear that efforts were made to design an entirely original type of monument, even though larger grave steles have not yet been discovered there. The graves of the Canosan necropolis are loosely scattered throughout the environs of the city. Between the isolated funerary monuments there are broad burial grounds with simple graves, so that at first glance it seems as though there were hardly any differences from the practices of earlier times. However, Canosa exhibits elaborate tombs with rock-hewn façades, such as the Lagrasta Tomb II. The latter was visible above ground on account of a freestanding series of eight columns above its façade. Cubic tombs have also been found, the various features of which include Doric friezes, friezes depicting gladiatorial scenes, and rounded shapes. It appears as though the aristocracy of Canosa sought to produce unique structures, the inspiration for which had likely been drawn from other cities in the area.

10 Central Italy

Recent studies have provided photographs of a series of small-scale funerary monuments in central Italy. A general spirit of change seems to have descended around the middle of the first century, in light of which small workshops were busy trying to fulfill many commissions for individualized sculptures. The panoramic results of this new trend were striking in their diversity. The geographic realities of different regions – such as waterways, streets, valleys and mountain ridges – played a prominent role in the cultural communication between the populations of different areas, and this was no less the case during the nearly 100 years of the later Republic and the early Empire, the epoch in which the majority of the small-scale monuments had been produced.

Regarding the inland territories of central Italy, a great deal of archaeological research has been devoted to the findings of the Iron Age and the Archaic Period, many of which had been discovered in necropolises (Capini and De Benedittis, 1984; Capini and Di Niro, 1991; Capelli, 2000). The second-to-first-century period of Romanization was the topic of a conference held in 1988, an event that occasioned the

Figure 4.2 Urn of Q. Minucius Saturninus, from Isernia. Source: Photo by Rossa, D-DAI-Rom 1975.2563, used with permission.

discussion of the central Italian funerary monuments of the Late Republic and the Early Empire (*La Romanisation*, 1991; Diebner, 1991). A group of decorated round altars, all similar in design, is known from the Marsi area in the Fucino basin, and some of these possess a cavity that seems to have been meant for the ashes of the dead. In fact, it is possible that the altars themselves had functioned as above-ground grave markers (Valdiserri Paoletti, 1980; Diebner, 1987). The areas of Isernia and Venafrum have received an archaeological investigation of their own, the efforts of which resulted in a thorough inventory of funerary monuments (Diebner, 1979). In neither of these areas, which had remained largely unfamiliar with Roman culture until their colonization, did the formal features of such monuments dovetail with those found in other Italic regions. Located at an important inland crossroad, Isernia is home to a great number of limestone monuments with Doric friezes that were used to honor local officials and also to serve as grave altars. A group of urns has survived whose form resembles that of money chests, and these have analogues in the Paelignian towns of Cocullo, Sulmona, and Corfino (Figure 4.2). Figurative motifs appear only occasionally, although they are often difficult to recognize as such.

In neighboring Venafrum (Venafro), which enjoyed a favorable location along the north–south axis between Rome and Campania, there are a great many vestiges of military activity. An impressive number of sculpted seats of honor (pulvini) were used for honorary purposes and for large grave altars, a practice that continued into Imperial years. It can also be surmised that local types of funerary monuments existed in Late Republican Latium, as was the case in other topographically confined regions. A collection of urns was discovered in a non-burial setting in the south of Latium, for instance, but their simplicity and lack of figural adornment cannot be said to betray anything of a local style (Diebner, 1983). Simply put, they are cubes of stone with simple recesses and conical lids, often bearing the inscription "ossa," or bones. Urns of this sort represented an affordable form of burial for a large portion of the local population; since their original location is unknown, it can only be guessed whether they might have been accompanied by any grave goods.

11 Picenum and Umbria

The series of historical events that led to the Romanization of Picenum is well documented. The region was ultimately subjected to Roman authority after the siege of Ascoli Piceno in 268, and the ratification of the Lex Flaminia, in 232, resulted in the administrative apportionment of Ager Gallicus Picenus. Before and during this period of Roman integration, the city of Ancona seems to have occupied a special position. The burial sites and grave steles discovered in this city and its immediate surroundings, which date from the fourth to the first century, consistently follow earlier models. For instance, the style and typology of the steles, none of which was found in connection with a gravesite, have close parallels in Rheneia. The rich Hellenistic tradition preserved in coastal Ancona, which is further evidenced by a great number of imports among the known grave goods, hardly left any mark on the interior of the region. Here the extant urns and steles, which were likewise discovered outside of their original contexts, date to the last years of the Republic and exhibit formal and decorative features that are unique to the area. In general, their differences from the finds in Ancona say much about the cultural exchange that took place between the mountainous inland and the Adriatic coast (Diebner, 1982, 2007).

The small funerary monuments known from Umbria, which was culturally heterogeneous, can be divided into a number of locally specific types. In the north-eastern part of the region, the predominant grave marker was the door-stele, the earliest examples of which can be dated to the middle of the first century (Verzár, 1976; Diebner, 1991–3). To the south – around Assisi, Bettona, Bevagna, Spello, Carsulae and Todi – the art of the urns and steles, which were produced by local workshops into the first century, displays the unmistakable influence of Etruscan precedents. The forms and motifs of these Umbrian urns, in turn, were adopted into the artistic repertoire that is on display in later Roman grave steles (Diebner, 1986). The funerary monuments found in the cities around Perugia are particularly indebted to Etruscan forms, motifs and craftsmanship. It remains unclear why certain places to the west of the Tiber had adhered to the traditional custom of urn burials, whereas other places preferred to designate graves with steles. In any case it is noteworthy that the variety of forms and motifs exhibited in the funerary monuments diminished considerably over time, so much so that tall rectangular steles became the common standard by the end of the first century CE (e.g. Figure 4.3). The remarkably imaginative architectural

Figure 4.3 Grave stone of L. Vesprius, from Assisi. Source: Photo by Rossa, D-DAI-Rom 1975.1755, used with permission.

and artistic combinations that characterized such monuments until the end of the first century ceased to be augmented, soon fell out of demand, and ultimately disappeared altogether.

It is interesting to note that the creative efforts of local workshops to synthesize traditional forms and motifs with the latest urban trends had been such that the original meaning of certain artistic concepts was frequently lost in the process (Diebner, 1986). In the area of Carsulae, a familiar example of such a synthesis is the so-called *cippo carsulano*, which was modeled after the Etruscan house-shaped urn but incorporated a wide variety of decorative motifs (Diebner, 1986–8, 2009). The extant urns of nearby Amelia, on the contrary, exhibit far more independence from Etruscan influences. Only after more material has been gathered and studied will it be possible to evaluate and contextualize this and other individual observations about the smaller funerary monuments of the Roman Republic.

Acknowledgments

This chapter was translated by Valentine A. Pakis.

FURTHER READING

The following titles, several of which appear in the body of the article, can be recommended as points of departure to interested readers: Steingräber (1981), Kockel (1983), Diebner (1986), Hesberg and Zanker (1987), Lippolis (1990), Hesberg (1992), Tirelli (1997), Pontrandolfo Greco (1998), Verzár-Bass (1998), Heinzelmann (2001) and Steingräber (2006b).

CHAPTER 5

Before Sigillata: Black-Gloss Pottery and Its Cultural Dimensions

Roman Roth

1 Introduction

Roman archaeologists, in general, consider each period to be represented by its own distinctive type of fine pottery. Thus, as far as Italy is concerned, Italian sigillata represents the Early Imperial period, succeeded by red-gloss wares produced in Gaul and Africa during the High and Later Empire. As the title of this chapter implies, black-gloss wares are usually seen to fit into this picture by virtue of having been the first widely distributed type of fine ware. Thus, they provide archaeologists with a "sign-post" artifact of the Republican period before the appearance of the first Arretine sigillata, dated conventionally to around 30.

As a general guideline, this model has certain merits, by, for instance, conveniently providing spot-dates for survey and excavation, even for non-specialists. Yet this chapter argues for a more complex approach to black-gloss wares: first, as established by Lamboglia (1952) six decades ago, black-gloss pottery was produced in different regions of Italy, and in a variety of shapes and fabrics. The number of known regional productions has grown exponentially over the years, but only a small number enjoyed a quantitatively significant and geographically wide distribution in antiquity. Second, black-gloss pottery by necessity forms an anachronistic analytical category since, in reality, its boundaries to other types of ceramic wares (such as coarse and plain table-wares) may have been much more fluid. In particular, this statement applies to the local and regional ceramics, the variability of which is still being discovered. Third, chronology imposes another factor of variability, involving significant changes in the morphological repertoires and the geographical locations of black-gloss production over three centuries. In addition to providing an overview of the geographical and chronological developments of the

A Companion to the Archaeology of the Roman Republic, First Edition. Edited by Jane DeRose Evans.

production, distribution and consumption of black-gloss pottery, this chapter also aims to map out different avenues of exploring those patterns of variability. For this reason, it is imperative to begin by looking at the basics involved in the creation of such variability.

2 Production Techniques and Technological Choices

Black-gloss pottery was produced in Italy from about the middle of the fourth into the second half of the first century, starting in Campania and the Etrusco-Latial area, before spreading to other parts of the peninsula as well as to neighboring areas in the western Mediterranean. While the technology involved in the production of black-gloss pottery may well go back in its origins to Athenian workshops of the Classical period, it had become firmly established in central Italy by the end of the fourth century, assuming a style that makes it easily distinguishable from contemporary examples produced in the Greek east. As with other classes of Roman pottery, this distinctive style has been predominantly associated with a specific morphological repertoire: this forms the subject of the next section. As the approach taken in this chapter is based on a more comprehensive understanding of style, however, this first section provides a summary of the different stages involved in creating a black-gloss vessel.

The basic techniques involved in creating the black-gloss effect are relatively straightforward: first, the potter has to coat his vessel with a partly evaporated and thus viscous suspension of levigated clay or slip; second, in order to produce a black surface, the atmosphere in the kiln needs to be oxygen-reduced (or reduction firing). The fineness of the slip and its content of ferrous minerals determine the smoothness and shine ("gloss") of the finished artifact, in conjunction with the extent to which the potter is capable of controlling the conditions of the firing process. The simplicity of the basic technology may go some way towards an explanation of why black-gloss wares were produced in so many regional varieties at so many different levels of quality. Potentially, the technology afforded each potter with a considerable range of technological choices affecting the style of the finished product. As noted earlier, this factor may justifiably lead us to question the extent to which we should consider it as a homogeneous class of material.

Technological choices do not, of course, enter the picture only at the stages of the slipping and firing of the vessels but begin with the selection of the clay. As shown by the considerable differences between, say, the clays used to make Campana A wares and those used to make pottery of the "Cerchia della Campana B" (Cibecchini and Principal, 2004; the terminology is explained below), not even the production of fine or very fine black-gloss vessels, in principle, required the use of specific types of clay, as long as the potters sufficiently refined it. However, the choice of clay more often than not determined the appearance of the finished products, which further serves to explain the considerable differences among the pottery produced in different parts of Italy across a three-hundred-or-so year span.

As all Roman fine wares, black-gloss pots were wheel-made. Although none of the surviving depictions of such wheels specifically relates to black-gloss pottery, one may assume that the standard tool was a simple pivot-set single wheel of the types that are still found in some parts of India (Peña and McCallum, 2009a: 69–52; cf. Miller, 1985). From the archaeologist's point of view, shaping represents a particularly important stage in the production process: as we shall see in the next section, the morphology of black-gloss pottery is central not only to our understanding of its chronology but also to wider debates concerning its evidential value for socio-economic history. The choices made by the potters when fashioning their clay into specific shapes (and not others) should, therefore, form a central part of our inquiry into widely debated issues such as the standardization of craft production during the Republican period.

While turning the vessel on the wheel, the potter, furthermore, had the choice of using simple tools to create patterns such as concentric circles around the center of the inside base; alternatively, he might apply stamped decorations showing stylized figures or objects or, more frequently, abstract or floral patterns. Both of these decorative techniques are commonly found throughout much of the period of black-gloss production. Stamps with the names of potters or, more probably, of the owners of workshops, are found only rarely, and exclusively on bowls made at Cales in north Campania during the third century (Morel, 1989: 480–2).

Applying the viscous slip, the next step after the thrown vessel had dried to leather-hardness involved a number of important choices. As noted above, the consistency and mineral composition of the slip critically affected the appearance of the finished pots. It is more than likely that most potters used the same clay to throw and to coat their black-gloss vessels (Peña and McCallum, 2009b: 171). However, the degree of levigation and, in particular, the addition of other, non-plastic ingredients to the suspension of levigated clay determined the gloss and smoothness of the finished pots' surfaces. An additional technological choice at this point concerns the way in which the potter applied the slip to the vessel. Holding the pot by its foot and dipping it in the suspension seems to have been the norm; this, however, left parts of the foot unslipped or covered by fingerprints, as is indeed shown by many surviving black-gloss pots. In other cases – usually these are particularly fine examples – the potter took care to apply slip to these and other imperfectly coated parts of a vessel (as, for instance, in the handle zones), using a brush, evidence for which survives in brush marks.

While firing his pottery in a reduced atmosphere did not, in itself, confront the potter with an insurmountable technological challenge, the degree to which he was able to control the temperature, achieve complete oxygen-reduction (as and when required), and manage both to separate pottery and to fire, significantly affected the quality of the final products. Scores of vessels, usually with orange or brown discoloration, attest to the fact that these challenges were not always met with success. Unfortunately, few black-gloss producing kilns have been excavated or, at any rate, published, while their existence and approximate locations are usually inferred from concentrations of wasters and misfired vessels on the surface or in dumps. One recently excavated and published example, however, confirms the fact that the production of black-gloss wares – at least outside the main centers of Campana A and the "Cerchia della Campana B" – was not necessarily a specialized activity. The complex of Marcianella has produced convincing evidence for the contemporary firing of different types of ceramics in the same kiln, possibly by the same people (Pucci and Mascione, 2003). Should this scenario be paralleled at other sites, it may have important implications for our understanding of the place of (regional) black-gloss pottery as an integral part of the ceramic spectrum, and not as a separate category (see also below, section 5).

To sum up: the relatively basic skills and facilities required for the production of black-gloss pottery provide ideal conditions for both the varying quality and the great geographic spread of black-gloss wares. While a high degree of standardization could be achieved, the requirements for creating wares with a black gloss were sufficiently low as to afford the potter significant freedom in his technological choices. As noted above, the choices which the potters made when fashioning their clays into specific shapes have made a particular impact on archaeological approaches to black-gloss wares and, together with the choice of clay (fabrics), form the main basis of any systematic approach to the material. The next section, therefore, focuses on these two aspects in more detail, by way of providing a critical account of the history of research into this class of material.

3 Shapes, Typologies and Chronologies

The study of black-gloss pottery as one class of material began relatively late. Therefore, much of what was written during the early stages of research into this class of material still affects its study to this day. For instance, the terminology reflects the fact that black-gloss wares have traditionally been associated with Campania, where several important productions were located. However, the term is misleading since it treats Campanian wares as a pars pro toto, just as the label "Arretine" detracts from the fact that much Italian sigillata was, in fact, produced outside Arezzo.

Based on his excavations at Albintimilium (Ventimiglia) on the French-Italian border in Liguria, Lamboglia developed his pioneering "classificazione preliminare della ceramica campana," a provisional system of organizing the large quantities of black-gloss wares which his teams were unearthing (Lamboglia, 1950, 1952). The fact that the systematic study of black-gloss wares was sparked in this way is significant: in responding to the requirements of a large-scale, stratigraphic excavation of an urban site, the "classificazione preliminare" was unique in its kind. At the time, excavated materials were usually classified according to systems that were based on connoisseurship and had been developed for the study of de-contextualized objects or with little concern for the needs of scientific archaeology in mind. Lamboglia's innovative, decidedly archaeological approach is very much reflected by the two principal aims of the "classificazione preliminare": the chronology of the material and the patterns of inter-regional trade on which it might shed some light. In realizing these aims – drawing not only on the material from Ventimiglia but also on comparanda from other parts of Italy – Lamboglia established a number of fundamental patterns. Despite the great progress of the subsequent six decades, these patterns still form the backbone of our current understanding of black-gloss pottery, particularly in as far as the situation of the second and first centuries is concerned, that is, the time-span covered by the excavations at Ventimiglia. Three basic conclusions arose from Lamboglia's studies:

1. Black-gloss wares occur in different fabrics, some of which are found across much of the peninsula, while others are of regional or local distribution. Lamboglia identified three main fabrics during the second and first centuries through visual analysis – Campana A (red clay matrix), B (beige clay matrix, gloss with a bluish sheen) and C (gray or reduction-fired clay matrix) – that were produced in Campania, coastal

Etruria and Sicily respectively. Campana D is, by contrast, a local variety found at Ventimiglia, which – as Lamboglia correctly postulated – finds corresponding types at almost every other Republican site. With the progress of stratified excavation and archaeometry, this picture has been greatly refined and, in places, corrected. Nevertheless, Lamboglia's basic observations regarding the distinction between black-gloss wares of inter-regional and regional distribution remain fundamental.

2. The three main fabrics occur in distinctive shapes, with a certain degree of cross-fabric sharing, especially between Campana B and C. At Ventimiglia, Campana D largely occurs in the shapes of Campana B, and could thus be considered a local imitation ware.
3. Black-gloss pottery provides an indicator of "Romanization," by signaling trade routes and exemplifying the increasing homogeneity of material culture. With some modifications, this has become the widely accepted view which this chapter subjects to critical scrutiny.

The typological system proposed by Lamboglia for a long time remained the standard among students of Republican pottery, and is, with modifications, still sometimes used today. The "classificazione preliminare" assigns a number to each shape, starting with Forma 1. As new shapes were discovered, this was steadily increased by Lamboglia himself and other scholars. As it was increasingly realized that many shapes could occur in more than one fabric and were not exclusively associated with just one of them – as Lamboglia had originally postulated – the two factors of fabric and size became dissociated from one another. In addition, a wave of studies focusing primarily on black-gloss wares in museum collections across Italy and the western Mediterranean during the 1960s and 1970s revealed the existence of more fabrics, some of which were widely traded, while the majority was confined to regional or local significance (see esp. Balland, 1969; also Morel, 1963, 1965, 1969; Pasquinucci, 1972).

This is not the place to discuss all of these fabrics in detail, but one of them merits the privilege of being looked at somewhat more closely here, in part because this umbrella of productions preceded the date of the main phases represented at Ventimiglia (and thus was not taken into account by Lamboglia), and because it sheds light on the early trends in black-gloss wares towards standardization and mass-exportation. Students of Roman Italy are highly likely to encounter this fabric not only in specialized reports but also in discussions of the Mid-Republican economy. This is the so-called "Atelier des petites estampilles," which J.-P. Morel identified in a ground-breaking paper published in 1969. Taking its name from a series of small bowls of the shape Lamboglia 27, which were typically decorated with four small stamps around the base, this "workshop" appears to have been the first producer of black-gloss pottery that was widely exported to other parts of Italy and the western Mediterranean – to areas as far-flung as Carthage and the south of France – throughout the third century. From the point of view of vessel shape, Morel rightly indicated that the emphasis on the stamped bowl Lamboglia 27 (the atelier appears to have produced other forms in smaller quantities) indicates a move towards standardization. A useful terminus post quem for the ending of the "Atelier des petites estampilles" is the votive deposit of Lucus Feroniae, sealed by the destruction layer of the Hannibalic sack in 211 (Livy XXVI.11; Stanco, 2009: 164).

While the "Atelier des petites estampilles" represents an early form of the trend towards morphological standardization which Morel identified, this phenomenon is exposed to its full extent only once the fully-developed repertoires of Campana A, Campana B and, to a lesser extent, Campana C, enter the picture. "To a lesser extent" reflects the fact that the reduction-fired, gray pottery produced in the eastern part of Sicily never enjoyed the same geographical spread as its two bigger brothers. One should add that black-gloss wares with a gray clay matrix are not confined to Sicily but – as established by Morel (1963; see also 1981a: 72 *et passim*) – are found throughout the "*aire punicisante*," that is, the western Mediterranean under the cultural influence of Carthage. This standardization began with Campana A during the 220s, if not even before then; the arguments in favor of raising the chronology are considered by Ferrandes (2006), with further references. However, it is important to note that, until quite recently, the early second century was the accepted terminus post quem for the production of this pottery, and thus this date looms large in archaeological and historical literature. The consequences of this for our understanding of black-gloss pottery as evidence for socio-economic processes are addressed in the next section, as are the implications of the fact that even the three "large workshops" (i.e. Campana A, B and C) were internally more heterogeneous – in terms of both their fabrics (i.e. places of production) and their shapes – than was realized until recently. First, however, it is necessary to provide a brief account of Morel's typological system found in *Céramique campanienne* (1981a).

On encountering a typological number in Morel (1981a) (henceforth "Morel numbers"), most readers will be struck by its complexity. The present chapter maintains that this complexity, in itself, betrays the heterogeneous nature of Republican black-gloss wares, which ultimately renders such a unified system, both in practice and heuristic approach, difficult to sustain. As Figure 5.1 illustrates, each Morel number is based on the basic formula of four digits plus minuscule letter plus digit, in addition to another minuscule (e.g. 2783a1a). In the case in Figure 5.1, we are dealing with a reference to one specific, archaeologically attested example of the generic bowl Lamboglia 27: this specificity to a real object is expressed by the final letter. As pointed out below (sections 4 to 5), this shape is ubiquitous in central Italy during the third and second centuries, and is thus not generally regarded as a reliable chronological indicator. Following quasi-Platonic principles, the typological system lists this specific vessel as a realization of an ideal type – in this case, 2783a, while the first four figures act as a descriptive code, preceding from the general ("2000" corresponds to the category of "shallow vessels without handles, lacking an out-turned rim" – there are eight other categories of similarly broad definition) to the increasingly specific.

By virtue of narrowing down specific vessel shapes through numerical capillarization, Morel numbers afford the most precise way of accurately describing an individual shape, to the point of inserting an individual artifact into what is effectively a generational system structuring the material world (Roth, 2007b; cf. Graepler, 1997: 78). With regard to entirely preserved vessels – and this applies to the vast majority of the individual, actual examples listed in *Céramique campanienne* – this typology works perfectly, provided that both the practitioner assigning a number and the reader confronted with such a reference have a clear understanding of the principles underlying Morel's typology. In contrast to Lamboglia's "classificazione preliminare," *Céramique campanienne* makes

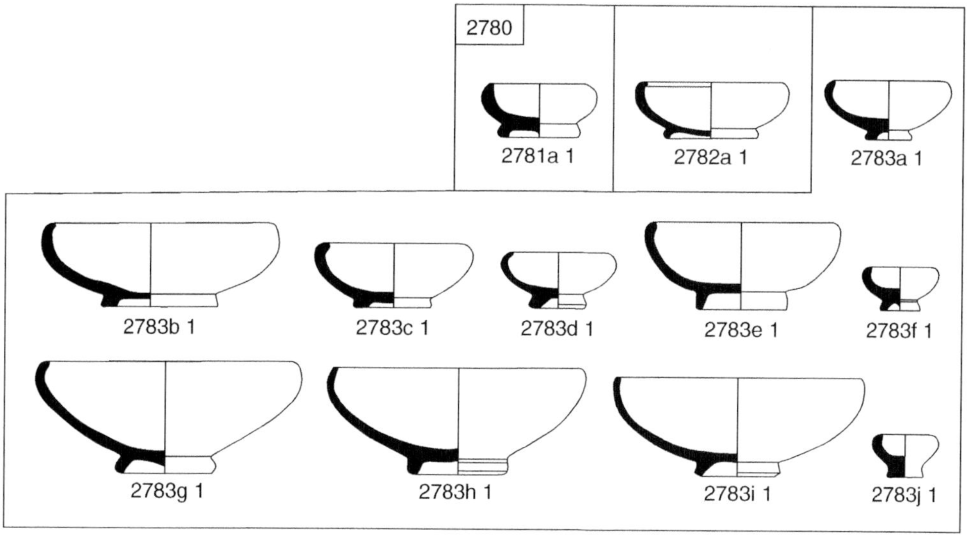

Figure 5.1 Typology of the widespread *forme* Morel 2783 and its closest cognates, progressing from broader characteristics described in the shaded boxes (2000: "shallow vessels without handles, lacking an out-turned rim"; 2700: "incurving rim"), to the morphological details of the specific examples illustrated by the line drawings. Source: Drawing by R. Roth, after Morel (1981a).

it possible for new vessel shapes to be inserted at a logical point within this generational structure (Morel, 1981a: 33). By contrast, newly discovered vessel shapes – which turned up much more frequently in the archaeological record during the first couple of decades following Lamboglia's "classificazione preliminare" than they do now – could only be added sequentially to Lamboglia's principle of assigning numbers, without bearing any logical relation to those that preceded or followed them. With regard to this aspect, the open system proposed by *Céramique campanienne* is clearly preferable to its closed Lamboglian predecessor, and indeed bears many parallels to other, widely accepted typologies like the *Conspectus* of Italian sigillata.

Yet, as so often, the devil is in the detail. First, it is difficult to impose such a complex numbering system without meticulously memorizing it, or without constantly referring to the volume of typological charts (found in vol. 2 of Morel, 1981a). As noted above, this is arduous enough in the case of vessels preserved in their entirety (and, more often than not, held in museum collections) but becomes a virtually insurmountable problem in the case of fragmentary evidence excavated in settlement contexts or recovered during field survey (cf. De Tommaso, 1987; cf. Morel, 2009: 479). Though Morel rightly maintains that the geometrical dimensions of telltale parts of a given vessel – such as its foot or rim – do allow it to be approximated to one of his numbers, even if only to the level of the second or third digit, this introduces a strong element of imprecision at the practical level – which is precisely what *Céramique campanienne* is designed to eliminate. Morel maintains that the small, morphological details described by each complete number are especially sensitive chronological markers; therefore, losing this detail through fragmentation in some cases questions the rationale for using the more complicated system in place of the comparatively simple and easily memorized typology of Lamboglia in the archaeological practice (Roth, 2007a: 109–15). Some scholars have resorted to using Lamboglia and Morel numbers alongside each other (e.g. Frontini, 1985; Palermo, 1998; Bianchini, 1996).

However, this is not merely a problem relating to the shortcomings of our fragmentary evidence: the drawback of using Morel numbers – whether or not the practitioner may be aware of it – is the underlying principle of coercing even minute variations in material culture into a historically interpretative framework. While ethno-archaeological inquiries have shown that potters repeatedly produce certain "ideas of a pot" – producing a variety of variations on a type – they also act as prolific innovators of ceramic form. Even by means of geometrical modeling we do not have a means of arriving at a realistic measure of cognitive action upon the material world in the past (see section 2 above). Of particular significance in this regard is Miller's study of rim morphology among vessels made at Dangwara (central India) (Miller, 1985: 42). It is this underlying, heuristic aspect of *Céramique campanienne*, which, in addition to the limitations imposed on it by archaeological practice, ultimately renders this typology not user-friendly, and, furthermore, unsatisfactory from the point of view of those who wish to study Republican ceramics as a source from a historical angle that does not coincide with the vision set out by Morel in *Céramique campanienne* (cf. Jolivet, 2009: 494; Morel, 2009: 481).

The inextricable link between typological organization and historical interpretation is, at the same time, the great merit and major practical disadvantage of the standard typology of black-gloss wares. The salient problems of this are discussed in the next section. To conclude the present discussion, however, it is necessary to move away from questions of morphology, and to turn towards issues of fabric, that is, the individual productions or – to follow Lamboglia's terminology – *classi*. By 1981, it had become evident that the basic scheme proposed by the late Italian archaeologist was in need of revision, in part because an increasing number of regionally and locally produced and distributed productions had come to light. In addition, the internal homogeneity of the so-called "grandes ateliers" – chief among them Campana A, B and C – was becoming a matter of debate. Since the mid-1990s, significant developments in, and the increasing accessibility of, archaeometric methodologies have rendered it possible to explore such issues of heterogeneity in detail. Morel's fundamental contribution on the question of internal heterogeneity within the "grandes ateliers" was followed by the publications of Frontini and Grassi (1998), and numerous other contributions reporting on the compositional analysis of regional assemblages (Morel, 1998a).

A key issue that emerged from Frontini and Grassi concerns the fabrics formerly known under the umbrella term of Campana B. This issue continues to be central to the understanding of internal, regionally based heterogeneity within this *classe*. Students of Republican archaeology will often be confronted by the terminology of *B-oïde* productions (esp. Morel, 1998a): this refers to wares that, generally speaking, conform to Lamboglia's description of Campana B – based on certain, dominant shapes and pale to light-beige clay matrices – but which, even to the naked eye, display unmistakable variations which compositional analysis has further helped to refine. While many points remain unclear, this chapter follows Cibecchini and Principal's suggestion that, for the time being, the most precise and economical way of referring to this group of productions is the "Cerchia della Campana B." The distinctive varieties of this group were produced in at least three (possibly four) regions of Italy, namely north Etruria, north Campania and the Adriatic (in addition to, possibly, the Roman suburbium) from the late third to the late first centuries (Cibecchini and Principal, 2004, refining some of the observations made by Morel, 1998a). Within and outside these regions, a number of larger productions – based in the areas of Arezzo (Morel, 1963; Schindler, 1967; and the later production in Ettlinger, 1990) and Volterra, Cales, Jesi and possibly Rome (see Stanco, 1999), respectively – enjoyed wide distribution. At the same time, many smaller productions that are represented exclusively at the regional and local levels display strong influences from their larger cousins, while also incorporating elements of specific regional productions, as well as of the "Cerchia della Campana B." In this way, for instance, regional productions of the Etrusco-Latial area of the second century tend to be dominated by the "fashions" of the "Cerchia della Campana B," while using Campana A shapes (such as the plate with hanging rim, Lamboglia 36). Older forms are inspired as well by other traditions, both of black-gloss (such as bowls of the type Lamboglia 27 that continues to be made in the Tiber Valley) and utilitarian wares (such as the closed shapes and *tegame* or shallow cooking pan found in Volterra and its territory) (Roth, 2006: 122, 2007a: 110–12, 118–20).

These and comparable patterns of variability appear to have peaked across central Italy during the second century, and thus impinge upon our understanding of the cultural history of Italy. As the next section argues, black-gloss pottery has the potential to provide pivotal evidence to explain patterns of cultural change that undoubtedly question many of the central tenets of older approaches to "Romanization" as a homogenizing phenomenon. Little noticed but of particular interest in this regard is the issue of how the productions of the "Cerchia della Campana B" were distributed. With transport by sea not providing a possible model in most cases (and in marked contrast to Campana A), it invites the suggestion that the importance of land-based and riverine movements of both materials and "fashions" needs to be reassessed (see Horden and Purcell, 2000; Coarelli and Patterson, 2009; and for sea-borne modes of transport of pottery, Morel, 1989). This, in fact, becomes evident as early as the third century, with the distribution of the "Atelier des petites estampilles" by both land and sea. This material forms the starting point of the following discussion of black-gloss wares in their cultural dimension.

4 Black-Gloss Wares and Cultural History

As outlined in the preceding section, the "Atelier des petites estampilles" is the first type of black-gloss pottery that not only enjoyed distribution far outside its region of origin but that also occurred in a relatively standardized repertoire of shapes. This alone justifies assigning privilege to this material in any discussion of the cultural history of Italy and the western Mediterranean during the pre-Hannibalic period even if, or especially if – as is argued here – the apparently homogenous appearance of this material can no longer be maintained.

Despite the wide geographical spread of pottery associated with the "Atelier des petites estampilles," the main concentrations of finds occur in Rome and its surroundings, including the suburbium, the Tiber Valley and Latium Vetus. This led Morel to argue that this pottery represents the first archaeologically visible example of a Roman business venture that successfully exported its products following in the wake of Rome's increasing expansion as the hegemon of the central and western Mediterranean. The ending of the production during the Hannibalic War may, then, reflect the negative repercussions for Rome's military and, as a consequence, her economic expansion (see esp. Morel, 1989: 480, 484–5, the only authoritative account to date in English of the historical significance of black-gloss wares). This standardization, Morel argued, could be construed as an indication of economic rationalism which, according to his influential, interpretative model, was to become characteristic of black-gloss production during the second and first centuries, with the emergence of Campana A and the wares known under the terminological umbrella of Campana B (Morel, 1981a: 491–2).

However, recent studies have shown that the "Atelier des petites estampilles" represents more than one workshop; in fact, we are dealing with a multitude of workshops. Some of these may indeed have been located within the suburbium while others appear to have operated across south Etruria and as far north as Cosa/Orbetello (esp. Stanco, 2009). The diversity among the stamps found on the vessels belonging to the circle of the "Atelier des petites estampilles" is, furthermore, worthy of note, as had already been recognized by Morel (1969). Thus, the "Atelier des petites estampilles" is not exclusively representative of the craft production of the emerging metropolis as such, nor does its distribution necessarily reflect cultural and economic trends that were driven by Rome (Morel, 1969, 1989: 480). What is more, the precise date of the pottery produced by these workshops cannot be resolved until more precise stratigraphic indicators are determined (see the implications of new interpretations for survey work in Chapter 12). Stanco (2009) proposes a stylistic approach to the dating of these wares in which they were successively influenced by Magna-Graecian, Etrusco-Latial and Roman tastes. Quite apart from the difficulties inherent in identifying such culturally specific traits – especially in the context of a *koine* (Chapter 27) – it should be abundantly clear that such an impressionistic approach cannot replace a stratigraphic one. However, Stanco's study remains extremely valuable by showing conclusively the existence of different productions under the umbrella of the "Atelier des petites estampilles," which future archaeometric work will no doubt help to refine further.

One important point is undeniable: however internally diverse a production as the "Atelier des petites estampilles" may turn out to have been, the fact remains that, compared to other, contemporary issues of black-gloss pottery, we can observe a level of standardization which goes beyond the merely incidental. Although several workshops were involved, we can observe an undeniable trend towards the assimilation and, as a consequence, the reduction of the morphological repertoire of black-gloss wares. This trend, of course, becomes even more pronounced with the emergence of Campana A and the "Cerchia della Campana B" during the second half of the third century. Yet an increasing number of regional and local wares also emerged at the same time as those, even if Morel dismissed the smaller productions on account of their insignificant economic and commercial impact (Morel, 1989: 499). These regional and local productions are distinguishable not merely by their fabrics but also by morphological eclecticism. The simple bowl Lamboglia 27, for example – being the "classic" shape associated with the "Atelier" – is widely produced across central Italy throughout the third and second centuries, to the point that most students of Roman pottery would not consider it to be chronologically sensitive. At the same time, local producers – while adhering to their own approaches to the fashioning of vessels – mix and match a variety of vessel shapes that no longer correspond to a single tradition but which mix influences of not only the large productions (as Campana A and the B group) but also incorporate the traditions of local utilitarian wares (see section 3 above). Herein lies the big challenge of the study of black-gloss pottery: namely that, with the advance of standardization by geographically widespread products, we are, at the same time, facing an extremely diverse body of material that, while technically representing black-gloss wares, bears little semblance to the global products (like Campana A). We begin to encounter this diverse body of material ubiquitously as Roman expansion continues, and it plausibly represents the flow of produce from agricultural powerhouses such as Campania and Etruria (Morel, 1989: 500–2). Yet this plausible conclusion does not in any way compel one to agree with Morel's model of the Republican economy, let alone to dismiss the importance of regional wares to the history of cultural history (Roth, 2007a: ch. 2).

In addition, continuing advances in the archaeometry of Roman fine wares have allowed us to gain an ever-improving level of resolution that has, in turn, exposed a far greater degree of heterogeneity even within those types of black-gloss pottery that were previously thought to represent individual centers of production (Morel, 1998a; Ferrandes, 2006). In addition to the points raised in the preceding section, one should also note that this is by no means limited to the Republican period but has come to be recognized as an important factor in the study of Roman ceramics of Italian origin in general, even if – as the case of Italian sigillata shows – this is not, on the whole, widely acknowledged or, at any rate, rendered accessible, by relevant publications. The same applies to the great advances in chronological studies. Based on the analysis of previously unpublished contexts, or of evidence re-examined in the light of newly gained insights, the traditional dates of the wares associated with Campana A and B have been modified, usually by raising their respective dates. However, since many publications, particularly those dealing with the broader patterns involved in the economic history of Republican Italy and the western Mediterranean, still adopt a framework that is, more or less, based on the traditional categories of the three main Campana productions, the orthodox view deserves to be expanded upon and criticized here. Once again, Morel's *Céramique campanienne: les formes* (1981a) stands out as the single most influential work in this area. For, in addition to providing students of Republican archaeology with an innovative, typological system accommodating the great majority of black-gloss shapes known at the time, this work also comprises a catalogue of all of the major regional fabrics known then, organized by their geographical origin but explicitly excluding lesser-known local fabrics. With regard to these, Morel openly acknowledged that they may not in all cases conform to the typological patterns set out in his work – which, of course, form the basis of his interpretative approach discussed here (Morel, 1981a: esp. 485). It remains problematic that many Roman archaeologists ignore these limitations of the typological system and end up applying it to locally and regionally specific productions. Rather than being overwhelmed with a detailed discussion here, the reader may be referred to De Marinis (1985), as well as to what is argued in Roth (2007a: esp. 49–50) and a critical response in Morel (2009).

Building on Lamboglia's work and on the data that had been examined by himself and other scholars since, Morel's *Céramique campanienne* espouses an interpretative approach that sees the emergence of the main three fabrics as an example of the fully-fledged type of Roman socio-economic imperialism. This had been foreshadowed by the distribution of the "Atelier des petites estampilles" but had been cut short by the setbacks experienced during the Second Punic War. Placing the date of the emergence of Campana A at the beginning of the second century, and that of the "Cerchia della Campana B" slightly later during the same century, this model viewed them as reflecting the expansion of the Roman agricultural regimen and a "mercantilistic economic system" that was based on both large slave-worked estates and substantial coastal emporia in Campania and Etruria respectively (Morel, 1981b; Carandini, 1981).

The problems associated with this interpretative approach have been addressed elsewhere, as have been the reasons why it is questionable that the study of ceramic evidence as espoused by those who favor such models makes historical sense. Therefore, it is not necessary to revisit these criticisms in any detail in this chapter of the *Companion*. Yet their continuing influence on the communis opinio necessitates not only that those views are acknowledged here but also that a far-reaching problem of Roman pottery studies be addressed: this is, the fact that ceramic chronologies, together with views concerning the heterogeneity and, especially, homogeneity of certain classes of pottery tend to be subjected to a text-based view of the history of Roman economy and society. On the whole, these approaches to using ceramics as an historical source assign privilege to highlighting certain trends which were brought about by specific events, such as wars and colonization, that are directly – or, in the case of Republican Italy, more often than not, indirectly – traceable in the historiographical record. Even recent, explicitly archaeological approaches ultimately remain prisoners of the chronological framework imposed by circumstantial historiographical evidence, especially insofar as the third century is concerned (for instance, Ferrandes, 2006; Stanco, 2009). Even the chronology imposed by the ceramic re-study of the Tiber Valley material betrays signs of chronological determinism (for instance, Patterson, Di Giuseppe and Witcher, 2004, although Witcher, 2009, later advised caution as to the chronological rigor with which periodization should be applied).

It is at this point that conventional ways of studying black-gloss wares as the main – albeit broad – category of widely produced and consumed example of Republican pottery reaches its limits. This problem is reflected by many approaches to the material culture of the period in general, and thus reveals much about the heuristic problems of the archaeology of Republican Italy as a whole. These limits are imposed by the very fact that students of Republican history and archaeology have tended to treat the material record unquestioningly as a source of confirmation for events that are more or (as in most cases) less well attested by historiographical and epigraphic sources. The rise of the "Atelier des petites estampilles," its supposed sudden demise and its replacement by other, increasingly "globally" distributed artifacts present us with a perfect case in point with regard to this procedure. The First Punic War – or, according to a more recent view – the even less well documented war with Pyrrhus, are portrayed as termini by which to explain ceramic distribution patterns, without ever discussing explicitly the relationship between the (perceived) history of events and the material record (Ferrandes, 2006).

Similarly, the wide acceptance – without any prima facie evidence, as is increasingly recognized – of the Second Punic War as a major turning point in the socio-economic history of Italy (and thus, implicitly, as possessing explanatory force as far as the distribution of black-gloss pottery is concerned) has led to a text-book view of the Campanian economy of the early second century, which more recent and nuanced studies struggle to rectify as far as non-archaeological readerships are concerned. Of particular note here is the realization that the production of Campana A wares – as the "signpost" of the new, post-Hannibalic order – in fact were made in the 220s, if not before; the Graeco-Italic amphoras which had previously been subjected to the same chronological straightjacket are now redated (see above and Chapter 6). While the cities around the Bay of Naples – and Naples, in particular – had begun to form a center of the production of black-gloss wares known as Campana A, during the latter half of the third century, this type of pottery did not immediately begin to wash the shores of the Tyrrhenian coast, nor to replace what had gone before. At a site such as the ancient settlement at modern Castiglioncello, for instance, Campana A accounts for just one – albeit a significant – portion of imported pottery, equaled in quantity by Volterran, south Etruscan and (to a lesser extent) north Campanian and even Iberian imports (see the contributions to Gambogi and Palladino, 1999).

Rather, we are increasingly dealing with a scenario in which Campana A is but a small part of the medley represented by the patterns of exchange that are exhibited by the material and, if one re-reads the sparse textual sources from a less Romano-centric point of view, overall historical evidence. If viewed as part of these, it indicates that it was regionalism, as opposed to what adherents to the unilateral, proto-rationalistic view (which is still fashionable in some quarters) would maintain. Archaeometric evidence, in particular, has begun to break down those neat correlations between supposed historical cause (lying in the realm of the history of events) and structural effect (evidenced by signpost-materials and embodying the – in itself – questionable notion of the "Roman economy"). In addition, it has been shown that even a class of pottery that is easily distinguished by the naked eye, and that was produced in a large but ultimately defined geographical region, could be significantly more diverse than previous studies may have suggested. We should also begin more seriously to turn our attention to how the difference in the materials, locations occupied and (one might add) specific technologies used might be significant in terms of our historical understanding of the workforce involved in the production of these wares.

While these important chronological and geographical distinctions are increasingly realized by archaeologists who specialize in Republican ceramics, it is unfortunate to see that relatively little progress has been made in realizing what they mean for the disciplines of Roman archaeology and history on the whole. Uncertainties are inconvenient, as far as text-led historians are concerned. Yet Campana A only represents the tip of the iceberg: even if the edifice of the Campanian economy is moved to accommodate a larger time-frame and to form part of a more diverse model of the central Italian and western Mediterranean economy of the late third/early second centuries, most historians will still accept it as conforming to the basic model of an agriculturally based, elite-run system that, for its expansion, relied on privileged access to the sea. These are, as has been pointed out elsewhere, the preconceptions shared by orthodox approaches to the economy of Roman Italy, no matter what interpretative quarter they might hail from. This supposedly neat sequence of events, and thus broader patterns – since, according to such views, material culture patterning will reveal the structural facts produced by events – should now be questioned, no matter how convenient its implications are. As shown above, this is best demonstrated with reference to the group of fabrics that were once referred to as Campana B. The final section of this chapter, therefore, seeks to outline an alternative route: that is, an approach to black-gloss pottery that assigns to it a space outside and beyond the realm of the patterns of the Mediterranean economy during the Republic, to which we usually subscribe.

5 The Future Scope of Ceramic Culture in the Archaeology of the Roman Republic

At the outset of this chapter, the production of black-gloss pottery was defined in terms of a series of choices on the part of the producer, which resulted in the stylistic properties of the finished vessel. As argued elsewhere, style does not just pertain to formal characteristics that would conventionally be associated with it – such as the shape of the rim, number and form of handles – but resides in all formal aspects of an object that come as the result of culturally informed choices (Roth, 2007a: 69–76). We may now add a second arena in which choices are made, and this concerns the consumption of pottery: how a certain vessel was put to use in a specific context, and how it related to other vessels and objects in this context, to name two obvious examples. As with any forms of interaction by human agents and the material world, the stylistic choices made at various stages during a vessel's life imbue it with meaning, not so much in the symbolic sense, as in one of contextual associations (Roth, 2007a: 65–85 discusses this within the context of the study of Roman pottery). We may see this in our own environments, in which, for instance, the choice to use a fine porcelain dinner set together with silver cutlery might reflect the high status attached to the occasion in question, those present, and the food consumed.

As the preceding discussion has shown, Republican black-gloss wares are characterized by a great degree of variability; and although an increasing trend towards standardization can be seen in some of the major productions throughout our period, diversity continues to flourish at the level of locally and regionally produced black-gloss wares that are usually found in the same contexts. This notion of variability extends beyond the level of fabric and – when stylistic aspects such as vessel shape, size, surface treatment and decoration are taken into account – characterizes the production and consumption of black-gloss wares as a whole.

Quite apart from the theoretical and methodological concerns expressed in the preceding discussion, it is against this empirical background of variability that both the dominant typological systems and mainstream interpretative approaches to black-gloss pottery were criticized. The key to the success of future work in this area lies in finding methodologically and heuristically sound ways of explaining stylistic variability, in all its aspects, in a culturally meaningful way.

To explore and understand this variability in more detail, therefore, remains a fundamental desideratum of the study of black-gloss pottery. In addition to archaeometric analysis, this effort needs to involve the chronological dimension. Too much of our knowledge still draws on un-stratified contexts like tombs, ship-wrecks and re-deposited assemblages, while too little is known about sequential changes in the patterns of consumption at settlement sites.

Last but not least, it may be time to question just how justifiable it is for us to single out black-gloss wares as an area of study in its own right. As was stated at the outset,

Roman archaeologists in particular take comfort in compartmentalizing their material evidence into supposedly neat categories. To a considerable extent, this is a sound approach in the case of black-gloss wares that, on account of the basic technologies involved in their production, their shared morphological repertoires, and their geographical and chronological distribution, in many ways represent a unified class of material. Yet there are arguments pertaining to all of those different aspects, which might lead to more context-based approaches integrating the study of black-gloss wares with that of other ceramic materials found in a specific context, be the latter defined by site, region or chronology. By isolating fine wares from other parts of the ceramic repertoire found in the material environment of human agents during the Republican period, we necessarily limit our potential to reconstruct and comprehend its cultural-historical significance. This is not to suggest that it is time to throw certain time-honored typological frameworks overboard. Rather, it is a call for fresh approaches to the study of the production, consumption and, of course, the distribution of black-gloss wares and Republican pottery in general. While it may remain practical to publish the ceramic assemblages from a particular site following conventional categories, this division is counter-productive at the level of cultural analysis (but see Jolivet, 2009: 494). Here, we must stop so as not to confuse the typologically useful with the heuristically sound, and start to move along interpretative avenues that may lead to a better understanding of ceramics as evidence for the cultural dynamics of the Republican period.

With black-gloss pottery, students of the archaeology of Republican Italy possess a uniquely rich corpus of evidence for the cultural history of this period. Much of this potential has already been tapped, leading to some fundamental conclusions about Roman economy and society, which have found their way into standard works of Roman history. It is now time to explore this material further, and to find ways in which it can help us address the new questions elicited by ongoing discoveries and significant shifts in the archaeological debate over the nature of "Roman Italy."

FURTHER READING

Significant parts of this chapter expand on the history of the study of black-gloss wares over the last six decades or so. Certain views on the chronology and geographical distribution of ceramic productions have recently been superseded by new discoveries, as well as by the increasing contribution of archaeometric studies. This way of proceeding is inevitable in a fast-developing field, particularly since much of what has been published (including standard works in the fields of archaeology and ancient history) is not in the first place concerned with ceramic studies. The reader should, furthermore, note that much of what is here is in the vein of views that were expressed in Roth (2007a, 2007b). This discussion is in respectful disagreement with the criticisms raised by Morel (2009), especially with regard to Roth (2007b: ch. 2), although where appropriate, every effort has been made to mitigate the shortcomings identified by Morel.

On Greek black-gloss technology, see Scheibler (1995: 71–106). With regard to some of the very early examples of black-gloss pottery found in south Italy, Lamboglia (1952) noted occasional difficulties in distinguishing these from contemporary Athenian imports. This is not to deny continuing mutual influences and inspirations between Italian and Greek pottery (or, indeed, between potters and craftsmen working in other media). However, Italian black-gloss wares are decidedly not imitation artifacts.

Early studies of some black-gloss wares include Pagenstecher (1909) and Beazley (1947). The typological systems proposed by Morel (1981a) and Lamboglia (1952) remain the standard way of describing black-gloss shapes today. Most readers of this *Companion* are likely to encounter these typological conventions in their studies and research. For a critical examination of black-gloss typologies, see Roth (2007b: esp. 63–4). The problem of regional manufacture and distribution was especially emphasized by Morel (1981a), who originally planned to produce a second volume on the subject of fabrics (as well as a third on stamps). Although neither volume has been realized to date, it should be noted that Morel continues to be at the forefront of research into the regional production of black-gloss wares and their distribution across Italy and the Mediterranean (Morel, 1998a, 1998b). The diverse wares of coastal north Etruria are discussed in Pasquinucci *et al.* (1998) and Gambogi and Palladino (1999). Regional variations are discussed by Scott (2008), which is the most recent publication of a substantial black-gloss deposit (Cosa); for contrasting reviews of her study, see Roth (2008) and Jolivet (2009).

The straightforward method of manufacturing black-gloss wares is emphasized by Olcese (2009), and somewhat qualifies the argument made in Roth (2007a: 134–9) regarding the relative value of black-gloss and other wares. This method contrasts with Italian sigillata (Schneider, 1990). Particular clays were used in the production of Campana A but many parts of Italy produced black-gloss wares of varying technical quality, employing a large number of different types of clay: even a cursory, visual comparison between the clay matrices of a Campana A and a Campana B (used here in the wider sense of the "Cerchia della Campana B") vessel will illustrate this point to the field archaeologist. For an overview of certain clay types (by geological origin) and their suitability for specific types of pottery and earthenware, both in general and with a particular focus on Pompeii and the Bay of Naples, see Olcese and Picon (1998) and Peña and McCallum (2009b: 167–8). On Ischian clays used in the manufacture of Campana A, as well as possible alternatives, see Morel (1998a: 11). In a similar way, observation (made in the field, E. Stanco, pers. comm.) demonstrates that vessels of the same shape that are produced in different parts of south Etruria show consistent variations in the ways in which their bases were constructed.

This ties in with the significance of culturally conditioned technological choices in creating material form, which are discussed by the contributions to Lemonnier (1993); Gosselain (1998) provides a particularly useful, ethnoarchaeological perspective for the impact of cultural knowledge on the manufacture of pottery and on the style of the finished product; see also van der Leeuw (1993). For a summary of the ethnoarchaeological debate on style and its relevance to the study of Roman and black-gloss pottery in particular, as well as for additional references, see Knappett (2005) and Roth (2007a: 69–76, and ch. 3 *passim*).

CHAPTER 6

Amphoras and Shipwrecks: Wine from the Tyrrhenian Coast at the End of the Republic and Its Distribution in Gaul

Fanette Laubenheimer

1 Introduction

Why will amphoras be the focal point of our discussion on commerce and the economy? Amphoras are marker artifacts which refer to economic life, the production of food products, exchange and markets. Indeed, "amphora" is a very old word; one finds it in Mycenaean documents and also in Homer, in the form "*amphoreus*," which can be translated as "a carrying vessel with two handles." But the word becomes "amphora" in Latin, and it remains a word that is well understood and often used by contemporary scholars and indeed by the general public. At the same time, the names of other food product vessels – such as "*kylix*," "*oinochoe*," "*lagynos*" or "*krater*" – remain in the vocabulary of the specialists, and outside of the use of the general public. If the word "amphora" is thus situated today in general use, it is surely not by chance. The amphora is distinguished from other ceramic vessels, and in particular from other storage vessels, primarily by its economic function: amphoras were never manufactured for anything other than utility, that is, what they were made for and they what contained and transported.

The idea of manufacturing a container with a large capacity and two handles to use for the transport of food products is very ancient. The shape originated in the Near East, probably with the Canaanites, the ancestors of the Phoenicians, on the coasts of Syria and Palestine, where amphoras appear at the end of the third millennium (Grace, 1979; Laubenheimer, 1990a). They were largely used in Egypt as well for the storage of wine or beer. Such containers are also known in ancient China: in the area of Shaanxi, the site of Banpo (near Xi'an) produced several examples of amphoras dating between 5000 and 3000 (Wang, 1994).

Simple packing containers, produced in a series and thus cheaply, amphoras were manufactured everywhere in the Mediterranean during antiquity. They were created to

A Companion to the Archaeology of the Roman Republic, First Edition. Edited by Jane DeRose Evans.

export food products – at first wine, but then also olive oil or fish sauce – to places far from their place of production. They were the means of marketing the surpluses of commercial farming. Sometimes the presence of workshops for the production of amphoras in rural areas is the only evidence for the economy of agricultural production.

Like our modern bottles, the forms of the amphoras change according to their contents, the country of their origin, and the era in which they were made. The identity of the product contained within can sometimes be judged by the presence of resin in the interior of the amphoras, if they contained wine or fish sauces; resin is not found in amphoras that contained oil. Sometimes stamps with the names or emblems of various types indicate the organization of production. Inscriptions painted on the top of the belly or the neck were used to announce the nature and quality of the contents, the quantity, the time or place of production, even the name of the estate or the name of the producer, as if they were control marks. The consumer would thus only have to read the painted inscriptions to understand the significance of the vessel; today, most of the inscriptions have disappeared and the situation is more complex.

Heavy and cumbersome, amphoras were transported primarily by boats, either by sea or by river barges; when the bulk shipment was broken up, they were transferred to carts or lashed to the backs of beasts of burden. Neither the weight nor the fragility of the container, the contents nor the distance carried were real obstacles to the distribution of amphoras, which were carried thousands of kilometers throughout the Roman world. Once at its destination, its contents were sold or exchanged as a whole or in part. When the amphora was empty, did it become a vessel without further utility? That is precisely what interests archaeologists: since terracotta is virtually indestructible, it is preserved in archaeological levels on sites. There is, indeed, no reason why we cannot find entire amphoras as accidental losses, placed, for instance, in rows in cellars or in shipwrecks. The material with which the archaeologist works in general consists of waste; here, the remains of amphoras have a whole history contained within themselves. We must try to understand why we find fragments of amphoras in any given context, before we launch into specific economic interpretations of the finds.

The study of amphoras was first undertaken by the erudite German epigraphist Heinrich Dressel at the end of the nineteenth century, following two exceptional discoveries in Rome. One, a space under a floor in the Castra Praetoria in Rome, was packed with dozens of discarded whole amphoras, with their painted inscriptions preserved. The second, the pile of discarded amphoras in the man-made mountain called today "Monte Testaccio," was built at the spot where ships were unloaded at the bank of the Tiber River; both sites held remarkable epigraphic material for Dressel's study. He first deciphered the painted inscriptions and the stamps, and then created a typology of the amphoras, speculating about the relationships between the shape, the origin of production, the date, and the contents of these objects. In his work, he discovered that the vessels could present evidence in the study of economic history. Yet this first leap into amphora studies was not followed up by scholars for more than half a century.

A return to amphora studies is, curiously, one result of the Second World War. It was then that techniques of undersea diving were developed. These allowed the excavation of shipwrecks, which made masses of material hitherto unknown available to researchers. Amphoras generated research of all kinds after scuba gear became commercially available, including the study of workshops and the production of the vessels, morphological classifications, and the creation of databases of the mineralogical, petrographic and chemical composition of the clays, analyses of their contents, lists of stamps and painted inscriptions found on them. In short, everything that could "speak" from these containers has been used in the context of contemporary archaeology. That is why I have chosen amphoras, as remarkably talkative tracers of economic history. Relations between the Tyrrhenian coast of Italy and Gaul are, thanks to them, of exceptional interest.

2 Italian Wines and the Vineyards of the Tyrrhenian Coast

By the second century, and especially at the end of the Republican period, vineyards were developed in Italy along the Tyrrhenian and Adriatic coasts. The production of amphoras is the best witness of this process of intensive viticulture, which is differentiated by the types of amphoras produced on either coast. From the eastern side comes the bulbous amphora type that is today called Lamboglia 2. They were exported to the eastern Mediterranean and in particular to the free port of Delos, home of the largest slave market in the ancient world. On the western coast of Italy, by contrast, the slender wine amphoras called Dressel 1 (following the typology established by Dressel) were produced (Figure 6.1). They were

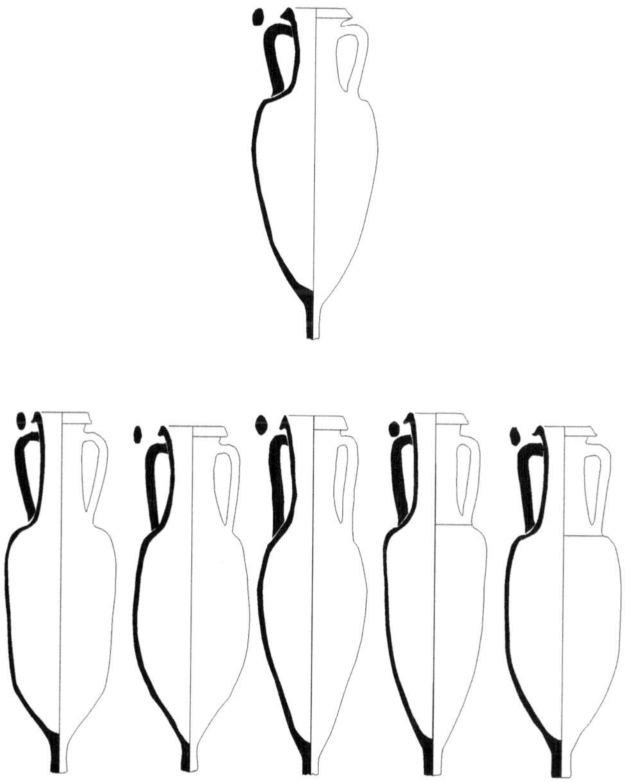

Figure 6.1 Typical profiles and sections of Dressel 1 Amphoras. Source: Drawing by F. Laubenheimer.

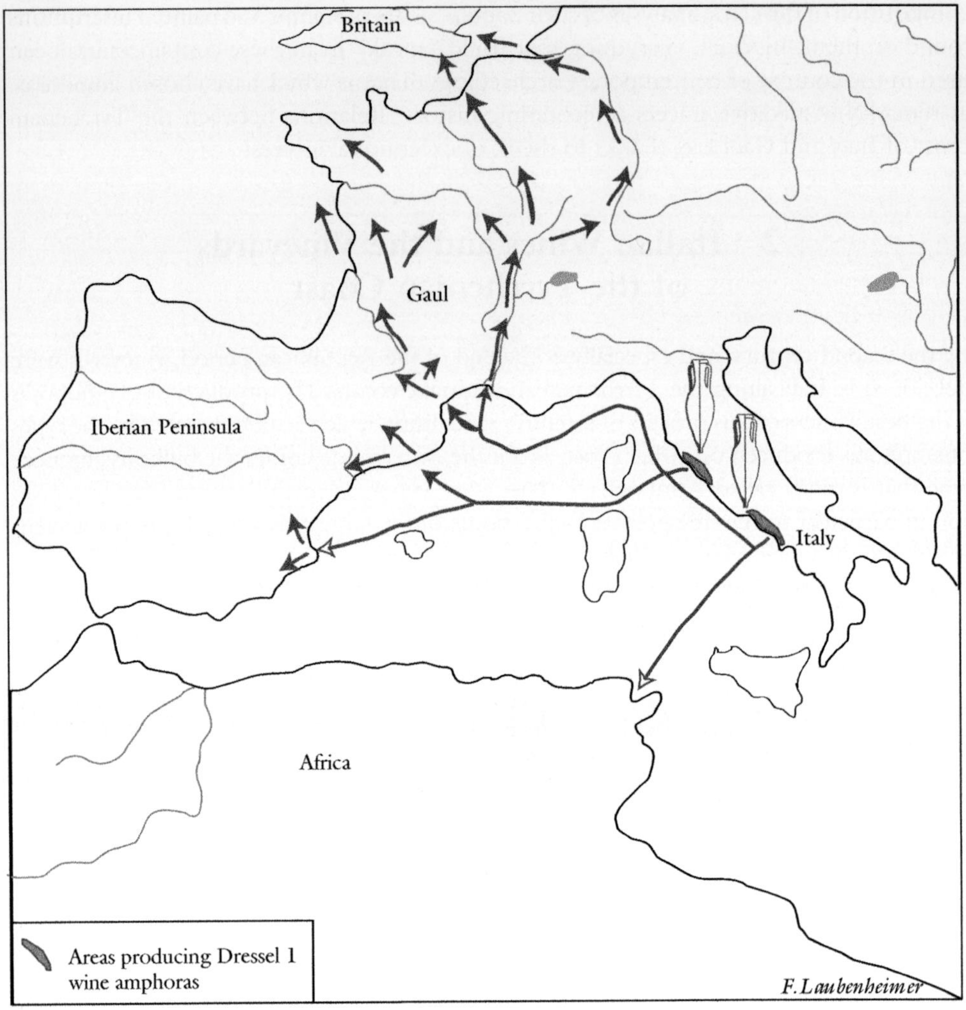

Figure 6.2 Trade routes of Dressel 1 Amphoras in the Republican era. Source: Drawing by F. Laubenheimer.

intended for trade in the entire western half of the Mediterranean basin, but are found primarily in Gaul (Olmer, 2003: 187 and fig. 49). Recent research allows us to estimate that these were made in about one hundred different workshops. I will concern myself with the production and trade of the Dressel 1 amphoras, since they are better known to us than the Lamboglia 2 types.

Two basic production areas are known today: in the north, Etruria, and in the south, Latium and Campania (Figure 6.2). We do not yet fully understand what was the primary cause of the trade – was it the development of Italian viticulture or the western demand for that wine? We can only note that in both the eastern and western halves of the Mediterranean a flourishing slave trade was essential in the exchange of goods for Italian wine.

In Etruria, large slave-staffed estates were the possessions of powerful senatorial families. The case of the Sestii, friends of Cicero, is a remarkable example. They owned vineyards, but also maintained an important manufactory of amphoras to market their wine. Their amphoras are stamped with their name (Manacorda, 1981). The estate of the Sestii is located in the territory of Cosa, in the ager Cosanus, and the amphora workshop is located beside the sea, near the Portus Cosanus. The significant production of this manufactory both provided the packaging needed for the produce of their own fields and allowed them to sell empty vessels to smaller producers without their own amphora workshops. The manufactory was not excavated, but underwater explorations in the port, led by A.M. McCann (1987), raised more than 700 amphoras, with 14 different stamps, all bearing the name of the Sestii. The name could be followed by various symbols such as a trident, anchor, caduceus, palm, double axe, star or arrow.

Alongside these stamps with names, other types of stamps are known from Etruria and in particular, stamps consisting of two letters, found at the base of the handle, are known from surveys; the clay of the amphoras seems to be characteristic of local workshops that have not yet been located. Manacorda suggested in 1981 that the two letters did not mark the initial letters of the praenomen and nomen of the producer, but were a code. And indeed, one of the letters varies in alphabetical sequence, so, for instance AA, AB and AC or AA, BA, BC. The systematic study of the particularities of the stamps was undertaken recently by Olmer (2003: 157–72). She offered four different interpretations for the meaning of each letter: the workshop and owner (*officinator*); the workshop and production cycle; the production cycle and workshop; alphabetic numeral and cycle of production. In the last case, the "large number of possible codes makes it possible to envision a simple and efficient counting of the number of amphoras coming from a lot of N individuals" (Olmer, 2003: 165). This seems to me a good solution, providing a clever method of counting that fits well with this type of amphora production, and different from the kind that displays the name of the famous owner, like the Sestii. The local distribution of these coded stamps shows either the location of the workshops in the vicinity of Cosa or a system of stamps relevant to the organization of different production areas.

A third center of production, altogether different but still in Etruria, is becoming better known. It is that of Albinia (Grosseto), which was in production in the last two centuries of the Republic. Over the past decade, exploration of the site was carried out under the supervision of D. Vitali, with a team from the Universitá di Bologna, where I worked as a specialist. Located not far from the sea, in a plain about 300 m from the mouth of the Albegna River, the workshop covered a huge area, about 200,000 m^2, as shown by fieldwalking and magnetic surveys. This is certainly the largest amphora workshop explored to date. Its placement near the coast may coincide with a likely small port (Calastri, 2007). The excavated part of the workshop includes, in its first phase, four large principal kilns, all of the same size, which were probably all in production at the same time (Vitali, 2007: fig. 21). One of the kilns, whose measurements at the floor of the stoking chamber were 3.20 by 5.30 m, can be reconstructed to have a 17 m^2 firing chamber, based on the premise that the height of the firing chamber is at least equal to the measurement of the longer side of the stoking chamber (Laubenheimer, 1990b: 70; Leenhardt, 2001). A very large batch of whole Dressel 1 amphoras (more than 100), which were discarded, were preserved by their reuse in a subfloor nearby (Vitali, 2007: figs. 10, 11). A paleomagnetic study of the position of objects in the kiln, in a study led by P. Lanos (Centre National de Recherche Scientifique, Rennes) and his team, was able to show that the amphoras were fired upside-down (Hill *et al.*, 2007). It is now possible to simulate the load in the kiln by placing the jars in a staggered fashion, with their feet in the air (Vitali, 2007: figs. 22, 23, 24). Because the amphoras have a maximum diameter of 30 cm, 170 vessels could be placed in one level. Because of their weight, the amphoras were stacked in three (rather than four) levels, allowing over 130 amphoras to be fired in one batch. We estimate that a firing would take one month, and the firing season was only nine months, excluding the winter and spring months (due to poor drainage in a land that often flooded) which gives a magnitude of nearly 3,900 amphoras per kiln per year. When one considers the shipwrecks that carried several thousand amphoras of the Dressel 1 type per load, the number looks appropriate. It also has the merit of helping us understand for the first time the quantity of the production of Tyrrhenian wine and the economic history of the region, for which we have been lacking evidence.

The excavations of Albinia have provided a large quantity of amphoras bearing a stamp – more than 500 – but not all are stamped (Laubenheimer, 2007). Only the late Greco-Italic amphoras (the older Dressel 1A and Dressel 1C) are thus marked. The Dressel 2/4 amphoras, produced at the end of the lifetime of the workshop, are not stamped. The stamps of Dressel 1A can be immediately differentiated from those on the Dressel 1C amphoras, since they are completely different. They illustrate two different types of organizations, undoubtedly linked, but their contents are not the same, as we will see. On the former, the names of about a dozen individuals appear. These names are of eastern origin or show a slave status: for example, AIAS, APPEL, NICIA, MAHE and SOTIC. Much less numerous are stamps which are codes of one, two or three letters. Eight different stamps appear on the Dressel 1C vessels, the majority of which are composed of initials followed by ".S," perhaps an abbreviation for "servus" (slave): for example, AR.S, DA.S and DI.S.

What was in those amphoras? Recent work in Baetica (south-east Spain) has shown that these amphoras contained fish sauces (Etienne and Mayet, 2000: 74–80). The same is true at Cosa (Will, 1987), where Dressel 1C amphoras are used in conjunction with the nearby fishery. However, in the case of Cosa, two Dressel 1C amphoras are stamped with the name of Sestius, the owner of vineyards, in stamps identical to those used on the wine amphoras. Thus, Sestius produced both wine and fish sauce (Laubenheimer, 2007: 77). Presumably, the workshop in Albinia produced amphoras in the same manner for both wine and fish sauce. The organization of the manufacture of vessels intended for one type of food or another was different, as illustrated by the stamps. But in the case of Albinia, we do not yet know if the workshop was owned by a large landowner. These few examples show the sophistication of workshops in Etruria, where we are only beginning to glimpse the complexity of the production of amphoras.

What is known of the wine of the great landowners in Etruria? Curiously, although we can see that production was important, there is no mention in ancient texts or elsewhere of the wines of the region, in contrast with those produced in Campania (Tchernia, 1986: 109). All indications are that this was a relatively recent production, or one favoring quantity over quality.

In contrast, remains of workshops for amphoras in Latium and Campania are known to archaeologists, but they have not yet been explored. Thus, archaeological evidence about the amphoras is frail, though our information about the wines of the region is robust. Strabo (V.4.3) writes, "This is where the Romans find their best wines."

The *grands crus* emerged at the end of the second century, and developed in the first century in tandem with the expansion and growing wealth of Italy (Tchernia, 1986: 63–5; Tchernia and Brun, 1999: 13–15). The most famous of these vintages were those of Falernia, Caecubus, the Alban hills, Mount Massicus and Cales, which were found on the best tables of Rome. A great wine was always aged from five to 25 years and sometimes much more. According to certain ancient texts (for which, see Tchernia, 1986: app. 2), these are white wines, and described as being strong, sweet, amber in color, strengthened with age, and expensive. For the aristocracy, it was important to drink great wines, but especially to have old vintages (Tchernia and Brun, 1999: 34). Large wine cellars of the oldest aristocratic families began to appear in the second to first centuries.

The density of the amphora workshops, the surveys of the wine-producing villas and the numerous references in ancient sources concerning vines leave no doubt about the importance and the fame of the southern vineyards. However, one wonders if such refined products were ever shipped to customers in the provinces of the western Mediterranean – Gaul, Spain and Africa. The painted inscriptions on the exported amphoras have mostly disappeared, or, if they survive, do not provide sufficient guidance about the quality of the transported wine. It seems that the genius of the Roman economy was to have differentiated among clientele. We must contrast the celebrated vineyards and the Roman elite customer to the vineyards for mass production, as in Etruria, which served the general trade of table wine.

3 What Can We Say about Shipwrecks?

Shipwrecks represent only the accidents of ship traffic which did not arrive safely in port, which must have been a small minority of all the ships carrying cargo, but even those tell us a tremendous amount. An inventory of shipwrecks in the Mediterranean gives us interesting insights into ship traffic (Parker, 1992). If we look at commerce between Italy and Gaul at the end of the Republic, there are 44 wrecked merchant ships known that were loaded with Dressel 1 amphoras that sank along the coasts of Gaul between the second and first century. Yet shortly before this era, at the end of the third to second century, only 13 ships carrying Greco-Italic amphoras are known. Regardless of the specific numbers of wrecks, the increase in numbers is significant in showing a boom of imports at the end of the Republic. What can be said about the quantity, in terms of the volume of vessels?

A few small coastal vessels, about 15 m long, which were capable of carrying amphoras around the western Mediterranean, are known wrecks, as the one off Cavalière (Var). But underwater excavations have revealed much larger vessels including those of Madrague de Giens, near Toulon (Var). Forty meters long, it was carrying about 7,000 amphoras. Another wreck, off Albenga on the Italian coast, was carrying even more – about 10,000. These *muriophoroi*, as they were called in ancient sources, were the largest merchant ships of their time. It was not until the fifteenth century CE that Genoese and then Venetian galleys exceeded those tonnages (Pomey, 1997: 89). The *muriophoroi* were built in order to transport enormous quantities of wine in Dressel 1 amphoras for trade, especially with Gaul.

The excavation of the wreck of Madrague de Giens, in water 21 m deep, in 1972–5 showed that the main cargo was wine (Tchernia, Pomey and Hesnard, 1978). The cargo of the boat was still intact, with three layers of amphoras carefully stowed so that the bases of the upper layers fit between the necks of the amphoras in the lower levels. The amphoras of the bottom-most layer were surrounded by branches of juniper, heather and rushes. Above, in the spaces between the amphoras, were the secondary cargo, consisting of coarse wares and Campanian black-gloss tablewares (Pomey, 1997: 149). The amphoras were still intact, with their cork lids sealed by pozzolana, a volcanic cement. Many of them carried stamps. The most common stamp was the name of Publius Veveius Papus, which located the production facility in the Latio-Campanian region of Terracina. The amphoras retained traces of their contents in the reddish-brown deposits at the bottom of the containers, which proved to be good red wine (Tchernia and Brun, 1999: 5). We would like to think that this wreck provides evidence for the massive export of red wine to Gaul. Thus, Campania and Latium had two categories of vineyards, supplying two kinds of wine: refined white wine drunk by the Romans; and red table wine for bulk export.

4 Tyrrhenian Wine Consumption in Gaul

Gaul was the primary market for Tyrrhenian wine. The first tentative imports were introduced in the late third and early second centuries, as we can tell by the small number of Greco-Italic amphoras found in excavations. They were found in sites in southern, central and northern Gaul (Laubenheimer and Marlière, 2010: fig. 4) and even in Britain (Laubenheimer, Ménez and Le Forestier, forthcoming). The influx of Italian wine reached its zenith during the second and first centuries, and we need to discuss why this is so. In part, the importation of wine was limited due to the Gallic preference for beer. In Greece and Rome, where wine production was controlled and regulated, wine-drinking was a refined art, a true measure of civilization. The Gauls drank only beer, which the Romans considered a drink for barbarians. This shocking custom on the part of the Gauls resulted in contempt and incomprehension on the part of the Romans, as we find in the ancient texts. Faced with this new beverage choice, the Gauls then adapted their own practices of drinking wine. Thus, Dionysius of Halicarnassus, a Greek intellectual in Rome in the first century, and known for his love of classical culture, wrote, "The Gauls at that time [s.v. the early fourth century] had no knowledge either of wine made from grapes or of oil such as is produced by our olive trees, but used for wine a foul-smelling liquor made from barley rotted in water" (*Ant. Rom.* 13.11, trans. E. Cary, 1950, Loeb). This is clearly the stunned and contemptuous reaction of a Greek, and is evidence of a clash of cultures. Faced with a drink he does not know – beer – he has no word to designate it, and does not hide the disgust it inspires in him.

Diodorus Siculus writes in the same era:

> Furthermore, since temperateness of climate is destroyed by the excessive cold, the land produces neither wine nor oil, and as a consequence those Gauls who are deprived of these fruits make a drink out of barley which they call *zythos* or beer, and they also drink the water with which they cleanse their honeycombs. The Gauls are exceedingly addicted

> to the use of wine and fill themselves with the wine which is brought into their country by merchants, drinking it unmixed, and since they partake of this drink without moderation by reason of their craving for it, when they are drunken they fall into a stupor or a state of madness. Consequently many of the Italian traders, induced by the love of money which characterizes them, believe that the love of wine of these Gauls is their own godsend. For these transport the wine on the navigable rivers by means of boats and through the level plain on wagons, and receive for it an incredible price; for in exchange for a jar of wine they receive a slave, getting a servant in return for the drink.
>
> (V.26, trans. E. Cary, 1939, Loeb)

For his part, when Posidonius of Apamea traveled to Gaul in the first century, he observed a Gallic banquet: "And the liquor which is drunk is, among the rich, wine brought from Italy or from the country about Marseilles; and this is drunk unmixed, but sometimes a little water is mixed with it. But among the poorer classes what is drunk is a beer made of wheat prepared with honey, and oftener still without any honey; and they call it *korma*" (quoted in Athenaeus, *Deipnosophistae* IV.36, trans. C. Gulick, 1929, Loeb).

A real market for wine did develop in Gaul. This trade was organized around the demands of the rich aristocrats and political leaders, as can be seen on the Arvernian coins, which feature amphoras, especially the coins of Vercingetorix. On the Italian side this expensive wine was traded for slaves, and probably metal (Laubenheimer, 1990a: 71–5; Tchernia, 2009). Archaeology and texts combine to show the specificity of the report of the Gauls in regard to wine. Recent work has focused on Gallic consumption, which has been well attested in the performance of religious rites, known through careful excavation (Poux, 2004a, 2004b).

The drink of eternity, the symbol of power, Italian wine was treated with respect, as can be seen by the transport amphoras that were placed in the graves of Gallic chieftains. There are numerous examples: in Trier (Metzler, Waringo, Bis and Metzler-Zens, 1991), among the populations of central Gaul (Ferdière and Villard-le-Tiec, 1993) and in southern Gaul (Garmy, Michelozzi and Py, 1981). Pits and wells were filled with amphoras, found in the hundreds in the area around Toulouse, in the region of Gers, at Agen or by the dozens in Rodez (Aveyron). These were originally placed with evidence of sacrificial offerings probably for a chthonic cult: animal bones and millstones, speaking of the fertility and fecundity of the land, and – by the hundreds – Italic wine amphoras, mostly broken, but nearly entire. So wine was ubiquitous, found, for example, in the pits of Rodez. It is here that the latest excavations have uncovered the remains of at least 270 Italian amphoras, enough to hold about 6,000 liters of wine. These amphoras date between 120 and 80 (Gruat, 2004). Amphoras arrived by the thousands in the market at Toulouse, which itself consisted of about 500 pits (Benquet, 2004).

Feasts were another important channel of consumption of wine in Gaul. The ancient texts describe the process that modern excavations have confirmed (Poux, 2004b). There was an orgy of food and drink, consumed in an ostentatious manner in order to reinforce the authority of the man who organized the feast, a display which had nothing to do with the refined symposium of the Greeks and Romans. The participants drank their wine unmixed, in order to achieve intoxication. A description of the political campaign waged by Luern of the Arverni is given by Posidonius (quoted in Athenaeus, *Deipnosophistae* IV.37): "to win the favor of the mob he rode in a chariot through the fields scattering gold and silver among the myriads of Celts who followed him; he also made an enclosure twelve stades square, in which he set up vats filled with expensive wine, and prepared a quantity of food so great that for several days all who wished might enter and enjoy what was set before them, being served continuously" (trans. E. Cary, 1928, Loeb). In Corent (Auvergne), the home of Luern, excavators have traced vast enclosure fences and large banqueting halls with lines in the soil showing the troughs for wine, along with fragments of amphoras and large quantities of animal bones, the remains of a feast (Poux, 2004b). We can visualize the elite warriors who gathered there at fixed dates for religious festivals, including political and electoral feasts.

Other examples of the remains of feasts and libations are provided by ditches filled with dozens of vessels and animal bones, for example in Lyon, on the plateau of La Saar (Mandy *et al.*, 1989; Poux, 2004a: 527–35) or in Aix-en-Provence (terrain Coq) (Guyon *et al.*, 1998: 206–7 and fig. 399). In one particular ritual, the mouth of the amphora was decapitated at the base of the neck, with the cork still in place (Poux, 2004a). Even if the wine was Roman, it was consumed in a very Gallic manner. Indeed, the consumption of wine in Gaul belongs to the religious, public, festive and political spheres. According to Herodotus, one feature of the barbarity of non-Greek-speaking peoples was their lack of manners in consuming wine (e.g. the Scythians, who drink it neat and in great quantities; Herodotus 6.84). The same sentiment existed in Rome, where one obeyed the proper code for drinking. In Gaul, there was no lack of ritual, but there were different ways of consuming wine; we cannot say that these rituals were barbaric.

5 The Means of Distribution of Italian Wine in Gaul

The distribution of Italic amphoras in Gaul has been studied sporadically over the last 30 years, with the result that distribution maps are still incomplete. But important progress has been made in recent scholarship by Olmer (2003) for Burgundy, Poux (2004a) for the area around Lyon, and in more regional surveys in the north-west of France (Laubenheimer and Marlière, 2010). In addition, a colloquium was held in early 2007 in the Musée archéologique de Lattes (Hérault); this will result in a rich publication (Olmer, forthcoming).

The most important era of importation of wine from Italy to Gaul dates from the late third century up until, and shortly after, the conquest of Caesar. The first imports were carried in the so-called Greco-Italic amphoras, which are known from 15 shipwrecks in the western Mediterranean (Cibecchini and Capelli, forthcoming). They entered Gaul through Pech Maho (at the mouth of the River Aude) or Lattes (Hérault) early in the second century, and towards the middle of the same century in Narbonne, or before the conquest of Narbonne, in Toulouse. Several amphoras in Toulouse bear inscriptions in the Iberian language spoken in the hillforts of western Languedoc. The Greco-Italic amphoras were also distributed in the interior; for example, they are found in Corent (Auvergne) between 160 and 150, at St-Gence (Limoges), Châteaumeillant (Cher), Lyon (Rhône), Meulan (Seine Valley), Paule (in the heart of Brittany), and even in the north-west of Gaul, where a dozen examples were discovered from the Eure to the Somme.

But the bulk of the imports of Italian wine occurred in Dressel 1 amphoras, beginning at the end of the second century through the first century. Strabo emphasizes the importance of the river system for importations into Gaul:

> Now the whole of this country is watered by rivers: some of them flow down from the Alps, the others from the Cemmenus and the Pyrenees; and some of them are discharged into the ocean, the others into "Our Sea." Further, the districts through which they flow are plains, for the most part, and hilly lands with navigable water-courses. The river-beds are by nature so well situated with reference to one another that there is transportation from either sea into the other; for the cargoes are transported only a short distance by land, with an easy transit through plains, but most of the way they are carried on the rivers – on some into the interior, on the others to the sea.
>
> (IV.1.2, trans. H. Jones, 1923, Loeb)

Two major lines of distribution have been uncovered by archaeological investigations (Figure 6.2). The first, on the axis of the Aude and Garonne Rivers, served the Volques and Ruthenian tribes, in the territory of Narbonne. A system of taxes and tolls levied abusively on wine amphoras by the governor of the province, Fonteius, was opposed by the Gallic populations. In the famous trial, Cicero himself defended Fonteius, who was accused, in part, of *crimen vinarium*, or illegal customs taxes on wine; we have fragments of Cicero's publication of his defense in *Pro Fonteio*.

The second axis, which runs north–south, supplied central and northern Gaul by means of the Rhône and Saône River valleys, but also by secondary routes such as the Hérault, Gard and Allier River valleys (Olmer and Maza, 2004: 209ff). The distribution was carried out through the major urban centers, the foremost of which are Bibracte and Toulouse. Besançon, Basel, Bourges, Chartres and the oppidum of Titelberg in Luxembourg have been less well studied, but certainly functioned as distribution points. What the recent research highlights is a widespread distribution of these amphoras which less regularly reaches into rural areas. The evidence is made clear along the Aude–Garonne axis, where recent studies have shown Dressel 1 amphoras on every small rural site reaching up into the north-west, into Auvergne and Burgundy (Loughton and Jones, 2000; Olmer and Maza, 2004; Loughton, 2005; Laubenheimer and Marlière, 2010). In the region around Paris, distribution in rural areas is uneven (Séguier, forthcoming; Barat and Laubenheimer, forthcoming). In the heart of Brittany, the aristocratic residence in Paule received hundreds of wine amphoras from the vineyards of Italy, while none have appeared in the south of that region (Laubenheimer, Ménez, and Le Forestier, forthcoming).

Given the new distribution maps, the fruit of systematic reviews, there arises the problem of identifying consumers, or, rather, buyers of Italian wine, if we assume the presence of amphoras can only signify the purchase and consumption of wine. Yet we know that the reuse of amphoras is common, especially for the heavily-built Dressel 1 amphoras. The question of the buyers of wine still needs to be researched. For some scholars, only the elite are able to buy and distribute wine, since it is an expensive commodity. For others, the presence of amphoras in small rural sites shows that the drinking of wine is widely popular. It is not possible to say with certainty that the discovery of amphora sherds at a site is evidence of the consumption, or even the purchase, of wine.

Italian wine was distributed throughout Gaul, except perhaps in northern Belgium, among the Nervian tribe, as is seen in a recent study (Laubenheimer and Marlière, 2010: fig. 4). Caesar observed (*B Gall.* I.1) that the merchants rarely penetrated into the territory of the Belgae, nor did they introduce "the commodities that make for effeminacy," that is, wine. In reality, a large part of that territory indeed received wine in Italic amphoras. A singular phenomenon also resulted from special conditions: the market was controlled by Italian merchants, perhaps with the cooperation of Iberian and Gallic merchants, in search of alluvial or mined gold, as suggested by Tchernia (2009: 102). This mechanism seems to draw Italian wine imported in Greco-Italic amphoras at the end of the third and into the second century. But in the last quarter of the second century, beginning with the creation of the province of Narbonne and lasting into the first half of the first century, imports of wine exploded, and we see amphoras arrive by the thousands. The strength of the Gallic demand was answered by Roman traders, in a fruitful exchange of wine for slaves and metals.

6 The End of the Wine Trade from Italy

The decline, and then the end, of the monopoly on the importation of Italian wines into Gaul after Caesar's conquest is not uniform throughout Gaul. In Picardy, at the Caesarian camp of Arras, we find almost exclusively amphoras for Italian wine. In Senones, Carnutes or Parisii the level of the consumption of Italian wine around the time of the conquest and even later remains very high (Séguier, forthcoming). However, in Lyon, imports had been in decline since the 40s (Desbat, 2004).

The strong drop in the import of Italian wine in the second half of the first century was coupled with the emergence of new wines from the east, from Spain and from Gaul, at the beginning of the Augustan period. This change of supply marked a profound shift in the economy and especially in society. Amphoras became quantitatively rarer, there was no longer an Italian hegemony in production, and the consumed products were packed in different containers; in some cases, barrels replaced amphoras. The loss of trading products available to the Gallic elites – first slaves, then metals from mines – largely explains the collapse of the Italian trade monopoly. But also, the elites lost the habit of drinking "like a Gaul." Wine was always present at aristocratic tables, including Italian wine, but in more modest quantities, and there was a great variety of wines consumed. Wine was now part of everyday life, sold in shops just as it was in Pompeii. And so we find beautiful artwork in the gravestones of vintners now in the archaeological museums of Dijon and Bourges. Being accompanied by wine in death is no longer reserved for princes, but becomes part of everyone's life: there are many gravestones that show persons of modest means drinking wine.

Finally, the vocabulary of images was enriched by associating wine with the entirely Gallic god, Sucellus, the distributor of drink. He was traditionally shown with the attributes of an olla or jar in his right hand and a mallet in his left, with a barrel – probably holding beer – at his feet. The novelty is that Sucellus can, during the Empire, be associated with wine amphoras, just as he is in a beautiful statue in Javols (Lozère).

Acknowledgments

I would like to thank Jane DeRose Evans for translating this chapter.

FURTHER READING

The best analyses of the production of amphoras and the use of wine and fish sauces in Gaul are Tchernia (1986), Laubenheimer (1990a), Olmer (2003) and Poux (2004a). A basic handbook in English on how amphoras are studied is Peacock and Williams (1991); on the wine industry in Italy (though without the benefit of information from recent excavations) see Purcell (1985); see also Laubenheimer (2010).

CHAPTER 7

Coins and the Archaeology of the Roman Republic

Jane DeRose Evans

1 Introduction

Ask an archaeologist why excavation coins can be useful, and s/he will probably answer that s/he expects the numismatist to help date the strata (see e.g. Rotroff, 1997; Walker, 1997). A few numismatists have begun the discussion on what coins mean on an archaeological site, but it is an area that is still greatly in need of development. Some of the silence is due to the specific language of numismatists, which can be daunting, even though there are now very good catalogs of coins minted in the Republican period to help the scholar of the ancient world. However, an area that is sorely lacking is still the publication of excavations, in part because the cost of publishing bodies of coin data is higher than other bodies of data, and in part because there are few numismatists who are also field archaeologists. I will begin by discussing the reasons why coins are found in strata, and uses and problems of using coins to date strata, and in so doing, explore the different ways that numismatists can define the date the coin was issued. Then I will explore how coins were used, including a particular method of use, in hoarding. Finally, I will end with a case study, the introduction of the Iberian denarius, to sketch out the ways excavated coins are combined with history, epigraphy and numismatic studies to provide both questions and answers about the Roman world during the Republican period.

2 Coins in Excavated Strata

Certain inferences about the nature of coin loss and excavation finds have been established for ancient sites. Excavation coins are differentiated into two overall categories, intentional

A Companion to the Archaeology of the Roman Republic, First Edition. Edited by Jane DeRose Evans.

and unintentional loss. Intentional losses (such as hoards or votive deposits) are rarer, and can be more closely tied to an individual's or community's behavior. Unintentional losses account for the huge bulk of the finds on any site. We expect that usually these are the coins that have been dropped as they were transferred from hand to pouch or hand to hand. The user of the coin may have been unaware of the loss, or he could have unsuccessfully looked for the coin after his loss. Almost all of these coins will be of low value.

There is less agreement on exactly how to analyze a body of excavation coins. Numismatists are well aware that there are many factors that influence the number of coins found on any site, and their comparisons of sites show that they are working with often unrecoverable factors to explain the data. For instance, the annual supply of coin is a figure that is estimated – sometimes by elaborate equations using the number of coin dies known – but the number of coins coming from the mint in any one year can never be precisely ascertained (see Buttrey, 1994; contra Callataÿ, 1995). However, it is generally agreed that coins that were issued in large numbers will be more commonly found on archaeological excavations than smaller issues. By contrast, smaller issues will be overrepresented in museum collections, since they are of more interest to collectors.

Unintentional coin loss is affected by initial coin supply to the site, the size and the material of lost coins, the value of the coin (which may be a highly personal issue to the owner), and the methods of excavation. The initial supply of coin varies by time and place. Coins are more commonly found in cities, towns and forts than in rural sites, where the initial supply would be much lower. In the sixth century, the only coins available to the Romans were those minted in Magna Graecia. It was not until the third century that mints in Magna Graecia, Etruria, the area around Rome, and Rome itself began to produce large numbers of coin; even so, rural sites and small towns show evidence of few unintentional coin losses. Thus, the initial coin supply is composed of two things: how many coins were in circulation and were available to be lost; and how many coins came to the particular site. The town of Cosa, founded in 273, is a good example of how these forces interplay. There are few third-century coins found in the excavations; the majority of the excavated coins were minted in Rome (Buttrey, 1980). We would expect some coin loss since this is an urban site, founded by people who were probably used to transacting trade in coin. The coins also reflect the supply of coin from the mint of Rome, the very city that sent out the colonists (see Chapter 30).

Size matters. Smaller coins more easily escape the detection of the owner's eyes. At Cosa, a coin of Massalia (Marseilles) weighing over 16 g was not unintentionally lost, but is likely to have been an intentional deposit – the numismatist suggested "a casual accumulation" (Buttrey, 1980: 32). It was found with other, smaller bronze coins and a large aes signatum weighing 12.97 g. Thus, this accumulation was very likely an intentional deposit, and not an unintentional loss, due to the size of the Massaliot coin and the aes signatum.

Material matters, as does the place where the coin was deposited. Gold and silver coins are more rarely found on sites because of the higher value of the coin; we expect the individual knew that he had lost such a coin and thus would make a more thorough search for it, as is most famously recorded in a parable concerning a missing silver coin (Luke 15:8). All of the unintentional losses at Cosa, save one, are bronzes, until the first century, when silver coins – denarii and quinarii – are added to the bronzes. A bronze coin is more easily camouflaged against the color of the earth on which it has fallen.

Coins are more commonly found in busy public spaces than in houses (where the owner has more time to look for an errant coin). They are also more commonly found in drains or ditches than on floors that are routinely swept. Deposits in these drains or ditches may contain more coins as water would sweep small coins into pipes, cisterns or sewers, especially if they were closed and thus not easily accessible to the person who lost the coin.

The rate at which coins are found by the archaeologist is determined by the methods used in excavation. Experiments in different excavation methods have shown that up to 25% of the coins in any stratum can be missed – even during careful excavation – unless the newly excavated dirt is also sieved. This number is extrapolated from Imperial period strata, which would include more numerous small bronzes than in the Republican period, but the point still holds (see Reece, 1996: 342). In the underwater excavations from the Liri River, small coins were obtained only when the vacuum hose was used; in some seasons, larger coins were simply picked from the river bed by hand (Houghtalin, 1985). The problem of non-retrieval is exacerbated by identification rates of the recovered coins. If the stratum in which the coins were deposited was sandy, saline or acidic, the number of coins that can be assigned to a given decade falls, sometimes dramatically. If there are a number of illegible coins from a site, then the overall picture may be skewed. The conclusion one should draw from pondering all these variables is that the unintended lost coins from the site, commonly called "random" or "scattered" finds, does not give one a picture of the circulation of coin at any given period.

But we begin to assemble the picture of the coins in circulation, and thus the coins that are available to be deposited on ancient sites, by combining our knowledge of the unintentional losses with that of intentional deposits. Hoards (a group of coins intentionally placed and unintentionally left) are the first type of intentional deposit that springs to mind, and these have received the lion's share of numismatic interest. But there are other coins that are deliberately left at a site – in graves, as votive deposits, as foundation deposits of buildings or even in mast-steps of boats. Some of these intentional deposits may be made of coins that are taken directly from circulation, the most famous instance being the coins carried by those who perished in the fumes from Vesuvius in 79 CE (Duncan-Jones, 2003). Others may have been accumulated over a period of time, such as a lifetime of savings in a hoard. It is more likely that precious-metal coins are found in this type of deposit than in unintentional losses.

Votive deposits are very likely to have been mixed with other materials such as terracottas and jewelry, and can be deposited over a very long period of time (as in the Liri River deposit; Houghtalin, 1985). The archaeologist cannot call any coin recovered from a sacred area a votive, as the coin may have been an unintentional loss when the visitor to the sacred area was purchasing an item. If there is specific behavior that marks the votive, such as defacement of the coin or placement in a collection box or area, the coin can be considered a votive deposit (Wigg-Wolf, 2005). But the placement of the votive deposit need not be confined to a box, pit or bag. Pliny notes that he saw coins given as offerings thrown into the source of the Clitumnus River in Umbria (*Ep.* VIII.8). There must be some evidence of intentionality for the votive deposit to be recognized as such – possibly by inclusion with other materials of a clearly votive nature. However, the story of Eutropia visiting Abraham's well near the oak of Mamre should be borne in mind. According to the fifth-century CE historian Sozomen, she watched as worshippers threw coins, incense, bread and wine into the well as they prayed (*Historia Ecclesiastica* 2.4). Only the coins would remain for the archaeologist to find. Thus, the type of deposit in which a coin is found should be analyzed by the numismatist in tandem with the archaeologist; in so doing, the numismatist may help clarify for the archaeologist what type of deposit was found.

3 Coins and Chronology

Although Republican coins can often be dated to within a decade or less of when they were struck, coins are not necessarily the best tool an archaeologist has to date the strata. Part of the problem of assessing the date of the deposit is the wear or corrosion of the coin. Wear can be due to the effects of the coin's use during its lifetime above ground, as it was passed from hand to hand or banged around in a bag. But the wear can also be caused by post-depositional effects on the coin, such as water or sand. Both kinds of wear can be detected by the flattening of the highest areas of relief. While the numismatist must carefully consider both lifetime wear and post-depositional wear, in reality there is no way to distinguish the two types. If the coin shows very little wear from handling, it may have been deposited soon after it was minted. If the relief is flattened, the wear could be caused by long use, or water erosion that results from a coin's placement in a ditch or drain. The coin's corrosion may be due to the matrix in which it was deposited. If it was embedded in a concrete subfloor, or recovered from sewers that funneled highly acidic liquids, the coin may have corroded quickly and thus become illegible; it might look more heavily used than another coin that was buried in a less corrosive deposit. Corrosion can be affected by the temperature and humidity of the stratum. The degree of corrosion or wear is also influenced by the type of alloy from which the coin is made. Gold is immune to corrosion; silver less than copper alloys. It is the impurities in the silver or the copper alloy that corrode first, but corrosion can even consume an entire coin. The combination of corrosion and wear, especially when paired with harsh cleaning methods, can result in a large number of illegible coins from a site, even up to 90% recorded by the numismatist at the Imperial baths at Hammat Gader in Israel (Hirschfeld, 1997: xi; for notes on the conservation of coins, see Andreau *et al.*, 2002: 129–33, 152–6).

But the greatest problem with using the coins as a basis of dating a layer is determining the year when the coin was placed in the deposit. Archaeologists agree that coins can only give us a terminus post quem for the deposit. What the coin cannot do, as is often forgotten, is tell us when the deposit was laid down – we do not know how many years passed after the coin was minted and before it was deposited. Sometimes a coin that is clearly much earlier than the rest of the coins in the deposit is called "residual." These coins are considered "obsolete" or not legal tender, but it can be that they were used long after their date of minting because of their fineness or their module, or even because their types remained somewhat familiar. The user believed that he could continue to pass the coin as legal tender, even if it was clearly "out of date." If such a coin is found, the numismatist must consider the deposit – the coin could be part of a disturbed or secondary deposit (a robbed-out wall, or even an animal burrowing in the soil), or the coin could be present due to the reuse of the material in the stratum. Or the residual coin could be an indication that the coin continued to be used for legal tender long after we would have expected it to have passed out of circulation. The numismatist must check whether there is any other residual material found in the stratum. For example, a coin of Philip II of Macedon (413–405) was found in the deposits of the Liri River, though the votive items probably only began to be thrown into the river after the foundation of the Roman colony Minturnae in 295 (see Figure 19.3; Houghtalin, 1985). This means that the coin of Philip II was in circulation at least 100 years after it was minted, in order to have been deposited in the river. The coin of Philip II, found by itself, would not give the archaeologist any understanding of the date of use of the votive deposit in the Liri River.

The easiest place to look for residual coins is in hoards, where coins that do not seem consistent with the bulk of the hoard are excluded as "extraneous"; sometimes the numismatist will explain that the patina does not match the rest of the hoard. One problem with publishing hoards that have not been excavated is that the numismatist must be aware of the potential of the seller to slip some extra coins into the hoard before presentation to the numismatist (see e.g. Crawford, 1969: 84, n. 172, and 105, n. 311). Hoards, then, must be carefully analyzed and the numismatist must note why s/he is admitting or excluding coins from the non-excavated hoard.

Although it is clear that some coins could circulate for many years beyond their initial minting, we still do not know how long, in general, a coin would remain in circulation. This lack of knowledge is partly due to our ignorance of where and how coins were introduced into circulation. Most numismatists work with the argument that silver coins were shipped to the armies and were used for administrative business within the city of Rome; hence, we would expect to find newer issues and more denarii on urban sites and in forts (gold coins are rarely minted in the Republic). Some numismatists also argue that silver coins were returned to Rome in the form of taxes, to be remelted and reissued. Yet this argument has been refuted by scholars who would see most taxes paid in kind, not in coin (Duncan-Jones, 1990; contra Hopkins, 1995/6). There is more debate on how and why bronze coins entered into circulation, an argument that is tempered by the long periods in the Republic when bronze coins are not minted. The one thing that is safe to say is that the numismatist cannot give the excavation coin a terminus ante quem or the date the stratum was laid down. In the publication of excavation coins, the numismatist should work closely with the pottery and other specialists in interpreting the date of the layer, as well as with the archaeologist to determine the nature of the deposit.

Numismatists of the Republican period weave together threads of coining techniques, metrology, epigraphy and iconography to interpret the coin. Coining techniques, such as overstrikes or the style of the die, can help place the coin in time and space. Overstrikes – the striking of a type over the top of another coin – allow us to see what types of coins were available to the minters of the second coin and help us to understand the chronology of certain issues. Thus, bronze coins of Acarnania and Oeniadae in Macedonia were used as coin blanks by the Romans after they had acquired the coins in the First Macedonian War, while they were simultaneously fighting Hannibal. The era was one when coins were commonly overstruck, to save the minting city one step in the production of coins with their own images and legends (Crawford, 1985: 58). This practice dies out by the end of the Second Punic War, as Rome becomes less stressed and other coinages gradually begin to fail. It is possible that after this date, coins taken as booty were remelted, not overstruck (see below, on the Iberian denarius).

In the late fourth and third centuries, mints in Etruria, Luceria, Sicily, Sardinia and elsewhere can be posited based on the needs of the army and the style and iconography of the die (see e.g. Chapter 24). For instance, the first issue made by (or more precisely, for) the Romans in the late fourth century was a bronze coin minted in Neapolis (Naples; Crawford, 1974: no. 1), not yet controlled by the Romans, as can be seen by the reverse type of the forepart of a man-headed bull walking right, an image found on coins of that city beginning *c.* 450. The obverse, the head of Apollo, is done in a Greek style; the Romans will not mint coins with heads of deities on them for another generation. The only thing that actually identifies this coin as belonging to the Romans is the inscription "Romaion," which is spelled out in Greek letters. The combination of style, iconography and epigraphy places the coin in the late fourth century, when Neapolis was allied with Rome. Didrachms – silver coins weighing just over 7 g, a denomination found in the Greek cities of Magna Graecia – with the head of Roma on the obverse and a victory on the reverse and the Latin inscription ROMANO (Crawford, 1974: no. 22) also have a control letter or letters on the reverse and a control symbol on the obverse. The sequence of letters follows closely the die recording system used on Ptolemiac decadrachms struck in honor of Arsinoe II after her death in 270. Thus, the Roman coins should date to just after the introduction of the Egyptian coins, which were available to the Romans (Thomsen, 1957–61, vol. III: 127–36).

Metrology, or the study of the weight of the average coin, can help us understand which coins were prevalent when a coin was first minted. It appears that only the weights of gold coins were tightly controlled; silver coins, and certainly bronze coins, were made al marco, that is, the batch of coins was made with a predetermined amount of metal, but each coin could vary somewhat from the theoretical weight of an individual coin. The weights of Roman bronze coins fell during the third century, especially during the Second Punic War, as did the weights of silver coins, although the correlation between the bronze and silver coins is still being discussed. Arguing from metrology quickly becomes extremely complex, as the numismatist tries to sort out the rate of the descent in weight, which is complicated by the restoration of weight standards which the Romans attempted from time to time, and the attempt to parallel the weight of coins from neighboring cities. The denarius was first struck around 4.5 g, but sank to 3.85 g after *c.* 157, when we are told by ancient authors it was revalued. The first weight is close to the Attic drachm of about 4.4 g, and about half the weight of the Carthaginian shekel, which also fell in weight during the course of the war. Bronze coins are much harder to trace, in part due to the sporadic nature of their minting. The severest period of reduction was in the Second Punic War. Yet even within issues, coins could come from the mint at full weight and as underweight (Crawford, 1974: 595–6).

4 Coins and Coin Use

Numismatists have made some tentative progress in explaining coin use in different deposits in order to explain or even predict behavior and thus coin loss. Normally only small bronzes are used for burials (S. Stevens, 1991: 249) and for coins placed beneath the mast-steps of ships (Carlson, 2007). Votive coins found in sanctuaries can be quite worn, or even counterfeits; precious-metal coins are rarely left as votives (Walker, 1988). I think we still need to examine if there are high percentages of "foreign" coins, or coins not minted in the city where they are placed, in votive deposits. Army camps tend to have higher percentages of gold and silver coins among the unintentional losses than do cities (Kemmers, 2006).

Counterfeit coins are a fascinating problem. Counterfeits – often called imitations or plated coins – routinely appear as unintentional (or are they intentionally discarded?) coin losses in southern France, Spain, the Balkans, and even Italy. The ancients considered them a problem, as there was a penalty for producing counterfeit coins, which would argue that there were no officially sanctioned imitations as some have suggested. Serrate denarii, or denarii with notches around the edge, have been explained as one official reaction to the problem of counterfeiting. We are not entirely sure how these serrate coins were produced, but they did not solve the problem of counterfeiting, as we know from plated serrati (see Kraft *et al.*, 2006). The first legislation about forging is the Lex Cornelia *de falsis*, which had a provision against counterfeiting gold and silver coinage. Counterfeiting bronze was probably also illegal, but at the time this law was passed, in 81, the Roman government was not coining bronze. The use of control marks on denarii to ensure proper production of the coin, or even extra-mint practices such as the use of "banker's marks," or small symbols stamped onto the face of the coin to ascertain if the coin was plated, show the ancient concern about plated coins. Unlike countermarks, banker's marks are not placed onto the coin to officially recognize a re-evaluation or reaffirm the currency, but are stamped or cut into the face of the coin by private users of the coin. Late in the Empire there was a suggestion that all mintworkers should be confined to an island so they could not pass forgeries of the coins they were making in official workshops (fourth century; Anon., *De rebus bellicus* 3.1–3). There was also legislation against stealing dies, which implies this was a circumstance that required the passing of such a law (*Dig.* 48.13.8). A sentence of exile was mandated for anyone housing counterfeit activities under his roof (*Codex Theodosius* 9.21.2). Some scholars have argued that Republican denarii that were struck from official dies were illegally produced in the mint itself; alternatively, they were produced outside the mint by making casts (or even dies) from genuine coins (Crawford, 1968).

After the Second Punic War, the denarius became the dominant coin used in peninsular Italy; to a lesser extent Roman bronze coins also replaced many local coinages. In Italy, the issuing of coin even became a mark of rebellion from the Roman state, as during the Social War, when the coins were modeled on the Roman denarii in terms of weight and iconography – and separated from them by the language used in the legends. Beyond Italy, denarii eventually replaced local silver coinages, especially the dominant silver units minted by Massalia (Marseilles). In the east, strong local coinages continued in their Hellenistic patterns and numismatists have explored the relationship between Roman control and local coinages in some issues, such as the large silver *cistophori* of Asia Minor, the *chalkoi* of Corinth, and the bronzes minted by Herod the Great (Crawford, 1985; Crawford and Burnett, 1987; cf. Callataÿ, 2011). The appearance of denarii and gold coins in ports in India have led to discussion on the shipping of coins, supply of the coins, trade in general, and the initial phases of that trade either in the second or first century (Berghaus, 1991; contra Gupta, 1991).

At the end of the Republic, the generals moving with their armies changed coin production and coin use from Spain to North Africa to Asia Minor. Gold was minted by the triumvirs at a level not previously seen. Mark Antony authorized vast numbers of denarii commemorating the legions serving with him in the east (the so-called "fleet coinage"), possibly with silver borrowed from Cleopatra. These denarii were issued in conjunction with an entire series of bronze divisions; they continued to circulate for many years; and are often found in very worn condition. They were rarely hoarded, most likely because their silver content was lower than the average Republican or Early Imperial denarius. The bronzes soon dropped out of sight, since they were smaller than bronze coins from the mints associated with Rome. But the coinage of this period shows that the mints outside of Italy were reacting to, if not conforming to, the coinages from Rome, and as such are another tool to help us define the changes that are called Romanization or acculturation, in both the east and the west (see e.g. Kroll, 1997).

5 Coins in Hoards

Hoards have retained a special interest to numismatists, despite problems such as hoards appearing on the market with no provenance information (nor any certainty that the hoard is complete), or incomplete or insufficient publication of the coins in the hoard. Scholars debate how much a hoard can reveal about coins in circulation: hoards may contain carefully selected coins of finer quality or heavier weight, skewing the average qualities of the body of coins in circulation (see Lockyear, 2007). The ordering of hoards allows the numismatist to date coins that are minted in a series like the Roman denarii; the work that was initially done here by Crawford has been refined and extended in the light of a few important hoards, such as the Mesagne Hoard (Crawford, 1969; Hersh and Walker, 1984). When a number of coins from different political systems circulate at the same time, hoards can help provide us with important clues about the relative chronologies of the issues, as well as the political and social ties between minting authorities. Some hoards show that coins remained in circulation for a long period of time, since they were available to the hoarder for placement in the hoard. And it is important to note that not all hoards are closed due to unrest, war or famine, but all are closed due to quite individual circumstances of loss. The 994 denarii found in an olla under a shop floor of Musarna (Viterbo) date between 211 and 67, although the bulk of them date to 120–80 (Andreau *et al.*, 2002: 160). No circumstance of war or famine can be attributed to central Italy at this date; instead, the hoard was closed and never retrieved due to a particular personal predicament.

Certain hoards that were closed due to war have provided dramatic testimony to the date of the introduction of the denarius. Numismatists who followed Pliny's (confused) account of the introduction of a silver coin in Rome in 269 or 268 argued that Pliny was noting the introduction of the denarius (*HN* 33.45), but when Mattingly and Robinson proposed a date of 187 for the introduction of the denarius, a storm arose that lasted until the early 1960s (Mattingly and Robinson, 1933). It was then that Thomsen (1957–61) finished his intricately argued volumes, comparing the denarius to other coins found on the peninsula by weight, style, iconography, epigraphy, placement in hoards and overstrikes, and, combining this information with the historical sources, proposing a date near 215 for the introduction of the denarius. Just as he was finishing his work, archaeologists found the destruction level of the city of Morgantina, captured by the Romans at the end of the Second Punic War, in 211. The destruction layer had a terminus post quem due to a coin of Hieronymous, minted in 215–214, and third-century coins found in the destruction level of the sanctuary of Demeter and Kore. Roman coins were found both in the destruction level and in hoards under the destruction level, which were hidden when the city was threatened. Most of these were early silver coinages, but there were some anonymous denarii mixed into the hoards. Thus, the hoard evidence and the unintentional loss evidence pointed to a date for the introduction of the Roman denarius just before the destruction of Morgantina in 211, in agreement with all of the other evidence that Thomsen had marshaled. The date of the introduction of the denarius is now securely set at *c.* 212, no earlier than 215 (Buttrey, 1989).

6 Case Study: The Introduction of the Iberian Denarius

In order to bring many of these threads of numismatics and archaeology together, we can look at the problem of the introduction of the Iberian denarius, a problem that has been approached from many angles. The argument is between the "high" date, or early in the second century, and a "low" date of about the mid-second century (see e.g. Knapp, 1977b; García-Bellido, 1993; contra Crawford, 1985: 84–102, 340–4). All of these arguments use the historical sources, metrology, iconography, overstrikes, hoards and evidence from archaeological strata, but the clinching evidence has yet to be brought forward (for a good introduction, see Ripollès, 2005). Numismatists have to await evidence from an excavation and/or a retrieval of a hoard, as happened in Morgantina.

The first issues of silver coins from cities on the Iberian Peninsula may be dated to slightly earlier than their counterparts in Latium, that is, to the mid-fifth and fourth century. The earliest Iberian coins are the silver fractions of Emporion and Rhode (see a map of Spain, Figure 33.1). Yet the bulk of the coinage used on the peninsula was minted in Massalia; coins minted in Asia Minor and Magna Graecia were also in circulation, as can be seen in the unintentional losses in excavated strata. All of these coins tended to be found only in the area along the Mediterranean coast and in the north-east of the peninsula, the areas that had more towns and cities than the rural interior. Bronze coins were minted by *c.* 300 and this is when the small silver fractions were abandoned in favor of larger silver drachmas. The impetus for widespread coin use and minting came with the Barcid control of Hispania; they probably began the exploitation of the silver resources. By the last decade of the third century no Hispano-Punic coins were made, and any that remained on the peninsula rapidly left circulation.

The silver coins minted under the Barcids were based on the Carthaginian shekel (at 7.20 g), but during the course of the First Punic War the coins were gradually reduced in weight to between 6.70 and 7.00 g. Bronzes were also minted, but after the war, they rose from 8–9 g to 11 g (at least some of which seem to date prior to the arrival of Scipio). Other smaller coins were minted by various cities during the war and these circulated with imitation drachmas, fractions attributed to Ampurias, and imitation Massaliot obols. Only some of the coins are epigraphic and thus can be assigned to minting cities; the identifications of many of the minting cities of the anepigraphic coins are no more than educated guesses. Coins arriving with the Roman soldiers included silver victoriati and quadrati (minted by Rome prior to the denarius), denarii, and some coins from Magna Graecia. In other words, in the third century, the Spanish Peninsula looked much like the Italian Peninsula, where individual cities minted bronze and silver coins and some of these issues were strongly influenced in terms of weight and iconography by the dominant political forces on the peninsula. These local issues circulated with coins from areas connected to the minting cities militarily or economically. In Spain, one can see the spheres of influence from the hoards and finds: Carthaginian influence is felt in the south, along the Mediterranean coast, while the Roman influence is felt in the Guadalquivir River Valley and the north-east of the peninsula (see Chapter 33).

At the end of the war, the Romans kept legions in Hispania, using them to put down native revolts, especially in the interior. When Cato's tour of duty expired in 195 he declared that the province had been pacified; Roman generals would again boast of pacifying the province in 194 and 180. In each of these victories, Livy relates that the generals were voted triumphs at which they displayed enormous amounts of booty taken from the peninsula, mostly in silver coins which Livy describes as "argentum oscense" (Livy XXXIV.10.4 and 6–7, XXXIV.46.2, XL.43.5). This term has occasioned much discussion, with some suggesting that it refers to any of the silver coins circulating in the province. In the third century, these would have been primarily Hispano-Carthaginian coins featuring either very Greek-looking profiles or facing heads of Herakles/Melqart, with Punic inscriptions, and the drachmas of Emporion and Ampurias, which featured Greek-style heads of the nymph Arethusa or Artemis on the obverse and Greek legends. None of these coins were minted in the second century, or the period when Livy notes the silver coins taken as booty. In 179, Tiberius Sempronius Gracchus, as praetor of Hispania, ordered the town of Certima to pay an indemnity of 24,000 nummi (Livy XL.47.2–10) – frustratingly, a word that does not give us clear evidence concerning the units in which the indemnity was paid (local coin, bullion, or denarii?). Soon after this, an embassy was sent to Rome to complain about Roman magistrates who were raising taxes in currency, again with no specification of the means of payment (Livy XLIII.2.12).

Since the historical accounts are problematic, archaeologists have proposed a date based on the stratigraphy of the camps used by the Roman legions during campaigns around Numantia (see Figure 14.4; Chapter 14). These are the only sites yet found that can be dated by marker artifacts within the early to mid-second century; they have bronze local issues and foreign coins that can be linked to the Iberian denarii in contexts that are known from historical circumstances. Included in the finds from the camps are bronzes from the Hispanic mints of Sesars and Sekaisa. The former minted bronzes with a diademed male head (sometimes bearded) on the obverse and a rider on a horse on the reverse, with a legend in Iberian. It is thought that these bronze coins were issued simultaneously with Iberian denarii with identical obverses and reverses, although it should be noted that no denarii were found in the camps. Bronzes of Kesse-Tarraco were also found in the camps, especially Renieblas Camp III; Kesse-Tarraco also produced denarii, as we know from the stylistic similarity of the diademed male head on the obverses on both the bronze and silver coins. Excavation of the camps at Renieblas was conducted by A. Schulten (1929) between 1909 and 1912, when he discovered five superimposed Roman legionary camps. The excavator dated these camps by a reconstruction of the history of the region, not by using the material remains or marker artifacts. His proposal that two of the camps belong to the same year is especially controversial; Schulten thought one level belonged to the summer camp, and another to stone construction for a winter camp in the same year. A study of the coins from the camps also remains controversial for its conclusions. These conclusions about the dates of the coins were argued by the numismatist Hildebrandt, using the wear of the coins and the weight of the bronze issues, which he maintained fell in tandem with the weight of the Roman bronzes (Hildebrandt, 1979). The real problem is the topmost camp, Camp V, which Schulten associated with Scipio's siege of Numantia in 133/2. Nine Roman coins were associated with this camp, along with seven bronze coins minted in Spain. The Roman coins were dated between 211 and 146, and six of the seven native coins were thought to be minted between 145 and 130. One Roman bronze coin in the same layer, an uncia, is now thought to have been minted in 108 or 107, though in Schulten's time the coin was thought to have been made at the beginning of the first century. Hildebrandt suggested that this was a residual coin from the higher level and suggested that Camp V should date to immediately after the siege of Scipio, due to the weight of and wear on the bronze coins. Interestingly, Schulten dated this very same layer to a camp built in 75/4, or the end of the Sertorian War, using the historical sources only. Hildebrandt's date is thus based on his suggestion concerning the residual nature of the late coin. Using the same coins but different arguments about metrology and residuality, other scholars suggest that the 108/107 date is a real terminus post quem (Dobson and Morales, 2008). Thus, the archaeological evidence remains problematic.

Hoards unfortunately give us little to go on. Roman and Iberian denarii only appear together in hoards in the last third of the second century, after both the "high" and the "low" date of the introduction of the Iberian denarius. And away from the coasts, only hoards of bronze coins from the late third and first half of the second century have survived, though these can include both Roman and Iberian coins. Numismatists have carefully considered the wear on the Roman and the Iberian coins to try and give them a relative order, but this reasoning has not won full acceptance in terms of dating the Iberian denarius.

And so, numismatists have suggested other ways of thinking about the introduction of the Iberian denarius: the denarii were created in order to make the indemnity payments to Rome by 195, and thus mimic Roman weight standards. Or they were created strictly for commercial purposes within the Spanish Peninsula, and can be explained by their being modeled on the dominant commercial coin of the second century. We do not know if they were created as exchange items used to cement alliances among the tribes, or even used as a profit-making device by the larger Iberian towns, which minted for their smaller neighbors. The coins may have been created to fund the insurgency against Rome, or the fact that the cities which emit denarii come from Roman-controlled areas implies that they were not used as a tool of resistance to Roman rule. They may have been created in order to pay Roman taxes, or the Romans may have supplied denarii to the Iberian Peninsula in order for the locals to pay taxes to Rome. If the latter, the creation of the Iberian denarius would be unrelated to the need to pay in specie. But taxes may have been largely paid in kind. The coins may have been used to pay Roman legions in Spain (and the Roman soldiers had them remade when they returned to Italy), or they may have been used to pay auxiliaries in Spain. We do not even know if the Roman soldier was paid in coin shipped from Italy or if he was paid in bronze or silver.

7 Conclusion

The case study has shown how important it is for archaeologists to work closely with numismatists on interpreting the excavation coins. The first thing the numismatist should do is work with the archaeologist to systematize the recording of the coins *in situ*, and to establish a careful method of conservation and preservation of the coins. Using this recording system, the numismatist, working closely with the pottery specialist, can help determine the nature of the deposit and decide if the coins are intentionally placed on the site or unintentionally lost. Only when these parameters are understood can the numismatist begin to help with the terminus post quem of the deposition of the stratum. It is expected that at the same time, the numismatist will pay attention to issues such as metrology, iconography, the use of the coin in the particular circumstance, and the overall picture of coin loss on the site and in the region. The result should be a final publication that is more than a simple list of the coins found in the excavation, but an interpretation of why the coins were found and what those coins can tell us about the site. Close collaboration between the archaeologist and the numismatist can uncover aspects of the ancient world that scholars of the ancient economy, ancient history, and even ancient language can share.

The coins of the Republican period still need a great deal of study, as can be seen by the case study of the Iberian denarius. The third century is a particularly contentious period, even if the picture begins to be sorted out at the end of the century with the introduction of the denarius. The numismatist must work closely with the archaeologist in order to interpret the evidence of the coins, using all of the tools available to her – from metrology to hoard evidence, iconography to prosopographical information, evidence from archaeological strata to the few references we have from ancient sources. Only by viewing this evidence holistically can we see how the Romans developed their monetary system, and eventually knit into it all of the provinces of the western Republican empire, and much later extending that impact to the eastern provinces of the Imperial world.

Acknowledgments

I extend my thanks to L. Houghtalin and the late J. MacIsaac, who read this chapter and gave me good suggestions for improvement.

FURTHER READING

There is no good handbook that discusses Republican coins and archaeology, and one older handbook, *Coins and the Archaeologist* by J. Casey and R. Reece (2nd rev. edn, 1988), is very largely concerned with Roman Britain and the Imperial period. Sheedy and Papageorgiadou-Banis (1997) is rather limited in terms of the theoretical issues addressed, and it concerns only Athens and Corinth. One can get an overview of archaeological thought in regards to coins in Noeske (1998) and in less detail in MacIsaac (1995). Reece (1996) (writing about Imperial sites) must be used carefully, as, for instance, Reece is sanguine that corrosion and retrieval methods cannot seriously disturb the overall picture of coin loss and excavation coins.

The important coin catalog for Roman Republican issues is Crawford's 1974 volumes, and his arguments on all aspects of Republican coinage must be taken seriously. His 1985 volume is a fundamental and important study of coins from Spain to Syria, from the fifth century until Augustus's reign. Thomsen's three volumes (1957–61) are still valuable for their exploration of monetary issues in the peninsula in the Republican period. The collected papers from a conference, edited by Crawford and Burnett (1987), provide insight into the coinages that were in use when the Roman denarius system arrived and dominated the expanding Republican Empire, from Germany to Spain, North Africa to Asia Minor and Greece. It should be used in conjunction with Burnett (1982), and Amandry, Burnett and Ripolles (1992), for local issues at the end of the Republic.

CHAPTER 8

Weapons and the Army

Andrew L. Goldman

1 Introduction

In a volume dedicated to the archaeology of the Roman Republic, a close examination of weapons and the army – to speak of *arma et viri*, as it were – is an appropriate subject, one touched upon only lightly within Erdkamp's *Companion to the Roman Army* (Erdkamp, 2007). The organizational reconfiguration and physical transformation that gradually swept through the Republican army is a topic of no minor significance, since the military and its institutions were organs both central and vital to the prolonged function and prosperity of Roman society. As the city-state ascended from its hard-won position of local dominance in central Italy to its role of uncontested Mediterranean heavyweight by the Late Republic, Rome's army – what was indisputably the most potent physical manifestation of its authority – underwent continuous and fundamental alterations, evidence for which is apparent in the adoption of new organizational principles and complementary sets of offensive and defensive weaponry. By the Late Republic, the changing nature of Rome's adversaries, the expanding diversity of encountered terrain, the widening variety of campaign types, and the increasing range and duration of Rome's military operations had all helped to compel Rome to modify its early army. The eventual result was a tempered, flexible and more homogenized force capable of astounding military feats, an army that would have been virtually unrecognizable to Rome's Republican founders. This chapter aims to explore this transformation through examining the legionary arms and armor of the Republican period, focusing in particular upon the nature of our current archaeological sources for aiding us in understanding the tremendous level of invention, modification and adaptation which took place between the sixth and first centuries.

A Companion to the Archaeology of the Roman Republic, First Edition. Edited by Jane DeRose Evans.

2 Archaeological Sources and Research

Discussions of evolution in the Republican army's structure and equipment have centered principally upon an examination of surviving literary sources and interpretation of their often contradictory or ambiguous content (Zhmodikov, 2000; Forsythe, 2007; Sekunda, 2007; Sage, 2008). In contrast, archaeological data relating to the Republican military has been employed to a far lesser degree to address this topic. This situation is primarily due to the quantitative and qualitative problems associated with this particular corpus of material. In comparison to the Imperial era, for which we possess an abundance of archaeological sources (e.g. permanent bases, caches of discarded weapons and armor, funerary monuments, etc.), physical evidence for the Republican army remains relatively scarce and its documentation disappointingly patchy. One obstacle has been the relatively fugitive character of the army's overall footprint, since the pre-Augustan army was rarely sedentary. Indeed, when Republican forces proverbially came, saw, and conquered, they also almost invariably marched off to the next engagement or turned homeward in short order, leaving little informative detritus behind in their wake. While investigation has taken place at a handful of formal camps and siege and garrison sites, much of the excavated material remains contextually problematic, having been recovered in the nineteenth and early twentieth centuries under circumstances which lack sufficient documentation and methodological rigor for conclusive analysis (see Chapter 14). Beyond merely assessing poorly excavated and stratified finds, scholars must also overcome numerous other complications which beleaguer the study of the Republican military and its equipment, including a lack of precision in analysis, delayed publication, a dearth of well-illustrated objects, the general absence of weapons in battle and funerary contexts, and uncertainty about the location of ancient camps and battlefields (Bishop and Coulston, 2006: 23–38). Furthermore, potentially useful archaeological resources that routinely aid our research prove disappointingly meager for the Republican era. There exist only a bare handful of useful visual monuments, funerary reliefs and numismatic representations that depict soldiers and their weaponry. In addition, epigraphic evidence for Republican soldiers of any rank is exceedingly thin and difficult to identify with certainty. In the face of such conditions, it is small wonder that in discussions of the Republican military, written sources – troublesome to interpret as they may be – have customarily gained greater consideration than archaeological ones.

Fortunately, over the past several decades this situation has begun to reverse itself, as steady forward progress has been made in the gathering and analysis of archaeological data relating to the Republican army. Leading the way is a quiet revolution in the area of military equipment studies, where new excavations, newly published finds, and the reappraisal of previously excavated materials have combined to provide an increasingly useful evidentiary corpus. This body of military equipment has begun to force scholars to reassess many issues once thought firmly settled by the literary evidence. Much of the credit for initiating this new movement can be traced to the efforts of H. Russell Robinson, whose authoritative work, *The Armour of Imperial Rome* (1975), ushered in an era of more exact methodological and analytical treatment of military equipment. Robinson's research drew particular attention to two critical issues: the need to treat with greater care empirical detail derived from iconographical representations of the army, and the potential capacity of experimental archaeology to expand our understanding of Roman weaponry. In response to this work and that of his colleague, P. Connelly, the investigation of Roman military equipment subsequently blossomed into a vibrant new subfield of army studies during the 1980s and 1990s. Important developments include a succession of international colloquia known as the Roman Military Equipment Research Conferences (ROMEC) and a spate of new publications dedicated explicitly to the study of Roman and Byzantine military equipment, including periodicals (*Arma*, the *Journal of Roman Military Equipment Studies*) and comprehensive surveys of Roman weaponry (Feugère, 2002; Bishop and Coulston, 2006). In terms of understanding Republican weaponry, a seminal meeting (ROMEC X) in Montpellier, France, produced promising results including valuable surveys of military equipment from regions throughout Europe (Feugère, 1997). Subsequent new fieldwork and research in Spain (Quesada Sanz, 1997a; Quesada Sanz and de Prado, 2006), France (Feugère, 1994, 2002: 66–9), Slovenia (Istenič, 2005) and, more slowly, the eastern Mediterranean (Oransay, 2001/2) are continuing to provide new data about Roman weapons on an almost annual basis.

An examination of archaeological evidence pertaining to Republican weapons and the army is thus not only a useful exercise, but perhaps now an obligatory one within any comprehensive overview of material culture during the Republican period. Of course, evidence culled from literary sources remains essential for understanding how the army equipped itself and fought during the Republic. Indeed, in spite of their inherent problems, those sources will always play a prominent role, providing us with key descriptions and military terms as well as evidence of perishable materials (i.e. wood, cloth, leather) that prove impossible to recover from most archaeological contexts. Yet armed with new archaeological data and avenues of related research (e.g. experimental archaeology and weapon reconstruction), it is increasingly possible to reach conclusions independent of the written records. This recent development not only has resulted in lively scholarly debate on the subject of arms and the army, but permits us to assess and challenge those reconstructions previously based solely or largely upon literary references.

3 The Early Republican Army and Its Weapons

Despite the advances in scholarship outlined above, relatively little can still be said with certainty concerning the composition and weaponry of Rome's army prior to the early third century. Nearly all of our information concerning the Republic's early army and its development is from literary sources, and virtually no aspect of those sources has remained undisputed. Filled with anachronistic accounts, often antiquarian in nature and rife with conflicting information, these accounts have proven tremendously problematic to use in addressing even the most fundamental questions on the army's size, organization, armament and tactics. Likewise, archaeological evidence for the Early Republican army remains problematic and elusive, particularly in the heartland of Rome and Latium, where finds of military equipment are rare due to a cessation among the Romans and Latins by the early sixth century of burying their dead with armor and weapons. Fortunately, more plentiful sources of archaeological material (e.g. funerary goods, votive deposits, tomb paintings, vase representations) are available from Campania, Etruria,

Lucania and Apulia for the sixth to fourth centuries (see Figure 24.3). Given the high level of cultural exchange and interaction which was taking place among the peoples of Italy during that period, such material has allowed scholars to extrapolate with some level of confidence the types of military weapons and organizational structure probably used by Early Republican Rome (Rawlings, 2007: 53–5; Sage, 2008: 64–5).

Both archaeological and literary sources are in general agreement on the fundamental character of Rome's early military and its weapons (Cornell, 1995: 184–6; Forsythe, 2007; Sage, 2008: 19–27). At its heart was an armed militia of heavy infantry (hoplites) who operated in a chiefly local, short-term capacity and probably in an ordered formation (phalanx). This model was likely inspired by the Etruscans, who had adopted the Greek hoplite panoply and phalanx fighting by the mid-seventh century, when such equipment begins to appear in Etruscan burials. When less formal needs arose, Rome also fielded smaller armed groups of a less organized nature to be used for raiding or retaliatory purposes. In either capacity, men were required to buy and maintain their own equipment, and those who could afford it used the basic weapons found in the standard Greek hoplite panoply, including an iron spear (hasta), a large round shield (clipeus), and a bronze helmet (galea). Variations of the panoply's main elements, found in graves and seen in artistic representations across the peninsula, were apparently in widespread use by 600. The hasta was the principal and largest offensive weapon of the period, and triangular- and quadrangular-shaped iron spearheads have been recovered in a broad range of sizes, indicating use in stabbing (large) as well as thrusting and throwing (mid-range to small). Some regional variation has been observed in this regard; while a heavier spear for both throwing and thrusting seems to have been popular in Samnium and western Lucania, Apulia and other parts of Lucania seem to have preferred lighter javelins for throwing (Small, 2000). In those cases where butts have also been found, spear length is estimated at *c.* 1.6–2.6 m. Less certain in terms of its widespread adoption and form is the iron sword. Numerous examples of short, double-edged blades (*c.* 60–70 cm long) of the type used by Greek hoplites have been found in non-Greek regions throughout Italy and the Adriatic region. It has thus been suggested that this sword type (or a close variation of it) was the one also used by the limited number of Romans who could afford one.

In terms of protective equipment, bronze helmets in Italy display a broad range of regional character in shape and decoration before the fourth century, when a greater level of homogenization took place. Popular fifth- and fourth-century helmet varieties include the Attic and Chalcidian types found in Etruria, the widely circulating pot-shaped Negau type in northern Italy (see Figure 23.1), the cap-like Apulo-Corinthian Greek-derivative type popular in southern Italy, and the open-faced Samno-Attic type found in Campania and Lucania (see Figure 24.3, worn on the head of the horseman). In central and northern Italy, most helmets belong to the general "pot-shaped" class, which feature a bowl-shaped helm with a broad rim and a variety of attachment knobs and crest holders riveted to it (Connolly, 1981: 102–4; Burns, 2003: 68–70). While less is known about shields, since they were largely made of perishable materials, pictorial representations in tombs and on vases indicated that they came as well in a wide variety of shapes (e.g. round, oval) and appear to reflect regional tastes. Bronze body armor – likely restricted only to the elites – is also well illustrated in surviving hoplite representations, the most common finds falling into two categories, the anatomically muscled cuirass and the chest protector (cardiophylax, literally "heart protector"), both of which were worn on top of a linen or wool undergarment. Although heavy and expensive, the former Greek-influenced cuirass gave the upper body full protection, with a cast bronze front and back that was hammered together. A variation of this type is the squared muscle cuirass, a much shorter version (*c.* 30 cm) covering only the upper abdomen (see Figure 28.3). Found in Etruria and southern Italy, this type is perhaps best known from the fourth-century tomb frieze of returning soldiers (probably Samnites) from Nola. In comparison, the cardiophylax was composed of a circular bronze disk (*c.* 17–35 cm in diameter) affixed to the chest by a harness of metal or leather straps over the shoulder, with a similar bronze plate on the back (see Figure 24.3). Although this arrangement did not give much protection, it was clearly a popular type, present across Italy from the seventh to the fourth century as well as in Corsica, southern France and Spain. It is probably the basis for the triple-disk breastplate of the fourth century closely associated with the Samnites, as seen on the bronze "Samnite Warrior" from Sicily, now in the Louvre. While the exact character of the Samnite panoply is disputed, the association of the triple-disk breastplate with them is probable, given the numerous examples represented on Campanian vases and recovered from the central highland and adjacent coastal contexts. Finally, there is plentiful evidence for bronze padded shin guards or greaves, examples of which have been recovered from tombs in Etruria, central and southern Italy (Connolly, 1981: 97, 102–9; Feugère, 2002: 73–4; Sage, 2008: 43–4, 67).

Literary descriptions of Rome's early army, such as those offered by Livy (I.43) and Dionysius of Halicarnassus (*Ant. Rom.* 4.16), support the use of these general weapon and armor types in the fifth and fourth centuries among the upper and lower classes of Roman legionaries. Wealthy elites in the heavy infantry (the classis) were expected to don the full bronze panoply: a round shield, a helmet, greaves, a breastplate, a heavy thrusting spear and a sword. Citizens lower down in the census were exempted from carrying more expensive items like the breastplate and greaves and were seemingly differentiated from the top class by the assignment of a different shield type, the oblong scutum. Further down still in the legion, the lowest census groups (the infra classem, literally "below the classis") operated as light-armed troops who carried little or no armor and used only a combination of thrusting spear, short throwing spear (verutum) or sling. The Early Republican army thus appears to have borne weapon types akin to those of their neighbors, and probably fought with similar (hoplite) tactics. This is hardly surprising, since by the end of the fourth century, the increased breadth and scale of conflict across Italy had brought about significant interaction between its peoples. Of course, all peoples of the peninsula did not uniformly adopt the same hoplite equipment and fighting tactics. Weapons were often adapted to local tastes and environments, and native types of equipment (e.g. axes) continued to be used alongside the Greek-influenced panoply. Indeed, it has been suggested that, using the hoplite panoply and their combination of light and heavy infantry, Etruscan and Italic forces might have fought in a far more open and flexible formation than has been assumed. However, while the archaeological record does bear out a high level of regional variation in Italy in regard to its weaponry, there is likewise a wide dispersal of similar armor and equipment designs, as noted above. Rome was at no time isolated from its enemies or its allies, and it seems highly probable that Rome's repeated, at times intense, contact with its neighbors would have an impact on their military practices (Burns, 2003; Rich, 2007: 16–18). Exactly to what degree this occurred is difficult to say with certainty, given the problematic nature of our early sources. But it seems fair to conclude that the development and character of Rome's early army and weapons were very much influenced by both local conditions and the wider, complex cultural continuum of sixth- and fifth-century Iron Age Italy.

4 The Manipular Army and Its Weapons

Significant changes in weapon and army configuration took place by the end of the fourth century, when new types of equipment and a new tactical structure – the maniple (manipulus, "handful") – had come to replace much of Rome's original military model. Fundamental to any discussion of the manipular army and its weaponry is the meticulous description preserved in Polybius (6.19–42), an oft-quoted source – written *c.* 160, but believed to describe the army of *c.* 220 to 170 – that contains an invaluable commentary on a broad range of military subjects. In arguably the most important set of passages (6.21.7–23.16), Polybius describes the legion as divided into five groups, four of infantry (4,200 men) and one of cavalry (300 men). The infantry fought in a three-line formation (triplex acies): a front line of younger men (hastati) heavily armed with javelins (pila) and swords (gladii), a central line of 1,200 more experienced men (principes) also bearing gladii and pila, and a rear line of 600 men (triarii) drawn from the oldest soldiers who fought with heavy spears (hastae). All of these men carried a standard oblong shield (scutum) and wore full armor, consisting of greaves, a helmet, and either a cardiophylax or, if they were wealthier (with a property qualification over 10,000 drachmae), a mail shirt (lorica; see Figure 8.1). Each of these three lines was divided into 10 maniples (30 in total), so that a maniple of hastati and principes contained 120 men, while a maniple of triarii had 60. The final infantry group was composed of 1,200 light-armed, unarmored skirmishers (velites) drawn from the youngest and poorest citizens, who were armed with a sword, javelins and a small circular shield (parma) in place of the scutum. Proportionally assigned to each maniple (40 men), the velites could be used as a screen in front of the hastati, to mop up small pockets of resistance, to badger the enemy before battle and to reinforce the cavalry, who were now armed with a stronger, longer spear and a round shield similar to that of the velites. Analysis of these passages has dominated all conversations about the Republican army and its weaponry, and for good reason. While other literary references to the Middle Republican army exist (e.g. Livy's discussion in VIII.8–10 of the army's organization in *c.* 340), questions abound about their veracity and the nature of their source material. In contrast, this Greek-born author is a rare contemporary source for the second-century manipular army, and in writing for a largely Greek audience, Polybius provides technical details about the military that many Roman writers tend to elide over, assuming familiarity among their audience. While the Polybian account is certainly not without its problems (e.g. he never mentions the basic legionary unit, the century (centuria) or names its tactical commander, the centurion), it does provides a useful framework by which to examine the character of legionary weapons (Connolly, 1989, 2000; Zhmodikov, 2000: 67–8; Sekunda, 2007: 349–50; Dobson, 2008: 47–121).

Exactly when and why the fundamental alterations to the army and its equipment came about is uncertain, although surely multiple factors were involved. The geo-political situation had shifted enormously since the Republic's foundation, when Rome is thought to have fielded an army of roughly 8,400 heavy and light infantry (or two legions). Rome

Figure 8.1 Late Republican soldiers, armed with scutum and mail, from the so-called Domitius Ahenobarbus Altar in the Louvre Museum. Source: Photo by Torben Retboll, used with permission.

was now the most powerful state in central Italy, with a vastly expanded territory and an unmatched body of manpower. Its army had doubled in size by 311 – from two to four legions – and it is estimated that by 225 Rome's available infantry – including contingents of allies (socii) – had come to number nearly three-quarters of a million men, with an additional *c.* 73,000 cavalry (Cornell, 1995: 181–3; Forsythe, 2007: 28–32, 36–8). Among the catalysts compelling Rome to restructure its defensive and offensive capabilities were a series of new external challenges which placed the army in situations (e.g. longer campaigns, mountain fighting against the Samnites, overseas expansion against states like Carthage) that demanded a more flexible formation than the previous model could or did provide. Moreover, continuous and increased exposure to Italic and non-Italic peoples familiarized the Romans with a broad range of different armaments and tactics, some of which Rome chose to adapt for its own use. While this transformation in weaponry and tactics was quite gradual, the end result was a far different, more complex and flexible force (Rawlings, 2007: 53–9).

Archaeological sources for Rome's manipular soldiers and their weapons are substantially better than for the preceding period, although the record is still not as free of fissures as one might wish. One particular gap is the lack of visual representations of Republican soldiers; with a few notable exceptions, such as the Aemilius Paullus monument from Delphi (*c.* 168), the so-called Altar of Domitius Ahenobarbus from Rome (*c.* 100; Figure 8.1), and the tombstone of the centurion Minucius from Padova (*c.* 44–42?), we cannot turn to sculptural monuments, as we can for the Imperial period, to aid us in reconstructing an accurate picture of legionary equipment and clothing (Keppie, 1997; Feugère, 2002: 64–6; Bishop and Coulston, 2006: 48–50).

The most valuable collection of material has been recovered from a series of weapon hoards, camps and siege sites in western and southern Europe (Figure 8.2). Covering a roughly 200-year span, the hoards have allowed scholars to examine aspects of equipment design, manufacture and regional variation during the Middle and Late Republic (Connolly, 1997: 41–4; Feugère, 2002: 66–9; Bishop and Coulston, 2006: 48). While some remain difficult to use, with unresolved questions concerning stratification and the identification of possible non-Roman elements, the dates of most are believed to be reasonably secure, related persuasively to key historical events known from the literary sources. The earliest hoard, from the Etruscan city of Telamon (Talamonaccio), has been tentatively linked to the defeat of the Gauls there in 225 (Polyb. 2.23–31; Luik, 2000). Since the material was found at various depths in the deposit, however, acceptance of this late third century date remains tentative. While the hoard recovered in 1891 at Šmihel near Grad (Slovenia) is also controversial, a recent review of this vast assemblage (over 450 weapons and other metal objects) has led to general agreement on a date during the first half of the second century (Horvat, 1997, 2002; Istenič, 2005). In north-west Greece, the burned Nekromanteion at Ephyra yielded a weapon hoard dated to the site's destruction by the Romans in 167 (Livy XLV.34; Baatz, 1982). Most of the hoards, however, belong to camps or siege sites in Spain and France that can be dated with more precision. For Spain, these include several sites involved in the sieges of Numantia, especially Camp III at Renieblas, *c.* 153–137; Castillejo, *c.* 137–133; and Peña Redonda, *c.* 134–133 (Dobson, 2008; Luik, 2010); the Sertorian wars, especially Cáceres el Viejo, *c.* 80 (Ulbert, 1984: 105–8) and Gracurris (Alfaro), *c.* 82–72 (Iriarte *et al.*, 1997); and the upheavals of the Late Republic, especially Osuna, *c.* 46 (Sievers, 1997). In France, there are important deposits from the siege of the Gallic fortified town at Entremont in modern Provence (*c.* 124–123), the destruction of the oppidum at La Cloche just north of Marseilles in 49 and the Caesarian circumvallation of Alésia in central France from *c.* 52 (Brouquier-Reddé, 1997; Sievers, 1997). For the latter site, new fieldwork and study of the weapons from the original excavations of 1861–5 have greatly advanced our knowledge about the Roman army and its Gallic opponents in the mid-first century. Other fruitful sources of military equipment include excavated settlements such as Delos, from *c.* 69 (Siebert, 1987), and Castellruf and Caminreal in north-eastern Spain, dated to the late third and late second century, respectively (Vicente, Punter and Ezquerra, 1997; Arza and Argente, 1999). Also from Spain comes evidence from *necropoleis*, which are often non-Roman but contain Roman equipment; battlefield contexts like Valencia (La Almoina); and deposits (votives?) recovered from underwater sites, which usually consist of single pieces of armor or arms found in rivers or wells. Much of the recovered equipment is in fragmentary condition, displaying pre-depositional damage that could have resulted from a variety of causes (e.g. abandonment, breakage in battle, or being stockpiled for later recycling and never reclaimed). Yet in spite of its gaps and dating problems, this combined assemblage has proved tremendously useful in clarifying the origins of and developmental patterns in fourth- to second-century Roman equipment.

The maniple's most important and widely discussed weapon is the pilum. The exact origin of this close-range javelin remains highly debated; ancient and modern scholars have variously suggested that the weapon was first developed among the Etruscans, the Samnites, the Spanish, or the Gauls. Yet a strong case has been made for an Italian origin,

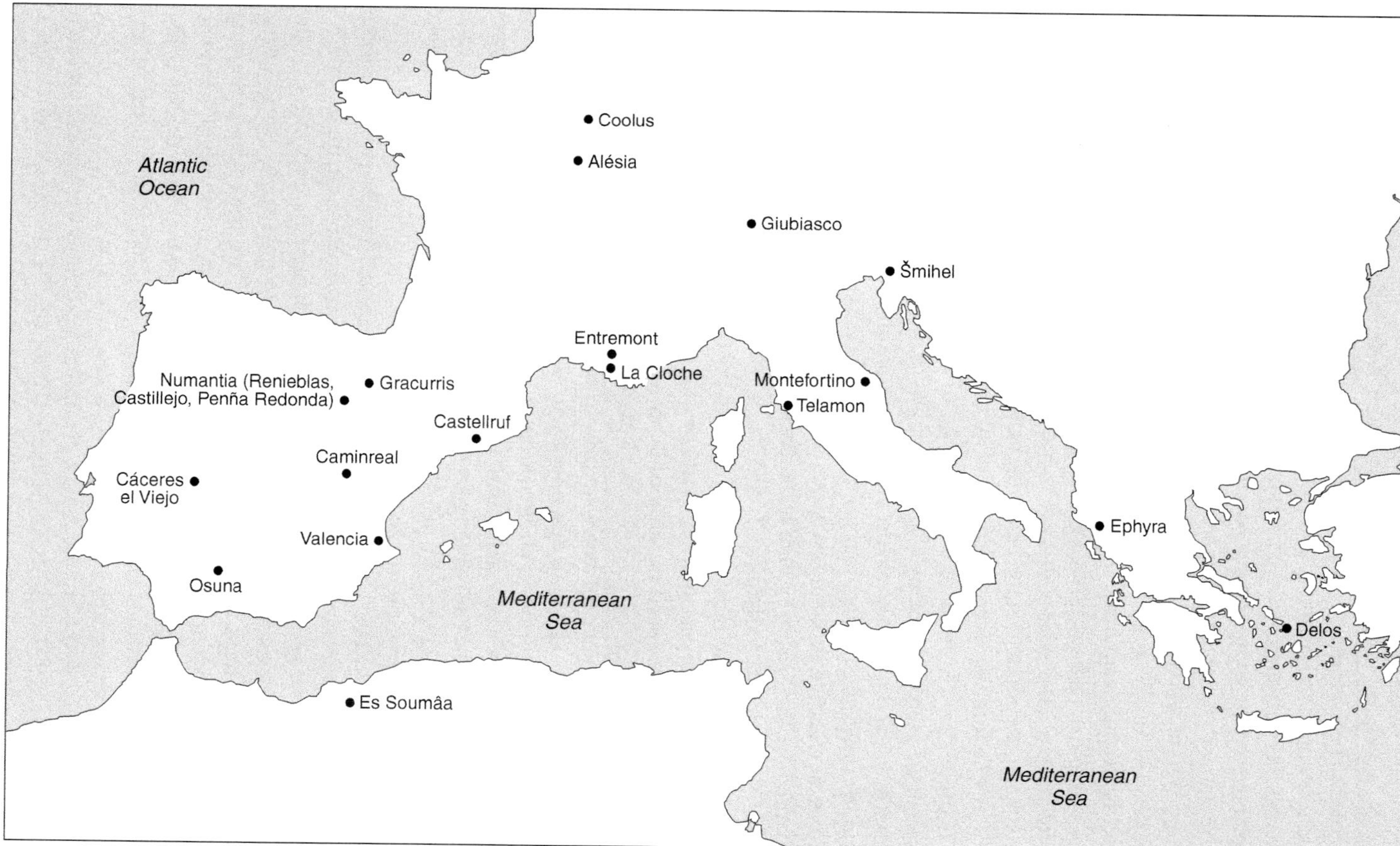

Figure 8.2 Location of major hoard, type- and find sites for Republican weapons. Source: Map by A. L. Goldman.

since early forms of pila were in common use there by the fifth century. Easily recognizable by its characteristic long iron shaft and barbed head, socketed versions are known from numerous northern Italian contexts including a possible fifth-century example (*c.* 120 cm long) from a tomb at Vulci, representations in the fourth-century frescoes from the Giglioli Tomb at Tarquinia, and fourth-century Celtic examples (*c.* 40–100 cm long) from sites like Montefortino. Alternatively, contemporary finds from southern Italy and Spain have lent support for a Samnite or Spanish prototype, perhaps adopted following contact with Carthaginian mercenaries of Spanish or Celtiberian origin during the First Punic War, at which time we have the first definite use of the pilum recorded by Polybius (1.40.11–13). Whatever the case, an earlier javelin of some type must have been in use, since the Republican army appears to have adopted the maniple by the fourth century and manipular tactics cannot work without the pilum or a similar, short-range weapon. While debate continues about whether one or two pila were carried into battle, the moment when they would have been discharged by lines of the maniple, and the frequency of missile fighting over the course of an encounter, the function of the pilum in combat is quite clear. Thrown at a close range of about 25 m, its heavily-weighted head was meant to pierce an enemy's shield and armor, for the purpose of killing the enemy outright, breaking an oncoming charge, entangling and forcibly disabling an enemy shield, or providing the legionary time to draw his sword. In a pinch, they could also be used as an effective thrusting weapon, and examples from Šmihel show that pila could be adapted for use as short-range incendiary missiles. Their overall design carried clear benefits, including the barbed heads that made them quite difficult to extract and an incidental tendency of the head to bend upon impact, rendering them useless to an enemy. Modern reconstructions and field tests of pila indicate that it was effective at a range of *c.* 25–30 m and produced the kind of bending on impact of the long iron shank found on excavated examples.

We possess plentiful evidence for both the heavy and light versions of the pilum mentioned by Polybius (6.23.9–12), who describes them as barb-headed with wooden shafts and heads each three cubits long (*c.* 1.38 m). Since a portion of the head's length was buried in the wooden shaft (to ensure that the metal would break before the head and shaft could separate), the overall length of Polybius's version was *c.* 2 m. Pila heads (cuspides) have been recovered from nearly a dozen sites and, contrary to Polybius's description, display wide variation in size, shape and profile. The oldest known type, the light pila, was normally socketed and varied greatly in length (*c.* 15 to 94 cm) during the Republic. In contrast, examples of the heavy pila (*c.* 25 to 95 cm long) have thicker shanks and show a preferred use of a flat or broad tang for attachment by two rivets to the wooden shaft (though socketed heavy versions have also been found). To reinforce the junction where the metal head was inserted into the wooden shaft, the edges of the tang were often hammered over to create a stronger bond. The design of the pilum, however, was hardly standardized, displaying constant changes in hafting and head shape. For example, in early second-century versions of the barbed heavy pila from Castellruf and Ephyra, the tang was secured to a wooden block atop the shaft, while slightly later examples from Renieblas and Caminreal bear pyramidal heads and were riveted directly to the shaft (Connolly, 1997, 2002; Zhmodikov, 2000: 68–74; Feugère, 2002: 80–3; Bishop and Coulston, 2006: 50–3; Sage, 2008: 79–84; D'Amato and Sumner, 2009: 6–7; for the reconstruction of pila, see Connolly, 2000).

A second weapon was the spear (hasta), the main armament of the light-armed velites and the triarii in the rear line, the latter likely a relic from the days of defensive phalanx fighting. The triarii used heavy fighting hastae, while the velites carried a lighter, more portable version (the hasta velitaris). On occasion, as at the Battle of Trebia in 218, velites also employed throwing spears (veruta) slightly shorter than the hasta velitaris. All remained simple in form, a wooden shaft with a socketed iron head and spiked butt on the end. While a large range of head sizes has been recovered from camps and other sites, no specific feature appears to distinguish Roman spearheads from non-Roman weapons during this period. For the most part, these hastae were designed to be grounded with the butt (which could serve as a secondary weapon) and used against swordsmen in battle. However, experimentation with reconstructions of the hasta velitaris has shown that when thrown, it has an effective range of *c.* 40 m (Connolly, 2000: 45; Bishop and Coulston, 2006: 53–4; Sekunda, 2007: 350–1).

Beyond a few literary references and carved depictions (e.g. on the Padova centurion), we knew little until very recently about the short iron sword of Republican legionaries, which Polybius (6.23.6) terms the "Spanish sword" (gladius hispaniensis). On the evidence of its name, most scholars have assumed that this weapon was of Spanish origin, taken up following extended contact with Celtiberian mercenaries during the First or (more probably) the Second Punic War. Recent scholarship has favored a more complex genesis, rejecting some earlier suggested prototypes (e.g. the falcata, a curved bladed saber found at Caminreal and Cáceres) in favor of the Romans' adoption of a Spanish adaption of a Celtic long sword. Finds of such Celtic long swords at Spanish sites appear to support this hypothesis. Happily, a growing number of actual Spanish swords have been positively identified, beginning in 1986 with the example from the House of the Sword on Delos (see Figure 8.3). This short weapon (76 cm long, 5 cm wide) with its straight double-edged blade (*c.* 60 cm long) and long, sharp point matches the literary descriptions of Polybius and others. Incredibly, the remains of the leather scabbard (*vagina*), its iron framework and four suspension rings – an upper and lower ring attached on each side of the upper scabbard's frame – were also recovered. More than 20 examples from sites in Spain (e.g. Renieblas, Osuna), France (from the Rhône River near Lyon, Mouriès), Switzerland (Giubiasco), Algeria (Es Soumâa) and Israel (Jericho) have subsequently been identified, the earliest at Šmihel (*c.* 175). All display a similar range in blade size (64–9 cm long, 4–5 cm wide), although a few are waisted in form, bearing a slight inward curve of the blade near the hilt, which would have been made of wood or bone. Exactly what kind of sword the Romans used prior to the gladius hispaniensis is not known, although it might have been similar to shorter (*c.* 50 cm long) types used by the Etruscans or Greeks. In any case, the Spanish sword would remain in use from the mid-third century BCE until the mid-first century CE, at which time the shorter, slimmer Mainz type replaced it. Useful for both slashing and thrusting, it was well suited for the close-quartered fighting of the manipular formation following the discharge of the pila (Connolly, 1997: 49–56; Quesada Sanz, 1997a, 1997c; Feugère, 2002: 79–80; Sage, 2008: 84–7).

Another weapon type with apparent Spanish ancestry is the dagger (pugio). It appears to derive from the Celtiberian double disc hilt dagger, and numerous Roman examples have been found in second- and first-century Roman camps (e.g. Caminreal and Cáceres), where they display little difference from native Spanish types. The double-edged blades

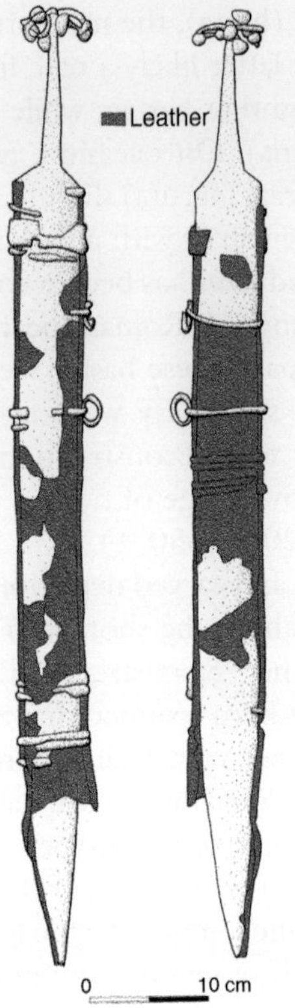

Figure 8.3 Spanish sword from Delos. Source: Drawing by A. Yandek, adapted from Bishop and Coulston (2006).

(*c.* 15–20 cm long) are generally waisted, with a long tip, a midrib and a bulbous handle with a riveted hand guard. This sidearm is not mentioned by Polybius, which has led scholars to question whether it was carried into combat or in widespread use before the first century. However, it has a long history in Italy and the Padova centurion wears it in a horizontal position below his waist belt. Thus it could be that the pugio was worn mainly by officers during the Republic and only became part of the standard combat weaponry during the Late Republic or Early Empire (Connolly, 1997: 56; Bishop and Coulston, 2006: 58; D'Amato and Sumner, 2009: 24).

The most important of the legionary's defensive arms was the scutum, the large curved, oval shield with central rib (spina) and round iron boss (umbo) seen carried by soldiers on the Aemilius Paullus monument and the so-called Altar of Domitius Ahenobarbus (Figure 8.1). Representations of the scutum or variations of it indicate that it was in use by Italic peoples by the mid-seventh century, and scholars generally accept an Italian origin for the shield, with possible later influences from oval Celtic or German shields brought into northern Italy. According to Livy (VIII.8), it replaced other types, such as the round hoplite shield, during the Latin wars of the mid-fourth century, as a far more suitable type for the maneuverability demanded by manipular tactics, sword play and the increased use of javelins. In terms of archaeological evidence, we possess fragments of iron bosses from the Spanish camps (e.g. Graccurris, Renieblas), and a single, remarkable example of an intact wooden scutum recovered in 1900 from Kasr al-Harit in Fayum (Egypt). While its date is uncertain (first century?), its preservation is outstanding, lacking only its original paint. Measuring 1.28 m in length and 0.635 m in width, it is composed of three layers of alternating vertical and horizontal plywood (birch) strips, a compositional technique that Varro (*Ling.* 5.115) credits with the genesis of its name ("secuta" is the act of cutting). The use of plywood both lightened and strengthened the weapon, as did the surface covering of lamb's wool felt. The exterior sports a round wooden umbo and a vertical spina, while the interior has a horizontal handgrip behind the umbo. It varies from Polybius's description (6.23.2–5) in several ways: it lacks metal edging (as do all visual representations of the scutum), the boss is not made of metal, and it is thinner at the rim than at the center, making it harder to penetrate. As the shield type was in use for several centuries, however, such minor modifications are hardly surprising. Polybius is certainly correct to assert that its large size, curved exterior surface and metal boss would have provided excellent protection against sword strokes and the impacts of stones and missiles. While it could also be used offensively, to unbalance and bring down an opponent, reconstructions suggest that its weight (*c.* 10 kg) made it too heavy to have been wielded freely with its single, horizontal handgrip. Rather, the scutum probably rested on the ground, held out by a straight arm, to provide frontal protection. This contrasted strongly with the parma carried by the velites, a circular ox-hide shield (*c.* 1 m in diameter) that allowed for more maneuverability but provided less protection (Polyb. 6.22; Treloar, 1971; Eichberg, 1987; Feugère, 2002: 76–8; Bishop and Coulston, 2006: 61–3; Sage, 2008: 76–7).

In terms of body armor, Polybius (6.23.14–15) describes two types in use by the manipular army, the cardiophylax or pectorale and the mail cuirass (lorica). Since the lines of heavy infantry were divided by age and experience and armor type was still closely linked to wealth, it seems that the manipular legion would have had a very non-uniform appearance. Unfortunately, Roman armor finds tend to be rare, and the armor of the less affluent soldiers – specifically the cardiophylax – is not represented on Roman figural monuments. However, plentiful examples of the pectorale, including rectangular, triple-disc and anatomical types, are known from Campania, Lucania and Apulia, where they became common by the mid-fourth century, as indicated by tomb finds and vase paintings. All consist of bronze breast- and back-plates attached by rings or shoulder- or side-plates, and their spread is likely a product of Italic and Samnite influence, inspired by earlier Greek and Etruscan armor. Bronze circular plates (*c.* 17–25 cm in diameter) with a raised central boss and attachment holes around the perimeter have been found in the camps at Numantia and probably represent a variation of the square (*c.* 23 cm) brass heart-protectors mentioned in Polybius. Also recovered from the camps (i.e. Renieblas) were copper-alloy fragments of ring mail (lorica hamata), a type of armor clearly adopted from Rome's Celtic neighbors. Composed of inter-linked rings, it was worn over leather or linen padding and belted in place to transfer weight from the shoulders to the hips and keep it stable. Soldiers wearing thigh-length examples of this mail type with belts and shoulder doubling for added protection from swords are seen on the Aemilius Paullus and Ahenobarbus monuments (Figure 8.1), as well as on a relief of a legionary from Osuna (probably Late Republican). Scale mail (lorica squamata), composed of small scales cut from iron or bronze sheeting and sewn on to a leather or cloth backing, might also have been in use, as a cheaper and more easily produced alternative. However, while well represented in the Imperial period, at present we lack both secure finds and artistic representations of this type for the Republican era. We likewise lack physical evidence for another type of armor described in the literary sources, the linothorax, a cuirass with shoulder guards composed of stiffened linen (or possibly leather) that would have been a more affordable alternative for soldiers. Finally, there is the muscle cuirass, which continued in use from earlier times and can be seen on the Ahenobarbus altar. It would have been rarely seen (if ever) on the battlefield, however, since its weight and inflexibility made it useful only for officers and cavalrymen (Feugère, 2002: 73–6; Burns, 2003: 71–3; Bishop and Coulston, 2006: 63–5; D'Amato and Sumner, 2009: 38–43).

Given the legionary's growing need for mobility in manipular tactics, one element of the defensive panoply that seemed destined to fail was the shin guard, or greave. Polybius briefly mentions their use (6.23.8), and it is the general scholarly consensus that the heavy infantry wore only one greave, on the left foot, which was thrust forward when in combat and thus needed greater protection. What type of greave might have been worn remains difficult to assess, since there are no known Republican examples, although a representation of one is visible on the Osuna relief. Their use in battle seems to have been abandoned in the late second or early first century, perhaps for reasons of cost and the restrictions it placed on movement, especially for marching (Feugère, 2002: 76; Bishop and Coulston, 2006: 64–5; Sage, 2008: 78–9).

The final piece of defensive armor was the helmet (galea or cassis), and evidence from a variety of contexts, including funerary, camp, well and river finds, reveals that the manipular army wore a variety of helmets. Indeed, studies of helmet distribution in the fourth century indicate that while regional preferences existed, styles tended to overlap and no one helmet type predominated. Many of the bronze and iron helmets belong to the family termed "jockey caps," with a strong resemblance of their bowl-shaped profile to the modern cap (though the helmets were worn in reverse, with the peak at the back to act as a neck-guard). By the mid-third century, the so-called Montefortino type – an Italo-Celtic hybrid that can be traced back to the fifth century in France and Austria – had become the most common jockey helmet in Italy, Spain and Gaul. It was thus arguably the one used by Roman infantry. Named after its central Italian find-site, this copper-alloy (and later iron) helmet consists of a hemispherical bowl beaten into shape, with a round knob at the top, an angled, stubby neck-guard, and wide, detachable cheek-pieces (in metal or leather) with fastenings for a chin-strap. This simple design perhaps enhanced its appeal, since it must have been comparatively cheap and easy to produce, a boon to soldiers required to purchase their own equipment. Polybius (6.23.13) describes infantrymen attaching plumes of black or purple feathers (*c.* 45 cm high) to the knob for increased visibility and identification. The protrusion might also have served to secure larger crests, as seen on the Montefortino helmets of the Ahenobarbus altar (Figure 8.1). Over time the neck-guard was enlarged in size, a modification that has led scholars to suggest that the legionary crouched in battle, placing his neck in slashing range of enemy swords. The cheek-pieces were also modified to a more protective form, with an earlier triangular shape (often decorated with three circular bosses) replaced by wider, scalloped-shaped pieces in the third century. A second variety of bronze helmet, the Coolus type, had a similar bowl-shape and projecting back-peak but no crest knob. Although this type dates back to the third century and shares a similar Celtic origin with the Montefortino type, it did not become popular until the mid-first century, when it is found at sites in northern Italy, Gaul and Germany. Neither type appears to have been adopted by officers, however, who are shown on monuments and in paintings as continuing to wear the Greek-influenced Italo-Corinthian and Attic types of earlier times (Robinson, 1975: 13–41; Quesada Sanz, 1997b; M. Burns, 2003: 67–70; Bishop and Coulston, 2006: 65–6). As in the pre-manipular period, then, variation continued in the form of armor, a fact that would have left Republican soldiers and their defensive armor looking oddly mismatched to the modern eye.

5 The Late Republican Army and Its Weapons

The final century of the Republic witnessed continued evolution in the tactical structure and weaponry of the Republican army, as the maniple gave way to a new organizational unit, the cohort. The former system found itself under rising pressure from an increasing number of internal and external factors, including a growing core of longer-serving, near professional soldiers, extended overseas warfare, the demands of long-distance supply, problems with recruitment, and unacceptable military setbacks against foes in Africa and northern Italy (e.g. Jugurtha and the Numidians, the Cimbri, the Teutones). The response of generals like Marius and Caesar was to move gradually towards a more homogenized system of organization, weaponry and training, with the general aim of producing a larger, better-coordinated and less expensive army.

The transition from maniple to cohort proceeded in irregular stages, with important ramifications for armament and tactical organization. The cohort had come into existence by the late third century; Polybius (11.23.1) first mentions its use in Spain at the Battle of Ilipa in 206. In this early context, the designation referred to a group of three maniples that were deployed together, perhaps in a temporary assignment. Such units seem to have been particularly well suited for the situation in Spain, where varying terrain and a mixture of guerilla tactics and frontal assault by the enemy necessitated a more flexible means of attack and defense. Even so, the maniple continued to be the legion's dominant tactical subdivision until the end of the second century or perhaps slightly later, as reflected in the layout plan of the camps around Numantia dating to the 130s (Dobson, 2008: 84–90; see Chapter 14). By the late second century, however, major conversions were taking place. Equipment seems to have become more standardized by the 120s (including the abandonment of the hasta, an unwieldy weapon within a cohort, in favor of the pilum), and shortly afterwards the light-armed velites disappear, a change probably caused by either the elimination of property qualifications under Gaius Marius or their replacement by rising numbers of allied troops who acted as slingers and archers. Allied and non-Italian cavalry likewise began to replace Rome's traditional citizen cavalry, which also soon disappeared. In addition, the legion was reorganized into 10 cohorts (*c.* 420 men) in place of the 30 maniples, a transition probably completed by Marius (although the older system seems to have continued in concurrent use to some degree, perhaps as a tactical alternative). While each cohort continued to consist of identically armed hastati, principes and triarii, the legion was now divided into lines of cohorts in a 4-3-3 formation. The advantages of the new system included a simplified command structure, units that could work effectively independently or in combination, and greater uniformity in training and equipment that enabled easier substitution and supply. Under

Marius's direction, a new level of regularity was reached, with men trained more thoroughly, each legion given an identical standard (the eagle) and soldiers made more mobile and self-sufficient by carrying both equipment and supplies (they thus became known as "Marius's Mules"; Frontin. *Str.* 4.1.7). The state also became more involved in equipping its soldiers, since the former property qualifications for legionaries had been dropped and those below the minimum census requirement (the capite censi) could not afford their own arms. With the new grants of citizenship following the Social War, former allies could now become legionaries and the allied units disappeared. Their replacements consisted of temporary units composed of native levies from the immediate area of operations as well as more permanent non-Italian, non-citizen companies (auxilia) with various tactical and weaponry specialties (e.g. Numidian cavalry, Cretan archers, Balearic slingers). Further standardization by the Caesarian period resulted in cohorts of *c.* 480 to 600 men armed in an identical manner, with legions composed of *c.* 4,800 to 6,000 men; this became the Imperial model (Cagniart, 2007: 80–8; Sekunda, 2007: 356–7; Sage, 2008: 199–208). The result was something of a trade-off, with Rome moving away from a purely Romano-Italic army constructed of independent field units having different arms and functions, towards a force with greater interchangeability among units yet saddled with a growing dependence on non-Roman troops for its tactical diversity.

Details from the archaeological record appear to support the growing trend of standardization within the military as well as reveal the effect that the expanding army had upon the production of military equipment. Changes in manufacture have been detected by the late second century, most notably in the production of Montefortino helmets, whose craftsmanship seems to have declined in quality, showing evidence of the use of scrap metal for the fittings and a lower standard of finishing on the cheek-pieces (or total lack thereof). Moreover, mass production seems to have been initiated by that time, as indicated by maker's stamps and, in the first century, the adoption of spinning lathes to create the bowl-shaped helmets more quickly and in larger numbers. A certain level of homogeneity in equipment types began to emerge from this large-scale production, and a resulting uniformity in outward appearance developed to a greater degree. Regional variation was still evident, however, as indicated by the wide variety of helmets in use during the second half of the first century, when the Montefortino helmet seems to have become less popular. Many helmets recovered from France and Germany belong to the increasingly popular bronze Coolus type mentioned above and a slightly heavier version of it known as the Mannheim type. There was also a simpler bowl-shaped version known as the Buggenum type, a rare variety popular around the region of modern Holland. Based on fragments found at Alésia, the Agen/Port type – another branch of the Gallic family of helmets – was also in use at this time. These iron helmets are characterized by a deep, rounded bowl with vertical sides that flare along the lower edge into a brim, one that is narrow in the front and almost twice as thick in the neck area. The Agen/Port type differs in that it has only a narrow front brim and is decorated in the brow area with a set of embossed, re-curved eyebrows. The cheek-pieces of both follow the line of the jaw and have semi-circular cut-outs along the front edge for the eye and mouth (Robinson, 1975: 26–43; Feugère, 2002: 70–2; Burns, 2003: 74; Bishop and Coulston, 2006: 65). The army of the Late Republic would thus have achieved a more uniform level of armament than its predecessors, yet still differed somewhat in the details of its equipment.

Some modification in offensive weapons also occurred as the cohort developed. Both socketed (light) and tanged (heavy) versions of the pilum – now the primary missile weapon of the infantry – continued in use, with square-sectioned versions of the former found at Alésia. In the case of the broad-tanged pilum, new hafting methods were adopted, with rectangular tangs locked into the shaft by two rivets, as seen at Caminreal and Valencia. Another first-century modification was the addition of a round or square iron collar at the juncture of the head and shaft to keep the latter from splitting on impact, a change necessary since the tang edges were no longer being bent around the outside of the haft for added reinforcement. A third type of pilum was apparently in use as well, differentiated by a spiked-tang that would become popular during the principate. Examples of this type from Alésia and Osuna, characterized by thick shanks and short heads (*c.* 29–39 cm long) attached by a single rivet, probably represent pila muralia, short-distance javelins used for defensive sieges. This evidence has proved helpful in addressing the famous modification in pilum design which Plutarch (*Mar.* 25.1) claims Marius made against the Teutones in 102, specifically that one rivet of the socketed pilum was replaced by a wooden pin in order to facilitate breakage upon impact and deny the enemy an opportunity to reuse the weapons. However, no example of a preserved pilum head with a double-holed socket has ever been found missing its second metal rivet. Recent experimentation has produced interesting results, however, indicating a pronounced tendency of the second metal rivet (closest to the juncture) to break on impact in any case, skewing the wood haft up to 90 degrees. It has thus been suggested that Plutarch is mistaken and is in fact not describing the double-riveted broad-tanged pilum, but rather the newly introduced spike-tanged pilum with its single rivet (Connolly, 1997, 2002). In any case, the pilum, the gladius hispaniensis and probably the pugio emerged as the main offensive weapons of the cohort and would continue to be so into the Imperial period.

6 Conclusion

While the Roman Republican army was hardly a perfect creation, it was indisputably one with the power to intimidate and effectively overwhelm its enemies, as it often did. Indeed, the army became so adept at meeting its challenges, in part through the reinvention of its military technology, that the mid-first-century writer Diodorus Siculus could justifiably claim that "the Romans as pupils always surpassed their teachers" (23.2.1). One should not fail to remember, however, that the evolutionary process from armed citizen militia to the near-professional army of the Caesarian era was a slow and uneven one. The emerging field of Roman equipment studies is helping to reinforce this fact, allowing us to examine the depth and complexity of this gradual transformation through the lens of archaeological sources. New evidence for weapons like the pilum and gladius is helping not only to clarify our understanding of their origin, use and manufacture, but also to reveal processual aspects of Roman state formation, with the development of Rome's military technology providing us with new insight into the character of intercultural contact and exchange between Republican Rome and its neighbors. Our overall reconstruction of Rome's substantial military transformation is admittedly far from complete, as critical data remain scarce for certain periods of

Republican history and for geographic regions like the eastern Mediterranean. Nevertheless, with a growing body of archaeological material to aid us, we are moving beyond the limitations of our literary evidence towards a more nuanced portrait of the Republic's army and its weapons.

FURTHER READING

For those who wish a general introduction to the Roman army and its weapons, the chapters in Parts I (Early Rome) and II (Mid- and Late Republic) of Erdkamp (2007) are an excellent place to begin, with current overviews of army organization throughout the Republic as well as general examinations of a broad range of related topics. For brief descriptions of military equipment see also Rich (2007: 16–18) (early Roman army), Forsythe (2007: 31–2) (centuriate army), Hoyos (2007: 69–70) (the manipular army) and Cagniart (2007: 88–91) (Late Republican army, cavalry); on contextual problems in Republican equipment hoards, see Connolly (1997: 41–2). For quality illustrations of the equipment of Rome, its neighbors and enemies (though lacking in bibliography and footnotes), see Connolly (1981). Essential for anyone wishing to look more closely at military equipment are the excellent recent surveys by Feugère (2002), Bishop and Coulston (2006) and D'Amato and Sumner (2009), the first such treatments to be published in well over half a century, since P. Couissin's venerable *Les Armes romaines* (1926). For those wishing to examine in greater detail specific regions or sites, the volume of *JRMES* 8 (Feugère, 1997) contains much of the research presented at ROMEC X on Republican military equipment. For reasons of space, this chapter focuses primarily on the legionary soldier and his arms. For the Republican cavalry, see McCall (2001); for supplementary military equipment: Feugère (2002: 173–82), Bishop and Coulston (2006: 67–70); and for artillery: Feugère (2002: 105–8). For the helmets found near the Egadi Islands, see S. Tusa and J. Royal, "The landscape of the naval battle at the Egadi Islands (241 BC)" JRA 25 2012: 7–48.

CHAPTER 9

Bodies of Evidence: Skeletal Analysis in Roman Greece and Cyprus

Susan Kirkpatrick Smith

1 Introduction

My job as a biological anthropologist is to bring the long-dead back to life by telling the stories written in their bones. History tells us about the notable people and events; it provides clues to how the people lived. But it does not bring common individuals' stories to us. With the analysis of human remains from archaeological sites, anthropologists are able to do this. We can document whether a woman's diet was based on fibrous or abrasive foods, or based more on starchy gruel. We can document whether a man engaged in strenuous physical daily activities. We can document whether a child suffered from periods of severe illness or nutritional stress. As biological anthropologists, we work to put the breath back into dry bones. I will present how biological anthropologists working in the Roman lands of the eastern Mediterranean have conducted their work and given new life to old bones.

Human skeletal remains can provide a wealth of information for classical era sites that is not available from other sources. Sadly, in much of the world of classical archaeology these analyses have not been as prevalent or as integrated into archaeological analyses as one would wish (e.g. Fox, 2005; Harper and Fox, 2008; Paine *et al.*, 2009). This lack of integration is changing (MacKinnon, 2007; Rife *et al.*, 2007; Ubelaker and Rife, 2011), but the need for collaboration between biological anthropologists and archaeologists is crucial from the planning stage. As MacKinnon notes, "Classical bioarchaeology should provide a truly interdisciplinary means to reconstruct the natural and cultural world of classical antiquity" (2007: 474). This interdisciplinary approach means that biological anthropologists should be involved in archaeological projects as early as possible, from the development of research proposals if possible. This will allow the biological anthropologist to contribute to the research questions and excavation plans so that the projects can be "truly interdisciplinary" and give us a more in-depth picture of the classical world.

A Companion to the Archaeology of the Roman Republic, First Edition. Edited by Jane DeRose Evans.

When biological anthropologists are involved in an excavation from the beginning, the project benefits from having an expert on the team who can begin the analysis before the skeletal remains are removed from the ground and valuable information is lost. For example, it can be difficult to determine the side of the body from which some finger and toe bones come, but if the bones are collected in the field by a biological anthropologist, the association with limb bones makes the job of "siding phalanges" possible. Or bones may be so fragile that they crumble as they are excavated. When this is the case, having a biological anthropologist on site during the excavation means that analysis can be conducted while the bones are still in the ground (Barnes, 2003: 438). Similarly, teeth are prone to falling out of the jaws. If this occurs in the field, the anthropologist can record exactly from where in the mouth the teeth came. While it is often possible accurately to identify loose teeth, it can be difficult or impossible if the teeth are heavily worn, damaged by caries, or broken. It is these types of teeth that provide some of the most useful dental information (see below for a discussion of dentition). A biological anthropologist present at the excavation would be able to determine whether teeth were lost before or after death, which may influence how the excavation should proceed.

Pathological bones that may not be recognized as human by a non-specialist can be evaluated in the field, and important contextual information recovered. Biological anthropologists are also better equipped to recover bone fragments that might be missed by others or considered too small or damaged to be useful. Barnes (2003), in her analysis of Frankish burials from ancient Corinth, shows the importance of an experienced biological anthropologist being present at an excavation. These human burials had some animal bones in with them. Barnes was able to recognize fetal bones within the pelvis of a female skeleton that otherwise might have been mistaken for animal bones. Even if they had been recognized as human fetal bones in the lab, the context of the bones would have been lost and the fetus's transverse position within the pelvis would have never been known (Barnes, 2003: 438). In this same excavation, Barnes discovered an unusual-looking bone that turned out to be several fused carpal (wrist) bones damaged by severe rheumatoid arthritis. By recognizing this in the field, she was able to direct the archaeologist she was working with to the possibility of finding other similarly afflicted bones. Additional bones were found in two adjacent graves, and the careful collection of the bones allowed her to determine that they all came from one individual.

Cremations, post-mortem disturbances, environmental conditions and other factors can result in poorly preserved or broken bones. It has sometimes been the case that these fragmentary remains were not kept by archaeologists. However, Rife's excavation of inhumed and cremated human remains from Kenchreai in the Corinthia shows the value of keeping and analyzing such small fragments (Rife *et al.*, 2007). Ubelaker and Rife were able to document mortuary practices, such as body placement of interments and the careful collection and placement of cremated remains in the tombs by combining careful archaeological recovery techniques of the skeletal remains under the supervision of a biological anthropologist (Ubelaker and Rife, 2011). The social activities surrounding the burial practices in Roman Kenchreai were complex and many people would have taken part in them. Members of the community had to engage in a significant amount of handling the dead. This understanding of the details of burial ritual would have been lost without the careful excavation and analysis of small amounts of bone.

The long history of occupation in Greece and Cyprus combined with intensive archaeological programs make these locations significant for an advantageous study of the effects of the Roman Empire on the biological and social characteristics of the inhabitants. The skeletal studies from eras prior to the coming of the Romans can serve as a source of comparison. The cultural changes brought about by the Romans included an influx of new people, bringing with them new diseases, new physical characteristics, new diets and new behaviors. Many of these changing cultural traits would have had an impact on people's lives that would be visible in the skeleton.

2 Biological Anthropology in the Archaeological Context

Analysis of human skeletal remains from archaeological sites follows a basic plan, with biological anthropologists assessing the minimum number of individuals present in a burial, establishing their demographic profiles and recording details of skeletal pathologies and anomalies. Buikstra and Ubelaker's (1994) volume provides standard methodologies for the collection of these data and is an excellent resource for non-specialists to understand the data available from human skeletons. What follows is a description of how data are collected, followed by examples from the literature that show how these data have been used to help us understand life in Greece and Cyprus during the time of Roman control.

3 Minimum Number of Individuals

Minimum number of individual (MNI) assessments must be made whenever skeletal remains are discovered. When the skeleton is recovered still in anatomical position, with little to no disturbance in the placement of the bones, it is easy to make the determination that we are dealing with one individual. However, many Roman tombs were used for more than one burial, and post-burial disturbances can cause remains to become comingled. When this is the case, the bones are often disarticulated, broken, fragile and incomplete. In these cases biological anthropologists must calculate MNI.

In order to determine the MNI present in a tomb or other burial, we can do a basic counting of each skeletal element present. If there are four left femurs (thigh bones), three right femurs, and three left patellas (knee caps), the MNI is four, based on the left femurs. However, if two of the left femurs are large and robust and two are small and gracile, while all three right femurs are large and robust, then there are three people who are large and robust, and two who are small and gracile, so our MNI goes up to five people. This process continues with all of the skeletal elements being considered. A seemingly small amount of skeletal material can sometimes represent a large number of people.

MNI assessments can indicate how tombs were used and reused in the past. The human skeletons from the Late Roman burials of Lerna Hollow in Corinth were in very poor condition, but the MNI information was able to show that grave type was associated with use practices (Wesolowsky, 1973). The more elaborate graves often had multiple individuals buried in them, while the simpler graves lined with tiles always had only one person. Based on this and other demographic information about the skeletal remains, Wesolowsky was able to show that grave type was based on such cultural factors as financial or social status or religion, and not demographics (age and sex) (Wesolowsky, 1973: 341).

The Roman burials from the north-east sector of ancient Corinth had a mix of primary and secondary burials. Secondary burials suggest that tombs may have been cleaned out to provide room for new burials, either as the tombs got full or as tombs changed ownership. The "Tomb with the Sarcophagi" contained an MNI of 98 individuals. All of the sarcophagi contained multiple burials except for one, which had fragmentary remains of two juveniles, and also a curse tablet. Walbank suggests that the curse tablet is the reason that this sarcophagus was never reused. In addition to the information from the curse tablet sarcophagus, an analysis of MNI over the 500-year history of the cemetery shows an increase in multiple burials over time (Walbank, 2005).

Finally, an example from a Late Roman church burial at Maroni Petrera in Cyprus demonstrates how an analysis of MNI contradicts what is suggested in the excavation (Fox, 2003). The bones coming from what the excavators called burials 2 and 3 turned out to contain a MNI of five individuals, all of whom were infants (Fox, 2003: 39–40). The difficulty of recognizing infant bones in the field, particularly by non-specialists, can result in this type of misconstrual of burial information.

4 Sex

One of the basic demographic assessments that biological anthropologists attempt with each set of skeletal remains is the determination of sex. Sex is best assessed on the skeleton in the same locations as in the living: the pelvis is the most reliable, followed by the skull. The other bones of the body often vary between males and females, but these differences only work for individuals at the extreme ends of the size spectrum; otherwise, the amount of overlap between males and females is too great.

Differences in the pelvis are due to adaptations in females for childbirth (Figure 9.1). The female pelvis has a larger pelvic inlet, and a wider flare to the top of the os coxa (hip bone). Even when only small parts of the pelvis are preserved, it is often possible to differentiate females from males. The pubis is square in females and more triangular in males; the greater sciatic notch is wider in females than in males. When the entire os coxa is present, an experienced biological anthropologist can have greater than 95% accuracy in determining sex (White and Folkens, 2000: 368).

When just the skull is present, an experienced biological anthropologist can accurately assess sex 80 to 90% of the time (T. White and Folkens, 2000: 362). Males have larger bony ridges over their eye sockets, squared-off chins, and a robust character to the skulls. Females develop more typically "masculine" characteristics on the skull as they age (Meindl *et al.*, 1985; Buikstra and Ubelaker, 1994; T. White and Folkens, 2000), so age and sex must be considered together. It is also important to be familiar with the population from which the skeletons came, as there are differences in how males and females look across different groups. Because there is currently no consistently reliable way to sex juveniles from their skeletal remains (Scheuer and Black, 2004), the discussions below of sex assessments will focus only on adults.

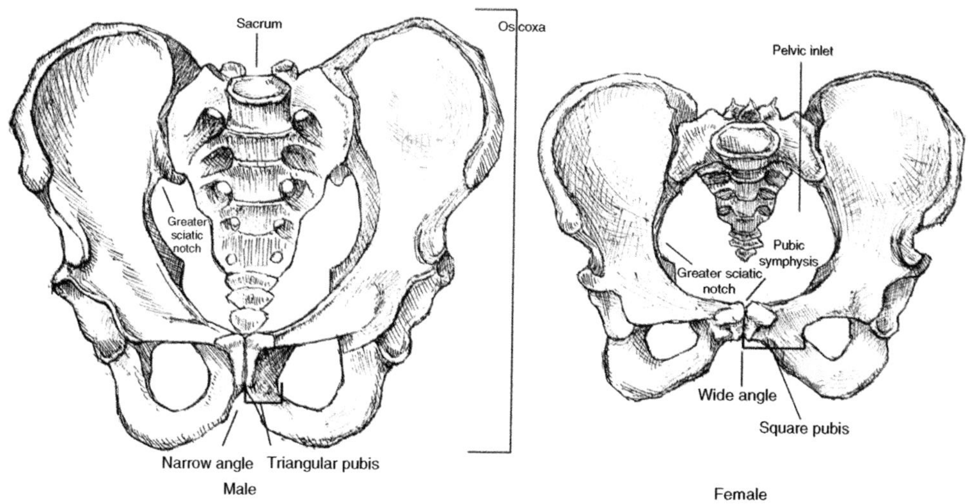

Figure 9.1 Male and female pelvises showing features used in sex determination. Source: Drawing by Danna Simonis.

Differences between males and females in health, average age at death, degenerative conditions and dental characteristics are often included in excavation reports. This can help to explain whether males and females were subjected to the same types of stresses, diets, physical activities or burial traditions. When there is variation between these factors and sex, it tells us something important about how the sexes lived in the past. For example, one of the major sex-related differences found in the archaeological record is that the average age at death for males is often much higher than for females (Bisel and Angel, 1985: 203; Bourbou, 2005; Fox, 2005). This contrasts with the greater longevity of females compared to males in modern populations (Mathers *et al.*, 2001).

Males and females received similar types of burial treatments in Roman Kenchreai, near Corinth, Greece (Ubelaker and Rife, 2011). While we cannot know all of the ritual events that occurred at the time of death and burial, we can see that the two options for burial, either inhumation or cremation, were used for both sexes. Wesolowsky (1973) also found at Lerna Hollow in Late Roman Corinth that the sexes seem to have equal access to the both rock-cut and tile-lined graves. Wesolowsky argues that grave choice was cultural, based on factors such as social status, financial status or perhaps religious practices, rather than revealing demographic characteristics. In Roman Greece and Cyprus there are not yet any data about health and diet differences between men and women. As more studies are conducted, these data will become available.

5 Age

It is much easier to assess the age of children than adults in skeletal populations. Childhood growth and development occur in regular and predictable patterns, while degenerative changes in adults are much less regular. We can often determine the age of children within

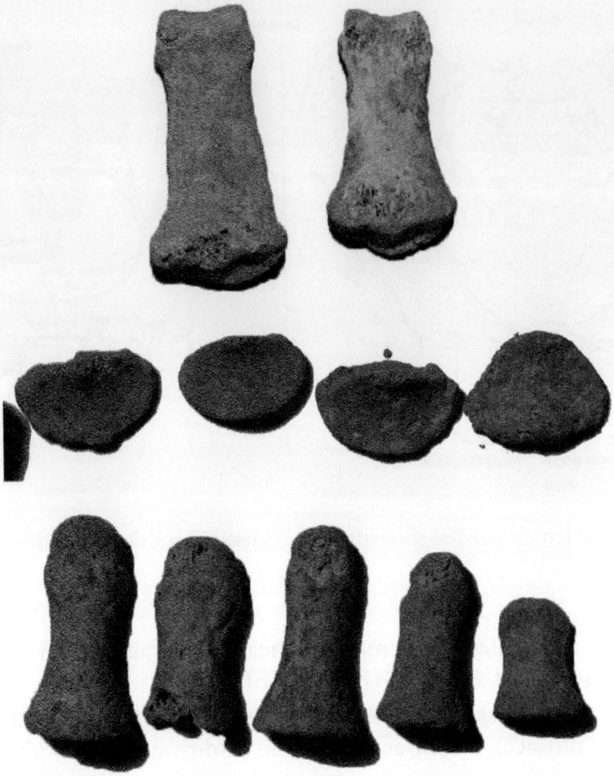

Figure 9.2 Adult phalanges (finger bones) (top) and juvenile phalanges (middle), with juvenile epiphyses (ends) (bottom). In the juvenile hand, the epiphyses are separated from the shaft by cartilage (which has disappeared). This lack of fusion is indicative of the immature state. Source: Photo by S.K. Smith.

a year or two, while with adults the age categories are much broader. Because the techniques used for assigning age to juveniles and adults vary, I will discuss them separately.

When children are born, their skeletons are composed of both bone and cartilage. As they grow, the bone replaces the cartilage until the adult stage of a completely bony skeleton is reached (see Figure 9.2). Long bones, for example, grow from multiple centers and eventually fuse into the complete bone with which we are familiar. During childhood, the size of bones and their stage of development provide very good estimates of age. Teeth also develop and erupt on a predictable timeline and can be used to augment data from the bones. Teeth are already beginning to develop in the jaws before a child is born, and the last teeth fully erupt in early adulthood. In the intervening years, the pattern of tooth development and eruption can be used to assign age.

Once body growth stops in early adulthood, aging must be assessed by degenerative changes to the skeleton. Two areas of the os coxa (Figure 9.1 and 9.3) are useful in the assessment of adult age. The pubic symphysis, where the two ox coxae meet in the front of the body, undergoes marked changes during the lifetime of an individual. In early adulthood, the face of the pubic symphysis is billowy and undulating (see Figure 9.3). As a person ages, the billowing pattern disappears, and the pubic symphysis becomes more

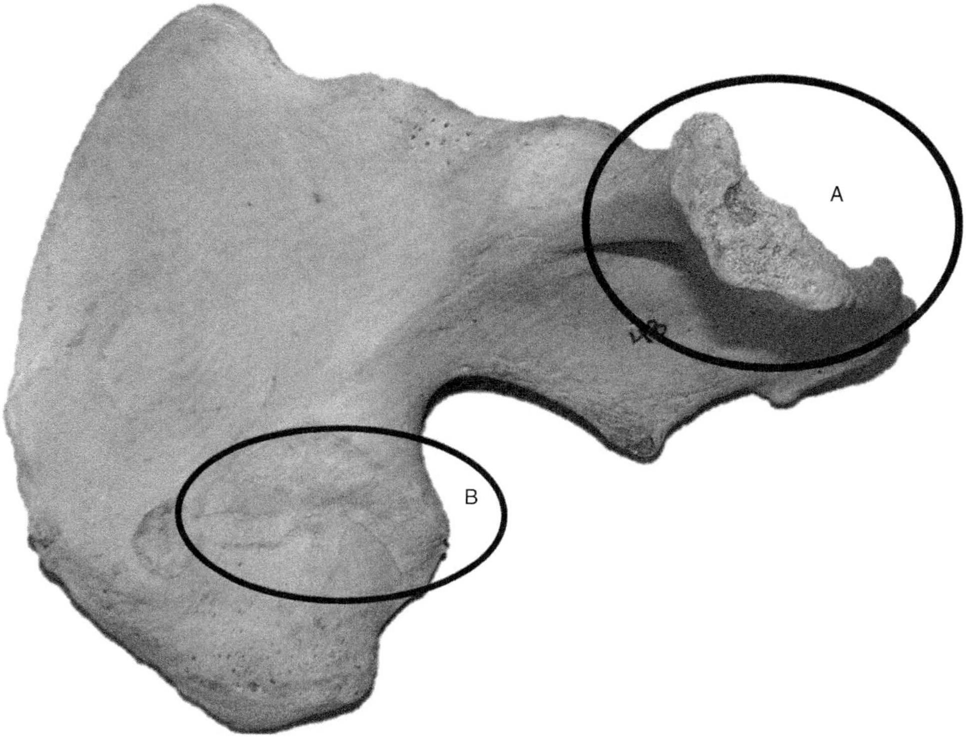

Figure 9.3 Left os coxa (hip bone) showing the pubic symphysis (A) and auricular surface (B) which can be used to age adults. Both of these surfaces have lost the youthful billowing appearance and have become more roughened and coarse. Source: Photo by S.K. Smith.

rugged looking. This pattern of change has been quantified (Todd, 1920; Suchey, 1979; Katz and Suchey, 1986) and tested for decades. The auricular surface, where the sacrum articulates with the os coxa (Figures 9.1 and 9.3), undergoes the same type of morphological change from a billowing appearance to a more coarse and rough appearance (Lovejoy *et al.*, 1985).

Another, less precise, method of assessing age involves the documentation of closure of cranial sutures (Figure 9.4). These sutures between the bones of the cranium are open in juveniles, and begin to close in adulthood. In some individuals the sutures become completely obliterated. By scoring the amount of closure on the cranial sutures, an age estimate can be produced (McKern and Stewart, 1957; Meindl *et al.*, 1985). It must be remembered that the age-related changes in the adult skeleton do not occur in the same regular ways as they do in children. It is much more difficult to accurately assess the age of an adult and it is particularly difficult to accurately assess the age of older adults.

Age-related burial practices reveal information about social position in the Roman world. A common view for the Roman period in Greece is that infants and children were not accorded the same burial rites as adults (Walbank, 2005: 271). However, a number of Roman sites in Greece have included the remains of adults and children of all ages (Carrington-Smith, 1982; Fox, 1996; Lagia, 2000; Walbank, 2005; Ubelaker and

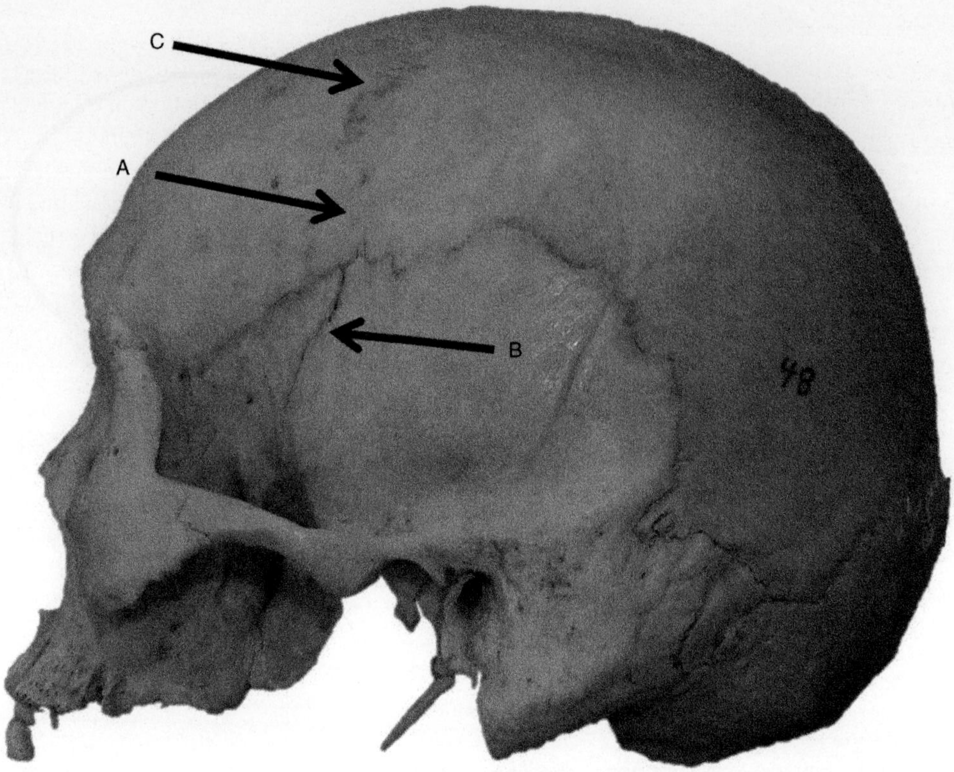

Figure 9.4 Skull sutures in various stages of fusion. The suture at (A) is completely obliterated. The suture at (B) has minimal closure. The suture at (C) has significant closure. Fusion of the sutures occurs with advancing age. Source: Photo by S.K. Smith.

Rife, 2011). Fewer sites have adults only (Bourbou, 2005) or too few children for the expected size of the community, suggesting that many children were most likely buried elsewhere (Wesolowsky, 1973). This variation in how children were buried suggests that the social position of children in the Roman east was not constant and unwavering across time and place.

Age at death data are a valuable window into the physical stress occurring in the lives of individuals and populations. Age at death of children can document if there were specific periods in childhood that were particularly dangerous. Fox recorded a high frequency of children from Late Roman Alassa (Ayia Mavri) in Cyprus who died around the age of weaning. This would suggest that this developmental stage was especially dangerous to children (Fox, 1996: 61). The average age at death for adults across time can document how culture change influences the lifespan and health of the people. The Roman period in Greece seems to have brought about a decline in health, both in populations on the mainland (Bisel and Angel, 1985: 203) and on Crete (Bourbou, 2005: 102), with a resulting decline in the average age at death from the Hellenistic to the Roman periods. Variation in age at death shows how Romanization affected the people in the conquered areas.

6 Stature

Stature can vary among genetically related populations and among individuals within a population based on a number of factors. Low protein and low calorie diets can stunt growth if the deficiencies persist throughout the growth period. Children who are frequently sick are also more likely to have stunted growth. These effects on stature persist into adulthood. Stature can be estimated from measurements of the long bones of the leg or arm. Measurements of the long bones are taken according to established protocols (e.g. Buikstra and Ubelaker, 1994) and the lengths are then plugged into regression formulae to give the stature estimate. For example, if a femur from a female were measured at 35 cm, the following formula would be used to calculate stature: 2.47(femur length) + 54.10 +/− 3.72 cm = stature (Trotter, 1970). According to the calculation of this specific female (2.47(35) + 54.10 = 140.55 cm +/− 3.72 cm), her stature was most likely between 138.83 and 144.27 cm. There are different formulae for each long bone and for each sex in different ethnic groups. While there are formulae that have been created based on modern Greeks (Eliakis, Eliakis and Iordanidis, 1966), those formulae are not used as frequently as are the formulae established by Trotter and Glesser (1952, 1958) and later revised by Trotter (1970), and hence the results may not be as reliably compared to stature estimates published in other studies.

In the Roman eastern Mediterranean, the number of individuals for whom stature has been calculated is small (Wesolowsky, 1973; Bisel and Angel, 1985; Fox, 1996; Lagia, 2000), with many large collections having no individual long bones complete enough to estimate stature (Bourbou, 2005; Ubelaker and Rife, 2011). It is necessary to have large samples of individuals for stature estimates in order for meaningful comparisons to be made between populations or segments of a population. The data so far suggest that the Roman period in Greece saw the highest stature for females from the Neolithic to the Byzantine period; the same is not true for males (Bisel and Angel, 1985: 203). This suggests that the lives of girls may have been improved under Roman rule. Further research into other collections will indicate whether this trend holds.

7 Dentition

Teeth are often better preserved than the rest of the skeleton due to the hardness of the outer enamel layer. This preservation is valuable because teeth are affected by many life events. Diets high in processed carbohydrates and sugars are cariogenic, so individuals who eat these foods are more susceptible to developing dental caries (cavities). Caries can become so large that they result in the loss of the tooth. When a tooth is lost before death, the alveolar bone surrounding the tooth begins to remodel, resulting in a smoothing over of the tooth socket (Figure 9.5). Therefore, information about dental health can be preserved in the bones even when the teeth themselves are not present.

Dental pathologies are reported in most osteological reports from archaeological remains, which results in an ability to compare pathology frequencies between regions and over time. The rates of caries in Greek samples have changed over time (Bisel and Angel, 1985). There are no statistically significant changes at the Hellenistic/Imperial

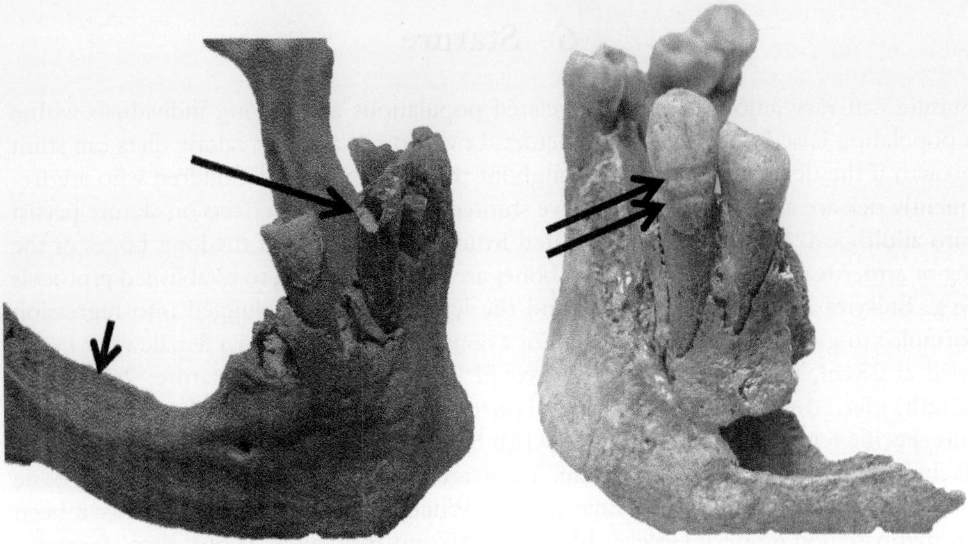

Figure 9.5 Mandibular incisor with a large carious lesion, marked by the long arrow (left). The smooth portion of the mandible, at the shorter arrow, shows where the mandible has receded after the molar teeth were lost prior to death. Mandibular canine tooth with two linear enamel hypoplasias, marked by the arrows (right). Source: Photo by S.K. Smith.

Roman and Imperial Roman/Byzantine transitions, although the Imperial Roman collections studied did have slightly higher rates of dental lesions than either the preceding or subsequent periods (Bisel and Angel, 1985: 203). Fox (2005: 78) did not find any differences between the caries rates of the Hellenistic and Roman skeletal remains from Paphos, Cyprus and the primarily Roman remains from Corinth. The combination of antemortem tooth loss and caries found in the Hellenistic site of Khania and the Roman site of Sfakaki, both in western Crete, are also similar (Bourbou, 2005: 104). Burns (1979) found a general trend toward increasing prevalence of caries in Corinth from the Late Roman period to the Byzantine period. The data are preliminary at this point, but in Greece and Cyprus, Romanization does not seem to have caused major dietary changes in the cariogenic quality of foods.

Teeth can also preserve evidence of childhood physical stresses. When teeth are growing, nutritional deficiencies and illnesses can cause a disruption in the formation of the enamel. This disruption results in visible lines on the teeth that persist throughout the life of the tooth (Figure 9.5). These lines, or linear enamel hypoplasias (LEH), are useful in documenting systemic stresses that occurred during childhood. Because the lines are only lost with the loss or severe attrition of a tooth, adults retain this information about their childhood health and diet throughout their lives. It is important to recognize that there is no way to know the precise nature of the stress that caused the development of the LEH, but high frequencies or sudden changes in frequencies of LEH in a population may indicate famines, influx of infectious disease or other changing cultural practices.

Fox (2005) compared the frequency of LEH in Roman Corinth with that of Hellenistic/Roman Paphos, in Cyprus. Corinth had a higher frequency of individuals with LEH than Paphos, indicating that the children of Corinth were subject to more nutritional stresses or illnesses. In addition, she found that there was greater infant mortality and shorter stature of adults in Corinth, as compared to Paphos. Clearly, the health of children in Corinth suffered in comparison to the health of children in Paphos. Fox suggests that this may be due to Corinth having been sacked by the Romans, while Paphos enjoyed a much smoother transition to the Empire. Harper and Fox (2008) compared the frequency of LEH in four Roman Cypriot sites and found a difference between urban and rural areas. Urban areas had higher rates of LEH than did rural areas. However, in this case, the frequency of LEH may indicate that the urban children suffered more from nutritional or disease stresses, but survived in greater numbers than the rural children, because the number of infants in the urban cemeteries is smaller than in the rural cemeteries. These data are perhaps examples of the "osteological paradox" (Wood *et al.*, 1992) which argues that sometimes populations exhibiting more instances of pathology may in fact have been healthier than populations exhibiting fewer instances of pathology. That is because the individuals with the osteological pathologies lived long enough with the illness or nutritional deficiency to develop bony evidence of the stressful condition, while those with fewer instances of pathologies died earlier in the course of the disease or nutritional stress, and before the condition had time to affect the skeleton. In these two studies (Fox, 2005; Harper and Fox, 2008), the higher rates of LEH may be associated with more frailty in childhood in one of the populations, but not in the other. LEH frequencies in Early Roman Agios Nikolaos in eastern Crete (S.K. Smith, unpublished) are even higher than either Corinth or urban Cyprus. The frequency is also higher than the earlier Hellenistic site of Khania (Bourbou, 2005) and the later Proto-Byzantine site of Eleutherna (Bourbou, 2003), both on Crete. This suggests that the transition to Roman rule on Crete was also physically stressful, which supports Fox's interpretation of the Corinthian population (Fox, 2005) suffering due to forced Romanization, which also occurred in Crete.

Often the number of teeth or individuals present at a site is so small that they do not provide any detailed information about the dental health of the people (Wesolowsky, 1973; Fox, 1996; Lagia, 2000). However, it is still beneficial to include the results so that these data can be combined with others into larger meta-analyses (e.g. Bisel and Angel, 1985).

8 Pathology

Many diseases, injuries, and congenital and degenerative disorders affect the skeleton. The manifestations of these pathologies can reflect the types of infectious agents, activities and genetic conditions that were present in an ancient population. Crowded living conditions, poor nutrition or cultural upheaval are often accompanied by an increase in disease. Warfare and the cultural acceptance of interpersonal violence result in higher rates of trauma. Degenerative disorders, such as arthritis, can vary according to the types of physical activities practiced in a given population. Genetic disorders may increase with population migration and the influx of new genetic conditions. Pathologies are sometimes presented as case studies of individual disease patterns (Gejvall and Henschen, 1968; Smith, 2010) and other times as population frequency data (Lagia, 2000; Walbank, 2005). Pathology data of both types are presented below.

Figure 9.6 The upper portion of the eye orbit (socket). The pin-prick holes are a sign that this person had suffered from anemia. Source: Photo by S.K. Smith.

The skeleton can react to pathogens, trauma or other disorders by exhibiting abnormal bone size, abnormal bone formation, abnormal bone destruction and abnormal bone shape (Ortner, 2003: 50). Because the range of skeletal responses to insult is limited, it is difficult to make a definitive diagnosis for conditions on archaeological bone, so researchers must be careful to present accurate and detailed descriptions of the conditions that they describe (Ortner, 2003: 112). A clear picture of pathologies in the past is also limited by the fact that the skeleton is rarely involved in the majority of pathologies, traumas and other disorders that affect us; therefore the people we see in archaeological populations with evidence of infectious disease, for example, are most likely those who suffered from chronic conditions (Ortner, 2003: 114).

The impact of anemia, and specifically thalassemia (an inherited disease that is now most common in the eastern Mediterranean), has been a concern for biological anthropologists ever since J. Lawrence Angel first began to publish on the topic (Angel, 1964). While for a long time the view prevailed that skeletal evidence of anemias in Greece and Cyprus was the result of thalassemia (which may give a degree of protection against malaria), it is now clear that other factors, such as nutritional deficits or other pathological conditions, are the cause of the anemia evident on the skeleton (Figure 9.6) (Lagia, Eliopoulos and Manolis, 2007; Bourbou, 2010; M. Lewis, 2010). Even though in Greece "no firm diagnosis of a genetic anemia has been reached to date" (Lagia, Eliopoulos and Manolis, 2007: 269), there is still great interest in skeletal evidence of anemias (Fox, 1996; Lagia, 2000; Bourbou, 2005; Harper and Fox, 2008; Lewis, 2010). Comparing anemia frequencies between populations can provide a measure of the relative health or physical stress met by each. Fox (2005) found similar rates of anemia in Paphos and Corinth in the Hellenistic and Roman periods. However, the physical manifestations of the anemias were different. She has therefore suggested that the cause of anemia differed between the two locations. Interestingly, while no definitive cases of thalassemia have been documented in Greek skeletal populations (Lagia, Eliopoulos and Manolis, 2007), there has been "one probable, and two possible cases of thalassemia" from a Roman cemetery in Dorset in Britain (Lewis, 2010). The presence of these individuals, all children, with possible thalassemia, contributes to our understanding of migrations during the Roman period. These children would have been too young to have been born outside of Britain, so they provide evidence of a second generation of immigrants in Poundbury, Dorset. Their burial in a mausoleum shows that their parents still practiced Roman (Mediterranean) burial customs, which implies a Roman self-identification.

While anemia has been one of the most systematically investigated pathological conditions found on skeletal remains in the Roman eastern Mediterranean, other pathologies have been reported as well. Gejvall and Henschen (1968) present a detailed analysis of two skeletons from Roman Corinth which exhibited pathological vertebral columns. The likely genetic root of the malformations suggests that these two individuals were brother and sister. Often burials of an adult male and female are assumed to be husband and wife, but in this case we see that life-long bonds between siblings were also a possibility in Roman Corinth.

Bourbou (2005) documented two cases of diffuse idiopathic skeletal hyperostosis (DISH) from Roman burials in western Crete. This disease is characterized by bony growth on the spine in the region of the anterior longitudinal ligament (Ortner, 2003: 559). Other parts of the body, such as regions of the pelvis, knee and heel, may also have ossified soft-tissue (Aufderheide and Rodríguez-Martín, 1998: 98). Both of the individuals with DISH were older, high status adults. The relationship between status and DISH has been suggested previously, but at this point it is not clear how or why status may be a factor in its etiology. Further studies of this condition in the Roman world could help in determining whether it is a condition more likely to be found in certain segments of the population.

Patterns of degenerative joint disease (DJD), commonly called arthritis, can be associated with physical activity. Lagia (2000) and Walbank (2005) discussed the high frequency of DJD in the arm bones of Roman skeletal remains and suggested that these patterns were likely the result of occupational activity. When the patterns are in an uncommon location such as the arm (Ortner, 2003: 547), and younger ages are involved, then occupational activity is more likely to be the cause, as Lagia (2000) noted in the skeletons from the Roman Kerameikos in Athens. Walbank (2005) was able to show that the Roman Corinthians may have been buried in tombs belonging to trade associations, based on the common pattern of DJD.

Trauma can provide similar information. The prevalence and type of fractures in a population can indicate whether a community was prone to interpersonal violence and which segments of a population were more at risk. Bourbou (2005) found that fracture rates in Roman western Crete were particularly low, although there were two fractures that may have been the result of violence. Harper and Fox (2008) documented a variation between urban and rural fracture prevalence in Roman Cyprus, with the rural population having fractures that were more likely caused by accident rather than violence.

9 Conclusions

The skeletal studies presented above provide a clear description of the types of data that can be gathered from human remains. What is important is that often the data provided are not available from any other sources, either historical or archaeological. In addition, the skeletal data combined with the historical and archaeological data provide a greater picture of life in the past than any of these three types of data could alone. The excavation at Kenchreai in Roman Corinth provides an excellent example of the benefits of this type of integrated approach (Rife *et al.*, 2007; Ubelaker and Rife, 2007, 2011).

The study of Roman skeletal collections from the eastern Mediterranean lags behind the study of Roman skeletal material from other parts of the Empire. Romano-British material has been particularly well studied and could provide a model for other Imperial era locations (i.e. Richards *et al.*, 1992; Mays and Faerman, 2001; Leach *et al.*, 2009; Redfern and DeWitte, 2011). Having multiple collections from the same geographic area allows us to see patterns in health and other biological and demographic characteristics through time. The eastern Mediterranean is rich with Roman sites with human burials. MacKinnon has called for us to "humanize the world of antiquity" (MacKinnon, 2007: 496). Learning about individual people, through their physical remains, provides an opportunity to heed his call.

FURTHER READING

There are several excellent textbooks that discuss the basics of human osteology and paleopathology. White, Black and Folkens (2011) present beautiful life-size photographs of each bone in the human body as well as a discussion of laboratory and field techniques for handling skeletal remains. Buikstra and Ubelaker (1994) has become the standard for the collection of systematic data from human skeletal remains from archaeological sites. Their discussion of how to collect various types of data is augmented by an excellent set of data sheets. Ortner (2003) is a wonderful choice for identifying pathological conditions. Many photographs accompany the descriptions and modern clinical cases are compared to archaeological examples. In order to see how all of this technical information is used to help us understand life in the past in the ancient world and especially ancient Greece, Schepartz, Fox and Bourbou (2009) provides work from different time periods, using multiple approaches. To expand from Greece, MacKinnon (2007), with its extended bibliography online, provides a comprehensive discussion of osteological research in and around the Mediterranean and in other parts of the Roman world. This work is particularly useful because it includes both human and animal studies.

CHAPTER 10

Population and Demographic Studies

Elio Lo Cascio

1 Introduction

One of the most significant developments that the social and economic history of the ancient world has undergone in the last 20 years is the great relevance finally being ascribed to investigating size, structure and dynamics of ancient populations. That such relevance went broadly unrecognized becomes especially bewildering if one considered the role demographic factors did and do play towards reconstructing the social and economic evolution of medieval and modern Europe. It is in this respect baffling to realize how much this lack of attention to demographic factors was shared by both Rostovtzeff's *Social and Economic History of the Roman Empire* (2nd edn, 1957) and Finley's *Ancient Economy* (1st edn, 1973), that is to say the most representative and influential works of the twentieth century in championing the two opposed interpretations of the Roman economy, the "modernist" and the "primitivist."

2 Ancient Evidence, Comparative Data and Demographic Models

The main rationale claimed by many a scholar to support such a disinterest was akin to an argument Hugo Jones had originally presented in his Inaugural Lecture at University College London in 1948: the lack of substantial ancient statistics would have made any demographic analysis worthless since any attempt at quantification was in fact precluded (Jones, 1948). Since the late 1960s and 1970s, however, pioneering studies by Hopkins and some other scholars such as Frier (Hopkins, 1966–7; Frier, 1982, 1983) stressed

A Companion to the Archaeology of the Roman Republic, First Edition. Edited by Jane DeRose Evans.

the potential of a clever combination of various types of ancient evidence – both textual and material (e.g. skeletal remains) – with comparative data and demographic models built in in order to produce estimates of the vital statistics (death rate, birth rate, fertility and reproduction) of contemporary populations, for which they are lacking (Coale and Demeny, 1966). Such an approach made it indeed possible to derive orders of magnitude about the vital statistics of ancient populations too, thus testing the demographic plausibility of data derived from specific sources of evidence.

Accordingly, a relevant case of plausibility is represented by that peculiar – albeit quantitatively modest – dataset which is made up of the papyri reporting the *kat'oikian apographai* (i.e. "declarations by household") that local authorities periodically required from the inhabitants of Roman Egypt as part of the census (Bagnall and Frier, 1994). Quite different, to the contrary, is the case of the thousands of funerary inscriptions recovered all over Rome's Mediterranean Empire, which report the age at death, peaking in numbers at Rome and, for example, in North Africa: derived patterns of life expectancy, age-/sex-distribution and age at death have proved totally implausible. Life tables built on the latter series of evidence present peculiar features indeed. Besides pointing to a life expectancy at birth in the order of 20–30 years, they highlight marked regional differences and a rather unusual age-/sex-distribution among the population that cannot find any correspondence with other reliably known population, past or present. Thus, funerary inscriptions imply that those who attained 10 years of age would have lived (on average) another 20 years if at Rome, but another 40 or more if in Africa. It is indeed possible that health conditions in a city like Rome (a metropolis with perhaps one million inhabitants, without comparisons in western Europe until nineteenth-century London) might have been so precarious as to determine higher mortality levels than in the rural areas, as it is more generally the case in the preindustrial world (by no means a foregone conclusion for ancient Rome: Lo Cascio, 2006). However, patterns from Africa are not matched anywhere else across the Empire and, in any case, they too are based on evidence derived from urban areas. Indeed, the reason behind the African anomaly rests not so much on a supposedly higher longevity of the African population as in the fact that those whose age was reported to be 90, 100 or more do constitute 7% of the sample in Africa, but less than 1% of the sample in Rome. Even the age-/sex-distribution, as evidenced by funerary inscriptions across the Empire as a whole, seems absolutely unusual and without comparisons anywhere else. Infant mortality is clearly lower than one would otherwise expect from a population featuring a life expectancy at birth of 20 years (very low). In a similar way, the levels of mortality among the young and the elderly, both in Rome and in the other regions of the Empire, are respectively much higher and much lower than in a population with a life expectancy of 20. There are various reasons why the age structure of this sample appears distorted. First of all, in terms of age reporting, there was a tendency to exaggerate figures and a tendency to round them off: the number of deaths at ages ending with 0 or 5 is hugely higher than the number of deaths at "intermediate" ages. Another reason might be found in the fact that parents tended to commemorate more often their sons and daughters when they died at a young age, something which explains the proportionally high number of deaths between 1 and 19 years of age. Another crucial factor of distortion is represented by the exceptional over-representation of men over women. A normal population for biological reasons has a slight prevalence of males at birth (1.05 males to every 1 female). In contemporary populations this prevalence is rapidly overcome, since men (till now at least) have experienced in general a lower life expectancy than women. The population structure as derived from Roman funerary inscriptions instead shows a strong prevalence of males over females of 135 to100. This ratio and therefore this kind of population structure, however, is highly unlikely, although it must be conceded that a quantitative male prevalence has been often reported among preindustrial European populations and it still is for some contemporary ones (e.g. in India and China).

This broad reliance on comparative evidence is therefore justified by scholarly awareness of the fact that "biological constraints" have determined a close relationship between different past populations which existed before the so-called "demographic transition." This transition took place at some point between the eighteenth and twentieth century, at a specific time which is different for each European and extra-European population. Accordingly, an increase in the health, sanitary and alimentary conditions led to a marked decrease in death rates, while birth rates remained, with only few exceptions, still very high. In turn, such a widening gap determined a rapid population growth at an annual rate that had been unthinkable during the "pre-transitional" period. One can appreciate the impact of this change by considering the fact that long-term growth of sufficiently large populations passed from 1–2 *per thousand* a year to 1–2 *per cent* a year: in other words, populations that doubled over a period of 350–700 years, started to do so in about 35–70 years.

3 Population Structure and Population Size

The renewed interest in the demography of ancient populations opened the way, in a first phase (especially during the 1970s and 1980s), for a new wave of studies that, by relying on an increasingly refined and sophisticated methodology, aimed at assessing specific aspects of the structure and dynamics of ancient populations: the normal mortality and fertility regime, the age at marriage, the diffusion of practices like infanticide or child-exposure, the diffusion of wet-nurses and their potential impact on fertility (Parkin, 1992). Much less popular, at least in the case of the Roman world, was the enquiry about the size of the populations across the different regions of Rome's Mediterranean Empire, although the first and most celebrated scientific treatment of the demography of the ancient world, K. Beloch's *Bevölkerung der griechisch-römischen Welt* (1886), aimed precisely at that by way of a philological appraisal of available evidence. This monumental work has for long been considered the final word on the subject; it is a study which was certainly amenable to further refinement, but nonetheless still broadly valid. As it became immediately clear to Beloch's followers and critics alike, however, his whole interpretation rested on a peculiar interpretation of a unique piece of evidence, which was in turn rightfully considered to be the most important and reliable source on the size of an ancient population we possess: the number of Roman citizens reported in the three censuses that Augustus had carried out in 28 BCE, 8 BCE and 14 CE. Indeed, once this interpretation had received wide support from the analysis presented in another masterpiece of modern historiography, Brunt's *Italian Manpower* (1971), it seemed unlikely that one could significantly improve on Beloch's estimates of the population of Augustan Italy and, on the basis of them, of the rest of the Empire.

This is not the case anymore and a lively debate has since spread among scholars specifically about size and regional distribution of ancient populations together with their evolution over time. Research has therefore focused, for example, on the potential impact of the great epidemics, some of which have been tentatively characterized as crucial turning points. This is the case of recent debates about the relationship between the Justinianic Plague and the end of the ancient world, a subject that has been at the center of a fundamental book which, edited by Little, not only includes a review of the broad array of literary (Greek, Latin and Syrian) and documentary sources, but also integrates material evidence by employing also, for example, molecular biology (Little, 2007). The same can be said of the impact of the "Antonine Plague," a likely smallpox epidemic that, starting in 165 CE, repeatedly affected many regions of the Empire over a period of 25 years, in relation to which the evidence provided by skeletal remains seems to show its own potential to identify local outbreaks (Lo Cascio, 2012). In a similar way, studies of Greek demography are undergoing a radical re-appraisal, especially following the most recent contributions by Hansen, who has raised the estimates that had been originally proposed by Beloch and then adopted by later scholars. Hansen now cogently argues in favor of a total population of the world of the Greek *poleis* in the fourth century in the order of 7.5–10 million, out of which 4–6 million people would have lived in Greece proper (Hansen, 2006). These figures are not only well above Beloch's estimates, but they are also much higher than the level that the Greek population had achieved at the end of the nineteenth century.

4 Roman Italy's Population: The Census Figures

This kind of re-appraisal appears especially radical in what pertains to the Roman world and, especially, Roman Italy, whose available evidence seems to provide a more solid support on which to base relevant estimates. Over the last 20 years Beloch's "orthodoxy" has been indeed cogently questioned as serious doubts have been cast on its very foundations, that is his interpretation of the results of the Augustan censuses as preserved in the eighth chapter of the *Res Gestae* in relationship to the results of the Republican censuses as reported by the literary tradition (annalistic, chronographic and antiquarian). We do indeed possess figures for more than 40 censuses carried out in Rome since the time of Servius Tullius, who, according to tradition, first instituted the census. Latin sources – especially Livy and the Livian tradition – present these figures by mentioning the names of the relevant magistrates and by employing the formula "censa sunt civium capita" followed by the number of the enumerated citizens. Given the aims of the census (military, political and fiscal), modern studies broadly agree on the fact that preserved Republican figures refer to all adult male citizens or a specific sub-group within them, including – according to some – the "cives sine suffragio" (non-voting citizens). The nature of this specific sub-group has been variedly interpreted as made up of the assidui (i.e. those belonging to the five census classes that were required to serve in the army), the iuniores (i.e. those of between 17 and 45 years of age who *actually* served in the army) and the "sui iuris" (i.e. the head of the household that made the formal declaration in front of the censor). Ancient authors seem to be, in any case, somewhat unanimous in presenting the census as having numbered adult males only or a sub-group within them: from Livy (who, following Fabius Pictor, considers Servius Tullius to have registered "qui arma ferre possent," or those eligible to carry arms, Livy I.44.2, cf. III.4.10) to

Dionysius of Halicarnassus (who, in reporting some of the fifth-century figures, presented them as referring to "*hoi en hebe*" ("those in their prime"), that is adult males, a group which would have included the sub-group of "*hoi echontes ten strateusimon helikian*" ("those of military age"), that is the iuniores as opposed to the seniores, adult males of 46 to 60 years of age; Dion. Hal. *Ant. Rom.* IV.15.6, V.20.1, etc.), to Festus (who explains the term "duicensus" as the one describing a citizen that is registered with his son: "Duicensus, dicebatur cum altero, id est cum filio census"; Festus p. 58L). Of course, the census required the head of the household to make his own professio in front of the censor (or his assistants, the iuratores) by declaring all the individuals "in sua potestate," free and slave, together with an assessment of his patrimony. However, it seems certain that the *enumeration* did involve all adult males or, as already mentioned, a sub-group within them. Scholars have considered also a passage from the second book of Polybius's *Historiae* as especially relevant (Polyb. 2.23–4). In this passage the historian reports figures from the lists (*katagraphai*) of infantry and cavalry forces that Rome had required its allies to provide on the eve of the Gallic invasion of 225. According to Polybius, these lists would have included *hoi en tais elikiais* ("those of military age"), a definition which is more likely to refer to iuniores only rather than all adult men, both iuniores and seniores (pace de Ligt, 2012: ch. 2). Besides reporting the number of already fielded armies, Polybius reports the number of potential recruits and sums it up as the total number of *hoi dynamenoi hopla bastazein*, "qui arma ferre possent," Romans and allies, that Rome could have employed in case of extreme danger. This passage, although not reporting the result of a census proper, is very important because it makes it possible to estimate the population of the so called Roman-Italian Confederacy towards the end of the third century, thus providing a comparison for the Italian population following the Social War.

5 The Augustan Census Figures and Their Interpretation

On the whole, the census figures, especially from the third century onwards, appear to constitute a trustworthy series, whatever the level of reliance we can place on each of them taken individually. They represent a plausible and coherent series, whose number of "civium capita" oscillates between 200,000 and 300,000. The increase following the Social War, especially as outlined by the census of 70/69, can be easily explained – although establishing whether or not the size of such an increase is credible is quite a different matter. What is very difficult to explain is the marked jump implied by the Augustan figures, themselves introduced by the expression "censa sunt civium capita," in relationship to the last Republican census: a more than fourfold growth from 900,000 (or 910,000) in 70/69 BCE, to 4,063,000 in 28 BCE, 4,233,000 in 8 BCE and 4,937,000 in 14 CE (*Res gest.* 8). A further (but much less problematic) increase is recorded by the Emperor Claudius's census of 5,984,072 citizens in 48 CE (as such preserved in Tac. *Ann.* 11.25, a figure which is to be preferred to the 6,941,691 reported by Eusebius's *Chronicon*, *sub Ol.* 206.2 or the 6,944,000 reported by Jerome *sub Ol.* 206.1).

The explanations that over the years have been put forward to justify this fourfold growth of the number of "civium capita" by the time of Augustus are basically two. The first one – so far the most popular – is the one put forward by Beloch himself in the *Bevölkerung*. According to Beloch, the census of 28, itself the first to be properly concluded with a lustrum after a lapse of 42 years as remarked by Augustus himself, would have enumerated *all* Roman citizens – including women and children – for the very first time. This explanation does not find any support in the available ancient evidence, but it is merely an argument from likelihood: the gap being so wide, the addition of adult male citizens from Transpadana (to whom Caesar had granted the status of civitas in 49) and the integration of new cives from Caesarian and post-Caesarian overseas colonies would not be enough to explain such a growth of "civium capita." Another argument from likelihood – already hinted by Beloch, but pursued especially by later authors – refers to the fact that if those four million "civium capita" in 28 were to be interpreted as adult males only, this would imply that the overall population of the Italian Peninsula had achieved a level in itself comparable to that of early modern Italy.

The second explanation links this gap to a marked decrease in the number of the incensi, that is, those who went un-registered, especially following the introduction of a decentralized census procedure in the time of Caesar as attested by the *Tabula Heracleensis*, an epigraphic text reproducing one or more Roman laws of Caesarian age (*Roman Statutes* 24, ll. 142ff.). Accordingly, following the extension of citizenship after the Social War, practical issues (e.g. the difficulties experienced by the heads of household in order to come to Rome to register) and hostility on the part of some members of the elite towards involving the new cives in the running of the res publica would have made the census highly inefficient, and thus it was discontinued after 70/69. Eventually, the situation would have improved once it became possible for the "sui iuris" (heads of households) living in the municipia, coloniae and praefecturae of Italy to register themselves and their families in their local communities and in front of local magistrates (Lo Cascio, 1994, 2001 and references there).

If we were to accept Beloch's theory, we would be required to believe that the census had been generally efficient during the whole of the Republican period and that the underlying numerical pattern of relevant figures indeed provides a reliable image of Roman demographic development. Accordingly, the supposed coeval decrease of Italian manpower precisely in the period of conquest would become even more significant, a decrease that, according to many, is in a way confirmed by the nature of the Gracchan reforms as reported by Appian and Plutarch. Furthermore, by accounting for the fact that those two centuries had witnessed an undeniable growth of urban centers in Italy, both in number and size, such a marked decrease of the population would have implied the almost total disappearance of the free population from the Italian countryside, in turn supplanted by a massive – albeit not quantifiable – input of slaves. This is the view of the demographic development of Italy as advocated by Toynbee (1965) and Brunt (1971), and which found its more elegant formulation in Hopkins (1978).

Of course, supporters of the alternative view would imply a growth pattern for the Italian population between the third century and the Augustan age. This increase would not have been determined, but only strengthened by the input of slaves, and would also explain the growth of the cities of Italy, and especially Rome (which becomes a metropolis with one million inhabitants or more), without having to assume a deserted countryside, a notion which seems to be disproved by the archaeological evidence produced by field survey projects in many parts of Italy (see Chapters 12 and 25). These investigations show that, even when allowing for the variety of regional situations, the overall density of settlement was on the rise: indeed, the vast majority of surveys would attest to the growth not only of larger sites (interpreted as villas), but of smaller ones (interpreted as farms) (Launaro, 2011). This scenario would allow also for a much larger population living in the Po Valley (Kron, 2005), thus putting the population more in line with what happened in later periods, whereas supporters of the opposed interpretation are obliged to reduce it to a rather modest presence. In a similar way, this scenario holds slavery to have played a much more limited role (in quantitative terms too) than that implied by the alternative view with few free rural inhabitants.

6 The Implications of the Alternative Explanations of Augustan Census Figures

As one can see, the controversy is not only – or not so much – over the issue of the size of the population of citizen status in the Augustan age as over the reconstruction of the most relevant phenomena related to the political, economic and social development of Italy in a period marked by the Roman unification of the Mediterranean under her rule. Even more than that, the argument is over the issue of how to view the demographic evolution of Roman Italy in comparison to the trend of the Italian population in the following two millennia, and therefore over the issue of how to frame the achievements of the Roman economy, in terms of per capita output and living standards. Supporters of Beloch's view envisage a long-term population growth in Italy, from the Roman to the medieval and modern period. Those who oppose Beloch's argument think of a Roman Italy that was as populous as in the fourteenth – or even the eighteenth – century (Lo Cascio and Malanima, 2005). Those who agree with Beloch believe that Roman agricultural productivity was generally not on the same level as that of Italy of the early modern period. Those who reject Beloch's view, on the other hand, believe that Roman agriculture was no less productive than in the early modern period: the main caesura would have been represented by the massive introduction of corn (maize) in the eighteenth century which would have substantially increased the "carrying capacity" of the peninsula (Lo Cascio and Malanima, 2005). In the same way, they believe the Roman economy to have been as productive as the economy of Elizabethan England (as recent attempts at estimating the GDP of the Roman Empire can be taken to suggest: Lo Cascio and Malanima, 2009 on Maddison, 2007; see also Scheidel and Friesen, 2009). In other words, at issue are two radically different ways of interpreting demographic, economic and social development in the long or very long term. And this explains the lively nature of the debate between those that within an extensive literature have been termed "high counters" and those who have been labeled "low counters."

It is important to stress the opposition between two radically different estimates: the "civium capita" amounted, between 28 BCE and 48 CE, to either 4–6 million or to 12–18. There seem to be no intermediate solutions, notwithstanding relevant attempts – even recent ones – in that direction. Accordingly, Hin has proposed that, during the whole Republican period, only adult male "sui iuris" (i.e. the heads of household) had been registered, whereas Augustus would have included women and children in the "sui iuris" category as well (i.e. widows, unmarried women, orphans) (Hin, 2008). Such a solution implies a higher estimate of the overall population in the Republican period (accounting for the fact that the "sui iuris" would have plausibly amounted to about half of the total males, whereas adult males would have represented much more than half of them), but would also bring the estimate for the total population in the Augustan period to about 10 million, or even less, rather than 12–14. It is nonetheless necessary to emphasize the fact that the already-mentioned passage about the duicensus by Festus could hardly be explained if we were to suppose that only adult males "sui iuris" had been registered. An intermediate estimate has been also put forward by those scholars who believe they can push survey evidence beyond a mere estimate of population trends, thus deriving a reliable figure for the number of inhabitants of a territory in a given time (Witcher, 2005, 2008, 2011; Fentress, 2009; see also Yntema, 2008, who finds, in his estimate of the third-century population of Messapia on the basis of the field surveys, a confirmation of the figure given by Polybius of the numerical strength of the Messapians in 225). These attempts are nonetheless undermined by unsurpassable difficulties: from the one of estimating an average number of residents for each settlement type (e.g. villa, farm) – a number which cannot but be arbitrary – to the difficulty of identifying well-defined categories of settlement in terms of size. The intensity of surveys, on the other hand, has been variable across the different areas and it is on the whole likely that many sites, especially the smallest ones, or even those not well integrated in the distribution network, might have eluded identification (see Chapter 12; Scheidel, 2008: 49–54; Rathbone, 2008; Launaro, 2011).

Leaving aside the possibility of achieving an intermediate estimate based on available evidence, there have been attempts to question the traditional view of the economic and social development of the last two centuries of the Republican period as outlined by Brunt and Hopkins without accepting the "high count" hypothesis. Accordingly, moving from the analysis of the Polybian figures and of the censuses of the second century, de Ligt has recently argued that the free population of Italy had experienced a moderate increase in that century (see now de Ligt, 2012).

An important set of implications of the two opposing estimates concerns the role played by slave labor and, therefore, the number of slaves. Of course, it must be stressed that there is no direct available evidence about this number and any estimate is unquestionably tied to the estimate of the number of the free population, hence the risk of circular reasoning (Lo Cascio, 2010). On the whole, "low counters" tend to propose higher figures for slaves and, conversely, those who believe that the slave population must have been high (e.g. most recently Andreau and Descat, 2006) should prefer a "low count" scenario, that is the only one leaving enough room for a significant slave population in the Italian countryside. On the contrary, "high counters" tend to envision the slave population as less relevant. It is nonetheless interesting to observe that among those who have argued in favor of much reduced estimates for the slave population of Late Republican Italy one can include a champion of the "low count," Scheidel (2005).

A lively debate has also developed about urbanization levels. The number of towns (which we know from the Plinian lists) as much as the size of Rome or even other urban centers like Ostia, Capua and Aquileia seem to represent a fundamental argument in support of the "high count." If the urban population was so high, how could it have been supported by such a low rural population as the one implied by the "low count"? Comparison with other preindustrial situations – for example, medieval and modern

Europe – seems to suggest that the percentage of the population unengaged in the production of primary goods could not have exceeded 15–20% of the total. To this argument the "low counters" have replied, on the one hand, by claiming that the vast majority of the Italian towns would have been very tiny (de Ligt, 2012: ch. 5). They also advance an hypothesis that could appear plausible for certain regions of the peninsula – as much as it seems plausible for other comparable situations, first of all that of the Greek *poleis* – that a good proportion of the towns of the peninsula would have been "agro-towns," that is urban centers whose population was largely involved in agricultural production (Scheidel, 2004, 2008). Even though such an explanation could well account for the nature and spread of urbanization in specific areas of the Apennine range, it is on the contrary hard to suppose that the population of cities like Ostia and Aquileia, let alone Rome, could have been made up of peasants who worked suburban lands. And this is true of many other urban centers of northern Italy, even if one is ready to accept the contention that their size must have been in general smaller than the size of their successors in medieval and early modern Italy (de Ligt, 2012).

7 A Way Out of the Impasse?

According to Beloch and his supporters, one of the strongest arguments against the plausibility of the straightforward interpretation of the Augustan census as reporting the number of adult males only would rest on the fact that it would imply too high a population – especially when considering later developments. It is not so much the fact that the last Augustan census (14 CE) would imply 14–15 million "cives Romani," but rather the fact that such a population would have kept growing, thus becoming 18 million in 48 CE and even more after that, since there would be no reason to believe that such a rising trend had slowed down or reversed before well into the second century CE (see now de Ligt, 2012). However, it is possible to counter this objection by devising an interpretive compromise. The "high counters" had already hinted at this solution (Lo Cascio, 2003), which has now been authoritatively reasserted in a recent contribution by Crawford (2008). We need always to remind ourselves that the census figures refer to the citizen population living in every region under Roman rule, even though, following the enfranchisement of the entire Italian Peninsula and then the Po Valley, the majority of "cives Romani" did certainly live in Italy. In the same way, we need to remember that, before and after the enfranchisement of the Italians, the citizen population was made up of those who were born citizens and those who had been made citizens, that is those who, for various reasons, had been granted citizenship – collectively or individually (e.g. participants in overseas colonial foundations; properly manumitted freedmen; "Latin" citizens that were granted Roman citizenship by having served as magistrates or members of their local senate). But there is more. The citizen body was also made up of those who had moved to the provinces, that is, those who emigrated there as part of a state-sponsored colonization (especially in the time of Caesar and then under the triumvirs and Augustus) or as individuals (in the framework of what is usually termed as "spontaneous" emigration), together with their descendants. A series of very detailed estimates of state-sponsored emigration can be found in Brunt's magnificent presentation. These estimates, however, are often too low because they were influenced by Brunt's own acceptance of Beloch's thesis. Moreover, spontaneous emigration, largely underestimated by Brunt, cannot be directly quantified. Its scale can only be indirectly appreciated by way, for example, of proxy indicators like the precocity and depth of the "Romanization" of those regions to which emigration was directed. This earliness and depth do characterize Cisalpine Gaul before 49, Sicily, southern Gaul and Spain. For example, it is not illegitimate, in the case of Spain, to consider the precocity of Romanization as an indicator of the relevance of the Italian immigration experienced by this region (Lo Cascio, 2003), a process that many years ago Gabba had already outlined for Baetica as much as for the Ebro Valley and the coastal areas (Gabba, 1954; for a slightly different view, see Chapter 33). Such a relevance is, for example, attested by the fact that when Metellus conquered the Balearic Islands in 122 and decided to found the new colonies of Palma and Pollentia, he employed the Romans of Spain (Strabo III.5.1).

Is it possible to give a quantified estimate of this emigration? Recently, Scheidel has based his important contribution on the mobility of the Roman world on Brunt's figures and has endeavored to estimate the NRM (i.e. Net Rate of Migration: the rate of migration in relationship to the population that experiences such migration in a given period of time) and the NROM (i.e. Net Rate of Out-migration: the rate of emigration in relation to the population that experiences such emigration in a given period of time) (Scheidel, 2004). He derived these indicators for those four periods that he deemed crucially important from this point of view: (i) between the last decades of the fourth and the first decades of the third century; (ii) the 25 years following the Hannibalic War; (iii) the half a century following Sulla; and (iv) the period of massive colonization under Caesar, the triumvirs and Augustus, between 48 BCE and 14 CE. It is this last period that is most relevant to us, when 96 colonies were created, the vast majority of which were either newly founded settlements or older settlements that were structured around Roman citizens of Italian origin, but include a consistent share of peregrini ("foreign" non-citizens). In this period the NROM, as reckoned by Scheidel, achieves a very high level, unparalleled by what happened before and after it. This level would imply, in turn, an implausibly high level of mobility, without parallels not only in the history of the Roman world, but also, for example, in the Greek Classical and Hellenistic periods. This is because the population out of which the migrants come – that of Italy – is set by Scheidel in accordance with the "low count." Therefore, this implausible level of mobility seems to constitute a further argument against the "low count."

8 Conclusion

In conclusion, one could assert that out of those 14–15 million envisaged by the "high count," a significant proportion (as high as half of it, according to Crawford) would have been "cives Romani" living in the provinces. It is possible that Crawford's estimate is an exaggeration. But it is also true that the higher the estimate of the proportion of "cives Romani" in the provinces in the Augustan period, the less populous Italy will be, especially in relationship with fourteenth- or eighteenth-century Italy.

The population of Italy, in this scenario, might have increased, even considerably, over the years of the long Augustan Principate. It might have increased even later and at plausible rates for a "pre-transitional" population. We must also account for the relevance, especially seen in the time of Claudius, of the increasingly generous granting of citizenship to non-Italians, that same granting of citizenship that Seneca ridiculed in his *Apocolocyntosis* and which made the proportion of "cives Romani" in the provinces grow progressively. One may conclude that, even according to a "high count" scenario, it is not necessary either to suppose that the Italian population had stopped growing in the course of the first century CE nor to suppose that it had achieved implausibly high levels in relationship to later developments.

Acknowledgments

I am indebted to Alessandro Launaro for penetrating comments and invaluable help in translating this chapter.

FURTHER READING

General overviews of the issues of Roman demography are Parkin (1992) and Scheidel (2001). On population size, structure and dynamics as determinants of economic performance in the ancient world, see especially Scheidel (2007). On the debate over the size of population of Roman Italy, see Lo Cascio (1994, 2000), Morley (2001), Kron (2005), Lo Cascio and Malanima (2005), Scheidel (2008), Launaro (2011) and de Ligt (2012). On survey evidence, and its limits, as a source of information, see Witcher (2008, 2011), Launaro (2011) and de Ligt (2012: ch. 6).

CHAPTER 11

Looking at Early Rome with Fresh Eyes: Transforming the Landscape

Albert J. Ammerman

1 Introduction

It is often said that the city of Rome was built on seven hills. For those who study the topography of the ancient city more closely, this old chestnut is, of course, stretching things. There are more than seven hills on the east bank of the Tiber, and sooner or later the Romans chose to occupy all of them. In addition, the list itself has varied over the ages. And in the eyes of the earth scientist and the geo-archaeologist, they are not really hills at all. Two of them, the Capitoline and Palatine, are isolated remnants of a low volcanic plateau at the foot of the Alban Hills volcano, and the rest are simply promontories on the edge of the same plateau (Alvarez *et al.*, 1996). The volcanic eruptions that gave rise to the formation of the plateau took place in the interval of time between 600,000 and 300,000 years ago. Then the Tiber River took over and played its hand in sculpting the plateau into the so-called hills. The river's handiwork entails a complex story that unfolds over the course of several ice-age cycles. In response to changes in sea level during the Pleistocene, it involves a combination of aggradation, down-cutting and lateral shifts in the drainage pattern. In the end, all of this geological activity came together to produce the familiar high places on the landscape. Among them was the one on the west side of the Palatine where a cluster of wattle-and-daub huts with thatched roofs was built in the eighth century. The purpose of this chapter is to introduce the recent studies in environmental archaeology that are throwing new light on the formation of the early city.

The low places between the hills have commonly attracted less attention, even though they constituted most of the area in the heart of the early city. This was clearly the case when I first went out to Rome in 1985 to conduct fieldwork at archaeological sites in and around the Forum. At the time, very little was known about the Forum basin itself

A Companion to the Archaeology of the Roman Republic, First Edition. Edited by Jane DeRose Evans.

and other nearby low places such as the Argiletum and the Velabrum. There was a good reason for this state of affairs: low places of this kind have all been deeply buried in the ground for many centuries. In the Renaissance (on the first steps toward archaeology in Rome at that time, see Schnapp, 1993), the situation was much the same as it is today. Only with great difficulty would the budding archaeologist at that time have had the chance to dig deeply enough in the earth to reach an archaeological layer dating to the first century or the closing years of the Republic. No one in the sixteenth century – with the possible exception of Michel de Montaigne, the French essayist – dreamed of digging even deeper in the ground: that is, first down through a sequence of archaeological levels dating to the Republic (the years from 509 to 27) and then further down through a sequence of levels going back to the regal period (753 to 509, according to literary tradition) in an attempt to reach the natural soil.

2 Getting Down to the Bottom of Things

To gain a better sense of the remarkable build-up of the ground level in Rome over the centuries, it is useful to consider a specific example from the Velabrum (see Figure 29.1). At the headquarters of the Vigili Urbani on the Via della Consolazione, the city police park their cars today on a street that stands at 23 m above sea level. At a depth of 8 m in the ground, one reaches the travertine podium of what most scholars in Rome consider to be the Templum Novum, the temple built in honor of the deified Augustus, the first emperor. If one then goes down another 9 m, one comes to an archaeological layer that dates to the seventh century. And at its base, one encounters the natural soil at an elevation of 6 m above sea level. The difference in elevation between the street level today and the top of the natural soil amounts to a total of 17 m (or 55 feet). Even for the archaeologist who has extensive field experience, it is a daunting task to think about a stratigraphic sequence that is so deep. In the context of the dense urban fabric in this part of Rome, there is, in fact, no practical way of making an excavation that will reach a depth of 17 m. And assuming that one somehow did manage to get down to a depth of 14 m, the archaeologist would then have to face another major challenge: that of excavating archaeological levels that occur below the water table of the modern city. In effect, one would have to dig the last 3 m in water-saturated soils – a very difficult task even when pumps are in operation. It should come as no surprise then that, over the wide area between the Forum and the Area Sacra di Sant'Omobono (a distance of 300 m), no excavation has ever been taken all of the way down to the natural soil. This can only be done by means of deep, machine-made cores. This explains why the Velabrum remained a terra incognita for so many years and why the situation only began to change when the first archaeological cores were drilled there in 1996.

3 Three Misconceptions

Based on the new evidence that has come to light since 1985, it is now possible to realize that much of what was said about the low areas between the hills in the previous literature was marred by misconceptions. In retrospect, this is fully understandable given the acute shortage of firsthand observations previously available on the environmental setting of the low places. By way of introduction, it may be instructive to consider three examples of such misconceptions. In the case of the Forum basin, scholars had wrongly assumed for years that all that was needed to make this low place a dry one was a drainage canal (Gjerstad, 1953: 49). While such a canal may have worked quite well most of the year, this was not the case on those rare occasions – perhaps only once or twice in a given year – when the Tiber rose in its banks and was in major flood (Ammerman, 1990a: 636–8; Aldrete, 2007). At such a time, the water would have flowed back through the Velabrum and flooded the Forum basin. In this case, a canal must be seen as working in two directions. During a major flood event, it would have actually facilitated the flow of water into the Forum basin. Remarkably, no one had picked up on this in the literature before (Ammerman, 1990a: 638). So drainage by itself was not enough; the ground level had to be built up. Indeed, it would take a major project of land reclamation to turn the seasonally wet Forum basin into the hub of the early city (Ammerman, 1990a). The second example concerns the Velabrum, the low area between the Forum basin and the Tiber. In the previous literature, there was the notion that much of this area was once covered by a large, standing swamp. When we had the chance to take deep, machine-made cores in the Velabrum, no evidence was found for a large swamp that dates to the regal period (Ammerman, 1998). The third example has to do with the position of the banks of the Tiber in the area of the early city known as the Forum Boarium. Maps showing the situation there in the sixth century have always placed the riverbank in its present position. Again, deep cores showed that things were quite different in the sixth century, when the riverbank once stood about 100 m back from where it stands today. Over the centuries, the Romans – just like the inhabitants of other great waterside cities around the world – have encroached on the banks of their river (Ammerman, 2006: 307). In short, what we see in Rome today does not provide a good guide to the situation in the remote past. Scholars writing about early Rome in previous generations often projected what they saw in their own day back on to early Rome. On the whole, they did not understand the changes that have taken place on the landscape over the last 3,000 years.

At this point, it may be useful to say a few words about the history of scholarship as it relates the second example. We can trace the idea of a large, standing swamp in the Velabrum back to a map that Giovanni Battista Brocchi drew in 1820 in an attempt to reconstruct the setting of primordial Rome (Ammerman, 2006: fig. 7). Brocchi's idea was then taken up by Rudolfo Lanciani in 1897 and further elaborated by Coarelli in the early 1980s (Lanciani, 1967; Coarelli, 1985, 1988b). Lanciani and Coarelli both chose to enlarge Brocchi's original swamp. The reconstructions that the three scholars put forward were based on what they could glean from reading the ancient sources and not on firsthand observations that they made in the field. It is worth noting here that the ancient sources that have come down to us on the early Velabrum are few in number, and they are limited in terms of what they have to say. There is the added problem that the authors who wrote the passages did not live in the first five centuries of the city's history but in later times. In other words, when they wrote, the low places in the Velabrum had all been filled in for some time. As in the more recent cases of Brocchi, Lanciani and Coarelli, the classical authors themselves were not in a position to make firsthand observations on the early Velabrum. Instead, they had to draw upon traditions handed down over many generations. Unfortunately, traditions of this kind are often less than reliable. Thus, in the study of the early Velabrum, the challenge has long been that of putting discourse on a more solid footing.

4 Coring in the Velabrum

In 1996, there was the chance to conduct the kind of fieldwork that was called for in the Velabrum. The City of Rome commissioned me to make a series of deep cores in the area between the Forum and the Tiber. Each of the machine-made cores began at modern street level and was taken down through a long sequence of archaeological levels well into the natural soil below. The first cycle of coring showed that the early environmental conditions there were quite different than scholars had previously thought (Ammerman, 1998). While we were able to document a swamp dating to a much earlier period of time (a swamp known as an alder carr, which occurs at an elevation of 3 m below sea level and dates to 7,500 years ago), there was no evidence for a swamp dating to the regal period. Two more cycles of cores were made in 1998 and 2003 in order to enlarge the coverage of the Velabrum, and they both produced results of considerable interest, as we shall see in the next paragraph. However, after taking a total of 24 cores, there was still no support for the notion of a large, standing swamp in the regal period. It was now clear that the early Velabrum should be viewed as part of a larger lateral valley that fed into the Tiber Valley. In the warm summer months, the Velabrum was essentially a dry place. In the winter months, it could be a wet one, especially on those rare occasions when the Tiber was in major flood. And flooding itself would have worked against the formation of a large permanent swamp on the valley bottom, since the deposition of Tiber flood sediments would have filled in the lowest places. Instead of envisioning the Velabrum as a static *palude* as scholars had previously done, it can now be seen as a more open and dynamic space that had its own seasonal rhythms (Ammerman, 2006: 308).

5 The Clay Beds in the Velabrum

The coring in the Velabrum led to another surprising discovery – the presence of beds of high-quality clay in a position just below the top of the natural soil (Figure 11.1). Previously, no source of clay of the quality and quantity required for the production of roof tiles was known on the east bank of the Tiber (Ammerman *et al.*, 2008). As documented in the cores, the clay beds turn out to be quite large in size. They measure some 3 m in thickness, and they extend for more than 100 m in length. Studies of the clay done by specialists in ceramic technology confirm that it is of good quality for the production of roof tiles, which made their first appearance in Rome around 625. The next step in the research was to compare the geological and chemical properties of the clay with those of the early roof tiles in Rome. Without going into the details of the characterization studies here, the two sets of fingerprints do match (Ammerman *et al.*, 2008). The inference to draw is that the clay beds in the Velabrum were the source of the raw material used for making the earliest roof tiles in Rome. Thus, the early Velabrum – a place that scholars

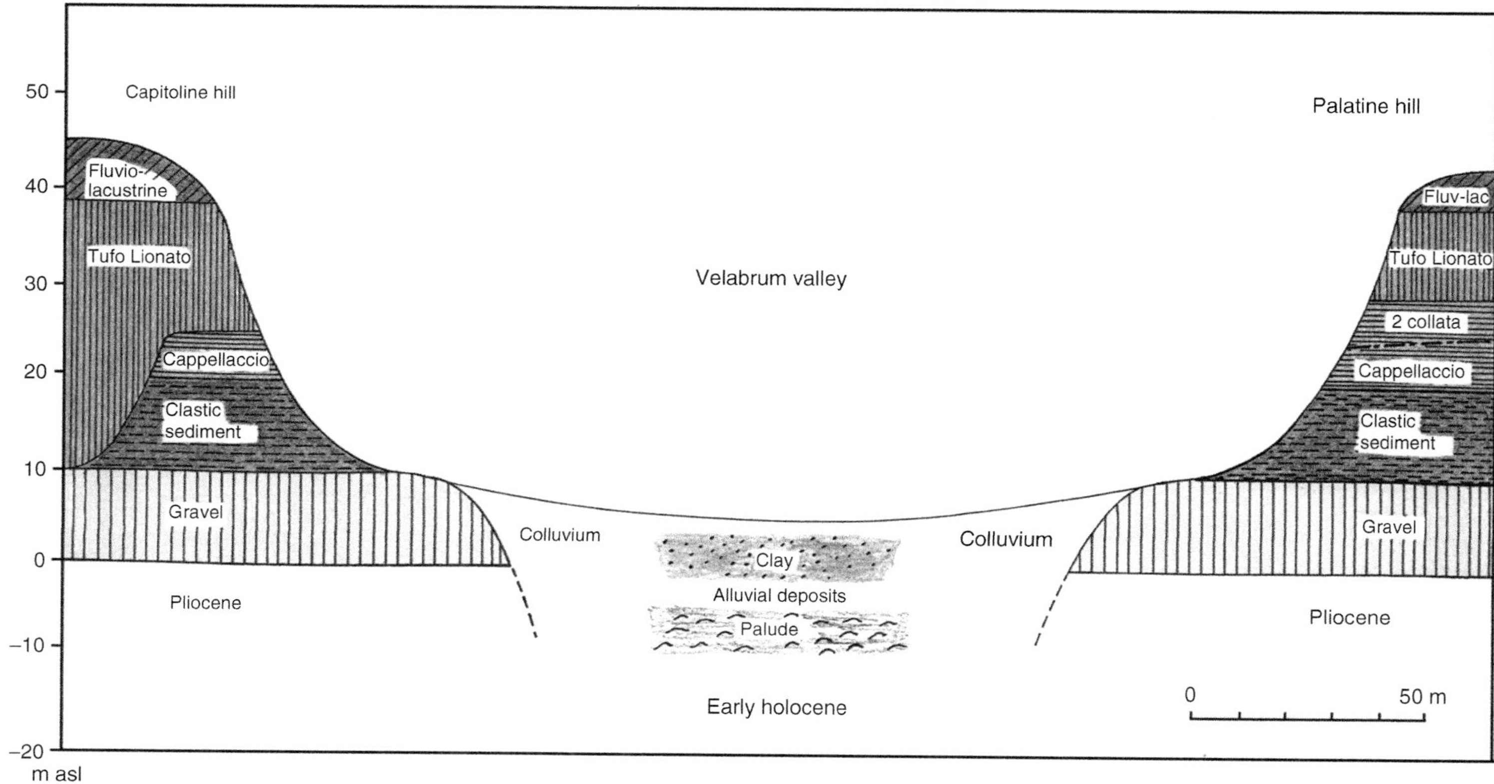

Figure 11.1 Cross-section showing the environmental settings of the Capitoline Hill, the Velabrum Valley and the Palatine Hill. Here the vertical scale is twice the horizontal scale. Source: Image by A.J. Ammerman.

for years had regarded as a swampy no-man's-land – can now be viewed in a different light. It was a busy place where clay was dug from the earth and where craftsmen made roof tiles. At the same time, the Velabrum and the adjacent Forum basin were low places that were not really suited for habitation, since they were flooded by the backwash of the Tiber from time to time. In short, the study of the Velabrum offers a good example of what it means to look at early Rome with fresh eyes.

6 Returning to the Aims of the Chapter

This is a good place to step back and say a few more words about the scope of this chapter. In a nutshell, it has two main aims. The first is to set the stage for the study of Rome in the years of the Republic by presenting what is currently known about two of the most important nodes in the city's formation: the Forum and the Capitoline Hill. They both go back to the time of the seven kings of Rome. According to Roman literary tradition, the regal period began in 753 with the foundation of the city by Romulus (on the legends concerning the first king and on those relating to his twin brother, Remus, see Wiseman, 1995), and it lasted until the fall of the monarchy in 509. One of the important aspects of the urban fabric of ancient Rome is the antiquity of several of its key nodes. For instance, many of the important buildings in and around the Forum were already in place by the third decade of the Republic (that is, by 480): the Regia, the Temple of Vesta and the Temple of Castor and Pollux on the east side and the original Senate House (the Curia Hostilia), the Comitium and the Temple of Saturn on the west side (Figure 11.2). In order to study the city of Rome in the Republic, one has to start with the footprint that it inherited from the regal period (see Chapter 26).

The second aim is to introduce the new and more interdisciplinary approach that is now being taken to the study of early Rome. In the literature, it goes under various names: environmental archaeology, geo-archaeology or even landscape archaeology. For the classicist and the common reader, the names of such fields of study are not household words. What they all share is a common interest in placing an archaeological site in its environmental context. Of course, each approach tries to give the treatment of the subject its own spin. Prior to 1985, studies of this kind were few and far between in the eternal city. In light of the significant gains that have been made since 1985, a question of interest is why work of this kind did not begin earlier in Rome. Again, space is not available in this short chapter to develop a full answer to this question. Only a few of the factors that may help to explain the late start will be introduced here. To begin with, the disciplines of archaeology and geology are both young ones in the academic world. It was only in the nineteenth century that they came into their own. Even as late as 1850, few universities around the world had a department of either geology or archaeology. And those archaeologists who excavated at classical sites in the nineteenth century already had their hands full: in moving large volumes of earth, in drawing monuments, in reading inscriptions, in studying coins, in identifying pieces of sculpture and in classifying pottery. They did not have the time to think about initiating environmental studies at their sites. Moreover, it was only in the second half of the twentieth century that the equipment and the methods required for such investigations became more widely available. If they had already existed a century ago, an archaeologist such as Giacomo

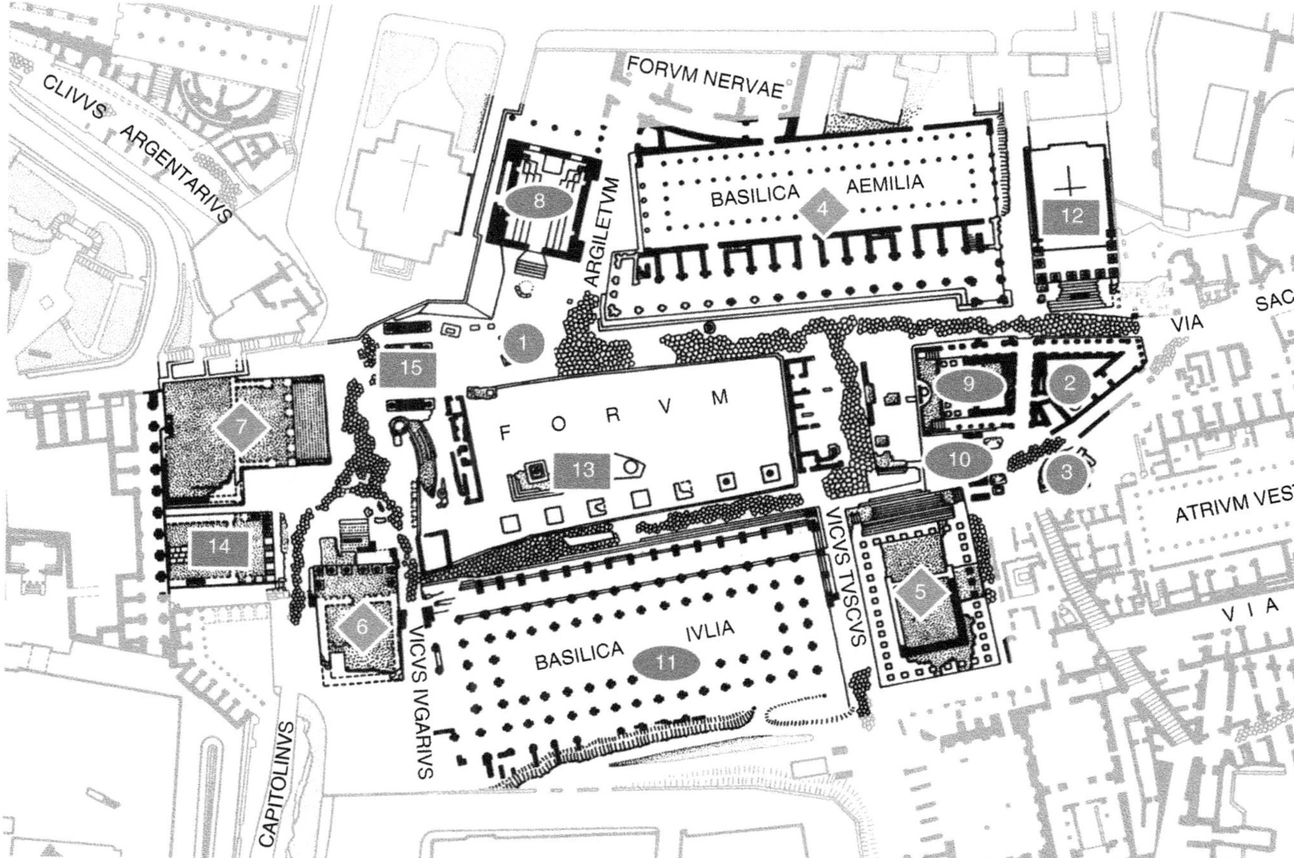

Figure 11.2 Map of the main sites located in and around the Forum. The construction of each monument was begun in one of four periods: (I) Regal (the time of the kings, circles); (II) Republic (diamonds); (III) the lifetimes of Julius Caesar and Augustus (the transition between periods II and IV, ovals); and (IV) Empire (squares). The monuments are the following: (1) Comitium; (2) Regia; (3) Temple of Vesta; (4) Basilica Aemilia; (5) Temple of Castor (Castors); (6) Temple of Saturn; (7) Temple of Concord (Concordia); (8) Curia Julia (Curia Iulia); (9) Temple of Deified Julius Caesar; (10) Arch of Augustus; (11) Basilica Julia (Basilica Iulia); (12) Temple of Antoninus and Faustina; (13) Column of Phocus; (14) Temple of Vespasian; (15) Arch of Septimius Severus. Source: Image by A.J. Ammerman.

Boni, who was trained as an architect and who carried out innovative excavations in the Forum, would have surely used them. And, up until quite recently, something else was missing as well. This was an interdisciplinary approach to doing archaeological research in the Mediterranean world. In Italy, in particular, there had long been a tradition that the humanities and the sciences were not supposed to mix. By the 1960s, taking an interdisciplinary approach to an archaeological project had become fairly common in northern Europe and North America. It could still raise eyebrows on the east bank of the Tiber in the 1980s.

7 Adding the Third Dimension in Rome

What all of this meant was that there was much work to do when I first went out to Rome in 1985 – sometimes even at the most basic level. For instance, there was still no profile of the natural relief of the Forum basin at the time. No contour map showing the natural relief of the Capitoline Hill had been produced even as late as 1990. What was missing from the study of early Rome, I soon realized, was the third dimension. Previously, there had been no systematic attempt to record the elevations of the natural soil even at some of the most important archaeological sites in the Forum and on the Capitoline Hill. Indeed, many of the previous excavations had not been taken down deeply enough to reach the natural soil. Given the previous lack of equipment for making the cores that could produce this information, knowledge of the third dimension turned out to be a scarce commodity. Thus, one of the first priorities that I set when we began to do environmental archaeology in Rome was to add the missing third dimension. On the face of things, this may sound like a straightforward task. In fact, it often represents a tall order. For example, it would take years of fieldwork on the Palatine before we could draw, even at the first level of approximation, a contour map of the hill's natural relief. In making a map of this kind, one has to work by steps of approximation, which means adding new observations to the map each year as they slowly come in. In retrospect, there was the good fortune that I had the opportunity to work at many different archaeological sites in Rome over a period of 20 years (1985–2004). Here it is worth adding that there was a remarkable figure in the sixteenth century, the French essayist Michel de Montaigne, who had the foresight to understand the nature of the challenge well before our time. This is why it is now called Montaigne's challenge (Ammerman, 2006: 308).

8 Montaigne's Challenge

Montaigne wrote a detailed account of the visit that he made to Rome in 1581. In his diary, he set down his impressions of the city in the third person, as was his habit. After a few days of initial enthusiasm, he had a clear sense of disappointment that the ruins of Rome, grand as they might be, did not really offer a satisfactory image of the ancient city. "He observed: that there is nothing to be seen of ancient Rome but the sky under which it had risen and stood, and the outline of its form; that the knowledge he had of it was altogether abstract and contemplative, no image of it remaining to satisfy the senses" (see Ammerman, 2006: 297). By the time of Montaigne's journey into Italy, there had been several recent attempts to reconstruct the entire ancient city, starting with the bird's-eye view map drawn by Pirro Ligorio in 1561. Montaigne had seen such "pictures" as he called them in his diary. However, when he now had the chance to examine the situation on the ground, he found serious problems with all of them. In the case of the Capitoline, it was difficult to reconcile the limited size of the hill as he now saw it with his own eyes with the grandiose vision of Ligorio, who crowded it with more than 25 temples and monuments. And Montaigne was no less critical in his thinking when it came to what he saw in the case of the Velabrum:

> But, in truth, many of the conjectures which one has formed from pictures of the ancient city are not at all borne out, when you get there, for even the site has undergone infinite changes; some of the valleys are filled up, even the deepest of them such, for instance, as the *Velabrum*, which on account of its lying so low, was selected as the main sewer of the city, and formed a watercourse, even this has now become as high as the other natural mountains that surround it, and this has solely been done by the gradual agglomeration of the ruins of old Rome.
>
> (Hazlitt, 1864, vol. 4, p. 308)

Montaigne was the first person in the modern era to look at ancient Rome with fresh eyes. Because of his critical thinking, he was able to move beyond what I have recently called the "syndrome of the eternal city" (Ammerman, 2006: 308) – the impulse to see ancient Rome as a highly idealized place that somehow has never changed over time. Instead, for Montaigne, the city of Rome was just the opposite: a place that over the course of its long life had witnessed infinite changes.

9 Landscape Transformations in Early Rome

In our time, the classical example of the syndrome of the eternal city is the notion that the Tiber bank in the area of the Forum Boarium has always stood where we see it today. This is, by the way, where it is shown on almost all of the maps of early Rome published in the twentieth century. As mentioned before, this is a misconception. Today the riverbank stands in a different position than it did in the sixth century. When the Romans in the mid-Republic built several temples in this place (Ziolkowski, 1992), they did so by encroaching on the space of the Tiber. The syndrome of the eternal city also conditions how we think about early Rome in less obvious ways. When we look out over Rome today, it is clear that there is a substantial difference in height between the tops of the Capitoline and Palatine and the low area between these two hills. Some of us may even begin to reflect on the problems that such a demanding site would have posed for those who once lived in the early city. However, the impression that we come away with does not capture the true difficulty of the original site. While the top of the natural soil on the Capitoline and Palatine varies in elevation to some extent from one point to the next on the two hilltops, it commonly stands at around 43 m above sea level (Figure 12.1). In contrast, the top of the natural soil in the middle of the Velabrum, as mentioned before, rests at an elevation of 6 m. This means that the variation in height between the highest and the lowest places on the landscape in the regal period was about 37 m (or 120 feet). In the same part of Rome today, the variation in elevation is much less: only about 27 m. In other words, the range of variation is now some 10 m less than it was 2,600 years ago. The reason for this change is connected with the thickness of the archaeological deposits in the two cases. While the build-up of archaeological levels in the Velabrum amounts to 17 m, it usually runs to less than 7 m on the tops of the Capitoline and the Palatine. The main point that I am trying to make here is that we tend to underestimate the rough and accentuated character of Rome's natural relief and the difficulty of building an early city there. One of the implications is that the early Romans had to learn how to modify the landscape. The purposeful transformation of the landscape has recently emerged as one of the leading new themes in the study of early Rome. Four places provide evidence for landscape transformations in the time before the Republic (Ammerman, 2006: 299–300): (i) the Forum Basin where a low, seasonally wet area was turned into the hub of the early city; (ii) the Capitoline Hill where the west side of the southern summit was terraced for the emplacement of the Temple of Jupiter Optimus Maximus; (iii) the deeply incised gulley at the foot of the North Slope of the Palatine which was filled in to create new urban space; and (iv) the Velabrum where the ground level was successively raised when new gravel pavements were laid down. In the last two sections, more will be said about the first two case studies.

10 On the Origins of the Forum

The Forum in Rome was at once a marketplace and a civic center. It would take a major project of landscape transformation to turn the Forum basin into the hub of the early city. Previously, there had been the idea that families living in wattle-and-daub huts had occupied the low central part of the basin in the time before the first gravel pavement of the Forum (Gjerstad, 1953). The fieldwork that we carried out in the Forum now raised serious questions about Gjerstad's interpretation and led to thinking about the origins of the Forum in a new way. The work on the profile of the relief of the Forum basin now showed that the natural soil in its center stood at a height of 7 m above sea level and that the environmental setting there was a seasonally wet and dry one. In 1990, in an article called "On the Origins of the Forum Romanum," I set out to do three things: (i) to review the literature on the origins of the Forum, which dates to around 625 (there is the need for a new cycle of excavations to improve the dating of the first gravel pavement of the Forum); (ii) to present the results of the recent work on profiling the Forum basin; and (iii) to put forward a new explanation for how the Forum began (Ammerman, 1990a). According to the new hypothesis, the creation of the Forum called for a major project of land reclamation. In order to bring the ground level up to an elevation of 9 m, it would take large amounts of both landfill (at least 10,000 cubic meters) and corvée labor.

Once the ground level was raised to this height, it was then covered with a gravel pavement. The gravel itself came from the gravel beds at the base of the Capitoline and Palatine Hills. As shown in Figure 11.1, the tops of the thick gravel beds, the foundations of the two hills, stand at 10 m in elevation (that is, at basically the same height as the first pavement of the Forum). It will be recalled that two of the most important lines of communication in Archaic Rome once ran across the tops of the gravel beds – the precursors of the streets that the Romans in time would call the Vicus Iugarius and the Vicus Tuscus. More recently, coring has brought to light the clay beds in the nearby Velabrum, as mentioned before. In addition, we know that the clay there was used to make the earliest roof tiles in Rome (Ammerman *et al.*, 2008). Indeed, it is now possible to think in terms of new connections between Tarquinius Priscus (traditionally the fifth king of Rome; Demaratus, his father, is said to have brought tile-making from Corinth to southern Etruria), the digging of the Cloaca in the Velabrum (again, the first drain there is attributed to the fifth king) and the beginnings of the Forum (it too is first mentioned in the ancient sources as a marketplace in the time of the fifth king). Needless to say, there is much work that remains to be done when it comes to the study of these connections (Ammerman, 2010: 220). What is important for our present purposes is that the Forum project demonstrates a noteworthy ability to transform the landscape in the time before the Republic.

11 The Siting of the Temple of Jupiter

The second example concerns the single most important temple in early Rome – the one dedicated to Jupiter Optimus Maximus on the Capitoline Hill. By making more than 40 deep cores at different points on the hill and by bringing together other lines of evidence (from excavations, exposed outcrops of bedrock and galleries quarried in the bedrock), it was possible to produce the first contour map of the hill's natural relief and to work out its geological structure (Alvarez *et al.*, 1996). In addition, we were able to date, by means of a new method of geo-chronology, the two main volcanic tuffs that make up the hill. As shown in Figure 11.1, they are the hard upper one with a reddish color known as Tufo Lionato (an ash-fall tuff dating to around 350,000 years ago, which breaks off naturally along cooling joints to form the hill's distinctive steep cliffs) and the lower and softer one called cappellaccio (a mud-flow tuff that dates to around 520,000 years ago). Here it is worth adding that the lower one, which is sometimes called a "pebble tuff," was used for making the oldest stone blocks in Rome, including those found in the foundations of the Temple of Jupiter itself (Cifani, 2008; see Chapter 17). What this means is that the three main building materials that were used in the regal period – cappellaccio from the slopes of the Capitoline and Palatine (for walling), clay from the clay beds in the Velabrum (for roof tiles and architectural terracottas) and gravel from the gravel beds at the feet of the Capitoline and Palatine (for pavements) – all had good local sources in the very heart of the early city.

The Capitoline is a comparatively small hill and it has two summits: the Arx on the northern side and the Capitolium on the southern one. Between them is a low saddle, now Piazza del Campidoglio. The new contour map now revealed how small the top part of the Arx was. It covered an area of only about 2 acres (less than 1 hectare). The southern summit, again in its original form, was not much more than twice that size. On maps of the Capitoline in the previous literature, the Temple of Jupiter was placed in the middle of the Capitolium. Where it actually stood was on the west side of the southern summit (Ammerman and Terrenato, 1996: fig. 5). In fact, the temple's north-west corner was built on an artificial terrace – formed of the soil removed from the temple's deep foundation trenches. Such trenches (up to 5 m deep in some places) were required because of the thick deposit of soft fluvio-lacustrine sediments on the hilltop (Figure 11.1), which could not support the weight of a large temple. Accordingly, it was necessary first to dig down to the top of solid Tufo Lionato and then to build back up again, using a dozen or more courses of cappellaccio blocks, so that the temple would rest on a solid foundation. At the same time, the emplacement of the temple on the west side of the southern summit solved, in an ingenious way, another problem as well. This was the need for ample space around the temple on its east side (that is, at the top of the Clivus Capitolinus) so that a victorious general could celebrate the conclusion of a triumph with proper pomp and ceremony (for the many ceremonies associated with the temple, especially the Roman triumph, Beard, 2007). Thus, the purposeful transformation of the landscape on the west side of the Capitolium served a twofold purpose: as the place to park the waste soil from the temple's foundation trenches and as a way to create more space on the temple's east side. As more work is done over the next 20 years at archaeological sites in Rome that date to the regal period and the Early Republic, further examples of landscape transformations will no doubt come to light. We are just at the start of responding to Montaigne's challenge and looking at early Rome in new ways.

FURTHER READING

Prior to 1990 there was not much in the way of literature on the topics covered in this chapter. Since that time, the specialist literature has grown considerably. A few of the main articles and chapters are indicated in this chapter. For the non-specialist, the best places to start would be Ammerman (2010, 2006). For a more detailed treatment of the new interpretation of how the Forum began, there is Ammerman (1990a). For the recent studies on the oldest roof tiles and architectural terracottas in Rome, one can read Ammerman *et al.* (2008). Since this volume is intended for students in the English-speaking part of the world, the bibliography in the Italian literature has been kept to a minimum, but see, for example, Ammerman (1990b), Ammerman and Filippi (2004) and Ammerman and Terrenato (1996).

PART II

Archaeology and the Landscape

CHAPTER 12

Survey, Settlement and Land Use in Republican Italy

Helena Fracchia

1 Introduction

A heated debate surrounds the contribution of field survey to our understanding of settlement development and land use. Archaeologists sing the praises of field survey and its uses but many historians of the Roman Republic remain skeptical about the usefulness of field survey results (de Ligt, 2006: 590–2; Rathbone, 2008: 305–7). Considerable debate has ensued about whether potsherds found on the ground can tell us something about changes in settlement type, about changes in settlement patterns, and about changes in the cultural identity of a given area. Two interconnected main themes in the debate emerge: was there settlement continuity or discontinuity in the creation of Roman Italy (Colivicchi, 2011); and can the ancient sources, rarely contemporary to the Republic itself, be more accurate than the archaeological evidence, itself variable and contradictory? (Rathbone, 1981; Bintliff, 1991; Bulliet, 1992; Mattingly and Witcher, 2004; Rathbone, 2008; Yntema, 2008). The time period of the Roman Republic, beginning around 500, includes the early and intense interaction of indigenous tribes of Italy with the growing Roman state and the Roman hegemony over all of Italy. The Republic demonstrates some of the most problematic issues of the Roman hegemony including colonization, the Gracchan land reforms, demographic development, the spread of the villa system and changes in land use during that time. The integration of archaeological evidence and ancient sources is not an easy task, especially in such a polarized debate (Dyson, 2003; Rathbone, 2008: 307–10; Yntema, 2008: 373–7; Patterson and Di Giuseppe, 2009). What contribution can field survey make to the understanding of these issues?

Archaeology is one of the historical disciplines, and field survey results – regardless of research goals or methodological approach – must, in the final analysis, contribute to the

A Companion to the Archaeology of the Roman Republic, First Edition. Edited by Jane DeRose Evans.

historical narrative of ancient Italy. Archaeology is the most visible and most immediate of the historical disciplines. Consequently, it is also the historical discipline most subject to manipulation. The variability in interpretation of the surface finds coupled with methodological differences is a serious issue in the debate about the historical implications that can be justifiably derived from field survey results.

2 Background

Originally field survey was simply a preliminary step to identifying a site for future excavation. After the Second World War, the increased use of mechanized agriculture in Italy brought to the surface, and progressively destroyed, vast numbers of archaeological sites. John Ward-Perkins, director of the British School at Rome, started an extensive field walking campaign in the 1950s and 1960s in southern Etruria. He postulated that the occupational history of entire landscapes could be reconstructed by identifying and dating pottery sherds found on the surface of fields where the remains of different types of ancient buildings, usually farms and villas or high-status sites, were also visible. Thus, information about the transformations in settlement and land use could be derived from the kind and types of fine-ware pottery found in the countryside. The intellectual framework for the original South Etruria Survey was the diachronic impact of the growth, success and decline of Rome on its hinterland in general (Potter, 1979; Patterson and Millet, 1998; Patterson, ed., 2004; Patterson, Di Giuseppe and Witcher, 2004; Patterson and Di Giuseppe, 2009).

In 1965, Arnold J. Toynbee suggested that the Second Punic War caused the total devastation of south-eastern Italy and, in particular, the decline of the Italian peasant class, due to the long military service required during the war years. After the war, the changed relationship between Rome and her Italic allies brought new economic opportunities and forms of land exploitation in that specific area of Italy (Toynbee, 1965). Field survey results began to be used to address demographic issues, land allotments associated with colonization, and land use.

In southern Italy, field survey became a tool by which to evaluate the decline of the fortified centers and their territories; the subsequent Hannibalic devastations and the depopulation of the Italian countryside; the impact of colonization; and the appearance of the villa as the defining element of the Roman presence in southern Italy. Information about possible desertification, abandoned agricultural land, the habitations where peasants, aristocrats and slaves lived, the kind of agricultural production (modest Mediterranean polyculture in contrast to large slave-run estates for wheat production), and demographic trends – all elements that Toynbee had attributed to the Second Punic War – could be gleaned from the diachronic incidence of farms and villa sites as determined by surface architectural and ceramic finds. These were first studied independently and then in relation to one another. As early as 1971, P.A. Brunt thought that Toynbee had overstated the effects of the Hannibalic War (Brunt, 1971). Brunt proposed instead that the events that followed the Hannibalic War were the major factor in the decline of the free Italian population, which could be seen in the lesser number of sites dated to the second century that could be identified by field survey. The decline in rural population was also interpreted as evidence that newly established villas absorbed part of the free population, paving the way for the slave mode of production and a recognizable rural Roman Italy (see Chapter 10).

In southern Italy, however, the problem of the second-century "crisis" cannot be separated, nor should it be, from the preceding period. Thus, the question is not just the second-century crisis, but the fate of the indigenous peoples throughout what had been Magna Graecia. The pre-Roman settlement pattern was characterized by the presence of fortified centers controlled by various elite groups known primarily from wealthy burials in the fifth and first half of the fourth century. By the second half of the fourth century, an intensive agricultural exploitation in the countryside and dense rural settlement had developed around the fortified centers (Bökönyi, Costantini and Fitt, 1993; see also Chapter 19). This rural expansion has been connected to the emergence of an intermediate social class. Nonetheless, by the mid-third century the fortified hilltop settlements generally declined as Roman hegemony spread across the southern parts of the peninsula. Following the Second Punic War, land in areas that had supported Hannibal was confiscated. Ager publicus, land belonging to the Roman people, was introduced into parts of Campania, Lucania and Apulia (Rathbone, 2008; Roselaar, 2008, 2009). The alleged concomitant depopulation of the countryside created the conditions in which, in some areas, large estates (or latifundia) using slaves for extensive agricultural use, could become established, although the degree to which this happened and why is still a matter of debate (see Chapter 13; Morley, 2001; de Ligt, 2006).

The low cost of field survey and its relatively rapid results soon gave rise to numerous survey projects throughout Italy. The raw material found on surveys had to be organized in a way that made it useful. The fine-ware pottery found in the survey was assigned to broad chronological periods. Sites with various indigenous ceramic wares were dated generally to the sixth to fourth centuries. The marker artifacts for the expansion of Rome and the Roman period were the presence of the highly standardized so-called Campanian black-gloss and its imitations (or *pasta grigia* pottery in parts of southern Italy), dated *c.* 200–50, and Italian terra sigillata wares dated 50 BCE–50 CE (see Chapter 5). The types of pottery found, in combination with the physical extension of the surface material and other surface elements (such as roof tiles, bricks, stone blocks, tesserae and glass), were then classified into settlement or habitation categories: for example, large or small farm, villa rustica or villa, hamlet, or vicus. Between survey projects, the criteria for what constitutes a large or small farm, villa or villa rustica can vary, as can the classification itself (Rathbone, 2008).

The interpretations of the raw material found in the fields in many areas of southern Italy and around Rome itself were often in direct contrast with what the ancient sources, writing centuries after the fact, said about the Hannibalic devastation. The contrast between the archaeological evidence and the ancient sources gave rise to a large body of source criticism (Patterson, Di Giuseppe and Witcher, 2004; Patterson and Di Giuseppe, 2009). Toynbee's assessment and Brunt's criticism further polarized the debate. Building on highly differentiated survey results, some scholars have attempted to identify ager publicus in the depopulated regions of southern Italy, while others have projected the developments delineated in specific areas across all of Italy, regardless of local geomorphology, topography, coastal or inland placement, suburban or rural site. Demographic

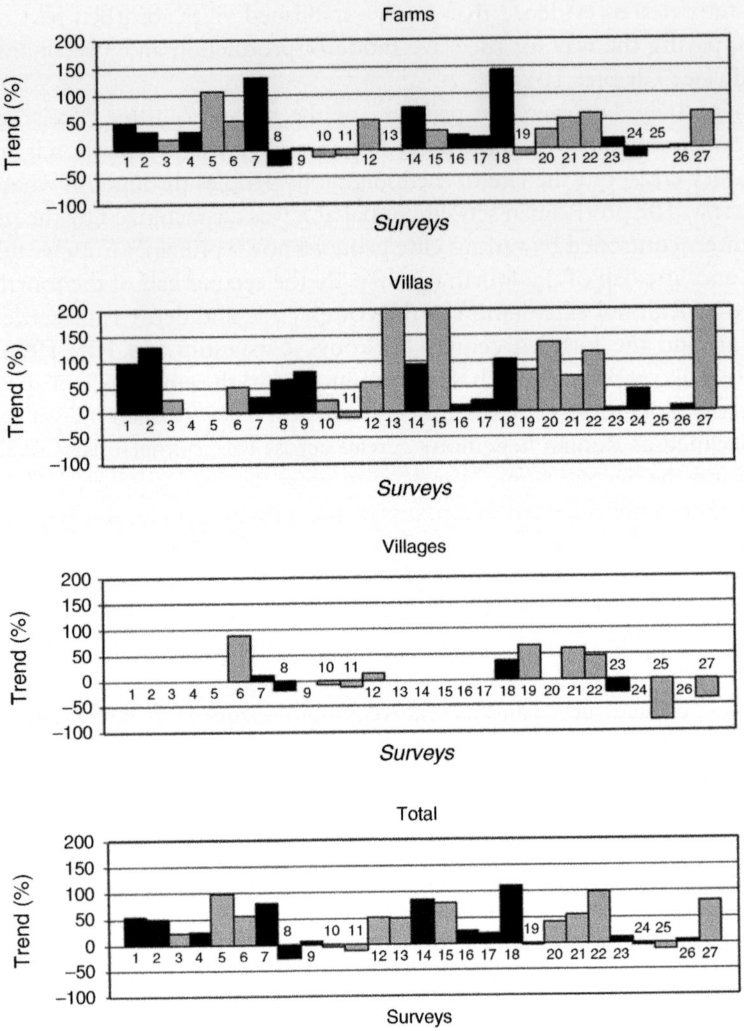

Figure 12.1 Comparative survey trends: settlement trends between the Late Republic and Early Empire, north to south (gray = calibrated; black = not calibrated). (1) Valli Grandi Veronesi; (2) Carta Archeologica del Comune di Poviglio; (3) Carta Archeologica del Comune di Modena; (4) Polcevere Valley; (5) Ager Lunensis; (6) Cecina Valley; (7) Potenza Valley; (8) Carta Archeologica della Provincia di Siena; (9) Scarlino; (10) Albegna Valley; (11) Tuscania; (12) Rieti Basin; (13) Marina di Montalto; (14) Ager Foronovanus; (15) Corese Survey; (16) Torrimpietra; (17) Ager Caeretanus; (18) Collatia; (19) Biferno Valley; (20) Fregellae; (21) Liri Valley; (22) Northern Campania; (23) Carta Archeologica della Campania; (24) Ager Venusinus; (25) Oria; (26) Carta Archeologica della Valle del Sinni; (27) Roccagloriosa. Source: Drawing by H. Fracchia, after Launaro (2011).

studies, based on survey evidence, have multiplied in recent years, with concomitant attempts to make survey results more comparative (see below) using statistical analyses and/or calibrated dates (Figure 12.1) (Alcock and Cherry, 2004; Launaro, 2011).

3 Field Survey: Methodological Problems

Numerous methodological problems abound when reconstructing a cultural and political history from fragments found in ploughsoil (Mattingly, 2000; Francovich and Patterson, 2000; Millett, 2000; van Dommelen, 2000; Terrenato, 2000, 2004; Attema *et al.*, 2002). Field surveys are research-specific, and to a large extent, the methodology employed is determined by the research goals, landscape and surface visibility. Funding and time are also major considerations. Recently, the ability to compare the data of field surveys has become important to scholars, particularly those working on demographic issues. Scholars debate whether valid comparisons depend on the same methodological application or if a similar methodology is more important than the research formulation (Bintliff and Sbonias, 2000: 253; Yntema, 2002: 3).

Methodological approaches to field walking are numerous and often mean that the results are not easily comparable between areas where, for example, complete coverage has been attempted by walking 5 m apart and areas where transects are laid. In order to make a transect within the selected survey area, strips or rectangles are laid out that represent the typical geophysical profile (such as the vegetation and the topography) of the entire area. The survey results from those delineated areas are then projected to provide a total number of sites within the entire survey area. The use of transects increases the speed of field survey but not necessarily the accuracy of the results. Intensive surveys (and the word "intensive" is directly proportional to the amount of manpower expended) carried out in a grid are particularly useful around excavated sites, or where large surface scatters of pottery have been identified. A combined pattern of field walking and intensive survey of identified sites is usually used in the wider territory. The necessity to survey areas repeatedly in different seasons when the degree of agricultural activity may have brought more material to the surface and over a number of years is a critical factor in increasing reliability, yet it also increases cost and time.

In order to organize what can be a vast amount of variable material, surveys are often carried out in a grid pattern, usually 25 by 25 m, but smaller units are also used. The smaller units, theoretically, allow a more accurate delineation of functional areas. The use of a grid is compromised in long-cultivated areas by sherd displacement, by plowing in particular, but also modern earth-moving and erosion. Factors in the interpretation of the survey include the number of times that a site has been visited; the visibility of the earth during each of those visits; the terminology used to define the sites; the classification of sites identified on field survey; and the basis on which those classifications are made. Consequently, the designations into site types have been regularly contested, and the judgment of the survey team and the narrative by which results are communicated become critical (Terrenato, 2007; Rathbone, 2008). The site designation is the most critical point of interpretation and creating the survey narrative. If a fourth-century site is underneath a Roman villa, the modern destruction necessary for material from the earlier site to appear on the surface is considerable. For example, the floor levels of the Roman villa would have to have been destroyed by modern plowing before sufficient ceramic material would appear on the surface to allow a field survey to recognize the existence of the earlier site and to determine the character of that site.

Sites are usually categorized according to the size (less than X, more than Y) and composition of the scatter. Isolated sherds, or less than three sherds, are sometimes considered simply evidence of field fertilization by manuring. A smaller scatter with little fine-ware pottery is usually designated as a villa rustica or farm although this may not necessarily be the case (Osborne, 1985, 1988; Rathbone, 1981, 2008: 321–6; Lohmann, 1992). The lack of archaeological visibility of the lower social groups is another problem in the assessment of field survey results. The conflation of archaeological descriptions (or size) with a historical interpretation (identifying a villa or farm) is influenced by the historiographical tradition which started with the Romans themselves (see Chapter 20; Rathbone, 2008: 306–7, 310–21).

Larger scatter sizes are usually identified as a richer or high-status establishment such as a villa or, if there are no luxury goods found, an agglomerated settlement or vicus, although the word vicus implies a juridical administrative qualification which many agglomerated settlements did not have. The designation as a villa/high-status site is determined by the presence of Roman cultural icons such as tesserae for mosaics, glass, or red-gloss pottery; the reasoning is decidedly circular and certainly fluid. By definition, all high-status sites are Roman villas. Leaving aside the question of production models, the definition of high-status sites means that there were no high-status sites before the introduction of the villa: this issue could certainly be debated. Etruscan aristocrats had more land than they could work: how is their property identified and defined by field survey? The same question has been posed for other areas. For example, in second-century Messapia, the sites around Oria identified by survey became fewer but the remaining sites grew much larger. At the same time, however, funerary evidence disappears, which probably indicates a substantial change in either status or custom. In Messapia (and parts of Daunia), the local elites lasted longer than elsewhere because they were supported by the Romans, yet we cannot discern if those larger establishments could have belonged to the local elites – on their way to becoming Roman – living in high-status villas; this was happening in Lucania, as indicated by brick stamps (Gualtieri, 2008a: 208, n. 7; Di Giuseppe, 2008). No sites dated before the second half of the second century have ever been designated high-status, but are rather generically classified as farms, yet the indigenous tribes in Italy all had an identifiable elite and the wealth in their tombs is well-known. But they, like the lower classes, cannot be recognized by field survey unless a necropolis is the survey subject.

The way in which the data have been chosen for collection and recorded is another critical element in creating the survey narrative. Was the recovery complete or were only diagnostic sherds (rims, bases, handles, decorated wall sherds) retained? The difference is critical, especially in light of ceramic chronologies and recent studies emphasizing more socio-economic applications of survey material (see Chapter 5; Roth, 2007b; Lippolis, 2008). An increased knowledge of the typological evolution and more precise chronology of imported fine wares have made chronological assignments more accurate since the 1950s and 1960s (Lamboglia, 1950, 1952; Morel, 1981a; Oxé, Comfort and Kenrick, 2000; Yntema, 2005). Numerous studies have concentrated on local ceramic production and chronology, with the result that local imitations, common wares and cooking wares can now be dated and are used to supplement and refine survey conclusions (Bats, 1996; Principal, 2006: 48–52; Roth, 2007b).

The question of the black-gloss pottery chronology (the marker artifact of the late fourth through the mid-first century) and its usefulness as an indication of Roman expansion is complex (Di Giuseppe, 2005, 2011; Roth, 2006: 139, 2007). The ware was originally imported from Greece, then imitated at many small and large centers in southern Italy, and eventually replaced by the Campana A productions which are a Roman innovation (see Chapter 5). Some forms have been dated by their associated excavated contexts. In typological compilations, the date is determined on the basis of what is said in a museum display without further documentation or on the basis of a recognizable proportion between the component parts of the pot. Whole pots are a rarity on field survey (Morel, 1981a: 30–2, 403). The biases in ceramic chronology, originally based on the ancient sources and Toynbee, posit a decline in the southern Italian indigenous centers to the end of the fourth or beginning of the third century (Yntema, 2006: 129, n. 49; Roth, 2007b; de Cazanove, 2008: 275–90; Lippolis, 2008). The number of indigenous centers that continued well into the second half of the third century establish that the original dating of decline was incorrect (de Cazanove, 2008: 275–90; Fracchia and Gualtieri, 2009). The existence of numerous local imitative productions of black-gloss, coarse ware and common ware pottery further complicate the issue (Patterson, Di Giuseppe and Witcher, 2004; Di Giuseppe, 2005; Patterson and Di Giuseppe, 2009). On small farm sites, local products and a reduced number of forms of pottery are most likely to constitute the bulk of the material retrieved. Rather than indicating a decline in population or economic level, the use of locally made pottery can indicate different sources for household pottery in different societal levels as well as different functional uses for individual sites (Roth, 2007b: 5–6, 60, 93–4, 169). Local productions are notorious for their lack of standardization, for the use of a few forms reproduced in all types of wares, and for a mixed and highly eclectic typology (Yntema, 2005; Roth, 2007b: 93–4). The conservatism of forms and thus an accurate chronological range are difficult to determine unless survey pottery can be compared to the pottery found in a stratified context or within a datable sequence. Conservatism, often a catchword for a reduced number of forms, can also be attributed to a decrease in regional cultural diversity, and a concomitant interest in expressing a cultural pertinence or cultural resistance.

With increasing frequency, calibrated dates are applied to the "life span" of pottery found on survey in order to restrict the variability and increase the comparability of the surface material by determining a median date for the individual ceramic forms. The distribution of the ceramic form through time, and thus the times of greater and lesser density, is based on the number of sherds and the "use life" of the forms. The life span of a given form is compared to the incidence of its frequency to provide a median date for the form itself. The use of a weighted average in order to calibrate the dates is, however, highly dependent on the established chronology of pottery types. The Morel typology of black-gloss pottery provides 2,174 datable forms for the fifth through first centuries. The chronological breakdown of the Morel typology is shown in Table 12.1.

But the use of calibrated dates is not without some problems. For example, if 80% of the black-gloss pottery found on a site was second-century forms, but only 5% of the diagnostic sherds were datable (meaning that 75% were body sherds) and the remaining 20% of the pottery was datable to the third century but 95% were diagnostic sherds, the

Table 12.1 Chronological breakdown of the Morel typology of black-gloss pottery.

Date	*No. of forms*	*Percentage*
Fourth century	383	17.6
Fourth/third century	367	16.8
Third/second century	174	8.0
Second century	306	14.0
Second/first century	157	7.2
First century	106	4.8
Totals	2,174	99.7

standard weighting application or calibrated data would skew the results heavily towards the third century. Nonetheless, calibrating survey results or trends has become extremely popular, particularly by scholars interested in estimating the population of ancient Italy and those trying to compare survey results that use differing methodologies, but it is necessary to be aware of the limitations of this method (Ikeguchi, 2006; Witcher, 2008; Di Giuseppe, 2011; Launaro, 2011).

The methodological structure of the field survey on the ground can vary enormously and is one of the critical elements in comparing survey results; the bibliography on the topic is vast (Cambi and Terrenato, 1994; Terrenato, 1998b, 2000, 2004; Ikeguchi, 1999/2000; Mattingly, 2000; Alcock and Cherry, 2004; Burgers and Recchia, 2009: 11–32; Launaro, 2011: ch. 3, esp. 3.3–3.5). Comparative survey results are notoriously difficult to evaluate, given the differences in methodology, terminology, recording and assessment of finds, with the result that in many cases, each individual survey often remains isolated within its own context. A simple reportage of site numbers is not very useful and does not advance a historical understanding, but site numbers have to be contextualized into the research framework that produced the numbers. Otherwise, the site numbers remain merely raw data. Calibrating survey data is one attempt to create a level playing field in terms of methodological differences. But how critical are these differences? Certainly comparing raw data, recorded differently, is difficult enough, but what are really being compared, in most cases, are the research narratives where biases can easily slip into interpretations. Data calibration may provide some refinements and a similar language, but the validity of data calibration depends on the validity of the ceramic dates and the way in which those dates have been determined.

Modeling constitutes a very important use of the information provided by field survey but modeling must be used with caution (Roth, 2007b: 35; Goodchild and Witcher, 2009). Site visibility, and hence site density, varies from year to year; in many cases, surveyors have estimated a site recovery at less than 50% of the actual number of ancient sites in a given territory. The number of variables for each survey area presents new configurations of the basic model, as is evident in the various interpretations provided for the areas near Rome. Nonetheless, various manipulations of partial evidence can illustrate how the dots on the map can become a working hypothesis that addresses settlement development, with a highly variable degree of accuracy (Witcher, 2008: 288–92; Goodchild and Witcher, 2009). The problem of site recovery is not insignificant and increases as any rural area is developed, either agriculturally or commercially.

Certain applications, such as Geographic Information Systems and Digital Elevation Models, can provide greater mapping precision as well as provide important environmental aspects for field survey. These applications are important in providing a common basis for discussion (Patterson, 2004; Coarelli and Patterson, 2009). Economic and geographic models have also been used to make sense out of a fragmentary picture. The Van Thunen model of geographic economy and Thiessen polygons for territorial spatial divisions are most frequently applied and can be useful in understanding the relative connections of sites to one another, for example, to market centers or sanctuaries. The Van Thunen model of geographic economy is based on the idea that a city is centrally located within a given area and has four rings or areas of activity surrounding that central city; each ring has a different type of productive activity, depending on such factors as distance to markets or cost of transport. Thiessen polygons are a formal method for exploring settlement patterns based on notional polygons constructed around a series of distributed points to form a lattice. The polygons provide a general approximation of the extent, shape, and orientation of the spheres of influence or territory around recorded settlements or other important sectors of a settlement. The use of models is risky although they can certainly help to understand various problems and to propose interpretive hypotheses, as long as the researcher takes into consideration all the regional political, geographical and social differences of the settlements. All of these approaches are an attempt to place a more scientific and therefore controlled result on what is partial evidence, in terms of material recovered and in the number of sites recovered.

Field survey reports have also come under criticism for providing very little real information about settlement development or about anything else connected with human activity. It is a fair criticism and recent appeals for a more humanistic approach to field survey have resulted in a number of thematic articles using the data from field surveys for such things as the placement of institutional, artisanal or market buildings, the significance of local ceramic productions, or for the placement of religious sanctuaries in a given area, another critical element of settlement development in a more humanistic and socio-economic optic (Witcher, 2006b: 47–64; Coarelli and Patterson, 2009; Rous, 2009; Stek, 2009; Stek and Burgers, in press).

Diachronic field survey results are more credible and can be used more substantially to address historical questions if the survey is part of an integrated archaeological methodology, using excavation, survey and geophysical exploration to ensure an accurate architectural and ceramic chronology for individual areas. Excavation examines intensively a small area while field survey can provide a wider contextualization for excavated remains. An excavation also provides a touchstone by which to evaluate survey material, which is always a partial picture of what is actually underground. Diachronic field survey is an excellent complementary tool in a structured research program focusing on settlement development and land use and it can provide the wider evolutionary settlement context for a given area and for an excavation. To a certain extent, the need to verify results by excavation or by geophysical exploration mitigates the low cost of field survey, but other methods, such as shovel testing, replicate the superficiality (in several ways) of the technique when used alone. Geophysical testing can provide an insight into the extent and overlay of architectural remains under the surface but the efficacy of geophysical testing can be compromised by soil type and soil conditions such as the amount of humidity in the soil. It is preferable to do at least two types of geophysical testing (magnetometer and electric resistivity, or electric resistivity and ground penetrating radar, for example) to create an overlay of results (Goodman, Nishimura and Rogers, 1995; Malagodi *et al.*, 1996; Conyers and Goodman, 1997).

4 Settlement Development and Land Use: The Second Century

With that background in mind, we can now look at a highly selective group of case studies from central and southern Italy to determine the extent of the second-century crisis. In some cases, at least one site has been excavated in the survey area and provides the parameters for chronological development. The results that have been obtained in terms of addressing settlement development and agrarian history are by no means uniform. A high level of regionalism is evident in settlement development and land use, but the word "regional" itself is ill-defined and does not necessarily contradict a wider and/or different contextualization, as rural areas are notoriously conservative. Areas closer to an intensive Roman urbanization may provide different results, for example in areas where colonies were founded (Laurence, 2001; Terrenato, 2001b: 2–3; Yntema, 2002: 3–4).

In the first instance, the landscape and the geography, the local geomorphology, the diachronic communications network, the settlement development that pre-dated the Roman hegemony, the role of a local aristocracy, colonial foundations and land grants, and the Roman urban revolution in general all play an enormous part in determining the highly differentiated settlement development and land use in Roman Italy (van Dommelen and Terrenato, 2007). Land use is closely tied to settlement development and, in particular, to site size in terms of production mode and actual production. A few large sites are thought to indicate intensive and extensive grain production, that is, the villa system and slave mode of production. Numerous small sites are thought to indicate a more restricted production or a Mediterranean polyculture of olives, vines and grain. The vici, and their relationship to other sites, and the social level of the inhabitants add another dimension to the problem. In turn these factors are tied, ultimately, to the demographic question (see Chapter 10; Terranato, 2007). The sampler of case studies below may help to outline the problems and define some of the questions that emerge from field surveys.

4.1 Central Italy

4.1.1 The Tiber River Valley

The South Etruria Survey Project and the Tiber River Valley Survey which has grown out of it have had a major impact on the development of and debate about survey, settlement and land use and the correlation of survey results to the comments found in the ancient sources (Potter, 1979; Patterson, ed., 2004; Patterson, Di Giuseppe and Witcher, 2004; Patterson and Di Giuseppe, 2009, where a history of the projects can be found) . The research program includes a number of individual survey projects throughout the entire area of southern Etruria, some of which are still in progress. The surface material has been collected at different times, with different criteria of collection and of recording the finds. Recently, calibrated dates have been used to compare the individual survey results.

The original goal of the South Etruria Survey was to delineate Rome's impact on the nearby territory. Some scholars used the original South Etruria Survey results to address the events of the Hannibalic War and the developments of the second century (Frederiksen, 1970/1). Recent research has focused on the restudy of the material of the earlier survey, and the integration of all the survey projects to create a more thematic approach to landscape, and on the discrepancies between the literary and archaeological evidence. The early survey results suggested that the events of this period had had little effect on existing settlement patterns in the area. Given the building activity in Rome itself in the late third to second century, the conclusion that the countryside around the capital city was well utilized and populated was not surprising (see Chapter 28; Welch, 2006a).

In order to (re)address the chronological framework and verify the initial results which contradicted the written sources, the black-gloss pottery found by the various survey projects that constitute the larger South Etruria Survey was assessed using the standard weighted or calibrated measure application. The results of this statistical approach indicate that there is a decline in the amount of black-gloss pottery sherds found in the second half of the third to the first half of the second century. Corresponding to what ancient authors said about the economic situation, a concomitant decline in rural settlement and the social status of the population seems to have occurred.

It has been suggested that the apparent decline in rural settlement could be the result of the fewer diagnostic ceramic types in circulation in this period. Concomitantly, it seems that the number of production centers for pottery in the middle Tiber River Valley diminished during the same time, indicating lessened market demand and confirming a reduced population (Patterson, Di Giuseppe and Witcher, 2004; Patterson and Di Giuseppe, 2009). On the other hand, the disappearance of local ceramic production centers should also be tied to the appearance of the Campana A products in the last decades of the third century. That the production centers decline is not surprising, as Campana A production is typically commercial production: this ware was mass-produced, standardized, simplified, anonymous and sold on a wide scale. Thus, a decline in small artisan centers may not solely indicate a lessened market demand and lessened population but may also indicate a progressively Roman landscape with a clearly Roman commercial profile. The conclusions are not uniform even within the defined study area. At Capena, a local, less refined, production of black-gloss pottery existed in the second century alongside the imported products. The local pottery was used for bowls and plates that were not produced in the finer ware; this production effectively questions substantially the notion of stagnation and decline proposed by the field survey (Roth, 2007b: 152–74).

New conclusions indicate that the small farms do not disappear entirely but the settlement emphasis or advantage is represented by villas (see Chapter 20). In the middle Tiber Valley the population in general grew discontinuously from the third century through the Augustan age. A discontinuity can be delineated, in particular, from *c.* 250 to 150. The type of mixed settlement now recognized in the middle Tiber Valley indicates a certain residual agricultural model in which farms and villas share the landscape, although the reduced population between 250 and 150 left the countryside open for the introduction of large landholdings associated with Roman villas.

4.1.2 *Northern and Central Etruria*

The conclusions of the South Etruria Survey are not replicated in northern Etruria, where numerous small farmsteads dotted the landscape from the fourth century onwards (Motta, Camin and Terrenato, 1993: 109–16; Terrenato, 1998b, 2007; Carandini and Cambi, 2002; Roth, 2006, 2007b). At Volterra, a recent study of the black-gloss pottery from the excavations has indicated two healthy and vivacious local ceramic traditions and production in the second century. One tradition was closely linked to the older fine ware production but a less refined fabric was used to produce vessels that were inspired by local plain and coarse wares. The existence of both types of ceramic production indicates a certain affluence and industry in central Italy during the second century. The number of small farms excavated or examined by geophysical testing in the area underlines a certain continuity and similarity of size and social class across eight centuries. The research project has made a significant contribution to our understanding of small land holdings because several of those small holdings have been excavated. The excavations have allowed the survey results to take on more significance. The composition and the extension of the scatters are now known; the interpretation of survey material becomes more certain, and in that certainty, the possible interpretations become more substantial. These results have caused some scholars to postulate an increase in population in the second century.

In parts of inland Etruria, such as the Val di Chiana, virtually no change in the distribution of sites is noted until the last decades of the first century, but several sites undergo a change in function earlier in the century when veterans of Sulla began to colonize the area. Those fourth- and early third-century farm sites or cemeteries remain such until some of the original sites are replaced with Roman villae rusticae which continue to be inhabited through the Late Antique period. The road systems remained the same and here, as elsewhere, the role of the local elite Etruscan or indigenous aristocracy is programmatic in the settlement pattern (Burgers, 2001; Gualtieri, 2008a, 2008b, 2009a, 2009b; Bandelli, 2009; Grelle, 2009). Similarly, agglomerated rural settlements were established in the fifth century along the major road system, and those same settlements continued, gradually taking on Roman cultural aspects until the early second century, when they died out, concomitant to a functional transformation in some of the nearby villa sites (Gualtieri, Fracchia and Ferrari, 2010: 131–9). The configuration in the Val di Chiana, a well-known grain producing area in antiquity, illustrates the importance of agglomerated settlements in the early transformation of the landscape through the late first to early second century CE.

4.2 *Southern Italy*

4.2.1 *Lucania*

Mediterranean polyculture is evident in this area from the locally produced amphoras, large storage vessels (dolia) and the number and types of grain mills found in the valley. The presence of wool washing basins dated to the Late Republican period suggests that transhumance was also an important part of the local economy. Studies of the region have divided the area into Grumentum and the upper Agri Valley, and the eastern and western coasts (Gualtieri, 1993; Roberto and Small, 1994; Fracchia, 2001a, 2001b; Gualtieri, 2003, 2008a, 2008b, 2009b; de Cazanove, 2005, 2008; Marchi, 2008; Fracchia and Gualtieri, 2011; Di Giuseppe, 2011; Di Lieto, 2011; La Torre, 2011).

In western Lucania, an extensively excavated fortified center at Roccagloriosa continued to exist through the third century, with a high level of urbanization already apparent in the fourth century. The countryside around the fortified settlement, densely inhabited through the first half of the third century, gradually became more rural. In the later third and second centuries, larger centers of population developed at the expense of the dispersed capillary settlement of the countryside. In the Mingardo-Bussento watershed, two farm sites have also been excavated and two sites have undergone geophysical testing. In this case, the actual impact of a Roman colony on the area can be delineated. The foundation at Buxentum in 194 had very little impact on the area because the pagus-vicus type of settlement was previously already well established (Capogrossi Colognesi, 2002: 190; Stek, 2009: 140–5). The introduction of Italian terra sigillata pottery in *c.* 50 at the rural agglomerated settlements and small sites found in the countryside establishes evidence for the renewed presence of farms which were probably still engaged in producing grapes, olives and wheat. When the sherds are associated with tesserae or glass, in combination with the extension of the scatter, rather more imposing structures such as villae rusticae have been postulated. At a certain point, the destiny of the countryside and that of the re-founded colony of Buxentum, the administrative center for the region, merged, but the foundation for the Roman landscape had its roots in structural changes within the indigenous world.

In eastern Lucania in the upper Bradano Valley, the evolution in the second and early first centuries is quite different. One villa site and one farm site have been excavated (Fracchia and Gualtieri, 1999, 2011; Gualtieri, 2008b). The fortified hilltop settlement at Oppido Lucano was abandoned by the mid-third century, but many of the Lucanian sites in the countryside, distinguished by the various black-gloss and *pasta grigia* wares, continued to be inhabited from the late third through the first centuries (Figure 12.2) (Gualtieri, 2008b; Fracchia and Gualtieri, 2011). The earliest villa-type structures in the upper Bradano Valley appear in the second half of the first century, usually overlaying earlier Lucanian farm sites, thereby raising the question of continued local ownership. In this case, the number of individual farms with earlier phases, and the relative scarcity of hamlets or villages, indicates that the Italian peasantry in this area were still on their land until the later first century CE when rural agglomerations and villas become the protagonists of the Roman landscape.

The upper Agri Valley, the theater of battles in 215 and 207 during the Second Punic War, makes the area critical for our understanding of the Hannibalic devastation and subsequent changes in settlement and land use (Munzi, Cifani and Fusco, 1999/2000; Gualtieri, 2009b). Originally an urbanized Lucanian settlement with a densely inhabited countryside in the late fourth to early third century, Grumentum was controlled in the late third or early second century by a praefectus iure dicundo: this fact has conditioned the Romano-centric optic through which the countryside is seen, although the actual foundation date of the colony, usually postulated for the third century, remains uncertain.

Figure 12.2 Sites with grey glaze pottery in the upper Bradano Valley: Large pin indicates more than 30 sherds; medium pin indicates 29–15 sherds; and small pin indicates less than 15 sherds. Source: Tanya Henderson, used by permission. Base map © 2012 Google.

The countryside was dotted with small farms through the third and second centuries, although some farms were temporarily abandoned. Field survey cannot tell us much about these changes. Only the chronological gap in the ceramic record can be identified. Nonetheless, in combination with the excavated sites in the upper Agri Valley, the extent to which the changes occurred can be contextualized more accurately. The third-century architectural and functional changes documented by the excavated farms indicate the emergence of an intermediate social group among the indigenous population. By the Late Republic, the presence of several senatorial families indicates a concentration of property in the hands of wealthy landowners, demonstrating an increased productive capacity of, most likely, grain. The survey evidence established an abandonment of some farms; the excavations have established both architectural and functional changes on farms in the later third century in those sites that did continue through the second century: certain continuity in ownership might be a warranted assumption. In this case, field survey has put the architectural developments revealed by excavation into the wider settlement context with the concomitant result that a scenario of gradual settlement development and changes in land use is more justified than it was several decades ago.

The settlement development in ancient Lucania was not homogenous, although the development probably does not matter in terms of Roman expansion. The indigenous western coastal areas of Lucania had a precocious political, administrative and social development which is not reflected in eastern Lucania, where there are pockets of change registered in the second century at places like San Giovanni di Ruoti and Satrianum (Roberto, Plambeck and Small, 1985; Roberto and Small, 1994; Osanna and Serio, 2009; Di Lieto, 2011). At Satrianum, the decline in the number of settlements and the appearance of Roman villas in the landscape is tied to the continuation of the well-known sanctuary at the site. In other areas, a kind of settlement continuity seems the norm, although the character of settlements may change. Yet the question remains, do the areas of Lucania that register a break in habitation at the end of the third to the first half of the second century represent ager publicus? And in particular, do they represent vast tracts of ager publicus? The answer depends on the topography and geomorphology of the individual sites as well as an accurate ceramic chronology based on an excavated site or sites (Di Giuseppe, 2011). Field survey results must be contextualized but a regional contextualization does not necessarily negate a different and more varied broader geographic contextualization.

4.2.2 Apulia

Apulia presents an interesting study for settlement and land use because ager publicus was instituted only in the second century and because parts of Apulia were precociously colonized (Sabbatini, 2001; Marchi, 2008; Grelle, 2009; Goffredo, 2010). Around the Latin colony of Venosa, more than 90% of the sites that existed in the pre-Roman phase disappeared when the territory fell under the influence of the new administrative center, which was constituted by a number of local elite incorporated into the original colonial cohort. In the fourth century, the majority of those sites were between 400 and 1,500 m^2. In correspondence to the colonial foundations farms became and remained much smaller (between 150 m^2 and 400 m^2) through the Middle Republic, when a number of farms became larger (400–800 m^2): these holdings may have belonged to a higher economic class. Under the triumvirate, land was confiscated and distributed to veterans. More and more farms of 400–800 m^2 can be attributed to this period but extensive slave-owning estates were not a characteristic of the area. Nor do the surface scatters of large extensions seem to be definable as villas. In this case it is clear that, in the first instance, the strong colonial foundations (whose numbers are still debated) destabilized the countryside and it was only at the end of the Republic that the farms disappeared, leaving the landscape open for the arrival of larger establishments. This development probably also signals the final break with the old local elite families and a new type of land use (Marchi and Sabbattini, 1996; Marchi, 1999, 2004; Sabbatini, 2001).

In the Messapian territory of the Salento Isthmus, two situations are evident. In the hinterland of Brindisi, a dense late fourth-century occupation in the countryside underwent some changes in the second century. At Oria, the number of sites found diminished, but the size of the individual sites increased. This is in complete contrast to the area of Valesio, which is closer to Brindisi and more coastal. There, the number of sites diminishes and no growth of existing sites has been documented. At Oria, then, a more specialized agricultural production has been postulated and, by analogy, the opposite would be the case for Valesio. The presence of the Via Appia seems to be programmatic in the rural expansion in one area and the population reduction in the other (see Chapter 19). One factor that distinguishes the Messapian (and Daunian) territory from elsewhere is the continued presence of aristocratic tombs in the second century near Oria, proof of the longevity of the indigenous presence (Burgers, 1998, 2001; Yntema, 1993a, 1993b, 2006, 2008; Gallone, 2007; Burgers and Recchia, 2009).

In general, the prevailing consensus is that the evolving settlement pattern in Apulia from the early third century onwards reflects a Roman adaptation to the pressures from the local elite landowners who managed to recuperate the land that they may have lost in the Second Punic War. As a consequence, the local pre-existing agricultural traditions of horse breeding and Mediterranean polyculture continued under a familial or clan ownership (Grelle, 2009; Goffredo, 2010). Pockets of a second-century decline have been noted in the border areas between Lucania and Apulia, at Botromagno, Gravina and Monte Irsi, and some areas of the lower Ofanto Valley (Small, 1992, 2001; Small *et al.*, 1998; Small and Small, 2002). On this evidence a crisis was perceived throughout the survey area and was considered to be a consequence either directly or indirectly of the Hannibalic War.

5 Conclusions

The mosaic that emerges from this highly selective sampler of field surveys in central and southern Italy is without any set pattern. Without wanting to over-emphasize regionalism, it is clear that different areas, including contiguous watersheds, may have had different developments, particularly in times of crises, which are never symmetrical. What then are the reasons? To a large extent, the force of the local aristocracy and the more or less direct rapport with Rome seems to have a considerable influence on both settlement development and land use. Apparently, areas closer to major road systems had a greater incidence of survival and those are the sites – villas or villages – that would flourish through time, in the service of nearby markets. A colonial settlement also carries with it potential changes in a countryside, although it depends in some areas on the structure of the pre-existing countryside, as in the presence of agglomerated settlements, and on the type of colony founded. Agglomerated settlements that existed in the Early and Middle Republic may or may not become vicus-type settlements which, in any case, have to be seen in relation to a villa/high-status site – either in terms of actual ownership or in terms of a labor source, even though the agglomerated settlement might well pre-date the Roman period.

The fertility of the soil, the hydrological resources, the proximity to an urban center and the presence of principal routes all are major factors in settlement development and land use, for the location of large-scale productive operations. But it is also evident that the appearance of villas or large productive establishments did not mean the end of traditional agricultural practices nor of small farms in large areas of central and southern Italy. Has field survey contributed to this debate? Yes it has, but the validity of the field survey narrative is increased exponentially when a site or sites have been excavated within the study area. The material from field survey has to be analyzed independently before it can be contextualized responsibly. Certainly, on the one hand, excavation and geophysical testing reinforce the independent conclusions of field surveys, and, on the other hand, those conclusions can then be more accurately contextualized within the historical framework. Sites found on survey did not exist by themselves or in a vacuum. The sites existed in relation to historical developments in the nearest city, in the wider territory, across the Italian Peninsula, but in the first instance the results need to be analyzed independently (and verified) before they can be applied to wide-scale historical developments delineated by the ancient sources or by modern scholars.

Field survey has made a major contribution to our knowledge in individual areas about settlement development. But settlement development should not be confused with land use, especially on the basis of field survey, and in the absence of an excavation. One problem in the interpretation of survey results is the application of historical chronological divisions using potsherds in a long-term and (at best) uneven process. In some cases, the optic of the research question has created a bias in the interpretation of the material retrieved. Field survey can honestly provide very little information about land use unless a site found on survey has been excavated and the paleobotanical record recovered. In the absence of a controlled excavated site, all the correlated hypotheses about size of land holdings remain just assumptions.

Field survey results can be compared to a kaleidoscope in which turning the cylinder creates different patterns from the same pieces: the pottery and site types are all common elements, but interpreting the changing patterns remains difficult. Understanding field survey results, settlement and land use would be much easier if the picture were uniform all across Italy but that is not the case. It would also be helpful to have surveys conducted and the material recorded in ways that could be used comparatively despite different original research formulations. Nonetheless, the major error lies in trying to establish a common historical development for a number of regions with a different geomorphology and different pre-Roman ethnic identities, as well as a variety of political, social and cultural situations. The strength of the pre-existing conditions and the tools of expansion employed are determining factors in the process of creating tota Italia.

Some areas will reflect changes and some continuity; some areas will reflect both change and continuity contemporaneously, depending on the regional history and local social, political, cultural, topographical and geomorphological conditions. A level playing field certainly creates a more comprehensible and general "story," but a simple story was not the reality of ancient Italy and does not reflect the variety of situations that constituted the Italian Peninsula. This fact should not be as problematic or troublesome to modern scholars: the end result was, in time, a landscape that can be recognized as Roman.

FURTHER READING

A number of recent volumes present varied points of view regarding survey, settlement and land use in the Republican period. Articles in Carlsen and Lo Cascio (2009) address the issues of survey, settlement and land use from the viewpoint of the literary evidence as well as the archaeological evidence. Articles in de Ligt and Northwood (2008) also present various points of view on the usefulness and correct application of survey data. An excellent discussion of comparative survey results and the problems inherent in surveys is found in Launaro (2011). Colivicchi (2011) provides an overview of the survey activity and settlement development in southern Italy in the Republican period. Attema and Schoerner (in press) contains a number of fundamental contributions regarding survey, settlement and land use. The use of models and spatial analysis is discussed comprehensibly in Hodder and Orton (1981).

CHAPTER 13

Agriculture and the Environment of Republican Italy

Helen Goodchild

1 Introduction

The traditional picture of Republican agriculture is one characterized by upheaval and warfare. The Punic Wars and the Gracchan agrarian reforms are key events in the history of Italian agriculture, but these crisis periods are not the full story. Sophisticated farming systems were already well-established in Italy by the time of the Early Republic, but these have often been overshadowed by the alleged disappearance of small-hold farmers and the rise of great slave-staffed estates. A reliance on ancient texts in the classic studies of Roman farming (particularly White, 1970) prioritized high-status investment farming and led to generalized narratives of agricultural production. However, the proliferation of field survey in Italy over the last 60 years, spearheaded by the South Etruria Surveys in the 1950s (Potter, 1979), has contributed to the realization that the rural landscapes of Roman Italy were incredibly diverse (see Chapter 12).

Many early regional studies of ancient rural landscapes addressed fundamental problems such as population numbers, the size of exploited areas, or the configuration of rural settlement. It was soon realized, however, that greater issues, such as larger economic and social systems, household organization, or regional interaction, could be investigated, thereby requiring more detailed data collection and a more nuanced interpretation (Alcock, 2007a). To fully comprehend rural agricultural systems, it is necessary to assess their impact on the people (and society as a whole) that produce, process, transport, administer, consume and profit from them. This study focuses on the landscapes and agricultural systems in place in this highly transitional period, concentrating primarily on the central and southern Italian regions. Using the wide variety of historical and archaeological evidence now available, it will be shown that there was greater regional and temporal variation in production and consumption than has been inferred previously.

A Companion to the Archaeology of the Roman Republic, First Edition. Edited by Jane DeRose Evans.

2 Landscape and Environment

Before discussing Republican agricultural practice, it is necessary to frame the environment in which rural activities were taking place. Landscape variation leads to differences in agricultural exploitation, due not just to topography, but also soil quality and drainage capability. Where people settled, and what farmers planted where, was influenced by many socio-economic factors, although agricultural fertility is constrained by the nature of the land, and has been argued to act as a significant driver in settlement patterns and socio-economic networks (e.g. Isayev, 2007a: 7).

Landscapes are highly dynamic entities, and areas with high rural populations and large-scale agricultural practice often correspond with geomorphological evidence for significantly accelerated rates of erosion (Lewin, Macklin and Woodward, 1995; Barker and Grant, 1999: 575). As such, the modern environment cannot simply be extrapolated back into the past. Vegetation studies in Italy have enjoyed a recent surge of interest due to the realization of their important contribution to landscape history in this area. Pollen is an excellent indicator of generalized patterns of vegetational change; activities such as pastoral activity and the management of woodland rely on palynological evidence or macroremains for their detection (e.g. Mazier *et al.*, 2006). For other systems, particularly crops, it is preferable to use samples derived from excavated contexts.

Environmental parameters can determine the likely location of certain plants and trees, but prolonged human activity can also have a major impact, making reconstruction complex. For example, all plants have an altitude limit whereby the growing cycle becomes so long it ceases to be agriculturally worthwhile. The limit for wheat is 1,000–1,200 m above sea level (Spurr, 1986: 21), for olives in Italy it is 500–700 m (Mattingly, 1996a: 215), but the altitude and temperature tolerance of grapes varies significantly depending on the variety (Creasy and Creasy, 2009: 65–72). It was noted by Pliny that beech trees grew along the Tiber River in the fourth century, but were considered mountain trees by the first century CE (Plin. *HN* 16.15); they now grow at around 1,000 m above sea level in the Apennines (Reale and Dirmeyer, 2000: 167). The tree line is also known to have varied considerably over time, as demonstrated by environmental work carried out in the Alps (Tinner and Theurillat, 2003). The Mediterranean region is a hugely variable and fragmented landscape and any attempts at simple zonation for plants and trees based on altitude are complicated by diversity in the bedrock, tectonic regime and climate of a region. In reality, Mediterranean landscapes are a complex mosaic of woodland, scrub and cultivated areas (Foxhall, Jones and Forbes, 2007: 95–6).

One of the greatest influences on production in Italy is the amount of water available for farming, and scholars have discussed "dry farming" (i.e. methods aiming to conserve as much water as possible) as a strategy within ancient Mediterranean environments (e.g. White, 1970: 157, 173; Horden and Purcell, 2000: 201–2). However, in antiquity, the Mediterranean climate was generally wetter than present, with a gradual warming occurring from the fifth century onwards (Lamb, 1995; Reale and Dirmeyer, 2000). Such analysis deals only with long-term trends, rather than seasonal variation, but given that even the modern annual rainfall in central Italy (Naval Intelligence Division, 1945: 533–4; Spurr, 1986: 20) is sufficient to render irrigation unnecessary for standard arable crops (as supported by research in Latium and south Etruria by Thomas and Wilson, 1994), assuming the wetter climate postulated by Lamb (1995), dry farming techniques would have been unnecessary in many areas of Italy. This is further supported by Pliny's implication that irrigation for arable crops was unusual (*HN* 17.250; see also Spurr, 1986: 20). Tree crops and vines were resistant to drought, and required little water, whereas horticulture and pastures needed extensive irrigation (Thomas and Wilson, 1994: 158–63).

Republican farmers worked under a variety of environmental conditions. Certain areas of the peninsula would have been more appropriate for particular kinds of systems (e.g. agriculture or pastoralism) but variation will always have occurred. Even in the same region, an elite investment farmer producing for the market will have farmed differently than a smallholder producing for family subsistence. Furthermore, though Roman agronomists had knowledge of soil quality and improvement (e.g. Cato *Agr.* 34–7), we cannot assume that such knowledge was universal, or that farmers always made the most "rational" use of their land.

The Middle Republic saw a series of victories by Rome over the various cultural groupings of the Italian mainland. During this time the character of both rural and urban settlement changed. The landmark South Etruria Surveys from the 1950s (Potter, 1979), along with other regional studies (e.g. Dyson, 1981; Attolini *et al.*, 1991; Barker and Rasmussen, 1991), have demonstrated a move away from large nucleated centers (as was characteristic of the Etruscan city-states, for example) with a dramatic increase in rural sites in the seventh and sixth centuries. This was followed a century later by an increase in the number of small farms, with major growth in the fourth to third century. This was originally thought to have been a consequence of Roman conquest (e.g. Potter, 1979: 93–101), but it has since been observed that such transformation was a wider trend across Italy and even parts of Greece (Terrenato, 2007: 142).

The move to dispersed sites was not universal, however. Nucleated settlement was a safer option, for example, when areas such as Latium or Etruria were frontline territory in the Middle Republic, and a number of sites from the third century have been interpreted as "agricultural villages" (Pelgrom, 2008: 352–60) from which farmers commuted to their land. The prevalence of small sites and villages demonstrates the importance of the small farmer at a time when a surplus was required to feed not only the army, but also the growing urban centers and new colonies.

Two major factors in the reorganization of exploitation and settlement around cities were roads and cuniculi. Road building is extremely important in the development of agriculture as it opens up greater areas within urban hinterlands, but also allows small-scale producers an outlet for exchange. Roads also played a major part in the domination of Italy by Rome, and the subjugation of certain cities (famously Veii; see Laurence, 1999, and Chapter 19 for the role of roads in Roman hegemony). The building of cuniculi – extensive underground shafts connected by tunnels – was a huge infrastructural change in south Etruria and the Alban Hills in the Etruscan period. These performed a number of functions: primarily the drainage of swamp and wetlands or land prone to flooding to create agricultural land, and also controlling lake levels. Rainfall in central Italy is more abundant and consistent than elsewhere in the Mediterranean, but shortages still occurred. These structures collected water for storage, and could be diverted to cisterns or other hydrological structures (Judson and Kahane, 1963; Potter, 1979: 79).

Cuniculi were hugely valuable in creating and retaining agricultural land, and recent studies have shown their continuing impact (Bersani, Canalini and Dragoni, 2010).

The control of water was important, particularly as the population of Rome grew. After the Roman conquest, the scale and impact of hydraulic works increased exponentially (Terrenato, 2007: 153). The Early Republic saw the draining of Lake Baccano for use as agricultural land (Potter, 1979: 21). The building of the first aqueduct (the Aqua Appia) was initiated in 312–310, followed in 272 by the Anio Vetus (Torelli, 2010: 92; Chapter 18). By draining lakes and diverting water, marginal land was made available to agriculture and settlement. This at first might seem beneficial, but it must be remembered that marginal land is not useless, and the removal of such micro-environments might have been detrimental to smaller-scale farmers. Such land reclamation creates an environment suitable for large-scale arable farming to feed a growing urban population, rather than for smallholders living from a variety of resources and foraging environments.

In the Middle Republic, the program of Roman and Latin colonization dramatically reshaped the landscape (Morel, 2007: 497). The development of infrastructure facilitated exploitation, and the colonization process is supposed to have increased rural cultivation due to land distributions, theoretically introducing a more sophisticated form of agriculture and husbandry (e.g. Dyson, 1992: 25). A recent re-evaluation of centuriation (i.e. allotments in regular gridded plots) has highlighted some interesting problems. Unlike later settlements, which follow the traditional model as outlined in the *Libri Coloniarium*, there is no convincing evidence for centuriated settlement in any Latin colony before the Punic Wars (see Chapter 16; Pelgrom, 2008: 365). Furthermore, land division is not simply the preserve of Roman colonization. Systematic field boundaries (*isomoiria* or "equal portions") have been identified at Metapontum in Magna Graecia, dating back to at least the early fifth century. Here also, land division might not have existed at the colony's foundation, instead appearing up to a century after the urban center was established (Carter, 2006: 6, 94–6). Meanwhile, possible pre-Roman grids based on an Oscan system have been identified in Campania (Castagnoli, 1953/5: 3; Chouquer *et al.*, 1987: 169–70; La Regina, 1999: 9), along with a potential Greek-influenced grid in Apulia (Pelgrom, 2008: 364–5; Volpe, 1990: 209–13).

A fragmentary cadastre excavated in Verona dating to the later first century shows three plots of varying size in a centuriated landscape (for sites mentioned in the text, see Figure 13.1). Rather than showing the allotment of surveyed land, the irregularity of these units was interpreted as the quantification of pre-existing properties within a centuriated framework (Cavalieri Manasse, 2000: 26). This supports an alternative theory that argues, based on the location of the grid in low-lying areas and function as drainage or irrigation systems, centuriation could simply represent land reclamation schemes rather than estate divisions (Franciosi, 1997). Excavation of boundary lines of the Metapontine system show that divisions served as delimiters, but were also a combination of major and minor roads and drainage channels (Carter, 2006: 104–12). Regardless of its format and function, centuriation still had a major impact on the landscape, despite probably being a later phenomenon. How rural sites were organized within them, however, cannot be simply extrapolated from their spatial organization.

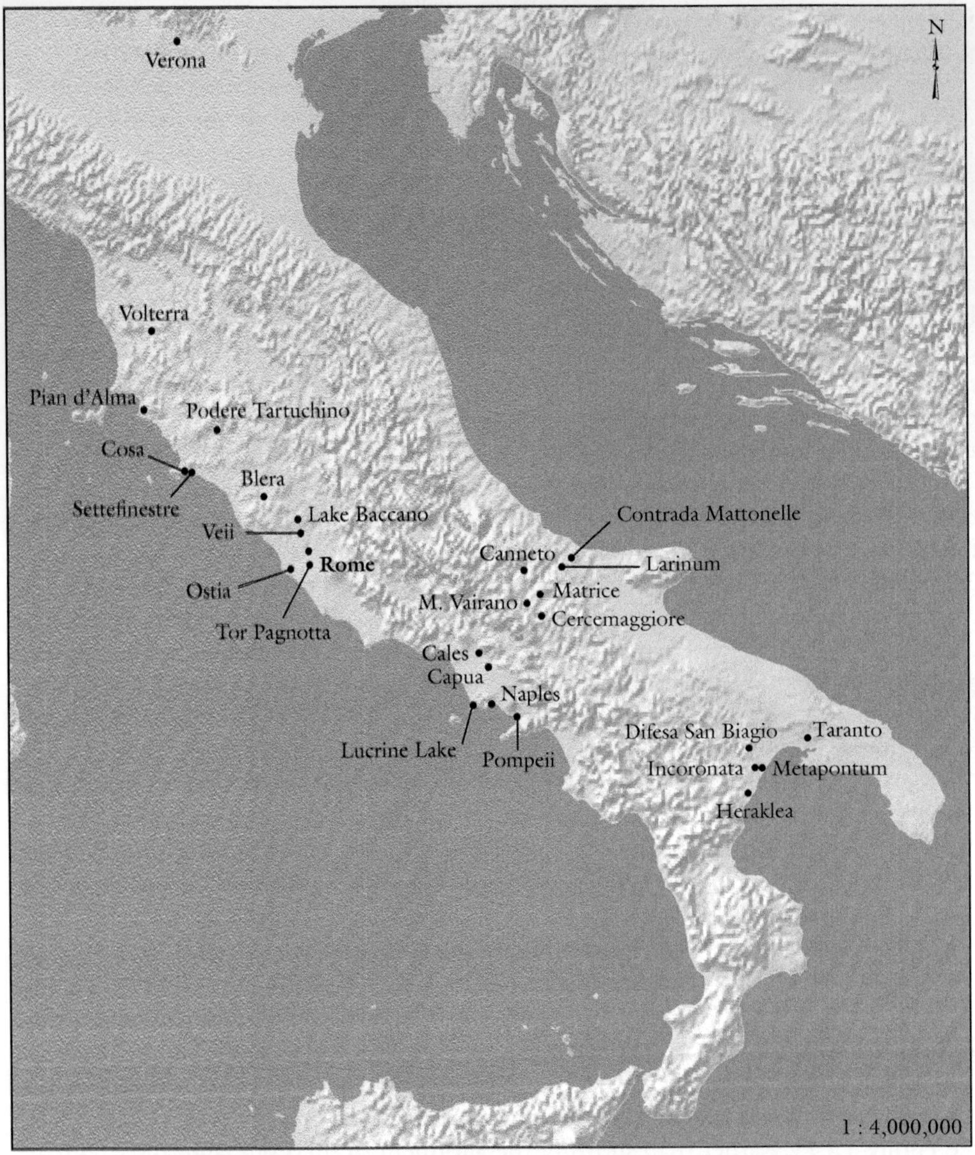

Figure 13.1 Map of Italy, showing sites mentioned in the text. Source: Drawing by H. Goodchild, with base data © Natural Earth. Free vector and raster map data at naturalearthdata.com.

3 Farming in Italy

In most studies of Italian agriculture, a differentiation is made between lowland arable agriculturalists and mountain-dwelling pastoralists. This stereotype has persisted within the literature, with a few exceptions where detailed study of environmental data from excavation has shown that, while regional specialisms were in evidence, most areas of Italy actually had a very diverse economy. Farming of cereals and legumes alongside the rearing of animals appeared in southern Italy in the sixth millennium (Barker, 1995b: 98). It has been argued, using faunal, botanical and artifactual data, that several components of the "traditional" Mediterranean landscape, consisting of pastoralism and seasonal transhumance, ox-drawn plowing, cultivated tree crops (possibly incorporating terracing), cereals and legumes, and forest management systems, were in place (though undeveloped) in various parts of the Mediterranean by the third millennium (Barker, 2004: 55–7). Combining textual sources with field survey and excavation data, however, can provide a clearer picture of how such systems spread across the peninsula throughout the Republic, and assist the interpretation of economic strategies at different types of rural site.

Later written sources state that emmer (*far*) was the earliest cultivated cereal in Rome: Pliny notes that in Rome's first 300 years, the people lived on emmer (often eaten as puls, or porridge) rather than other wheats or barley (*HN* 18.62). During expansion in the fifth century, conflict with their Latin and Etruscan neighbors combined with a series of famines meant their reliance on supplies from more distant areas of Etruria, Campania and Magna Graecia, making available a wider variety of grains such as breadwheats and barley (Livy II.34, IV.25, IV.52; Dion. Hal. *Ant. Rom.* 7.1.3, 12.1). This is alleged to have caused tension, as these new grains came from areas beyond Roman control, and were representative of "un-Roman indulgence" (Purcell, 2003: 334–5).

Analysis of plant remains from pre-urban deposits in Rome shows a different picture. Samples from the Forum and Palatine demonstrate a considerable increase in grain varieties, plus legumes, olives and grapes, up to two centuries before the events described in the sources (Costantini and Giorgi, 2001: 240–3; Motta, 2002: 74–6). This implies that the tension was a rhetorical device rather than representative of the true farming situation.

Botanical data provide evidence for a similar variety of cereals and legumes at Etruscan farms of the fifth to fourth century (Ricciardi, 1987; Perkins and Attolini, 1992; Mariotti Lippi *et al.*, 2003). In southern Italy, though, barley was the grain most identified with Greek colonists. This was due to its higher tolerance of drought, and the alkaline and slightly saline soils found in the south. Wheat was considered a luxury cereal in many Greek-colonized areas of the Ionian coast, due to the difficulties of cultivating it in such soils (Foxhall, Jones and Forbes, 2007: 102–3). Conversely, barley may not have been as popular in central Italy given its lower tolerance of damp soils (Costantini and Giorgi, 2001: 244). However, excavation data from the *chora* (territory) of Metapontum indicate a diverse economy, with intensification and diversification of agriculture from the sixth century onwards. This colony was famous for its barley production (as shown on their coinage), but wheat, legume and forage crops were also present (Carter, 2006: 13–24). By the late fourth century, however, there was a dramatic move to a virtual monoculture of wheat. This is a risky strategy, due to the vulnerability of a single crop to failure, and could feasibly be linked to the economic decline of the *chora* and *asty* (urban center) witnessed after 300. This may be related to events in the neighboring area of Heraklea, where epigraphic evidence from the late fourth-century "Heraklea tablets" indicated that the price of grain was a public concern, and outlined rental payments using barley, a more reliable crop (Carter, 2006: 26–7).

Similar diversity in crops is found in the "pastoral" region of Samnium, which was actually made up of a number of cultural groups (Chapter 25). Two of these located in the Biferno Valley – the Pentri, an upland group, and the Frentani in the lowlands – both demonstrated crop production alongside pastoral pursuits. In the upland region, though good arable land was scarce, excavation at the late Samnite farmhouse of Matrice has shown the production of cereals alongside sheep and pig pasturage from the third century. Breadwheat was found nearby at Monte Vairano, where there were also examples of agricultural tools such as billhooks, mattocks and spades. Meanwhile, Larinum in the coastal lowlands had access to the best agricultural land of the region, and grain was likely to have been the main crop of the farms in the valley (Lloyd, 1995: 204–5; see also Polyb. 3.101, 107).

Cereals and legumes were almost ubiquitous across Italy from the early Republic. Despite the regional specialization of certain products, cereals were very much a part of a mixed system, with legumes an important component, used for "green manure" (White, 1970: 135–6) as well as food and fodder (Foxhall, 1998). Even in southern Italy, when a virtual wheat monoculture was emerging in the late fourth century, the Greeks were not reliant on a single wheat species, and other crops were present, though a minority.

Vines and olives form a major part of the Mediterranean diet, are easily incorporated into a mixed farming system with other crops, and as such are common in archaeological assemblages. Grapevines were not a human introduction to Italy (contra, Plin. *HN* 14.1), but were likely to have been the result of natural movement due to changing environmental conditions (Kelly and Huntley, 1991: 219) and have been found in central Italy from the Bronze Age (Mariotti Lippi, Bellini and Mori Secci, 2010). Vines can tolerate a wide range of environments: Columella (*Rust.* 1.1) quotes the first-century agriculturalist Saserna when he expresses surprise at the variety of locations that vines grow, attributing their spread to climatic change (see also Plin. *HN* 14.4.2). It has been inferred from amphoras that Etruscan wine production operated on a large scale from the seventh century, and exported widely by the sixth century (Barker, 1988: 781–3; Brun, 2004: 172–3; Dietler, 2005: 43; Riva, 2010: 211; Chapter 6).

The olive's domestication in Italy is less clear. Pliny (*HN* 15.1) states that olives did not exist in Italy in the time of Tarquinius Priscus (late seventh to early sixth century), but that they were cultivated and traded by the mid-third century. It is known from pollen evidence that wild olives were present since at least the eighth to sixth century (e.g. Bonatti, 1963, 1970: 29–31; Kelly and Huntley, 1991: 254), and olive wood was used for fuel at the sixth-century Etruscan farm of Pian d'Alma (Mariotti Lippi *et al.*, 2003). It is difficult to prove cultivation from such evidence, but the disappearance of oil imports in Etruscan contexts and the manufacture of local containers has been used to argue that oil was produced in the late seventh century (Brun, 2004: 174), and amphoras found in a sixth-century shipwreck off the Tuscan coast contained numerous olive pits (Bound, 1991).

Olive pits and grape seeds appear frequently in excavated contexts, but evidence for production can be problematic. Processing facilities are only required for medium- to large-scale production, as smaller-scale farms were more likely to utilize portable (probably shared) equipment or rely on treading (Rossiter, 1981: 348). Nevertheless, examples of vine trenches and pressing equipment were found at two Etruscan farms (Ricciardi, 1987; Perkins and Attolini, 1992: 117–23), indicating medium-scale production at non-elite sites in the fifth to fourth centuries, while very large-scale vine trenches have also been found dating from the early third century (Tor Pagnotta; see Brun, 2004: 172–8). For olives, early evidence for processing comes from the Villa dell'Auditorium in the suburbs of Rome, where an olive press was installed as early as 500 (Terrenato, 2001a).

Campania has arguably produced the greatest amount of evidence for villas dedicated to wine and oil production, with elaborate processing equipment and storage facilities found at numerous sites (see Rossiter, 1981: 348–9). It is also now known, from study of Dressel I and Greco-Italic amphoras, that a flourishing wine trade was in existence in Campania prior to the Second Punic War (Tchernia, 1986: 85–7, 2009), and a close study of the ancient literature implies wine production between the fifth and third century in Campania and northern Latium (see Chapter 6; Brun, 2004: 176–7).

Contrary to the traditional picture of the mountain pastoralist, excavation in Samnium has shown a wide range of agricultural activities (Curti, Dench and Patterson, 1996: 180). Olive growing was restricted in certain areas, as many parts of the Samnite uplands lay above the maximum altitude for successful cultivation, though vines could be grown in most areas (Lloyd, 1995: 194–205). The large-scale lowland rural site of the fourth century onwards (interpreted either as a villa or village) at Contrada Mattonelle produced two oil and wine installations, along with buried dolia containing olive pits (Ceglia, 2008: 193–6). This site is of particular interest as, if the village interpretation is correct, it would seem to demonstrate the sharing of resources between a number of smallholders. Excavation at smaller rural sites in the region also indicate a mixed economy from the third to second centuries, with finds including millstones, plant remains, dolia, animal bones, loomweights and a possible treading floor. The Samnite farm at Cercemaggiore shows a similar enterprise, with several dolia and a mortar-lined basin, along with other containers (Lloyd, 1995: 194–6). Larger-scale production, meanwhile, is evident from the neighboring Trigno Valley at Canneto, where excavation of a late Samnite and Roman villa uncovered large wine and oil installations to rival those of Campania (Di Niro, 1982; Lloyd, 1995: 201).

Until recently it was thought that wine production in southern Italy was limited. The area was previously omitted from studies of the wine trade due to the lack of legible stamps on amphoras, but field survey, excavation and in-depth study of the forms has enabled the re-evaluation of their contribution to Italian viticulture (Vandermersch, 1994). Olive production in Magna Graecia is also problematic to assess as, although olive pits have appeared in botanical assemblages (Carter, 2006), there is currently little convincing evidence for oil production or processing in the Republic (Rossiter, 1981: 361 mentions oil processing sites at Salapia (Apulia) and Scalea (Calabria), but provides no dates). However, fragments of olive presses of the fourth to first centuries have been recently found in Apulia and the Metapontine Plain (Goffredo, 2010: 25; Mercuri *et al.*, 2010: 890–1), and the texts refer to intensive and specialized oil production in this area (Varro *Rust.* 1.8.2; Dion. Hal. *Ant. Rom.* 1.37.2).

The size of estates dedicated to wine or olives is extremely hard to establish. The northern Etrurian villa at Settefinestre, a classic example of a slave-staffed, market-oriented villa (Carandini, 1985a), has been estimated to have had vineyards of between 50 and 100 iugera (approximately 13–26 hectares), based on storage facilities and number of presses found during excavation. This compares well with Cato's model vineyard of 100 iugera (*Agr.* 11). This seems a relatively small area, particularly in comparison with the overall size of giant estates known from sources (e.g. Brunt, 1975). However, it has been noted that even small vineyards could produce capacities equivalent to modern French estates (Potter, 1987: 108). It has also been argued that the prominence of viticulture in Virgil's *Georgics* demonstrates that wine production was an activity common among small-town landowners, rather than senatorial villas or large-scale, slave-run latifundia (Purcell, 1985: 8).

There is limited information on the size of olive groves, but olives and vines are often integrated with other crops using intercultivation techniques (see for example the olive tree spacings suggested by Columella: *De Arboribus* 17.3; *Rust.* 5.9.7). This system (also called intercropping, polycropping or *coltura promiscua*) was possibly in use since the Etruscan period (Barker and Rasmussen, 1998: 184–5, 193). It not only takes advantage of the intensification of land, effectively allowing more to be produced from the same amount of land, but is also an efficient form of risk avoidance. Intercultivation was considered the normal strategy for both large and small farmers, certainly by the Late Republic.

Oil and wine are typical products for market engagement, and much evidence for consumption and distribution comes from ceramics linked to transport. This only works for long-distance trade, however, and it must be remembered that the importation of amphoras does not imply the absence of local production (Brun, 2004: 160). Much information regarding local exchange is lost due to the perishable nature of the skins or barrels used. Many areas of Republican Italy engaged in such production, but olives and vines were not ubiquitous. Palynological research indicates that olive groves were not established on the Lepine hill slopes in the Ager Pomptinus until the third century (Attema, Delvigne and Haagsma, 1999), despite their early cultivation near Rome, while some areas were simply unable to cultivate them (e.g. upland Samnium).

Animals are an extremely useful resource: they act as stored food, providing meat, leather, bone and horn at the end of their lives, as well as secondary products such as wool, milk/cheese, manure and, importantly, energy for traction. Pastoralism is often depicted wrongly as a symbol of economic underdevelopment, and animals could also be well integrated into arable systems (Horden and Purcell, 2000: 198–200). Transhumance systems were probably extremely important in certain regions, but are unlikely to have left many archaeological traces, given their mobility, ephemeral shelters and minimal material culture (Barker and Grant, 1999: 584). However, transhumance into high altitude zones can be inferred from pollen analysis (e.g. Lowe *et al.*, 1994). There is some evidence that elite-owned transhumant flocks were an important component of the Etruscan agricultural economy, although large-scale long-distance transhumance was probably not characteristic until the Roman period (Barker and Rasmussen, 1998: 194–8). Elsewhere, the Samnites utilized long-distance transhumance between Apennine summer pastures and wintering in the Apulian lowlands, near wool working centers such as Tarentum (Taranto) (Bispham, 2007b: 184). Such routes are hypothesized to have persisted, and reconstruction has been attempted using computer modeling to find optimum paths (Bell, Wilson and Wickham, 2002). Varro (*Rust.* 2.6.5) described specialized transhumance for southern Italy in the first century, but anything beyond short-distance transhumance was unlikely to have been common in Magna Graecia (Foxhall, Jones and Forbes, 2007: 104–5).

Evidence of domestic, stall-fed animals integrated into the agriculture of the homestead is found in the Republic. Foxhall (1998: 35) has argued that small-scale integrated animal husbandry would have been heavily dependent on arable by-products for fodder.

These include chaff, straw and stubble of cereals, legume stalks and husks, and weeds, along with the residue of more specialized production such as olive skins and lees or vine trimmings (Cato *Agr.* 25, 37.1.2, 54, 91.6; Varro *Rust.* 1.55.7, 2.2.16–19). The Samnites were considered to be predominantly pastoralists, and faunal samples from a variety of contexts suggest sheep were the main stock, kept primarily for their milk and wool; pigs were kept for meat, and a few cattle maintained for traction (Barker, 1977: 24). The Samnite farm at Matrice showed that sheep or goats and pigs were the dominant species with a small number of cattle, with evidence for butchery on site. Surplus was produced for the local and regional sanctuaries (Lloyd, 1995: 202–3; Bispham, 2007b: 207). Other sites in the region also showed a reliance on secondary products, with loomweights common, and the occasional specialized object, such as a ceramic vessel probably for cheese-making found at Monte Vairano (Lloyd, 1995: 205).

At Metapontum cattle were dominant in faunal assemblages from the sixth through the second century indicating a reliance on traction and consequently an arable-dominated economy. Pigs were also common, while the nearby site of Incoronata contained the earliest known chicken remains in Italy (Carter, 2006: 78–80). However, in the nearby Bradano Valley (at Difesa San Biagio), the Late Hellenistic landscape was likely to have been dedicated to sheep/goat pastoralism, as indicated by the presence of fungal spores that live in animal dung, and pollen from grazing plants (Mercuri *et al.*, 2010).

Silvicultural pig keeping is also likely to have been an ancient Italic practice (Purcell, 2003: 340). Though they can be raised on meager scraps, or by woodland foraging (pannage), pigs were not just the preserve of the low-status and pigsties are known both from excavations and the agronomists (Columella *Rust.* 7.9.9–10; Varro *Rust.* 2.4.13–15; Carandini, 1985a: 182–4). In the Later Republic and Imperial period, pork dominated the faunal samples from towns including Capua, Naples, Pompeii, Ostia and Rome (Barker and Grant, 1999: 590). Indeed, pigs have even been described as "cash crops" given their relatively simple conversion of a rich natural resource into a high-value product for redistribution (Horden and Purcell, 2000: 199; see also MacKinnon, 2001; Varro *Rust.* 2.4). Animals added important nutrition to the diet of Republican Italy. While one can survive solely on cereals and legumes (Sanders, 1984), animal products offer important variety, and also food during times of shortage. Archaeological and textual evidence has shown animals were well integrated into both agricultural and trade systems.

While the Mediterranean triad of wheat, grapes and olives dominated production, there were many other food sources available to the Roman farmer. Fruit and vegetables could be obtained via either specialized horticulture or foraging. Horticulture is often associated with denser, urban populations, but can also play a significant role in risk reduction, representing a different kind of cultivation, often operating within the market (Horden and Purcell, 2000: 210–21). Even low-status farmers with small plots of land could potentially take part. In the pseudo-Virgilian poem *Moretum*, from the late first century, the peasant farmer Simylus grows vegetables, not for his own consumption, but in exchange for staples, thereby maximizing the production of his limited land.

Some economies were very much the preserve of the elite, due to the high levels of expenditure required. The production of luxury foodstuffs for the market known as *pastio villatica* is not likely to have become widespread until the Imperial period. However Varro's third book is dedicated to the subject, and Pliny dates its development to before the late second or early first century, when C. Sergius Orata raised oysters in the Lucrine

Lake in Campania (*HN* 9.79.168). *Pastio villatica* was essentially divided into three categories: the breeding of birds, hares and fish, although other specialist activities included beekeeping, dormouse breeding and game reserves. It was, however, only in those areas with good transport access to an urban center that it would have been profitable to follow such a market-oriented regime (Morley, 1996; following Thünen, 1826).

Elite market-oriented production would not have fed the vast majority of the population. Nutritional diversity for lower-status farmers was likely to have come from activities such as foraging. Garnsey argued that "peasants have always been systematic foragers on uncultivated land" (1988: 53, 46). In foraging, many areas traditionally dismissed as marginal are likely to have played an important role. Marsh, mountain, forest or sea are able to provide a diverse, productive opportunity (Horden and Purcell, 2000: 182). Etruscan and Samnite assemblages have established wild food on domestic sites, including fruits, nuts and game (Barker and Rasmussen, 1998: 200; Bispham, 2007b: 185; Sadori, Mercuri and Mariotti Lippi, 2010: 945). Hunting was not restricted to the elite (see Kron, 2008a), and woodland was also exploited for fuel and building materials. Other woodland and scrub products include honey, willow props for viticulture and, importantly, items such as acorns that provide forage for animals – particularly pigs (Columella *Rust.* 3.3.2, 7.9.6).

Wetlands are diverse environments varying in size and salinity, including pools or lakes, and seasonal or perennial marshlands. They can provide winter grazing (but malaria would be a risk in the summer; see Sallares, 2002), and offer opportunities for fishing, fowling, hunting and edible vegetation. At drier times of year, the margins could be used for traditional cultivation. Additionally, the supply of salt from natural pans at marsh edges was hugely important in food preservation (Horden and Purcell, 2000: 187–96), and was the root of conflict between Rome and the Etruscans on the Tiber delta (Morelli *et al.*, 2008: 215).

While cultivation is clearly the focus of farming, it would be worthless if there were no means of storage to last throughout periods of non-production (i.e. winter in the short term, or poor harvests in the long term). Storage was an "economic necessity" and facilities are regularly unearthed at archaeological sites (Garnsey, 1988: 53; Forbes and Foxhall, 1995: 69). Comparative evidence from later societies implies that it was desirable to have up to two years' supply of food (or four of olive oil), to guard against bad harvests and shortages (Forbes, 1989: 93; Gallant, 1991: 96). In the sources we have little direct information regarding suggested storage amounts. Cato (*Agr.* 3.2) simply advises that a farmer should have "plentiful storage" for an elite estate, reflecting his desire for villa self-sufficiency as well as allowing an estate owner to take advantage of market prices.

It is evident that, across Italy, the same basic crops and animals were exploited, with local conditions determining the concentration on certain products. Samnium, for example, was known for its pastoralism, yet the region still had large-scale agricultural establishments in areas capable of producing wheat, wine and oil. Conversely, even villas dedicated to investment agriculture were likely to have included animals. Cato includes sheep among his recommended "equipment" for an olive grove (*Agr.* 10). Furthermore, a concentration on pastoralism does not automatically equate to a backward economy and a lack of involvement in trade. All the regions discussed have been shown to have acted within the market to some extent, some geared towards Rome, others remaining independent until much later (e.g. Samnium's trade with Magna Graecia).

Though these systems did not all develop simultaneously, or to the same scale, the evidence points to diverse and active economic systems (not all reliant on Roman conquest), which formed the basis of urban development built upon an agricultural or pastoral surplus. Despite the upheaval during Rome's rise to supremacy, and the conflict of the Punic Wars, these systems either persisted or adapted to the changing conditions of the time, forming the backdrop for the events of the Late Republic.

4 Agrarian Crisis

The traditional view of agriculture in the Late Republic is that a combination of aristocratic land hunger and extended absence of peasant soldiers from their land during the Punic Wars contributed to a decline of the smallholder. This led to their replacement by slaves, encouraging mass migration to urban centers, with the effects felt particularly in the south. The heavy military losses were thought to have caused a severe reduction in the rural community, with those returning finding their farms neglected, war-torn, and often reduced to debt. This was argued to have precipitated a change in the system of land organization into larger units capable of producing a surplus using less labor. This larger-scale method of production also meant that smallholders were increasingly unable to compete in the marketplace (Toynbee, 1965: 105; Hopkins, 1978: 2–3; De Neeve, 1984b: 9–10). Though aspects of this were challenged early on (e.g. Spurr, 1986), recent studies have led to a number of reinterpretations of this traditional paradigm (see de Ligt and Northwood, 2008; Bowman and Wilson, 2009; Carlsen and Lo Cascio, 2009), and the new orthodoxy is becoming one of a variety of regional responses (Chapter 12).

For the Punic Wars, it has been argued that the effects may not have been as long-lasting as scholars such as Toynbee (1965) and, more recently, Cornell (1996) have suggested. Dyson (1992: 28) argued for a quick recovery: new colonies were established at this time (requiring large-scale mobilization of manpower), and it is possible that the population recovered within a generation (see Chapter 10). Rosenstein (2004: 9–10) followed this theory by reducing the scale of analysis from global trends to the household perspective, arguing that military service would have had less of an impact than originally thought. The assumption that women, children and the elderly were incapable of supporting themselves through farming is an outdated concept. Even if men were absent for extended periods (or permanently), analysis shows that most farms were robust enough to survive, and that families had the potential to recover rapidly (Rosenstein, 2004: 104–5). Additionally, it has been argued that legislation in the later second century, endowing land to increase the numbers eligible for military service, was not a sign of the decline in eligible men, but of a general lack of enthusiasm for service (Patterson, 2006: 620, n. 73). Decline and expansion were not due solely to the effects of warfare, therefore, but also the natural ebb and flow of population dynamics.

Toynbee (1965) assumed that southern Italy was dominated by small-scale agriculture and pastoralism before Hannibal's arrival, allowing investment villas to fill a vacuum. It has been shown above, however, that between the Greeks of Magna Graecia and other local cultures, a more varied and sophisticated system was in place. Likewise, regional field surveys have shown a number of responses to crisis: decline, persistence and even increase in rural populations in some areas of Italy (Chapter 12). While some areas are likely to have conformed to Toynbee's view of large-scale, slave-run systems, this was not a universal trend (see Carter, 2006: 239–43, for evidence of recovery in Magna Graecia).

Small and medium-sized farms contributed substantially to the growing economy, but we cannot completely refute the traditional model of land reorganization and slave-led agriculture. Simply because small rural properties persisted and potentially profited in some areas in the second and first centuries does not mean that they became the norm. It is clear from a range of evidence that substantial numbers of slaves were brought into Italy in the second century, and that large numbers may already have been brought into Rome from Italy from the late fourth century onwards (Patterson, 2006: 617). It cannot, however, be assumed that the villa boom of the first century was a conversion to slave-led investment agriculture. There is no real evidence to support slave movement to great estates rather than urban or industrial locations, or even non-elite farms (slave ownership is argued to occur well down the Roman social scale), and any new slaves probably served to replenish losses made during the war rather than increase numbers dramatically (Rosenstein, 2004: 9–10). The supposed incompatibility of slaves and peasants on the land has also been exaggerated, as the Roman writers do suggest supplementing the slave workforce with casual free labor at times of heavy workload (Cato *Agr.* 1.4; Columella *Rust.* 1.7). Even in Magna Graecia – an area traditionally held to have been dominated by a servile workforce – the move to fewer, larger sites in the post-Hannibalic period does not automatically prove the existence of a slave-led economy, particularly as material culture recovered from excavation was of higher quality than would be expected for such low-status occupants (Carter, 2006: 244–5).

There are, nevertheless, examples of slave-staffed villas producing for the market in the Late Republic. The villa of Settefinestre has been used to exemplify the practices advised by the Roman agronomists, and was said to be "the best example of Varro's *villa perfecta*" (Carandini, 1985a: 194). The faunal evidence indicates a clear separation between the quality of food eaten by the senatorial family and the slaves working the estate (King *et al.*, 1985; Barker and Grant, 1999: 590). However, even here it is unlikely that the slave staff could have fulfilled the full labor needs of the estate without relying on day laborers (Rathbone, 1981: 12).

Diversity is key: villas in coastal Etruria and northern Campania may have acted as intensive crop producers for the market, relying on a predominantly servile workforce, but all villas cannot be assumed to have operated in a uniform fashion. In the Ager Cosanus (Etruria) and Ager Falernus (Campania) such villas were mostly a phenomenon of the second half of the second century to the early first century (Patterson, 2006: 619–20). Meanwhile, areas such as the Albegna Valley were more likely to exploit more marginal land for the local market with a more varied workforce (including tenants and slaves) – Carandini's "peripheral villas" (Carandini, 1985a). In Volterra, further north, there is a striking continuity in the landscape from the pre-Roman period, with very few villas, while Magna Graecia, as discussed above, saw a move to smaller numbers of larger sites. Terrenato (2007: 145–7) argues that it is "unwise to consider the remarkable diffusion of Late Republican villas as evidence of a new and all-pervasive mode of production." This implies that it was the combined production of small and medium-sized farms, and not simply investment agriculture, which was crucial in supporting the increasing urbanization and expansion of Rome throughout the Mediterranean.

The simplistic division of investment villas or small subsistence farms is flawed for many reasons. Not all villas produced for the market, and many small farms were likely to have produced surplus to allow them to engage in trade. Along with this is the range of rural settlement types which co-exist between these two stereotypes: medium-sized farms and estates existed in a number of regions, and there were also nucleated settlements within the rural landscape which could act as foci for their hinterland, bases for farmers to commute from, a center for shared resources, or even as markets themselves (for example, the villages of Samnium above).

The process of intensification and specialization is not merely the preserve of the larger establishments, but could also be achieved by the small farmer carrying out mixed farming, which had the potential to be highly productive (see Goodchild, 2009; Kron, 2008b). Indeed, it is dangerous to assume a strict dichotomy between small subsistence farmers isolated from the larger market and surplus-producing villas heavily integrated into the economy. Likewise, the presence of infrastructure for on-site production does not preclude the importation of goods at either end of the scale (Witcher, 2006a: 342).

Various studies have attempted to gauge the profitability of agriculture using indirect approaches, such as calculating the yield of a vineyard, farm or orchard of a given size and subtracting expenditure such as labor, land costs and equipment (e.g. Duncan-Jones, 1982: 33–59; Morley, 1996: 124–36). This may be a flawed approach, as the farm or estate is not considered to be operating within an active market with competition and fluctuations of price, thus assuming an unlimited market for goods (Rosenstein, 2009: 244). Despite the proliferation of high-status villas at this time, not all were necessarily involved in agricultural production for the market, as there was unlikely to have been enough market capacity to support it.

The investigation of market activity on smaller rural sites can also be problematic. A related economic aspect is the emergence of tenancy (the "colonate"). Though some have argued this was an Imperial phenomenon (e.g. De Neeve, 1984a; Rosafio, 1994: 155), others have argued for its role within Republican rural systems (Foxhall, 1990). Areas such as Capua in Campania that did not succumb to investment villas have long been argued to represent tenant farmers following the Punic Wars (Toynbee, 1965: 123–6, though he sees this as an "aberration"). Tenancy is difficult to prove based on archaeological evidence, but Foxhall (1990) has suggested that limited material culture in combination with sophisticated buildings may be indicative of such a system. While an attractive proposition, the effect of taphonomic processes on the recovery of field survey material cannot be underestimated (Witcher, 2006a: 344). The lack of recovered material culture does not mean it was not there – a classic problem in the interpretation of field survey results (Chapter 12).

We must therefore look at indirect methods for investigating the emergence of tenancy in this period. One example analyzed farm and villa productivity in the Late Republic using a combination of field survey data and computer modeling (Goodchild and Witcher, 2009). The idea of carrying capacity is not new, but its application in a small region of the Ager Veientanus showed that villas in continued occupation from the Late Republic to Early Imperial period were situated on better land than those newly founded in the Early Imperial period. This is to be expected as good quality land becomes scarcer, though new villa foundations may not have valued agricultural productivity as an important factor in site choice, perhaps valuing aspects such as scenic views or proximity to Rome. Surprisingly, however, the opposite was true for small farms, with new farms being situated on relatively better land than the new villas. This may demonstrate the emergence of tenancy, as the elite allowed their land to be fragmented and let to small farmers, with both owners and tenants reaping the benefits of increased production and socio-economic ties of patronage (Goodchild and Witcher, 2009: 214–15).

Understanding the social status of farmers is important to studies of Republican agriculture, as economic strategies are likely to have differed if producing simply for one's own household, or providing for the family as well as rent. The difficulties in such a study, however, and the lack of archaeological indicators for social status, mean that this is a poorly understood aspect of Republican rural life.

5 Conclusion

There was clearly great variety in production and landscape exploitation in Italy throughout the five centuries of the Republic, and in preceding and contemporary cultures. Agriculture was not a uniform activity across Italy, or even within regions, contrary to the impression given by the ancient texts and studies founded upon their work. Different patterns of urbanization led to different patterns of exploitation (or vice versa) and detailed reconstruction of the agricultural economy is consequently a complex undertaking.

Long-term, regional multidisciplinary projects have been influential in shifting the focus from highly generalized large-scale analyses to local-scale variability. Controversial debates intrinsically tied to agriculture, such as the high/low count of Roman population, can now be firmly rooted in particular regions of the country (see Chapter 10). However, while field survey has been invaluable in providing a more detailed and nuanced view of Italy, disproportionate emphasis on local and regional variety of developments can be to the detriment of the interpretation of larger-scale processes (Lo Cascio, 2009: 10). The answer therefore lies somewhere between the local and national, and could potentially be achieved by analyzing at the local scale, then re-aggregating multiple small-scale analyses in order to investigate larger processes (Goodchild and Witcher, 2009: 192–3).

From the Early to Late Republic, Italy employed a huge variety of exploitation strategies adapted to unique local environments and socio-economic conditions. Indeed, it has been stated that "there is no absolute quality of land anywhere: its value and potential depend on the choices and perceptions of those who make use of it" (Horden and Purcell, 2000: 231). While this is true to a certain extent, the role of the landscape and environment must not be undervalued. There is an element of choice and adaptation, but production is ultimately constrained by the environmental conditions in which agriculturalists operate. As such, a detailed reconstruction of agricultural and pastoral production, and detailed figures for how many people could be supported by such produce, is simply not possible. The landscape itself has altered, both in shape and fertility due to both natural processes and human activity, and so, despite our increasing knowledge of the ancient landscape, it is still difficult to model absolute production levels effectively. Nevertheless, with the increasing ruralization of Roman archaeology, we are no longer reliant on the urban-centric, elite sources, which originally encouraged generalized statements about agricultural systems based on limited, biased information. Local areas must not be studied in isolation, however, and determining how they fit into the larger economic processes remains the challenge.

Acknowledgments

I would like to thank Jane DeRose Evans for inviting me to contribute to this volume. I am also extremely grateful to Rob Witcher, Kevin Walsh, Fay Glinister, Guy Bradley, Ben Croxford and Nick Daffern for their enormously useful feedback on the various aspects of this broad subject, and for the editing skills of Liam Herringshaw.

FURTHER READING

Studies of Roman agriculture should begin with the texts of the Roman agronomists: Cato *De Agricultura*, Varro *Rerum Rusticarum* and Columella *De Re Rustica* and *De Arboribus*. A wealth of further information can also be found in Pliny's *Natural History* (particularly Books 15–18). The original classic work is White (1970), which is based heavily on the works of the agronomists. This was followed up by Frayn (1979) and Spurr (1986), who attempted to use archaeological and ethnographical evidence alongside the ancient texts, though these are now dated.

Horden and Purcell, in their broad study of the Mediterranean (2000), placed an emphasis on micro-environments and regionality. While criticized, it forms a good antithesis to studies which have neglected regional variability. An excellent overview of evidence for oil and wine production is Brun (2004), though there are no equivalent publications in the English language. Vegetation studies are currently undergoing something of a renaissance, with a good introduction to current work being a themed volume of *Plant Biosystems* journal (2010, issue 4) which showcases a number of environmental studies in the Mediterranean and demonstrates the range of applications. For the current state of play on the agrarian crisis, and Late Republican rural systems, there have been a few key recent volumes. In English there is de Ligt and Northwood (2008), and Lo Cascio and Storchi Marino (2001). Most recently, Carlsen and Lo Cascio (2009) have articles in English, Italian and French on a wide variety of subjects related to Late Republican agriculture, rural settlement and trade from archaeological and historical evidence.

CHAPTER 14

No Holiday Camp: The Roman Republican Army Camp as a Fine-Tuned Instrument of War

Michael Dobson

1 Introduction

All ancient armies required camps, but the Roman army made them a speciality, even a science, and definitely an art. Scenes of Roman camp building, with tents inside and defended entrances, are shown several times on Trajan's column, clearly indicating the importance of "castrametation" to the Romans. The camp was a second home to the Roman soldier – the ramparts his city walls and his tent his house (Livy XLIV.39.5). This expertise in encampment started and was perfected in the Republic and continued without significant change throughout the Empire. For the soldiers themselves, regularized encampment was put into practice on a near-daily basis, and so common was this military "art" that Roman writers could refer to camps as an everyday object known to all their audience, rather like today, when no one has to explain what a railway station is.

Some Roman authors took their armies' camps so seriously that they even theorized about them in treatises. For example, Vegetius's *Epitoma rei militaris* provides many details about camps and the treatise now commonly called *De metatione/munitionibus castrorum* and often (wrongly) attributed to Hyginus (Birley, 1982; Dobson, 2008: 5) is solely about the subject. These two examples date to the Empire, but Vegetius is known to have used Republican sources, as he refers to some that he consulted (*Mil.* 1.8; Milner, 1993: xiii ff.), and it is likely that *De metatione castrorum* also made use of a collection of earlier sources which could go back to the Republic (Birley, 1982; Dobson, 2008: 5). Consequently, both can potentially provide information about Republican encampment.

For the Republican period itself, military treatises clearly existed, as, for example, Lucius Cincius Alimentus, an early historian (pr. 210), is known through a quote by Gellius (*NA* 16.4) to have written a book on military matters – *De Re Militari.* Marcus

A Companion to the Archaeology of the Roman Republic, First Edition. Edited by Jane DeRose Evans.

Porcius Cato (cens. 184) also wrote a military treatise, *De Re Militari* or *De Disciplina Militari* (Veg. *Mil.* 1.8; Tatum, 1992: 639). It seems likely that camps would have formed part of Cincius's and Cato's discussions, but since neither work is extant, the amount of such discussion is unknown. These specific military works are separate from military handbooks (commentarii) that seem to have been supplied to senior army officers and must have included information about camp construction (Fraccaro, 1934: 158; Rawson, 1971: 15; Dobson, 2008: 54).

It was not just Roman authors who felt moved to write about camps. Such was the impact of the Roman camp that the second-century Greek historian Polybius devoted a long section of his work to precisely this subject (6.19.1–6.42.6). Much later, Josephus felt it important to provide quite a detailed description of Roman camps to his Jewish readers during a digression on the Roman army (*BJ* 3.79–92; cf. 1.3, 6). Ironically it is Polybius, rather than a Roman author, who remains our main source for the nature of the Roman Republican camp.

These earthwork "weapons" clearly also made a great impression on those being invaded or fought against, regardless of whether during the Republic or Empire. One of the main reasons for this was because the highly ordered Roman camps contrasted so markedly with the relatively crude practices of the non-Roman writers' own armies. Enemy commanders could also be very impressed by the ordered and disciplined nature of Roman camps. In 199, Philip V of Macedon (Livy XXXI.34.8), once he had seen the regimented Roman camp, is credited with declaring that clearly he was not facing a "barbarian" army. The same comment is attributed to Pyrrhus in 280 by Plutarch (*Pyrrh.* 16.5), but this may be a case of Plutarch making use of the later statement (Briscoe, 1973: 141).

2 History of Research into Republican Camps

The interest in Roman camps did not stop with the end of the Empire. Polybius's long digression on camps formed one of the important elements of military studies from the Renaissance period on, with analysis of the text beginning at least as early as the fifteenth century (Nissen, 1869: 23, n. 1; Fabricius, 1932). It also featured heavily in the highly influential work of the late nineteenth- and early twentieth-century German school of research into the Roman Republican army (e.g. Marquardt, 1881; Mommsen, 1893, 1900; Stolle, 1912; Fischer, 1914; Kromayer and Veith, 1928; Fabricius, 1932; Oxé, 1909, 1939; Schulten, 1905, 1914, 1926, 1927, 1928, 1929, 1931, 1933), which formed the start of the modern study of all things Roman military.

Sadly, after this German impetus of research into the Republican army and its camps, little else followed for much of the twentieth century, as attention soon became focused on the installations and army of the Empire. This was largely because an increasing amount of archaeological evidence for this later period was quickly emerging, thanks to the large-scale excavations, fieldwork and associated research carried out from the late nineteenth century in Germany and Austria, particularly the state-funded work on the *Limes*, and in Britain. Testimony to the longevity of the early German research on the Republican army is that Fabricius's 1932 interpretation of the Polybian camp remained the standard and unquestioned version reproduced in any publication about the army of this period, until very recently (Dobson, 2008: 3). Similarly, Schulten's own summary publication of the excavations he carried out before the First World War at the highly important complex of Republican camps at Numantia and nearby at Renieblas in north-eastern Spain was even republished in full and translated into Spanish as recently as 2004 (Schulten, 1933; Wulff, 2004).

Until the latter part of the twentieth century there were a few notable exceptions to the lack of research on the Republican army, with some including discussion of camps. These were the studies of Bell (1965), Harmand (1967), Sumner (1970), Connolly (1978, 1981) and Keppie (1984). Interest in the Republican army, including its camps, has steadily increased since the 1990s. The aspect of Republican equipment was included in Bishop and Coulston's work (2006) and has seen frequent articles in the *Journal of Roman Military Studies*, with even an entire volume devoted to the Republic (see Chapter 8; Feugère, 1997). A new translation of Polybius's account of the Roman camp, together with *De metatione castrorum*, was published in 1994 (Miller and DeVoto, 1994). Pamment Salvatore (1996) published a discussion on the layout and planning of Republican camps which has become frequently cited. The issue of the development of the cohort, and hence how camp layout was affected, was raised both by Pamment Salvatore (1999) and by Cadiou (2001). Changes to camp layout as could be traced in developments in army organization were also caused by the disappearance of the legionary cavalry in the first century. The history of this unit was examined by McCall (2001), and I agree that its disappearance was not due to it proving ineffective in battle, as was the usual theory, but due to changes in the career paths of its members. Richardson (2003, 2004) included the Polybian camp in his discussion of space allocation in camps. The organization and development of the Republican army, and its system of encampment, with a detailed reassessment of both Polybius and the complex of camps at Numantia and Renieblas, was published by the current author (Dobson, 2008).

In fact, 2008 saw several significant publications on the Republican army. Cadiou (2008) published a thorough discussion of the army in the Iberian Peninsula which included much material about camps there. A convenient collection of army-related classical texts, translated into English, was published by Sage (2008). Johnston (2008) published the first detailed discussion of the military consilium, the commander's advisory council, whose location in the Polybian camp is unclear, though members of it are referred to in connection with guard duties (Polyb. 6.35.4; Dobson, 2008: 80). Phang (2008) discussed aspects of military service during the Late Republic. More recently, the Late Republican army, and particularly the credit given to Marius for significant reforms, has been looked at by Sampson (2010) and the evidence of Livy for Republican infantry combat has been studied by Koon (2010).

3 The Form of Republican Camps

Little can be established about the form of Roman camps, and even the army itself, prior to the late third century. References in ancient literary sources to earlier camps are either unreliable or too vague to allow us to derive anything meaningful. Archaeological evidence cannot help either, as the earliest confidently identified camps date from the second century, in Spain. Clearly thousands of camps must have been constructed around the Mediterranean as Rome expanded its sphere of control. The absence of identified camps in these areas is the result of lack of fieldwork and topographical and climate conditions being unsuited to showing remains of camps through aerial photography. In contrast, the sites of many battlefields in this area have been conjectured or confirmed, but no associated camps have been located (e.g. Connolly, 1981; Cowan, 2009; Matyszak, 2009; Fields, 2010).

Polybius wrote his account of the Roman army and its form of encampment probably *c.*160, but his wording suggests that the system dates to an earlier period (6.19.1–6.42.6). It may derive from a written source which was also created before his time, possibly a commentarius for a military tribune (Fraccaro, 1934: 158; Rawson, 1971: 13ff.). He is possibly referring to things as they stood at the beginning of the Second Punic War, which is where the digression is placed within his *History*, at the end of events for 216 (Dobson, 2008: 54ff.). The regularized form of camp that Polybius describes, or at least some aspects of it, may pre-date this period, as some of the practices he describes originate in augury and agricultural land management which could easily date long before the late third century (Dobson, 2008:71ff.). Hence, it is uncertain when the Roman camp started to have a regularized form.

It can be claimed that one of Polybius's main reasons for giving such a detailed account of the Roman system to his Greek audience was to show just how different and inferior their own practice was. Hence he comments that the Greeks adapted their encampment and its defenses to suit local conditions, resulting in little conformity among camps and uncertainty within the army about the relative location of units (Polyb. 6.42). The result must have caused logistical and communication problems and delays in deployment. In contrast, Polybius emphasizes that the Romans took a "mechanical" approach to create a camp with the convenience of an unvarying single form in both layout and defenses (Polyb. 6.41). Admirable as this ethos was, the reality was that local topography did inevitably lead to alterations from the theoretical ideal, as shown by surviving archaeological evidence, but at least all armies had a known and coherent starting point.

The nature of the camp described by Polybius has been much discussed, as it is felt by some that there is ambiguity about whether the camp is for four manipular-organized legions with allies and commanded by two consuls (a double-consular army) or for two legions, allies and one consul (a single-consular army), because of potentially conflicting statements of Polybius 6.32.6 and 6.32.8 (Dobson, 2008: 68ff.). As already stated, the most influential interpretation has been that of Fabricius (1932), who favored the single-consular camp theory. A strong argument can be proposed to counter the single-consular camp, since the location of the digression in Polybius occurs in a period when double-consular armies were the norm (Dobson, 2008: 47ff.). As a result, simply for convenience, Polybius describes one half of a camp (which half would have contained two legions, allies and one consul) that is a mirror-image across its long axis (Dobson, 2008: 68ff.).

The overall shape of this double-consular camp was rectangular (Figures 14.1, 14. 2). This shape remained the ideal for camps, as indicated by *De metatione castrorum* and can be seen in many surviving Roman camps from all periods. From the Republic, there are good examples of rectangular camps in Spain at Renieblas (Lager IV and V; Figure 14.3), Almazán and Cáceres el Viejo, and in France at Mauchamp and Faux-Vésigneul (which may be Caesarian; Reddé *et al.*, 2006: 27, 277ff.). There are, however, examples particularly

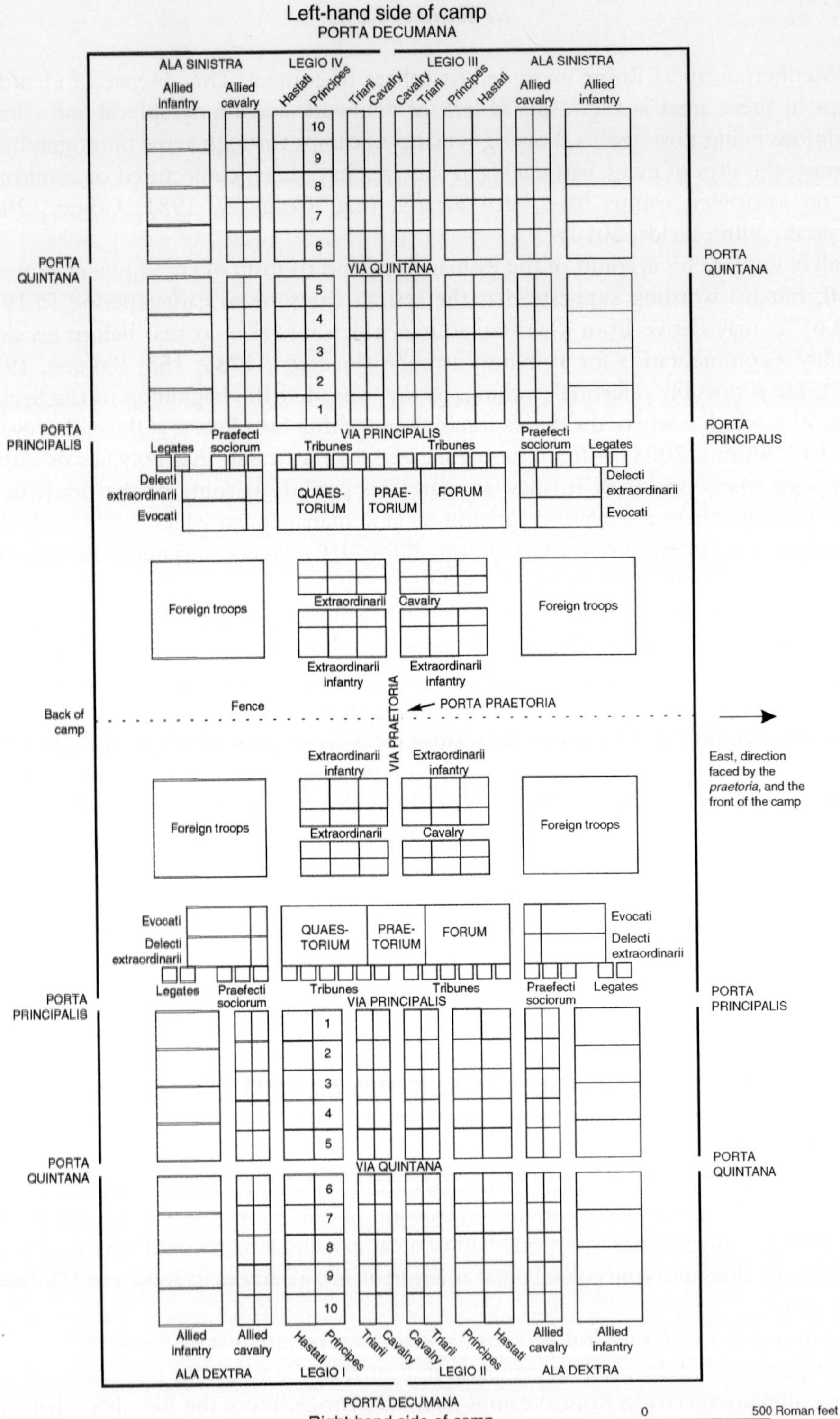

Figure 14.1 Reconstruction of the Polybian manipular double-consular camp. Source: Drawing by M. Dobson, from Dobson (2008).

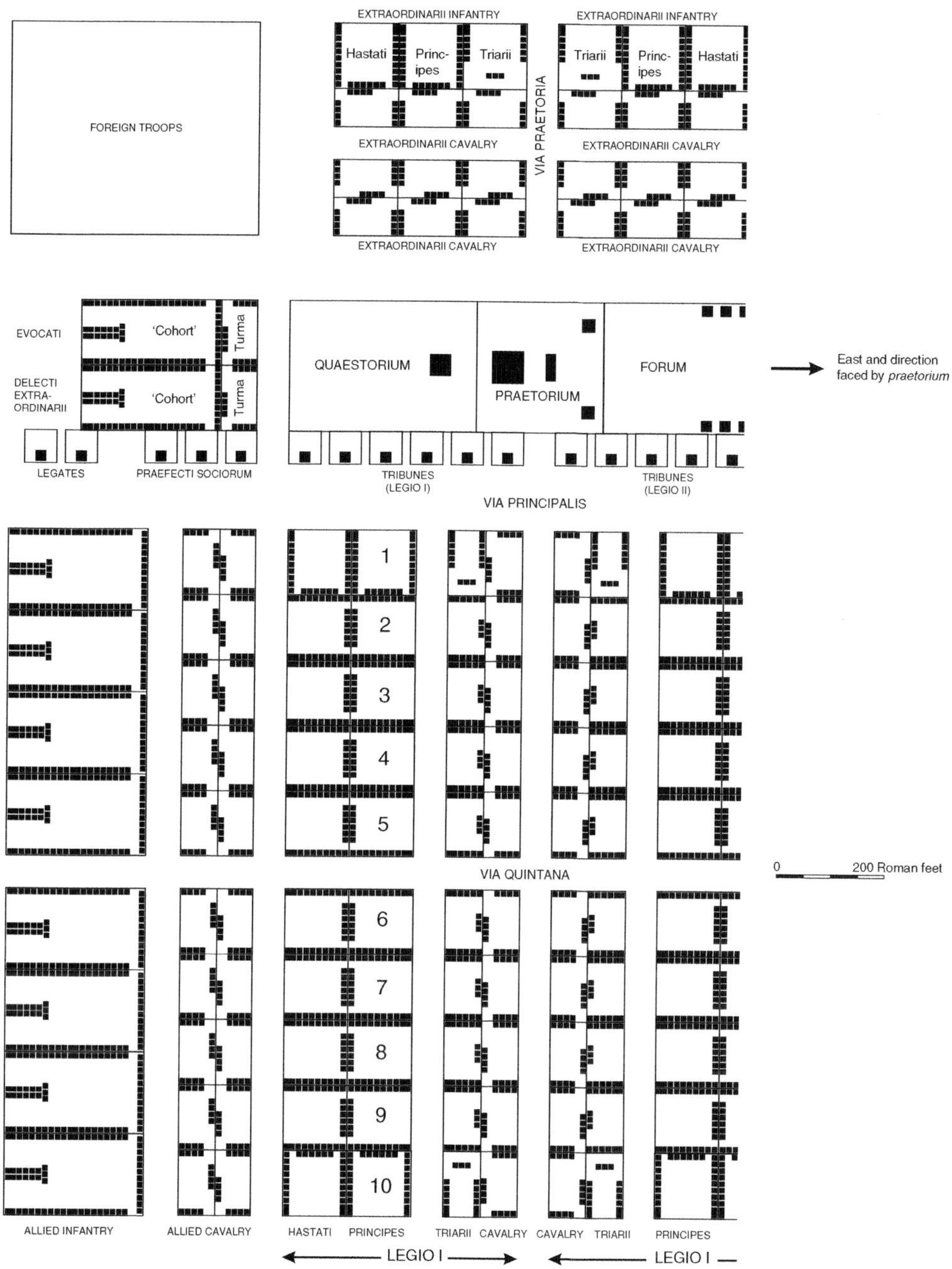

Figure 14.2 Detailed reconstruction of the "south-western quarter" of the Polybian manipular double-consular camp. Source: Drawing by M. Dobson, from Dobson (2008).

from the Republic and Early Empire of irregular shapes. For example, from the Republic they occur at Numantia (Figure 14.4), Renieblas (Figure 14.3) and Alésia. Topographical reasons do not always explain why these camps diverged from the rectangle. From the mid-first century CE the rectangular shape seems to have become the norm even when

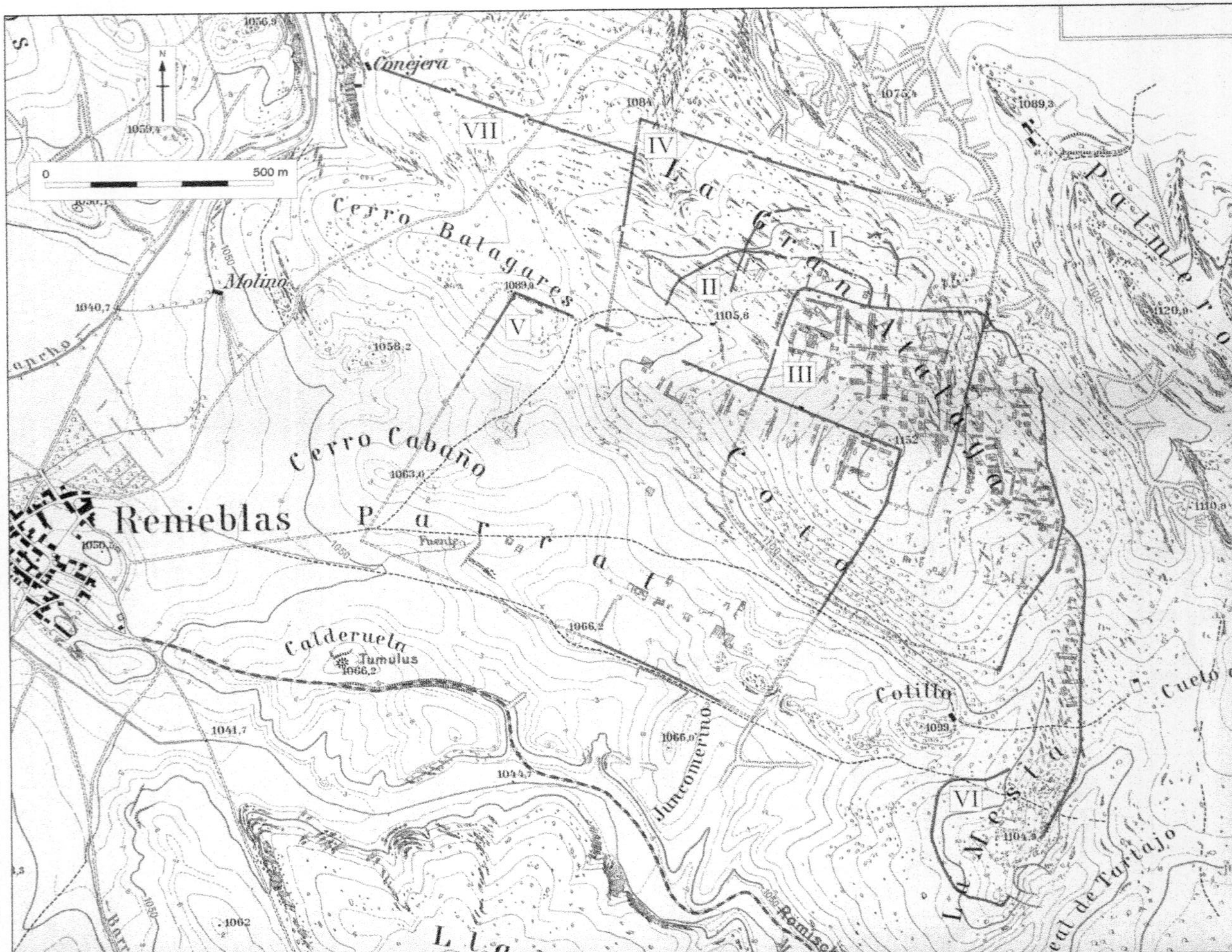

Figure 14.3 The camps at Renieblas, Spain. Source: Drawing by M. Dobson, from Dobson, (2008).

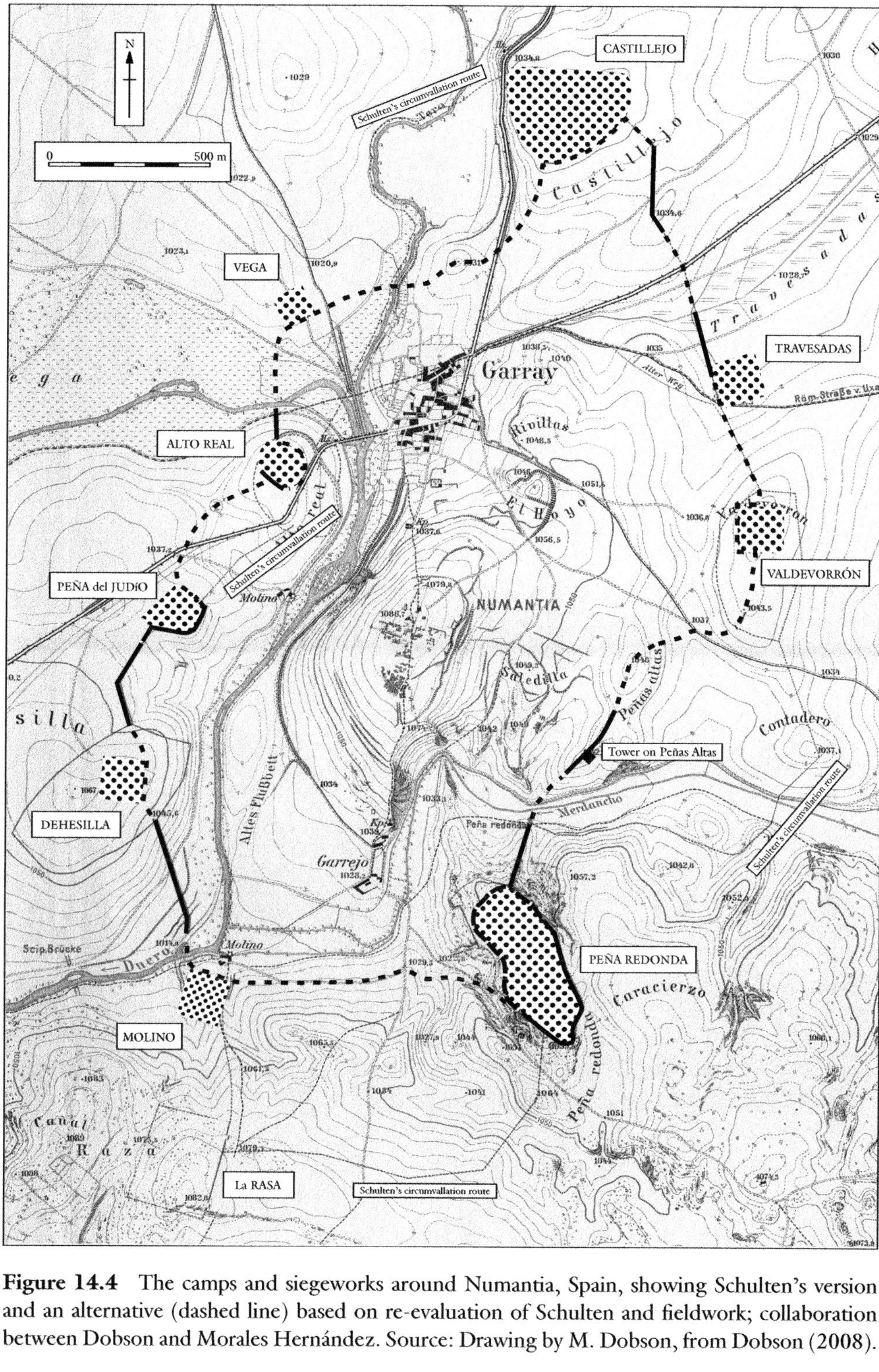

Figure 14.4 The camps and siegeworks around Numantia, Spain, showing Schulten's version and an alternative (dashed line) based on re-evaluation of Schulten and fieldwork; collaboration between Dobson and Morales Hernández. Source: Drawing by M. Dobson, from Dobson (2008).

topography could have reasonably allowed divergence. The reasons for this development are not clear (Schnurbein, 2001: 510, 2006: 71ff.; Reddé, 2008).

The ideal orientation of the Polybian camp was to the east, because of the requirements of augury and religious practices (Dobson, 2008: 70). Since these activities were ancient in origin, it is probable that the eastward-facing layout of camps would have been observed from long before the period Polybius described. In practice, topographical or logistical restrictions probably prevented such a rigid orientation at all camps.

The camp's interior was based on a regular grid-layout. This grid system was also derived from agricultural land arrangement practices (see Chapter 16; Dobson, 2008: 71ff.). It is thus very likely that the Polybian form of both rectangular and grid-based camp pre-dated the late third century, but by how much is uncertain. A grid-based layout, in appropriately amended form, continued in use throughout the Roman period for camps, fortresses and forts. Polybius provides details about the size of the grid. The area unit he specifies is Greek (*plethron*), which coincidentally matches the equivalent Roman area (*heredium*). For dimensions of the grid he uses feet. This is invariably and unthinkingly taken by modern scholars to mean Roman feet, but the Greeks used the same-sounding term for measurement. There is good reason to claim that Polybius is in fact using Greek feet, and indeed this would have been the obvious thing for him to use for his Greek audience (Oxé, 1939; Dobson, 2008: 71). Hence, converting the Greek feet to Roman (100 = 120), the camp grid system can be seen to have been based around the Roman agricultural field grid of 120 Roman feet (35.52 m) – the actus quadratus (Dobson, 2008: 71ff.).

The grid included intersecting streets (Figure 14.1). The spacing of these again ties in with Roman agricultural practice, and relates to how roads were ordered within field systems (Dobson, 2008: 83). Polybius indicates that most of the streets were 60 Roman feet wide, with exceptions being the two important streets on either side of the headquarters range, which were 120 feet wide. Remains from Numantia and Renieblas suggest that such widths were generous and streets were narrower in practice, presumably to economize on space (Dobson, 2008).

It can be established from Polybius and other sources that several of the streets had names and led to named entrances (Figure 14.1) (Dobson, 2008: 80ff.). Such terminology appears to have continued throughout the Roman period, though the location of each, relative to the front of the camp, changed after the move away from double-consular armies (Figure 14.5) (Dobson, 2008: 110ff.). Concerning street names, the one of "via decumana," used frequently in modern discussions to describe a street leading to the rear entrance, should not be used, as it has no literary evidence and was fabricated by Schulten, as he freely admits (Schulten, 1927: 107; Dobson, 2008: 83).

Polybius's description concerns the general layout of camps. Even at this level, the amount of information is variable and at times non-existent. For example, he notes the size of each area occupied by the different types of legionary infantry maniple (triarii, principes, hastati) and of the cavalry turmae, but provides no such information for the equivalent allied units, instead remarking only in general terms about them (6.28.1ff.). The nature of the areas occupied by the select allied troops, the *extraordinarii*, who were placed by the headquarters range, is even vaguer and the information is confusing (6.31.1f.; Dobson, 2008: 94ff.). As for the layout of the headquarters range itself, consisting of forum, praetorium and quaestorium (see below), no details are provided

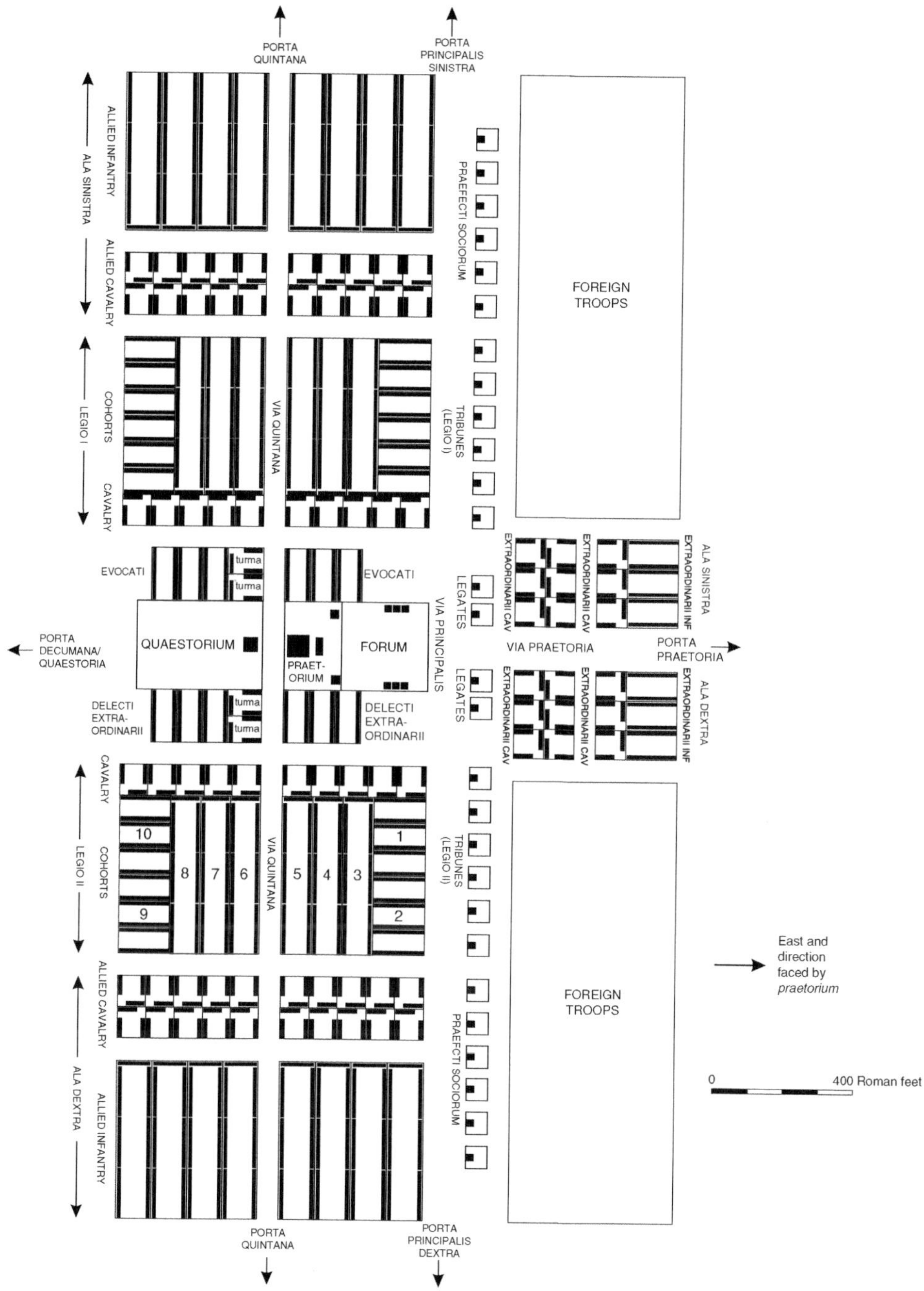

Figure 14.5 Reconstruction of the cohort-organized, single-consular camp of the late second century. Source: Drawing by M. Dobson, from Dobson (2008).

other than the relative siting and overall size of the praetorium (6.27.1ff.). Given the relative importance of these areas for an army, this lack of information is surprising. One exception to the amount of detail that Polybius provides concerns the layout of the tribunes' quarters (6.27.4ff.) and indeed about the activities of the tribunes in camp in general. This adds weight to Polybius's source being a tribune's commentarius (Rawson, 1971).

Despite Polybius's shortcomings, reconstruction of the camp is possible by using the information he provided, in combination with other literary sources and the archaeological evidence, particularly from Numantia and Renieblas (Figures 14.3 and 14.4). These camps are the earliest known and mostly date from the second half of the second century. Their publication (Schulten, 1927, 1929) has become one of the main sources for the organization of Republican camps, as the internal layout of many of them is indicated by stone structures, since the usual tented accommodation (which would leave no archaeological traces) was unsuitable in Spain because of weather and siege conditions. Also, no other camps of this period have been the subject of such large-scale excavation. Schulten's interpretations should, however, not be accepted uncritically, as they all too often have been, as recent reassessment of his results shows that different reconstructions are possible (Dobson, 2008). Caution should also be exercised in using this material for aiding the (retrospective) reconstruction of late third-century camps, and of creating potentially circular arguments that Polybius helps interpret Numantia, which helps to interpret Polybius.

The layout of the individual areas within Polybius's account is uncertain, as it largely lacks such details. For example, the dimensions for the legionary areas are given, but there are few details about internal layout. No details at all are provided about the cavalry areas. Polybius does say that the first tents in each of the legionary maniples are occupied by the centurions (6.30.5), but how the tents of each maniple are arranged is confusing and the number and size of the tents are not stated. It is consequently fortunate that *De metatione castrorum* (1) has a good description of how a century's tents are arranged and even states that they were 10 Roman feet square and provided accommodation for eight men – a contubernium.

The physical form of tents has been particularly studied by Driel-Murray, based on depictions on Trajan's Column (which shows different sizes and types of tents) and surviving leather fragments from military contexts, usually Imperial, but one Republican example exists, from Alésia (Driel-Murray, 1990a, 1990b, 1991, 1998, 2001; Richmond, 1982). Driel-Murray concludes that the similarity in form of the fragments, regardless of period, shows that once tent design had been perfected in the Republic it needed little subsequent change (Driel-Murray, 2001: 366). This adds weight to the idea that what *De metatione castrorum* describes has origins in the Republic and hence reflects elements of accommodation practices in the Polybian army (Dobson, 2008: 86). The evidence can also be added to the Polybian-related archaeological remains from Numantia and Renieblas to help with the reconstruction of such areas (Figure 14.3). The work of Morel (1991), which looks at archaeological evidence from early Imperial installations, and literary evidence for the size and layout of tents and the relationship between tents and barracks should also be noted here, as should that of Pamment Salvatore (1993).

By Polybius's own period, armies had reduced to being single-consular (Dobson, 2008: 56ff.). Hence, Polybius refers to the camps of such armies (6.32.8). Army organization was also moving away from larger units called maniples towards smaller cohorts. Polybius was aware of this, as he explains what a cohort was (11.23.1). The chronology of the transition to cohorts has been much debated (e.g. Bell, 1965; Pamment Salvatore, 1999; Cadiou, 2001; Dobson, 2008: 58ff.), but there is growing consensus that it was a fluid development with both types of unit being in use throughout the second and into the early first centuries, before the cohort became the standard tactical and operational unit in the first century and thereafter.

Polybius is vague about how the single- and double-consular camp differed, other than to indicate that the central command and administrative areas were located between the two legions (Dobson, 2008: 68). A reconstruction is possible if evidence from Numantia, Lager III at Renieblas in particular, and other literary sources are used (Figure 14.3) (Dobson, 2008: 101ff.). This type of camp is highly significant as it displays the main features of all future camp, fortress and fort layouts of the Republic and Empire in embryonic form. These are the basic T-shaped main streets of *via principalis* and *via praetoria*, with the headquarters at their intersection; this central area is given a protective surrounding screen by troops.

Polybius provides no discussion about the layout of the early cohort-based camps. They were presumably based on the preceding tried-and-tested manipular-based ones. This is borne out by elements of the manipular camps still being apparent in installations of the Empire. Radical change was also probably unlikely as armies are generally conservative and prefer gradual changes based on successful, established principles rather than a sudden adoption of totally new and potentially risky practices. The evidence from Numantia and Renieblas can help reconstruct the early cohort-form of camp, as remains from both manipular and cohort-organized armies appear to be present at these installations (Figures 14.3 and 14.4) (Dobson, 2008).

4 The Use of Camps

During the campaign season, when the army was regularly on the move, it constructed a camp at the end of each day's march, as attested by many literary references throughout the Roman period. Consequently, these camps are often referred to today as "marching camps." Such camps may have been occupied for just a single night, within a sequence of daily maneuvers. Some may have been in use for a few days, providing, for example, a temporary base for foraging for supplies or from which to send out patrols to collect information to prepare for the next stages of the campaign. They also formed good temporary bases from which to carry out local offensive operations. There are many references to the army working in this way to pacify an area or to regain a rebellious area.

The "temporary" aspect of camps was actually very flexible. Occupation of these summer camps (castra aestiva, usually just referred to as castra) could be extended to provide accommodation for over-wintering, literally becoming winter camps (castra hiberna, or simply hiberna). Alternatively, specific winter camps could be established in areas more convenient for over-wintering, for example, if the campaigning season ended in an area where adequate supplies could not be provided. Sometimes the more comfortable facility of being quartered in a town, rather than a hiberna, was adopted. There is growing archaeological evidence of the army occupying native defended settlements or oppida, perhaps after the previous inhabitants were removed. For example, at La Chaussée-Tirancourt and Petrisberg (Trèves) in France and on the Titelberg (Pétange) in Luxembourg, finds of military equipment suggest occupation in the period 30–20; the context for the latter two sites may have been a consequence of a revolt of the Treveri in 30–29 (Reddé, 2006: 27, 249ff.).

For hiberna, the tents used for summer accommodation were replaced by timber or stone structures in the same locations; there was no good reason to move the relative position of troops between camp and winter quarters, and such a practice would in fact have had a negative impact on communications and logistics. The timber and stone structures clearly provided warmer and drier accommodation. There is also increasing evidence to suggest that sometimes the tents remained, but were given improved insulation and water-proofing by being thatched over and even built around with low walls (Morel, 1991; Dobson, 2008: 254ff.). This seems to have been done by Caesar (*B Gall.* 8.5) and there is evidence for it at Numantia (Dobson, 2008: 254ff., 389; Dobson, forthcoming). The provision of more permanent accommodation in the hiberna naturally developed into the much longer-lasting timber and Imperial stone forts and fortresses.

A camp was always established before a battle. This had obvious advantages. Troops could prepare for the forthcoming fight within a protected enclosure. The commander could use this strategically strong position to take his time planning when to deploy, in consultation with his consilium (e.g. Livy XXII.3.8, XXVII.46.5; Johnston, 2008: 39). This would be an efficient site as the relative location of troops within the camp related directly to their fighting order; they could simply assemble in the streets and the space immediately around the inside of the defenses and march out ready to fight without significant (and potentially vulnerable) field maneuvers in front of the enemy (Polyb. 6.31.12).

Important for the future survival of the army, as well as its morale when fighting, the baggage and support activities were protected in the camp (Caes. *B Civ.* 1.41f.; *De metatione castrorum* 18; Polyb. 6.31.13). The consul Lucius Aemilius Paullus points out that protecting the baggage was not a job for cowardly troops; up to a quarter of the army could be given this role (Livy XLIV.38.6). Such a large proportion of the army being away from the battlefield seems surprising, but makes good sense, since what remained in the camp was vital to the future functioning of the army. Also, if the unexpected happened and the army was defeated in battle, the camp provided a nearby and very defensible refuge; the troops left behind inside could readily start to defend while the retreating troops came into the camp and were taking up positions along the defenses.

The camp could provide the setting for a defeated enemy to negotiate with the Romans for their surrender (Johnston, 2008: 50). Similarly it could form the reception area for the captive defeated enemy leaders (Johnston, 2008: 54). This, no doubt, had the advantage of raising Roman morale as it allowed the troops to ridicule the captives as they passed along one of the main streets towards the central command area of the praetorium and its associated forum, where the captives could be displayed and harangued. But the ritual also had parallels to the punishments used on the Roman soldiers themselves. The condemned went from the judgment performed in the praetorium to the place of punishment outside the camp, along a street through the main troop encampment, allowing the troops to show their "displeasure" to the condemned as they passed by (Dobson, 2008: 100).

The Romans never intended to fight from a camp – it was not like a medieval castle – but always sought battle in the open, where their advantages of maneuverability and firepower could be fully exercised. For example, Livy (X.35.1ff.) reports a consul encouraging his troops to leave camp to fight because of the impracticalities of fighting in camp.

This did not mean that troops never fought in the camps, since camps were attacked or even besieged; there are many accounts of this happening. But there is no record of a camp ever having been taken by the enemy, unless the Romans had been previously defeated in open battle. This is clearly a solid indication of the excellent military qualities acquired by the Roman camp. The reports of two attacks and one siege coincidentally provide an important source of information about the location of the quaestorium, the quarters of the quaestor (the officer in charge of the army's finances, who could act as deputy commander; Johnston, 2008: 8ff.), suggesting it was near the rear entrance in single-consular camps (Livy X.32.3ff., XXXIV.46.4ff., XL.27.1ff.; Dobson, 2008: 105ff.).

The power of camps meant that one of them, or several forming a circuit, with associated siegeworks, could bring about the capitulation of an enemy city. The camps provided a conveniently located and defensible base for all types of siege-related operations. The visible presence of a disciplined army in a highly ordered camp, carefully preparing and working towards the intended destruction of the city, sometimes through expending huge amounts of labor on constructing earthworks intended to strangle the besieged city, must have had a very significant demoralizing impact on the besieged onlookers. It is no surprise that Roman siegecraft became a very skilled operation (Campbell, 2005; G. Davies, 2006).

There are many literary references to camps being used in the context of sieges during the Republic. Caesar, for example, repeatedly used this technique in attempts (not always successful) to overcome Gallic oppida. Alésia and Gergovia are notable instances of his sieges, to which both literary and archaeological evidence can be brought to bear (*B Gall.* 7.69ff., 7.36ff.; Deberge and Guichard, 2000; Reddé and Schnurbein, 2001). In Spain, the famous example is the siege of Numantia by Scipio in 134–133, where there is also both literary and archaeological evidence (Figure 14.3; App. *Hisp.* 76ff.; Dobson, 2008).

The high regard the Roman army had for camps and several of the main aspects of how they used them are well summarized in a speech of 168 by the consul Lucius Aemilius Paullus to his consilium, defending his decision to stay in camp rather than give battle:

> … can we possibly consider it a small advantage that we shall go out, with the good help of the gods, from this camp in which we have spent last night, to do battle either today or at least tomorrow if we so determine? Does it make no difference whether you bid a soldier take up arms and lead him into battle when he is wearied neither by the toil of a journey nor by construction on that day, when he is rested and refreshed in his own tent, when he is full of strength and vigorous in body and in spirit? Or whether he is wearied by a long march …
>
> … you say, we should have had a battle-line undeployed and disorderly; but our camp was fortified, our water-supply found, the way to it secured by placing guards, and reconnaissance made all around … Your ancestors regarded a fortified camp as a haven against all the mischances of an army, from where they might go out to fight, and to which they might find shelter from the storm of battle. Therefore when they had walled in their camp with fortifications, they used also to strengthen it with a powerful garrison, because one who had been stripped of his camp, even though he had won the battle, would be considered the loser. Camp is the shelter of the conqueror, the refuge of the conquered. How many armies meeting with none too favorable fortunes in battle, after being driven

within their rampart, have in their own good time, sometimes after only a moment, sallied forth and routed the victorious enemy? This abode is a second home for the soldier, its rampart takes the place of city walls and his own tent is the soldier's dwelling and hearthside. Should we have fought like nomads with no abode, so that, whether conquered or conquerors, we might return – where?

(Livy XLIV.38.6–44.39.5; after A. Schlesinger, 1951, Loeb)

5 Creating the Camp

Roman encampment was a very efficient operation. Specialist troops commanded by a tribune went on ahead of the main force to locate a suitable site. They then used a system of colored flags and spears to mark out the parts of the camp, starting with the commander's area, the praetorium (Polyb. 6.41.1ff.). The system for laying out camps meant that their size could easily be adjusted to suit the requirements of each force, as shown by the very wide range of sizes of surviving camps. When the main army arrived, Polybius says that the surveying system was so well-known to the troops that they could very quickly and straightforwardly go to their allotted place, itself usually in the same relative position, very much like "the return of an army to its native city ... and everyone ... reaches his own house without fail, as he knows both the quarter and the exact spot where his residence is situated" (6.41.11; W. Paton, 1923, Loeb).

6 Camp Defenses

The general modern view about Roman camps is that they had a standard defensive provision of an encircling ditch, with the excavated turves, soil and stones heaped on the inner side to form a rampart, on top of which was some form of stockade. The Romans also clearly placed extreme value on man-made defenses for the security of those inside, as they sought to prevent any damage to them by imposing the death penalty, or at the very least dismissal from the army, on any Roman troops who crossed rampart or ditch (*Dig.* 39.16.3, 39.16.18; Zonar. 7.3).

It seems though, that defenses were not always required and not as universal as modern discussions imply. Gilliver (1993) points out that there are a number of literary sources which indicate that defenses could be completely dispensed with, or either ditch or rampart not constructed, if encamping in friendly territory or if circumstances meant the army was not under threat. Nevertheless, the numerous references in the sources to camp defenses implies that their construction was very frequent, though since most of the references are to conflict situations, we would expect that defenses would always be provided in such circumstances.

All soldiers carried entrenching equipment to facilitate ready creation of the ditch and rampart (Fuentes, 1991). Polybius relates how responsibility for the construction of each side of the camp's defenses fell to the troops located closest to that side (6.34.1). Each maniple was required to build a specific section, supervised by their centurions, with overall command by the tribunes (6.34.2). Juvenal makes a telling comment about how forceful the centurions could be in getting the men to dig: "a centurion's knotty staff would be broken over his head if his pick were slow and sluggish in the camp's trench" (*Sat.* 8.247f.; G. Ramsay, 1918, Loeb).

It seems to have been standard practice to provide a guard in front of those involved in the defense construction when hostile forces threatened. Caesar, for example, specifies the number of legions he detailed to stand guard against nearby forces while the remaining legions constructed the camp ditch (*B Civ.* 1.41ff.). Vegetius comments that the cavalry and the infantry excused from fatigues guarded those involved in defense construction (*Mil.* 1.25, 3.8).

Polybius indicates that the rampart was 240 Roman feet (71 m) from the accommodation (6.31.11), creating an area known as the intervallum (Isid. *Etym.* 15.9.2). This space was to provide a sufficiently wide area for assembling troops before and after deployment, for keeping cattle (for food) and for booty, but most importantly to prevent enemy missiles reaching the tents (Polyb. 6.31.11ff.). Polybius's dimension for the intervallum may have been only a theoretical ideal, as the remains at the camps of Numantia, Renieblas and Cáceres el Viejo, for example, indicate the area was much narrower than this and could also vary in width around the camp (Dobson, 2008; Ulbert, 1984).

Polybius provides very little information about the physical form of the defenses and entrances. This is possibly because his source concentrated on the activities of the tribunes (see above), and these officers did not need to know construction details. Such things could have been left to the supervising centurions. Other literary sources help, but detailed information is generally poor, with the exception of Vegetius and *De metatione castrorum*. Caesar actually gives some dimensions for defenses. For example, in operations against Afranius, who was close by, one camp is reported as having a ditch 15 feet (4.4 m) wide (*B Civ.* 1.41; Frontin. *Str.* 1.5.9). When fighting near Bibrax, Caesar constructed a camp with a ditch 18 feet (5.3 m) wide and a rampart 12 feet (3.6 m) high (*B Gall.* 2.5). This camp may well be the one found at Mauchamp (see below). Such ditches seem very wide and would have required a great deal of physical effort to dig. Caesar may have been exaggerating, but Vegetius (1.24, 3.8) states that where a large enemy force threatened, ditches up to 17 feet (5 m) and 9 feet (2.7 m) deep should be dug. In contrast, Vegetius (3.8) states that for short-occupied camps the ditch could be 5 feet (1.5 m) wide and 3 feet (0.9 m) deep, the same as given by *De metatione castrorum* (49) for all camps. For longer-occupied camps, Vegetius (1.24; 3.8) prefers ditches between 9 (2.7 m) and 13 feet (3.9 m) wide and 7 feet (2.1 m) deep. *De metatione castrorum* (49) also states that the ditches should have a "V"-shaped profile or a Punic form, in which the outer face is vertical and the inner sloping.

There is some archaeological evidence for Republican camp ditches. Recent excavations at Caesar's camps at Alésia revealed several V-shaped ditches up to 4.2 m wide and 1.5 m deep and at Gergovia a ditch survived that was 3.5 m wide and 1.8 m deep (Deberge and Guichard, 2000; Reddé and Schnurbein, 2001: 176ff.). These dimensions indicate that those given by Caesar may not have been exaggerated. Less substantial ditches were nevertheless also created, as found at Almazán (Spain), for example, where the ditch was only 2.8 m wide and 1.6 m deep (Gamer and Frías, 1969: 177). The example of Gergovia also shows that ditches could be dispensed with if the local topography presented a sufficiently strong natural defense. Here, no traces of a ditch could be found at the small camp on the very steeply sloping southern and western sides (Deberge and Guichard, 2000).

Sometimes it seems that the provision of ditches had to be ignored, even though encampment was in hostile territory. For example, there is no evidence for any at Numantia and Renieblas. This could be because the early twentieth-century excavation techniques failed to see them, but more likely, the absence was probably genuine as the ground conditions were simply too rocky to allow their creation (Dobson, 2008: 123ff.).

If a ditch was present, it would clearly have made sense for the excavated material to have been heaped up on the inner side to form a rampart. The size of this would obviously have related to the amount of material dug out of the ditch (i.e. the size of the ditch). *De metatione castrorum* (50) advises that the rampart should be 8 feet wide and 6 feet high. The profile would have depended on how friable the excavated turves were and if the soil was able to retain stability when heaped, as is pointed out in *De metatione castrorum* (51). The remains of such ramparts were found at Alésia, for example, where they survived as mounds between 2 and 5.5 m wide and up to 0.4 m high, consisting of earth and stones dug from the ditches (Reddé and Schnurbein, 2001: 178, 236ff., 280), though slumping and erosion will presumably have altered the surviving widths and profiles from the originals.

At Numantia and Renieblas are examples of how ramparts could be constructed without material being provided from a ditch. Here was a ready supply of suitable building stones. These were used to construct an inner and outer revetment wall, with an infill of earth and stones, producing a rampart generally between 2.5 m and 3 m wide and surviving up to 1 m high (Dobson, 2008: 401ff.). This width compares favorably to that stated by Appian (*Hisp.* 90) for the Scipionic siegeworks and so the height for the rampart, equivalent to 3.55 m, may also be plausible (Dobson, 2008: 46). Similar stone rampart construction technique and width were found at Cáceres el Viejo (Ulbert, 1984: 17ff.).

Evidence for the nature of the stockade placed on top of the earthen ramparts, (or that could be constructed where a ditch was absent) is available in literary sources. Livy (XXXIII.5.5ff.) compares Greek and Roman practice in this respect. He is critical of the Greeks because they cut large trees and leave too many branches on them. These were not only difficult to carry at the same time as weapons, but they had to be laid relatively far apart, resulting in them being easily pulled out by a very small number of the attacking troops. In contrast, the Romans cut light trees or branches with only a few, sharpened, offshoots, allowing several to be carried at once by one man, together with his weapons. Livy emphasizes their portability by saying that soldiers actually carried a stockade with them on a short march, so a camp could be fortified in any place (XXXIII.6.1). These prepared branches could be assembled so closely together that they formed a dense and prickly hedge that could not easily be parted when attacked (XXXIII.5.10ff.; cf. Polyb. 19.18.1ff.; *De metatione castrorum* 51).

Although other authors also refer to stakes being carried (Fuentes, 1991), Caesar, for example, indicates that they were not always. He notes that on one occasion the troops could not go out and collect them near the camp, due to deforestation by the enemy. Consequently he built what he terms a novel form of rampart, a fairly complex structure of a pair of parallel stone walls, roofed over and strengthened with cross-timbers (*B Civ.* 2.15). At another location he comments that material for a stockade could only be procured at some distance from the camp (*B Civ.* 1.42). Perhaps the branches were only carried when circumstances advised it. This would make good practical sense as it kept the usual marching load to a minimum.

A long-standing belief among scholars was that a stockade was created by pila muralia, objects referred to, for example, by Caesar (*B Gall.* 5.40; 7.82; Beeser, 1979). Examples of what were thought to be pila muralia have been found in sites dating to the Empire (Beeser, 1979; Bennett, 1982), but none so far from the Republic. These are lengths of timber, generally 1.5–2 m long, square-sectioned, with points at both ends and a short reduced-width section in the middle. These are not, however, pila muralia. A pilum murale was probably a short-range heavy weapon thrown downwards from the rampart (Connolly, 1997: 44, 2002: 2). Such weapons may be represented in the hoard from Talamonaccio, probably dating to 225, and also at Alésia (Connolly, 1997, 2002: 2).

The exact way in which the double-pointed stakes were used is uncertain (Bennett, 1982; Gilliver, 1993; Davies and Jones, 2006: 26ff.). It seems unlikely that sufficient numbers could have been carried to form a continuous fence-like palisade either on top of or instead of a rampart. Experiments have shown that such a fence structure can be pushed over easily and the upwards-facing pointed end would have been blunted when the stake was hit into the ground (Gilliver, 1993; Junkelmann, 1986: 206ff.). Despite this negative evidence, reconstructions of camps can often be seen showing the stakes used like this. A viable suggestion is that groups of three such stakes could have been tied together to form a series of large, interlocking caltrops along the top of the rampart (Davies and Jones, 2006: 26ff.; Gilliver, 1993, 2001). Hence perhaps these objects should be known as tribuli, referred to, for example, by Vegetius (*Mil.* 3.8). Assembled in this way they would in effect have formed a more "machine-produced" version of the prickly hedge-like stockade described in the sources.

7 Entrances

Polybius includes only two references to entrances, both concerning the guards, the light-armed skirmishers or velites (6.35.5, 6.36.2). Neither reference gives any information about how many entrances there were, what their physical form was or what their names may have been. An idea about the number, relative location and names of the entrances can be gained from other literary sources (Figures 14.1 and 14.3).

The physical form of entrances can be established from archaeological evidence. The actual entrance passageway was created simply by interrupting the main camp defenses. Possibly some form of movable timber barrier was placed across the entrance passageway to act as a gate in the very short-lived marching camps (Gilliver, 1993: 52ff.). At camps occupied for some time, such as over winter or under siege conditions, there is evidence suggestive of proper timber or stone gateway structures, even with flanking towers, as at Alésia, Cáceres el Viejo, Numantia and Renieblas. Some of those at Alésia and Lager III at Renieblas are quite elaborate (Ulbert, 1984: 21ff.; Reddé and Schnurbein, 2001: 179ff.; Dobson, 2008).

To prevent a direct assault on the entrance, a common device was to place a short length of defenses directly in front of the entrance passageway – a titulus. Where a camp was given both rampart and ditch, the titulus would have consisted of the same, with the ditch being on the outer side (Figure 14.6, bottom), as in the southern entrance of

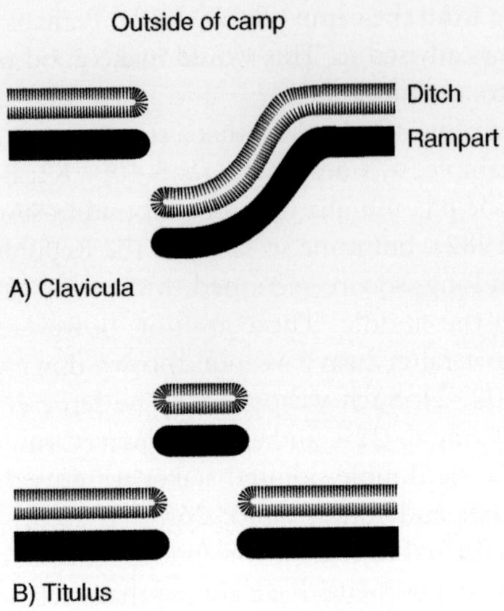

Figure 14.6 Types of Roman defended camp entrances (schematic, not to scale). Source: Drawing by M. Dobson.

Camp A at Alésia (Reddé and Schnurbein, 2001: 276ff.) and Almazán, where the rampart sections of the tituli were seen by Schulten in 1911. They presumably had a ditch in front as the camp itself had a ditch (Gamer and Frías, 1969). At Numantia and Renieblas, where ditches are absent, only the rampart aspect of the titulus occurred: in Numantia – Peña Redonda, at the eastern entrance; in La Rasa; and in Renieblas – Lager IV, VI, VIII (Dobson, 2008).

Another form of defensive entrance feature was the clavicula. This was a curved length of defenses extending from the main ones, across the rear of the entrance (Figure 14.6, top). Until relatively recently, such features were thought to date from the Flavian period, based on the examples known. They have now been demonstrated to go back to the Republic, having been found at Alésia (Camps A and C), here also with a narrow timber-revetted titulus in front of the entrance (Reddé and Schnurbein, 2001: 179ff., 259), and at Mauchamp (see below).

8 Camp-Construction Training

The creation of a Roman camp was a well-honed art. Archaeological evidence from the Empire shows that troops actually practiced camp building, as shown by groups of small practice camps in Germany and Wales (Bogaers and Rüger, 1974; Davies and Jones, 2006: 163ff.). These concentrated on the difficult aspect of the corners, which had to be rounded to build a strong enough structure in earth, and the defended entrances. Vegetius mentions camp construction as part of troop training (*Mil.* 1.25). It can be presumed that such practice activities took place in the Republic. When Scipio was given command of the army for the Numantine Wars, he is recorded as having put his troops through a rigorous training routine, which included daily fortifying new camps, demolishing them, digging deep ditches and backfilling them (App. *Hisp.* 86). Metellus did something similar when preparing to campaign (Sall. *Iug.* 45.2). Such resulting efficiency meant that a Roman army could quickly create the vital security of an ordered camp with good weather-resistant accommodation and protected supplies within, that was easily defensible if required. The overall time required from surveying a site to completion of the camp is clearly uncertain, but one estimate, based on Caesar (*B Gall.* 2.17–28) suggests only nine hours were needed (Peddie, 1994: 75).

9 The Archaeological Evidence

As stated above, the earliest known camps that certainly date to the second century are found in the Iberian Peninsula. In this area there should be a wealth of archaeological evidence for the Republican army, as conquest took over 170 years, but relatively little is known, for after the work of Schulten there was little interest in Roman military archaeology there for many years. Where archaeological work did occur, the results were rarely or only briefly published (Morillo and García-Marcos, 2002: 779; Morillo, 2006a: 7, 2006b: 92ff.).

No new Roman military sites were discovered in the Iberian Peninsula until the 1960s, with developments in scientific archaeology and the start of aerial photography there, but more attention was paid to Imperial camps than the Republican ones (Morillo and García-Marcos, 2002: 779; Mora, 2006; Morillo, 2006a: 7ff.). More recently, there has been a very marked increase in activity, sparked by a series of conferences beginning in 1998 (Morillo, 2002) and continuing since (e.g. Cadiou *et al.*, 2008; cf. Dobson, forthcoming). This has involved fieldwork, excavation and aerial photography revealing several new military sites (e.g. Cerro de las Fuentes de Archivel, Cerro del Trigo, Gandul, La Cabañeta de Burgo de Ebro, La Poza, Los Cascajos de Sangüesa, Muro de Agreda, Segeda, Villa Joiosa, Zalbeta) as well as more information about Numantia and Renieblas (Didierjean, 2008; Didierjean and Petit-Aupert, forthcoming) Not all sites have been securely dated, however, and some dates are surmised on the basis of appropriate surface finds of weaponry and coins rather than the form of the defenses. Of particular interest are sites where a Republican camp, nearby battlefield and associated earthworks are emerging (e.g. Andagoste, El Pedrosillo and Las Albahacas, the last the possible location of the battle of Baecula, but only the battlefield has been found).

The picture in Gaul is very similar. Much interest in Caesarian camps was created by the fieldwork and excavations sponsored by Napoleon III during the mid-nineteenth century, especially at Alésia, Gergovia and Mauchamp, but most research moved on to sites of the Empire. From the mid-twentieth century things improved, as in Iberia, with increased use of aerial photography (e.g. by Agache, Braasch, Dassié and Goguey) leading to discoveries and sometimes subsequent fieldwork and excavations of several probable Republican military sites (e.g. Le Catelet (Vendeuil-Caply), Faux-Vésigneul, Folleville-Quiry-le-Sec (Le Blanc Mont, sometimes called Breteuil-sur-Noye) and Mirebeau-sur-Bèze (La Fenotte); see Reddé *et al.*, 2006). Very important excavations, fieldwork and surveying also took place in the 1990s at Alésia and Gergovia (Deberge and Guichard, 2000; Reddé and Schnurbein, 2001).

Nothing is known archaeologically about pre-Caesarian Roman military activity in Gaul. Curiously there is nothing known for certain after him either, until the period of

Augustus, even though there is literary evidence for military activity in the intervening period in Gaul (Reddé *et al.*, 2006: 25). The earliest potential camp in France is Mauchamp (Berry-au-Bac). This has often been associated with Caesar's activities at Bibrax in 57 (*B Gall.* 2.6ff.), but until recently, the camp was dated to the Flavian period because of the presence of claviculae. Recent coin evidence from the nearby oppidum at Vieux-Reims strongly supports this being Bibrax, so the camp could well be Caesar's (Lambot, 1997; Reddé and Schnurbein, 2001: 536ff.; Reddé *et al.*, 2006: 24, 139, 225ff.). Compared to the number of camps known from the Empire, the list of Republican ones is still very short. It can only be hoped that archaeological research activity into the Republic will increase. It is an area of Roman military study that has far too long been ignored.

FURTHER READING

Keppie (1984) remains one of the main modern discussions about the Republican army in general and includes discussion of camps. Connolly (1981) provides an excellent and very readable discussion of army organization, weaponry, camps and battles, with many wonderful reconstruction drawings. The period of the second century in particular, both the literary and archaeological evidence for army organization and camps, is discussed in detail by Dobson (2008). Another thorough discussion, focusing on the army in Spain, is offered by Cadiou (2008). Good bibliographies on the Republican army can be found in Cadiou (2008) and Dobson (2008).

General discussions about Roman Spain and Gaul, including military sites, are provided by Keay (1988; see also Chapter 33) and King (1990) respectively, with Knight (2001) producing a convenient field guide to sites in France. Useful resumés of archaeological work in the Iberian Peninsula, with bibliographies, are in Luik (1997) and Morillo and Aurrecoechea (2006). A very good summary of work carried out in Gaul, with an excellent gazetteer of sites and bibliographies for each, is provided by Reddé *et al.* (2006).

CHAPTER 15

Reconstructing Religious Ritual in Italy

Alison B. Griffith

1 Introduction

In what way do archaeological remains inform the study of religious ritual? This issue challenges every archaeologist simply because all human cultures believe in gods and engage in rituals that honor, placate, entreat, thank or communicate with them in some way. For an archaeologist, an investigation of ritual behavior is limited by the nature of the evidence and by this inescapable fact: archaeological remains may attest the occurrence of a ritual action by revealing the environment and spaces where such actions occurred, or the equipment that was used, but they do not generally inform us about the nature of ritual actions themselves. It is therefore not the purpose of this chapter to substantiate descriptions of ritual from written sources merely by drawing attention to ritual equipment recovered by excavation. Rather, the discussion herein provides an overview of physical evidence for religious activities and experiences that do not unequivocally fit strict social-scientific definitions of "ritual," but which were nonetheless part of ritual and, therefore, of religious experience. In other words, this study adopts a broad definition of ritual that includes a range of activities occurring in spaces intentionally set aside for religious purposes. It follows Renfrew and Bahn's list of indicators of ritual: focusing of attention (with place, equipment and symbols); boundary zone; presence of the deity; and participation and offering (Renfrew and Bahn, 1991: 359–60).

Interest in religion in ancient Italy has skyrocketed in the past 20 years, as is evident from the number of publications and conferences devoted to the subject. There are several new historical studies of Roman religion based primarily on texts and inscriptions, but the diversity of religious practices among other groups in the Italian Peninsula, as evidenced by archaeological data, has also received considerable attention. Scholars have

A Companion to the Archaeology of the Roman Republic, First Edition. Edited by Jane DeRose Evans.

confronted the dearth of knowledge about practices beyond Latium and have questioned and revised their views on the spread of Roman religious practices throughout Italy as a result of Roman conquest. In addition to local or indigenous practices, many inhabitants of south Italy, Sicily and Sardinia arrived from elsewhere and brought their own traditions. Trade and political interactions around the Mediterranean were also a source of new deities (e.g. Magna Mater, Isis, Melqart and Astarte) and diverse forms of ritual observance. Among the most important developments are the systematic publication of the thousands of votives, in-depth study of the religious practices of sub-groups (e.g. the Etruscans and Samnites), re-examination of well-known sanctuaries and publication of lesser-known examples outside Latium, and the appearance of regional studies that incorporate field survey data.

Given this exponential growth, it is difficult to do justice to the subject of religious ritual in Italy within the scope of a brief survey. The discussion herein focuses on physical evidence for ritual, which is invaluable for reconstructing the ritual environment. It will touch on evidence from visual representations; votives; the phenomenon of large, multi-functional sanctuaries in Italy; the importance of natural settings; and, finally, theaters. As will become apparent, archaeological remains reveal rituals as diverse, multi-sensory experiences, with sounds, smells and sights that were not necessarily part of daily life.

2 Evidence from Visual Representations

Religious observance in ancient Italy occurred in many forms. In addition to sacrifice, libations and other offerings, there were processions, prayers (petitions, vows, oaths and thanksgiving), hymns, dramatic ritual, dance and music. From the third century onwards, theatrical performances (ludi scaenici) and competitions in the circus (ludi circenses) were also added to the festivals of certain deities, although these are not always classified as "ritual" activities. From an archaeologist's point of view, this variety of practices shows that religious observance involved a prodigious quantity of equipment including altars (permanent or portable), statues of the deity, tools for killing and butchering sacrificial victims, crockery for cooking and consumption of food and drink, vessels for liquids and solids (storage, transportation, use in sacrifices), musical instruments and lamps.

Ceramic remains (coarse ware, fine ware and lamps) and items in stone (altars and statues, and some votive objects) have been recovered from many sites and are more consistently and completely published now than in the past (Hermann, 1961; *Thesaurus Cultus et Rituum Antiquorum* 1.2a "Sacrifices"; 1.2b "Libation"; 1.2d "Dedications"; and 5.2b "Cult Instruments"). In addition to votive items, paterae (plates) and containers for liquid and other substances used as votives or sacrifices are well represented in the archaeological record. Material remains do not tell the full story, however, for we know that Latin has a rich vocabulary for sacrificial clothing and equipment, as well as for attendants at sacrifices. Visual representations, mostly relief in stone or terracotta, help to fill this gap in our knowledge and constitute an important source for understanding ritual. For it is here, and sometimes only here, that we see religious equipment made of perishable material such as wood, leather and wool, and especially metal tools for stunning, killing and butchering sacrificial victims. For instance, we know of the sacerdos or flamen (priest), who wore a toga, who can be shown capite velato, or with the toga pulled over his head; some had special items as well, such as the *apex* cap of the flamines, made of soft leather with a wooden spindle and piece of wool at the top of the head. The camillus (attendant) might hold a pitcher (urceus, guttus) or cloth (mantele), an incense box (acerra) or some other platter or container for sacrificial items such as a lanx for first fruits or situla for cooking the exta, or entrails. The popa, the attendant who stunned the victim with a hammer (malleus), can be shown with his implement; the victimarius, the attendant who slayed and butchered the victim, was often depicted with his knife (culter), but he is sometimes shown with an axe (dolabra or securis). Attendants are also distinguished by their clothing: camilli and popae wore the toga at first, but when the role was filled by slaves, the costume switched to a tunic. Victimarii wore the limus, an apron-like garment that was girded at the waist, leaving the upper torso bare. Even the victims wore a fillet and covering on the back (dorsualis).

Most evocative in the reliefs, but elusive in the ground, are numerous musical instruments. Representations of sacrificial scenes often include a lyre player (lyricen or fidicen) and tibicen, the player of pipes (tibiae). In scenes with sacrifices related to triumphal processions, there are sometimes players of straight trumpets (*tubae*), and in Etruscan funerary processions, curved horns (the cornu and lituus) (*Thesaurus Cultus et Rituum Antiquorum* II, s.v. "Musique romain" nos. 1, 2 and 16). The variety of instruments and their frequent depiction emphasizes their importance: human voices were stilled at sacrifices, but the repeated representation of musicians suggests that these were far from silent observances. In addition to sacrifices, we know that music accompanied hymns (carmina) and dance, and that processions sometimes moved according to a particular rhythm (Fless and Moede, 2007). There is, however, no evidence for percussive instruments until the introduction of the tambourine (tympanum) and cymbals (cymbala) with Magna Mater in the late third and early second century and, somewhat later, the sistrum (rattle) with Isis.

The Imperial date of most of the evidence, both physical and iconographic, requires a cautious approach. There is the obvious concern that iconographic representations of sacrifice, libation and processions dated to the Imperial period may not provide reliable evidence of Republican practices. Moreover, we have no actual examples of musical instruments from Italy (except a sistrum from Pompeii), let alone from the Republican period, although representations in stone and bronze corroborate the use of tibiae, *tubae* and the lituus (Ryberg, 1955: 6–37). In considering the value of Imperial examples we must also acknowledge bias in the evidence. Emperors and their families are represented more frequently, as are particular rituals such as the suovetaurilia (Ryberg, 1955: 190–1), whereas in the Republican period sacrifices related to triumph are more commonly represented and occasionally funerary rites. Nevertheless, while the relief images themselves are mute, their subject matter prompts us to recall that ritual activities were by no means silent affairs.

3 Votive Material

The most prolific form of physical evidence for ritual activity is tens of thousands of votives in terracotta, stone, bronze and, sometimes, other metals. These are being systematically published in the series *Corpus delle Stipi Votive in Italia* according to

13 typological categories, but a useful overview of this Italian material can be found in the *Thesaurus Cultus et Rituum Antiquorum* (I: 328–91). The sheer number of these objects has inspired the term "votive religion" since votives are tangible evidence for the principle of reciprocity – *do ut des* (I give so that you will give) – and the notion that a petition made to or granted by the gods required an offering (Turfa, 2006: 90). Despite their quantity, and occasional inscriptions that recorded a vow fulfilled, we have no idea of the precise ritual by which one dedicated a votive object. Whatever this entailed, it must certainly have required words, either according to a set formula for vowing an object, or in a prayer appealing to the deity for help or conveying thanks for a request fulfilled. Inscriptions, mostly in Latin, give us some clues as to the formula (Turfa, 2006: 91–2), but these are succinct and rarely state the reasons for the dedication. Any object could be vowed, but generally votives consisted of:

- perishable items (grains and plants, milk, wine, honey, cakes);
- personal items (toys, amulets, jewelry);
- practical items (fish hooks, loom weights, tools, utensils, incense burners);
- statuettes/statues (gods, men, women, swaddled babies, animals);
- body parts (miniaturized models of every description);
- altars, cippi and bases;
- ceramic items (miniaturized pottery, lamps, temple models);
- coins.

These items were left on display in the open in sanctuaries, placed in temples, sprinkled over altars, cast into water or sometimes ritually buried in pits, as at the Temple of Jupiter Anxur (Borsari, 1894).

Apart from demonstrating the quantity and variety of votive types, the systematic publication of votives has revealed how widespread this habit was in terms of context and chronology. Votives have been recovered from all types of sites, both urban and rural, with architectural improvements or without, and the habit is not only common throughout Italy, but also among other cultural groups in the Mediterranean. The practice of dedicating votives is an old one, and is more likely to be evidence for Mediterranean *koine* rather than the influence of a specific foreign culture in Italy. While the fourth to first centuries marked a peak in votive dedications at Italian sites, there are older deposits, for example those at Sant'Omobono and the Lapis Niger in Rome. It has even been argued that grave goods in Villanovan burials were a form of votive dedication (Turfa, 2006: 95).

The ability to explore the social, historical and political issues relating to votives is another benefit of publication, and the state of this question is well summarized by Glinister (2006). As she rightly observes, anatomical votives constitute the most common votive type by far, especially during the fourth to first centuries. Most of these were mold-made and were almost certainly purchased from stock in shops rather than specially commissioned, although there are exceptions to this rule. These are, by and large, the easiest type of votive to interpret – they appear to be either requests to be healed (examples with actual pathologies are rare, however) or thank-offerings for a cure. Glinister has suggested other interpretations for certain types, which may be dependent on context. Genitalia, for example, might represent the request or gratitude for a cure from venereal disease, but they might equally express a petition for fertility, or mark a rite of passage into puberty. Other body parts might represent function rather than malfunction: feet could represent a journey or pilgrimage, hands a prayer, and so on (Glinister, 2006: 11–12).

The vast quantity of examples and difficulty in dating them continue to challenge the study of votives. Initially, votive objects were placed in geographic categories: Italic, Magna Graecian/Sicilian and Etrusco-Latial-Campanian (Comella, 1981: 758). The latter group consisted mostly of anatomical votives, a practice that is thought to derive, ultimately, from those offered in the sanctuary to Asclepius at Corinth. The distribution of these has been closely connected with Rome and the foundation of Latin colonies, or the process of Romanization (Torelli, 1999; de Cazanove, 2000). However, more detailed information about provenance shows that anatomical votives had a wider distribution in Italy – many were found at Etruscan sites and, as ongoing excavation is starting to reveal, sites in the Apennine and Abruzzo regions (Colonna, 1985; Campanelli and Faustoferri, 1997; de Cazanove and Scheid, 2003). Therefore, the practice of dedicating anatomical votives may not necessarily be the result of the spread of Roman power (Glinister, 2006: 18, 23–6). Dating also remains a problem simply because it has proven difficult to identify stylistic categories. Moreover, the fact that votives were openly displayed and, sometimes, ritually buried in pits means that they are difficult to date precisely. The arrival of Asclepius in Rome in 291 *may* be a result of the increasing popularity of anatomical votives in the previous century, but it has given rise to circularity in dating, whereby anatomical votives are automatically dated to the fourth to first centuries (when they were most common), and their presence at a site is used as a basis for dating that site or archaeological context. Increased awareness of the potential difficulties of studying votives and the growing number of publications about them should, in the future, result in a more nuanced understanding of this aspect of ancient religion.

4 Ritual Practice and Architectural Environments: The Sanctuary Phenomenon

That ritual practices in antiquity took diverse forms and utilized a range of "equipment" is further supported by the architectural remains of sanctuaries, insofar as these attest to the complex environments where some of these activities occurred. One could reasonably argue that architectural remains provide only general evidence for ritual activities by demonstrating *where* these occurred. But in many definitions of ritual certain characteristics recur: intentionality, a prescribed sequence of meaningful actions, and the necessity of a controlled environment. It is the last of these – controlled environment – that is best demonstrated by architectural remains precisely because ritual actions "are concerned with the world as it should be and not as it is" (Ekstrom and Hecht, 2007: 246). Altars, shrines, temples, groves and sanctuaries attest not merely where religious activity took place, but how varied were the spaces in which ritual was performed, and how diverse the approaches to the divine (Renfrew, 1994: 51–2). The importance of such spaces has been well articulated by Jones, who argued that sacred buildings are not merely the back-drop for ritual activities, but sites where sacred architecture, ritual and individual experience interact in "ritual-architectural events, wherein people engage with works of art and architecture in a kind of dialogical exchange" (Jones, 2007: 252). Put in the context of religious observance in Roman Italy, Jones encourages us to ask whether visitors' experience of the architectural contexts within a sanctuary constitutes one or more "ritual-architectural" events.

The most elaborate and frequently studied sanctuaries in Latium (e.g. Praeneste, Gabii and Nemi) are a hallmark of Late Republican religious architecture, but excavation and field survey projects of the last three decades have demonstrated a much wider geographical and chronological distribution of this phenomenon. The growing number of known sanctuary sites in Etruria, Latium and (increasingly) in the Apennines and southern Italy can in some cases support sophisticated investigations concerning spatial distribution, spheres of political influence, and manifestation of or resistance to Roman control (see Chapters 12, 24, 22, 25; Colonna, 1985; Campanelli and Faustoferri, 1997). In a recent study, Stek exposed the shortcomings of previous studies and addressed the challenge of understanding whether sanctuaries were a tool or result of Romanization, or a cultural expression of indigenous groups in Italy. Using field survey data, he adopted a new approach by examining the relationship between the S. Galdo sanctuary and features of the surrounding environment such as settlements and settlement patterns, *necropoleis* and roads (Stek, 2009). Stek was primarily interested in the pagus-vicus system of land organization and the location and role of sanctuaries within this system, but this approach is a promising one that can be applied to other lines of inquiry.

The antiquity of some of the sanctuaries in Latium is worthy of note. Many were established in honor of archaic indigenous deities and were in use long before the second-century improvements that have made them the focus of architectural studies (Castagnoli, 1977, 1980b). For example, small finds suggest a date of the eighth or seventh century for activity at the sanctuary of Diana at Nemi; architectural terracottas indicate a sixth-century date for Juno Sospita at Lanuvium; and there are indications that the site of the sanctuary of Juno at Gabii was in use from at least the sixth century. There is little reason to doubt that the sanctuaries to Hercules at Tibur and Fortuna Primigenia at Praeneste were similarly archaic, even if that is difficult to establish on the basis of the extant remains. Activity dating to the sixth and fifth centuries has also been attested at Rome (Sant'Omobono) and at other centers in Etruria (Pyrgi, Veii, Murlo and Orvieto) and Latium (Ardea and Satricum), which suggests that the establishment of some sanctuaries is closely connected to urbanism in Italy. Sanctuaries existed in demarcated (sacrosanct) space that served a specific community but, more importantly, the ability to site, build and host a sanctuary was a significant undertaking that required sufficient civic and social organization (Cornell, 1995: 108–12). As religious and (often) socio-political entities, sanctuaries were highly evolved; the deity or deities, motivations and facilities of a particular sanctuary depended on its location, for example at political or territorial boundaries, in or near towns, or out in the countryside (de Cazanove, 2000, 2007; Edlund-Berry, 2006a: 124–7; Stek, 2009: 39–42). In some cases sanctuaries became a significant source of wealth or a monumental expression of political clout or territorial control.

It is generally assumed that deities at the various Italian sanctuaries were closely connected to or embodied a natural feature of the local environment, and that this association is reflected in the location of the sanctuary itself. Groves, or more specifically the clearings within them, were an important site of religious activity. Early worship at many sites almost certainly occurred out in the open, as seems to be the case with the open-air altars at Lavinium but also at Pietrabbondante in Samnium and, perhaps, the rocky outcropping at Terracina, prior to second-century improvements. Bodies or sources of water were important in two ways: as a natural feature embodying divine power (numen) and as a source of water for rituals. So, for example, the sanctuary of Fortuna Primigenia at Praeneste was located at a spring, while the sanctuary of Juno at Gabii overlooked the now dry Lacus Gabinus (Lago di Castiglione) and that of Diana at Nemi overlooked the crater-lake but was also close to the spring that fed it. Sacred sites at springs are also common in Etruria, for example at the Aquae Cutiliae and Aquae Taurini. Outside Latium the noxious sulfuric gasses and boiling mud pools produced by volcanic activity were the inspiration for the sanctuary of Mefitis in the Lacus Ampsanctus (Valle d'Ansanto). Other aspects of the land might have influenced the choice of location; the Sanctuary of Hercules at Tibur is located on the site of an ancient sheep market, a source of sustenance arising from the land (Coarelli, 1987b: 85).

The power of nature to create and set the mood, and to inspire a sense of reverence and awe, cannot be overstated. While the examples listed above related to specific features of local geography, they were also located on hilltops or hillsides that afforded a spectacular view. This feature of sanctuary sites is reasonably consistent throughout Italy and must have served a religious function. Some rituals, such as augury, obviously required open and elevated space, but it may be that the elevated location itself served to mark physical and sacred boundaries (Edlund-Berry, 2006a: 120). No doubt the climb to the top and the outlook upon arrival assisted visitors to make a transition from the profane to the sacred and to put them in a religious frame of mind.

Some hillside locations afforded settings for oracles or divination by lots, or so it is thought. For example, two different features at the Temple of Jupiter Anxur have been consistently interpreted in this way. The first was directly east of the temple and consisted of a spur of natural rock that had been surrounded by a base of concrete faced in opus incertum and measuring 6.9 by 5.96 m. In and of itself this feature might have marked a place hit by lightning, as would be appropriate for Jupiter, but the rock was pierced with a hole that communicated with an underground grotto some 7 m below, and the whole structure was surmounted with an aedicula resting on brick columns with Ionic capitals in travertine (Lugli, 1926: 172; Borsari, 1894: 103–4; Coarelli, 1987b: 119). There was a second grotto accessible from the inner corridor under the temple platform, which is also generally accepted as a site of oracular activity (Lugli, 1926: 166; Borsari, 1894: 105; Coarelli, 1987b: 117) As Coarelli observed, this installation is reminiscent of the terrace of the hemicycles at Praeneste, where, in front of the eastern hemicycle, there was a well similarly adorned with an aedicula and beside it an altar (Coarelli, 1987b: 48–50, 67–70, 119). Not surprisingly, underground corridors at Tibur and Gabii have also been tentatively identified as oracular seats.

The problem in identifying sites of oracular activity is one of physical evidence, which is hardly surprising. The few surviving examples of lots are of bronze, lead or stone and have succinct inscriptions, although there is some doubt about their identification as actual lots. Of these, most were recovered from Etruscan sites, and thus we must rely on visual representations (e.g. on Etruscan cinerary urns) and information from texts, such as Cicero's *De divinatione* (II.85–7), where he recounts the foundation-myth of the oracle at Praeneste (Klingshirn, 2006: 140–7). While it may prove difficult to identify actual oracular seats conclusively, there is sufficient evidence from other sources to attest to the popularity of oracles in the Republican period, and so the possibility that underground caves and passages at sanctuaries were used in this way is plausible. Moreover, it demonstrates another way in which religious activity was integrated into and exploited the natural environment of hillside sanctuaries.

Nothing attests the continuing importance of religious observance in the Late Republic better than the second-century improvements to sanctuaries in Latium and beyond (Coarelli, 1987b). In several locations around Rome individuals or communities amassed the money, labor, materials and skills to transform older sanctuaries into feats of architectural engineering that are still astonishing to behold and, in their time, were made aesthetically pleasing through the use of Greek architectural facades, axial symmetry and mathematical ratios. The influence of the great Hellenistic-era sanctuaries such as those at Cos, Lindos and Delos is undeniable, and yet each site reveals evidence of distinctly Roman practices both in terms of fulfillment of local religious requirements and construction (Lauter, 1976; Pollitt, 1986). All are based on a combination of basic elements adapted to suit the site in question: temple, portico, ramps and steps. For our purposes, the interest in this phase of building lies in the incorporation of much older functions of the site into a larger, unified whole.

Although it seems an unexpected example, the theater-temple-portico complex of Pompey in Rome is useful for understanding the underlying religious functions of Republican sanctuaries. In purely structural terms it contains all the elements of a Republican hillside sanctuary apart from the hill: porticoes, theater and temple. The use of harmonious mathematical proportions throughout the entire complex has been canvassed by Gleason, who further observed that the view from the "Curia of Pompey" in the portico up to the temple of Venus Victix, with the large mass of the theater substituting for a steep hillside, must have been as awe-inspiring as the view from any of the sanctuaries in Latium (see the problem of the existence of the scaenae frons, in Chapter 29; Gleason, 1994: 19–24). Yet strikingly different from sanctuaries outside Rome is the relative proportion between these elements in Pompey's version. Here the triporticus is a full quadriporticus (there was no landscape view to obstruct) and the theater was as wide as the portico. The Temple to Venus Victrix was aligned with the central axis, but was effectively cut off from the porticoes by the theater. The inescapable conclusion is *not* that the deity had dwindled in importance – nothing really supports this idea – but that context, fashion and the competitive political climate in Rome created a much greater demand for a large theater in this location. I will discuss the incorporation of theaters into sanctuaries in more detail later. What is of additional interest here is the presence of a grove.

Sources for Pompey's porticoes – the Severan Marble Plan fragments, recommendations by Vitruvius concerning porticoes, and other literary references – attest and even emphasize shade from trees planted at the center of the quadriporticus (Vitr. *De arch.* V.9.5; Ov. *Ars am.* 1.67–8, 3.387–8; Prop. 2.32.11–16; Mart. 2.14.9–10; *NTDAR*; Gleason, 1994). These trees suggest the idea of a sacred grove, or nemus, which was the primordial abode of the deity, in this case Venus. The idea of a simulated sacred grove has also been argued for the sanctuary of Juno at Gabii by Coarelli (1987b: 16–17; also

Almagro-Gorbea, 1982: 587–90). Here the remains of 34 evenly spaced pits (each 1.5 by 1.6 m) and 70 smaller pits (1.2 by 1.3 m) from two successive phases attest the planting of trees to represent a grove. This feature may occur elsewhere, such as at the sanctuary of Hercules, whose plan was similar to that at Gabii, and even at Lanuvium where, it has been suggested, the western area of the hill of San Lorenzo, which is otherwise inexplicably devoid of structures, may have been left as a grove (Chiarucci, 1983: 255; Coarelli, 1987b: 90). These intentionally planted groves may be understood to fulfill two functions. First, they constitute a formal, albeit somewhat artificial attempt to recreate and therefore preserve the natural domain of the deity. Second, the shade provided by the grove enhanced the ambience of the sanctuary, whose appeal derived in part from dramatic contrasts (Pollitt, 1986: 235–6).

As Coarelli rightly reminds us, the nemus (sacred grove) must be distinguished from the lucus, a clearing in the grove that was an inaugurated zone corresponding to the templum (the area freed from other attachments and made sacred to a god) (Coarelli, 1987b: 16–17). The sanctuary of Diana at Nemi was perhaps the best Republican example of monumental improvements to a lucus created from the sacred grove. But it takes little stretch of the imagination to comprehend many other hilltop and hillside sanctuaries as inaugurated templa carved from surrounding groves. The porticoes, temple and other monumental improvements to the sanctuaries were, effectively, situated in the lucus, the land specially cleared for worship of the deity. The actual status of these groves in the Roman mind is difficult to determine. A simulated sacred grove is, on one level, a contradiction in terms because it could not possibly reproduce the awe-inspiring power of a natural grove or the peculiar aura that instilled reverence and even fear. And yet the existence of groves at Gabii and Pompey's portico, and perhaps elsewhere, suggests that this was an important component of the sacred space within the sanctuary.

If the nemus and lucus were necessary, water was even more indispensable for ritual performance. Many rituals, and especially pre-ritual purification, required "living" water (flowing water from springs, rivers or the sea) rather than water delivered by pipes. For this reason many temples and sanctuaries were located close to natural sources, or had a fountain, basin or well near the entrance or at the foot of the stairs leading to the temple (Scheid, 1991: 209–12; Edlund-Berry, 2006b: 162–80). Natural springs at some sites marked out the home of a deity for centuries, as at Nemi and Praeneste, but we know that water from particular sources was required in certain rites, especially in Rome. One striking example of this was the Fons Camenae at the Porta Capena, which was used by the Vestal Virgins (*LTUR* 1: 216 [Rodríguez Almeida]). Where there was no spring we sometimes find architectural remains that attest the collection and storage of rainwater, as at Gabii and Terracina. At Gabii rainwater was channeled into a cistern that supplied two wells accessible from the portico, and at Terracina it was collected in cisterns that fed wells accessible from the main platform supporting the temple (Borsari, 1894: 104; Lugli, 1926; Guaitoli, 1981; Coarelli, 1987b: 16, 119). Transportation of water from its source to its point of use is an equally important issue, but it is attested only indirectly, for example through the presence of miniature ollae (round storage jars) at votive deposits (*Thesaurus Cultus et Rituum Antiquorum* 1: 368–79, esp. 373, and 2: 63–80).

Apart from fulfilling the requirements of ritual purification, some sources of water were highly regarded for their therapeutic properties, whether or not they supplied a temple to a deity associated with healing powers. On the advice of physicians, patients might partake of a particular source of water by drinking it, immersing themselves in it, or sitting in its vapor and inhaling it. The list of diseases thought to be remedied or at least relieved by partaking of or bathing in water was a long one, and the recovery of those who had suffered from various complaints is attested by numerous anatomical and other votives that have been recovered from sites with therapeutic waters (D'Amato, 1989; Jackson, 1990, 1999). The diseases thought to be best treated by water include diseases of the joints, muscles, nervous system, digestion, urinary tract and kidneys, skin, eyes and lungs, as well as fevers and infertility. Despite this widespread and frequent habit of dedicating anatomical votives to thank a deity for a safe recovery, the coincidence of therapeutic bathing with religious sites and ritual observance occurs only in isolated locations in Italy during the Republican period (see Chapter 1). So, for example, the numerous bathing establishments at Baiae took full advantage of the mineral waters there, but bathing at this site seems to have been a strictly secular activity. No votives have been recovered there, nor any sanctuaries to obvious deities such as Asclepius or Hygieia (Yegül, 1996; Medri, Soricelli and Benini, 1999). By contrast, in south Etruria several ancient authors refer to numerous hot and cold springs, whose therapeutic capacity is also attested by votive dedications (Gasperini, 1988: 27, nn. 1–11; Chellini, 2002: 236–40). At the Aquae Caeretanae (Sasso di Furbara) the recovery of two votive dedications to Jupiter and Fons (or Fontes), attested by inscriptions, raises the possibility of a small shrine annexed to the site of the bathing establishment (Gasperini, 1988: 31–2; Consentino and Sabbatini Tumolesi, 1989: 109–12). Numerous votives, part of a sanctuary and a bathing establishment have been found at the Aquae Apollinares (Stigliano), and a "treasure" of gold, silver and bronze objects and coins was found at the Aquae Apollinares Novae (Vicarello) in the mid-nineteenth century, including 400 kg of *aes rude* and some 5,250 coins, of which around 1,400 are bronze *aes grave* and 50 or so are gold. Silver and bronze objects were also recovered, including several open vessels inscribed to Apollo, suitable for holding liquids (Colini, 1967/8; Künzl and Künzl, 1992; Chellini, 2002: nos. 4.11.1, 4.12). The coins, thrown into the water itself, span seven centuries (the fourth century BCE to the fourth century CE), but the vessels in precious metal date only to the Imperial period. No temple is attested by the architectural remains, although one is inferred on the basis of the number and value of the votives recovered (Künzl and Künzl, 1992: 274). Apollo is thought to be the tutelary deity of the waters, as seems to be indicated by a statue identified as a copy of an original by Leochares, a votive altar to Apollo, and an inscription to Apollo, Silvanus, Asclepius and the Nymphs (*CIL* XI.3294, XI.3296).

The sanctuary at the Ponte di Nona, just east of Rome on the road to Praeneste, was a thriving religious center until the late second century, consisting of a temple, a bath suite, a possible pool and other buildings. Modern farming and quarrying in the 1960s destroyed most of the buildings, but a rescue excavation in 1975–6 established a general picture of the chronology and activity at the site, which was in use from the fourth century on (Potter and Wells, 1985: 23–5). Nearly 8,400 terracotta votives, consisting mostly of body parts, but also some animal models, were recovered from pits where they had been intentionally buried. A statistical study of the anatomical votives revealed a higher than average number of eyes. Hands, heads (whole and half heads), limbs and hands were also common, but genitalia (male and female) far less so. The excavators concluded that this sanctuary specialized in the cure of certain ailments, particularly afflictions of the eyes, and that the proportions of votive types (hands and feet) reflect the injuries most likely to afflict a farming community. The deity of the sanctuary is unknown, and the precise date is difficult to ascertain. The pottery mixed in with the votives dates to the third and second centuries, but the process of filling the pits happened much later, during the fifth century CE (Potter and Wells, 1985).

Some caution is required when interpreting the dedication of anatomical votives as a ritual act and in identifying sanctuaries as "healing sanctuaries," even when a particular sanctuary yields evidence for votives and sacred springs. Anatomical votives recovered from the sanctuary of a particular deity do not, in and of themselves, attest the capacity to heal as a primary function of that deity unless there is corroboration of this ability from another source. Rather, a god might heal by virtue of his or her power, not necessarily by *healing* powers (Scheid, 1991; Glinister, 2006: 13, n. 11). Similarly, a sanctuary with a source of sacred water thought to have therapeutic properties was not a necessarily a healing sanctuary; nor were baths at a sanctuary necessarily curative baths. According to the strictest criteria, a hydrotherapeutic sanctuary was one in which the source of water was captured in a basin, which effectively made it sacrosanct, as is attested at the sanctuary of Clitumnus at Hispellum and the baths at Forum Traiani (Fordongianus) in Sardinia (Scheid, 1991: 207–8). Otherwise, the principal function of fountains, baths and other sources of water at sanctuaries was merely to provide for lustrations and ablutions. This is thought to be the case, for example, with the baths at the sanctuary of Dea Dia at La Magliana, where the Arval Brethren prepared for ritual activities. Such bath suites are more common at extra-urban sanctuaries, where the distance from the city was such that bathing had to take place on the spot (Scheid, 1991: 209–11).

The baths at the sanctuary of Diana at Nemi present an interesting case. Certainly the rural location of the sanctuary made baths a requirement, but the only evidence attesting baths in the Republican period is an Imperial inscription that commemorated a restoration of these (*CIL* XIV.419). Magnesium-bearing springs have been discovered nearby, but it is uncertain whether these fed the baths (Morpurgo, 1931: 297–8; Green, 2007: 63–6). On the other hand, anatomical votives have been recovered from this and other sanctuaries, and Diana had a long association with healing, especially of war wounds and skin diseases (Green, 2007: 235–55). It is thus a reasonable possibility that healing, either through physicians or the therapeutic waters in the baths, was a function of the sanctuary even as early as the Republican period.

It has been argued that the sanctuaries of Mefitis may also exemplify the connection between therapeutic waters and religious observance (Edlund-Berry, 2006b: 175), but this argument is not entirely convincing. Mefitis was a divine personification of the Oscan word for the suffocating sulfurous gas from volcanic emissions and, like Febris (fever) and Robigo (mildew), she was a goddess to be placated, one whose ill effects were to be warded off (Poccetti, 1982; Ranini, 1985). Her sanctuaries, attested mostly by inscriptions and literary texts, were located in south Italy and Sicily, but also around Latium and Rome, and as far north as Cremona (Edlund-Berry, 2006b: 176 nn. 94–5). Excavations at temples in ancient Apulia at Valle d'Ansanto and in Lucania at Rossano di Vaglio have revealed architectural remains, ceramics and several dedicatory inscriptions dating from the fourth century to the first century CE, which constitute the major body of evidence for the cult (Adamesteanu and Lejeune, 1971; Lejeune, 1990).

The temple at Valle d'Ansanto was located at the Lacus Ampsanctus, a lake famously described by Virgil and Pliny the Elder (among others) that emitted a gas so toxic that inhalation was (and still can be) fatal (Plin. *HN* 2.95; Verg. *Aen*. 7.563–7). To date, no archaeological or literary evidence indicates that bathing occurred in sulfurous waters associated with Mefitis at this sanctuary. Ceramic material recovered from the site of the sanctuary included a quantity of black-gloss ware (small cups, plates and *skyphoi*), unguentaria, and unpainted coarse ware for domestic use, but no anatomical votives (Ranini, 1985). Other votives recovered during the unpublished excavations of 1953–4 were concentrated in a creek bed in the so-called "Vado Mortale" (Death Valley) (Ranini, 1985: 8–9). Not all sanctuaries to Mefitis were located at sulfurous springs, however. It is of interest that a spring lying to the north-west of the Rossano di Vaglio sanctuary was redirected to fountains in an open, central court, where there was a large altar (Adamesteanu and Lejeune, 1971: 45–6). The architectural remains of cult buildings do not permit us to draw more specific conclusions about the role of water in the cult, but it has been suggested that there was a healing cult utilizing the sulfurous waters associated with Mefitis here and at other sites (Edlund-Berry, 2006b: 176–8, 2006c: 87). The votives from Rossano di Vaglio included a few fragmentary draped bronze figurines, but mostly comprised objects with a military theme (greaves, belts, helmets, spear tips, a model war chariot and horse) and thus do not explicitly demonstrate an association with healing (Adamesteanu and Lejeune, 1971: 44–5). Rather, a number of inscriptions recovered from this site reinforce the idea that Mefitis was a chthonic deity, that her tutelary associations extended to flocks and springs as well as sulfurous waters, and that she was sometimes associated with other deities, especially Jupiter and Ceres, but also Venus, Mars and Juno Lucina (Adamesteanu and Lejeune, 1971: 81–3; Lejeune, 1990: 56–63; Coarelli, 1998b).

On the basis of the current and ever-increasing body of evidence, it would appear that sanctuaries played a central role in facilitating a range of ritual activities and interactions. They fulfilled a basic need to engage or communicate with the gods through ceremonies, festivals, oracles and the dedication of votives. They also provided space for such encounters that was distinguished by a boundary but also by ambience, and within a natural setting or specially built environment that emphasized dramatic contrasts and sensory stimulation. In their most developed form sanctuaries also facilitated dramatic performance.

5 Theaters

There is a long history of sanctuaries with space for dramatic performances in Italy (Nielsen, 2002; Sear, 2006: 195). While the phenomenon is frequently associated with the second-century sanctuaries around Rome, some examples fall outside these geographic or chronological parameters, most notably the theater and temple at Caralis (Cagliari) in Sardinia (dated anywhere from 300 and 50, on the basis of associated finds), the gymnasium theater at Syracuse (late second century), and even the steps overlooking the open space in front of the sanctuary of the Magna Mater in Rome (later augmented by a platform supported by a terrace wall).

J.A. Hanson coined the term "theater-temple" in the first comprehensive study of sanctuaries with theaters in Italy, arguing that the foremost examples close to Rome (at Gabii, Tibur and Praeneste) influenced Pompey's temple-theater-portico complex, and that the type derived from the contemporary development of the contiguous curia-comitium complex known from Cosa, Rome and Paestum (Hanson, 1959). But the phenomenon of temples with theaters is more widely distributed in Italy (Tagliamonte, 2007). In addition to the example at Caralis mentioned above, there are other examples near Rome at Teanum (second century), as well as in Etruria at Castelsecco (mid-second century) and in Samnium at the second-century sites Iuvanum and Pietrabbondante (Sear, 2006: 139, 152–3, 165). One might even include Sarno and the first-century theater opposite the Temple to Hercules at Alba Fucens (Sear, 2006: 138, 149–50). In a recent study of theaters and ritual drama, Nielsen distinguished theaters located within a sanctuary and directly related to the temple of a deity as "cultic" (Nielsen, 2002: 171). Cultic theaters provided space for ritual drama and featured a temple at the back of the cavea, so that the deity was part of the audience and the performance occurred "in the sight of the god" (Nielsen, 2002: 149). These were small theaters, normally consisting of an orchestra and cavea with seating for around 1,000–1,200 people, but lacking a permanent scaenae building. The absence of the scaenae suggests either that it was not always deemed necessary, or that these were erected only as needed. The lack of formal theaters during the early phases of sanctuaries suggests that dramatic performance – whatever form it took – did not require a stage, but merely an open area sufficiently large for the purpose, and space for seating. The Temple of Magna Mater provides a good example, since its steps and those of the adjacent Temple of Victory were used as seating for performances in the open space in front of the temples. In such examples it is easy to identify the location of dramatic ritual or ritual drama, but we must bear in mind that many potential locations for ritual drama can no longer be recovered archaeologically.

The development of theaters as architectural types, with or without temples, was closely connected to the development of dramatic performance in Italy, although there is considerable debate on this subject. While literary drama (ludi scaenici) is generally recognized as first occurring from the late third century, there is no agreement on the form that dramatic performances took prior to this. Hanson's investigation focused primarily on theater-temples in central Italy and examined the variety of Roman ludi alluded to by ancient authors and the possible locations for the performance of these, without making any distinction between ludi and other forms of ritual drama (Hanson, 1959: 9–26). But, as Nielsen shows, the phenomenon is older and more widespread in Italy (Nielsen, 2002: 191–2). Terracotta masks recovered from Tharros, Cagliari and Motya are similar to those recovered from Sparta, Samos and Cyprus, and in great quantities from Carthage in the seventh century (Nielsen, 2002: 150–1). In her view, the need for cultic theaters arose from pre-literary dramatic forms consisting of "dramatic ritual" down to 364 and ritual drama from 364 to 240, which in the case of some gods co-existed with ludi scaenici that became part of the festival activities later on (Nielsen, 2002: 156–65; 172–80). Dramatic ritual related to seasonal festivals in which satirical poems were recited, rather than a narrative with a plot. Ritual drama consisted of dramatizations of myth, or myths based on Roman historical events, and possibly also hymns (carmina).

Greek drama and theaters in the colonies of Magna Graecia undoubtedly influenced similar developments elsewhere in Italy, as is evident from representations in wall and vase painting in Etruria. Phoenician and Punic colonies also played a role, especially from the seventh to fourth centuries, when there was considerable contact through trade.

A case has been made for ritual drama at the temple to Astarte/Uni at Pyrgi, but also at temples with other manifestations of Astarte, such as at Gravisca (Aphrodite-Adonis) and the sixth-century temples at Rome (Fortuna and Mater Matuta), and of Melqart, such as the Ara Maxima of Hercules and the Temple of Saturn, also in Rome (Nielsen, 2002: 151–6; but compare Mingazzini, 1949: 223–4; Hanson, 1959: 32–3). While it has been argued that the axially symmetrical plan and use of porticoes and terraces derives from Hellenistic cities in Asia Minor, Nielsen focused on Phoenician models for *function*, citing the Punic theater-temple complex at Caralis, which she dates more specifically to 200–150 (Nielsen, 2002: 151–5).

Roughly contemporary with the large sanctuaries in Latium are the temple-theater groups at Castelsecco in Etruria, and at Sulmo and Iuvanum in Samnium, that do not conform to any architectural type, except in the arrangement of the cavea set out in front of the temple. This would also appear to be the case at the sanctuary of Diana at Nemi, which underwent a major refurbishment in the Late Republic including the addition of a theater. This was not symmetrically arranged with the temple and terrace and thus did not conform to the schema of the other Latial sanctuaries. At one extreme, the non-axial alignment of the theater and its small size (*c.* 300 places) have been interpreted as evidence of private ownership; the theater is therefore completely unconnected to the sanctuary (Coarelli, 1987b: 180–2). On the other hand, the small size and adjacent location of the theater are cited as reasons for its inclusion in the sanctuary and use from the second century (Morpurgo, 1903: 291–2, 1931: 237–51; Nielsen, 2002: 234–6; Sear, 2006: 127–8; Green, 2007: 24, 63–6).

Clearly, provision for drama in a ritual context was widespread in both time and space. In taking a broad view, the famous theater-temples of the second century – at Gabii, Tibur, Praeneste, Pietrabbondante and elsewhere – appear exceptional in their centralized, axially symmetrical and integrative design. They can be understood both as the culmination of a long developmental process and as a new architectural type that responded to the accretion of multiple modes of worship and incorporated these diverse activities into a symmetrical and unified whole. In each case the remains primarily represent the latest and most elaborate phase of these sanctuaries, but excavation has also yielded evidence of activity on these sites in the preceding centuries, as discussed above. In the second century these were flashy new sanctuaries for indigenous gods. Scholarship has teased out the multiple cultural influences (Greek, Punic and Hellenistic) for the architectural form and its underlying spatial conceptualization – rectilinear but terraced space that gathered the constituent parts into a regularized unity that was both aesthetically pleasing and functionally suitable. Here, as we have seen, worship occurred in different forms (sacrifice, ritual drama, oracles and votive dedications, among others) and utilized equally varied components (porticoes, groves, cavea seating and performance space, oracular caves, altars and temples).

6 Summary

The foregoing survey of archaeological evidence for ritual shows that this area of study is one that is rapidly expanding. Not only is much older material finally being made accessible through publication, it is increasingly the focus of new studies and approaches.

Ongoing excavation has revealed that sanctuaries dotted the Italian landscape, and that these often exploited aspects of their natural surroundings such as springs, groves, geothermal activity and other features of the landscape to create a stimulating environment for a variety of ritual observances. The material finds from these sites show both commonalities and differences in ritual practices in ancient Italy, but mostly they attest to the many different ways in which people engaged with the gods. The staggering number of votives, which continued to grow throughout the Imperial period, reminds us that ordinary people communicated with the gods, and that they did so frequently and in great numbers. Taken as a whole, the rich material record reveals not just that religious practices were embedded in all facets of life, but that they were, in modern parlance, "highly interactive" and stimulating to the senses.

FURTHER READING

The literature on this subject is copious and includes both archaeological reports and secondary interpretations of them. Important issues are canvassed in Bispham and Smith (2000), Schultz and Harvey (2006) and de Grummond and Simon (2006), all of which have extensive bibliographies. Stek (2009) summarizes past scholarship and articulates the current state of the question very well: the influence of Roman religious practices outside Rome is not as extensive as once thought, and is better attested in urban centers than in the countryside.

One of the most exciting recent developments is the compilation and publication of evidence for religious practices in the Greco-Roman world, the *Thesaurus Cultus et Rituum Antiquorum*, vols. I–VI. Here, the evidence for a range of practices is presented according to its type: literary sources; epigraphic sources; iconographical representations; archaeological documentation; and physical remains.

PART III

Archaeology and Ancient Technology

CHAPTER 16

The Orientation of Towns and Centuriation

David Gilman Romano

1 Introduction

The establishment of Roman colonies in Italy was a fairly common occurrence from the late fourth century but it was only later, in the second half of the second century, that colonies were founded outside of the Italian Peninsula. Eventually colonies were established in many parts of the Roman world and the foundation of each colony included a decision or decisions about its spatial organization and the orientation of the city. Although there were differing types of colonies, a colony was essentially a city-state that included an urban as well as a rural area. In Italy, Latin colonies (coloniae Latinae) were settled by individuals who did not possess Roman citizenship, and Citizen colonies (coloniae civium Romanorum) were settled by individuals who did have Roman citizenship. There was a third category, that of municipium, which was an annexed foreign community that, following its incorporation into the Roman state, was allowed to keep its own governing structure. Eventually there were also coloniae militares, settled originally for the veteran soldiers of Sulla and Julius Caesar (Salmon, 1970).

The decision about when and where to found a colony was approved by the Roman people and during the Republic colonization was preceded by a law (Lex coloniae) passed by the Plebeian Assembly (Gargola, 1995). Commissioners, often including ex-consuls, and usually three in number, were appointed to oversee and supervise all aspects of the foundation. They defined the boundaries of the new colony, assigned the allotments, settled any disputes, wrote the charter of foundation and appointed the first magistrates and priests of the colony (Dilke, 1985, 1987a). The formal act of colonization was known as deductio and it was this practice that included the laying out of the area of the colony together with the necessary elements of an urban and rural structure, either on new land or on land that had been occupied previously. Deductio also included

A Companion to the Archaeology of the Roman Republic, First Edition. Edited by Jane DeRose Evans.

religious augural rituals of foundation. According to Virgil (*Aen.* 5.755–6) Aeneas founded a second city Troy in Sicily by plowing, and then he distributed the land by a lottery. Tacitus (*Ann.* 12.24) mentions the Forum Boarium as the location of the starting point where the founder of Rome began digging his furrow, "sulcus primigenius"; this was the original furrow and the outer boundary of a newly founded city, made by a plow pulled by a bull and heifer. Appian and Plutarch give a good deal of information about the foundation of Iunonia at Carthage in 123, a colony whose charter was later revoked. Appian (*B Civ.* 1.24–5) says that the triumvirs oversaw the location of the city, the centuriation of the land and the recruitment of the colonists. Plutarch (*C. Gracch.* 11) mentions that formal ceremonies were carried out despite prohibitory signs from the gods, including probably the laying out of boundary markers that were set upon by wolves and carried away.

2 Roman City Planning

Organized and regular city and land planning was not invented by the Romans. There are a number of earlier civilizations that had practiced this kind of interest, including the Etruscans, the Greeks, and civilizations of the Near East (Ward-Perkins, 1974). Varro noted that the origins of centuriation come from the Etruscan world and that limites (paths or roads) between plots of land originated with the Etruscans (extract in Frontin. *De limitibus*; see Campbell, 2000). Probably the best excavated example of Etruscan town and landscape planning is at Marzabotto, on the banks of the Reno River, 25km south-west of modern Bologna; the city dates to the late sixth or early fifth century. The plan shows a regular organization of the roads and streets into a grid with divisions approximately 165m long and 40–60 m wide. The orientation of the city is close to true north–south (Ward-Perkins, 1974; Grimal, 1983). At Marzabotto, the mundus was discovered; it was a central feature of the organization, a sacrificial pit and a link between the city of the living and the dead. Also found at Marzabotto were cippi that marked the intersection of the major roads and, at what presumably would have been the center of the planning organization, was the cippus that marked where the gromatic survey would have originated.

Colonial Greek examples of regular city and landscape planning in southern Italy and Sicily exist, for instance, at Paestum and Syracuse by the sixth century. There is even evidence of regular city planning in Megara Hyblaea as early as the late eighth century (Ward-Perkins, 1974). Aristotle (*Pol.* 7.1330a), in describing the optimum city, notes how a city is to be divided and laid out. But it was the Romans who took the idea of the organized and well-planned city and landscape and applied it to the cities of the ancient world with common standards and objectives. The underlying reason that the Romans were so keen on urban and rural organization is that it tied in very closely with their system of taxation and the generation of revenue. With an organized system of land in the urban and rural areas of their cities, taxes could be easily calculated based on the area of land owned. It was an early form of what we would call in the present day a GIS (geographic information system), linking the spatial data of the urban insulae and rural plots with information about ownership, agricultural production and taxes. The importance of this information cannot be underestimated, since the tax revenue from the produce of the agricultural fields of rural plots of land would be one of the driving forces of the politics of the Late Republic.

Yet even by the beginning of the Republican period, Roman colonies and cities in Italy, and eventually all over the Empire, were characterized as having regular and predictable organization for both their urban and rural aspects (Salmon, 1970). There were streets and roads in the urban center of the city that were parallel and that met perpendicular cross roads at regular intervals, making the orthogonal "Roman grid." Of course the cities were not necessarily oriented to exact north, or exact east, and in reality there were many different orientations for Roman colonies and cities. The layout of the roadways was critical to the successful life of a city. The cardo maximus (the main east–west street) and the decumanus maximus (the main north–south street) would have been the first roadways to be surveyed and laid out. Pseudo-Hyginus (*De limitibus constituendis*) mentions that Augustus had specific recommendations for the widths of these two major roadways: 40-foot width for the decumanus maximus and 20-foot width for the cardo maximus, with 12 feet and 8 feet for the lesser roadways (Campbell, 2000). It is known that centuriation was introduced to the Latin colonies after 338 and spread throughout the Roman Empire. Regularly planned Roman cities have been discovered all over the ancient world and it is the archaeological evidence, together with the literary and historical texts, that give us many of the details of the planning and organization of the urban and rural aspects of Roman cities.

Within the city, space for the forum was set out from the planning stage and it would have a specific shape and area (Rykwert, 1988). Typically, the forum would be close to the middle of the colony or city and it could be the place where the cardo maximus and the decumanus maximus intersected. Vitruvius (I.6.1) tells us that the proportions of the forum should be at a relationship of three times for the length and two times for the width, and he ties the size and shape of the forum directly to the fact that in Italy the spectacles were held in the forum.

Insulae were often of a uniform size; buildings and structures would be built within them. The amphitheater and circus could be set out within the new plan in fairly predictable places and the same would be true, generally speaking, for the markets, the baths, the basilicas, the temples and the areas of public assembly, including the theater and *odeion*. Areas of the city would be designated for houses and the domestic region; walls would commonly enclose the city, with gates opening to major streets (Rykwert, 1988). The agricultural fields that surrounded the Roman city or colony would have been divided up into regular units. These fields would characteristically be divided into squares or rectangles (centuriae) and they would be divided by paths and roads (limites) at regular intervals. Centuriation was typically applied to public land (ager publicus), land acquired by the state as a result of conquest, which was usually cultivated by individual landowners. Boundary stones were set out throughout the centuriated land to indicate where the territorium of the colony began and ended, and more detailed boundary markers were set into place to mark the specific locations of the agricultural fields with respect to the overall plan of the colony. Centuries were identified as being to the left (sinistra) or right (dextra) of the decumanus maximus and either beyond (*ultra*) the cardo maximus or on the near side of it (citra). The implication of this kind of organization meant that the surveyor was facing east for these descriptions to be true. The viewer would be able to learn from a single boundary marker in the landscape how far he was away from the cardo maximus or the decumanus maximus and how far from the origin of the system.

The map of the colony (forma), engraved in bronze, was set up in the forum of the colony to show the centuriated land. Although no forma in bronze has survived, there are multiple examples of formae in marble and on papyrus that show aspects of the city and land planning of a colony. The best examples of surviving ancient tax maps are the stone cadastral maps from Aurasio (Orange) in France (Piganiol, 1962; Salviat, 1977). These are remarkable documents that illustrate the location of fields around the Roman city and give the details of ownership and taxes. The maps also show areas of the landscape that are reserved for sanctuaries, or areas that are unable to be divided due to the presence of swamps or rivers. Some of the most interesting aspects of these marble maps are that they represent several different time periods and that there were land divisions that occurred one after the other in the same area; the latter represent surveys of land that were at times superimposed on one another. By the Imperial period, there appear to be seven or eight different orientations of land division in the areas that surrounded the city. These different orientations appear to be the result of differing phases of colonial settlement, from the original Gallic tribe occupying the area, the Tricastini, to the Roman colony for legionary veterans Colonia Iulia Firma Secundanorum, which was founded on their territory around 35. Another phase of centuriation likely developed in the Flavian period, when the town name was changed to Colonia Flavia Tricastinorum; the result was a change in land distribution that included allotments for the Tricastini tribe.

The stone maps depict fields divided by lines with no great width, probably indicating narrow paths, but the major roads have a width in the landscape. This reflected the practice of reserving land in the landscape for the most important of the roadways: an example of "iter populo non debetur" (see below). In their original location, these ancient maps were located near the theater of the city, to be used as a reference for magistrates, tax officials and other inhabitants of the city.

3 Roman Agrimensores

When a new Roman colony was built, the plan for the city would be provided and the measurement of the city would be carried out by the agrimensores, the land surveyors. We know a good deal about this professional group, which was a part of the Roman army, from a compilation of ancient texts that is known as the *Corpus agrimensorum Romanorum* (Dilke, 1971). The *Corpus* was originally compiled in the fourth or fifth century CE but includes texts as early as the first century CE. These texts give us information about the training of the agrimensores and their day-to-day activities, as well as some of the practical issues that they faced in the field. A number of hand-drawn illustrations, likely based on earlier Roman drawings, accompany the text and relate to the topics covered there.

The agrimensores used as their principal surveying instrument the groma (Grimal, 1983). This was a very simple instrument that consisted of a vertical staff with two horizontal cross bars connected by a bracket. Each of the ends of the crossbars supported a cord that was held vertical by a plumb bob. The agrimensor was able to use the groma to sight straight lines and right angles with great accuracy and was able to use this instrument in conjunction with decempeda or sighting rods. There would have been multiple assistants holding decempeda, working together with a surveyor and a single groma.

At the measuring out of a Roman colony, the groma would have been set up first in a location where much of the planned urban and rural areas of the colony or city could have been seen. The major cardo or decumanus would have been first sighted, the other major roadway would have been created at a right angle, and then the subsidiary roads would have been added. The agrimensores had a sophisticated method of checking the accuracy of the orientation of roads whereby they were able to make a visual check every fifth road (quintarius) and if an error was discovered it could be identified and swiftly corrected. The quintarius was usually a wider road in the Roman city.

Frontinus (*De limitibus* 37–43) gives us specific information about the question of orientation of colonies (Campbell, 2000). Frontinus says that colonies may take as their orientation the rising and setting of the sun, although he indicates that other solutions are possible as well. For instance, he cites the case where a reversal of orientation is possible when two colonies are close to one another, to prevent the same orientation for both of the colonies. Pliny (*HN* 18.76) has a passage on orientation in general in which he describes the method for an individual to determine east from sunrise and west from sunset and thereby locate south and north. Pseudo-Hyginus (*De limitibus constituendis*) indicates that a correct south bearing for the sun is necessary for centuriation of a colony. The correct south bearing that could be ascertained with the assistance of a gnomon would be the first step in being able to determine true east or true west, which were the cardinal directions for surveyors.

In preparation for a colonial foundation, the agrimensores would typically have worked under the supervision of the commissioners who were in charge of the foundation of the colony. The commissioners would have had several responsibilities: to delimit the boundaries of the territory of the colony (the pertica); to measure the limites of the city; to survey and to subdivide into sections the urban areas; and to do the same for the territorium of the colony. The survey of the urban as well as of the rural area of the city would be accomplished before the arrival of the colonists, since the division of urban and rural land would have been necessary before the allotment of plots of land could be carried out. Only after this had been accomplished could the actual allotments of land, both urban and rural, be measured out and assigned to the individual colonists.

The basic linear measure that was used by the agrimensores was the Roman foot (0.295+ m) and multiples of the Roman foot were calculated in terms of an actus, 120 linear feet. A square actus was 14,400 square feet and 2 square actus were known as one iugerum (thus Varro, as quoted in Frontin. *De limitibus*). Allotments of land for urban as well as for rural subdivision were calculated in iugera and in heredia, or 2 iugera. As an example in the Caesarian colony of Corinth the urban insulae were measured 1 actus wide and 2 or 4 actus in length, producing insulae of 1 or 2 iugera. Cardines, which could vary in width from 8 to 24 feet, divided the insulae. The major roadways were typically the widest, 25 feet or wider, and there were a variety of smaller roadway widths.

The roadways in the urban center could be planned and built so as to fully respect the entire calculated area of the insula according to a legal formula "iter populo non debetur," meaning that the widths of the roads were added outside of the regular iugera measure of the insulae (Gargola, 1995). This would result in an urban area where there would be a multitude of roadways, cardines and decumani, but the areas of the insulae, since they were calculated in regular numbers of iugera, would be preserved. In the rural landscape there was a different solution, "iter populo debetur," which meant that roadways could be added over land that was divided into iugera for farming purposes. This would mean that portions of the assigned rural land would in fact be utilized as paths or cart roads through the agricultural fields. The reason for this was related to the fact that the rural roadways between large tracts of agricultural land were not always major thoroughfares and often could be narrow tracks. However, there were exceptions to this scenario, primarily for the widths of the cardo maximus and decumanus maximus in the territorium. These were in fact planned in the manner of the urban roadways as "iter populo non debetur," meaning that the land for these major roadways in the rural territorium would be reserved, since they were important thoroughfares.

The objective of urban planning was to provide space for all of the many civic functions of a city. It would have been essential to start with a drawing that would have been the basis of the survey to be undertaken by the agrimensores. If the colony were to be created de novo, this drawing would have located every major building and structure in the city, including the forum, theater, temples, sanctuaries, baths, markets, basilica(s), tabularium, and possibly an amphitheater and circus. As the size of a colony or city would grow over time, there would be the need for increased numbers of buildings and structures, but it is likely that there would have been a reservation of land for the major buildings of a city from the beginning of the planning process. There would also have been a decision made as to the number of colonists and the necessary amount of domestic space in the urban area of the colony. Clearly the population could enlarge over time and the area for domestic buildings and services could have increased.

4 Rural Planning as Centuriation

The rural organization of a city or colony followed the survey by the agrimensores of the territorium (see also Chapter 13). The process of surveying and subdividing the land was known as limitatio. Once the surveyors had set out the borders of the colony, then the rural land could be subdivided. Originally the land was divided into centuriae that were equal to 100 units of land. The most common measurement of a centuria in the Augustan period was 20 by 20 actus, producing an area measuring about 200 iugera (Dilke, 1987b). Much of the area of the Po Valley was divided in this way. Limites were the essential framework of the centuriation system. Surveyed straight lines of no width were called *rigors* and surveyed lines with widths were called limites; the width of the limites could have varying dimensions. It was important for some of the rural limites to have a width to allow for passing vehicles. The centuriae were not always squares of 20 by 20 actus; they could also be 12 by 12 actus rectangles at Forum Iulii (Fréjus), or 21 by 20 at Cremona, or 25 by 16 at Beneventum (Benevento), or other measurements, as indicated by the ancient authors. The allocation of land varied from colony to colony and it is known from literary and historical sources that allotments of land came in various sizes depending on a variety of factors. Allotments for individual settlers ranged from 2 to 50 iugera. The smallest division of land was 2 iugera, also known as a heredium, a heritable plot of land. Typical allotments were in the range of 5, 8, 10 or 12 iugera, but there were also larger allotments of 20 iugera or more; in exceptional cases, as many as 100, 140 or 200 iugera were allotted to a colonist.

In the rural landscape there would be areas that were not attractive for agricultural purposes; for instance, swamps or gullies or lakes would likely be excluded from the measurement of centuriation, as at Aurasio. In addition, there could be rivers and streams that could overflow their banks at times and could diminish the amount of land within a particular surveyed area. There could also be previously existing sanctuaries or cemeteries or roadways that would not be surveyed. Land that was not surveyed by the agrimensores for any reason was called subseciva. The distribution of land to settlers utilized the method of a lottery that is clearly stipulated by Pseudo-Hyginus (*De limitibus constituendis*) and it is probably very likely that over time a single allocation of land may have been divided and subdivided based on inheritance over many generations.

It is fairly typical that the orientation of the centuriated land that is associated with a colony or city was in keeping with the orientation of the urban colony so that either the cardo maximus or the decumanus maximus or both would be represented in the territorium as extensions of the urban roadways and at the same orientation. This is clearly documented in many of the illustrations that accompany the *Corpus agrimensorum Romanorum* as well as by scientific study of the Roman cities in the modern day. It is also true that there are systems of centuriation in the landscape surrounding a Roman city, the orientations of which do not necessarily correspond with the planning of the Roman city or colony. Certainly the orientation of some systems that are associated with a Roman colony are organized with respect to an existing roadway or to a physical feature of the environment: a coastline, a mountain range, or a river valley. Such is the case with Tarracina (Terracina), on the west coast of Italy midway between Rome and Naples, which was founded as a colony in 329 (Gasparri, 2000). Here the Via Appia, a major regional roadway extending south-east from Rome, serves as the decumanus maximus for the colony, as is explained by Pseudo-Hyginus (*De limitibus constituendis*). The roadway runs north-west–south-east and most of the centuriated land was to the north-east of the roadway towards the mountains. The centuriation appears to be laid out with respect to the decumanus maximus which intersects the coastline near the urban colony. A hand-drawn illustration from the *Corpus agrimensorum Romanorum* illustrates the layout of Tarracina with the Via Appia running through it. From Livy (VIII.21.11) we know that after defeating the Volscians Rome sent 300 settlers to the new colony, each receiving 2 iugera of land. This was a fairly small colonial settlement, lasting only three centuries, although additional settlers may have been added at a later time.

5 Roman Corinth

From Roman Achaea, we know a good deal about the planning of the successive Roman colonies founded at Corinth, the first under Julius Caesar in 44 as the Colonia Laus Iulia Corinthiensis, and the second during the Flavian period as the Colonia Iulia Flavia Augusta Corinthiensis (see Chapters 35 and 37). Archaeological excavation and research, including computerized topographical survey with the methods of digital cartography, GIS and remote sensing, have provided a great deal of information about the successive Roman cities (Romano, 2003, 2005, 2010).

Following the sack of Corinth in 146 by the consul Lucius Mummius, who opposed the Corinthian leadership in the Achaean League against the coming of Rome, the Greek city was deprived of its civic and political identity. The men were killed while the women, children, freedmen and freedwomen were sold into slavery. From the archaeological record it is clear that there was a partial and selective destruction of Greek structures and the city walls. Following the defeat of Corinth, the Roman Senate sent ten commissioners to assist Lucius Mummius in the settlement of Greece. This work included the sale of property confiscated from those who had been prominent in the fight against Rome (Livy *Per.* 51; Paus. 7.16.7–9).

During the 102 years from the capture of Corinth to the colonization by Julius Caesar, the land that had been under its control became largely ager publicus, although several ancient authors mention that Sikyon had taken over the care of a part of the Corinthian land. Evidence for the work of the Roman agrimensores at Corinth can be divided into three phases of activity: (i) an interim period lasting from the sack of Corinth in 146 until the foundation of the Caesarian colony in 44; (ii) the foundation of the Caesarian colony Colonia Laus Iulia Corinthiensis in 44; and (iii) the foundation of the Flavian colony of Colonia Iulia Flavia Augusta Corinthiensis.

The Lex agraria of 111 includes information that some parts of the Corinthia were measured out for sale or rent and boundary stones were erected. This epigraphical evidence matches evidence of a Roman roadway, in use between 146 and 44, which was found to have broken through the old wall of the Greek city of Corinth, near the Asklepieion. Both the epigraphy and the topography suggest that the activity mentioned in the Lex agraria was the same as the division of land to the north of the city and between the long walls of Corinth. A series of roadways extending from the coastline of the Corinthian Gulf to the city of Corinth have been identified as 8 actus apart from each other; one of these roads is the same width as the road that breaks through the Greek wall at the Asklepieion. There also exists a reservation of one actus in the landscape to the north of the city, interrupting the 8 actus units, that corresponds with the location of the cardo maximus within the Roman city. If these preparations had been made in the area to the north of the city between 146 and 44, it seems very likely that preparations for the colony were in progress long before the colonists were on their way. The process of centuriation, preceding the actual colonization, was one way in which the groundwork was laid for the eventual colonization process.

The orientation of these north–south roadways that extend from the Corinthian Gulf to the Greek walls of the city are uniformly 3 degrees west of north and they agree in their orientation with the later evidence of the roadways and insulae of Roman Corinth, taken from the curbstones of the principal roadways of the city. There are also several interim period roadways from the 146–44 phase that have been discovered within the limits of the former city, indicating that a new system of roads was extended throughout the former Greek city. This orientation was utilized by those who founded the colony of Julius Caesar in 44, as the urban organization of the colony is largely based on this direction. Three degrees west of north was close to true north or close to true east and was possibly related to the measurement of sunrise from the east. More importantly, it provided for the division of land north of the city and south of the Corinthian Gulf, land divisions that were consistently parallel to the coastline.

There is some evidence to suggest that the general orientation of the earlier Greek city of Corinth was more north-westerly, in the vicinity of 13 degrees to the west of north. This suggestion is based on the presumed orientation of the Greek Lechaion Road within the Greek city and several other Greek roadways, as well as Greek buildings and structures within the walls of Corinth. It is also likely that the Greek port at Corinth was not in the same location as the Roman port but was located several kilometers to the west. When the Roman surveyors came in during the second century to lay out the centuriation to the north of the former Greek city of Corinth, their general north–south orientation of 3 degrees west of north was not the same as the earlier Greek orientation.

5.1 *Colonia Laus Iulia Corinthiensis*

There is considerable literary, historical, epigraphical and numismatic evidence concerning the foundation of Corinth by Julius Caesar. Although we do not know from the historical record how many colonists arrived to settle the colony, we do know from Strabo (8.6, 23) that most were freedmen, and Plutarch (*Caes.* 57. 5) implies that many were veterans. The Roman agrimensores' plan for Roman Corinth included a measured design for the overall urban limits as well as the location and dimensions for other specialized areas of the city. From archaeological evidence and the study of topographical maps, air photographs and satellite images, we know that the "drawing board" plan of the urban center included 4 centuries each of 32 by 15 actus or 240 iugera. Each of the four centuries of the urban colony was characterized as having the capacity for 29 cardines and 29 one-actus-wide insulae at the orientation of 3 degrees west of north, although the implementation of the colonial design may not have been completed in all areas of the city. The width of the excavated roadways varies from 8 to 24 feet. Each century was also designed to have the capacity for six decumani. The urban centuries of the colony were organized according to the formula "iter populo non debitur" meaning that the road widths were added outside of the regular insula measure. The overall size of the planned urban colony was 240 hectares or 593 acres and it fit largely within the Greek city walls. The cardo maximus of the urban colony was the Lechaion Road that extended the north–south length of the colony to the Corinthian Gulf, some 3,150 m from the rostra of the forum.

The evidence for centuriation of the territorium of Corinth outside of the urban center dating to the Caesarian colony indicates a division of land into large units of 16 by 24 actus, at the same orientation as the orthogonal grid of the city, 3 degrees west of north, or 3 degrees north of east (Figure 16.1). The large units were subdivided into smaller sections. The overall area of centuriation in the Corinthia associated with the colony of Julius Caesar at Corinth is *c.* 100 km.^{2}

Within the area of the urban colony of Julius Caesar the forum area as a whole occupied 24 actus^2 or 12 iugera in its topographical center, measuring 6 actus east–west and 4 actus north–south (Figure 16.2). This large area included important aspects of the former Greek city including several large buildings, notably the South Stoa that was 164 m long. Many of the most important civic, political and religious buildings of the Roman colony would eventually be constructed within this space. Interestingly, the most important Augustan buildings that were built within the area of the forum were at an orientation other than the principal orientation of the colony. These buildings included the Julian Basilica and the South-east Building at the east end of the forum, Temple F (likely dedicated to Venus Genetrix) and the Fountain of Poseidon. Along the northern border of the area was built what is known as the North-west Stoa and towards the

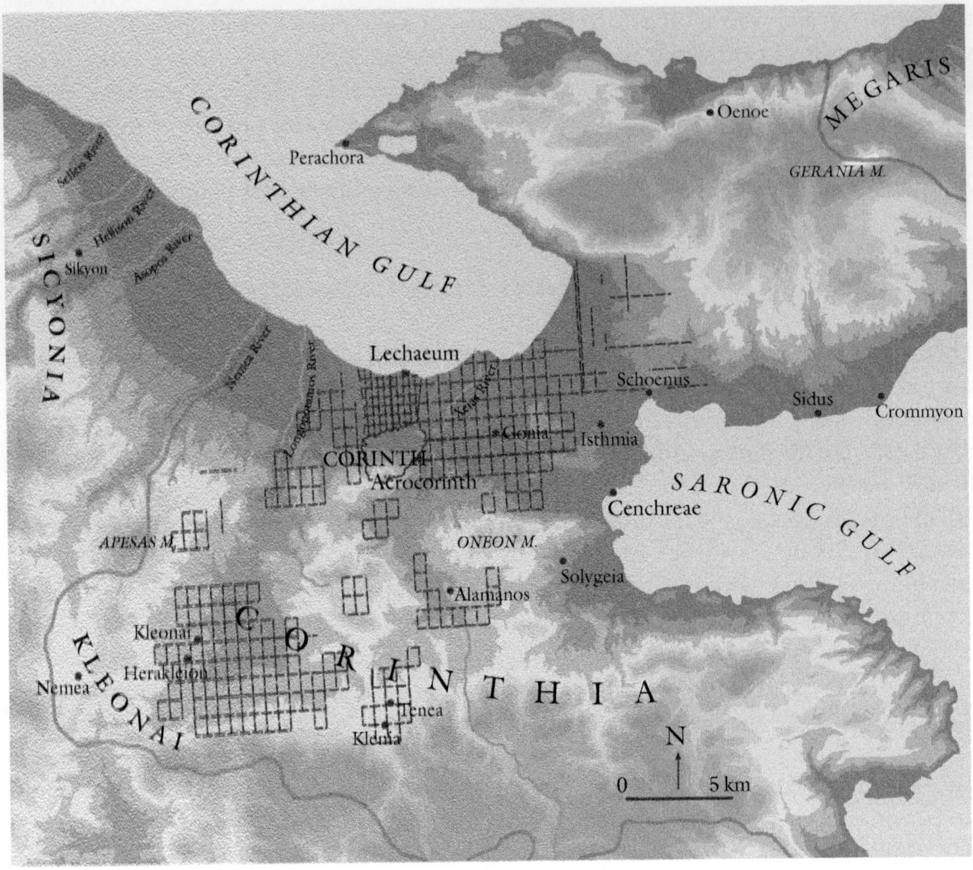

Figure 16.1 Centuriation of the territorium of Corinth, dating to the time of the Caesarian colony. Source: David Gilman Romano, Corinth Computer Project.

middle of the forum area was constructed the Rostra and a large altar. In the Late Augustan period or slightly later, Temple E, Temple C and the West Shops were built in what appears to have been a western appendage to the forum, 2 by 4 actus in area (Figure 16.2).

It is perfectly clear to anyone looking at a plan of the Roman forum at Corinth that the Roman surveyors picked an orientation for many of these Augustan and later Roman buildings and structures that was not in keeping with the urban insulae grid of the new Roman colony of 3 degrees west of north. Although the overall space reserved for the forum was rectilinear, and designed with respect to the city plan, most of the new Roman buildings and structures added inside this space were not oriented with respect to the urban grid. The orientation of many of the Augustan buildings within the area of the forum was based on an earlier Greek orientation that was known from a Greek racecourse in the area as well as the Greek North Building to the north of the forum that was slightly different from the orientation of two other major Greek buildings of the area, the Temple of Apollo on Temple Hill and the South Stoa. It is not completely clear why the Roman agrimensores chose this orientation for the most important buildings of the Roman colony, but they did.

Figure 16.2 Centuriation of the urban center of Corinth, dating to the time of the Caesarian colony. Source: David Gilman Romano, Corinth. Computer Project.

Table 16.1 The north–south orientation and unit size for parcels around Corinth in the Flavian era

Unit	*Geographical region*	*Orientation*	*Area (km²)*
A00	Sikyon north-west	N20°20'22"E	10
A0	Sikyon north	N34°22'32"E	12
A1	Sikyon	N62°26'52"E	40
A2	Sikyon coastal region	N48°24'42"E	58
A3	Nemea River area	N34°22'32"E	108
A4	Longopotamos River area	N20°20'22"E	27
A5	Corinth, Lechaion to Kenchreai	N6°18'12"E	142
A6	Corinth to Kenchreai, south corridor	N20°20'22"E	12
A7	Xerias River area	N7°43'58"W	30
A8*	West of Isthmus	N34°22'32"E	11
A9	West of Isthmus	N21°46'8"W	15
A10	East of Isthmus	N35°48'18"W	16
A11*	Area of canal (parallel to canal)	N48°20'02"W	17

5.2 *Colonia Iulia Flavia Augusta Corinthiensis*

The refoundation and renaming of Roman Corinth by the Flavians is known from epigraphical and numismatic evidence. The physical vestiges within both the city and the surrounding rural area attest to a second Roman land division that may be equated historically with the establishment of the Colonia Iulia Flavia Augusta Corinthiensis by Vespasian, although we do not know the exact date of the refoundation. The Flavian system of centuriation is characterized by a fan-shaped grid that is divided into ten differently oriented units in the plain immediately to the south of the Gulf of Corinth. Each of the units corresponds to a specific area of the coastal plain and all of the units, except one, are linked to each other by the simple ratio of the arctangent of 1:4, equal to the angle of 14° 2' 10". Each of the ten rectilinear units is roughly parallel to the coastline and to the river of the area. The length of the entire stretch of Flavian centuriation along the south coast of the Corinthian Gulf is *c.* 30 km. This kind of linking of centuriated units of land is known from other regions of the Roman Empire and is specifically a technique of Roman land division. Of course, the Roman agrimensores did not use degrees, minutes and seconds to determine the linking of these individual units of land. They used the simple ratio of 1:4 in a right triangle that produced this exact angle (see Table 16.1).

The reason for the linked orientation of these units of land along the southern coast of the Corinthian Gulf must be related to several geographical factors. First, the rivers drain towards the Gulf and many of the units seem to follow the general course of the rivers. Second, each of the units seems to be perpendicular to the coastline, creating rectilinear plots of land. Some of the units were surveyed on top of earlier surveyed areas of the Corinthia that are associated with the colony of Julius Caesar but there are several areas in the Corinthia that had not been centuriated previously. Along the coast these areas are in and around the city of Sikyon. The total amount of newly centuriated land

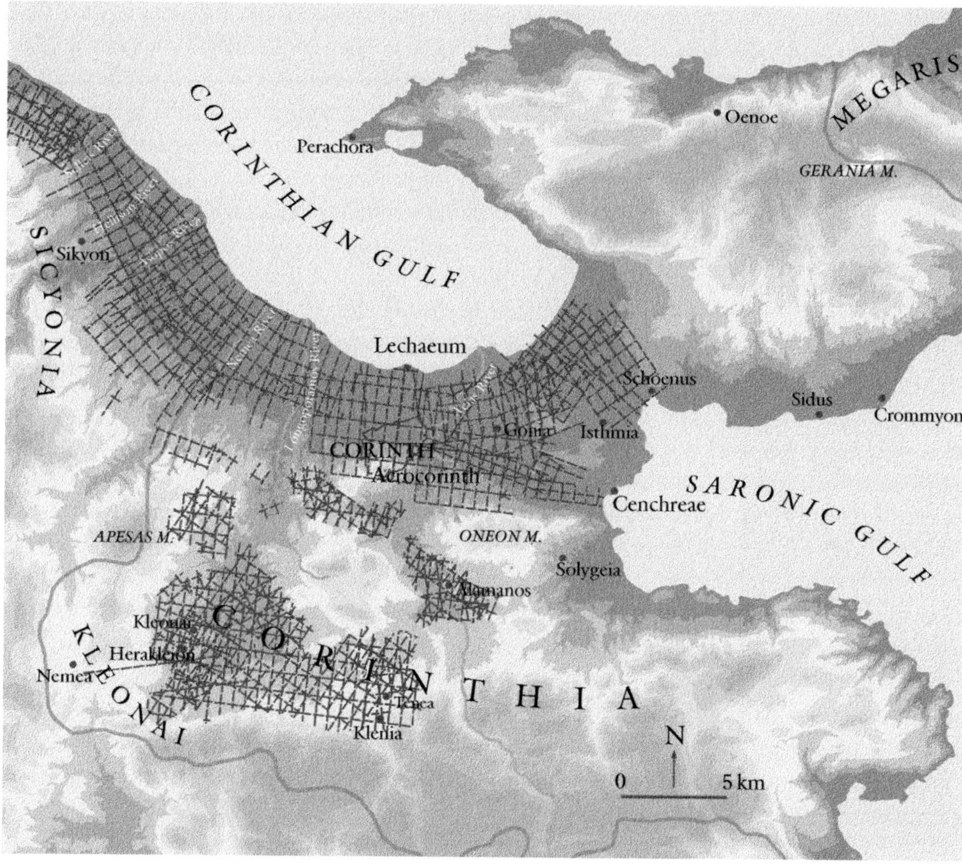

Figure 16.3 Centuriation of the territorium of Corinth, dating to the time of the Flavian colony. Source: David Gilman Romano, Corinth Computer Project.

in the Corinthia associated with the Flavian colony, including a large area of the central and southern Corinthia, is approximately 300 km^2 or roughly three times the area of the land centuriated in the Caesarian colony (Figure 16.3).

The proof that the large system of centuriation along the southern coast of the Gulf of Corinth is datable to the Flavian colonial refoundation at Corinth is found in the area of the Sanctuary of Demeter on the lower slopes of Akrocorinth. Three parallel Ionic prostyle Roman temples were built on the Upper Terrace of the Demeter Sanctuary, all with similar orientation. The temples are identified with Demeter (west), Kore (central) and possibly the Morai (east) and are datable to the period following the earthquake of the 1970s. A setting line on the top surface of the eastern wall of the central temple was measured to be only 3.5' of 1 degree from the orientation of the Flavian centuriation in this area, determined by independent means. It would appear that the temples were constructed with the specific orientation of the centuriated land. Their orientation is very different from the other temples that were built within the forum of Corinth and they were also different from that of the insular grid of the colony. Another building in the heart of the Roman city should be mentioned with respect to the Flavian surveyors and the Flavian system of centuriation. In the Forum South-west, immediately to the west of the South Stoa, is a long narrow structure that is identified as the "Long Rectangular Building." This structure has been dated by ceramic evidence to the period of Nero. The orientation of the building and the adjacent monumental arch are less than 2 degrees, measured at foundation levels, different than the Flavian centuriation orientation, suggesting a possible connection.

The implications of the new system of centuriation in and around Corinth were that the areas of centuriated land were greatly enlarged during the Flavian period and there was, in great likelihood, an increased number of colonists. I have made the suggestion that this increase in population may have been related to an account in Josephus (*BJ* 3.540) that in 67 CE, Vespasian called for 6,000 strong male Jewish slaves, who were assembled in the stadium at Tiberias, to be sent to the Isthmus at Corinth, presumably in connection with the construction of the canal that had been initiated by Nero. Although the canal was not finished by Nero or by Vespasian, it is possible that the importation of these Jewish slaves to the Corinthia may have been related to the refoundation of Corinth as a Flavian colony. It is possible that the slaves may have been given their freedom together with plots of land in the Corinthia, or possibly the slaves may have become tenant farmers of the refounded colony. There may also have been the relocation of citizens of Sikyon or of other cities nearby.

With respect to the question of the size of the allotments that were given to the Corinthian colonists, based on the area of centuriation, I have estimated that between 12 and 24 iugera was the original allocation, to between 1,500 and 3,000 colonists for the Caesarian colony. For the Flavian colony the newly centuriated area is approximately three times the size and, if the allotments were the same, this would suggest an increase of between 4,500 and 9,000 colonists.

6 Conclusions

The study of urban and rural city planning bears on many different aspects of ancient history and archaeology. By means of modern techniques of computerized survey, digital cartography, GIS and remote sensing, together with the traditional literary, historical and archaeological evidence, aspects of the ancient city and landscape can be documented and interpreted in new ways. With respect specifically to the studies of the orientation of towns and centuriation, precise mapping on both the small and the large scale can provide unexpected and important discoveries.

FURTHER READING

Archaeological excavation and research in the area in and around Corinth was conducted under the auspices of the Corinth Excavations of the American School of Classical Studies at Athens; it was combined with a study of the planning of the Roman colony in the Corinth Computer Project from the University of Pennsylvania Museum of Archaeology and Anthropology. The results can be found at the Corinth Computer Project (http://corinthcomputerproject.org).

For an excellent study of historical, literary and practical aspects of Roman colonization see Salmon (1970). Rykwert (1988) brings together many subjects, civic as well as religious, which relate to early examples of city and town planning. Dilke has written several important books and articles on the work of the agrimensores (including Dilke, 1971, 1985, 1987a, 1987b). An excellent translation of the *Corpus agrimensorum Romanorum* has been made by Campbell (2000). The study of Roman centuriation across Europe has been undertaken by the COST G2 Commission of the European Community, and a series of publications have appeared edited by Clavel-Lévêque and Orejas (1998, 2002), with a catalogue of case studies of documented systems of centuriation, often with discussion of the methods used to determine the results. Ward-Perkins (1974) provides an excellent introduction to the salient aspects of Greek and Roman city and landscape planning. Specific studies of ancient cities have produced many articles and monographs that deal with city and landscape planning. For Corinth, see Williams and Bookidis (2003) for a recent summary of current research at the site including the study of Roman centuriation and land planning.

CHAPTER 17

Scientia in Republican Era Stone and Concrete Masonry

Marie D. Jackson and Cynthia K. Kosso

1 Republican Era Construction Engineering

Monumental buildings constructed in Rome during the nearly 500 years of the Republican era record a progressive evolution in builders' refinements of diverse construction methods with local volcanic building materials, and their innovations in developing the complex stone and concrete masonry of the first century. Descriptions of extant examples of these masonry elements integrated with analyses of Vitruvius's *De architectura* (written *c.* 30–20) reveal that these builders were astute engineers. They recognized the material characteristics of the volcanic and sedimentary deposits of the Roman landscape (Figure 17.1) and selected specific rocks to enhance the structural stability and long-term durability of monumental architecture in Rome (Frank, 1924; Blake, 1947; Lugli, 1957; DeLaine, 1995; Jackson *et al.*, 2005; Jackson and Marra, 2006; Jackson *et al.*, 2006; Jackson *et al.*, 2007; Jackson *et al.*, 2010).

The central objective of this chapter is to investigate the hypothesis that Republican era builders integrated a long-standing tradition of thoughtful reasoning and invention with skilled workmanship to develop highly sustainable masonry materials, construction technologies, and engineering solutions (Vitr. *De arch.* II.1.2, I.1.1). Indeed, Vitruvius writes that the plan, elevation and perspective of architectural design:

> are born from the power of reason (cogitatione) and invention (inventione). Reason is painstaking effort (cura), diligent, zealous and watchful attentiveness (studii plena e industriae vigilantiaeque), effecting a plan with pleasing characteristics. Invention, on the

A Companion to the Archaeology of the Roman Republic, First Edition. Edited by Jane DeRose Evans.

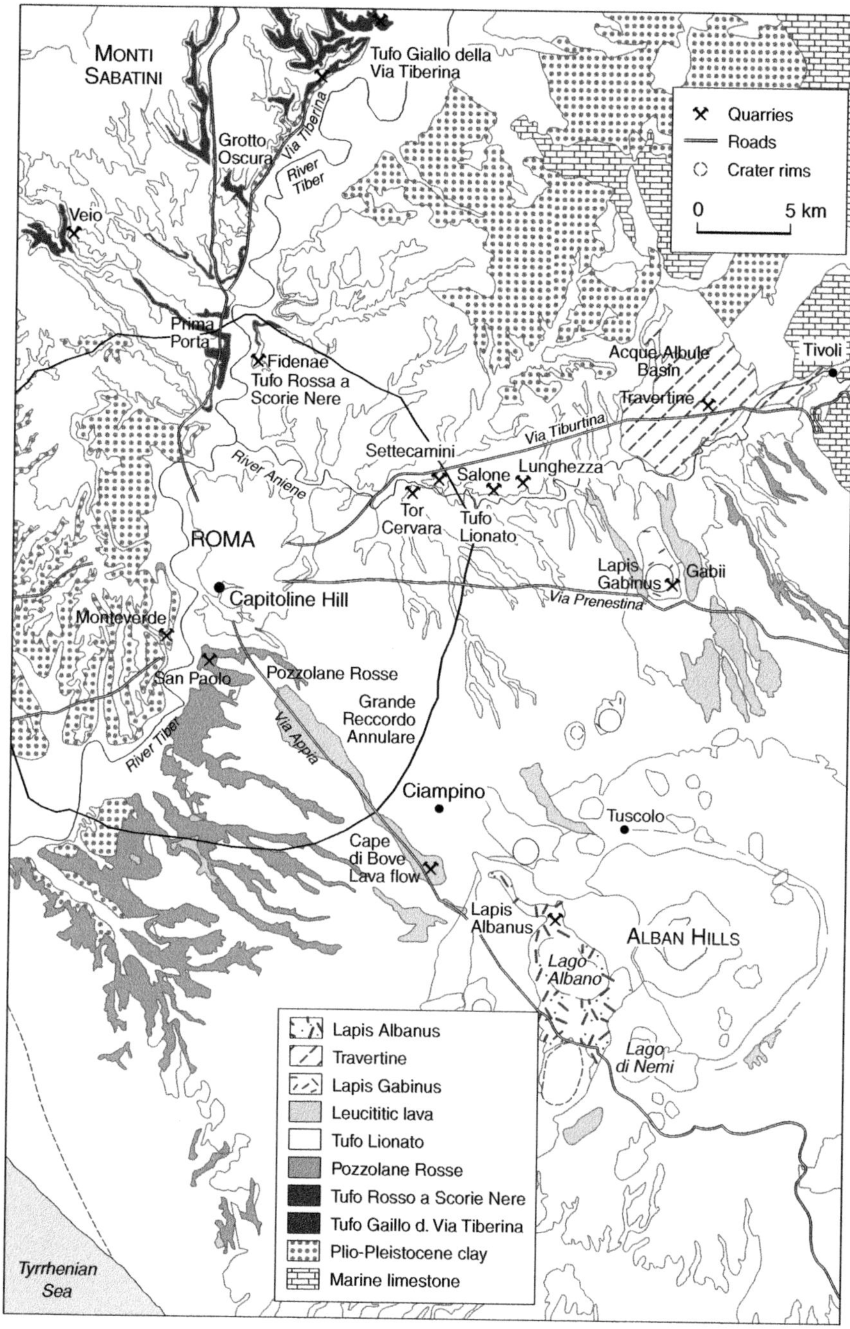

Figure 17.1 Quarries in the vicinity of Rome that were active during the Republican era. Pozzolane Rosse is grouped with Tufo del Palatino and Pozzolane Nere, and Tufo Lionato is grouped with Pozzolanelle; see Jackson *et al.* (2005), Jackson and Marra (2006) for a complete stratigraphy. Source: After Jackson *et al.* (2005), adapted, in part, from De Rita, Funiciello and Parotto (1988) and De Rita *et al.* (1993).

other hand, is the unfolding of intricate matters (obscurarum quaestionum) and a procedure (*ratio*) for discovering (reperta) new things through mental activity (vigore) and flexibility (mobili).

(*De arch.* I.2.2, trans. Jackson and Kosso, based on the text in Callebat, Gros and Jacquemard, 1999)

Descriptions of stone and concrete masonry presented in support of this hypothesis show how builders accumulated the knowledge and expertise (scientia) needed to create the durable pozzolanic concretes and increasingly complex architectural designs that would establish the city of Rome as the emerging capital of the Mediterranean area (see Chapter 28).

The earliest Republican era builders quarried soft volcanic tuffs close to building sites, such as for the massive dimension stone or opus quadratum podium of the Temple of Jupiter Optimus Maximus on the Capitoline Hill, dedicated in 509 (*LTUR* 3: 148–53 [De Angeli]; Chapter 11). By the first century, builders used numerous tuffs quarried from the Alban Hills and Monti Sabatini volcanic districts and durable travertine limestone quarried near Tivoli (Figure 17.1) as cut stone elements integrated with innovative concrete masonry in intricate architectural designs, as at the so-called Tabularium, *c.* 78–65(?) (perhaps an archival records office), the Theater of Pompey (*c.* 61–55/52), the Forum of Caesar (*c.* 46–44) and the Tomb of Caecilia Metella (*c.* 30). Moreover, the extant monuments and the text of *De architectura* are explicit evidence of Europe's first sustainable building practices, which were standardized during the Augustan age and form the foundation of the magnificent architectural achievements of the Imperial era.

Republican era builders owed a great deal of their success to a series of voluminous, mid-Pleistocene epoch, explosive eruptions from Rome's neighboring volcanoes. The diverse compositions of these pyroclastic deposits, and the varied processes of weathering, alteration and lithification that affected them on the earth's surface, provided an exceptionally wide array of volcanic building materials. Tuffs are lithified volcanic conglomerates composed of consolidated volcanic ash: the conglomerate consists of glass and pumice particles, crystals, and lithic fragments of lava. Roman tuffs range from glassy, pumiceous and lightweight to compact, cohesive and packed with lava rock fragments. Pozzolans are inorganic materials that react with calcium hydroxide (hydrated lime) to form binding cementitious compounds in hardened mortars. The Roman pozzolane are deposits of unconsolidated volcanic ash. The stability of Roman concrete constructions depended on cohesive volcanic ash pozzolan-hydrated lime mortars. In fact, the highest quality mortars were contingent on ash from a single explosive eruption dated about 456,000 years ago from the Alban Hills Volcano – the Pozzolane Rosse pyroclastic flow, and its fortuitous alteration on the ground surface during a humid interglacial period (Karner, Marra and Renne, 2001; Jackson *et al.*, 2010).

Erratic natural disasters and human catastrophes punctuated urban life in Rome. Unpredictable floods (Bencivenga, Di Loreto and Liperi 1995; Aldrete, 2007), earthquakes (Molin *et al.*, 1995) and fires damaged buildings and required constant reconstruction within the city. These events also motivated builders' innovations in materials and architectural design. Livy described such disasters in 193 and 192:

> There was an enormous deluge that year, and the Tiber inundated the low-lying parts of the city. Around the Porta Flumentana certain [buildings] even collapsed and fell in ruins ... The Tiber attacked the city more violently than formerly and destroyed two bridges and many buildings, especially in the area of the Porta Flumentana. A huge boulder, either undermined by the heavy rains or by a small earthquake which was not felt at the time, fell from the Capitol onto the houses in the area of Jugarius (vicum Iugarium) and crushed a great number of people.
>
> (XXXV.9.2, XXXV.21.5–6, trans. Jackson and Kosso)

These types of destructive events necessitated the development of stable, water-resistant and fire-resistant building materials. At the same time, an influx of unprecedented wealth resulting mainly from the spoils of military conquests reinforced a tradition of patronage and benefaction by generals and other magistrates, and often encouraged innovation and rivalries of ostentation in construction and reconstruction projects, ranging from Appius Claudius's aqueduct and Via Appia (*c.* 312) to Quintus Catulus's restoration of the Temple of the Capitoline Jupiter (dedicated in 69 after the fire of 83), to Pompey's theater in the Campus Martius (Diod. Sic. XX.36.1–2; Plut. *Pomp.* 40.5; Livy *Per.* XCVI.40.11–12, XCVIII.69.5). Many utilitarian buildings imitated Greek design practices, but with uniquely Roman tuff and concrete masonry (see Chapter 28).

1.1 The Knowledge and Expertise of Republican Era Architects

Vitruvius records Roman builders' technical expertise in developing an architecture of durability (firmitatis), utility (utilitatis) and grace (venustatis) (*De arch.* I.3.2). He discusses an ethos of Roman building that incorporated a long tradition of observation, display and exhibition, imitation, thoughtful reflection, skill, and the ability to learn, in order to improve construction durability and architectural design:

> Hence, after meeting together, the [early builders] began: some to make shelters of leaves, some to dig caves under the hills, some to make places for shelter with mud and wattles, imitating the nests of swallows and their methods for building. Then, observing the houses of others, and through thoughtful reflection adding new things daily, they produced better kinds of shelters. Since men are by nature imitative and easily teachable, they daily exhibit or display to one another the execution of [their] structures and other glorious inventions and so, they brought about, through their natural disposition for rivalries, better judgment every day.
>
> (*De arch.* II.1.2)

> ... Then, however, preparing themselves in spirit and looking out and forward with greater imagination born from the variety of their skill, they began to bring to completion not only the shelters but now, too, houses on foundations with brick walls and stone, and with roofs of wood and tile. After that, having observed through study, they understood rich material for building to be a lavish and abundant part of nature from which they drew; they attended to and prepared this material, extending through skill satisfaction with the elegant things of life.
>
> (*De arch.* II.1.7)

Vitruvius's emphasis on the importance of knowledge and expertise (scientia) in the development of architectural craftsmanship and technical skill (ars) seems to correspond to the modern concept of construction engineering: the discipline, art and profession of acquiring and applying technical, scientific and mathematical knowledge to design and to implement materials and structures that safely realize a desired objective or invention. He writes:

> The knowledge and expertise (scientia) of the architect is from many disciplines and [is] equipped with many kinds of learning; his work is accomplished by his judgment of all the other arts. This [knowledge] is produced from theory (ratiocinatione) and skillful effort (fabrica). The fabrica is the continuous and frequent exercise of practice (usus meditatio), where manual work is performed out of whatever material [according] to the design (propositum) of a representation (deformationis). Theory, on the other hand, is the ingenuity (ratione) to demonstrate the ability to explain those things made from calculated proportions (ratione proportionis).
>
> (*De arch.* I.1.1)

He further specifies:

> Therefore, those architects who aim at manual skill (manibus) without learning (sine litteris) have not had the power to effect a prestige corresponding to their labors, likewise, those who have trusted (confisi) to theory and literature only (ratiocinationibus et litteris solis) obviously have chased after a shadow (persecuti umbram) and not reality (rem). But those who have learned both thoroughly, like everyone furnished with [the proper] tools, have soon effected influence and authority, and attained their purpose.
>
> (*De arch.* I.1.2)

Vitruvius made an evident effort to integrate these approaches and present both practical and theoretical knowledge in *De architectura*.

For instance, he describes many "hands-on" construction practices to enhance the durability and service life of buildings. These include procedures for preparing and installing tuff in well-consolidated stone and concrete foundations and walls (*De arch.* II.7.1–5, II.8.1–5, VI.8.5); guidelines for minimizing compressive and tensile loads that could lead to the deformation and failure of masonry structures (II.8.16–20, VI.8.1–4); instructions for the preparation and application of stucco coatings (tectoria) (II.3.2, VII.2.1–2, VII.3.3–11, VII.4.1–3); criteria for the selection of volcanic aggregates, harenae fossiciae or pulvis puteolanus, to optimize the cohesion of mortars (materia) in buildings or harbor constructions (II.24.1–3, II.6.1–6, V.12.2–6); and methods for the selection, calcination and hydration of lime for enduring concrete masonry (II.5.1–3). He bases his recommendations on practical experience (ars and manus) and rational precepts (scientia and *ratio*): the tendency of soft tuff building stone to crumble and fall apart when directly exposed to weather, or frost, or sea air (II.7.2); the effects of water absorption in brick, clay, earth substructures, wood and tuff (rubro saxo) (II.3.2, VI.1.8, VI.8.5, VII.3.11, VIII.1.2); the effects of daily fluctuations in relative humidity (VIII.2.3–4); and the measurement of specific gravity of materials (VII.8.3, IX.Pref.10–12).

He also gives great attention to explaining the physical principles underlying these practices. Some of these explanations are well-grounded in empirical observations. For example, rock testing experiments substantiate the soft and yielding character of the Roman tuff building stones and the greater durability and strength of Roman travertine (Figure 17.2; Vitr. *De arch.* II.7.2; Jackson *et al.*, 2005). Other explanations are presented as hypothesis-driven trials (VIII.1.4–5, VIII.6.13, VIII.3.3, VIII.3.18), with valid specific conclusions. Significantly, Vitruvius proposes a well-controlled experiment to test the hypothesis that there is a cause and effect relationship between hot or damp climates and their human inhabitants' physical and mental characteristics: "We can observe by experiment that heavier and deeper effects arise in damp places and lighter and shriller effects occur in hot places" (VI.1.8). This hypothesis seems to derive, in part, from Hippocrates' theory of Airs, Waters and Places (Wallace-Hadrill, 2008: 149). Vitruvius proposes:

> Suppose two cups of equal weight have been baked in the same oven and equally, and when rattled [together] they produce one sound. From these [two cups] lower one into water, and after removing it from the water, then strike them both. Thus, this being done, there will be a difference in their sounds, and the cups will no longer be of equal weight.
> (*De arch.* VI.1.8)

Here, the specific conclusion is correct: the water-soaked terracotta cup changes in weight, just as in modern experiments for water sorption (Jackson *et al.*, 2005), and it resonates differently from the dry cup. However, when Vitruvius applies these experimental results to the more generalized theory, his deduction that the heat or humidity of climate produces shrill or deep utterances in humans "born of the same shape and conjunction of the heavens" seems illogical to the modern scientific mind.

Vitruvius had a high regard for Greek theoretical constructs (*De arch.* IX.Pref.15; Wallace-Hadrill, 2008: 145–7), yet he admits that he struggled to convey these accurately (V.4.1). His limitations arise, in part, from the difficulty of translating into Latin the scientific principles and sophisticated astronomical discoveries of Hellenistic scientists (*De arch.* IX.1.1–15; Russo, 2004: 238–9, 378–9). Furthermore, his explanations that rely on the earlier Empedoclean Theory of the four classical elements – earth, air, fire and water (VIII.Pref.1) – present rather obscure illustrations of physical processes. For example, Vitruvius theorizes that the fire element is the fundamental force that causes the mixture of lime, powdery volcanic ash pozzolan from Baia, and tuff caementa to solidify in pozzolanic hydraulic concretes, since the violent heat of the fire (ignis vehementi) in the volcanic environment "seizes moisture from the tuff (tofu) and the earthy ash (terraque) in the same manner which in the furnace, liquid is seized from lime" (II.6.3). When these three were hydrated, and "by suddenly accepting liquid, cohered into one [substance] and quickly solidified due to the moisture and were made solid," they formed very durable concrete piers in seawater (II.6.1). Vitruvius then remarks that it remains unclear why the powdery type of volcanic ash, or pulvis puteolanus, of Pozzuoli in the Bay of Naples, does not occur in the area of Etruria, north of Rome, since both areas have closely spaced springs of hot water (II.6.4), presumably associated with the "far distant fire and flame" of deep subterranean heat and the "violent force" of the fire element (II.6.1–6; Jackson *et al.*, 2007). He questions why the carbunculus type of excavated volcanic sand (harenae fossiciae) from Etruria and the pulvis puteolanus volcanic ash from Campania produce concretes with different material characteristics. Although carbunculus has advantages for buildings on land and pulvis for piers in the sea (II.6.4, II.6.6),

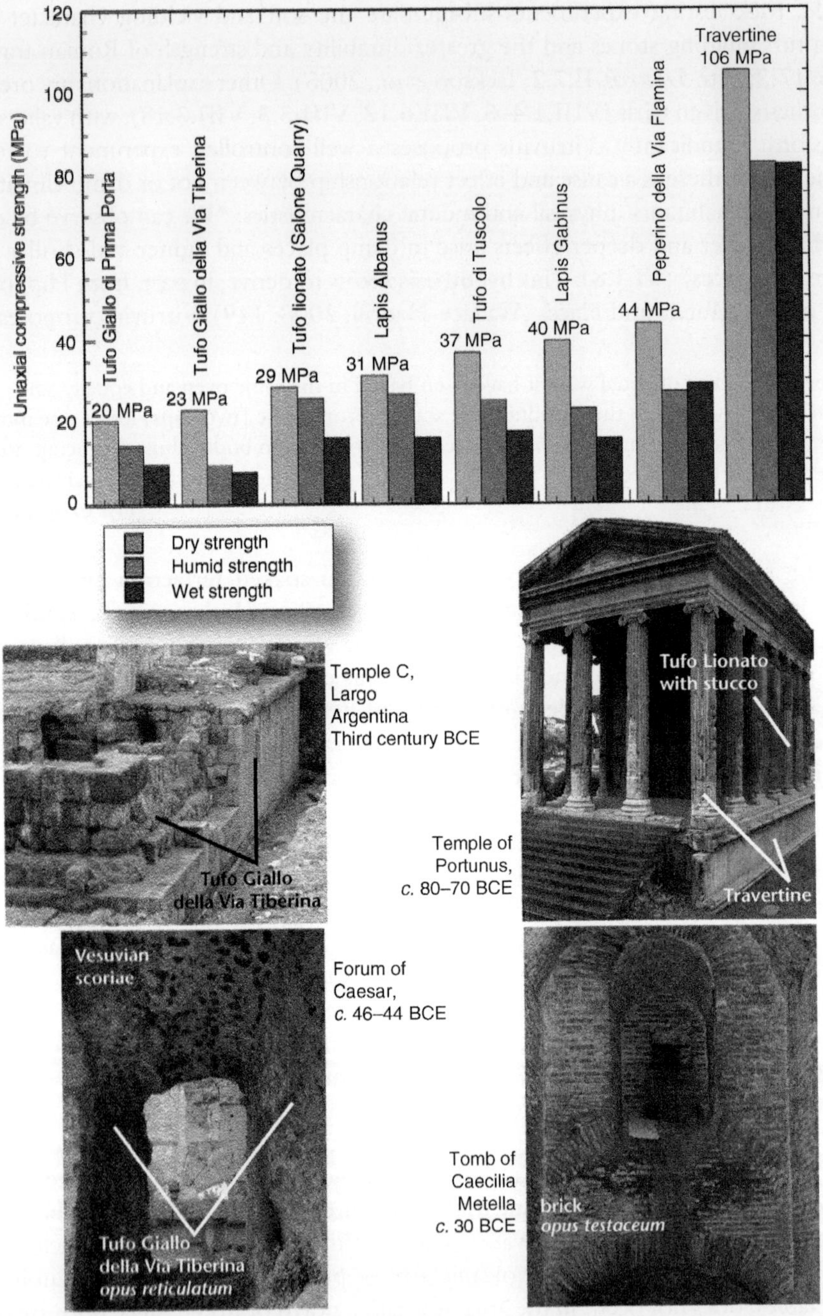

Figure 17.2 Uniaxial compressive strengths of the Roman tuffs and travertine, from oven-dry, water-soaked, humid and water-soaked experimental tests, and photographs of these building stones in Republican era monuments. Notably, two tuffs from the Via Flaminia – Peperino della Via Flaminia, a hardened facies of Tufo del Palatino, and glassy, brittle Tufo Giallo di Prima Porta – were not apparently used in Rome. Tufo di Tuscolo was used during first century CE. Source: After Jackson *et al.* (2005).

both are associated with geothermal environments and the "hot lack of moisture" of the fire element, which also causes concretes with pulvis pozzolan to harden underwater (II.6.1–4). He thus infers, obliquely, that there may be a discrepancy in understanding the role of the fire element in producing the differing pozzolanic characteristics of the carbunculus and pulvis materials and the processes of consolidation of concrete structures (II.6.4). Vitruvius strives to teach "true explanations in the precepts of physicists" (II.1.9), yet he shows awareness of some of the difficulties of integrating theory current at the time with practice and his own observations. He seems to present a Late Republican era perspective that grapples with integrating accurate empirical science with what we now see as untenable components of natural philosophy.

2 Geologic Foundations of Rome

About 560,000 years ago, explosive eruptions from Monti Sabatini and Alban Hills volcanoes began to deposit large volumes of volcanic ash in Rome (Figure 17.1). The tuffs were commonly emplaced as pyroclastic flows of hot gasses and incandescent particles that covered topographic plateaus or filled ancient valleys (Marra and Rosa, 1995; Karner, Marra and Renne, 2001). Travertine sedimentary rocks were deposited by warm, calcium carbonate-rich waters in the Acque Albule Basin, near modern Tivoli. Vitruvius correctly described the material characteristics of these rocks:

> Now order demands that I explain about quarries, from which both squared blocks and the supplies of rough unhewn stone for building are obtained and readied. These, in turn, will be found to have unequal and dissimilar qualities. Some are soft (tenera) and yielding (molles) around the city itself, in the manners of the Rubrae stones, the Pallenses stones, the Fidenates stones and the Albanae stones. Some are of moderate strength, like the Tibur stones, the Amiternae stones, and the Soracte stones, and others of this type. Some are hard, like lavas (siliceae).
>
> (*De arch.* II.7.1)

The soft and yielding stones are tuffs: Tufo Lionato (Rubrae), Tufo Giallo della Via Tiberina (Pallenses), Tufo Rosso a Scorie Nere (Fidenates) and Lapis Albanus (Albanae) (Figures 17.1 and 17.2). "Yielding" (molles) may describe Vitruvius's understanding of the tuffs' low yield strengths, and their tendency to rupture under heavy loads, especially when water-soaked (Jackson *et al.*, 2005; Jackson *et al.*, 2006). The Tibur travertine, and Amiternae and Soracte marine limestones are more durable, and have greater strength when wet. The siliceae stones are indurated lava flows. Builders' accurate empirical knowledge of these various material characteristics is confirmed by their selections of building stones in the extant Republican era monuments.

2.1 *Volcanic Tuff Deposits*

The diverse strengths and durabilities of the tuff building stones are a direct consequence of their relative abundances of porous glass or pumice fragments, crystals and lava rock particles (Nappi, De Casa and Volponi, 1979; De Casa *et al.*, 1999; Jackson *et al.*, 2005;

De Casa and Lombardi, 2007). For example, Lapis Albanus, the product of an explosive eruption and debris flow at the Lago Albano crater, and Lapis Gabinus, the product of a pyroclastic surge eruption at the Valle Castiglione crater, are commonly referred to as peperino. They have relatively high proportions of rock and crystal fragments, up to 30–50 volume percent, respectively. Zeolite mineral cements firmly adhere to the grain-supported framework of hard, dense lava fragments, which provide good bulk specific gravity and uniaxial compressive strength to the tuffs, about 23 to 26 MPa (megapascal, a measurement of force per unit area) in the humid ambient conditions of Rome. The geographer Strabo comments, "Gabii lies on the via Praeneste, it has a stone-quarry, [quite] serviceable for Rome [and] better than others, it is equally distant, around 100 stadia, from Rome and Praeneste" (V.3.10, trans. H. Jones, 1923, Loeb). In contrast, the coherent facies of tawny orange, glassy Tufo Lionato quarried near the Aniene River, known as Anio or litoide, has intermediate proportions of lava rock fragments, about 8–12 volume percent. It has compressive strength of about 25 MPa in humid conditions, but its glassy fabric is vulnerable to decay.

Pumiceous Tufo Giallo della Via Tiberina, known as grotta oscura, has low compressive strength, generally less than 9 MPa in humid conditions, which decreases further when water-soaked. The high water sorption of this tuff and Tufo Rosso a Scorie Nere, about 20–28 weight percent, encourages the weathering of the glass and pumice components to a clay mineral, reducing the adherence of natural zeolite cements and thus decreasing coherence and compressive strength. These porous tuffs are particularly susceptible to disaggregation, but as Vitruvius notes, "as long as they are used in covered areas (locis tectis), they will sustain loads (sustineant laborem), but if they are put in open uncovered places, then, once they have been saturated with ice and frost they crumble apart and dissolve" (*De arch.* II.7.1–2). Absorption of water occurs mainly through saturation of tuff foundations and walls with groundwater and inundations of Tiber River floodwater, and adsorption of water vapor occurs during large daily fluctuations in relative humidity (VIII.2.3–4). The Republican era builders coated most tuff masonry with protective layers of stucco to provide a decorative surface and to protect soft, porous stone from water sorption. The thermal expansion of all the Roman tuffs is low, and so they are rather fire-resistant building materials (Jackson *et al.*, 2005).

2.2 *Travertine Deposits and Marine Limestones*

Roman travertine, or Tibur stone, is exceptionally coherent and has far greater compressive strength than the Roman tuffs, about 105 MPa, with very low water absorption, 0.6 weight percent. This gives a long-lasting durability even in humid conditions. However, limestone building stones are susceptible to fracture at urban fire temperatures: "Travertine, on the other hand, and all stones of the same type, withstands heavy loads and damage incurred during stormy, wet seasons, but they cannot be safeguarded against fire. As soon as they come into contact with it, they crack apart and fall to pieces" (Vitr. *De arch.* II.7.2). Travertine, marine limestone and marble (metamorphosed, recrystallized limestone) are composed of calcite crystals. When heated, calcite lengthens along one crystallographic axis but contracts perpendicular to it. The contacts among calcite crystals then rupture and the stone abruptly fractures, as Vitruvius describes

(Jackson *et al.*, 2005). The Amiternae stones and the Soracte stones are the marine limestone bedrock of the Apennine Mountains. When calcined, limestone from the Monte Soratte quarries, about 50 km north-east of Rome, produced nearly 95% pure lime for the mortars of pozzolanic concretes (Jackson *et al.*, 2007).

2.3 Sedimentary Bedrock and Epiclastic Sediments

Plio-Pleistocene epoch marine clays and sandstones form the sedimentary bedrock of Rome. Builders used some of these clay horizons for brick manufacture (Ammerman *et al.*, 2008). A highly irregular and interlayered pattern of volcanic deposits covers most of this sedimentary bedrock in Rome. During periods of mid-Pleistocene sea level fall, the ancient Tiber River and its tributaries eroded valleys into these pyroclastic rocks. During periods of a rise in sea levels, the river system deposited accumulations of fluvially reworked epiclastic sediments, eroded from the volcanic landscape into these low-lying areas (Marra and Rosa, 1995; Alvarez *et al.*, 1996; Corazza, Lombardi and Marra, 2004). A distinct topography of hills and valleys eventually emerged in Rome, with a complex sequence of interlayered pyroclastic flow deposits, airfall ash deposits, and reworked epiclastic sediments.

2.4 Granular Ash of Alban Hills Pyroclastic Flows

Widespread unconsolidated ash deposits of the voluminous Pozzolane Rosse, Pozzolane Nere and Pozzolanelle pyroclastic flows surround the Alban Hills Volcano. During a humid interglacial period, a soil formed on Pozzolane Rosse that produced an underlying horizon of reddened scoriae with surface coatings of opaline silica and clay mineral. A deeper horizon has dark gray scoriae with zeolite mineral surface coatings. These alteration components have good pozzolanic reactivity with hydrated lime (Jackson *et al.*, 2010; Jackson, Scheetz and Marra, 2010). The scoriaeous Pozzolane Rosse ash, quarried in the San Paolo area about two-and-a-half kilometers south of the Capitoline Hill corresponds to the harenae fossiciae described by Vitruvius as reddened (rubra) and dark gray or black (nigra), which is "the type that makes a harsh, raspy, grating noise when rubbed vigorously in the hand [and] is best (optima). That which is earthy (terrosa) in texture is not rough enough. Likewise, if it is wrapped in a white cloth that is afterwards shaken out, if it neither stains the cloth nor has [fine] earthy particles that settle within it, then it is suitable (idonea)" (*De arch.* II.4.1). The cana sands seem to correspond to white, pale yellow or ashy-gray harenae fossiciae excavated from epiclastic sediments in Rome, with an earthy fabric and a fine grain-size fraction (Van Deman, 1912; Blake, 1947; Jackson *et al.*, 2007).

3 Examples of Dimension Stone Masonry

During the sixth and fifth centuries, builders used rather soft tuffs quarried within the city. Tufo del Palatino, or cappellaccio, is a friable, moderate olive gray tuff erupted from the Alban Hills Volcano and deposited as a mud flow in Rome (Alvarez *et al.*, 1996; Ammerman and Montanari, 2000). It forms the massive podium of the Temple of Jupiter Capitolinus;

the podium of the Regia, which was destroyed by fire and reconstructed in the sixth century; and the podia of the archaic temples at Sant'Omobono built in the same era. The clay-rich fabric of Tufo del Palatino is susceptible to decay when water-soaked, but it is more durable than the sun-dried mud-brick walls and foundations that collapsed when saturated with floodwater, as described by Cassius Dio for the Tiber flood in 54 (39.61).

The earliest extant wall around the Forum, which dates to the eighth century, is constructed of large, rough, rubble chunks of Tufo Lionato and Tufo del Palatino, quarried from the flanks of the Palatine Hill (Ammerman and Montanari, 2000). In contrast, Tufo Lionato dimension stone forms the foundation of the Temple of Apollo Medicus (dedicated in 431), which has soft Tufo del Palatino in its internal structure (*LTUR* 1: 52 [Viscoglioso]). The soft Tufo Lionato may have come from the Aventine Hill. Quarries in the Monteverde area on the right bank of the Tiber River were an important source of the tuff until the second century, when quarries for the more durable Tufo Lionato opened along the Aniene River.

After the conquests of Fidenae (435/426) and Veii in 396 (Livy IV.22, IV.35), builders began to quarry porous, glassy tuffs from the Monti Sabatini volcanic district. For example, an early structure made of soft Tufo del Palatino in the Aracoeli garden is surrounded with large blocks of light brown Tufo Rosso a Scorie Nere, the Fidenae stone of Vitruvius (Tucci, 2005). If this corresponds to the original podium of the Temple of Juno Moneta (345/344), then it suggests that by the fourth century, builders had observed that the glassy tuff was the more durable material, even though it was more difficult to quarry and had to be transported much longer distances. On the other hand, builders used porous, pumiceous Tufo Giallo della Via Tiberina, or Pallenses stone throughout Republican Rome. The soft, pale yellow tuff of the fourth-century "Servian" city wall is still exposed at the Termini train station and the foot of the Aventine Hill, weathered and decayed after centuries in the open air (Panei, 2011). The Tufo del Palatino blocks of the wall protected underground in the lower levels of the modern station, however, remain well-preserved. The four-meter-tall inner podium of the early third-century Temple C at Largo Argentina is composed of a highly pumiceous facies of the tuff, which is deeply weathered and degraded (Figure 17.2). Significantly, builders selected a far more compact and cohesive facies for the outer corniced perimeter wall, demonstrating a deliberate effort to protect the porous tuff of the podium interior from floodwaters which regularly submerged the Campus Martius. The overhanging eaves and the alae of the Italic-type temple removed tuff masonry surfaces from direct contact with rainwater (*De arch.* IV.7.2; see Chapter 26). Vitruvius's recommendations to place the soft tuff stones under cover (locis tectis) clearly had their origin in much earlier traditions.

Builders' regular use of dark olive gray, coherent Lapis Albanus tuff from blocks that were transported on oxcarts 25 km to Rome coincided with the construction of the Via Appia Antica in *c.* 312 over the surface of the Capo di Bove lava flow. They selected the cohesive tuff for the fourth-century foundation and walls of the Tullianum (Karner, Marra and Renne, 2001), for example, and the column drums, capitals and bases of the Ionic colonnade of the northernmost of the victory temples at San Nicola in Carcere in the Forum Holitorium, possibly the Temple of Janus (*c.* 260), and for the ornately carved architrave of the Temple of Spes (possibly 212). The fluted Lapis Albanus columns with plaster coatings at Via delle Botteghe Obscure (possibly first century) are reinforced with travertine capitals and bases. The compact fabric of the Lapis Albanus stone was more suitable for imitating the decorative elements of Greek marble architecture (Davies, in press a), as compared to the more porous and fragile Monti Sabatini tuffs. By the mid-second century, the Roman fascination with the elegance and refinement of the Hellenic world had opened a private trade in imported Pentelic marble for ornamental stonework (Davies, in press b; Bernard, 2010).

The trio of travertine, Lapis Gabinus and Tufo Lionato building stones quarried near the Aniene River (Figure 17.1) were described by Strabo: "From there, the river goes out through the fertile plain past the Tiburtine stone quarries and the Gabii [quarries], and of those called 'redstone,' so export from the quarries and ferrying across [the river] are perfectly convenient – [the stones for] most of the works being constructed at Rome are from there" (5.3.11, trans. H. Jones, 1923, Loeb). Builders selected moderate gray Lapis Gabinus, which they apparently considered the most durable of the tuffs (Strabo 5.3.10), for structural elements exposed to water-soaked conditions. Arched sections of the Aqua Marcia aqueduct (144–140) have Lapis Gabinus voussoirs, and long slabs of Lapis Gabinus form the floor and roof of the water channel (Frank, 1924: 137–9; Van Deman, 1934: 7–9; see Chapter 18). The reconstruction of the Milvian Bridge in 110/109 and Pons Fabricius in 62 have external piers and vaults of massive Lapis Gabinus dimension stone faced with travertine, which protected the internal blockwork of Tufo Giallo della Via Tiberina and Tufo Lionato, respectively (Frank, 1924: 141–3). Lapis Gabinus forms the facing of the concrete foundation of the "Tabularium," the facing of the Tufo Lionato piers of the tabernae of the Forum of Caesar and the fire wall of the Forum of Augustus, the last dedicated in 2 CE. Travertine keystone blocks and imposts reinforced the Lapis Gabinus voussoirs of arches at the "Tabularium" and the Forum of Caesar, increasing the durability and weight-bearing strength of the vaults.

Refined Tufo Lionato masonry forms the fluted column drums of the Temple B in the Largo Argentina (*c.* 100), the reconstruction of the Temple of Portunus (*c.* 80–70) in the active Tiber River flood plain (Figure 17.2), and the enlargement of the podium of the Temple of Apollo Medicus (*c.* 30). Travertine bases and capitals reinforce the Tufo Lionato columns, which still retain more recent coatings of plaster. Thick travertine panels protect the Tufo Lionato blockwork of the six-meter-tall podium of the Temple of Portunus from direct sorption of floodwaters, and the fluted columns of the *pronaos* are travertine. Travertine from Tivoli seems to have been introduced in the mid-to-late second century (Lugli, 1957: 319–36). It seems to have been augmented with early first-century decorative elements of Carrara marble, which were shipped about 300 km southward from the Luni quarries along the Ligurian coast. In the reconstruction of the Temple of Apollo Medicus, builders used Luni marble for fluted columns and their Corinthian capitals. Cleverly, they constructed much of the entablature of local travertine overlain with carved marble veneers, and they faced the Tufo Lionato cella with travertine overlain with molded decorative stucco (*LTUR* 1: 52–3 [Viscogliosi]).

4 Examples of Concrete Masonry

Durable concrete constructions in Rome apparently date to the second century. The Navalia, or ship-shed of Rome, is an extensive system of concrete walls and barrel vaults on the left bank of the Tiber River, long considered to represent the 192/174 construction of a Porticus Aemilia (Livy XXXV.10.12, XXXXI.27.8–9). The construction has now been re-dated to mid-second century (Cozza and Tucci, 2006; see also Chapter 21 for a different date). The concrete walls have opus incertum tuff facings and conglomeratic cores with various soft tuffs as coarse aggregate, or caementa. Narrow voussoirs of Tufo Lionato form the intrados of the concrete barrel vaults, indicating a rather skilled installation. The mortars are in some places dominated by Pozzolanelle ash aggregate, but generally formulated with epiclastic sediments. They are coherent, with a significant proportion of lime, and a somewhat well-developed cementitious matrix (Jackson *et al.*, 2007). Experimental rustic concrete vaults occur, for example, at the Aqua Marcia (144–140), where much of the above-ground roof is Lapis Gabinus dimension stone. At a few points, however, there is a roughly built, rounded roof of "crude tuff voussoirs held into place by a mass of coarse early mortar" (Van Deman, 1934: 8).

Concrete elements in the second-century monuments show a great diversity in mortar and coarse aggregate compositions. The friable concrete podium and walls of the 121 construction phase of the Temple of Concord in the Roman Forum have Vitruvius's cana mortar aggregate. These are earthy, pale yellow-gray, pumiceous epiclastic sediments apparently excavated *in situ* from the Capitoline Hill. The nearby Temple of the Castors, originally built of Tufo del Palatino (*c.* 484), has Tufo del Palatino caementa, perhaps derived from the weathered dimension stone, and a rather porous mortar with epiclastic ash, also apparently from the Capitoline Hill. Significantly, after a fire in 14 or 9, the temples of Concord and the Castors were rebuilt in 10 CE and 6 CE, respectively, with the newly standardized Augustan-age concrete compositions: harenae fossiciae of the reddened scoriaceous ash quarried from the Pozzolane Rosse pyroclastic flow and Tufo Lionato caementa (Jackson *et al.*, 2007) .The imposing podium of the Temple of Magna Mater (205–191) on the Palatine Hill, may have originally been constructed mainly of Tufo Giallo della Via Tiberina. It was rebuilt in heterogeneous concrete after a fire in 111, apparently with a friable, light ashy-gray mortar and Tufo Giallo della Via Tiberina and Lapis Albanus opus incertum facings and caementa, and repaired yet again in 3 CE after another fire, this time in Augustan-age concrete materials (Frank, 1924: 96–7; Blake, 1947: 330; *LTUR* 3: 206–8 [Pensabene]).

Petrographic examination of the mortars of the massive concrete structural elements of the "Tabularium" indicate that they have epiclastic aggregate composed of diverse reworked ashes, apparently excavated from the building site. Van Deman (1912: 247) describes this mortar as "of the ashy gray type, but it is slightly less friable than that of the preceding period." By contrast, the highly innovative mortars of the Forum of Caesar contain highly weathered pumiceous epiclastic sediments, but in variable proportions with Pozzolane Rosse and diverse fresh, glassy pumices, indicating that builders at the worksite were actively experimenting with aggregate design mixtures from several sources (Jackson *et al.*, in press a). They also experimented with the caementa of the concretes, using Tufo Giallo della Via Tiberina in the opus reticulatum and caementa of the tabernae walls, but glassy scoriae from a Vesuvian eruption for the caementa of the barrel-vaulted ceilings (Figure 17.2) (Amici, 1991; Lancaster *et al.*, 2011; Jackson *et al.*, in press a). Nevertheless, the concrete is poorly compacted, and pervasive spherical voids in the mortars suggest a free water porosity, which may indicate difficulties in regulating the water content of the Pozzolane Rosse-pumice-hydrated lime wet mix. Significantly, the durable and compact mortars of the Tomb of Caecilia Metella (*c.* 30) and the Theater of Marcellus (27–13/11), which are "grayish red and almost free of the friability of the earlier type" (Van Deman, 1912: 250; cf. Jackson *et al.*, 2007; Jackson *et al.*, 2011), have the reddened Pozzolane Rosse aggregate of the Augustan age.

Some question remains as to when builders first began to excavate the rubra and nigra harenae fossiciae from the middle and lower horizons of the Pozzolane Rosse pyroclastic flow. Further analytical work is needed to understand the diverse fabrics and sources of the variable compositions of the predominantly Pozzolane Rosse mortar aggregate of, for example, the irregular opus reticulatum at the Theater of Pompey (Gagliardo and Packer, 2006; Packer, Burge and Gagliardo, 2007); the Caesarian phase of the Circus Maximus, initiated in 46 (Ciancio Rossetto, 1995); the foundations of the Curia Julia (*c.* 44–29); and the podium of the Temple of Saturn (*c.* 42) (Van Deman, 1912: 248–51). In this regard, the Sanctuary of Fortuna Primigenia constructed above Praeneste (Palestrina) provides a reference point for late second-century builders' daring ingenuity and proficiency with pozzolanic concrete construction. The concretes have limestone opus incertum facings and coarse aggregate, quarried from the local bedrock. A complex series of ascending concrete terraces, gigantic ramps and staircases, and porticoes with coffered ceilings, culminates in an open theater and round temple with an expansive view to the sea (see Chapter 15). The builders' skilled expertise with cut-stone construction is shown by the Tufo Lionato voussoirs placed in the intrados of concrete barrel vaults and the integrated Tufo Lionato half-columns with travertine capital and bases. Careful macroscopic inspections of the coherent, pale grayish-red mortars indicate that the scoriaceous aggregate is Pozzolane Rosse from the deeper alteration facies. Exposures of Tufo Lionato and Pozzolane Rosse occur in the stream drainages leading directly southeast from the Sanctuary, and modern quarries for Pozzolane Rosse exist at nearby Zagarolo and Cave (De Rita, Funiciello and Parotto, 1988; Marra *et al.*, 2009). Apparently, builders had discovered the raw materials of standardized Imperial era mortars for highly cohesive concrete work nearly 100 years before the Augustan age, at the foot of the Apennines.

5 Historical Context of Construction Innovations

Observations of the compositions and material characteristics of the dimension stone of fifth- to second-century monumental constructions support Vitruvius's remarks about the development of Roman construction engineering (*De arch.* II.1.1). They indicate that Early-to-Mid-Republican builders actively strove to employ a correct treatment of building materials according to their functional qualities, to expand the supplies of various dimension stones from the Monti Sabatini and Alban Hills volcanic districts and, through attentive observation, to imitate effective Greek and southern Italian building practices (see Chapter 28) for enhancing the durability and utility of porous tuff masonry in the often humid and water-soaked environment of Rome.

During the second century, masonry construction in Rome was influenced by distant military activities and Mediterranean conquests, architectural developments in bridge constructions and fortified camps (as well as siege equipment, roads, walls and defensive towers) by engineering units led by architecti (Polyb. 6.26–32; Vitr. *De arch.* X.1; Shirley, 2001). Aristocratic competition stimulated the construction of many public buildings in

Rome. The movement of tens of thousands of people into Rome stressed infrastructure systems and required advances in the civil engineering of aqueducts, drainage canals and mills (see Chapter 18; Baker and Baker, 1994; Hodge, 2002). Builders' practices and manual skill with dimension stone masonry, acquired during the earlier Republican era, and their inventions of cohesive concretes installed in massive and innovative civil engineering structures such as the Navalia and the Aqua Marcia, reflect these new commercial and infrastructure demands. An inflow of wealth derived from military conquests encouraged the pre-existing system of architectural patronage and building projects that enhanced the power and prestige of victorious generals (Wallace-Hadrill, 1989; Favro, 1996; Lomas and Cornell, 2003; Taylor, 2003). The Milvian Bridge, for instance, was originally constructed in *c.* 206 after the battle of the Metaurus, during the Second Punic War, but was demolished and rebuilt in stone in 110/109 during the censorship of Marcus Aemilius Scaurus. Temple B at Largo Argentina was vowed on July 30, 101 as Aedes Fortuna Huisce Dei, after the Battle of Vercellae. The Temple of the Castors was renovated by Caecilius Metellus Delmaticus in 117 after his triumphal celebration over the Dalmatians (Cic. *Verr.*[2] 1.154; Asc. *Comm. Scaur.* 24, in Lewis, 2007: xxii, 203, 227–8). Architecturally, these monuments were rather conservative, indicative of "gradual adaptation and cautious advance" (Ward-Perkins, 1977: 52). By contrast, the concretes of the Sanctuary of Fortuna Primigenia in Praeneste are "remarkably advanced in the principles of their composition and in the virtuosity of their construction techniques" (Ward-Perkins, 1977: 35–9).

Repairs to buildings damaged by fire were apparently commonly carried out in concrete, as at the Temple of Magna Mater, and construction in response to political violence and civil disturbances opened space for architectural innovation (Nippel, 1984; Patterson, 1992). For instance, the Temple of Concord was constructed (or rebuilt) in 121 after the murder of Gaius Gracchus; the senate would meet here in a symbolic gesture of concordia when discussing controversial issues, such as land reform, which were likely to result in public dissonance (Plut. *C. Gracch.* 18.6; Scullard, 1982; Stamper, 2005). Scipio Nasica demolished the stone theater begun by Lucius Cassius (*c.* 154), because he considered this a likely source of new riots and civil disturbances (App. *B Civ.* 1.4.28). Marius captured and plundered Ostia in 87, one of only several episodes of destruction and consequent rebuilding of that city in the first century (App. *B Civ.* 1.8.67). According to Plutarch, Crassus would buy the lots in Rome where houses had been set afire; he was said to have had 500 builders ready to rebuild these properties (*Crass.* 2). P. Clodius Pulcher had Cicero's villas and house burnt down in 63 (Cic. *Red. sen.* 17), and after Clodius's murder in 52, fires associated with riots burned the Basilica Porcia (built in 184) and Curia Cornelia (dated to *c.* 90–80), as we read in Cicero's *Pro Milone* (37, 41). Later replacements for some of these buildings – the Theater of Pompey and the Curia Julia, for example – were highly inventive in their concrete materials and architectural designs.

It was in this unstable environment of civil unrest that early first-century builders initiated innovative architectural designs with cut-stone masonry, including improvements to the corbel, lintel arch and tie (DeLaine, 1990: 422). During this same period, builders perfected methods for the production and installation of durable seawater concretes with pulvis puteolanous aggregate in harbor constructions (Oleson *et al.*, 2004; Jackson *et al.*, in press b). In Rome, they integrated dimension stone and concrete masonry elements in far more architecturally sophisticated monuments, as at the "Tabularium," initiated after the 83 fire that consumed much of the Capitoline Hill; the Pons Fabricius, built to replace an earlier wooden bridge destroyed by a flood (Cass. Dio 53.33); and the Theater of Pompey, constructed by Pompey to win support from the people in his rivalry for power with Julius Caesar. The "Tabularium" is an example of the transformational use of dimension stone and somewhat rustic conglomeratic concretes into complex vaulted designs that might have seemed "startlingly modern, though the modernity was carefully tempered to public taste behind a conventional façade" (Ward-Perkins, 1977: 68). As we have noted above, the builders transported large quantities of dimension stone from the Aniene River quarries to the Capitoline Hill, but formulated concretes with mortar aggregates excavated from epiclastic ash at the building site.

The extant monuments of the mid-first century record the high degree of mental acuity and flexibility (*De arch.* I.2.2) that Republican builders brought to the task of enhancing the strength and durability of monumental construction. The tabernae of the Forum of Caesar provide an especially clear illustration of the builders' excellent empirical knowledge of the diverse material properties of the Roman tuffs and travertine and their skill in installing these rocks in complex structural elements (Jackson *et al.*, 2005: 506). Robust Lapis Gabinus voussoirs with good compressive strength (40 MPa) and bulk specific gravity (1.81) form the flat arches of the tabernae. These are securely reinforced with travertine blocks and keystone voussoirs, with high compressive strength (105 MPa) and specific gravity (2.58). The upper façade is formed of round arches of Tufo Lionato with moderate compressive strength (29 MPa) and lower specific gravity (1.73), which rest on lower, internal walls of Tufo Lionato that are protected behind an exterior course of more durable Lapis Gabinus dimension stone. The concrete barrel vaults within the interiors of the tabernae have walls with opus reticulatum facings and coarse aggregate of soft but easily worked Tufo Giallo della Via Tiberina, with low compressive strength (23 MPa) and specific gravity (1.52). The weight of the vaulted ceiling concrete was deliberately lightened by vesicular glassy scoriae caementa from Vesuvius (Lancaster *et al.*, 2011), and the incorporation in the mortar of additional, diverse pumices as harenae fossiciae imported from distant volcanic districts, based on their crystal assemblages (Jackson *et al.*, 2010). Significantly, vesicular basalts that are similar to the glassy scoriae have relatively high compressive strengths, about 40–50 MPa (Al-Harthi, Al-Amri and Shehata, 1999), suggesting that the glassy scoriae lend strength to the concrete despite their porous texture (Bianchi *et al.*, 2011). Builders' brilliant selections of Tufo Lionato, Lapis Gabinus and travertine dimension stone according to unit weight and durability illustrates they had thoroughly integrated manual skill with theoretical relationships regarding rock specific gravity, compressive strength and, quite possibly, yield strength to achieve extraordinarily innovative materials and structural designs (see Vitr. *De arch.* I.1.2; Jackson *et al.*, 2005; Jackson *et al.*, 2006: 1692–4). Their versatile installations of these diverse dimension stones and concrete formulations record the long traditions of observation, imitation and thoughtful attention that earlier Republican builders brought to the development of architectural craftsmanship and technical skill (Vitr. *De arch.* II.17).

The diverse fabrics and aggregate compositions of mortars formulated with epiclastic ash sediments or various Pozzolane Rosse mixtures in second- through mid-first century concretes, indicate that the very high quality Pozzolane Rosse pozzolan did not become firmly established in Rome until the initiation of the Augustan age. At this point, the collective accumulation of builders' geological knowledge and their identification of thick Pozzolane Rosse deposits close to Rome, perhaps in the San Paolo area, seem to have permeated and then permanently transformed construction practices. The highly cohesive and durable concretes and the complex internal spaces that emerged from the Tomb of Caecilia Metella and the Theater of Marcellus were constructed by builders trained in Republican era methods and materials, but reflect a new conceptual framework and collaborative approach in construction engineering practices (Gerding, 2002; Jackson *et al.*, 2011), which form the foundation of systematized construction technologies and standards of Imperial Rome.

FURTHER READING

For a more complete picture of the practical knowledge and capable engineering practices (scientia) of Republican builders, read *De architectura* while integrating Vitruvius's remarks regarding an empirical scientific method in construction engineering with the archaeological and historical context of the Republican era monuments (Chapters 1, 18, 26, 28; Davies, in press a, 2012) and the material characteristics of stone and concrete masonry of the Republican period (Jackson and Marra, 2006; Jackson *et al.*, 2010; Jackson *et al.*, 2011). Discussions of time-scale of constructions and the correlation of events of the geological, archaeological and historical record by Alvarez, Claeys and Montanari (2009) and Christian (2010) provide a broad conceptual framework with which to periodize the historical intervals of accelerated Republican era innovations in construction engineering. Russo (2004) describes the contributions of Hellenistic scientists, which form the basis of the scientific revolution of the Renaissance and who are cited by Vitruvius: Archimedes (IX.Pref.9–12), Ctesibius (IX.8.2–14), Eratothenes (IX.Pref.13–14) and Hipparchus (IX.6.3), as well as earlier Classical mathematicians (IX.Pref.19.7–8), and others (IX.6.1–3). For more information on pyroclastic volcanism see Fisher and Schmincke (1984), and for the Roman marble trade in general see Fant (1988) and Waelkens, de Paepe and Moens (1988).

CHAPTER 18

Aqueducts and Water Supply

A. Trevor Hodge

1 Introduction

The study of Roman aqueducts, as has been noted by Gunther Garbrecht, a leading authority, can impinge on a wide range of disciplines: "history, political administration, legal aspects, sociology, water demand, engineering, hydrology, hydraulics, stability analysis, construction materials, literary evidence, archaeology, and ecology" (pers. comm.). This is comprehensive indeed, yet aqueducts, though today spectacular monuments, were by no means indispensable features of ancient cities. In fact, they were often an expensive luxury. Moreover, though they are often thought of as distinctively Roman, this demands clarification.

As Roman power expanded overseas, in many of the new provinces, particularly in Hellenistic Greece, there were cities that now came under Roman control and which already had functioning aqueducts. Their water supply commonly came in pipes rather than the masonry conduits favored by the Romans, and terminated in some central fountain-house. Athens, with its Enneakrounos fountain, is an example of such a system. Though we normally consider them rather in the context of their Greek origin, fountain-houses certainly did form part of the Roman Empire and were in daily use by its inhabitants during the Republic. They were familiar also to the Romans and in particular to Roman army engineers, who must often have seen them in operation while serving in the provinces, and were no doubt impressed by such technical *tours de force* as the sixth-century kilometer-long tunnel on Samos or the tremendous inverted siphons at Pergamon. Closer to home, the Etruscans also had extensive waterworks, though usually in the form of tunnels cut under hills to drain away unwanted water from one valley to the next one. It is generally true that drains, such as Rome's Cloaca Maxima, come much

A Companion to the Archaeology of the Roman Republic, First Edition. Edited by Jane DeRose Evans.

earlier than water supply, since their purpose, to increase arable land by draining swamps, was seen as more important (see Chapter 28). Thus even if they were not currently engaged in constructing aqueducts at Rome, many Romans profited from an acquaintance with this background material on hydraulics.

Aqueducts were very costly to construct. So why and how did the Romans do it? Under the Empire, aqueducts were usually built and financed by the emperor, often to serve one of the monster bath complexes, themselves a product of imperial beneficence. Under the Republic these criteria did not exist and the decision to build an aqueduct rested with the Senate, apparently spurred on either by the personal initiative of an individual senator or by the ready availability of abundant funding; maintenance was under the care of censors or aediles. As an authority on financing, Ph. Leveau, has put it:

> Aqueducts, being public works serving the public, did not attract private investment, and private citizens, contributing to their construction, did not in principle look for any profit in return. It has therefore been pointed out, and rightly, that the utility of an aqueduct did not lie in the field of economics, for while it promoted the health (salubritas) and quality of life (amoenitas) in the city, it did nothing for its commercial development.
>
> (Leveau, 2001: 86)

A further motivation was civic pride. Water was a necessity, but a city did not have to have an aqueduct to provide it. Many did not, relying instead on wells and cisterns. However, once a city did have the abundant water that an aqueduct supplied, its further growth and expansion could be thereby facilitated or promoted, so giving the aqueduct a utilitarian function as it supported the increasing population. Moreover, an aqueduct publicly provided unlimited water. If it had arches and arcades that was an added spectacular demonstration of the prestige and prosperity of the city that possessed it. What marked aqueducts as a particularly Roman phenomenon was the breadth and grand scale of the vision that inspired them.

Aqueducts built by the Romans mostly date to the Imperial period, though metropolitan Rome did acquire four under the Republic. Our knowledge of them comes from two sources. One is archaeology, the other literary. Archaeology is centered chiefly on the great bridges and arcades that form so prominent a part of the aqueduct network. For a full understanding, two qualifications must be noted. First, most of the length of the aqueduct conduits, being underground, remain invisible and hence can be found and visited only with the assistance and guidance of specialists. Second, within the city of Rome itself, the details of the urban distribution network, which would otherwise be of great interest, remain uncertain and hard to follow, largely because of the modern city built on top. As for literary evidence, Vitruvius does give us an account, sometimes confusing and erratic, of aqueduct building, but our chief source is Frontinus. Sextus Julius Frontinus (*c.* 30–104 CE) was appointed in 97 CE by Nerva as Commissioner of the Waterworks. He then subsequently published his *De aquaeductu urbis Romae*, describing the network and its history. Though of later date, he is a prime source also for the Republican aqueducts, which were still running and under his control. An administrator rather than an engineer, he yet expounds in detail the volume of water carried by each aqueduct and its distribution, as also the legal history and issues raised by their construction (e.g. *Aq.* 106–11). On their building and maintenance he is less comprehensive.

2 General Principles

In the building of the Roman aqueducts some general characteristics may be noted. They usually take their water from springs, from a river, or from an aquifer, a semi-porous rock into which tunnels can be driven to collect the water seeping through. Unlike modern practice there were rarely any reservoirs or storage tanks. The aqueducts ran on the principle of continuous off-take, so there were no peak hours or slack hours. The water was used as it came (like a modern electricity grid); most of it, indeed, was unused and went straight into the drains. This at least did keep them clean and diluted all solid waste, which helped the city to dispose of it without any sewage treatment arrangements. Since the Romans lacked any effective pumps the aqueduct ran only by gravity, so the downhill slope was continuous, though not uniform. A city such as Rome might have multiple aqueducts, for when more water was needed the normal answer would be to build a new channel, completely independent, rather than to build tributaries feeding the extra flow into an existing one. As different sources might provide water of differing quality (some clear, some with sediment), the principle was that they should be kept separate throughout and not mixed, hence the number of separate aqueducts. Nor, upon arrival at the city, was there any way that an aqueduct could be shut off when not needed, say at night, any more than a river can be shut off. The water would keep coming and any attempt to close a sluice would simply lead to overflow and flooding of the town. Since the water could not be stopped, only diverted, if a section of the conduit had to be closed for repairs or maintenance its discharge would be diverted by a crossover channel into another neighboring aqueduct (which had to be built at a lower level since the water would not flow uphill into it, and which also involved temporarily mixing their waters). Indeed, aqueducts may well be conceived rather as artificially channeled rivers, with all that that implies.

3 Construction

For its first four centuries, Frontinus (*Aq.* 4) tells us, Rome depended on river water, supplemented by springs, wells and cisterns. At Rome the subterranean water table is relatively close to the surface, facilitating the digging of wells, while their water, coming from the Tiber, had percolated through a layer of sand that filtered all the impurities that in antiquity, as now, lent the river its distinctive yellow (flavus Tiberis) color. As for cisterns, they depended on rain water, which is why so many Roman houses have roofs sloping inwards, towards a central atrium or courtyard; the function of the roof being not to get rid of the rain but to collect it and deliver it to an underground cistern. Domestic cisterns and wells were at all times common in Rome and other cities of the Empire, although once an aqueduct was built it would usually furnish a supply both more convenient and, perhaps, more reliable. Rome's first aqueduct, the Aqua Appia, was built in 312, followed in 273 by the Anio Vetus, in 144 by the Aqua Marcia, and in 125 by the Aqua Tepula. All other aqueducts are of Imperial date.

Before tapping a source there arose the question of purity and how to test for it. Local inhabitants and animals which regularly drank from it could be examined to see if they actually looked healthy, or a pot of water could simply be heated to see if, when it had all boiled away, any sediment or impurities were left (Vitr. *De arch.* VIII.4.1). The quality of the water in the different aqueducts varied, and the Aqua Marcia was rightly famed as the best, very cool and fresh, not to mention voluminous.

There was also the question of financing and motivation. Aqueducts were expensive. Whole cities might, and sometimes actually did, go bankrupt in attempting to build one. The army might be used for labor, which would then come free, or perhaps as an accounting modification in the budget, but supplies, expropriation and other incidentals could still come dear. The Anio Vetus was financed by booty from the war with Pyrrhus (Frontin. *Aq.* 6), a period when financial misgivings seemed to outweigh civic pride, for the Aqua Anio was considered to provide low quality water. The chosen source also had to be high enough to maintain a suitable aqueduct gradient all the way to the city, thus excluding sources in the flat Campagna, while the further away it was, the higher it had to be so as to maintain the slope. From the source the water ran in a masonry conduit (specus), with flat bottom and sides lined with waterproof cement, and a vaulted roof. It was usually buried about 1 m below ground level, to avoid pollution. Dimensions were roughly standard for all aqueducts, around 70 cm wide by 1.20 m high. This uniformity reflected the governing factor, human access for cleaning, which was facilitated by inspection manholes every 70 m or so (Vitr. *De arch.* VIII.6.3 specifies 35 m), irrespective of the actual volume of water carried.

Cleaning was important, for in almost all aqueducts the water had a high calcium carbonate content which, separating out, would leave a calcareous deposit (usually referred to by the German term, sinter) in the form of a lining inside the channel (Frontin. *Aq.* 122). The build-up was slow, 1 mm per year, but if left untouched would accumulate to deform and reduce the free space in the conduit. In time the sinter could result in the water overflowing, though this stage would not be reached until, in the decline of Rome, there was no maintenance while the water kept coming.

Most conduits, in fact, usually ran about one-third full, for unlike a piped supply, they were not entirely watertight. True, they might have a roof or cover, but its purpose was to stop earth or pollutants getting in rather than to stop the water getting out. This is turn meant that while a piped supply could be laid out to follow a straight line, rising and falling over minor inequalities of the terrain (much like a garden hose), an aqueduct, relying entirely on gravity flow throughout, often had to follow a more circuitous, not to say serpentine, route, something like a contour line, in order to maintain an approximately constant gradient. When there was an unavoidable drop in the terrain it might be negotiated by cascades, a sort of internal waterfall built into the conduit, or by simply slanting the aqueduct sideways across the hillside, losing height gradually. At Rome nearly all aqueducts start in the Sabine Hills, in the valley of the River Anio. On leaving the hills at Tibur (modern Tivoli) the Anio itself plunges straight down to the Campagna in waterfalls, while the aqueducts, to preserve a more uniform gradient, curve away to the south in a wide loop. The increased length meant the water took longer to get to Rome. That did not matter; what did was the extra cost in building, servicing and maintaining the extra length.

Intervening mountains would be pierced by tunnels, which were of two different types. In one the tunnel could be dug by two crews, one starting from each end and meeting in the middle, though sometimes the meeting was not very precise, resulting in a kink or twist at the tunnel's mid-point to bring the two halves together. The other way was to sink a series of vertical shafts from the surface ground down to the tunnel level, and then to dig out a tunnel from the bottom of each one until these sections all joined up, forming the completed tunnel. The first method could be very slow for it involved only two work crews doing the digging, with space at the workface for only two men in each; there was also the problem of keeping the work on the correct alignment. The second way, with many work crews working at once on the shafts and on the horizontal gallery at the bottom, could be quicker but needed a big and cheap labor force, such as the Roman army. Moreover, since the shafts visibly began on the surface, at ground level, there would be no problem of keeping them, and the underground tunnel below, properly aligned, and upon completion they would offer convenient access, as did the manholes, for maintenance crews. However, this choice might be impractical if the hill rose too high, so that the vertical shafts would have to go too deep.

When valleys intervened they would be spanned by bridges. Their arched structures are not to be confused with the arcades that carry the conduit across the flat Campagna, and we should perhaps note that the most familiar and spectacular of bridges, the Pont du Gard in France, was not built in the Republican period. At Rome, bridges are common along the route of the aqueducts, ranging from minor culverts over streams to massive masonry structures such as the Ponte Lupo, 115 m long and 30 m high, on the Aqua Marcia. The principle of the inverted siphon was presumably known to Roman engineers from their acquaintance with Hellenistic work, and siphons do often occur in Roman aqueducts of the Imperial period, but there are none preserved in the Rome metropolitan network, either Republican or Imperial, though a few probably did exist. There certainly was one at Alatri, in the Alban Hills, built 130–120, known primarily through an inscription erected for L. Betilienus L.f. Vaarus (*CIL* I^2.1529). Along the route water might be diverted into side branches, either openly and legally, or illegally and surreptitiously, the object usually being agricultural irrigation or to serve a nearby villa, but this was not too common.

On approaching the city, the channel might be carried on a long continuous arcade of arches, even over a completely flat plain. The purpose of this was to ensure that on arrival the water was still at a high enough level to provide service to hilltops (and Rome was, traditionally, built on seven hills) or to the upper stories of houses. It is these arcades that today are prominent in illustrations and are readily seen from the Rome–Naples railway, leaving most of the aqueduct's actual length invisible. The arches thus may give a false impression of what a Roman aqueduct was like. Where possible the conduit was always buried underground, and the great arched structures that we think of as typical were only resorted to as an unavoidable and costly engineering necessity.

The aqueducts of Rome generally were built to a gradient of around 3.0 m per km. This was not uniform throughout their length, reflecting local exigencies. The degree of slope had various results. The steeper it was, the greater the erosion on the conduit from the increased speed of the water, while the slower speed on a flatter section permitted sediment to settle on the bottom, clogging the channel. The velocity of flow would accordingly vary, but this was automatically compensated by a simple law of hydraulics: the current ran faster but shallower on the steeper sections and slower but deeper on the flatter ones. The result of this was that at all points on a given aqueduct the discharge, the amount of water actually passing through, was the same. The actual velocity was usually in the range of around 3.5 to 5.5 km per hour, so that in the Aqua Marcia, for example, the water would have taken, say, 15 to 20 hours to get to the city.

4 Urban Distribution

On arriving at the city the water would enter a distribution tank (castellum). The best example is at Nîmes, and though of Imperial date, probably represents Republican practice as well (Hodge, 2002: 284–8). Thence it would enter the piped supply of the urban network, for which the best evidence is found at Pompeii. The pipes might be of terracotta but were usually of lead, in spite of the possibility of lead poisoning. The ancients did indeed understand this – Vitruvius advises against lead pipes because of it (*De arch.* VIII.6.10–11) – but lead was nevertheless regularly used, and, it would seem, apparently without any ill effects. There are two reasons why lead poisoning did not occur. First, the sinter, previously noted in the aqueducts, was also in the pipes, there forming an inner insulating sleeve that effectually stopped the water from ever actually touching the lead. Second, given the principle of continuous off-take, the water was in constant flow through the pipes and so would never be in contact with the lead for long enough to absorb anything from it. Indeed, given that while in the aqueduct it was running through a masonry conduit and was only in a lead pipe in the last stretch, from the city distribution tank to the house, the water was probably in the lead pipes only for some ten minutes or so, in any event not long enough to become contaminated. It would thus seem clear that while the Romans might indeed suffer from lead poisoning, they did not get it from their pipes: for example, the lead-based glazes on drinking cups or in lead cooking vessels may be a more probable source. Moreover, lead had various advantages. It was cheap and abundant, being formed as an unwanted byproduct in the refining of silver, and it could easily be bent, facilitating getting the pipe round corners. Its chief disadvantage was its great weight, adding to transport costs for installation if it had to come from any distance.

In distribution the water would be divided up among three principal uses: baths; street fountains; and domestic supply. The monstrous bath complexes of Imperial Rome, such as those of Caracalla or Diocletian, had no obvious parallel in the Republican era, but, given the frequent absence of any washing facilities in modest houses and tenements, small local establishments existed to fill the gap (see Chapter 1). They did not necessarily need an aqueduct to feed them, as did the great Imperial bath complexes. For example, during the Republic, the Stabian Baths at Pompeii were supplied from a well, supplemented by a bucket-chain from an aquifer below. Street fountains were aqueduct-supplied, and were the main source of water for most residents. Pompeii again offers the best example: the fountains were not only conveniently located at crossroads and street corners, but were numerous enough that residents never had to walk much further than 50 m to find one, or carry away its water. In design, each fountain consisted of a spout, often decorated with a masonry lion's head, delivering a constant flow into an open stone tank. Residents could thus either hold their jars under the spout or wait for them to fill, or simply dip them into the water in the tank, depending on how fresh they wanted their water to be; that no doubt depended on what it was to be used for – drinking, cooking, or washing. Given the constant supply, the tank would often be overflowing, especially at night, thus helping to keep the street clean, before running into the drains or sewer.

As for domestic use, supply to individual houses could be, and often was, achieved by fixing into the side of the masonry channel a bronze nozzle, or adjutage (*calix*), through which the water would pass into a lead pipe supplying the house. The Romans had no meters for measuring water use, so it was controlled by the nozzle, which, being of a fixed, approved, size governed the amount of water that could go to the house. As Frontinus (*Aq.* 39–45) tells us, there was a whole range of nozzles of different sizes, to correspond to varying circumstances, and they were made of bronze to stop fraudulent misuse, for a softer lead nozzle could easily be enlarged by a crowbar, increasing the supply passing through. However, this did not obviate improper installation of the nozzle by the workmen, depending on whether they had or had not been adequately bribed. For example, a nozzle angled to face the current in the masonry channel would deliver more than the authorized supply; one angled away from it would withdraw less water, resulting in a surplus in the main channel that could then be illegally sold elsewhere.

Not every house, let alone apartments in the insulae, would have had running water piped in. That was a privilege, or added luxury, that depended on the issue of an official permit, and probably the payment of a "water tax" (Frontin. *Aq.* 94). Part of the trouble here is that for the administration of the urban distribution we are greatly dependent on Frontinus, who describes in detail the system at Rome but, naturally, says nothing of how things were run in the provinces, and for them we have no other equally comprehensive source. Internal domestic use was of two kinds. First, there was the kitchen. Not all houses, and few apartments, had one, hence the multiplicity of thermopolia, hot-food shops. Toilets also might be connected to the water supply, but again this should not be taken for granted (Hobson, 2009). Many toilets simply consisted of a cess-pit, and in any case the common provision of public multi-seat toilets in the streets would seem to indicate that many houses were without one. Second, if the house, or villa, had a garden it might be adorned with decorative fountains where, in the absence of air-conditioning, the constant sound of falling or trickling water perhaps gave a welcome feeling of cool freshness to an otherwise hot environment. Since the water supply available was governed by the adjutage, there might not be enough to keep all such outlets running at once. There thus arises the question of taps, to regulate the distribution. There was no "mains tap," in our sense of the term, that would cut off the entire supply at its entry into the house, but taps might be built in at individual locations indoors.

Taps, though not as common as in modern practice, certainly did exist. They were made of bronze, and, unlike the modern screw taps, were of the type known as rotary plug. This consists of a circular hollow casing, inside which turns a cylindrical solid metal plug pierced by a horizontal hole (Hodge, 2002: 322–6). When the hole is aligned with the supply pipe the water flows through, while, given a quarter-turn, the hole is set crossways to the pipe, cutting off the water flow. The tap could not be turned by hand. Internal friction made it very stiff to move, so the top of the tap carried a handle in the form of a square loop into which could be inserted a portable iron bar or wrench to give the necessary leverage to turn it. This design of tap could not resist any high static pressure in the water supply and was liable to burst. Thus it was the taps, not the pipes, which were the weak links in the Roman urban network.

The next step in the completion of this hydrologic cycle is the disposal of waste or used water. Drains and sewers might be provided under the roadway to carry away storm water or overflow from the street fountains, and human waste. The most familiar example is the enormous Cloaca Maxima, built by Tarquinius Priscus, supposedly king of Rome from 617 to 578 (Chapter 28). In the beginning an open ditch, later covered by masonry, its original purpose was to drain swampland around the Forum. Discharging into the Tiber, where its arched-over exit into the river may be seen under the Lungotevere embankment (see Figure 28.2), close by the round temple by the Tiber (the Temple of "Vesta"), it is still in existence, and, in some sections, actually in current use and carrying sewage. Its size is attested by the unlikely tradition, recorded by Strabo (V.3.8) that a wagonload of hay could pass through it, and Pliny (*HN* 36.104) tells us that Agrippa, in his capacity as Water Commissioner, really did go through it in a boat. At Pompeii there was nothing so monumental, but there was a network of drains beneath the sidewalks. However, at the time of Pompeii's destruction by the volcano in 79 CE, the drains under the sidewalk still served only a small portion of the city. Likewise in Rome, despite the often sophisticated planning of the aqueducts, much of the city was still served by open drains down the middle of the street. Suetonius (*Ner.* 26) records the emperor's playful habit of knocking passers-by into them during his incognito nocturnal escapades. Our admiration of the Romans' hydraulic achievements should not blind us to the fact that many of their cities were served still by open sewers. This may seem strange in view of their devotion to baths and washing, but, as has been well remarked, what attracted the Romans was not so much hygiene as comfort.

5 The Four Republican Aqueducts of Rome

5.1 *Aqua Appia*

Built in 312, this was the first aqueduct to serve the city. It was 16 km long, with an estimated discharge of 73,000 m^3/24 hrs. As it ran almost entirely underground, it has left few identifiable remains. Within the city its relatively low level prevented it from serving any of the hills; it apparently ended at the Forum Boarium.

5.2 *Aqua Anio Vetus*

Built 272–269, its length was 81 km, with an estimated discharge of 176,000 m^3/24 hrs. Drawing its water from the River Anio, it followed the river to Tivoli, thence a circuitous course down to the level of the Campagna, setting the pattern for subsequent aqueducts. It is again almost entirely subterranean; its most striking feature, the Ponte San Gregorio, is a Hadrianic rebuild. It entered the city at the Porta Maggiore. Its distribution within the city cannot readily be established, though it may have broken up into a few branches. Its water was cloudy and unpleasant, and Frontinus (*Aq.* 92) says it could only be used for watering gardens, and "sordid uses."

5.3 *Aqua Marcia*

Built in 144–140 by the praetor Q. Marcius Rex (hence its name), it was 91 km long, with an estimated discharge of 187,000 m^3/24 hrs. Drawing water from springs beside the River Anio, it followed much of the same course as the Anio Vetus to the city. Approaching the Campagna it crossed two valleys by bridges, the Ponte San Pietro and the Ponte Lupo. The Lupo is a massive monument, 30 m high and 115 m long, built in 144, but later subjected to so many repairs and additions, including a road bridge, that it now appears an amorphous and disorganized patchwork. For the last 10 km or so the channel ran on top of substructures and arches, entering the city at Porta Maggiore and ending on the Viminal, whence several branches distributed its water further. Its water was well known for its excellent purity and freshness.

5.4 *Aqua Tepula*

Built in 125, it was 18 km long, from its source near the Alban Hills, with an estimated discharge of 17,800 m^3/24 hrs. The original routing was independent and does not now remain, but in 33 its channel was moved and placed on top of the Marcia's arches, which, for the last 8 km or so, it rode piggy-back to the city. Its water was tepid (hence the name) and unpleasant; because of this and its low discharge it was the least important of Rome's Republican aqueducts.

6 Conclusion

It will be seen from the above figures that at the end of the Republic Rome benefited from a daily supply of some 453,000 m^3. It must be emphasized that these figures are both approximate and subject greatly to variation, for the discharge, since it depends on the rainfall, will not only change with the weather and seasons, but at any given time may be different from one aqueduct to another, reflecting their different catchment areas. The total volume of water provided, the discharge, was nevertheless very high, though most went straight into the drains (Hodge, 2002: 347). This may not at first appear as useful water use, but one must always remember that the aqueducts operated on the principle of continuous off-take and continuous disposal, storage tanks or reservoirs being virtually unknown.

We may also note that in the cities, as well as the domestic uses noted above, water was also needed for industry. By modern standards this was small-scale, but such workers as potters, dyers and fullers needed it. As for the thermopolia that are so common a feature of Pompeii streets, they also presumably depended on a ready supply, whether for drinking or cooking. These aspects have been studied rather less than the domestic, as has the other possible rural use, irrigation. Agricultural irrigation seems to have been based mainly on various water-lifting devices, such as the Archimedean screw and animal-propelled water-lifting wheels that lifted water from a nearby stream or source rather than direct channels from the aqueduct. Perhaps the best-known example of agricultural irrigation by an aqueduct is the Crabra channel, near Tusculum. Among the properties that it served was one owned by Cicero (*Fam.* 16.18.3), though it is unclear whether we are there dealing with an aqueduct proper or a natural stream (Hodge, 2002: 250).

Given, then, that the great period of aqueduct construction came under the Empire, what attitudes and skills did its builders inherit from their Republican predecessors, and in turn, what did the Republicans lack? Their greatest single lack was undoubtedly concrete, which only came into extensive use with the Augustan builders, and in so doing revolutionized all construction, architectural as well as aqueducts. Apart from that, the aqueducts saw little technical improvement or innovation. All the main principles of surveying, tunneling, bridging and the like were already well known, and though the sophisticated technology of the inverted siphon was apparently not used in any of the Republican aqueducts at Rome, it was surely familiar from elsewhere. This known and established corpus of techniques and their employment was the great legacy bequeathed by the Republican aqueduct engineers to their successors, a basis for them to build on. A further legacy was in the field of legislation, though our main source for this information is the sixth-century CE *Digesta* of Justinian, which includes and describes legal principles established in the Early Republic. According to this source, as far back as the Twelve Tables (*c.* 450), the law defined the difference between public and private water. Water in the rivers and the city aqueducts was public, while water in springs, streams and wells was private, belonging to the relevant property. What this meant, notably in the city, was that if a resident had a permit to take water from the aqueduct, this was a permit valid only for him personally, and if he died or sold the house it would no longer be supplied. If, on the other hand, there was a spring or a brook on a farm, the water from it belonged to the property, not to the owner or occupier, and if the land changed hands, by law the water supply went with it. Sometimes, moreover, a property that was itself dry might also possess the legal right of access and use of water located on a neighboring property – a prolific source of litigation (Bannon, 2009: esp. 10–14). Frontinus (*Aq.* 94f.) also describes several of the relevant laws, particularly as regarding aqua caduca, water overflowing from public street fountains and other locations. His comments have been described as possibly a nostalgic longing for "the good old days"; they are certainly recognized as a clear factual outline of Republican practice (for which see also De Kleijn, 2001: 93–106).

Also established was the idea that a reasonable municipal water supply was to be accepted as a standard feature of urban living, so that as new towns spread throughout the Empire they were often designed from afresh to have an aqueduct to serve them, while in existing cities, such as Rome itself, the copious supply afforded by the aqueducts enabled and greatly facilitated their subsequent growth in population (Bruun, 1991: 104, 110–14; De Kleijn, 2001: 68–74). Indeed, we may perhaps conclude with the observation of Edward Gibbon in his *Decline and Fall of the Roman Empire* (1837): "The boldness of the enterprise, the solidity of the execution, and the uses to which they were subservient, rank the aqueducts among the noblest monuments of Roman genius and power" (Gibbon, 1897, vol. 1, p. 47).

FURTHER READING

The standard introduction to the field is Hodge (2002): a general and comprehensive survey, it shows how a Roman waterworks worked (including an explanation of the hydraulics involved). Ashby's (1935) detailed study of the ruined remains of the aqueducts of Rome (including the four

Republican ones) is itself monumental, while Aicher (1995) is a practical handbook for those seeking the ruins on the site. To set aqueducts into the overall context of ancient living, the relevant chapters Oleson (2008) and Wikander (2000) are recommended. As their titles indicate, Bannon (2009) and Hobson (2009) both offer specialist studies of limited topics that are often neglected. Of ancient sources, Frontinus stands alone, especially on legal and administrative aspects; Blackman and Hodge (2001) contains a critique (pp. 137–50) of the strong and weak points in his aims and methodology. Vitruvius (*De arch.* VIII) gives an overview of aqueducts but is sometimes confused and confusing on the technology, which he may not have fully understood. For individual aqueducts, studies are usually to be found in articles in bibliographies in the general works cited above: in particular is the German contribution to aqueduct research, centered on the work of the Frontinus-Gesellschaft, a learned society for the history of waterworks: their journal is the *Schriftenreihe der Frontinus-Gesellschaft.*

CHAPTER 19

Roads and Bridges

Ray Laurence

1 Introduction

There is a fundamental problem for the archaeology of Rome and its territories that needs to be recognized at the outset: often things built in the Republican period tended in many cases not to fall out of use for the best part of half a millennium but were subject to restoration, rebuilding and redesign. The roads of Rome with their paved surfaces that have been excavated and preserved are rather later than the surfaces of roads utilized in the Republic. The cutting of sections through later road surfaces reveals earlier surfaces of compacted gravel and a build-up of repairs over time (Quilici Gigli, 1990). Certainly, in the Republic, few roads were paved rather than being formed of compacted gravel or glarea. The Via Appia was, of course, an important exception that was worthy of note by the Augustan historian Livy (X.23, X.47, XXXVIII.28), who suggests sections were paved close to Rome, and it is possible that the process was extended to other sections of the road (*ILLRP* 448; Coarelli, 1988a: 37). Where we do find secure dating material in roads with paving, these mostly point to the first century CE or later.

The precise chronology of road construction has for the most part to be determined from literary sources and, where they survive, milestones dating from the time of construction. In most cases a precise chronology of construction is not possible (see Wiseman, 1970), with the result, for example, that we cannot know when the Via Appia was extended from Capua to Brindisi. Bridges are a little easier, but often there is a reliance on stylistic dating to identify a bridge as Republican; for instance, the Ponte di Nona (discussed below) is logically dated either to the second century or to the Sullan period. These problems are exacerbated by the fact that in the Imperial period roads and bridges were restored with inscriptions recording this action (e.g. *CIL* X.6853).

A Companion to the Archaeology of the Roman Republic, First Edition. Edited by Jane DeRose Evans.

Figure 19.1 Ponte di Nona, one of the best preserved Republican bridges on the Via Praenestina. Source: Photo by R. Laurence.

There is a long tradition of the study of the topography of Roman roads, the identification mapping and photography/sketching of standing remains along their course. At the end of the nineteenth and into the early twentieth century, Thomas Ashby, the first Director of the British School at Rome, followed by bicycle or on foot the routes of the major roads leaving Rome (Figure 19.1). His camera lens recorded many things that can no longer be observed (Le Pera Buranelli and Turchetti, 2003). It is still worth returning to these early publications (see Hodges, 2000 for full listings), not least because during the Second World War numerous Roman bridges were destroyed and, in more recent times, urban development has expanded into what had been in the early twentieth century a landscape that was far emptier (see Wallace-Hadrill, 2001 on changing landscapes and priorities of fieldwork).

2 Road Building in the Roman Republic

A statue of Appius Claudius stood in the Forum of Augustus (completed in 2), alongside other famous men from the Republic. Its inscription recorded the man's great deeds for the Roman state; it ended these with a statement that as censor he had built the Via Appia and brought the Aqua Appia to the city (see *CIL* XI.1827 for a copy of the inscription from Arezzo). These actions took place 310 years prior to the completion of Augustus's Forum, but are perhaps underrated by modern historians studying ancient Rome. The building of the Via Appia marked a fundamental step in the development of

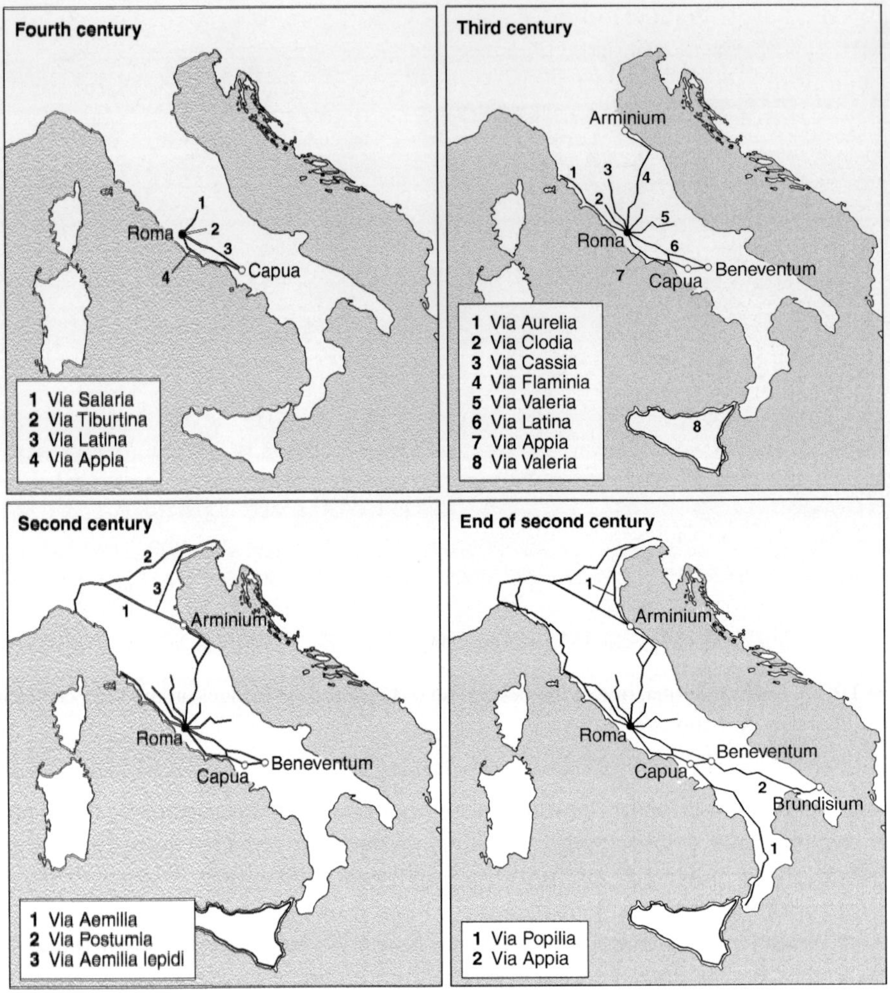

Figure 19.2 Map of the development of the public roads in the Republic. Source: Drawing by L. Bosworth.

the Roman state. It was the biggest public project ever undertaken, and connected Rome to Capua some 132 Roman miles away (Figure 19.2). The work was an impetus for Rome to mint its own currency to pay for the new work (Crawford, 1985: 29; Humm, 1996). This was, it would seem, an addition to the traditional salt road, the Via Salaria, leading from the coast to the Sabina and the Via Latina that connected Rome to the cities of Latium. Importantly, it was a long-distance road along which colonies were to be established with migration to the area from Rome and its allied cities (Coarelli, 1988a).

Road building and town foundation went together. This can be seen most clearly from an inscription known as the Elogium of Polla. Frustratingly, the first lines containing the name of the patron are broken off from the stone. The remaining lines tell us that the unknown man built the road from Capua to Rhegium, built bridges on it and set up milestones; he also built a forum and a temple at this place. The presence of a forum points to a new type of civic settlement (see Laurence, 1999: 29–32 for discussion). It can be dated from two other events mentioned in the Elogium, which may be related to the revolt of slaves in Sicily: the return of slaves to their masters and (since slave labor was now available) the conversion of land from pastoralism to arable farming (*ILLRP* 454). Hence, the inscription may be dated to the second century. A key aspect of road building was the creation of measured distance marked by milestones (see *ILLRP* 448–66 for examples). There was another aspect to road building though, and that was the transportation of agricultural equipment and produce from estates, which is a fundamental part of Cato's treatise *De agricultura*. This work stipulates the advantages of being close to a road and the origins of equipment for the estate – all of which points to a developed geography of movement and transportation by the second century (Laurence, 1999: 95–108).

Over the course of the third and second centuries, the Roman road system expanded. However, this was not a smooth process. There was no sudden rush to build further roads after the construction of the Via Appia and it was not until the late third century with the building of the Via Aurelia in 241 and the Via Flaminia in 220 that further expansion of the road network on a similar scale occurred. Over the course of the second century, the road system expanded to the north and to the south to form a network that integrated places on the road, while at the same time not penetrating into all parts of the Italian Peninsula. At a macro-level, those parts outside the road system were the least integrated into the state formed by Rome and its allies and it was precisely in these areas that the allies rebelled against Rome in the Social War. The roads effectively created an uneven geography of development with towns and cities being concentrated on these lines of communication. It is perhaps no coincidence that the greatest densities of cities and surviving inscriptions are found in the regions associated with the routes of the Via Appia and the Via Flaminia.

3 Agricultural Development, Road Building and Town Foundation: The Via Appia

Fortunately, for more than 200 years, Popes, and Kings of Naples and the Italian Republic planned to create an archaeological park along sections or the entire route of the Via Appia (Quilici, 2002). Hence, preservation of monuments and regulation of buildings along the route of the road has been controlled. As a result, there is a far greater degree of preservation than would have otherwise been achieved. In addition, sections that ran through the Pomptine plain have been preserved due to the less-than-healthy environment of the marshes in the past. Not surprisingly, the Via Appia has been the focus of a number of archaeological projects, the results of which are discussed here with a view to understanding the relationship of road building to changes in settlement patterns in the fourth and third centuries.

The straightness of the Via Appia is something that never fails to impress us (Figure 19.2). The road is aligned with reference to two points: one in Rome, specifically at the Temple of Diana on the Aventine, and the other in Terracina. Through the use of the new Pythagorean geometry a straight line was maintained (Humm, 1996). The cost of straightness is revealed from an examination of the engineering of the viaduct that leads the road into Ariccia. The masonry of this viaduct provides a second-century date for the structure (Lilli, 2002). The viaduct carrying the road crossed a landscape that included other roads that ran at a right angle to the Via Appia. The latter duly passed over one of these roads with a vaulted passageway through the viaduct to allow for the road to pass under the viaduct. This structure, probably named the pons Aricinus in antiquity, was over 231 m in length, rising to a height of 13 m and built from peperino blocks, and carries the road at a 1 in 4 gradient causing the traveler to climb up some 50 m. This structure might be connected not to the initial building of the road but to a restoration in the second century – perhaps even part of the project by Gaius Gracchus to construct better roads (Plut. *C. Gracch.* 8; cf. Muzzioli, 1996).

In contrast to the climb to Ariccia, the Via Appia in the Pomptine plain is totally horizontal or flat with no change in its level; it continues in a straight line to Feroniae, a sanctuary outside the colony of Terracina (founded in 299) over a distance of some 60 km. It ran alongside the Decemnovius or the 19 miles of canal made famous by Horace (*Sat.* 1.5). This area could only be utilized after land reclamation and full control of drainage was achieved; if the drainage channels became blocked or were neglected the region returned to marshland (Attema, 1993; see Purcell, 1996 on conquest and the control of water). The only settlement of any size, Forum Appii, was located at a bridging point of one of the few rivers in the region and associated also with a sanctuary (Bruckner, 1995). The landscape has been the subject of considerable archaeological investigation, notably a survey led by Peter Attema from the University of Groningen (Attema, 1993). The Dutch team investigated transects at right angles to the Via Appia to obtain an understanding of the landscape associated with this major road. The results demonstrate a major landscape transformation in the fourth century. The marshes were turned from a natural resource with few inhabitants into an agricultural resource for the supply of arable crops to Rome as part of the ager publicus or public land (Attema, 1993: 233–8). However, what is fascinating is that the land reclamation of the region preceded the building of the road (Attema, 1993: 236). As a consequence, the road with its new bridges (Lilli, 2006) not only linked Rome to the south, but also connected a developing agricultural region created by land reclamation with Rome. This demonstrates how the Dutch project altered a view based on literary sources that the road was built entirely for reasons of military expansion or control of territory at a distance (e.g. Uggeri, 1990: 21–2; Purcell, 1990: 12–13).

The development of towns in association with both the Via Appia and land reclamation can also be identified archaeologically. The colony of Terracina was founded in 299 at the point where the Via Appia met the sea. This settlement was dominated by the sanctuary of Jupiter Anxur overlooking the city at a height of 220 m above sea level. The colony underwent considerable development during the Republic, but it is worth noting that the well-preserved paved forum adjacent to the route of the Via Appia was not constructed until the Augustan period. This might suggest that the majority of construction was focused on the sanctuary (see Coarelli, 1987b for details). The route of the Via Appia on its ascent from the gate of the city ran along the walls that connected the colony to the sanctuary. This stretch of road was lined with tombs constructed from opus incertum that faced onto the road (Böethius and Ward-Perkins, 1970: 178). Further to the south of Terracina, Minturnae was founded in 296 (Livy X.27.7–10). This city had been founded as a small Roman colony of 300 citizens. There was no need for a forum, since

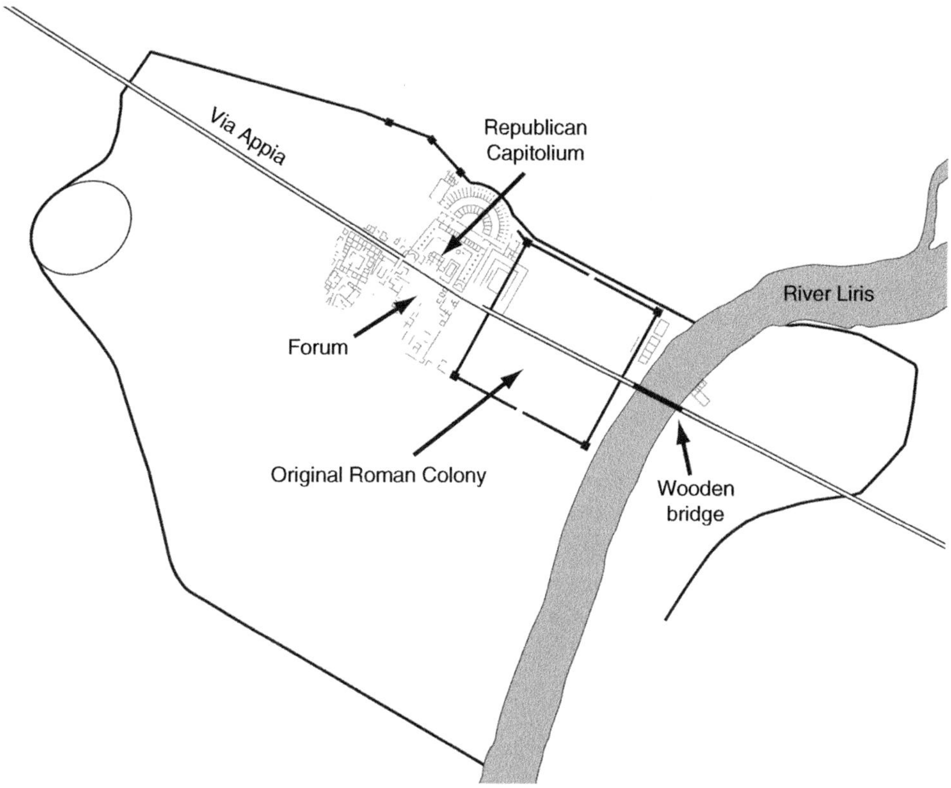

Figure 19.3 Plan of the expansion of the Roman colony of Minturnae: ribbon development along the course of the Via Appia. Source: Drawing by L. Bosworth.

the settlers were all citizens and thus their forum was in Rome. However, the town expanded from its original walled area with ribbon development along the Via Appia (Johnson, 1935). It was outside the original walls of the colony that Republican temples were constructed, and they were aligned to front onto the Via Appia. On the other side of the road, an open space was provided as a forum (Figure 19.3). On the other side of the colony, the Via Appia crossed the River Liris on a wooden bridge (Frier, 1969; Brookes, 1974; Ruegg, 1995). It is also located close to a sanctuary. Hence, this Roman colony, like the location of Forum Appii, was situated at the intersection of the road and the river – two complementary transport arteries.

4 The Extension of Rome's Hinterland: The Via Flaminia

The discussion of the Via Appia focused on the development of the road in the fourth century in relationship to the alteration of the landscape associated with former marshland. Our discussion now turns to roads in the uplands in connection with the bold move to construct a new road that crossed the Apennines from Rome to Rimini on the Adriatic coast (Figure 19.2). In this case we are looking at the development of the landscape of central Italy from the middle of the third century, delineated by the foundation of the colony at Rimini in 268, but looking back to the foundation of the colony at Narni in 299. Hence, the construction of the Via Flaminia in 220, as a continuous road connecting colonies, should not be seen to be completely disconnected with the development of the Via Appia. In both regions there is clear evidence in Livy that, as in the Pontine region, Central Italy supplied Rome with food via the Tiber (Livy II.34, IV.12–13, IV.52) and was subject to a number of schemes to drain the land and increase the cultivable area (Livy *Per.* 11; Vell. Pat. 1.14; Cic. *Att.* 4.15). The Via Flaminia defined a routeway as a road or single element leading across the landscape.

The Via Flaminia north of Rome follows what may be described as natural routes in order to connect existing towns (Ashby and Fell, 1921). Hence, it may have followed existing routes of communication. What is important is that in 220, Gaius Flaminius created the road as a single entity as a public thing to be owned by the Roman state. New settlements similar to that of Forum Appii were established on the Via Flaminia in part due to the traffic on the road at Forum Flaminii, Nuceria and Forum Semproni (Strabo V.2.10). Bridges were an important feature of the new road with the Pons Mulvius (see Livy XXVII.51 for its existence in 207) being an important development upstream from the city of Rome. This was the key route to the north linked not just to the Via Flaminia but also to the Via Cassia, Via Clodia and Via Amerina leading to other cities in Etruria. Thomas Ashby (Ashby and Fell, 1921: 140–4) makes a very important point with regard to the route of the road: it did not – as a modern road would – follow the valley floor of the Tiber, but instead climbed up to the higher ground. Being built over the volcanic rock known as tufa (known to geologists as tuff), there are relatively few river valleys to cross for the first 35 miles of the road. The road bypasses nearly all major settlements in the area, including Veii, so that no major settlement or city was encountered until a traveler reached Ocriculum over 40 miles from Rome. There are a whole range of minor roads leading from the Via Flaminia across this region that were the subject of considerable study after the advent of aerial photography (see Potter, 1979: 101–9). Interestingly, the Via Flaminia, like other roads leading north from Rome, did not pass through the Etruscan cities, and in a way redefined the geography of the region as primarily connected to Rome, while extending the influence of the capital city across the region (Witcher, 2006c). Interestingly, however, from studies of the number of sites in the Tiber Valley identified by field surveys we find that the expansion of rural settlement occurred prior to 220 and that the period in which the Via Flaminia was established was one of settlement contraction (see Patterson, Di Giuseppe and Witcher, 2004: 10 for data; see Chapter 12). This has also been considered to be the case with the Pontine region and the period following the building of the Via Appia (Attema, 1993: 233–8). Although it might be tempting to trace the contraction of the settlement pattern to the Hannibalic War as a cause (Patterson, Di Giuseppe and Witcher, 2004: 13–17), there is a case for seeing this contraction of the settlement pattern as an aspect of the extension of Roman territory and the definition of the Via Flaminia. Put simply, the road facilitated communications with regions at a greater distance from Rome and the expansion of settlement in the north of Italy caused a migration from this region as new land became available (on the Via Flaminia and Ager Gallicus, see papers in Luni, 2002).

The crossing of river valleys was not avoided by the course of the new road. The Treia River – that even in the 1920s could be in such rapid flow that it could take away its bridge – presented a major obstacle: a valley about 1,300 m wide involving a descent of 80 m to the river's bridge and then an ascent of 50 m (see Ashby and Fell, 1921: 158–9 for full description and illustration). Today, it is possible to see part of the viaduct (the Muro del Peccato) ascending on the northern side of the valley. It is not unlike the viaduct at Ariccia. The road continues on level ground again until it descends to cross the Tiber at Ocriculum. From Ocriculum, it is possible to see the Roman colony of Narni a mere eight Roman miles away, and from Narni it was only ten Roman miles to Interamna (Terni). From there, the road took the traveler a further ten miles to the colony of Spoletium (Spoleto), founded in 241. The settlement pattern is a contrast to the first 40 or so miles of the road in which no cities were encountered. The close spacing of cities ends once Forum Flaminii (Foligno) is passed and the road rises into the high Apennines. The constructions that were Republican are difficult to identify, since there were tunnels built by the Flavians and the whole of the Via Flaminia, including its bridges, was restored by Augustus (see Ghedini and Rosada, 1997 for tunnels).

5 Transhumance, Water Resources and Settlement: The Via Tiburtina

The Via Appia and Via Flaminia were seen in our discussion as key routes for the development of long-distance transportation and the integration of new agricultural resources with the city of Rome. The recent Swedish project on the Via Tiburtina points to this road not as a new route but as a road that develops from a form of agriculture based on the movement of flocks of sheep or transhumance (Figure 19.2) (Santillo Frizell, 2009). Movement in antiquity was often associated with the labors of Hercules (including the driving of the flocks of Geryon). Along the Via Tiburtina were a series of shrines dedicated to Hercules. Included among these was the major sanctuary of Hercules on the outskirts of Tivoli (Tibur). This large Hellenistic sanctuary is built on a raised platform under which the Via Tiburtina passed in a covered tunnel (Santillo Frizell, 2009: 50–3). The tunnel allowed for the counting of sheep to ensure tax was paid upon them and was equipped with storage facilities for the sale of commodities to the shepherds – including salt, a state monopoly (see Bonetto, 1999 for analysis). In terms of transportation, flocks would have been moved from their winter pastures near Rome (after shearing) along the Via Tiburtina up to the mountain pastures for the summer. The Via Tiburtina was ultimately based on a traditional transhumance route that was intensified in response to the need to supply Rome both with food and with wool for textile production.

The Via Tiburtina is not unlike other roads, such as the Via Salaria, leading to the north-east and east of Rome. These roads cross areas that geologically are underlain by limestone, rather than volcanic rocks (see Figure 17.1), and as a result water is scarce with far fewer rivers. As a result, settlement in the region is, in part at least, determined by the availability of water sources. These have been shown to map onto the locations of sanctuaries in the region that were places of healing not just for humans but also for animals (Santillo Frizell, 2009). Moreover, it can be suggested that in the case of towns in the Apennine Mountains that these were cattle markets (as is suggested by Santillo Frizell, 2009: 56–7), as in the case of the colony at Alba Fucens, founded in 303 and presided over by the god Hercules. The importance of moving animals should not be underestimated when discussing Roman roads. Regions beyond the immediate vicinity of Rome became famous for the production of specialized animals: for example, Umbria on the Via Flaminia was known for its white bulls to sacrifice to Jupiter; Reate in the Sabina on the Via Salaria was famous for the breeding of mules (Laurence, 1999: 123–35). The latter were increasingly used for pulling carriages of the elite and were bred for this purpose, with the female mule becoming the preferred beast for this purpose during the Republic (Adams, 1993 delineates the linguistic developments associated with this change). All these animals needed to be transported to Rome and its large urban/state market. In response to this need, the state and/or cities on these routes developed an infrastructure of supply of water, salt and other products to ensure that the transportation of animals was achieved.

6 A Bridge and a Healing Sanctuary for Travelers

The Via Praenestina led from Rome to Praeneste (Palestrina) – a city that was a rival and is now justifiably famous for the surviving sanctuary of Fortuna from the Republican period (Figure 19.2). The route of the road is very well known and passes through Gabii with its well-preserved sanctuary close to the Via Praenestina. Just prior to reaching Gabii, at the ninth milestone, is the Ponte di Nona, one of the best preserved Roman bridges dating back to the late second century. The bridge crosses a river valley some 70 m in width and is raised on seven arches more than 12 m above the foot of the valley (Quilici, 1974: 373–81). It creates a level surface for the road and replaces an earlier bridge that can still be seen at the foot of the valley. Beside the bridge was the stopping point known in antiquity as Ad Nonas – a village or vicus. This area had been a focus of archaeological research from the beginning of the twentieth century, with excavations defining a sanctuary dating back to the third to second centuries (see Mancini, 1912 for the excavation report; Quilici, 1974: 364–73 for an overview; see Chapter 15).

A new impetus to work on the site began in the 1960s, when a pozzolana quarry was opened, and was followed up by further rescue excavations in the 1970s to ensure the site was not pillaged for the antiquities market (Potter and Wells, 1985; Potter, 1989; see Chapter 37). The latter was focused on a dump into which clay votives had been placed at a much later date, but coins and pottery in the dump point to a collection of material dating to the second or third century. Many of these clay votives resembled human body parts – most prominently, eyes, ears, faces, half-faces, hands, legs, genitalia and wombs – or representations of animals (see Potter and Wells, 1985 for analysis). These terracotta votives are assumed to be dedicated to the god or goddess to ensure that the person or animal might be cured (compare Santillo Frizell, 2009: 48–9 on *CIL* XIV.3911). It has to be remembered when reading the report from the excavation that during the 1920s, with the advent of deep plowing, much material had been previously unearthed (Quilici, 1974: 369). However, the later excavations in the 1970s with more than 6,000 examples excavated provide a guide to the overall nature of this assemblage (Potter and Wells, 1985: 28–9): the greatest number of these (39%) were feet; with the next highest type being heads (22%); followed by limbs (15%); hands (10%); and eyes (6%). The rest of the sample was made up of genitals, ears, uteri, breast and intestines, as well as a small number of animal figurines (in all cases each type represents less than 3%). Earlier excavations at the site had revealed a similar pattern, with the exception of the number of votives in the form of eyes, which were of a much greater number and come in part from a cavity at the north-east corner of the temple – 290 examples were located here (Potter and Wells, 1985: 27). Votive deposits of this nature have been found at more than 90 sites in Italy with the majority of them concentrated in central Italy to the north and south of Rome (Fenelli, 1975). The composition of these assemblages varies from sanctuary site to sanctuary site, which would suggest that each may have been associated with specialized cures. The large number of feet found at Ponte di Nona, which are the size of adult feet, would suggest that injuries to feet may have been a specialty and suggest that the injured were carried to the site (Potter and Wells, 1985: Appendix). However, a roadside sanctuary with a thermal spring would have attracted passing trade going to or from Rome – people walking the road who may have undertaken journeys over considerable distances with a degree of wear and tear on feet and legs, and even on other parts of the body. This factor is confirmed by the fact that a votive assemblage from a non-roadside sanctuary (Campetti, near Veii) did not feature nearly as many feet or limbs.

7 Road Building in the Provinces: The Via Egnatia and the Via Domitia

In discussion of the Via Appia and the Via Flaminia, it was argued that long-distance roads such as these were developed across settled landscapes and utilized existing routes; yet, once designated as a road or single geographical unity, these roads did not develop the landscape of the region but instead facilitated movement through that landscape rather than to that landscape (Figure 19.2). The same was true of roads in the provinces. Cicero (*Prov. cons.* 2.4; *Pis.* 40) speaks of the Romans' ("our") military road through Macedonia to the Hellespont. This was the Via Egnatia, a single geographical space stretching from Dyrrachium across the Adriatic from Brindisi (the end of the Via Appia) to what would become Constantinople. Strabo (VII.7.3) makes it clear that the road and Roman rule associated with it did not result in the development of a local population – quite the opposite. By the end of the Republic, Strabo could enumerate depopulation. The road was built by a Gnaeus Egnatius, as indicated on a milestone, but his precise identity is difficult to pin down (Collart, 1976: no. 1; Fasolo, 2003). However, it seems certain that the road was constructed in the second century. There is good evidence for Macedonian road building in the region, with milestones and measured distance being a feature going back to the third or even fourth century (Adams, 1982: 272–8, 301–2), and this explains the rather unusual milestone from the Via Egnatia discovered in the 1970s (Romiopoulou, 1974, now in the Archaeological Museum of Thessaloniki; see Collart, 1976 for a review of later milestones). The bilingual inscription on a grey marble column, 1.3 m tall, marks distance on the road and the space of the road as that of the Proconsul Gnaeus Egnatius, with its Latin text preceding the Greek. The latter integrates the space with a Macedonian tradition of measuring distance, whereas the Latin integrates it with a Roman measurement of distance found in second-century inscriptions in Italy (e.g. *CIL* I^2.617–20 on the Via Aemilia; *CIL* X.6838) and made much of by Plutarch (*C. Gracch.* 7). The designation of Gnaeus Egnatius is spelled out in the Greek as not just a proconsul, but a *Roman* proconsul (Romiopoulou, 1974), which certainly marks the space of the road as Roman, or, in Cicero's language, "our road."

The Via Domitia in southern Gaul (Gallia Narbonensis) is very similar to the Via Egnatia in its imposition of a Roman space of transport across the landscape of this new province (Purcell, 1990: 14). The former followed what was known to Polybius (3.39) in the second century as the road of Hercules. Like the Via Egnatia, it was equipped with milestones, one of which was discovered in 1949 (Campardou, 1949). It is almost 2m in height and bears the inscription "Gnaeus, son of Gnaeus, Domitius Ahenobarbus" with the statement that he was an imperator, as well as the number XX to indicate the measurement of distance. Gnaeus Domitius Ahenobarbus was consul in 122. The establishment of the road, like that of the Via Appia, was associated with the foundation of the colony at Narbo (Narbonne) in 118/117, but a precise chronology is almost impossible for knowing when Domitius Ahenobarbus was in Gaul (see Duval, 1949 for discussion). The expansion into the new territory of Narbonensis included road building that defined the territory as that of Rome, which can also be seen occurring in the establishment of the province of Asia in 129 with the establishment of roads with milestones to measure distances, all done in the name of M. Aquilius as consul and translated into Greek for the new subjects to read (*CIL* I^2.646–51; see French, 1988: 428 for others). Interestingly, by the end of the Republic, the translation into Greek was dropped in the province of Asia with the text solely given in Latin (*CIL* III.462, a statement of the restoration of a road in 46 by P. Servilius Isauricus). Even in the Greek-speaking part of the Empire, the language of the state and thus also of the Roman roads was Latin, with the distance carefully measured in Roman miles and articulated in Latin numerals. This occurred in a period when the roads were in need of repair, as is stated in this inscription from Asia and confirmed by Cicero (*Font.* 8.18).

8 Movement and Space in the Roman Republic

Road building quite literally remade the geography under Rome's control and, as we have seen, was an almost instant reaction to the conquest or acquisition of new territory. Yet, strangely, in the study of the impact of Rome on its subject states (provinces) this macro-scale reconfiguration of geography hardly receives a mention, unlike (for instance) the foundation of colonies (Alcock, 2002: 44–8; but compare Alcock, 2007b: 690–1, in which road construction is recognized as a factor; Purcell, 1990 is an exception to this trend). The juxtaposition of the power of road building and the measurement of distance found on Roman milestones can be contrasted to the use of the Greek language locally in the new provinces of the east. The road can be seen as a memorial to Roman power and its construction coincided (in the case of the province of Asia) with the definition of a road as a single entity and the measurement of distance between two places. These roads owned by the state were monuments and a means of simplifying the conquered landscape. Like the roads built by colonial powers in the modern empires, these roads were built by and for the imperial power to obtain a resource from a distance or connect with a distant point. The places in between the start and finish points of the road might benefit from traffic, but there was no scheme to develop the economy of the wider landscape alongside the road. The road actually has the potential to cause the extant landscape to become irrelevant, since it was the distant point that was the object of the state's attention. Thus, as we have seen, the Via Flaminia was more about the development of northern Italy than about the economic well-being of the Tiber Valley. The link between road building and landscape change, due to the drainage of the land, is often stated, but, as has been discussed above, this action could precede the road being built, as in the case of the Via Appia. Equally, we have seen the restructuring of both the western and eastern Mediterranean in quite different ways by the construction of the Via Domitia and the Via Egnatia. The latter linked Roman Italy to Asia Minor and, in some ways, caused "Old Greece" to cease to be relevant except as a cultural resource (perhaps mostly for a Roman/Romanized elite) founded on a historical memory of a past, rather than as an integrated province similar to Africa or Macedonia (Alcock, 1997, 2002; see Chapter 35).

The argument put forward here is that the road system that emerged in the Roman Republic enabled greater economic development on the line of a road and in the distant places to which the road led. Further development of the regions which a road traversed only occurred with the building of feeder roads or local roads that integrated the surrounding landscape with the line of communications. Hence, the new geography founded on road transport in Italy and its adjacent provinces, which emerged by the end of the second century, provided the infrastructure to enable a flow of goods, taxes and services to the city of Rome (fundamental to K. Hopkins's 1980 model of taxes and trade; compare Morel, 2007: 499). This was a new geography that altered the economic *longue durée* of the Mediterranean by establishing a land-based connectivity that expanded the range of places integrated with both the Mediterranean and more importantly with the expanding city of Rome (Harris, 2007: 535). It is important to realize that the alteration of the transport system by its expansion had an effect upon all other parts of that system in some way. Hence, the expansion of the road system had a negative effect on existing settlements. For example, the Latin colony of Cosa was re-founded at the beginning of the second century, but failed to develop successfully (Fentress, 2003; see Chapter 30). This situation, perhaps, was repeated across the network of Latin cities closer to Rome including those on major roads, for example Narnia, and causing these cities to enact legislation to prevent migration and preserve their population (Broadhead, 2001, 2003, 2004). The need for such legislation to enable settlements to remain unchanged points to the effect of road construction as an enabling feature of migration and movement. However, we should perhaps recognize that settlement change is an inherent feature of the Roman Republic and we should not expect continuity over time. At the end of the Republic, Cicero (*Balb.* 29–31; see also Broadhead, 2001: 85) could comment that every state was connected to Rome by a road and these routes were used in the past for migration to Rome. Road building had created a movement economy that included the migration of people and the creation of a tradition of migration as part of Rome's cultural memory of its past. This is a sign of a major change in *mentalité* that is as impressive as the scale of the undertaking of building these monuments that crossed space and created surfaces for travel.

FURTHER READING

Key to further reading is the *Barrington Atlas of the Greek and Roman World* (R. Talbert, ed., 2000), since so much of our understanding of Roman roads and bridges involves the study of the landscape through which they pass. The following listing refers only to books in English; there is a far greater amount of material available in other languages, particularly Italian, some of which I have noted. A general treatment of the issue of roads, mobility and cultural change can be found in Laurence (1999). Purcell (1990) provides an overview of landscape transformation and road building with a greater focus than here on textual sources. Books on Roman roads are few and far between: Chevallier (1976) remains a standard work on the study of the archaeology of roads; Staccioli (2003) has provided an illustrated overview; while Van Tilburg (2007) examines the issues of traffic and congestion. For visiting roads near Rome, Coarelli (2007b) provides guidance on some of the major sites on the roads. For the Via Flaminia, Ashby and Fell (1921) continues to be the most extensive description in English. O'Connor (1993) provides a survey of Roman bridges and viaducts; there are also 177 images of Roman bridges at http://www.signaromanorum.org/sr_lista_pon.htm.

CHAPTER 20

Villas and Agriculture in Republican Italy

Jeffrey A. Becker

1 Introduction

The iconicity of the "Roman villa" affords it a rare status in that its appeal easily cuts across the boundaries of multiple disciplines. This is perhaps because the villa has always stimulated our imagination about the ancient world and cultivated a longing for that realm of convivial, pastoral bliss that the villa conjures up for us. Just as Seneca contemplated Roman virtues in the context of the villa of Scipio Africanus (Sen. *Ep.* 86), modern (and post-modern) thinkers continue to privilege the villa both as place and space, often using its realm as one in which to generate reconstructions and visions of the ancient past. In the nineteenth century the Roman villa appealed to the Romantics and the exploration of Vesuvian sites, in particular, fueled a growing scholarly interest in the architecture and aesthetics of the Roman villa (most recently, Mattusch, 2008). Often guided by ancient texts, villas were divided into typological groups, as were the interior appointments from wall paintings to floor mosaics. Villas seemed to be a homogeneous type, representative of a "Roman" cultural norm. The fascination with villa life began in antiquity, not only with the likes of Seneca but also poets and scholars such as Virgil and Varro. In spite of the iconic status of the villa from antiquity to modernity, a good deal of uncertainty remains with respect to the archaeology of Roman villas of the latter half of the first millennium.

The scholarly approach to the Roman villa finds itself at something of a crossroads, particularly with respect to the villas of the Republican period in Italy. This state of affairs has been brought about by a myriad of factors including a reassessment of received wisdom and the introduction of new data by way of archaeological excavation and field survey. New and critical approaches to Republican villas find, in some cases, that the

A Companion to the Archaeology of the Roman Republic, First Edition. Edited by Jane DeRose Evans.

mainstream of Republican villa studies stemming from post-Second World War approaches are in need of updating, if not more radical intervention. At the heart of these reassessments are questions about the utility of traditional approaches to the villa as well as to the intertwined threads of influence in the Mediterranean – for instance, are the villa's roots Italic or a co-opted form from Greek or Punic spheres? This sort of argument can easily become a "chicken or egg" sort of diatribe, but nonetheless a reappraisal of the archaeological evidence in Italy is instructive, especially as a way to situate the better understood phenomena of villas in Late Republican and Early Imperial Italy, and, indeed, in the Roman Empire writ large. This contribution explores the archaeology of villas in Republican Italy, highlighting debates that center on questions related to the origins of villa architecture, the morphology of villas, and also their agricultural functions. To that end Republican agriculture will also be discussed, with the overarching goal of contextualizing the villa and its function within the framework of Republican Italy. The reader must also keep in mind that in many cases earlier villas have little to do with the rarefied notion of villas that developed in the nineteenth and twentieth centuries, but that the development of villas (and their functions) is an important part of the development of the Republican period.

2 Current State of Villa Scholarship

Any useful *Companion* chapter should address the state of the field with respect to its topic; although such an assessment is needed for Republican villas, the task is a tall order. It will prove useful to highlight several recent and pivotal developments in the study of villas in Republican Italy in order to emphasize the degree to which the field of villa studies has moved forward in recent years. Traditional approaches to the Roman villa (e.g. McKay, 1975) tended to focus only on the most opulent sites, especially the so-called villae maritimae (Lafon, 2001), or on the aesthetics of the Roman villa. While these aesthetic and elite-centered scholarly approaches still retain their value, they do not contribute substantially to the exploration of the roots of the villa phenomenon or on the phenomena of monumental elite architecture of the Italian Peninsula. This topic has been treated quite extensively in Etruria (Stopponi, 1985), but only recently has the topic begun to receive similar attention in Rome's sphere (Andrén, 1939; Cifani, 2008; Winter, 2009). While one could argue that such a focus is simply another scholarly agenda, it is, in fact, fundamental to understand the means by which the Late Republican villa became a part of the rarefied elite way of life. At the heart of this question is when the villa originated; are the origins to be assigned to the Archaic period? Mid-Republican? Late Republican? Older approaches relied on the chronology of domestic architecture which was founded on the dates that scholars maintained for Pompeian sites, dates that are increasingly unreliable as chronologies have been falling ever lower. The result is a sort of architectural gulf in the Middle Republic, with a number of "grandfather" prestige sites that exist in the landscape and a huge proliferation of "classic" villas dating to after the time of the Gracchi. The implication is that the number of "grandfather" sites is quite small, but nevertheless provided powerful examples to novi homines seeking to establish themselves alongside their hoary patrician colleagues in a time when the Roman world was experiencing radical changes (Pollitt, 1978).

Another significant approach situates villas in their cultural contexts; in this methodology, villas are a locus connected to memory and identity (Boatwright *et al.*, 1982; Bergmann, 1995; Bodel, 1997). The Late Republican villa functioned as a venue in which to cultivate and perpetuate memory and identity, from aesthetic schemes that reflected social standing to the display of specific images and texts that connected the present to the past by way of ancestor worship.

Other scholars emphasize how the villa allowed elite Romans to create a platform upon which to situate their ideas and ideologies with respect to the agrarian life and their own political agendas. This *topos* is reflected in the famous agrarian treatises, notably those of Cato, Varro and Columella. While these texts are being used in a fairly straightforward way by some scholars who still view them as handbooks or guides to Roman agriculture, the most exciting (and potentially revelatory) approaches to these treatises are those that push beyond the treatment of the Late Republican treatise as simply a handbook to ask deeper questions about the function of such writing in contemporary society at Rome. Ancient writers sought to classify the villa as well as to describe its function(s), in direct and paradigmatic ways. In some cases the villa was simply a setting, though the setting ranged from the literal to the metaphorical. Cicero and Pliny the Younger, for instance, described the life and duties of the villa owner in various ways, while philosophers and poets used the villa and its setting for their own purposes (Boatwright *et al.*, 1982; Bodel, 1997). Others used the villa and the limning of "agrarian life" to frame their own ideological platforms, predicated upon the notion that the purest form of Roman citizenship was the yeoman farmer who could, like Cincinnatus, leave the field to take up arms and then return directly to the field when the threat had been averted. Poets, too, weighed in on this issue, with works like the *Georgics* of Virgil both extolling the agrarian ideals and also positioning the farmer (and his farm) within the loss of the integrity of the Roman Republic (*G.* 1.506–8, 2.533–4). It is in light of these motives that philologists have produced new and creative approaches to the Latin agrarian treatises that have focused on the centrality of farming in the world of Rome's Republican aristocracy (Green, 1997, 2012; Reay, 2005; Terrenato, 2012). For instance, Cato's *De agricultura* has traditionally been read as a literal guide to rural living, home cooking and folk medicine, yet some of his advice hardly seems useful in light of the practicalities of actual farming. Scholars such as Reay examine the value of the text as a sort of position piece, an agrarian bulwark for Cato's conservatism. Likewise, Green's approaches to Varro explore the allegorical nature of his narrative, even finding that the carefully chosen names of the characters in the text are a part of the Varronian narrative (Green, 1997, 2012). Historians such as Marzano also considered the villa from the view of social history, engaging the ways in which the villa – as a center of production and an essential part of Italy's settled landscape – helped to cultivate individual identity and status (Marzano, 2007). Interpretations of these Late Republican treatises remain debated, but it is of remarkable interest that in the face of a rapidly Hellenizing world, a crisis of the social orders and a tidal wave of wealth in the hands of relative newcomers on the social scene, conservative politicians built their platform upon a mythologized agrarian past. Yet this was the ancient argument about the meaning of what it was to be a Roman.

The archaeology of villas has also advanced by fits and starts over the course of the modern era. With the re-discovery of Pompeii in the eighteenth century, the more opulent villas in the landscape became the targets of wholesale looting, as collectors and dealers sought portable art and wall paintings for sale and show. Thankfully they remained more or less ignorant of more humble rural villas (see Chapter 37). These rural sites came closer to the forefront following extensive field surveys of the mid-twentieth century. In the 1970s Marxist scholarship that focused on the issue of slavery had obvious implications for the investigation and interpretation of productive sites (see Chapter 12). Now in the twenty-first century, villa studies find themselves in a healthy place, but also a controversial one in terms of site classification and the debate over the adequacy of the word "villa" as a useful archaeological term. The range of sites and site typologies concerned is enormous – and this is only a consideration of the Italian Peninsula. In general terms the rich textual and aesthetic corpora for villas need to be augmented still by more extensive investigations of Republican villa sites, investigations that are not guided by *a priori* assumptions gleaned from texts. This is particularly true for villa sites of the Early and Middle Republican periods, where the possibility exists to map more comprehensively the emergence of perhaps the most iconic of Roman architectural types and thus better understand the phenomenon by which the "classic" villa proliferated in Italy and across the Mediterranean basin from the final decades of the Republic onwards.

3 What Is the Republican Villa?

While this question might seem innocuous and overly simplistic, it hinges on a number of current (and entrenched) scholarly debates about the topic at hand. From the traditional standpoint, the Roman villa was a country residence, the pendant, yet apposite, partner of the domus, a house in town. The villa could be situated either in the interior or at the coast, and extant textual testimonials in Latin describe such places. Pliny the Younger, for instance, described a number of his villas that include one by the seaside as well as the more working-class villas that he leased to tenant farmers (Plin. *Ep.* 3.19). Thus, even from the start the Roman villa had an identity crisis in that it could be a working farm or a rich man's pleasure palace, and yet still be called by the same name. Ancient authors further subdivided villas into two classifications: the villa urbana and the villa rustica. These labels have entered modern scholarship and have, at times, been applied in a rather uncritical fashion. The villa rustica is presumed to be a more humble place, connected often with production and slaves, as well as with store-rooms for produce, stalls for livestock, and the like. The villa urbana is presumed to be more refined, evocative of the comforts and luxury that the villa owner would also enjoy in his domus. Modern scholars have continued to construct typologies beyond the villa urbana and the villa rustica, as both Rostovtzeff and Carrington did (Rostovtzeff, 1957: 550, nn. 25, 26, but the typology was first proposed in the 1926 edition; Carrington, 1931). The former created a rubric for defining the villa rustica while the latter collected a sample of such villas from the area around the Bay of Naples. Carrington attempted to apply Rostovtzeff's scheme for classification to archaeologically attested villas, many of them connected, by way of inscriptions, with known historical personages. In terms of today's scholarship, most researchers would be reluctant to approach the classification of sites precisely according to this system, but we may still observe the continued classificatory impulse in the works of Romizzi (2001) and De Franchescini (2005). Marzano's villa volume represents the most comprehensive and nuanced recent scholarship, as it expertly combines the archaeology of villa sites with economic and social history (Marzano, 2007). The impulse to break down a vast archaeological corpus into smaller, more digestible segments is understandable but is only advisable when (and if) the cataloger is prepared to admit the limitations attached to the classificatory scheme. This seems especially true (and pressing) with respect to the villa corpus since the terminological insufficiency of the word "villa" not only impacts our ability to agree upon site classification but also extends to the interpretative universe in explaining the role of the "villa" in the landscapes of Roman Italy.

Within the Roman domestic sphere itself, yet another subdivision was created by modern scholars that mirrors the aforementioned, since the villa tends to be segregated into two distinct zones, the pars urbana and the pars rustica. The former quarter is considered as the more refined part of the house, the seat of the paterfamilias and the place to display the trappings of his status and ancestry. The latter is the part of the house connected to domestic service, slaves, and productive activities. This dichotomous nature, which also exists for the domus, has been well explored in recent scholarship (see Chapter 3).

By the Late Republican period we tend to speak of a "classic" villa plan, one that is perhaps inspired by extant villas that have been deemed canonical. The Villa of the Mysteries at Pompeii often factors in such discussions, as it demonstrates both the general tendencies of internal division discussed above as well as the centrality of an atrium space in the villa's layout. Further, for those most comfortable with a reading of the architecture informed directly by the ancient sources, the Villa of the Mysteries also conforms to the scenographic description given by the younger Pliny of his villa at Laurentum (*Ep.* 2.17). Pliny's description stresses the relationship between the natural and the built environment, the indoors and the outdoors, architecture and vegetation, and, perhaps above all, lines of sight and perspectives. As important as Pliny's description is, it is worth noting that despite some morphological trends, great variation exists within the corpus of villa sites and it is difficult to speak in terms of standardization of villa architecture.

4 Early Rural Architecture in Central Italy

The search for the beginnings of the "Roman villa" usefully ought to begin not with structures such as the ones described by Cicero and the younger Pliny but rather with rural buildings from across the social spectrum. Ancient authors, including Pliny the Elder, sought to explain villa origins, and the villa emerged as a central *topos* in the treatises of Late Republican and Early Imperial authors. In archaeological terms, rural architecture in central Italy has been the focus of scholarly attention since the advent of large-scale field surveys that began in the wake of the Second World War and have influenced dramatically the excavation of sites in peninsular Italy (see Chapter 12; Potter, 1979; Patterson, ed., 2004). The identification of a multitude of previously unknown rural sites provided not only an impetus for the examination and reconstruction of ancient landscapes, but also for the potential to excavate selected sites. The 1970s witnessed a spate of "villa archaeology," from the investigations in the Ager Cosanus to the archaeology of rural sites in Rome's immediate hinterland (see Chapters 13 and 30). In this latter category, Bedini's work along the Via Laurentina continues to prove seminal for the discussion of early rural architecture in central Italy. At a series of sites, emblematized by the sites of Torrino and Acqua Acetosa Laurentina, Bedini documented non-elite rural architecture of the archaic period (Bedini, 1984). These humble farm sites demonstrate the crude construction that was the vernacular of the non-elite architecture of central Italy, including the use of wattle-and-daub and a marked absence of rectilinear plans. Other archaic structures in central Italy, for instance, the multi-phase farm at Podere Tartuchino in Etruria, show a tendency for agglomerative construction in the Italian Peninsula (Attolini and Perkins, 1992). These rural sites show a tendency in their ground plan that is interesting in light of later morphological development, that is, the construction of a semi-enclosed corral area.

The work of Bedini and that of the excavators of Podere Tartuchino is made all the more interesting when the humble, non-elite sites are set alongside their contemporary elite sites. In Etruria and Latium, evidence for palatial architecture exists in the Late Archaic period, in the Etruscan palaces documented at Acquarossa and Poggio Civitate (Murlo), among others (see Chapter 23; Stopponi, 1985). High-quality construction materials and techniques characterize these sites, as does the material culture evidence for elite occupation and activity. These elite sites reach a monumental threshold with the late sixth-century building at Poggio Civitate, which is over 1,500 m^2. The architectural tendencies that link these sites include the use of colonnaded porticoes that both define and frame central spaces; this is true of the central, enclosed courtyard at Poggio Civitate as well as the L-shaped portico of zone F at Acquarossa.

The Poggio Civitate structure still awaits comprehensive publication, yet it can still be singled out as perhaps the largest structure in peninsular Italy in the third quarter of the sixth century. The destruction of the site notwithstanding, it set important elite architectural precedents both in terms of monumental scale and quality of construction. The nascent monumentality at Rome and in its hinterland took cues from the archaic elites who secured and maintained social position in order to display economic and political power by way of material wealth and monumental architecture. Yet one can reasonably raise the question as to what connection may be (or should be) drawn between the patrimony of Archaic Italian elites and the emergence of the architectural form that we call the villa.

5 The "Early Villa"

When and where can we identify the "first villa" in Italy? This search finds analogy in an epidemiologist's search for a "patient zero" and is no less complicated, especially given bifurcation of the issue: was the idea of a villa imported or indigenous? Scholarly rationale exists for both lines of thought and the argument is extremely complex and, at times, convoluted. Those who support exogenous origins point to structures like the Dema House in Attica, a fifth-century rural property whose ground plan evokes the Italian villa (Jones, Sackett and Graham, 1962: 75–114). In this case, the argument is that its parallel in Italy is seen in sites like the one at Moltone di Tolve in Basilicata, a case advanced by Torelli (2012; for the site, see Russo, 1992). But support has also been given to an argument for indigenous origins for the architectural form. The architectural morphology of several recently discovered sites, including the Villa dell'Auditorium near Rome (see below) and the structure at Prato-Gonfienti (Cifani, 2008: 275), lend credence to a claim for an Italian origin for the "classic" villa form. More than 20 years ago Carandini christened the Republican villa at Selvasecca di Blera as the "first villa in Italy" (Carandini,

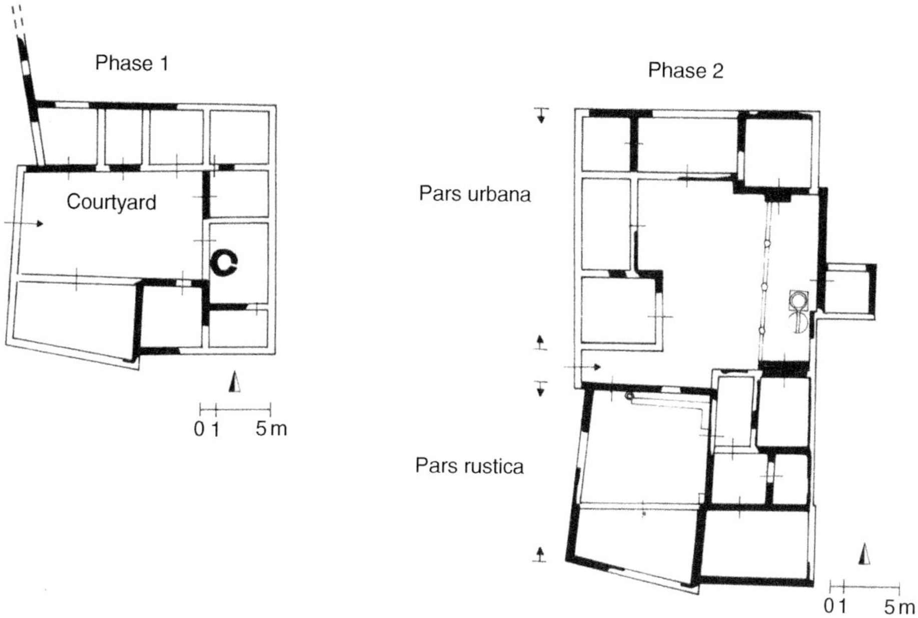

Figure 20.1 Plan of the Auditorium site. Source: Drawing by J.A. Becker, adapted from Carandini *et al.* (1997).

1989: 101–200). Dating the villa at Selvasecca is problematic as the site has a long history, as do other prestige sites that we shall now discuss.

A major development in early villa studies came in 1997 with the discovery of the Villa dell'Auditorium, near Rome, a monumental multi-phase elite residential complex uncovered during salvage excavations in Rome's Parioli district. The Villa dell'Auditorium is clearly an elite structure even from its earliest phase, with high-quality construction seen in the use of ashlar masonry. The complex underwent a number of stages of development with fairly radical alterations to the ground plan; occupation continued into the middle of the Roman Imperial period. An early appraisal by Terrenato identified the Villa dell'Auditorium as an early architectural precursor of the "classic" villa form in central Italy (Terrenato, 2001a).

Carandini, as the principal investigator, wove a narrative that adds to the mix components that include ritual usage and slave quarters (Carandini, D'Alessio and Di Giuseppe, 2006). The Villa dell'Auditorium (which I will refer to as "the Auditorium site" to maintain some neutrality as to its function) finds itself at the heart of a terminological debate at the ground level – should it be termed a villa, palace or workshop (*fattoria*)? The title of Carandini's final report – *La fattoria e la villa dell'Auditorium* – belies the biases of the principal investigator and his reckoning of the site's role in a slave-based economy. The first phase of the Auditorium site dates to the final third of the sixth century and covers some 300 m^2 (Figure 20.1) (Carandini *et al.*, 1997: 120–1). Phase II at the Auditorium site enters the same category as the archaic complex at Poggio Civitate, in that both exceed 500 m^2 in terms of built surface area, unusual in pre-Hellenistic Italy (Terrenato, 2001a: 15). This second phase oriented itself around an enlarged central courtyard that provided the focus of the plan and that, in later phases, became the atrium proper. The use of ashlar blocks of cappellaccio demonstrates elite status, as this is effectively the same building technique employed in contemporary monumental structures in the center of Rome, for example, the podium of the Temple of Jupiter Optimus Maximus (see Chapters 17 and 26; Cifani, 2008: 80–108).

Its less well-preserved neighbor, the Villa delle Grotte at Grottarossa on the Via Flaminia, is the Auditorium site's likely architectural peer. This villa, excavated in 1926, demonstrates a primary and secondary phase akin to those of the Auditorium site and it, too, remains in use across a broad time frame (Becker, 2003). The Villa delle Grotte, along with the Auditorium site, represent potential archetypes for the villa; their longevity in the landscape allowed them to serve as models for later villa builders and owners (Becker and Terrenato, 2012).

These two remarkable sites are not alone in Rome's hinterland, as we may safely add other potential peer sites to this group. Among these is the Casale Ghella villa site on the Via Cassia (Messineo, Petracca and Vigna, 1985; Messineo, 1987–8). This villa, excavated in the 1980s, is poorly preserved, but the dimensions of its early phase and the presence of a cistern connected with a series of rock-cut cuniculi suggest an elite residence similar to the early phases of the Villa delle Grotte. The Villa delle Grotte has as well a system of cuniculi cut into the tuff bedrock beneath the structure. These cuniculi are likely connected with a system of cisterns for water collection (Becker, 2003: 17–19). Also like the Villa delle Grotte, the Casale Ghella villa underwent substantial Late Republican renovation, thus complicating attempts to arrive at accurate chronologies for the earliest phases. However, it would seem that this constellation of prestige "grandfather" villas in Rome's suburbium establish an important precedent both in terms of their early monumentality and their long-range occupation, thus setting them apart from the vast majority of suburban villas.

Taken together these long-lived residential elite complexes in the urban periphery suggest that the landed, rural elites maintained a literal basis of their power outside the pomerium of the city, an interpretation that calls to mind both the scholarship of L. Ross Taylor (1960) who focused on voting districts as a way to examine the political landscape of Republican Rome, as well as more recent approaches to the gens as a means to devise a state formation model for Rome (Smith, 2006; Terrenato, 2011). These sites also are most appropriately compared to the archaic palaces of Etruria, both in terms of built surface area and in terms of aesthetic adornments. For instance, the eaves tile depicting the river god Acheloös discovered at the Auditorium site clearly demonstrates the presence of elite occupants, as it is now widely accepted that architectural terracotta decoration is a mark of elite status (La Rocca, 2006: 485–500). Another mark of distinction for the sites in this category is the fact that their early origins mark them as prestige sites in the landscape and they enjoy a long and unbroken history of occupation and status. Their early origin sets them apart from the vast majority of villa sites that emerge during the Middle and Late Republican periods.

6 Villas and the Middle Republic

Recent scholarly trends in both archaeology and ancient history have focused more attention (and deservedly so) on the Middle Republican period. It was during this time that Rome's hegemonic expansion was fully under way with colonial foundations placed across the peninsula (see Chapters 19 and 30). An important question in this case relates to the appearance of the villa in the landscape of peninsular Italy. It is important to keep in mind that numerous Late Archaic and Early Republican prestige sites like the Villa delle Grotte and the Auditorium site continued in use, perhaps still projecting the local power of the clan chiefs who may have inhabited them. It does not seem, however, that the onset of Roman expansion translated directly to a reproduction of these prestige sites in other parts of the Italian Peninsula. Some sites with still unclear chronologies, such as the villa at Selvasecca di Blera, may have undergone restorations during this period, but even Selvasecca seems to pre-date pronounced Roman expansionism. The chronological problems of the Selvasecca villa remain essentially unresolved, but it is clear that the interpretations offered in the preliminary report have not found widespread support, prompting more recent reappraisals (Berggren and Andrén, 1969; Klynne, 2006–7). The villa is large (*c.* 1,100 m^2 in area) and is constructed of local tufo using opus quadratum masonry. The preliminary report gives a second-century date for the construction of the villa, but this attribution is not consistent with the numerous fictile fragments that may be dated to the fifth or fourth centuries (Terrenato, 2001a: 23). Unfortunately, the villa remains enigmatic owing to incomplete excavation and publication, yet the combination of domestic space (pars urbana) and production space (pars rustica) suggests that the building had elite status as the pars rustica includes facilities for the production of architectural terracottas.

The Middle Republic sees the proliferation of small farms in central Italy (Figure 20.2). Among these are the villa sites on the Via Gabina, identified by field survey and subsequently excavated (Ward-Perkins and Kahane, 1972: 91–126; Oliver-Smith and Widrig, 1982: 99–114; Widrig, 1986: 141–82, 1987, http://viagabina.rice.edu). These sites were located in Rome's eastern suburbium in the area of Tor Bella Monaca. The two villas that were excavated represent a long period of occupation stretching from the Republican period to Late Antiquity. The Republican phases demonstrate household industry with evidence for presses and vats and the production of wine and processing of wool that took place at the site. The Via Gabina villas are representative of the small-scale farms common to the Mediterranean basin that predominate in the Italian landscape in the Middle Republic (Volpe, 1990; Terrenato, 2007).

7 Late Republican Villas

The Late Republican period witnessed a veritable explosion of villa architecture across the Italian landscape. Many of these villas belong to the first century and their construction radically changed the settled landscape of peninsular Italy. The classic villa plan was replicated again and again, becoming a widely distributed architectural type. The Francolise villas, located in the Ager Falernus, are good representatives of villa construction in this period. These villas at Posto and San Rocco are modest in size and connected to productive activities. The first phase of the Posto villa belongs to the late second century and its plan centers on a courtyard (Cotton, 1979: 17). The San Rocco villa, which has a slightly more complicated plan than Posto, is a terraced structure with the lower terrace given over to the pars rustica. The pars urbana occupies the higher terrace (Cotton and Métraux, 1985: 11–12). The Francolise sites show a range in terms of

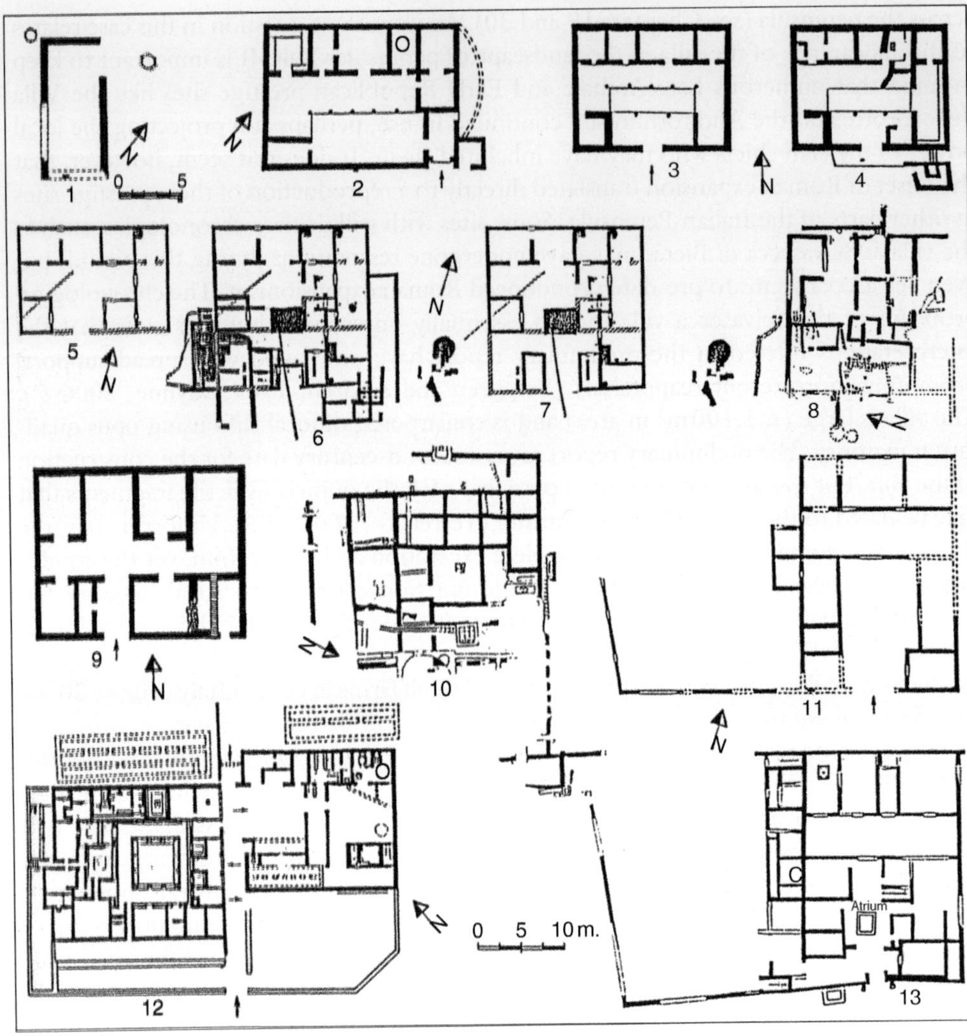

Figure 20.2 Typological sampling of Hellenistic farmhouses and villas. (1) farmhouse at Monte Forco near Lucus Feroniae (second half of first century BCE to second century CE); (2) Dema House in Attica (end of fifth century BCE); (3) Vari House in Attica (first phase, fourth century BCE); (4) Vari House (second phase, late fourth century BCE); (5–7) farmhouse at Moltone di Tolve (phases 1–3, end of fourth to third century BCE); (8) farmhouse at Giardino Vecchio near Capalbio (early second to late first century BCE); (9) farmhouse at Sambuco near Blera (second half of second to mid-first century BCE); (10) farmhouse at San Giovanni di Ruoti (Augustan period); (11) farmhouse on the Via Gabina near Rome (first phase, late third to late first century BCE); (12) San Rocco villa at Francolise (second phase, late first century BCE to mid-first century CE); (13) Via Gabina villa (second phase, Julio-Claudian period). Source: Drawing by J.A. Becker, adapted from Volpe (1990).

appointments and remind us of the combination, in the space of the villa, of the productive activities alongside other features.

The so-called Villa of the Volusii Saturnini at Lucus Feroniae on the Via Salaria is also an important Late Republican case study, one that occupies a different end of the archi-

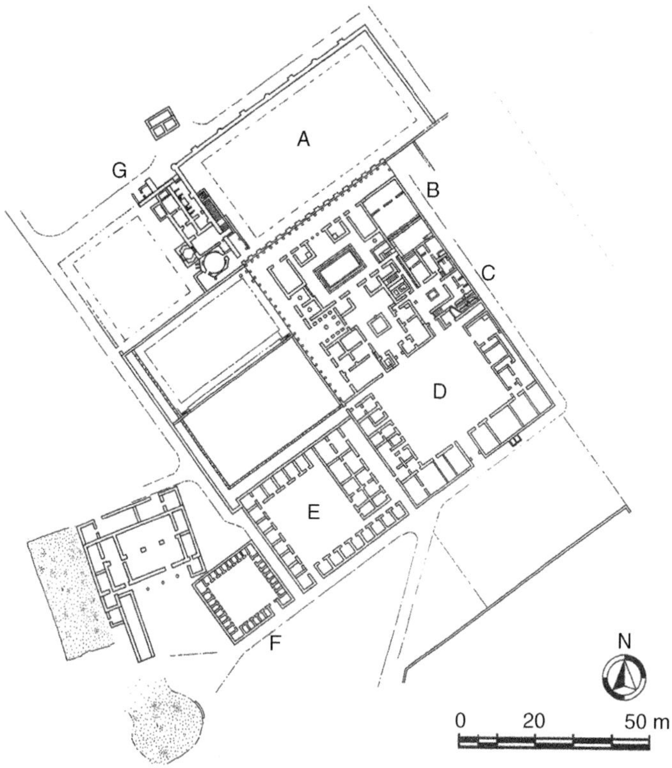

Figure 20.3 Plan of the villa at Settefinestre. A, D, E and F comprise the pars rustica; B, C and G are the pars urbana of the villa. Source: Marzano (2007: 654, fig. T3). Used by permission of Brill.

tectural spectrum than the Francolise sites. Dating to *c.* 50, the villa is associated with Q. Volusius Saturninus, whose family may have served as patrons of the nearby town of Lucus Feroniae. The family flourished under the principate of Augustus and, in turn, their villa was beautified and monumentalized. As Marzano notes, the villa itself was a visible projection of status and surely attracted attention (Marzano, 2007). What is extremely interesting is that the model here hearkened back to its earlier Republican patrician aristocrats, such as the gens that surely occupied the Auditorium site. Late Republican aristocrats relied upon the entrenched villa culture of Republican elites in order to convey and consolidate their own authority on the local level. With the family of the Volusii Saturnini achieving the status of patroni coloniae, their ability to project their status at the local level became all the more important.

The villa at Settefinestre, located in the Ager Cosanus between Orbetello and Capalbio, looms as one of the most important Late Republican villas (Figure 20.3). Excavated between 1976 and 1981 by Carandini, the villa has become the archetype of a large-scale, slave-run estate in Republican Italy, and the embodiment of Varro's notion of the villa perfecta (*Rust.* 3.1.10). Such an interpretation is not without controversy (Carandini, 1985a; see Kuziščin, 1984). Latifundia, for some ancient authors, represented the ruination of Italy and the Roman Empire (e.g. Plin. *HN* 18.7.35); in Carandini's reconstruction, the Settefinestre villa is an example of the Republican sources of that decline.

The villa has been linked to the Volusii family; the first phase is assigned to the middle of the first century, with modifications in the first century CE. It was briefly abandoned in the later second century CE, but later reoccupied. The villa is a sprawling complex, resting on a great cryptoporticus. Yet it is seen not as a luxury villa like those from the Bay of Naples region, but as a center of Roman agribusiness.

In architectural terms the Settefinestre villa looks forward as much as it looks backward, reflecting the heritage of prestige architecture in the Republican landscapes of Italy as well as establishing a long-lived architectural trajectory that would spawn the massive suburban villas of the Middle and Later Imperial periods, from Tivoli to Piazza Armerina. Indeed such villa sites also mark a pronounced connection between the Roman world and its successors in the medieval period.

8 Villas and Agriculture

In recent years ancient historians have focused a good deal of attention on farming and farmers in the Republican period (see Chapter 13). The rationale behind this scholarly trend is that the issue of farming is connected to larger research questions, especially those connected to hegemonic expansion and demography, both of which continue to enjoy robust scholarly interest. Additionally, manpower studies remain in vogue, following the work of P.A. Brunt, which still is discussed as vigorously as it was when it was introduced (see Chapter 10; Brunt, 1975; Rosenstein, 2004). Important contributions, like those of Rosenstein, help to contextualize agriculture within a larger socio-political framework (Rosenstein, 2004, 2008). Reconstruction of the rural economy and demographic trends continues to be important in the reconstruction of the history of first millennium Italy (Laurano, 2011).

In general, studies of agriculture in Roman Italy are at something like a scholarly turning point, due in large part to the growing importance of archaeobotanical and paleoenvironmental research (e.g. Motta, 2011). A paradigmatic example of these changes is Foxhall's 2007 monograph dealing with olive cultivation in the Greek world. Foxhall's work, although not directly applicable to the Italian sphere, is a worthy example of this new trend as her research uses archaeological material to challenge long-held preconceptions and assumptions about cultivation in the ancient world. Roman farming practices and the scholarship surrounding those practices also show a textual dependency, as is evident from the work of White (1970, 1975).

At a lower level, the combination of impulses such as those embodied by White's volumes with material evidence has produced, in some cases, useful reconstructions of the fundamental Roman farm implement, namely the plow. Working from a figural model of a plowman and team of oxen, reconstructions of Roman plows help aid our visualization of the Roman farmer and his work (Manning, 1971; see also Edlund-Berry, 2006a: 116–17; and, in general, Aitken, 1956). The plow is perhaps the most fundamental Roman farm implement and, as a result, the image and idea of plowing was of great interest to ancient artists. The farmer tilling the earth may be connected both with the fundamental ideological role played by farming in the Roman mindset and also with the ritual practices connected with inauguration. The imagery of the plowman remains one of the most vivid images connected with Roman agriculture that survive from antiquity.

Increasingly, scientific work on animal husbandry and fish farming, for example, continue to inform us about the intricacies of the Roman diet and the economy of production connected to villa sites (Higginbotham, 1997; MacKinnon, 2001). Compendia like Rossiter's collection of evidence for presses in peninsular Italy are useful for extending White's original work (Rossiter, 1981; Margaritis and Jones, 2008). It should be noted, however, that the discipline remains without any comprehensive spatial collation of evidence for Republican productive sites, although the work of the Comune di Roma, plotting villa sites and putative villa sites in the suburbium, represents an important step forward (Volpe, 2012).

In terms of archaeological evidence for cultivation and processing at villa sites, there is an abundance of information that has been collected, but a thorough and systematic compilation of these data still remains to be done. For instance, the Via Gabina sites show evidence for small-scale production – from harvesting to processing – with evidence for pressing and processing of the products of the fields and storage facilities associated with the villa sites. These are humble sites, with relatively small plots of land. Larger affairs, such as the villa sites at Centocelle, show evidence for viticulture in the form of planting trenches for vineyards (Gioia and Volpe, 2004; Volpe, 2009, 2012). This evidence comes in the form of vineyard trenches, of which archaeological traces are still evident (Carandini, 1980; Volpe, 2009).

The other side of the rural economy and production question is the market, a topic that has garnered a great deal of recent attention. For the Italian Peninsula, De Ruyt's work on the macellum remains fundamental (1983), but increasingly the collection of site compendia is made stronger when combined with theoretical frameworks for understanding the ancient economy, as Oxford's Roman Economy Project is seeking to do (http://oxrep.classics.ox.ac.uk). More collation of archaeological data for production and storage sites, as well as for markets and periodic markets, is badly needed in the archaeology of peninsular Italy.

9 Conclusions

The world of Republican villas is one that has been transformed thanks to archaeological fieldwork over the course of the past 35 years. From the perch of an early twenty-first-century observer, it is challenging to imagine the scholarly landscape of Republican villa studies without the excavations of villas such as the Villa of the Volusii at Settefinestre or the Auditorium site in Rome. These discoveries over the last four decades have been transformational. Taken together with constant revision and down-dating of the architectural remains from the Vesuvian sites of Pompeii and Herculaneum, the arena of Early and Middle Republican villas is now one in which a great deal is at stake. Yet the importance of archaeology can be felt especially in terms of the way in which we view and understand the Late Republican cultural milieu. This period is the one which we think we know best, as it is contemporary to (and embraced by) ancient authors; and its archaeology is more abundant and better preserved. However, Late Republican villa culture does not spring on to the scene *ex nihilo* – it originated somewhere (and sometime) earlier. This debate involves most every facet of Republican archaeology and philology and the potential for a comprehensive reassessment of the archaeology and culture of Republican villas has never been greater.

FURTHER READING

The bibliography for Republican villas is a vast one. A copious bibliography appears in Torelli (1990) as well as in Gros (1996–2001). Earlier studies on Roman villas are numerous, yet the scholars who wrote them tend to be more interested in later Republican and Imperial villas (e.g. Painter, 1980; McKay, 1975; Percival, 1976; Mielsch, 1987). Rostovtzeff essentially created in 1926 (see 1957) a three-part division of the villa rustica corpus, emblematic of late nineteenth- and early twentieth-century scholarly propensities for systems of neat classification. Such reasoning also informs Lugli's first volume of the *Forma Italiae* series (1926), where platforms in rural landscapes are referred to as *basis villae* (J. Becker, 2007: 215–31; Bodel, 2012). Numerous compendia enumerating villa sites have appeared in the last decade (e.g. Lafon, 2001; Romizzi, 2001; De Franceschini, 2005). These texts, in some cases, establish typological schemes for the sites that they collect. More recently, social historians have become involved in the study of villas of the Late Republic and Early Empire (e.g. Marzano, 2007); however, such an approach has not been realized for villas of the Early or Middle Republic. The 1970s produced a tradition of Marxist-driven Italian scholarship on villas as the loci of slave-oriented production (e.g. Giardina and Schiavone, 1981).

For the earliest Republican villas, the publications on the Auditorium site near Rome should be consulted (Carandini *et al.*, 1997; Terrenato, 2001a; Carandini, D'Alessio and Di Giuseppe, 2006), as should the work on the villa at Selvasecca di Blera (Berggren and Andrén, 1969; Klynne, 2006–7). Work on archaic period rural sites is also helpful (e.g. Bedini, 1984; Cifani, 1995, 1998, 2002; Attolini and Perkins, 1992). Important publications have resulted from the excavations at Centocelle (Gioia and Volpe, 2004; Volpe, 2007) and from the *Suburbium* congresses hosted by the École francaise de Rome in 2003 and 2005 (Pergola, Volpe and Santangeli Valenzani, 2003; Jolivet *et al.*, 2009).

The advent of bioarchaeological and environmental archaeological approaches has influenced the field to a great extent, with an increased interest in diet and production (MacKinnon, 2001; Motta, 2011). These studies of diet and distribution also extend to scholars whose interests lie in human skeletal pathology (see Chapter 9); in the not-so-distant future, a profitable convergence of scholarship on agriculture should occur with scholarship addressing nutrition and pathology of disease by way of human osteological data. Closely linked to this is the growth area of demography studies (e.g. de Ligt and Northwood, 2008; also Chapter 10) and a growing interest in new approaches to the rural populace of Roman Italy (Launaro, 2011) and slaves (Rathbone, 1983; Bradley, 1984, 1994; Bradley and Cartledge, 2011). Garden archaeology has also proven influential in Roman villa studies (MacDougall and Jashemski, 1981; Jashemski and Meyer, 2002).

CHAPTER 21

Ports

Steven L. Tuck

1 Introduction

The subject of this chapter is the development of ports in the Roman Republican world. In Roman port complexes we see the largest, most technically challenging and, arguably, most vital works of Roman infrastructure in terms of holding together the entire Roman Empire as well as supplying the capital. Ports also reflect the changing trends in technology as it is in their installations that we see the earliest evidence for Roman concrete and the adoption of Greek models for construction of lighthouses and harbor facilities.

Ports were vital to the city of Rome, to the growth and flourishing of the Roman Republic and to the success of its system throughout the Republican period. The location of Rome, off the coast yet on a navigable river, meant that construction of port facilities was an inevitability. These installations provided spaces for ships to dock and to berth in shipsheds, for them to be constructed and maintained, and for the bulk commodities to be offloaded, stored and sold, making life possible in the dense urban development of Rome. This is not only the modern scholarly perspective. We have testimony from the ancient world as well, the best provided by Cicero:

> Romulus was admirably successful in achieving all the benefits that could belong to maritime cities, without incurring the dangers to which they are exposed. He built Rome on the bank of an inexhaustible river, whose equal current discharges itself into the sea by a vast embouchure, so that our city can receive all it wants from the sea, and discharge its superabundant commodities by the same channel. It finds, in the same river, a communication by which it receives from the sea all the productions necessary to the

A Companion to the Archaeology of the Roman Republic, First Edition. Edited by Jane DeRose Evans.

> conveniences and elegancies of life, and possesses an inland territory beside, which furnishes it with an exuberant supply of provisions.
>
> (Cic. *Rep.* 2, trans. F. Barham, 1841, London)

As this quotation makes clear, by the Late Republican period, some Romans at least were aware of the central role of the Tiber in the development of Rome including its key role as a transportation conduit allowing the flow of commodities into and out of the city.

At 409 km, the Tiber is the third longest river in Italy and remarkable for its navigability, which made the eventual size of Rome possible. Ships in antiquity could sail up and downstream along at least 80 km of the lower Tiber, to and from the Tyrrhenian Sea. This fact, coupled with the high cost of overland transport, especially of bulk commodities, meant that the Tiber fulfilled the key role of providing transport for Rome's important bulk commodities including food (notably grain), olive oil and wine, as well as stone and timber – the raw materials of Roman urban life. Rome's eventual size and dense urban population were made possible by exploiting the benefits of the Tiber, as Cicero eloquently stated in *De republica*. It provided all of the positive aspects of a coastal community with the protections of an inland foundation, ensuring that Rome developed with adequate resources yet free from the dangers of attack. The port installations and other facilities along its banks made all of this possible. Rome may have been built on seven hills, but it was founded and kept fed and supplied because of the Tiber. That supply relied on the port facilities along the banks. This pattern is also critical to the larger picture of Republican archaeology because as the Romans spread their culture and way of life through colonies, they relied on the model of ports established at Rome, later also informed by Alexandria, to create port complexes throughout the Mediterranean.

A word needs to be said about our sources and the necessary interpretation of the ancient literary references to Rome and its relationship to the sea. The Roman sources clearly exaggerated Rome's rejection of the sea and naval matters in order to emphasize (or create the perception of) differences between Rome and Carthage. Yet studies of the archaeology, coinage, history and art make clear the use of the sea and its place in the Roman system. Roman treaties with Carthage dating back to the sixth century refer specifically to Roman ships, which indicates the need for such treaties in the Roman worldview. Some of our evidence for early ports in Roman territory is indirect, coming from epigraphical and historical sources. For example, in 338 the captured ships of Antium were put on display in the navalia of Rome. In 313 the Pontine islands were occupied; by 311 Rome had created the office of duumviri navales "to repair and equip a fleet." In 273 the coloniae maritimae of Cosa and Paestum were founded. In 267 were added the officers, quaestores classici, "quaestors of the fleet," stationed in a number of coastal towns including Ostia (Cornell, 1995: 388). These sorts of bare historical facts are on the surface very illuminating, and what they, in aggregate, support is the notion of Rome in the Republic as a maritime as well as terrestrial power. Once we accept that premise, the notion that the Roman system required a naval infrastructure, notably port complexes, follows naturally.

2 The Fourth Century

The Tiber is the final corridor and its banks the terminus for all of the bulk goods that made their way to Rome, including building materials such as imported stone and timber, living cargo (notably slaves and animals for the games, including elephants), and food (especially the major commodities of wine, olives and olive oil, and grain; see Aldrete and Mattingly, 1999). Since the tremendous ships of the grain fleet could not navigate the Tiber, their role in transporting the half a million tons of grain that came into Rome annually was taken by scores of smaller ships and boats of a variety of sizes and styles, all fulfilling a myriad of functions along the river including transport of commodities, passengers, ferrying across the river, guiding ships into dock, transferring cargo and other tasks.

Apparently always separate from the docks and facilities for commerce were the navalia, the shipsheds of the Roman war fleet. These were more developed than simple docks, as the fleet required dry dock facilities between deployments to preserve the ships and allow for maintenance. The navalia should be understood as roofed sheds set behind docking areas that allowed ships to be drawn up out of the waters of the Tiber. These were built initially on the left bank of the Tiber adjacent to the Campus Martius and according to Livy were in place at least from 338 (Livy VIII.14.12), who refers later to a rebuilding or extension in 179 (Livy XL.51.6) and 167 (Livy XLV.35.3). The structures were in use for their original purpose at least through the end of the Republic. If the account of Procopius can be accepted, they continued to survive into the sixth century CE, although their use had undoubtedly changed, as the Roman war fleet was no longer stationed at the capital itself after the end of the first century BCE (Blackman, 2008: 23). Their role throughout the period of their use seems to have been at least partially symbolic, since they represented Rome's naval power in a tremendous building complex along the banks of otherwise undeveloped space until the time of Augustus. Some references also suggest a museum purpose as the navalia were used to display captured ships and the "ship of Aeneas." There are no remains of the navalia, as there are none from the Republican period river port at Ostia, where navalia are documented by an inscription dated to the Republican period (Meiggs, 1973: 126).

There are no Roman lighthouses, as we currently use the term, which pre-date the lighthouse at Alexandria. There are, however, a number of Greek and Phoenician watchtowers and signaling towers on and just off the coast of Italy, beginning perhaps as early as the fourth century. Examples include the Greek tower on the Straits of Messina and the Phoenician one at Mozia. Even the Etruscans, if we can follow the testimony of Rutilius Namatianus, navigated by means of their city towers on occasion (Giardina, 2010: 5–7). True lighthouses and dedicated navigational towers and monumental landmarks post-date the Colossus of Rhodes, built in 292, and the Pharos at Alexandria, in 280 (Giardina, 2010: 9–12).

3 The Third Century

Cosa and Paestum were founded in 273 as maritime colonies. Each was designed to project Roman power along the Tyrrhenian coast into recently disputed territory.

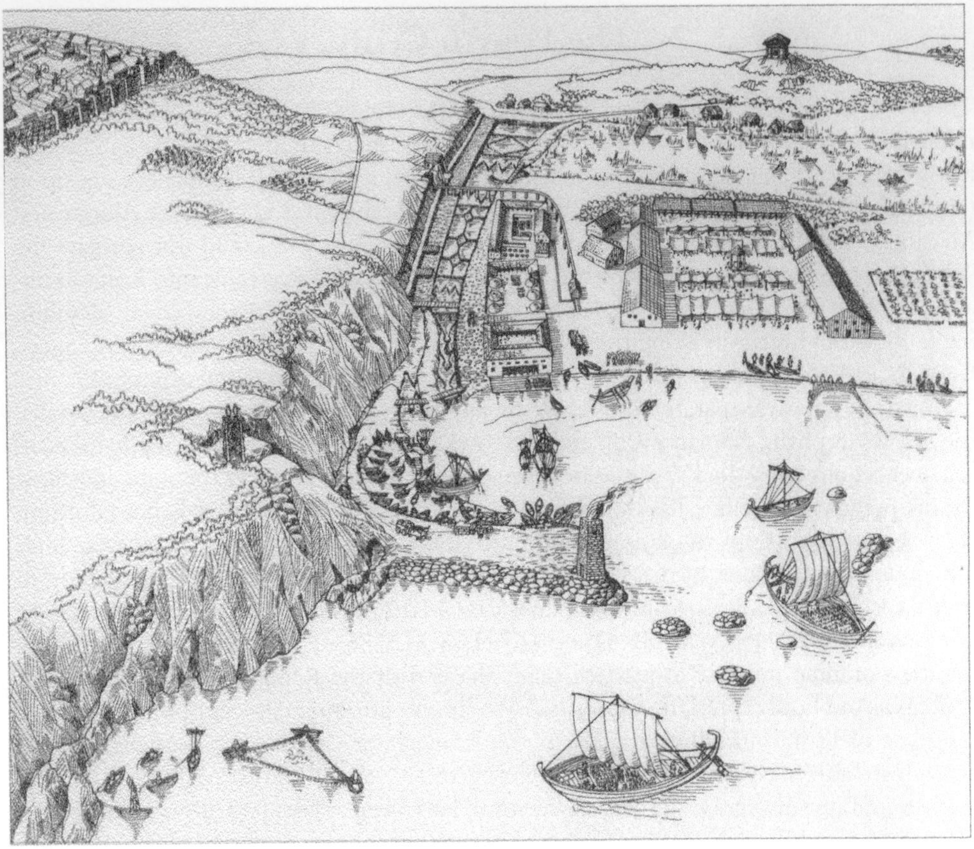

Figure 21.1 Hypothetical reconstruction of the port at Cosa in the late second and first centuries. Source: Drawing by Kathleen K. Borowick, adapted from McCann (2002). Used by permission.

Paestum was placed in the land of the Greek colonies and Lucanian peoples of southern Italy and Cosa was wedged into Etruscan lands to the north of Rome (see Chapters 24 and 30). Each represents the furthest extent of Roman power at the time. The Portus Cosanus, the port at Cosa, provides in many respects the best-preserved, excavated and studied port complex of the Republican period (Figure 21.1). Cosa is important for this chapter as the site of the earliest Roman port so far known and explored. Its location as a colony was probably because of the potential for protected anchorage. It has the only high promontory with a protected anchorage for hundreds of kilometers. Its ideal form for a coastal colony echoes Cumae or Agropoli in southern Italy. In fact, the earliest known coinage from Cosa depicts a dolphin as part of the reverse design, an indicator of its connection to the sea. The first phase in the construction of the port probably occurred shortly after the colony was founded (see Chapter 6). The remains of a breakwater can still be seen, made of loosely packed limestone blocks, quarried from the cliffs to the north-east of the port. This feature was originally 110 m in length and up to 70 m in width. The resulting irregular face of the breakwater and its gaps would serve the twin purposes of disrupting incoming waves while allowing silt to drift back out to sea to avoid silting up the harbor basin of the port facility (contra, Chapter 30). The only known antecedent for this design is the Etruscan port of Populonia (McCann, Bourgeois and Will, 1977), suggesting an Italic rather than Greek tradition at work here. The harbor basin at this period is already approximately 25,000 m^2, making it comparable in size to the later basins at Antium or Kenchreai. The entrance channel was at least 33 m, while the depth of the channel was 6 m. Evidence that the port was in full use by the second half of the third century is provided by Livy (XXII.11.6), who relates that in 217 a re-supply fleet from Ostia was captured off the port of Cosa by the Carthaginians. More direct evidence is found in the port itself where amphoras from this period are recognized among the many fragments found and analyzed (McCann, 2002: 18).

During this period we also see the rise of what would become one of the most important Roman ports in Italy, Puteoli. Conquered by the Romans in 338, the Greek city of Dicearchia was renamed Puteoli, and during the Second Punic War it served as a critical Roman base with a garrison and as a maritime, commercial and military outpost within the Bay of Naples. As the best natural harbor on the west coast of Italy it was a desirable location and it was apparently enhanced with constructed port installations very early in its Roman phase. The exact date of the earliest port facilities is the subject of great scholarly debate, but it seems clear that the major construction took place at the end of the second century and the first century, well after Puteoli was named a Roman colony in 195. This colonization and expansion occurred during the building and economic boom that followed the end of the Second Punic War and was probably also fueled by Rome's expansion into Greek lands to the east during the first half of the second century. Puteoli eventually became the western terminus for the Alexandrian grain fleet, leading to a tremendous expansion of port facilities in the first century BCE and first century CE (Ostrow, 1981).

The Romans may have quarried the raw material for their hydraulic concrete, opus caementicium, that revolutionized all Roman construction including that in port facilities from the vicinity of Puteoli (see Chapter 17). The raw material is the pozzolana mortar, a discovery of the late third century (Lancaster, 2005: 3). This mortar, whose name is related to and perhaps etymologically derived from Puteoli, is a volcanic ash found in large quantities in and around Puteoli and its territory on the north side of the Bay of Naples. Mixed with lime in a set ratio, with water added, it becomes the basis of Roman hydraulic concrete, that is, concrete with the critical property that it can set up underwater. To this is added any number of a variety of aggregates such as tufa, brick, broken amphoras, pumice or other forms of stone to make the mass of a concrete structure. Given its discovery here and the early importance of Puteoli as the Roman port in the region, scholars have suggested that the earliest example of Roman hydraulic concrete should be found here. Owing, however, to overbuilding and lack of large-scale excavation, that is not the case. Instead, we must turn to Cosa in the second century for the earliest documented opus caementicium.

The earliest port facilities known in Rome excepting the navalia were the Emporium and Portus Tiberinus, thought to have been constructed at the end of the third century along the left bank of the Tiber near the Temple of Portunus, which is located very near the banks of the Tiber in the Forum Boarium, facing the road that ran to the Pons Aemilius (Figure 28.4) (Colini, 1980). During the Middle and Late Republic, literary accounts suggest that sea-going vessels would regularly sail up the Tiber without stopping at Ostia to transfer cargo to smaller vessels. The description of the arrival of the cult stone of Magna Mater in Rome implies a long-range Mediterranean vessel in use on that occasion. Certainly the largest vessel ever attested to have sailed up the Tiber was that which carried Aemilius Paullus to Rome after his victory over the Macedonians in 167. Livy (XLV.35.3) describes it as the personal ship of King Perseus of Macedon, an enormous naval vessel powered by either eight or 16 banks of oarsmen. This was before the explosion of bridge building in the Late Republic and early Principate, which probably restricted the channel and subsequently led to the use of river boats rather than sea-going vessels to make the trip up the Tiber. The overall picture we get is of a very crowded river with constant traffic, perhaps hundreds of vessels a day, of which eventually one hundred at any one time might be from the grain fleet (Tac. *Ann.* 15.18.2). To understand the importance of the port, one has to imagine the volume of boats on the river – in motion up and down stream, ferrying passengers or goods intermittently across the current, and coming in and out of the docking. The liminal space of the banks of Tiber, between water and city, thus deserve some extended consideration.

The docking facilities that enabled the transfer of bulk commodities into Rome were not a simple arrangement. They served the purposes of flood control, docking ships, and allowing for loading and unloading by the use of cranes, among other purposes. These facilities represented a vast set of buildings and works of which the sprawling Emporium district on the south side of Rome is probably the best understood. They are certainly not the only example of docking facilities used by the city, and are probably later than the Portus Tiberinus, north of the Pons Aemilius and south of the theaters in the southern Campus Martius (see Coarelli, 1968a). The docks probably developed early to supply Rome. Since Ostia was not founded until the second half of the fourth century, all of Rome's long-distance trade until then probably traveled directly up the Tiber and docked at the city at facilities for which evidence no longer exists, due to later overbuilding. After the development of Ostia, larger and larger vessels over time stopped at Ostia (and then Portus) with smaller ships and boats supplying the city. These changes led to alterations in the dock facilities at Rome, but never allowed for their elimination. Port facilities began to develop at Rome in the Middle and Late Republic until they filled much of the left bank of the Tiber with stone and concrete embankments. These were constructed in a sporadic manner over the centuries with no overall plan, leading to a variety of designs, building materials and locations. Their major purpose was to allow for the docking of ships coming up river, but secondarily they included buildings for storage of commodities as well as perhaps shipsheds for dry docking ships for repair and maintenance. Ship construction is another matter and probably took place elsewhere than at Rome.

4 The Second Century

The great river port in Rome was constructed in the second century with a complex of buildings, grouped together under the term emporium, which referred to its use as the place where goods bound for Rome were offloaded, warehoused and sold. The area leading up from the actual docks to the warehouses was paved and steps were made from the river bank up to the horrea, or warehouses (Rickman, 1971: 149). The architecture for horrea, dry docks and shipsheds can be interchangeable in design, leading to uncertainty of the date and use of a particular facility. However, the general development of docks and their support systems is clear from the literary and epigraphic record.

The earliest large-scale construction projects we know of along the Tiber at Rome occurred in the early second century. At this period, the facilities at Ostia were restricted to sets of river docks, as at Rome, and large merchant ships were offloaded in open water, a risky procedure for ship, crew and cargo (Strabo V.3.5). Thus, the need for docks at Rome was very great. According to Livy (XXXV.10.12), M. Aemilius Lepidus and L. Aemilius Paullus constructed a number of works, including a portico and quay along the Tiber in 193. Livy further records (XL.51) that the quays were updated in 179 by the censors M. Aemilius Lepidus and M. Fulvius Nobilior. The censors of 174, Q. Fulvius Flaccus and A. Postumius Albinus, extended the quays and repaired the horrea, again according to Livy (XLI.27). The forms of the facilities are not precisely known, but sections of the embankments excavated in the preparation of the 1876–1910 embankment projects revealed paired flights of steps leading from the Tiber up along low walls; each embankment culminated in a connecting walkway along the top. Ships moored using projecting mooring rings carved from large stone blocks. The entire works were constructed of cut tufa and travertine and the form indicates that ships docked broadside to the embankment and were unloaded fore and aft, perhaps simultaneously. The locations of all these works are not securely identified, but based on the descriptions and extant remains, the areas near the Forum Boarium, alongside the Aventine near the Pons Sublicius and downstream at the Emporium area seem to be the primary focus of development during this period. Eventually the port and warehouse facilities stretched along the left bank of the Tiber from the area of the Pons Sublicius south past the Emporium region, across most of the banks of the Campus Martius and finally along the right bank in the area of Transtiberim. Although the eventual result is of a solid line of port facilities, these were never planned in concert and instead seem to have grown almost organically in response to Rome's needs for imported goods, notably grain, wine and olive oil.

The generic nature of Roman port installations and architectural forms and, perhaps, deliberate use of standard, modular forms in a variety of building types have led to debate in terms of some of the earliest extant port buildings. The major example of this at Rome is the so-called Porticus Aemilia, which represents perhaps the largest and most contentious component of any Republican port in terms of its identification and date (see Chapter 17). The building is a vast warehouse, parallel to the Tiber, located in the Emporium complex south-west of the Aventine. It is constructed of Roman concrete faced with opus incertum. In shape it resembles an enormous shed measuring at least 487 m long and 60 m wide. The interior is divided into seven longitudinal naves and 50 transverse naves by 294 pilasters. The structure is covered by a series of concrete vaults perpendicular to the facade. The traditional date and identification of the building are based on the location of the building and a fragmentary inscription on the Severan Forma Urbis that shows this structure, but only preserves the ending of a word in the name:]LIA. This has been expanded to read [PORTICVS AEMI]LIA and thus to identify the building as the one mentioned by Livy. Since its structure is clearly not that of a porticus, a number of other identities have been suggested.

The size and location of the building, coupled with the lack of remains for any of Rome's navalia, has led to the proposal that this building is the missing shipsheds. Coarelli (2007a) revised the identity of the building while retaining the date, arguing that the broken inscription on the Marble Plan of Rome should be read [NAVA]LIA, not [AEMI]LIA. That identification is accepted by Cozza and Tucci (2006), although the lack of a sloped floor and the distance from the river, and the many intervening structures between it and the river, would make this unique among shipsheds (Blackman, 2008). Such an identification also contradicts the literary sources that place the navalia further north on the edge of the Campus Martius, nearer the middle of the city (*NTDAR* 266). Both identification and the date are rejected by the present author, who sees the building as Sullan and a horrea, not shipsheds (Tuck, 2000).

A number of historical events and political decisions of the second century created a context that accelerated and encouraged port construction away from Rome. Protective legislation of the mid-second century prevented the Gauls from planting vines, ensuring a continuing market for Italian wine and a thriving export industry (see Chapter 6). Delos opened as a free port in 166; the aftermath of the Second Punic War created a context for both Roman expansion and Rome's extended naval activities. It is assumed that these events also had a direct impact on Ostia, leading to the expansion of the river port facilities there to create space for additional goods to be transshipped up to Rome. The banks of the Tiber at Ostia have revealed later constructions but the depth and extent of the explorations have not yet uncovered the proposed second-century phases thought to be buried under later buildings and the sands of successive floods (Meiggs, 1973: 31). The best evidence for the location of the Tiber port facilities at Ostia is provided by five second-century travertine cippi serving as boundary stones with inscriptions that preserve the space along the banks of the Tiber for the Roman people. These seem to have preserved an area at least 600 m in length west from a point just outside the eastern Sullan gate.

At Cosa in the mid-second century, the port facilities developed into a far more complex installation. The enclosed lagoon to the north of the port developed during this phase, with a temple probably dedicated to Neptune that overlooked a thriving port, fishery and perhaps amphora factory (see Chapter 30). The port became a suburb of the urban community and a thriving economic and religious center. McCann, in the excavation report of Cosa, argues that key elements of the port there date to the last quarter of the second century, making it one of the earliest uses of Roman hydraulic concrete in a port facility (McCann, 2002). This is not to say that the date of concrete construction at Cosa is without dissenters. Gazda (2008) has argued that the extant piers date from no earlier than the 70s, and may date perhaps as late as the third quarter of the first century. That argument is outlined below with the first-century developments. Here we trace the actual port facility remains using the original excavator's published dates.

The port at Cosa expanded tremendously, beginning in the final quarter of the second century. From this phase date the concrete remains of the port including the harbor piers, a wall, bridge and platform, the fish tanks in the lagoon, the platform for the Spring House and an aqueduct, the latter three elements supporting a fishing facility centered at the lagoon, probably producing garum (McCann, 2002: 22). The concrete utilized pozzolana mortar combined with limestone, tufa and amphora fragments. The large concrete piers inside the port allowed for the anchorage of small boats and ships within the harbor basin yet off the beach or breakwater. In addition to the concrete components, a pier at the end of the breakwater, called by the excavation team "pier 5," served at this period as the foundation for a lighthouse. The existence of a lighthouse at this time is supported by the remains of the port, by a votive model from the sanctuary of Vulci, and by a stamp on a local amphora of the Sestius family that dates as early as the late second century but before 75 (McCann, 2002: 28). While its size, particularly its height, cannot be determined with any certainty, it probably measured 5 m on a side, based on the dimensions of pier 5, and rose to about 30 m. It was certainly a multi-story tower, perhaps square in profile if the votive can be taken as a guide (Zancani Montuoro, 1979).

As the Romans moved around the Mediterranean, they brought port and lighthouse construction with them. Both Pomponius Mela and Strabo mention a tower, which can only be read as a lighthouse, in the port of Menesteo in Spain at the mouth of the Betis-Guadalquivir River. The tower, built in 108 by Q. Servilius Caepio, is compared by Strabo (III.1.9) to the Pharos of Alexandria in structure and the purpose of navigation (Giardina, 2010: 12). C. Marius is credited with building a lighthouse in Massalia, modern Marseilles, probably in *c.* 101. It marked the end of a navigation canal that Marius caused to be dredged at the mouth of the Rhône (Strabo IV.1.8), a canal that Strabo says became quite lucrative for the inhabitants of the city.

5 The First Century

In addition to extraordinary constructions, the docks, embankments and quays, and bridges required continual as well as periodic maintenance, which was supervised by officials who oversaw the clearing of the channel, the supervision of the banks and the removal of debris along the river. In the Late Republic, these works, like the constructions in the second century noted above, seem to have been undertaken by censors. The censors of 55, P. Servilius Vatia Isauricus and M. Valerius Messalla Niger, are referred to as curatores riparum on a series of cippi erected along the banks of the Tiber. The extent of their work is unclear, but at least part of it seems to have been designed to delimit the boundaries between public and private land, probably to prevent or reverse encroachment on public land, a constant problem in ancient Rome. The assertion of public rights in the first century suggests that the port facilities were still in use and perhaps vital to Rome, since they were not allowed to lapse into private hands.

Two of the major buildings in the history of Roman concrete and port construction have been re-dated to this period, the Porticus Aemilia and the port facilities at Cosa (contra, Chapter 17). In terms of the history of concrete, the date of the Porticus Aemilia is perhaps the most important example we have. It is traditionally dated almost exactly 100 years earlier than the next extant work in opus caementicium. Two attempts have been made to bring it down to the early first century, one based on the concrete work, the form of the building and epigraphical evidence (Tuck, 2000) and the other on parallels provided by other concrete structures (*NTDAR* 143). Neither attempt has gained much scholarly support. Any final solution to the problem of date will likely rely on new information, perhaps from radiocarbon dating of the concrete at Cosa.

An analysis and re-dating of Cosa's port complex has also been proposed, suggesting that the buildings dated to the second century in the original excavation report were actually built *c.* 70. The support for this comes from a radiocarbon date of 57 BCE to 33 CE from a wood sample in one of the piers, along with the date of the Dressel 1 amphora fragment of Will Type 4b, datable after 80, also embedded in a pier (Gazda, 2008).

This dating would remove Cosa as the earliest evidence for Roman hydraulic concrete in a port and would require an alteration in the thinking about the history of concrete and the economy of Cosa. The date would revise its great economic expansion to the first century rather than the second century.

The largest extant remains of the port of Puteoli date from this period. This is the famous mole that defined the harbor basin below and to the north-west of the acropolis. It currently consists of the foundations for what was originally a series of six piers which supported an arcade of opus caementicium with tufa as the aggregate, faced with opus reticulatum. The form of this arcaded breakwater is notable. It seems to have been designed to address the problems of silting that plagued so many Roman ports. With the construction of an arcade, the strength of the waves was broken, but the sand and silt that they carried were not trapped in the harbor basin as they would have been by a solid breakwater. Consequently, Puteoli avoided the silting problems and the need for constant dredging. These factors may have contributed to its use as an economic and commercial center into the sixth century CE. By contrast, Cosa was abandoned by the end of the third century CE because of silting.

That there was a lighthouse at Puteoli is supported by literary as well as artistic testimony. The most important evidence comes from a wall painting found in a villa at Stabiae, now in the National Archaeological Museum of Naples, which shows the entire port at Puteoli including the characteristic arcaded breakwater and a number of other elements including piers, porticoes and a lighthouse (Giardina, 2010: pl. 70). The lighthouse is placed on a small island or, more likely, independent breakwater separate from the arcaded one. Additional artistic evidence comes from a series of etched glass vases that show the port and monuments visible from it (Ostrow, 1981). One of these shows a building labeled "FAROS," a variant of the Greek *pharos*, adjacent to a personification of the port itself (Giardina, 2010: pl. 69). The representation shows a tall tower with what appears to be a platform built over the pitched roof. The placement parallels the lighthouse at Cosa and suggests a Late Republican date for its foundation.

The location of lighthouses in Republican harbors such as Cosa is consistently either on the left mole or breakwater from the perspective of approaching ships or on a free-standing platform defining the left edge of the entrance channel. Starting with Augustus (and in all later Imperial harbors), lighthouses are always placed on the right from the perspective of ships entering the harbor, a consistent element that seems to be one of the many changes that occur under the Principate, including the revision of these spaces into outlets for imperial ideology and display of dynastic and triumphal imagery.

What may be considered the final port planned and constructed during the Roman Republican period is at Forum Iulii (Fréjus). The Roman community here, including both the town and the port complex, was founded by Julius Caesar in 49. Its creation is linked to the need for a Roman military and naval presence along this stretch of the Mediterranean (Gébara and Morhange, 2010: 12). That the design of the town and harbor belong to Caesar and not to Augustus is supported by the evidence of the cadastral system, which differs from Caesar to Augustus, and even now can be recognized in the layout of the community (Gébara and Morhange, 2010: 14). The port was certainly functional by 31 when at least a portion of the captured fleet from Actium was brought here. For a port described by the excavators as a military base, the lack of navalia is surprising although they might still be located in one of the unexplored areas of the port.

Figure 21.2 Reconstruction of the port of Forum Iulii in the late first century. Source: Drawing by J.-M. Gassend, IRAA/CNRS, adapted from Gébara and Morhange (2010). Used by permission.

Although the site of the port of Forum Iulii has not been fully explored, the survey and excavations that have taken place reveal a harbor basin of *c.* 110,000 m^2 with an entrance channel of *c.* 100 m in width (Figure 21.2). This means that the basin was almost exactly four times the size of Cosa's harbor, with an entrance that was three times as wide, perhaps arguing for the use of Forum Iulii by much larger ships, probably the warships believed to be stationed here. The basin is surrounded by installations in opus caementicium including quays, a jetty that projected out into the basin for additional anchorage, a port bath complex, and the remains of walls and building foundations that need additional exploration. Although not yet subject to analysis, the mortar for this concrete may have been pozzolana, which was shipped by the Romans to port construction sites in the first century, including ones as far away as Caesarea Maritima (modern Israel). At present, the only shipwreck containing pozzolana in its hold from the area around the Bay of Naples is a large Roman merchant vessel which sank between 75 and 60, off Madrague de Giens, a short distance away from Forum Iulii (Brandon *et al.*, 2005).

The foundations of two towers were found at the entrance flanking the outer end of the channel leading into the harbor basin. The excavators have tentatively restored the one on the left (from the perspective of those sailing into the harbor) as a lighthouse (Gébara and Morhange, 2010: 139). The remains of a structure on the right might be a lighthouse but its only certain feature is a fragmentary colossal limestone sculpture of a triton, originally probably designed to decorate a tower or other building at the harbor entrance. The triton monument seems based on Greek Hellenistic antecedents such as the Colossus at Rhodes and the decorative sculpture that adorned the lighthouse at

Alexandria, perhaps demonstrating the Roman reliance on Alexandrian and Greek port models even after hundreds of years of experience constructing their own port facilities across the expanding Roman world.

FURTHER READING

The last attempt at a comprehensive survey of Roman port construction was published in 1923 by Lehmann-Hartleben. An excellent update is Blackman (1982), but this is now over 20 years old. The best sources are those that either concentrate on a particular site, such as excavation reports, or those that address the construction and technology of port architecture. For the latter, the critical studies are those on the history of Roman concrete (Lancaster, 2005) now supplemented by the work of the Roman Maritime Concrete Study (ROMACONS) whose work on Roman port construction around the Mediterranean is critical. For their work see Hohlfelder (2008) and the group's bibliography at http://web.uvic.ca/~jpoleson/#romacons.

Site-specific studies are many, but the most critical for this period include those at Cosa (McCann *et al.*, 1987; McCann, 1987, 2002; Gazda, 2008), probably the best excavated and published port site. The port at Puteoli is also important but owing to its continual occupation is not as well studied or published. Ostrow (1981) gives a useful survey of information to that point. Conclusions about the Republican remains at Ostia are collected by Meiggs (1973). For Rome itself, a useful survey is Colini (1980), with excavation reports in Castagnoli (1980a). The date and identity of the Porticus Aemilia, the largest extant component of the emporium at Rome, have yielded a number of studies: Rickman (1971), Tuck (2000), Cozza and Tucci (2006) and Coarelli (2007a). Finally, Giardina (2010) is a survey, analysis and catalogue of Roman lighthouses. These printed sources are supplemented by some digital ones. The most important of these is the web site on Ostia, http://www.ostia-antica.org, which continues to provide useful information on the entire city in surveys of literature, essays, photos, primary sources in translation and bibliography on Ostian topics including port construction.

CHAPTER 22

Material Culture, Italic Identities and the Romanization of Italy

Tesse D. Stek

1 Introduction

Far from Italy, in the foothills of the Sierra Morena in Baetica in southern Spain, a remarkable monumental cult complex was built during the reign of Vespasian (Figure 22.1). Laid out on a hill dominating the forum area, the complex formed literally the culmination of a general architectural makeover of the tiny urban center of Munigua in this period. The sanctuary was laid out over three terraces, the first of which was reached by two opposing ramps, visually forming a triangular base. From here, lateral stairs moving away from the central axis led to the next terrace. This level supported a podium with the temple building in the center, and was visually framed on three sides by a pi-shaped portico. In front of the actual temple building, which was elevated on a high podium, was a hemicyclic exedra, which formed part of the entire visual composition of the complex in that it directed the view from the opposing ramps at the base to the cella at the summit.

In its scenographical layout on a hill-side, its extensive application of substructures for terracing, and more in particular in its access ramps and hemicyclic space in front of the temple, the sanctuary closely echoes the sanctuary of Fortuna Primigenia at Praeneste in Latium, Italy (Figure 22.2). The similarities are so precise that there can be no doubt that its commissioners must have chosen deliberately to model the new Flavian sanctuary after the renowned Republican Latial cult place. In its composition of the upper levels, the rectangular temple building integrated in the raised pi-shaped portico as well as the hemicycle in front of it, the sanctuary at Munigua moreover reflects a similar Late Republican building complex in Latium, the sanctuary of Hercules Victor at Tibur (Figure 22.3). The otherwise unexpected choice to include these particular references to

A Companion to the Archaeology of the Roman Republic, First Edition. Edited by Jane DeRose Evans.

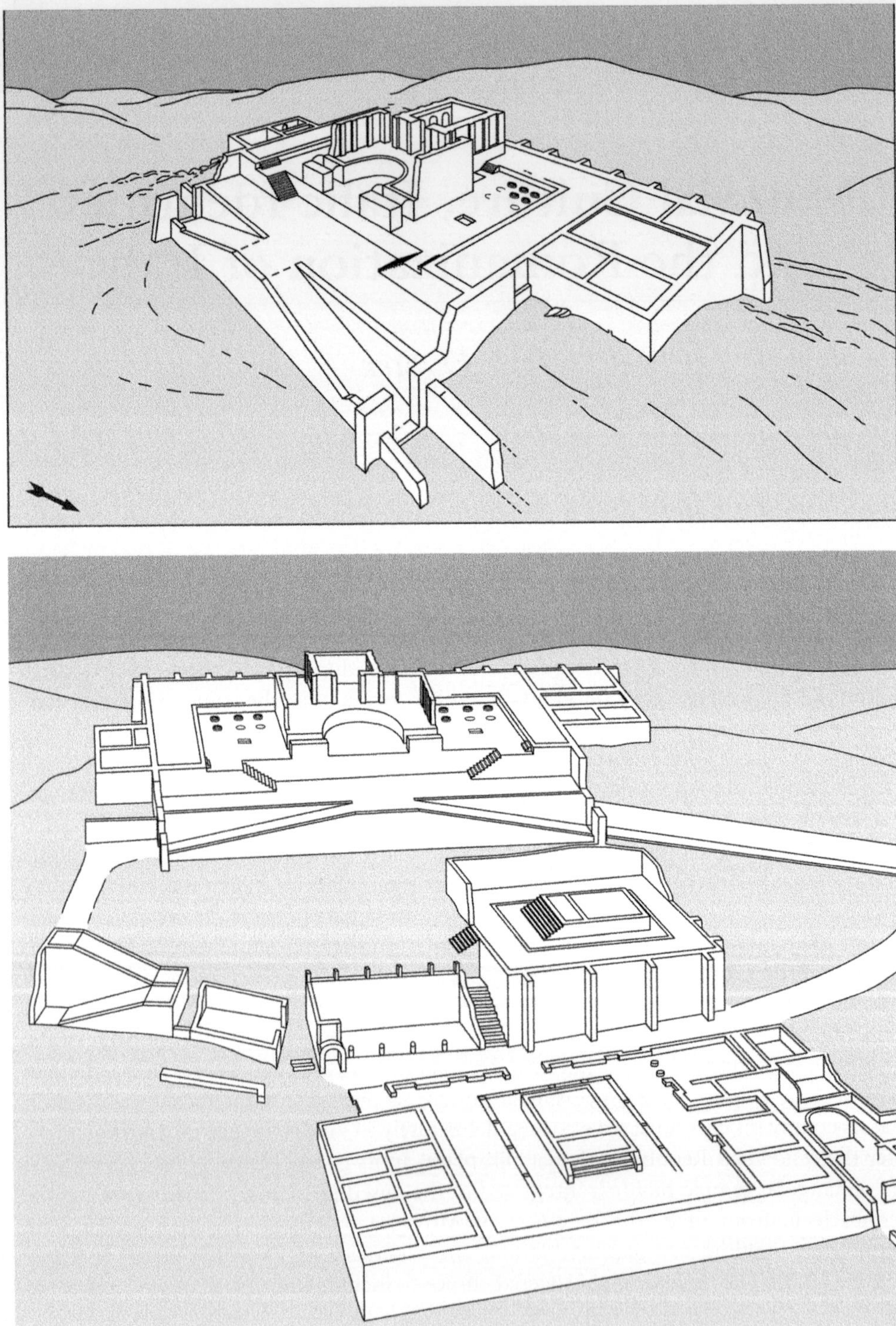

Figure 22.1 The terrace sanctuary at Munigua, two views. Source: Illustration by T. Stek, after Grünhagen (1959: fig. 5) and Hauschild (1968: fig. 3).

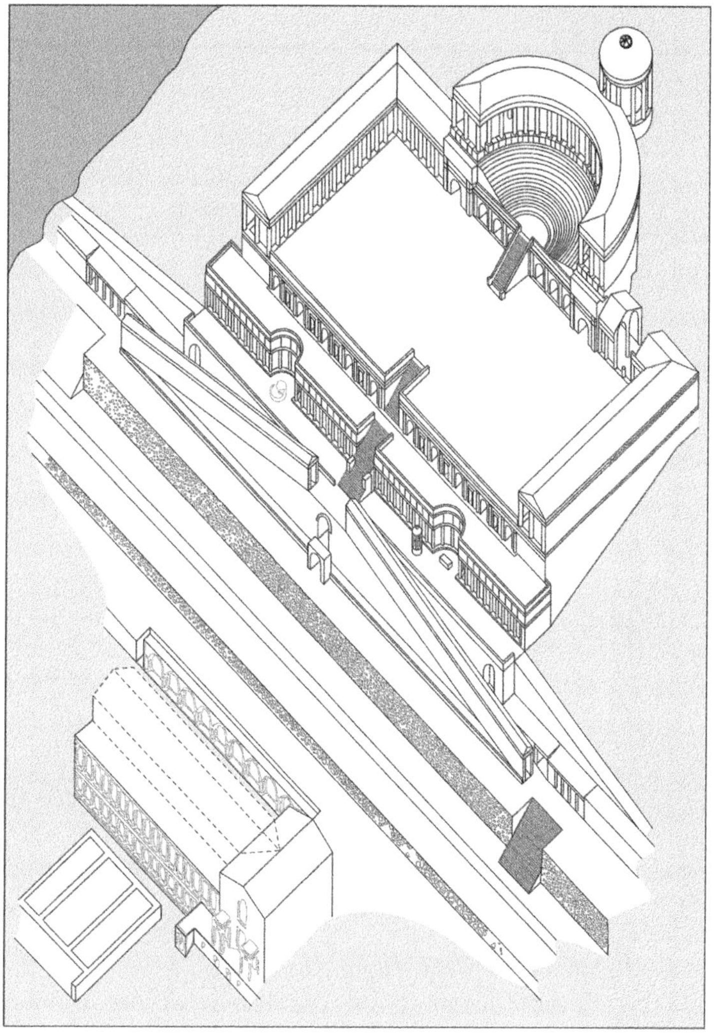

Figure 22.2 The sanctuary of Fortuna Primigenia at Praeneste. Source: Illustration by T. Stek, adapted from Kähler (1958).

antique religious architecture from Latium in the new building project is best explained by the promotion of the civic community of Munigua to Latin status at precisely that time (Coarelli, 1987a). The Latin right (ius Latii) was a particular legal status within the Roman imperial hierarchy. It denoted a relatively favorable set of rights which had initially been bestowed on the Latin peoples near Rome, before evolving into a legal category detached from any ethnic connotations, as its application to new communities outside Latium as far as Spain exemplifies. The new sanctuary of Munigua thus consciously symbolized its newly acquired Latin status in a Roman world.

How could an ancient local Italic ethnic develop into an indicator of privileged status in relation to Rome, and eventually be evoked in a province far from Italy by citing the architectural lay-out of a Hellenistic sanctuary? The perplexing case of Munigua pushes

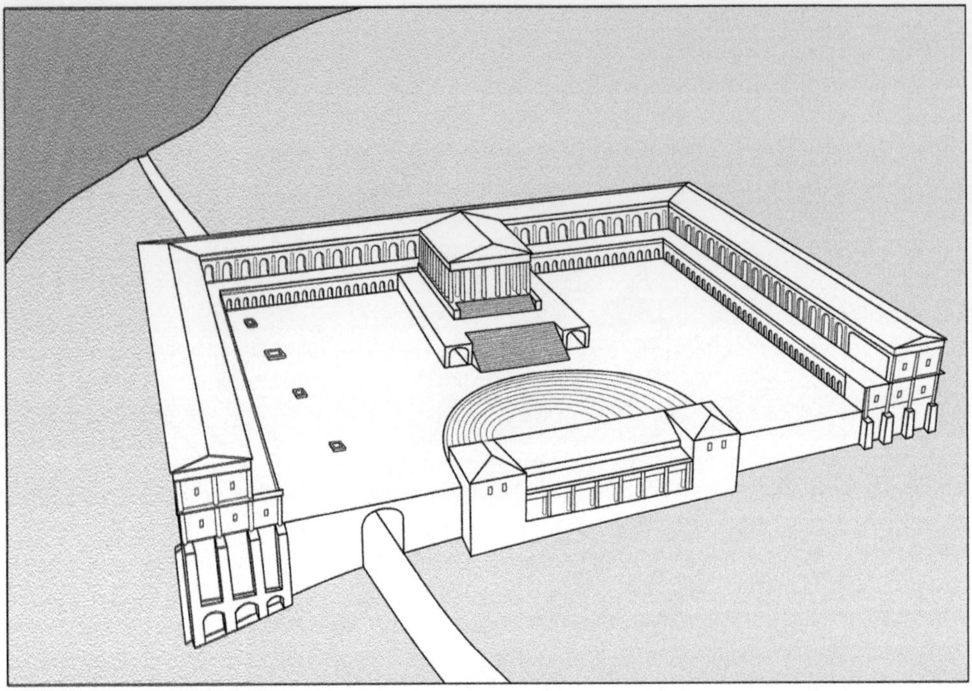

Figure 22.3 The sanctuary of Hercules Victor at Tibur. Source: Illustration by T. Stek, adapted from Giuliani (2004).

the boundaries – at least in geographical and chronological terms. But it goes to the heart of important questions about the intricate connections between conceptualizations of non-Roman Italian identities, Roman imperial ideologies, and its material expressions, questions that will be central to this chapter on cultural change in Republican Italy. Intended as an introduction to the ensuing contributions on specific Italian regions or localities and their (material) interactions with Rome, this chapter focuses on some recent trends in thinking about the impact of Rome in peninsular Italy, the significance and expressions of local or regional identities in these areas, and the relationships between the two.

2 The Terms of the Debate: De-centering Rome and Italic Identities

The inhabitants of the geographically diverse areas that make up peninsular Italy were implicated in momentous changes during the Republican period. Often within one or two generation's time, these transformations in the political, social, economic and cultural realms will have changed the living environment, including its material appearance, significantly and in some cases beyond recognition (Chapter 13). Much less clear, though, is how these developments, material and otherwise, connect to collective self-perceptions ("cultural," "ethnic," or more generally "group" or "aggregative" identities) in this period, and how they were, if at all, related to the Roman conquest and incorporation of these areas – the two main terms in which the debate has historically been posed. Any narrative is predicated on the questions and terms of the debate that are chosen, and this is no different for the study of ancient Italy. It is therefore important to emphasize at the outset that focusing on the specific issues of "Roman expansion" and "Italic identities/peoples" significantly structures the way in which we perceive cultural change in the Italian Peninsula during the time that the city-state of Rome was a republic. It almost naturally leads to an overly bipolar model of Roman and native, traditionally highlighting Roman impact, at the cost of broader contextualization (cf. Woolf, 1996–7).

In fact, many changes, cultural and otherwise, which occurred in the Italian Peninsula during the last four or so centuries BCE are probably better understood without *direct* reference to Rome. In most cases, our understanding of developments in material culture and settlement organization in specific areas of the Italian Peninsula benefits from regarding these areas as implicated in the dynamics of the Hellenistic world at large, a world in which the aggressive city-state of Rome formed merely one – though increasingly important – factor among many others (Purcell, 1994; Curti, Dench and Patterson, 1996; Keay and Terrenato, 2001; Dench, 2003; van Dommelen and Terrenato, 2007; on interpreting Hellenistic material culture cf. esp. Gallini, 1973; Zanker, 1976; Wallace-Hadrill, 2008). Such a shift of perspective is, at the least, a salutary counterbalance to overly Romano-centric approaches which long dominated the debate. Strong views of Romanization presented Rome as bringer of civilization in the Italic areas, promoting the formation of a politically, technologically and culturally advanced and unified Italy. In the case of Italy, they had their origins not so much in ancient Roman literary discourse or modern colonialism (particularly significant in other parts of the Roman Empire), as in the Romantic nationalism of nineteenth-century Italy and Germany (Terrenato, 1998a; Mouritsen, 1998). Some of the most influential historical works on "Roman Italy" were written in this period, paramount among which is Mommsen's *Römische Geschichte* (1854–6). In a much more explicit way than has long been appreciated in scholarship building on them, these works reflect specific contemporary preoccupations with the unification of highly diverse areas in Europe into nation-states, in which context the history of ancient Italy could be explored as a precursor or historical example. The most important strand of this discourse for our concerns here is the concomitant conceptualization of cultural change in Italy, which was presented as a gradual process of convergence under Roman guidance, predating but eventually leading to a politically and culturally unified whole. The general narrative thus created posited a neat sequence, in which the dissolution of the Iron Age "Italic peoples" preceded the emergence of a Romano-Italic culture promoted by Rome and her colonies, preparing the smooth establishment of Roman rule in the centuries thereafter. Such a conception and its chronology have perhaps been supported in part by the supposed fifth-century "crisis" of Italic cultures and the increased standardization of material culture during the mid-Republican period. But the framework itself is unmistakably of historiographic derivation, and observed developments in material culture have as a rule contributed little to it. The other way around, however, this general framework logically led to the attribution of many different social, economic and cultural developments in Italy in the last four centuries BCE to the rise of Rome in this period.

Thanks to important advances in both historical and archaeological research this state of affairs is now rapidly being corrected. Historiographical studies exposed the ideological roots and pervasiveness of the convergence model, which is a precondition for looking afresh at cultural developments in this period (Mouritsen, 1998; Bradley, 2002). Combined with the careful reconsideration of the development of Rome itself as part of the Italic Hellenistic world (on the "Romanization of Rome," see Curti, 2000: 90–1; Chapter 26), the self-evidence of Roman cultural dominance in Italy before the Social War has been fundamentally undermined. These developments tap into two broadly contemporary trends in archaeological work. First, archaeological field work especially over the last three decades has yielded an enormous amount of new data allowing better contextualization of developments, both regionally and in a wider Mediterranean perspective. Second, theoretical studies have increasingly pointed out the malleability of material culture, styles, and the meanings and associations attached to it, compromising any attempt straightforwardly to link certain types of material culture with specific and static groups or ideologies.

3 De-centering Rome

The correctives following from these and related insights have, over the last two decades, generally tended to de-center Rome from many developments in the Republican period. For instance, previous generations have insisted on Roman expansion and colonization as the cause of some of the most radical changes in rural settlement organization, for the Republican period notably the explosive spread of small farms in fourth- and third-century landscapes (see Chapter 13). But field survey data increasingly show that this development is actually part of a much wider Mediterranean phenomenon already in gear, that is, irrespective of Roman conquest (Terrenato, 2007; Attema, Burgers and van Leusen, 2011; see Chapter 12). Also, phenomena which have commonly been interpreted as indirect consequences of, or reactions to, Roman expansion can benefit from a broader perspective. The widespread construction of massive hillforts, for example, has in many Italian areas been related to specific moments of conflict with Rome (e.g. Oakley, 1995). This is not in itself improbable in particular cases, but the fact that the construction of such walled sites appears to be part of a much wider Mediterranean phenomenon related to general changes in modes and intensity of warfare puts it into perspective. Similar arguments can be made for the spread of other technologies and categories of material culture. At the moment that black-gloss pottery, originally conceived in Athens, is produced in several different areas, notably Campania and Latium, but equally in virtually every other region of Italy in locally specific forms, it requires special pleading to see the appearance of such crockery in general as an indicator of "Romanization" (Roth, 2007a; Chapter 5). Even villas, traditionally perceived as symbols of the Roman way of life *par excellence*, served very specific local and traditional functions, sometimes in fact perpetuating long-existing power structures dominated by the same elite families, as has notably been demonstrated for the Caecinae in Volterra (Terrenato, 1998b). Such findings concur with recent work on areas outside Italy, in which the Roman conquest is increasingly being regarded as much less decisive than other distinct socio-economic and cultural convulsions that were, in part, not causally related to Roman hegemony, and may indeed have enabled its spread in the first place (Wells, 1999; Keay and Terrenato, 2001; Terrenato, 2008).

In sum, many of the socio-economic and cultural developments in Italy in the (Mid-) Republican period can be characterized as momentous, dynamic, and often coupled with increased interaction and exchange, but not necessarily causally related to or centered on Rome. However, necessary and salutary as this revisionist line of reasoning is, there is a risk in turn of overly minimizing Roman impact. As regards historiography, demonstrating that the way by which previous scholars supported a certain view of historical developments was wrong does not automatically imply that that view itself is wrong, and must be turned on its head. Similarly, the multi-interpretability of material culture does not preclude that in certain historical contexts specific objects, styles and ways of living indeed did relate to Roman agency, and were meant to signal association with the new dominant power, or even document its blunt imposition. The most important realization, therefore, is that Roman political and cultural dominance is not self-evident – not that it was non-existent. A new perspective for analyzing specific historical situations has thus been opened up, which has suggested different emphases in research. To illustrate these points, we will briefly consider two issues; first, changed ideas on the actual mechanisms of cultural interaction between Italic and Roman people, and second, the creation of centrality in Rome itself by seizing, and internalizing, powerful symbols.

4 Cultural Interaction in Practice

One logical consequence of the re-thinking of the self-evidence of Rome as a dominant cultural factor is increased attention to the practicalities of cultural change. How and where did cultural interaction take place? Migrations, in the form of colonization, and, to a lesser extent, the army, have traditionally been regarded as key factors in the Romanization of Italy. Both fields are currently undergoing thorough revisions, in part stimulated by the above-mentioned intellectual developments. Manpower from the Italian allies was seminal for Rome's military success (Brunt, 1971). But the image of the army as a melting pot promoting allied identification with Rome turns out to be problematic (Pfeilschifter, 2007). To begin with, the levy and swearing-in of the soldiers, potentially integrative events, proceeded according to local community standards, not according to Roman standards, for Latin and allied Italic soldiers alike. Moreover, for reasons of practical administration and communication, the allied units were usually composed according to ethnic or regional criteria (thus, we know of a cohors praenestina, paeligna, etc.), and presumably usually addressed in their own language. Even later acknowledgements of allied contributions advertised in their home towns were in Oscan, as for instance a dedication by the Roman commander Mummius, captor of Corinth in 146, to the allied community of Pompeii (Martelli, 2002). Although shared experiences of Roman citizen soldiers and the soldiers from different Latin and Italic communities in victory and defeat in areas often far from Italy undoubtedly contributed to enhanced interaction and mutual awareness at some levels, this particular constellation of the army will equally have reinforced and shaped ethnic or regional sentiments. At the same time, it is quite possible that the logistical requirements imposed on the Italic allies led to increased or changed patterns in centralization in terms of settlements or other foci of social communication. That is to say, Roman agency had presumably considerable impact on various aspects of the lives of part of the non-Roman Italic population, but not in a traditional integrative "Romanizing" sense.

An example with a somewhat clearer material correlate regards colonization. Colonies established by Rome from especially the fourth century onwards have traditionally been viewed as key instruments in the spread of Roman ideas, models, and associated material culture *par excellence*. But this view is to an extent predicated on the assumption that colonies were small replicas of the city of Rome (Gell. *NA* 16.13.9), a view that is, as it now turns out, difficult to maintain even for the Triumviral and Imperial periods (e.g. Beard, North and Price, 1998: 331–4; see also Chapter 30). The character and rationale behind colonies in the Mid-Republican period is currently intensely debated (Bradley and Wilson, 2006; Stek and Pelgrom, forthcoming). The degree of their Roman-ness has been questioned, and a much more heterogeneous image emerges when the local archaeological and epigraphic evidence is taken as a starting point rather than later literary images (Bispham, 2006). Moreover, the indigenous contribution to colonial communities, though still debated, might have been considerable (Bradley, 2006). In terms of identity, locally constructed civic identities, as expressed in coinage, epigraphy and cults, will in most situations have been more relevant than their status as a colony related to Rome. In this respect they do not differ much from other civic communities in Italy.

Even their colonial persona was not necessarily centered directly on Rome. The most numerous and largest colonies had Latin status, and in certain historical circumstances these preferred to focus on an imagined shared Latin, rather than Roman, descent (people from Latium were often part of the new colonial populations, but underlining the Latin element has probably more to do with the new colonial configuration). Specific emphases in colonial panthea may be explained in this sense, in all their local variegation. For instance, a dedication at the Latin sanctuary of Diana at Nemi by a magistrate of the Latin colony of Ariminum on behalf of his community must be interpreted in this vein (*CIL* I^2.40; Cicala, 1995). If the role of colonies in the urbanization of Italy on a Roman model can also be contested for at least the Early and Mid-Republican periods, little seems to remain of the image of proud new Roman cities irradiating Roman culture over the peninsula. Many colonies were actually established in pre-existing cities, with many of the urbanistic details derived from contemporary Greek, not Roman, models (Sewell, 2010), and some new foundations may only have developed a proper urban aspect later in their history (Pelgrom, 2008). This is not to deny the significance of colonies in cultural and other developments in Italy altogether. But it shows that framing their role in terms of straightforward Romanization does not do justice to the various modes of cultural interaction at play. For instance, whereas the role of urbanism in the spread of Roman culture is being redressed, recent research points at the establishment of small rural communities integrated in the Roman administrative system (Tarpin, 2002; Capogrossi Colognesi, 2002). Interestingly, these communities tapped into new Roman ideologies at least as early as the colonial centers (Stek, 2009: 123–70). Such developments undermine the center–periphery models of cultural change normally underpinning narratives on the spread of Roman influence, and confound the neat contiguous geographical and territorial conceptualizations of Roman and non-Roman.

Another example related to colonization regards the debate on a special category of terracottas, the so-called Etrusco-Latial-Campanian group. As the name suggests, the group originates in these areas, but its subsequent appearance in other parts of Italy has traditionally been related to Roman colonization. It seems now that their static interpretation as "guide fossils" of Roman colonists and associated notions of the progression of superior Roman material culture are problematic and overstated (Gentili, 2005; Glinister, 2006). Yet it remains to be seen if their association with the newly established communities, closely connected in a dynamic network of urban and other communities in Italy and the wider Mediterranean, as such is to be rejected. It is rather their direct association with Roman ideologies that is presumably anachronistic. The complexity of such issues on the ground is only beginning to emerge, as a recent study on votive heads from a deposit in the territory of Vulci shows (Söderlind, 2002). Here, local Etruscan workshops deliberately adapted existing types, invariably uncovered heads, by applying hand-made veils. The secondary creation of a capite velato type must reflect the adoption of new ritual traditions in the region alongside the existing uncovered ritual, as uncovered types continued to be produced too. Whether this complex situation must be connected with the influx of colonists or broader changes or diversification in ritual practice is hard to prove, but the possibility that different social or cultural identities were played out in such a way remains distinct.

5 Seizing Symbols of Power

Rome itself was in this period developing fast in conceptualizing its newly acquired central position in the known world, a position created both physically, in the network of roads centered on Rome (Coarelli, 1988a; Laurence, 1999; for discussion on the dating, Pékary, 1968; Chapter 19) and mentally, in growing proprietorial attitudes to Italy and beyond (Purcell, 1990; Dench, 2005). This new position and ambition was increasingly reflected in the shaping of the material environment (Purcell, 2007a). In identifying such expressions, one should naturally not be looking for primordial and intrinsically Roman forms. But that does not mean that those implicated in the power game at Rome were not industriously looking for appropriate ways to express their superiority. For this purpose, common languages and symbols of success, victory and dominion available in the Hellenistic Mediterranean were mined. Perceived origins do not matter much at that point, or can, conversely, even be actively used to accentuate the power of incorporation. For instance, it would be perverse to argue that the Capitolium-type temple (three cellae, high podium) cannot have been a symbol of Rome, because Roman writers construed it as based on an Etruscan tradition. What seems to happen here is the successful adaptation of a widely existing, yet powerful, symbol and its subsequent promotion as an icon of Roman authority. How and when the model developed and spread is a different matter. Recent studies have rightly questioned the importance, and very existence, of Capitolia in the earliest colonies – another case of the projecting back of later literary images to earlier colonization (Bispham, 2006). But just as significant is the actual establishment of the model, beginning at least in the early second century, and its gain in popularity in the latter half of the first century and beyond, in line with different historical contexts within which both allegiance to, and centralizing ambitions from, Rome were growing (Purcell, forthcoming).

Just as Rome, other Italic communities were in search of new and powerful ways to express themselves, and this explains competing claims and shared symbolic languages. In general, the models were ultimately derived from Greek Hellenistic examples, but could be recast to serve diametrically opposed purposes. This process can be seen already before Rome became a factor of significance, famously in the Lucanian tombs of Paestum (cf. Chapter 24 and Figure 24.3). Here, the Greek-Barbarian dichotomy is effectively turned around, with a Lucanian warrior cast as the victor (Rouveret and Greco Pontrandolfo, 1983: 121; Greco Pontrandolfo and Rouveret, 1983; Curti, Dench and Patterson, 1996: 184). Powerful as they were, these south Italian pictorial traditions were to have an important influence on Roman visualizations of victory, and perhaps not incidentally started to appear in our record after the capture of these areas (Nicolet, 1962; Holliday, 2002). For the process in the mid-Republican period, the concept of Victoria might be emblematic. Ultimately inspired by the Greek notion of Nike, it was a relatively new cult in Rome, receiving a temple only in 294 after a victory over the Samnites. After that, Victoria and related concepts of success and valor became very popular and quite certainly associated with Roman power, as Victoria's appearance on mid-Republican silver coins, victoriati, underlines (Hölscher, 1967; Fears, 1981). In all its architectural distinctiveness and cosmopolitan eclecticism, it is quite clear in what discourse the "resistant" Samnites, the Pentri placed themselves with their late second-century temple at Pietrabbondante, with its three-cellae, high podium temple and dedication to *Vikturrai*. As with the Capitolium temple, the layers of meanings attached to these cultural models is a given, but that does not make them less powerful when applied at the right moment at the right place.

6 Italic Identities: Beyond Ethnic Regionalism?

De-centering Rome has thus been an important and productive shift in approach for Republican Italy, as long as care is taken not to throw the baby out with the bathwater. Less clear, however, is if one should instead focus on Italic ethnic groups or regions associated with them as the basic units of analysis. Regionalism along ethnic lines has a long history in research on ancient Italy, which can in part be connected with the fertile ground it has found in contemporary societal developments – most recently in modern discourses of regional particularism, in reaction to EU or central Italian government policies, before that in the nineteenth- and twentieth-century erudite humanism of especially south Italy, while its modern roots reach further back to the Etruscologia of the seventeenth century (Torelli, 1999: 1–2; for the analogous issue of methodological nationalism for the provinces cf. Woolf, 2004; Chapter 27). Traditionally, this separate branch of studies has had limited impact on conceptualizations of the Romanization of Italy, because their chronological focus was, not incidentally, usually placed before the advent of Roman expansion. Simply put, the process of subsequent change has thus traditionally been rendered as an equation in which Italic sentiments and customs are supposed to diminish proportionally to the adoption of new ways of life, that is, the substitution of peculiar Italic identities with a Roman one (Brunt, 1965; Toynbee, 1965; Keaveney, 1987: 21). That such a view of (ethnic) identity is not merely modern is proven by the observations of the Cappadocian Strabo, who wrote a historical geography in the Augustan period:

> But the [Leucani], and the Brettii, and the Samnites themselves (the progenitors of these peoples) have so utterly deteriorated that it is difficult even to distinguish their several settlements; and the reason is that no common organization longer endures in any one of the separate tribes; and their characteristic differences in language, armor, dress, and the like, have completely disappeared; and, besides, their settlements, severally and in detail, are wholly without repute ... The Leucani are Samnite in race ... But now they are Romans.
> (VI.1.2–3, trans. H. Jones, 1929, Loeb)

There are similar accounts for other peoples throughout Strabo's work. The problem here is the unitary, and therefore untenable, conception of identity, presenting the process as that of communicating vessels. Whereas we must be careful not to confuse fluidity and flexibility with total freedom in choosing one's role with respect to others, the situational nature of identity is indubitable. Ironically, one usually has to look hard for unambiguous expressions of ethnic identities in the diverse Italic areas precisely until the start of Roman interference in these areas. References to Italic ethnic designations can be found well into the Imperial period. Any temptation to link this with the recent reconsideration of Roman impact to suggest that little changed must, however, be suppressed. Indeed, emphasizing in turn the continued importance of Italic identities after Roman incorporation poses its own problems, especially when presented in terms of persistence, as will be seen below (cf. section 9 below).

7 Archaeology and Ethnic Identity

To start with, the notion and usefulness of ethnicity has, even for the perceived apogee of the Italic peoples in the Iron Age and classical period, met with major theoretical and methodological challenges. Anthropological, historical and archaeological research over the last few decades has shown how deeply problematic is the connection between peoples mentioned in the (mostly later) Greek and Roman sources and archaeological cultures distinguished by modern scholars (cf. Chapter 25). This regards its culture-historical theoretical premises, failing to acknowledge that certain assemblages of material culture are but one of many indicia possibly signaling ethnic feelings (e.g. Shennan, 1989; Hall, 1997, 2002). They may equally relate to other expressions and historical processes, such as other levels of congregation, patterns of trade, fashion, and so on, which blur ethnic distinctions (see Torelli, 1995 for examples). Objects and styles can start a new life, and can be imbued with new significance in different contexts. The localized process can be described as cultural *bricolage* (Terrenato, 1998a); that is, the changing roles of the object can only be understood by studying its cultural biography (Kopytoff, 1986). Related to these problems is the high risk of circularity in linking material assemblages and ethnic groups, as identification of a specific material assemblage is subjective and highly dependent on what types of evidence are privileged. Even arguably "emblematic" types or styles raise considerable problems of interpretation, as an example from south Italy illustrates. In the so-called Messapian area, female burials were characterized by the positioning of a specific pot with high handles decorated with wheels or *trozzella* in the grave. Originally an Iron Age form developed in the eighth century, it continued to be made until at least the third century, exclusively for funerary use. But whether it signaled a "Messapian" identity, or rather relates to a localized form of gender or other social identity, or combinations of all these facets, can only be guessed. Incidentally, it is worth noting that the pot itself is the result of *bricolage*, as it draws on Greek technologies (the use of the wheel; slip) and fashion (decorative motives), which are blended with a native form and arrangement of decoration (Yntema, 1974).

Further complicating our recognition of ethnic expressions is its profound volatility. Ethnic identity is highly political, and therefore extremely sensitive to changing historical circumstances. By definition (or by most definitions, at least), a shared past is essential in the construction of ethnic identity, but the way in which it is presented is profoundly more telling about contemporaneous concerns than that evoked past. This means that accepting the long lines of continuity drawn through time and space as intentionally presented in ethnic expressions must be avoided. For instance, in the second century a "Samnite" (*safinim*) group identity is being asserted in modern Molise (Vetter, 1953: 149; Chapter 25). It apparently builds upon the prior existence of a *safinim* identity that is attested epigraphically already by the sixth century in modern Abruzzo further to the north. But rather than indicating the stable and everlasting presence of an ethnic group over such distant geographical areas, the second-century referencing relates to contemporaneous claims to a distant and heroic past, quite possibly as part of the antagonizing political climate in reaction to Roman dominion in this period (Dench, 1995: 198–203; Tagliamonte, 1996: 7–13). These points show that, at least from an archaeological point of view, ethnic frameworks are not exactly the most self-evident path to follow, unless we are willing to abandon now commonly accepted definitions of ethnicity (e.g. Morgan, 2009; Antonaccio, 2009). In light of the essential fluidity of ethnic identities over time and space, ethnic regionalism is something of a contradictio.

8 Asserting Local Distinctiveness in New Forms

More important than noting difficulties in recognition, however, is asking how important ethnic identity was *vis-à-vis* other levels of organization or congregation. Posing our historical narratives in ethnic terms, thereby properly following the ancient authors, sometimes obscures the fact that most evidence–epigraphic, numismatic and archaeological–actually points at smaller (and occasionally at larger) communal identities as worthy of monumental expression as the ethnic groups known from literary descriptions (for Etruria, see Chapter 23). Thus, urban communities from south to north generally signal their civic identities by displaying their name on coinage, by worshipping polyadic deities, and sometimes claiming their territories by *necropoleis* or extra-urban sanctuaries (e.g. Zifferero, 2002). Although there is evidence for coinage being used for larger, presumably ethnic groups at specific historical moments, such as, for instance, in the Lucanian case (*loukanom*, discussed by Isayev, 2007b: 9–54, 2009; *safinim* and *italia/víteliú* on coins, discussed by Pobjoy, 2000a), we do not normally find "boundary markers" of ethnic or tribal territories, neither in urban nor in non-urbanized areas. Rather than a result of modern inabilities to recognize such boundaries, this questions the very existence of, and surely the importance attached to, expressions of fixed boundaries in these terms. As with Roman colonial territories and other administrative categories, the maps indicating peoples and statuses to which we are so accustomed appear to represent profoundly anachronistic ways of thinking.

This does not mean that ethnic identities were insignificant. These clearly emerged in various periods in relation to internal structuration processes, their ultimate causes being increased population densities and increased contact with other groupings, be they fellow Oscan speakers, Greeks in the south or Romans in the west. But we most certainly risk overemphasizing the importance of a level of aggregation that presumably only in certain circumstances came to the fore, at the cost of other, more evident levels of communal organization and related identities (Bradley, 1997; Bradley, Isayev and Riva, 2007). The ways in which emphasis on the ethnic or "state" level has obscured our understanding of other levels of socio-political organization and its cultural significance are several, both in less urbanized and urbanized areas. One example of the former, with quite penetrating historical consequences, regards the internal areas of central-southern Italy, or "Samnium," itself an ethno-geographical concept presumably developed first in a Roman context (*Samnio* appears in the elogium on the Tomb of the Scipiones, *CIL* I^2.6–7, dated to the mid-third century; see discussion on the probably adjectival form *safinim* in Rix, 2002). Whereas historical studies are rapidly dilapidating previous conceptions of the Samnites as a unitary state (in different ways; Letta, 1994; Cornell, 2004; Senatore, 2006), with all its consequences for our understanding of the Roman conquest of Italy, archaeological research increasingly points towards local variegation in material culture and practices (for instance in burial practices: Scopacasa, 2009; Chapter 25). This warrants more attention to local communities in their own right – all the more so since the relationship between such communities and higher levels of political organization is much less evident now than it was before the rejection of the main model of Italic political organization, the pagus-vicus system (Capogrossi Colognesi, 2002; Tarpin, 2002; F. Russo, 2003; Stek, 2009: 107–21). The evidence may show that different communities only organized themselves in the face of external threat.

In the more urbanized areas the importance of local civic identities is even more evident. The case of Latium is illustrative. The area around and east of Rome inhabited by Latin speakers consisted in the first place of different urban communities. These shared a common material culture nowadays called *cultura laziale*. Yet there seems to be no reason to assume that any political organization existed along ethnic lines (Cornell, 1997, 2000b). The shared rituals at the common sanctuary of Jupiter Latiaris on the Alban Mount rather served to establish power relationships between the different urban communities, which are proudly listed (Plin. *HN* 3.69; Livy XXXII.1.9, XXXVII.3.4). Difference in unity seems also to be expressed in the 13 different altars in the sanctuary of Lavinium, probably reflecting different Latin communities. Presumably, Latin speakers could identify with each other in the face of external threat (at least on one occasion apparently against Rome: Cato *Orig.* 58 P) and it is precisely such processes that allowed Rome to gain a dominant position in the first place (Cornell, 1995: 293–301).

Otherwise, however, the archaeological and historical evidence for the communities in Latium primarily points at the expression of local civic pride, all the more so in the Late Republican period (Wallace-Hadrill, 2007, 2008). The famous urban and extra-urban sanctuaries built in this period by different cities in Latium, but equally in other parts of Italy, are best interpreted as efforts by local communities, and their leading elites, to put themselves on the map (cf. e.g. the local senate and censors mentioned for the sanctuary of Fortuna Primigenia at Praeneste: Degrassi, 1969, or the quattorviri involved in the construction of the Temple of Hercules Victor at Tibur, *CIL* I^2.1492). The relationships between such building projects must therefore surely be seen in terms of peer polity interaction, and particularly competition, rather than as expressions of a common Latin identity. In terms of material culture, these sanctuaries are radically innovative, both in the adoption of new construction techniques (notably opus caementicium) and the application of architectural notions and models freely inspired by contemporary architecture in the Mediterranean east. At Praeneste, this is visible in the scenographical layout with terraces and ramps, creating an imposing sacred landscape over the town, as well as in many of its architectural decorative details (see Chapter 17). It probably incorporates an older cult place, as is suggested by the use of opus quadratum for the oracular pit on the hemicycle level as well as its position (Coarelli, 1987b: 50, 72–4). The same mechanism met throughout this chapter, that of expressing local difference in newly acquired material forms, incorporating new techniques as well as real or perceived traditions, applies here. For the contemporary Praenestini (who, incidentally, did not have Latin status) and presumably most other observers, the new magnificent sanctuary will have stirred up many associations, but most of all civic pride and achievement. If the sanctuary ever evoked a shared Latin identity, it must have been later, in a Roman Imperial context.

9 Questioning Continuity: Roman Imperialism and Italic Identities

After the Social War, all free inhabitants of the peninsula became Roman citizens, and were now also officially politically integrated into Rome. The first century has, in many ways, been seen as crucial in the integration of Italy, in socio-economic, cultural and notional respects. It is now that a notion of "Italy" as a whole, which had been developing since the third century, becomes fully articulated and indeed seems to culminate with Augustus's tota Italia. It has been observed, however, that the traditionally envisaged "end of the Italic peoples" in this period must be reconsidered, pointing to the vitality of Italic ethnic designations far into the Roman Imperial period (e.g. Giardina, 1997; Bradley, 2007; Chapter 23). The question is what such evocations in a still more uneven power balance mean, in terms of self-perceptions and cultural strategies by both those in power and those not in power. The earlier noted flexibility and topicality, in both temporal and spatial terms, of ethnic designations gains special significance in Roman Imperial discourse.

Certainly, outsiders' designations of ethnic groups had of old played a role in the formation and shaping of ethnic identities, not only in the common oppositional mechanism but also in terms of auto-identification with etic labels (e.g. Hill, 1996). Such processes formed an essential part of ethnogenesis in Italy, where ethnic identities were often modeled on dominant Greek ways of seeing, and the same mechanism applies to the period when Rome became dominant. In the Transpadane area, for instance, Roman interference may have generated the first instance when the people living there saw themselves as Galli. Indeed, the ethnic categories of Insubres and Cenomanni, which persisted until the late first century, presumably developed through a process of auto-identification with what were essentially Roman categories (Williams, 2001: 214–18). An early example of Rome's role in the formation and/or expression of ethnicity may be the Volscians of the Pontine area (Gnade, 2002), and, as discussed above, in second-century Molise political mobilization against Roman pressure may have catalyzed a process of ethnic construction. Also in the case of Umbria, the Roman conquest may have enhanced, rather than weakened, expressions of regional or ethnic identity (Bradley, 1997, 2000a: 190–269). But Roman agency may actually have gone much further, as notably Dench has shown (1995, 2005). The key issue is that specific and variegated images of the incorporated peoples of Italy were of direct importance to developing Roman policies and Roman constructions of identity in relation to others. In this way, peoples previously depicted as dangerous and anti-Roman, such as Samnites and other central Apennine peoples, could evolve, after their incorporation and useful employment in the Roman armies, into brave and frugal role-models for a Rome that was perceived as morally degenerate. In the process, a whole new ethnic was coined, Sabellians, to evade negative connotations and forge a connection with the ancient morally upright Sabines (Dench, 1995). Actually, the "Samnite" did survive, but in a profoundly perverted way – the adjective became a specific type of gladiator, attested already under the Republic (Lucil. 149–52 M). This could lead to the interesting situation reflected in an inscription from Cádiz (*Inscripciones Latinas de la España Romana* 5690), where a man of Greek origin (natione Graeca) with a Latin name (Germanus) defines himself as "Samnis" (Caldelli, 2001).

Pushing this line of enquiry further, the continued importance of evoking Italic ethnic identities throughout Late Republican and Imperial times can be related to the development of a plural notion of Roman identity, in which Italic difference could offer an attractive model for later incorporations and related mechanisms of self-positioning (Dench, 2005). The variegation of Italy provided, in other words, a point of reference for further imperial projects and this may be one of the key reasons for its continued existence. The question is, then, how such discourses relate to ethnic or other local identities on the ground. Most self-references to Italic ethnicities or localities come from individuals thoroughly entangled in the Roman power game, in Rome itself. This not only accounts for the well-known evocations of their non-Roman Italian roots by Late Republican and Early Imperial writers and poets (Gasser, 1999), but also for the numerous ethnic names adopted or promoted by influential families (Farney, 2007). The specific class context of such evidence can hardly be overstated, and it is good to remember that the often-evoked meditations on the conciliation of different loyalties, such as in Cicero (*Leg.* 2.5), or earlier indeed in Ennius's "tria corda" (Gell. *NA* 17.17.1), are quite exceptional. Imperial period references to ethnic identities pitched at local Italian audiences bear, perhaps not surprisingly, no recognizable relationship to continued archaeological regional variability, but they do point to the dominance of Rome in the negotiation. An inscription put up by the inhabitants of Superaequum, a small community in what is nowadays Abruzzo, seems to encapsulate some of the processes at work: Q. Varius Geminus, apparently native to the area, is being honored by the Superaequani for his achievements in Rome. Notably, he is "the first of all Paelignians" to have become senator. The phrasing demonstrates firm reference to a Rome-dominated power structure, the existence of something of a Paelignian identity as a relevant frame of reference still in the first century CE, but equally competition between different Paelignian communities and indeed local Superaequan pride (*CIL* IX.3306). Only from a different perspective can conscious Roman strategies be recognized in the antiquarian revivals of ancient Italic festivals, notably under Augustus. The ludi saeculares revived in 17, for instance, highlighted the incorporation of the Latins (and not other or all Italians) because these, subjugated and incorporated successfully already in the fourth century, provided an appropriate role model for the empire as a whole (Cooley, 2006: 230–7). The Rescript from Spello documents the celebration of "ethnic" festivals in Umbria and Etruria as late as the early fourth century CE, probably in a form conceived under Augustus (Coarelli, 2001). These examples show that the extent to which Italic ethnic designations can be confounded and indeed manipulated cannot be underestimated, and that the creation of the Italian regiones by Augustus with its probably unofficial yet suggestive ethnic labels is merely one of the clearest examples of the forms that such reconfigurations could take (Nicolet, 1991; Laurence, 1998). From this perspective, the transformation of the ethnic Latin to a legal status within the Roman administrative organization, detached from any geographical and ethnic reference points, can be seen as just one particular expression of a much broader phenomenon in which ethnic designations are employed to conceptualize the new order in terms both of geography and different sorts of statuses. The shift from an ethnic to a legal definition of Latinity was only possible in a Roman imperial framework – and so was its signaling through the quotation of the material forms of the civic sanctuaries of Praeneste and Tibur in the Baetican cult place of Munigua.

Acknowledgments

This research has been supported by the NWO (Netherlands Organization for Scientific Research) and Brasenose College, Oxford. I am grateful to Ed Bispham, Marijke Gnade, Eric Moormann, Jeremia Pelgrom and Miguel-John Versluys for discussion and comments on previous versions of the text. I should like especially to thank Marijke Gnade and Eric Moormann for their incessant confidence and support.

FURTHER READING

Historical overviews of Italy in the Republican period in relation to Roman history can be found in the *Cambridge Ancient History* (esp. Purcell, 1994; Cornell, 1989; Crawford, 1996; on the Samnite wars see now Cornell, 2004). Important contributions/introductions to the debate on the Romanization of Italy in English include Curti, Dench and Patterson (1996), Keay and Terrenato (2001), Dench (2003) and Bradley (2007), and part of Torelli's fundamental work is accessible in English (1995, 1999). For the general theoretical framework, see, for example, Hingley (2005, 2009). The historiographical background of the convergence model of culture change is discussed in detail in Mouritsen (1998). On Hellenization, its relation to Roman hegemony, and its local meaning, see Gallini (1973), Zanker (1976) and Wallace-Hadrill's work (most recently Wallace-Hadrill, 2008). Further key contributions on socio-cultural change in this period include *Modes* (1983), Mertens and Lambrechts (1991), Herring and Lomas (2000), Jehne and Pfeilschifter (2006), van Dommelen and Terrenato (2007) and Roth and Keller (2007).

More specifically on changes in settlement patterns, see, for example, Barker and Lloyd (1991), Lo Cascio and Storchi Marino (2001), Attema, Burgers and van Leusen (2011); on the revision of the pagus-vicus system, see Tarpin (2002), Capogrossi Colognesi (2002), discussed in Russo (2003) and Stek (2009); on colonization, see Crawford (1995), Fentress (2000), Bispham (2000), Bradley and Wilson (2006), Sewell (2010) and Stek and Pelgrom (forthcoming). On migration and demography, see, for example, Scheidel (2004) and de Ligt and Northwood (2008); on the socio-cultural effects of migration, see, for example, Pina Polo (2006). On important linguistic aspects, see J.N. Adams (2003, 2007); on religion and cult places, see, for example, Torelli (1999), Bispham and Smith (2000), Schultz and Harvey (2006), de Cazanove (2007) and Stek (2009, in press). On the municipalization of Italy, see Bispham (2007a); on the relationship between Italic identities and Roman imperialism, see Dench (1995, 2005).

For regional studies see, in Italian, the *Biblioteca di Archeologia* series of Longanesi (e.g. Cerchiai, 1995, Tagliamonte, 1996), and, in English, for example, Harris (1971), Dench (1995), Bradley (2000a), Isayev (2007b), Lomas (1993), Holloway (1994), now complemented geographically by the essays in Bradley, Isayev and Riva (2007), and, of course, the chapters in this section and their Further Reading sections.

PART IV

The Archaeology of Identity

CHAPTER 23

The Importance of Being Elite: The Archaeology of Identity in Etruria (500–200)

P. Gregory Warden

1 Introduction

How do we define identity, especially in terms of a population, the Etruscans, that flourished for almost a thousand years, a group whose identity has been so often interpreted and reinterpreted by others in the succeeding two millennia? The study of identity is now in vogue, and the term is freely used, but the search for identity has its own dangers if that term is not used carefully and with rigor, and the term is especially precarious when used in the singular. In the case of the Etruscans, can we really speak of a single identity? Identity is a construct that is both mutable and fluid, that reacts to both internal and external sources, highly subject to change. Identity is by nature complex: "What is important is that there is now wide agreement in theoretical literature on the constructed and fluid nature of ethnic groups throughout human history … It is increasingly apparent that any attempt to identify a particular historical point at which an ethnic group becomes definitively formed is misguided" (Bradley, 2000b: 113). At any particular point in time and in any specific place multiple identities will exist. This is especially the case with the Etruscans, a culture that is archaeologically and historically traceable for almost a millennium. In addition, Etruscan material culture is highly varied, seemingly very regional in nature, testifying to a complex interaction of the Etruscans with other groups (Greeks, Romans and Celts, to name just the more obvious ones). The notion of a singular Etruscan identity must give way to interpretations that factor multiplicity, hybridity and complexity. As Isayev (2009: 223) asked in a different context, "Were they or did they think of themselves as Lucanian? The simple answer to this would be: some of them, at certain times, and depending on who was asking." And to take this scenario one step further, whose point of view is most useful in ascertaining identity, the person who answers the question, or the person who asks it? Identity is made up of both a group's

A Companion to the Archaeology of the Roman Republic, First Edition. Edited by Jane DeRose Evans.

perception of itself, the emic, and a group's perception by others, the etic. These two aspects are not unconnected, and rather than seeing them as opposites, we might think of them as a continuum, where the emic perception of self is communicated through actions or through objects that then define etic identities as perceived by others. In this sense identity is not always defined by conscious acts (Hodos, 2009: 3), an important consideration in trying to define the identity of a population in the remote past, especially in the case of the Etruscans. Identity as perceived by others may be no less useful, perhaps even no less connected to cultural realities, than a group's own perception of self: "the field of experience should include identities that might contradict what the actors would have *said* about themselves, sometimes called the emic view" (Antonaccio, 2009: 50). The evidence for what the Etruscan elite said about themselves is scattered, but it does point to an identity that changed over time. These changes are reflected in the elite Etruscans who ruled Rome, and who were in turn changed by the Romans they encountered.

2 Etruscan Identity, Material Culture and Social Ritual

External sources have dominated the historiographic view of the Etruscans, and both Greek and Roman sources are by now well documented, especially in the areas of religion and origins, even if the etic nature of the evidence is not always fully acknowledged. The reliability of these sources is subject to the same issues as were outlined by Herring (2000: 48–9) for southern Italy. The question of origin is by now moot, at least archaeologically; the scholarly consensus is that Etruscan culture was clearly formed on Italian soil. But questions of origin also connect to issues of ethnicity that are now being brought up again through the study of DNA (Perkins, 2009: 95–111) and especially, and perhaps even more interestingly, in the question of how ethnic identity might have been determined by language. Both language and literacy in early Etruria have received scholarly attention of late (Wilkins, 1990: 53–72), as have other important questions such as the relationship of Etruscan to other Italic languages (Penney, 2009: 88–94), the transition from Etruscan to Latin at sites like Tarquinia (Kaimio, 2010), or the disappearance of the Etruscan language in the later Hellenistic period (Adams, 2003).

Because of the scarcity of historical and epigraphic evidence, reliance on material culture and visual representation is fundamental to gaining an understanding of the Etruscans. Material culture expands tremendously in the seventh and sixth centuries and is continuously abundant through the second century. Most material culture is known from funerary contexts, with a much smaller amount from sanctuaries or temples. Although increasing evidence is now emerging from settlements, the amount of information given by settlements is still relatively minor, albeit perhaps more important because of the light shed by the material culture of the settlement on hitherto less visible segments of Etruscan society. Our understanding is therefore highly skewed, and the nature of this evidence has created an inordinate focus on "princely" culture, an ill-defined term that focuses more on objects than contexts. This terminology is especially problematic in trying to interpret identity: "What most clearly defined identity was what was done with the artefacts, not what artefacts an individual possessed" (Mattingly, 2009: 287). Aspects of patronage, production, artistic identity, or even the diffusion and perception of style are important but not always clear, and related to all of these is the question of how these objects might have functioned in the self-definition of the Etruscan elites that used and displayed them. Part of the problem may be that traditional archaeological taxonomies have failed to address questions of function and process, and a productive approach might be to look at the way types or classes of objects would have reinforced status and rank in the context of elite display. Jewelry, personal ornament and costume, as imaged in Etruscan art and found in material culture, have been productive areas of research, especially the characteristic types of Etruscan objects that had wide cultural diffusion, for instance bullae (Warden, 1982: 69–75; Palmer, 1996: 12–28), textiles (Gleba, 2008), or specific types of jewelry. In the latter case Etruscan diversity seems to have given way by the third century to uniformity based upon pan-Mediterranean models (Castor, 2010: 42).

Etruscan archaeology's reliance on material culture is tainted by the processes of selection, processes that we really do not fully understand yet: "burial evidence also can be, by its very nature, idealized, conservative (with inclusion of heirlooms), and not necessarily representative of the real population" (Isayev, 2007a: 8). Tomb assemblages were not intended to replicate life exactly. They reflect choices made by either the deceased or the family or clan, choices that on one level may have been idiosyncratic but that must also have been culturally normative. Some of these norms can be easily identified and remain unchanged throughout Etruscan history: the symbolic preservation of the elite body; the recreation in funerary contexts of elite domestic settings, the house or palace, especially its interior; and the imaging of social rituals that reinforce status and rank, most notably the banquet. This interest in the elite individual in the context of a broader social landscape continues throughout Etruscan history. In this sense Etruscan funerary custom remains conservative, even though the type and style of objects included in tombs may change over time, and even if the mortuary contexts themselves may be quite differently configured. The body, for instance, is first imaged or "preserved" through urns that are treated as bodies (covered with helmets, clothed in fabrics, or even adorned), just as the elite body in later periods is expressly imaged on sarcophagi and urns that bear the image of the deceased. The house or home is imaged first as a hut and then more elaborately in later periods by tomb interiors in the form of houses or palaces, as in the Tomb of the Reliefs at Cerveteri with its palatial layout and walls festooned with varieties of domestic and personal possessions. The banquet is ubiquitous and is first represented by the inclusion of banqueting paraphernalia in tombs or in actual depictions of banqueting (Tuck, 1994). From then on banquets are imaged in sculpture, in tomb painting, even in the way that tables with food are sometimes placed in front of the deceased. When imagery is this rampant and diachronically robust, it is fair to conclude that it is an important indicator of identity; the overarching theme is of the perpetuation of the rank and status of the elite Etruscan as imaged through body, building and social ritual.

3 National, Ethnic and Civic Identities

The highly regional nature of Etruscan culture as reflected in art and material culture may have resulted from Etruscan political organization, or lack thereof. The development of identity at levels broader than clan or tribe is usually linked to the processes of urbanization and state formation, to the creation of city states. But even though we do have good evidence for the physical aspects of Etruscan cities, what do we know about the way that they functioned? To begin, we have scant evidence for what constituted citizenship. There doubtless would have been some sense of civic identity; the material evidence itself makes this evident, but how exactly did the urban or state identity intersect with other identities, the self-awareness of individuals, clans, tribes, or even the Etruscans as a nation or ethnic group? State formation and identity may often be connected, but "it is not inevitable that any society, which is subject to the conditions which might foster state formation, will develop the features of a state society" (Herring, 2000: 73, n. 4). The Etruscans, at least as reflected in their later, admittedly loose political organization, were an urban culture or at least a culture organized in urban groups. Ritual definition of Etruscan sacred space would have included urban spaces (Briquel, 2008). The vestiges of Etruscan cities, strongly walled and surrounded by the vast cities of the dead, are still among the most impressive physical remains, even if we know little about their size and actual organization, both factors that would have changed over time. There have been debates about early Etruscan urbanization, well summarized by Riva (2009b: 1–13), and there is clear consensus that later Etruscan cities may have functioned, at least from an etic, Roman point of view, as city-states, but much work needs to be done to determine the nature of these city-states. The Etruscan examples (Torelli, 2000) will need to be studied in a wider Mediterranean context, as has been pointed out by Riva (2009a: 191). But what did civic identity entail, and how did it relate to other negotiated identities? Material culture certainly suggests regional variation (as was first convincingly articulated by Banti, 1973), and the intense regional quality of Etruscan art is now a topic even in introductory texts. The differences in the material culture of major urban centers in southern Etruria has even been proposed recently as an argument for the distinct identities of cities like Vulci, Tarquinia and Caere (Torelli and Moretti Sgubini, 2008), but this is a case of a very loose definition of identity. That different urban centers would have their own typologies of material production is not especially surprising, and even so, different types of objects do not necessarily translate into different identities. This kind of typology-based definition of identity ignores the issues of how intense interaction between distinct groups would have been negotiated by different interests, individual, family, community and city-state (Häussler, 2000: 151).

Better evidence for civic identity is provided by Etruscan coinage (Rutter, 2001: 23–42), which is quite late in date, mostly third century or later, and whose attribution to specific centers is primarily the result of analysis of archaeological find-spot and hoard data. These coins were minted in extremely limited quantities, with the exception of the issues of Populonia, a city that produced silver coins in the Hellenistic period on a scale not otherwise seen in Etruria. In fact, the majority of coins circulating in Etruria were non-Etruscan coins, mainly from Rome or south Italy (Catalli, 1988). What is notable is that even though these were civic issues, civic identification either by legend or by characteristic image type is rare. Circulation also seems to have been extremely limited. The inference is that the Etruscans did not adopt a fully developed monetary economy until quite late in their history, so that "foreign" coins were sufficient to meet economic needs until the period of Roman domination. Judging from the types and legends, there does not seem to have been a compelling interest in civic identification, a situation that is quite different from Rome, where civic identification is clearly articulated in coins from the third century onward. All this could be attributed to an economic and social

Figure 23.1 Helmet of Negau type from Vetulonia, ritually crushed. Source: Photo courtesy of the Florence Archaeological Museum, by permission.

organization where family and clan identity were more important than urban or civic identities, a hypothesis supported by other evidence. Etruscan military organization, for instance, may have been more familial than civic; armies were often organized by clan affiliation (H. Becker, 2007). Prime evidence would be the 125 helmets of Negau type inscribed with the name of the Hapnas clan, ritually mutilated and interred outside the walls of Vetulonia in the fifth century (Figure 23.1) (Warden, ed., 2008: 274). The dominance of Etruscan clans in the life of the city would also have been mirrored programmatically in sacred architecture, for instance on the Talamone pediment, discussed below. This makes sense as temples are primarily urban phenomena and important economic agents (H. Becker, 2007). The temple/sanctuary, like the city, is humanly defined and is a consecrated space that reflects a sacred order. Other forms of urban architecture, such as the city gates at Volterra and Perugia, combine civic symbolism with clan identity by including sculptures portraying leading citizens. On entering Perugia through its second-century gate, the Porta Marzia, a visitor would literally have been greeted by images of citizens frontally placed and gazing down from a colonnaded parapet (Figure 23.2). The scenographic message is quite clear, as individual citizens, plausibly representing some of the major clans of the city, were placed high above in high relief in a place of privilege, serving as almost heraldic symbols of civic identity. There is an ambiguity of vision, an uncertainty about who the spectator is and who is being viewed, alongside an absolute certainty about the hierarchies of power, as the Etruscan worthies gaze down on the throngs below. It is tempting to hypothesize that these relationships might have in fact represented the negotiated boundaries of Etruscan civic identities.

Figure 23.2 Porta Marzia, Perugia. Source: Photo by P.G. Warden.

There is the also the question of national identity. Did the political structure of the Etruscans, based on a loose organization of urban centers that were often at war with each other but also connected by reciprocal agreements and even ties of clan and family, produce a sense of national identity? The Etruscans as a whole, if not as a nation, are better defined by external sources, for instance, the Greeks, who saw them as political and commercial rivals and from whom we gain a picture of Etruscan mercantile and maritime interests that is rarely explicit in the archaeological record. Roman sources are more plentiful and produce a more complex picture, and the Etruscans can be characterized as a group in cultural terms, in Livy's famous reference to the most religious of peoples (V.1.6), or in Seneca's summary of the causality of Etruscan belief (*Q Nat.* II.32.2), but not in political terms. Insight into what may have mattered most to Etruscans in terms of identity, at least in the Archaic period, is provided by Livy's account of Etruscan intervention into Roman politics at the bequest of Tarquinius Superbus, whose "appeal to the people of Veii and Tarquinia was that he was 'of the same blood' (*eiusdem sanguinis* – Livy 2.6.2); the Tarquinians were moved by the parading of his 'name and kinship' (*nomen et cognatio* – 2.6.4); to Lars Porsenna his claim is one of 'name and blood' (2.9.1)" (Rasmussen, 1997: 29). Kinship seems to have been a primary factor, but the reference to blood ties suggests a possibly greater identity. Roman sources, however, often refer to the Etruscans as a whole as the "peoples" of Etruria, reflecting the loose political organization of the Etruscans as a federation of city-states connected more by religious and cultural, rather than political, ties. This organization, which in the later period manifested itself as the duodecim populi (Torelli, 1985), was in fact reproduced by the Etruscans when they expanded outside their original homeland into Etruria Padana. Nor is there much evidence for an Etruscan sense of homeland, of a geographical identity. Admittedly Roman sources (Strabo V.2.1–9; Plin. *HN* 3.50–5) described Etruria quite explicitly, but they describe Etruria of the Roman period rather than a demonstrably Etruscan homeland. Also questionable is to what extent the Etruscans conceived of themselves as a single entity. Even if the Etruscans were characterized as a group by Greeks and Romans, did they have a conscious group identity? There are more modern ethnographic examples of peoples that do not have names for themselves (Renfrew and Cherry, 1986: 156), and this may even hold true for the Etruscans, even if we assume that they called themselves Rasna or Rasenna, a term that may simply denote "us" rather than being a specific ethnic determinant. In any case, language is not of much help: "Nor is it possible to establish what part language played in the Etruscans' conceptions of their own ethnicity: there is simply no evidence" (Penney, 2009: 88).

4 Modern Mythologies of "-ization," Greek, Roman or Otherwise

At issue is the problem that was posed by Mattingly: "what makes Romanization and Hellenization particularly unhelpful constructs is that the terms are used to describe both *process* and *outcome*, so that the terms have become their own explanation" (Mattingly, 2009: 285–6). Often we have little evidence for the social and ritual contexts of objects found in tombs, objects that because of their provenience or style are used to define cultural periods. These objects then come to define cultural "processes" like Orientalization, Hellenization or Romanization that are then turned into outcomes defined by external phenomena, as in the terminology we use for the Etruscans: Orientalizing, Classical or Etrusco-Roman. The question of influence and diffusion is connected to the phenomenon of imported objects, imported styles and even, in some cases, imported artisans. The importation of skilled artisans and artists has been well studied in the earlier periods in Etruria, but probably would have continued into the Hellenistic period, just as Etruscan artists would have traveled elsewhere, for instance to Rome at the end of the sixth century to work on the Capitoline temple. But if identity is culturally constructed rather than ethnically determined, then artifacts "produced through the interactions of humans, and also productive in relationships with themselves – generate a hybridity that may be defined as cultural, rather than ethnic" (Antonaccio, 2009: 50). This complexity would certainly apply to those objects that are tied to the so-called "Hellenization" of Etruria, such as the ubiquitous Greek vases, and the Etruscan and Italic copies, that are found virtually at every Etruscan site and in every Etruscan context, whether domestic, religious or funerary (Reusser, 2002). The multitude of Greek imports is often interpreted as cultural dominance: "Etruria remained permanently 'colonized' by this civilization, which it absorbed with unflagging enthusiasm" (Braudel, 2001: 154). That these objects had a high social value to the Etruscans, in spite of whatever their value might have been to the Greeks, is clear from the way that they were privileged as dedications or prominently displayed at Etruscan banquets, as depicted in Etruscan wall painting. But if they were certainly valued, how were they perceived?

Would an Etruscan patron have cared whether a vase was genuinely Greek or a good Etruscan copy? Would these objects have been perceived as Greek or as foreign? Did the Greek inscriptions on the vases matter? The concept of "Hellenization" would suggest so and indicate an Etruscan interest in trying to be like the Greeks, in accepting and aping Greek culture. But the very ubiquity of these ceramics in Etruscan contexts can also be used to argue that their "Greekness" really did not matter all that much. To the Etruscans what might have been more important would have been the novelty and visual quality of the ceramics, much in the way that Chinese export porcelain might have been perceived in eighteenth-century Europe. The latter may have spawned all sorts of "orientalisms" in the decorative arts, but no one would argue for the "Orientalization" of Europe in this period. In colonialist terms, quite the opposite is argued for the eighteenth century, and this may hold true for the Etruscans as well. Greek vases may well have reinforced Etruscan concepts of power and trade rather than an Etruscan interest in being Greek. A Greek artist working at Cerveteri producing Campana *dinoi* (Hemelrijk and Den Boer, 2007; Warden, 2008b) or Caeretan *hydriai* (Bonaudo, 2004) would have been working for Etruscans patrons, possibly even as a slave, just one of many artisans involved in producing objects for elite consumption and display. Recent scholarship that situates the interpretation of the visual culture of these objects within the Etruscan context of their patronage (Bonaudo, 2004) has met with resistance from scholars who insist on a strictly Greek, hermetic reading of the imagery that ignores the complexity of Etruscan culture in all periods (Hemelrijk, 2007; Hemelrijk and De Boer, 2007). A good example of a more nuanced approach to the hybridity of material culture is provided by Riva in her study of an early bucchero vase (Riva, 2009b), a close reading of an Etruscan version of a Greek myth which is methodologically appealing because it attempts to explain the way that Etruscans might have perceived or received Greek myth. It suggests that Etruscan reception might have been sophisticated and discerning, even in the earlier periods, and the choices that were made might allow us to perceive how foreign narratives were harnessed to express Etruscan interests, possibly even Etruscan values. This distinction is important in a cultural and artistic context where anything of quality or complexity is immediately attributed to foreign influence, even if much of Etruscan visualization can be shown to be firmly rooted in the Etruscan cultural systems. Other examples of Greek myth imaged in Etruscan manner for Etruscan patrons are the birth of Athena (Kennedy-Quigley, 2001), Herakles and Hera/Uni (Bonfante and Swaddling, 2006: 35), and even the Actaeon myth. A recent analysis of Etruscan visual culture from the Orientalizing into the Hellenistic period suggests that the Etruscan use of Greek mythology is neither ignorant nor unsophisticated: "Hellenization was not a process of unthinking acceptance of a somehow preferable repertoire; it was (if it can be said to have existed at all) a process of exchange and interaction" (Izzet, 2007a: 122).

5 Religious Identity

The visual language of Etruria in the Classical period is best interpreted through the lens of the Etruscan belief system. Religion and ritual, as shown by theoretical studies (Hermanowicz and Morgan, 1999: 197–214), are fundamental to the formation of collective identity. A contextual approach helps explain an Etruscan interest during the Late

Classical and Hellenistic periods in particular scenes from Greek mythology (J.P. Small, 1983: 87–92). Even more specifically, there are types of iconography that can be linked precisely to the Etruscan religious belief, for instance the death/sacrifice of Actaeon (Warden, 2008a), a scene connected to Etruscan interest in ancestor veneration (Damgaard Andersen, 1993; Camporeale, 2008). This forms part of a greater argument that there is an "interpretatio etrusca," an inherent Etruscan/Italic mode of visualization that is connected to Etruscan cultural attitudes. While some of these inherent attitudes would have been cultural and social, many others would have been connected to the Etruscan belief system; even so, social and religious contexts are not often easy to differentiate in Etruria. This should not come as a surprise in a theocratic context where the ruling elite exercised both political and religious power, if we are to believe Roman sources, especially Livy and Seneca, which describe intense Etruscan religiosity. There really is no reason to distrust these sources in the light of the heavy Etruscan influence on Roman religious doctrine, and recent archaeological work (Warden, 2009, 2011a), as well as intense analyses of visual sources that have produced a new synthetic understanding of religion and "sacred history" (de Grummond, 2006; de Grummond and Simon, 2006), which bears out the historical view. Indeed, religiosity in many forms seems to be a fundamental aspect of Etruscan culture that recurrently manifests itself in the archaeological and visual record. One manifestation would be the Etruscan interest in bounded and defined sacred space (Izzet, 2001), in ritually consecrated and sometimes even deconsecrated space (Edlund-Berry, 1994; Warden, 2011a), that is connected to another Etruscan development, the so-called Tuscan temple. The definition of this distinct form and its surrounding space is intimately connected to Etruscan belief and ritual. Even the characteristic moldings of this type of building are quite different in form from their Greek counterparts, and Meritt has argued convincingly that in form and in the way that they are used in a building, these molded elements define a characteristically Etruscan aesthetic that can be documented from the Archaic period into Roman times (Meritt and Edlund-Berry, 2000). Admittedly the Tuscan temples of the sort described by Vitruvius are widespread in Italy and are probably better dubbed Etrusco-Italic. Here once again it is difficult to differentiate between what is Etruscan and what is Italic (Pallottino, 1985: 46–53), and even though canonical temples were still being constructed in the third century, for instance at Volterra, by the second and first centuries temples of this type were mainly being erected in the Roman and Latin colonies (Colonna, 2006: 164).

6 Metalwork as Material Production in the Fifth through Second Centuries

Another characteristic of Etruscan material culture is the abundance of metal objects of every type and of high quality in Etruscan graves of all periods. This phenomenon was connected to Etruscan traditions of mineral exploitation and correspondingly high competence in metallurgy. This view is also supported by historical evidence praising the quality of Etruscan metalworking, extensive evidence for Etruscan exploitation of mineral sources, and the wealth of metal objects and sculpture from ritual as well as funerary contexts, especially in northern Etruria where votive deposits are almost exclusively made of metal. Metal production is a hallmark of Etruscan material culture, and indeed even part of civic fabric, as at Marzabotto where foundries and smithies are not uncommonly found within the residential areas of the city. The production of metal objects – not just figurines but also utensils and utilitarian objects of every type – from the fifth century onward suggests an expansion of the breadth of Etruscan patronage. Combined with evidence for increased numbers of graves, along with a decrease of the wealth of these graves, this broader patronage has been used as evidence for the emergence of the Etruscan middle class. The evidence, however, could also be explained by population growth and changing funerary custom, and even the poorer depositions of the Hellenistic period. The last consist of ash urns of modest type accompanied by more limited material assemblages, and can be attributed to constitute the elite stratum (or strata?) of the Etruscan social landscape.

Metal production from the fourth century onward also raises interesting questions about what is Etruscan about Etruscan art, as important centers of production can now be identified outside of Etruria proper (such as Latin Praeneste) that produced objects that seem typologically and stylistically Etruscan. This site produced large quantities of objects for elite, mostly female consumption, for instance mirrors, toilet implements, and the characteristic cistae (Bordenache Battaglia, 1979). Our difficulty in interpreting this large corpus of material comes from the lack of contextual evidence, for the cemeteries of Praeneste were excavated badly or at least not systematically. There is no question, however, that by the fourth century, metalworkers at Praeneste produced objects that reflected a visual *koine* that connected Etruria, Latium and south Italy. The Ficoroni cista is a case in point. Commissioned by an elite Etruscan woman, Dindia, as a gift for her daughter, it exemplifies the cosmopolitan and stylistically syncretic aspects of metal production that had strong connections to Etruscan metalworking traditions combined with Greek mythological imagery depicted with strong links to Greek/south Italian modes of representation. The signature of an artist, Novios Plautios, shows he is of probable Greek origin. The signature itself is a Greek custom, although Etruscan signatures are known (Sassatelli, 2004: 179–80), but the Latin inscription mentions Rome, linking the production of the cista to that city. As extraordinary as the Ficoroni cista may be – and it is also exceptional in the extraordinarily high quality of its decoration – it is but one of many objects that are part of a vast localized production at a non-Etruscan site, objects whose style is characteristic of the complexity and hybridity of Italic (we should now use that term rather than Etruscan) material culture of the fourth century. Praeneste is but one of many examples of material culture that suggests increased Italic identity (Massa Pairault, 1992: 100), and once again it is difficult in any period to distinguish between what is Etruscan and what is Italic (Pallottino, 1985: 46–53).

The so-called crisis of the fifth century and the radical changes in the political power structure of Etruria and Latium in the fourth century, through the agency of both Rome and the continued intrusions of the Gauls, created a period of turmoil that seems to have been ironically fertile for the production of art and material culture of quality in central Italy. From this period come funerary sculpture and painting, fictile architectural decoration, and even large-scale bronze sculpture like the fourth-century Chimaera of Arezzo. The latter is only the finest example of what must have been common large-scale dedications of sculpture at Etruscan sanctuaries, if Roman historical evidence of the spoils from the sacking of Etruscan sanctuaries like the Fanum Voltumnae can be trusted. The Chimaera, apart from its quality and scale, is of special interest for the question of display that it raises, as it possibly was part of a much larger group that could have included Bellerophon, or even Pegasus. Its unusual discovery in a votive stips is also of great interest, as it might have been ritually "sacrificed." And even though the Etruscan inscription (made as part of the casting of the object), which is a dedication to Tinia, proves that it is Etruscan, the hybridity of its style, combining Etruscan and what are arguably Greek elements, continues to stimulate discussion (Cohen, 2010; contra, Warden, 2011b).

These pan-Italic elements can also be documented in terracotta production of the same period, and again the hybridity of style produces the same issues of interpretation. The horses from a pediment of the Ara della Regina temple offer the same challenge: a soft naturalism that connects them to late fifth- and fourth-century Greek art, and even if the medium is indicative of production to local coroplasts, the group has been called pan-Hellenic and linked to Olympia, Athens and Phidias; we even have the suggestion that the horses are a copy of a group dedicated at a major Greek sanctuary (Massa Pairault, 1992: 101). Again, the interpretation of this kind of object through the lens of "Hellenization" produces what Mattingly has called a "self-fulfilling paradigm" (Mattingly, 2010: 285–6). Admittedly, when fictile decoration is fragmentary, as is the case with the Ara della Regina group, the context can be elusive, but there are instances that can allow us to glimpse a more localized sensibility. The spectacular second-century pediment from Talamone dates to a period when Etruria was under Roman dominion, and it is tempting to read into its complex visualization of the myth of the Seven Against Thebes a political message the reminds us of the fragility of the city. Two aspects of its message must certainly have resonated with a second-century Etruscan viewer. The first is the connection of family identity to civic identity, for the Theban cycle, if nothing else, is a paradigm for a city torn asunder by interfamily rivalry. The civic is thus equated with the familial, a message that would not have been lost on what remained of the elite Etruscan clans that had dominated urban life for centuries. The second message of the Talamone pediment is firmly rooted in Etruscan religious traditions, for in the center of the pediment, directly behind the place where an officiating priest would have stood to survey space and divine the portents according to the disciplina etrusca, stands Oedipus, facing forward as both aggrieved patriarch and blind seer. The combined message of the importance of familial loyalty and the role of an elite Etruscan as link between past and future, human and divine, is conservative and characteristic of the Etruscans, ironically so in this period of the twilight of their power.

7 Funerary Sculpture, Gender and Family

The essential conservatism of Etruscan visual representation from the fifth to the second century can be documented in funerary sculpture. A few examples will suffice. Sarcophagi or urns of elite family members often depict the deceased on the lid, usually reclining as in the act of dining. This kind of iconography, even if the figures are represented in up-to-date style, continues long-held traditions of elite afterlife, as in the images of the deceased at banquet that portray or at least symbolize the body of the deceased. Both men and women are depicted. More important than the sex of the individual may be the elite status of the deceased, but certainly the women are imaged in as privileged a position as their male counterparts. The overall picture is well known and well discussed for the Etruscans, a social fabric where women have agency and rank comparable to men. In this sense the high social role of women is usually thought to be a characteristic, possibly even exclusive feature, of Etruscan society, even if there is some evidence that this exclusivity may be overstated (Riva, 2010: 86). Thus gender may be connected to identity, and particularly relevant to this issue are Etruscan urns that depict married couples. Although not common, at least 36 urn or sarcophagus lids of this type can be documented. The tradition of imaging couples on urns and sarcophagi can be traced back to the famous terracotta sarcophagi in the Villa Giulia and Louvre Museums. Two very grand, much later, examples from Vulci depicting members of the Tetnies family (Rowland, 2008: 151–64), now in the Museum of Fine Arts in Boston, attest to the enduring aspects of this kind of imagery. The context of the more banal examples, the cinerary urns studied by Nielsen (1992), is not known, although in the case of later urns where the couple's urn is placed in a larger context that includes urns of several generations, as for instance in the Tomb of the Calisna Sepu at Monteriggioni (Cristofani *et al.*, eds, 1975–7: 168), one might suspect that the couple is being imaged for the glorification of family and ancestry (Figure 23.3). The married couple symbolizes family union and establishes the importance of descent on both the father's and the mother's side, a visual equivalent of the Etruscan custom of patrilineal and matrilineal inscriptions.

Figure 23.3 Urn from the Tomb of the Calisna Sepu, Monteriggioni, now in the Florence Archaeological Museum. Source: Photo by P.G. Warden.

Especially noteworthy is the remarkable terracotta urn of an older couple in Volterra, analyzed in detail by Nielsen (1992), whose realistic, pseudo-veristic style raises interesting but perhaps unsolvable questions about realism in portraiture in third- and second-century Etruria and its relation to Roman realism.

8 Tomb Painting in the Fourth and Third Centuries

At first glance later Etruscan wall painting seems a radical departure, and there are obvious differences of style and tenor. But the themes have not really changed; they are just represented in a new language. For instance, the elite banquet continues, even if it now takes place in what is either a different or more specific setting. Banquets now take place in the underworld, in the presence of underworld figures. The afterlife is thus clearly visualized, and death itself is now represented as a voyage to that other place, the "*al di là*," usually accompanied by daemons that seem to specialize in this kind of process. The imagery is so specific that it has even been convincingly argued that there is a kind of topography of death that is carefully depicted in both funerary sculpture and painting (Briquel, 2008: 27–47). Even though the imagery is quite different from earlier funerary iconography, it need not be simplistically attributed to a new "pessimistic" view of the world caused by Etruscan political decline (Ridgway, 2004–6: 127–41). This interest may in fact result from a new need to explain identity at all different levels, in all places. As part of the process of moving into this other world the deceased is now greeted by a retinue of high-ranking individuals that include previously deceased members of the clan. There is now an overt sense of spectacle, and the scenes resemble a Roman processus magistratualis. The emphasis on status is nothing new in Etruscan funerary settings, but earlier it was done in different ways, through tomb paraphernalia or by modes of representation that were more self-assured, not nearly as overtly self-conscious and self-displaying. The earliest example of a processus magistratualis in an Etruscan mortuary context, as has been recently argued (Vincenti, 2009), dates to the fourth century and is found in the Bruschi Tomb at Tarquinia, where, interestingly, the painters atavistically used hierarchical scale to denote rank. In the Bruschi Tomb, and probably in the François Tomb as well, material from earlier tombs was moved into the tomb, a kind of ceremonial parentatio (Vincenti, 2009: 135) where remains of ancestors would have been moved and reinterred within the new tomb. The overall picture is one of single-minded interest in depicting the status of the clan, whether through ancestry or political status as defined through new kinds of ceremonial contexts. The imagery reflects a kind of insecurity, an insistence on status as defined by others whose self-consciousness may well be indicative of the changing political demography of the Hellenistic period in Italy, of radically altered power structures resulting from Roman expansion.

The new approaches of later Etruscan tomb painting were not a radical departure. The thematic continuities of depicting the elite family, the banquet, and the afterlife housed in a palatial or domestic context all flow from an iconographic continuum that developed over many centuries. Already by the fourth century some of the new iconography was creeping into funerary representation. The fourth-century Golini I at Orvieto (Feruglio, 1982), for instance, re-imagines the traditional banquet by explicitly placing it in the underworld while adding a wealth of detail about the preparation of the food and drink provided to both the immortals of the underworld and the elite Etruscans that have joined them in eternal revelry. In this case, however, we get a rare glimpse of non-elite Etruscans, the minions that prepare the food, each of them now identified by name in individual inscriptions. This is a unique instance of depiction of non-elite Etruscans, albeit hardly a "middle class," for they are servile both in what they do and in their costume, or lack thereof. The recently excavated Tomb of the Infernal Quadriga at Sarteano is another example of changes in Etruscan funerary visualization as early as the fourth century, with its combination of elite banquet and dramatic otherworldly imagery (Minetti, 2006; Steingräber, 2006a: 228–30). What characterizes late Etruscan painting is its diversity and originality, its ability to incorporate new means of expression for values that are at their heart traditional (in terms of the themes of the imagery) and conservative (in terms of the articulation of values connected to individual and family identities). The François Tomb at Vulci is a case in point, for whatever one makes of the nature of the much-discussed historical elements that are represented (Buranelli, ed., 1987; Rasmussen, 1997: 25–7), the programmatic contrast of Greek and Etrusco-Roman history/legend in the "tablinum" area (Steingräber, 2006a: 238) is both ambitious and highly original. Also noteworthy is the depiction of specific individuals such as Vel Saties, with all the appropriate symbols of rank (Steingräber, 2006a: 184). Despite the novel mechanics of representation, we once again have the elite individual in the context of the elite family, buried and represented in a tomb that will serve as a palace for the dead, by now a trope of Etruscan mortuary custom. Once again the identity that is expressed spatially and visually is connected to the rank of the individual and family.

A major problem with using material culture to reconstruct identity is the issue of selection; the surviving evidence is selective because of cultural processes, of the vagaries of archaeological preservation, or even because it is chronologically or geographically specific. This is the case with late Etruscan funerary painting; it is a southern phenomenon, localized at Tarquinia, which, with its close proximity and interaction with Rome, is a place where elite Etruscans (Torelli, 1975) would probably have evolved different concepts of identity than, say, in the far northern limits of Etruria, at Fiesole or on the Apennines near Bologna. In these areas, as in many other parts of Etruria, archaeological evidence and different kinds of material culture must be brought to bear. In northern Etruria, even the major sites are less historically and archaeologically visible, yet compelling evidence is emerging for Etruscan urbanization at sites like Gonfienti (Poggesi, 2000: 61–9) and Marzabotto (Sassatelli and Govi, 2005), or for complex interaction of Etruscans, Gauls and Romans at sites like Monterenzio/Monte Bibele (Vitali, 1990) and Poggio Colla (Thomas, 2000). Etruscans in this part of Italy faced very different pressures than their southern counterparts. A site like Poggio Colla, with a fortified acropolis sanctuary and extensive surrounding settlement, is providing evidence for more than just elite culture. The discovery of more contexts connected to artisans and other Etruscan groups leads to questions of whether any of the conclusions relating to Etruscan elites have much bearing on "everyday" Etruscans. Would Etruscans working in the kilns at Poggio Colla, or in the artisan workshops at Poggio Civitate or Cetamura (de Grummond, 2009), or anywhere else in Etruria have had an Etruscan identity, a local identity, or would they have lived and worked much as any other workers in the Mediterranean, then or even in later periods?

9 Conclusion

What is visible is the elite Etruscan, and here too we might ask if elite identity might have transcended the boundaries of place, language or ethnic group. Elite Etruscans at Tarquinia may have had more in common with the elite inhabitants of Latium or Campania than with the non-elites of southern Etruria. Social rank may transcend what we perceive as ethnic identity. It seems clear, however, that elite Etruscan identity changed and even adapted over time. In the later periods, when highly unstable environments would have stimulated the emergence of shared identities (Rousseau and van der Veen, 2005: 286), Etruscan elites continued to rely on imagery that emphasized their clan rather than city or nation, even if they did so in a new visual language. From the earliest times Etruscan identity had been rooted in the status of the individual as connected to a physical environment (house, home or palace) that reflected "family values" in every sense of that phrase. As Etruscan power crumbled, Etruscan elites seem to have clung to this image of self. Even those individuals who successfully negotiated the changes of the Hellenistic period, the ones who successfully transformed themselves into the leaders of a new Italy under Roman rule, for instance Aule Metele, the "Arringatore" (Colonna, 1991), or even the new Latin ruling families of Roman Tarquinia (Torelli, 1975: 191), continued to define themselves through Etruscan traditions of rank, family and ancestry.

FURTHER READING

For discussion of general theoretical approaches to identity, see Hodos (2009) or Antonaccio (2009). Izzet (2007b) provides a good summary of the current issues in Etruscan studies while the book in which it is found, Bradley, Isayev and Riva (2007), is a collection of essays whose title intimates the difficulties of defining too closely questions of identity and ethnicity. The fundamental importance of myth, religion and belief system – so important to understanding the Etruscan social landscape – is surveyed broadly by de Grummond (2006), and Tuck (1994) provides a good case study of a fundamental Etruscan cultural trope, banqueting.

The relation of ethnicity to gender is covered comprehensively by the series of articles collected by Cornell and Lomas (1997). Nation state, and civic identities are elusive concepts in Etruscan Italy, but the question of city-states is investigated by Torelli (2000), and nationhood in relation to linguistic evidence by Wilkins (1990). Torelli and Sgubini's (2008) exhibition catalogue on southern Etruscan cities takes a very old-fashioned approach by defining civic identity through well-studied classes of elite artifacts. More methodologically satisfying is Bradley's (2000b) essay on southern Italy; the approaches discussed therein are applicable more broadly to other parts of Italy. Shifting identities under Roman dominion emerge from Adams's (2003) study of bilingualism, and Torelli's (1975) classic investigation of the *Elogia Tarquinensia*. The new evidence provided by recent DNA studies is well evaluated by Perkins (2009).

Identity as interpreted recently through material culture is considered in diverse and interesting ways by Castor (2010) in relation to jewelry and female adornment, by Edlund-Berry (2008) in relation to the Etrusco-Italic temple, by Gleba (2008) with respect to textile manufacture, and by Riva (2009b) in her discussion of early imagery on Etruscan elite pottery. An earlier study by J.P. Small (1983) about patronage and choice of subject on Etruscan urns raised fundamental questions that are still pertinent today.

CHAPTER 24

Greeks, Lucanians and Romans at Poseidonia/Paestum (South Italy)

Maurizio Gualtieri

1 Background

Some major aspects of the most recent debate on the archaeology of identity have already been discussed in Chapter 22. The aim of the following pages is to present a specific aspect of the problem: the definition of cultural identities in the southern part of the Italian Peninsula which developed concurrently with the emergence and rise of the Roman Republic, and eventually became part of the Augustan tota Italia (Nicolet, 1991: 194), an "imagined community" (Hobsbawm, 1990: 48–9) in which all the disparate local histories and geographies could be unified (Laurence, 1998: 109). Of major relevance to the subject matter of our discussion is the fact that in recent decades there has been a substantial reassessment of the problems posed by the Roman Republic's expansion in Italy and in the western provinces. This reassessment is certainly a result of our increased knowledge of the broader context and long-term developments, made possible by a fast-growing body of archaeological data. Already by 1971 P.A. Brunt expressed the view that "it is from archaeology that we can best hope to extend and deepen our understanding of social and economic conditions in ancient Italy" (Brunt, 1971: viii). This has been matched by the new emphasis placed on a contextual "reading" of the evidence and a comparative anthropological approach in the interpretation of the archaeological data. It is no exaggeration to say that the archaeology of Magna Graecia or Greek south Italy, the focus of this chapter, and the analysis of the methodological problems posed by a comparative anthropological approach to the phenomena of "acculturation" and self-definition is almost four decades old (Gallini, 1973; Prontera, 1988: 1–5). In more recent years there has been a marked upsurge of interest in the manifold aspects of interaction among ancient societies which has, in most cases,

A Companion to the Archaeology of the Roman Republic, First Edition. Edited by Jane DeRose Evans.

stimulated a more nuanced approach to the study of cultural contact and cultural change. Indeed, the dominant descriptive vocabulary – integration, identity, continuity and innovation – used today in the analysis of such phenomena (Hitchner, 2009: 651) and the concurrently less frequent use of more "loaded" terms such as Hellenization and Romanization, clearly betrays a change in perspective and an adoption of a new conceptual framework. With specific regard to the problem of Greek/native interaction in Magna Graecia the new approaches have been very critical of what may be characterized as a monolithic concept of culture still embedded in pioneer archaeological studies of the past century (as in Childe, 1956: 8) and, concurrently, have come to adopt with growing frequency a more fluid concept of identity (Coarelli, 1971; Antonaccio, 2005: 108).

It was the norm for generations of archaeologists of the first half of the twentieth century (the so-called cultural-historical school of thought) to define boundaries of cultural groups with some kind of ethnic identity on the basis of assemblages of material culture. Such definitions obviously relied on a unitary concept of identity. If we assume that a group or individual had only one identity, then it must express that identity according to a distinctive pattern of material culture (Jones, 1996: 63, 1997). This is the conceptual framework which, in many ways, underlies the traditional definition of Hellenization and Romanization, two commonly used labels that, as already hinted, have undergone radical questioning in recent decades. That Greek colonization in the west and the Roman conquest of Italy spread new cultural models and new assemblages of material culture is a truism. However, to interpret the progressive adoption of a new language, different social habits and the spread of new types of artifacts as an expression of giving up an identity to adopt a new one – as the terms Hellenization and Romanization might imply (and indeed have often come to imply) – is a fairly bold and objectionable inference, for the very fact that such reasoning is based on the belief that only a single identity of a group or individual is possible.

In fact, a well-known example from southern Italy reported by the literary sources may provide a salutary note of caution and seems to play directly against such a belief. The epic poet Ennius, author of the early second-century *Annales*, who had a key role in giving expression to a national Roman identity, used to define himself as having "tria corda" (i.e. three hearts), for he spoke Greek, Oscan and Latin. He was born in Rudiae, in Messapia on the heel of Italy and therefore belonged to all of Hellenized Magna Graecia: Tarentum, the major city in the region, was one of the foremost centers of Hellenistic culture and for this very reason he may certainly be labeled in all respects as "Greek." The native tongue of the region, however, was Messapic: within the complex mosaic of cultures and languages of fourth- and third-century Magna Graecia it is easy to understand that "his family defined itself as belonging to the broader group of dialects stretching southwards from Samnite central Italy, known to the Greeks generically as Oscan" (Wallace-Hadrill, 2007: 355; cf. De Simone, 1992; Lomas, 1993: 163). He spent a long time in the company of the foremost Roman generals of the day, including Cato, who brought him to Rome (Musti, 2005: 371). Above all, he was adopted in the familia of the Scipios, in whose sepulchral monument he was commemorated (Coarelli, 1972). What made this Greek/Oscan intellectual also a Roman? In this particular case, it might easily be pointed out that he was a Roman citizen by legal status, since we happen to know that he gained Roman citizenship from his inclusion in a Roman colony (either Pisaurum or Potentia). Thus, the effective metaphor of the "tria corda," which

Ennius himself formulated, is most appropriate and, as opportunely underlined in a recent account of the creation and expression of identity in the Roman world, "We must leave room for Ennius to be a model Roman, without excising either his Greek or his Oscan hearts" (Wallace-Hadrill, 2007: 356). Still, tension between local and Roman identities no doubt resulted in a number of instances (Smith, 2008: 387).

The very eloquent piece of textual information regarding Ennius seems to reflect a much more fluid and less exclusive concept of identity and is no doubt emblematic of vast areas of Magna Graecia. It is, then, no surprise that in the wake of post-colonial theory, scholars have come to adopt a conceptual framework which relies on a much more nuanced terminology, such as "hybridization" or "creolization," to define the varied identities of (to mention the most recent contributions) the Hellenized people of Sicily (Antonaccio, 2003, 2005) or the mosaic of cultures of the Italian Peninsula in the course of the first millennium (Riva, forthcoming). One of these contributions, in particular, strongly emphasizes "local and regional hybrid identities" as a "more appropriate focus of future discussion" (Antonaccio, 2005: 111–12). Similarly, for a different geographic context, a study in the late 1990s highlighted the complex and variegated "creole" communities of the Roman west (Webster, 2001).

2 The Poseidonia/Paestum Case Study

In the light of the most recent debate on cultural interaction and identity in the archaeological record, the site of Poseidonia/Paestum provides us with an excellent case study on phenomena of cultural contact and cultural change in the southern part of the Italian Peninsula during the second half of the first millennium. Founded at the beginning of the sixth century in one of the most fertile plains along the Tyrrhenian coast of south Italy, Poseidonia is no doubt emblematic of the key role played by the western Greeks in the early history of the Italian Peninsula (Pedley, 1990: 30–5; Osborne, 1998: 255–8). At the same time, the rich and varied archaeological record for the Lucanian and Roman phases provides first-hand documentation of the cultural and socio-political changes occurring in south Italy during the period which saw the emergence of Rome as a new protagonist on the political stage of the Mediterranean. Not surprisingly, when we consider the monumentality of its impressive temple architecture and its almost unique state of preservation, earlier research at the site privileged its importance within the context of the more general discourse of Greek art and architecture in the west and its distinctive "local" flavor (Pedley, 1990: 81–8; Holloway, 2006). A new wave of archaeological research that started in the late 1970s has placed greater emphasis on the study of the phenomena of cultural interaction with the native peoples in the hinterland of the colonial settlements in keeping with the conceptual framework indicated above (Gualtieri, 1987; Pedley, 1994: 321–2). This research has provided a welcome counterpart to the overt Hellenocentric bias of the earlier studies and has set in motion what may justifiably be labeled as the "new archaeology" of Magna Graecia. Archaeological research at the site during the last two or three decades has entailed, on the one hand, a contextual reading of the rich funerary evidence, followed by a careful analysis and systematic publication of the remarkable group of figured wall paintings of the fourth-century tombs of the Lucanian elites (Corrigan, 1979; Pontrandolfo, 1987a; Pontrandolfo and Rouveret, 1992). On the other hand, the extensive exploration, started in the mid-1970s inside the ancient city (Greco and Theodorescu, 1990, 1996) and, not least, the more recent systematic excavation of a major extra-urban sanctuary (Pedley and Torelli, 1993), have provided important parallel information on the transformations of the urban texture of a major city of Magna Graecia as a result of the phenomenon of "Samnitization" (or, more appropriately, "Lucanization") of the later fifth/early fourth centuries and under the impact of Roman colonization of the third century (Curti, Dench and Patterson, 1996: 188). Not least, the rich and unbroken epigraphic record found at the site and extending through the Late Republican and Early Imperial periods provides very direct and detailed documentation of those socio-cultural changes, while at the same time highlighting the lasting vitality of its Greek roots (Lomas, 1993: 110–14, 1996: 108).

Finally, we should not overlook a more general aspect of the evidence for the area and time frame considered. Rome's encounter with the cities of Magna Graecia, in the early third century, meant its first direct contact with Greek culture and art. Such an event marked the beginning of an active and creative process which, in the authoritative catalogue of a recent exhibition titled "L'arte greca conquista l'Italia," has been significantly defined as "a phenomenon that finds no analogies in history: a society appropriates the culture of the defeated enemy (the Greeks) to such a degree that the latter becomes an integral part of its identity" (Zanker, 2008: 165, translated from Italian by Gualtieri).

3 Greek Poseidonia

The establishment of the sanctuary at Foce del Sele just opposite the Etruscan settlement of Pontecagnano, a few miles north of the Sele River (a major divide in antiquity between Campania and Lucania according to Strabo VI.1.1), defined the northward extent of the new city's territory. The city itself was located not far from the springs of the Capodifiume River, in a commanding position over the southern part of the plain (Pedley, 1990: 61–2). Recent archaeological discoveries in the southern sector of ancient Campania have provided the detailed picture of a culturally homogeneous milieu extending as far south as the Sele River (Greco and Pontrandolfo, 1990; Cerchiai, 1995: 117–20). They have also contributed to clarify one of the key functions of the sanctuary at Foce del Sele as a sort of cultural and spiritual stronghold against the non-Greek communities on the opposite (north) bank of the river, the material expression of a frontier between Greek and Etruscan territory. This limit, however, did not entail a sharp separation between two mutually incompatible worlds. On the contrary, it was the visible sign of a complex reality, in which the coexistence of diverse cultural and socio-economic groups stimulated cultural contacts as well as a push toward cultural homogeneity. Thus, without an adequate consideration of the archaeological evidence from southern Campania (especially Fratte and Pontecagnano) and a proper reading of its cultural significance, it would be easy to underestimate the influence on early Poseidonia of the Etruscan or "Etruscanized" communities of the region around modern Salerno. Conversely, the role that Poseidonia came to play within the surrounding region is well attested by a growing body of epigraphic texts from Foce del Sele (Colonna, 1976: 153–5, 1983). Between the sixth and fifth centuries both the city and the major extra-urban sanctuary had become a center of attraction for the non-Greek communities settled in surrounding territories.

The monumental architectural remains of the urban and extra-urban sanctuaries are a clear indication that Poseidonia prospered during the second half of the sixth and the first half of the fifth century. A process of growth was already underway by the mid-sixth century, as is shown by the important group of early sculptural decoration (relief metopes, pertaining to a still unidentified early temple) at the major extra-urban sanctuary at Foce del Sele (de La Genière and Greco, 1996). The myths represented are meant to reassert the cultural identity of the colonists while also dealing with issues critical to the community. In particular the group of metopes showing Herakles fighting several antagonists is a likely reflection of Greek contacts with the non-Greek peoples of the western Mediterranean and they functioned as metaphors of the world beyond the boundaries of the territory of Poseidonia. In their unique blend of models or sources of inspiration from various areas of the Greek world and in their "enthusiastic experimentation with technique" (Pedley, 1990: 70), the metopes at Foce del Sele stand as a cultural counterpart to the architecture of the urban sanctuaries. On the other hand, the striking absence of any sculptural decoration from the urban sanctuaries may not be a mere coincidence. At the Foce del Sele site, which marked Poseidonia's territorial boundary, surprising emphasis was placed on the sculptural decoration of a temple, at a time when figural and especially narrative architectural sculpture was in its infancy in the Greek world at large. This emphasis is most probably to be explained as a strong signal of cultural identity by the early Greek settlers, not by chance to be found at the site of a major extra-urban sanctuary, which also functioned as a sort of interface between Greek and the "other" (Pedley, 1990: 62–7; De La Geniére and Greco, 1996; Nafissi, 2001).

The regular layout of the city, including the two main north–south and east–west streets that cross at what was to become the south-west corner of the later Roman forum (Figure 24.1), within a massive defensive system, dates to *c.* 500: the ashlar walls still visible today, however, are dated to the Lucanian period with some later re-patching. Together with the Temple of Athena of *c.* 500 at the north end of the new urban plan and the Temple of Hera II (the so-called Temple of Neptune) of *c.* 460 added to the southern urban sanctuary, the walls are an important indication of the scale of building activity between the Late Archaic and Early Classical periods and testify to the success of a growing community.

Although in some respects the urban layout of Poseidonia may appear rather anomalous as compared to the standard plan of a mainland Greek city, there is no question that by the first half of the fifth century it had developed many of the characteristic physical features of the classical Greek polis, such as a clear-cut division of private and public space and a number of buildings which were the expression of the values and socio-political organization of its citizen community. The most significant among these is the recently identified public space ***par excellence*** of a Greek city – the agora (the public square or market-place) in the area between the Temple of Athena to the north and the urban sanctuary dedicated to Hera at the south end of the city. There was as well the recently excavated circular building for public assemblies (an ***ekklesiasterion*** or ***bouleuterion***) at its north-east side (Pedley, 1990: 78–80). As a result of the impressive urban growth and its particular geographical setting, the city became an important intermediary between the Greek colonies of Magna Graecia, the Etruscan area of Campania immediately to the north and the Italic hinterland. If earlier studies, in the wake of a lasting Hellenocentric tradition, had looked at Poseidonia almost exclusively as a sort of "bastion" of Greek culture in the west, a much more nuanced and variegated picture of the nature of colonial

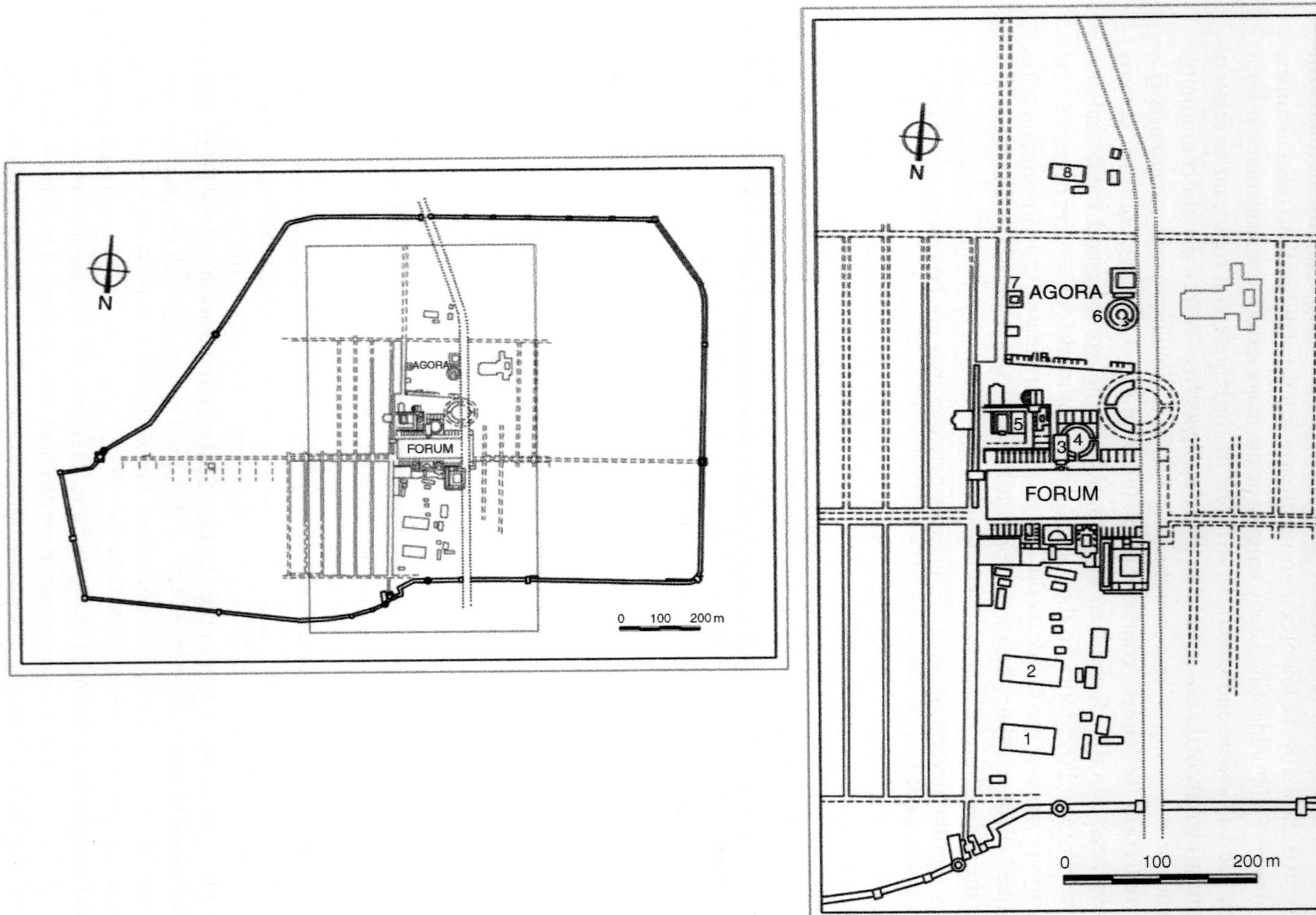

Figure 24.1 Plan of Poseidonia-Paestum. (1) Archaic Temple of Hera; (2) Temple of Hera II; (3) Republican Temple; (4) Comitium-Curia; (5) "Piscina"-gymnasium; (6) *Ekklesiasterion*; (7) "Heroon"; (8) Temple of Athena. Source: Illustration by S. Ferrari.

Greek settlements privileged by recent research is best exemplified by the lively debate about the cultural significance of the Tomba del Tuffatore and the still-open question of the identity of its owner (Osborne, 1998). Against the initial claim of the discovery of a precious and almost unique testimony of the lost Greek painting (Napoli, 1970), careful analysis of the entire context has come up with a rather different interpretation. The whole conception of the tomb (the plastered interior of the stone cist decorated with figured scenes of a symposium and the representation of a diver symbolically plunging into the other world on the inside of the lid) is a definite anomaly in the context of a Greek city. The custom of decorating the plastered surface of the interior of the tombs is rather to be found, as early as the Archaic period, in peripheral areas of the Greek world and especially in Etruria. Also, it has been appropriately underlined that Greek culture keeps strictly separate, as contrasting if not altogether antithetic categories, the world of death and the world of the symposium (Murray, 1988: 248–9). Thus, we may be closer to the truth in looking at those paintings of *c.* 480/470 as an explicit document of close contacts between the colonial Greeks and the elites of the many variegated neighboring communities. As Murray (1988: 249) points out, "If we want to offer a social explanation of this monument we would have to connect it with Etruscan or Lucanian attitudes to death." Furthermore, as indicated by a most recent overview of the iconography, the anomalous and somewhat peculiar shape of the so-called *krater* set at the center of the symposium scene is rather a faithful rendering of a traditional Italic wine container, the *krater-kantharos* (Colivicchi, forthcoming; contra, Holloway, 2006). A striking cultural pendant to the imagery found in the Tomba del Tuffatore and its mixed cultural aspects is provided by a remarkable inscription found on a black glaze jug (*olpe*) manufactured in Poseidonia and deposited in a tomb dated to *c.* 460 in the Etrusco-Campanian site at Fratte (Pontrandolfo, 1987b; Greco and Pontrandolfo, eds, 1990). It is a long metric inscription in two lines running all around the body of the vase impressed on the unfired clay (that is, it was written in the workshop in Poseidonia). It celebrates, in the form of a poetic verse, presumably sung in the context of the symposium, erotic games among individuals of Etruscan, Greek and Italic origin, as indicated by the names of Vulca, Apollodoros, Ksylla, Ybrikos and Parmynius. One of the many interesting aspects of this text, which makes it also an important piece of evidence of the social and cultural history of fifth-century Poseidonia, is no doubt its close connection to the sphere of the symposium, where personal relationships were established among the participants (in this particular case elite members of diverse ethnic communities), that often had more complex social and political implications (Cerchiai, 1995: 103–4; Pontrandolfo, 1995).

Both of these documents concur to delineate a scenario of intense interaction (and perhaps even "osmosis") between the colonial Greeks and neighboring communities, featuring Poseidonia as a "cosmopolitan" community which entertained exchanges with the Etruscan world to the north and, at the same time, was in close contact with the Italic populations of its immediate hinterland (Pontrandolfo and D'Agostino, 1990: 110–12; Ampolo, 1992). The few available bits of epigraphic and literary evidence confirm this role of the city over its regional surroundings. Most relevant is no doubt the well-known inscription on a bronze plaque dedicated at the Sanctuary of Olympia (Meiggs and Lewis, 1969: 18–19), dated toward the end of the sixth century: the text refers to a treaty of friendship between Sybaris and an Italic population called the Serdaioi, in which Poseidonia stands as a guarantor of the treaty, presumably on account of the already-mentioned credit the city enjoyed among the Italic peoples. Likewise, Poseidonia maintained its international standing within the Greek world at large, to judge from Herodotus's report that an athlete from Poseidonia, a certain Parmenides, won the first prize at the competition of the stadion (a race of *c.* 200 m) in the seventy-eighth Panhellenic games at Olympia, that is, in 468.

4 Lucanian Paestum

Between the fifth and fourth centuries, signs of change are indicated by the funerary record. The ostentation of wealth, through a careful selection of the grave goods and the multiplication of objects displayed, is in sharp contrast with standard Greek funerary practices (Pontrandolfo, 1987a: 242–4). Above all, a group of burials in the north cemetery, just outside the city walls, include weapons and defensive armor: most significant is the presence of the three-disk cuirass, a characteristic item of Samnitic burials, featured on a well-known Italic bronze statuette, the so-called Mars now in the Louvre, dated to the latter part of the fifth century (Colonna, 1970: n. 402; Cipriani, 1996, with a note of caution by Crawford, 2006a: 61; see also Polito, 1996). Understandably, such elements of evident discontinuity in the archaeological record have been explained by several scholars as early signs of the take-over of Poseidonia by the Lucanians reported by Strabo (V.4.13, VI.1.3) (Pontrandolfo, 1987a: 250–60; contra, Musti, 2005: 294–7).

The Lucanians were the southern branch of the Oscan-speaking Samnites, a major group of the Italic populations settled in the Apennine hinterland (often collectively labeled as Osco-Sabellians) that were responsible for a number of major changes in the cultural map of southern Italy during the fourth and early third centuries (Gualtieri, 1993: 21–5: see Chapter 25). They are best known to the ancient authors for their military confrontations with the expanding Roman power, such as the so-called "Samnitic Wars" of the second half of the fourth century (Cornell, 2004).

With specific reference to Lucanian Paestum, it should be pointed out that a commonly used term such as "take-over" might be an oversimplified and somewhat schematic label for the transformations shown in the archaeological record, which must have implied long-term phenomena of interaction. It is also likely that demographic factors may have been at play (Gualtieri, 1993: 232, 328; Musti, 2005: 296; Lombardo, 2006). The Greek historians, ethnographers and geographers say little about the Lucanians, except for recording military conflicts between these Sabellians and the Italiotes (the settlers of the Greek colonies). What is more, the surviving record is only to be found in late authors: it is therefore difficult to determine their sources and hence the reliability of their information, although it would be fair to surmise that the Greek literary tradition looks at the Lucanians and other Oscan-speaking peoples through Hellenic eyes. It is evident, for instance, that Strabo, our major source on the matter, presents the unquestionable phenomena of expansion of the Samnite peoples in the model of Greek colonization. Recent analyses of the problem and a growing reliance on the new archaeological data concur to indicate that, "The Lucani did not migrate from the Apennine regions of Central Italy as a single block as if the Samnites dispatched them to colonize Italy's southern regions" (Chiranky, 1982: 98–9). Actually, there are indications that the Lucanians or other Italic groups were on the road to settlement already in the sixth century: an epigraphic text from Castelluccio di Lao (Rix, 1997) at the border between ancient Lucania and Bruttium and the recently discovered lex sacra from Tortora, at the

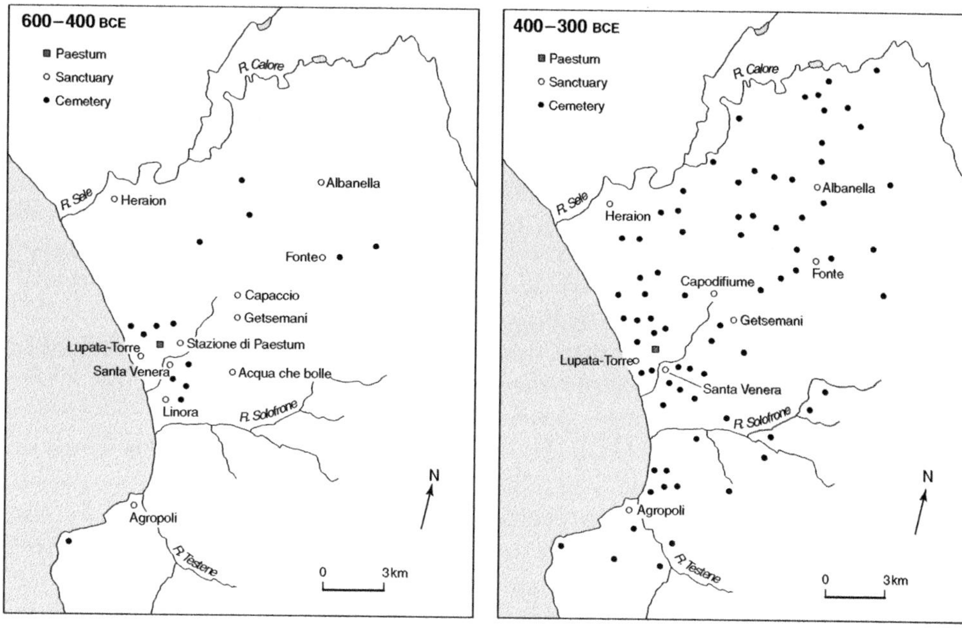

Figure 24.2 Sanctuary and cemetery sites around Paestum, 600–300 BCE. Source: Illustration by M. Gualtieri.

extreme south-west of the region (Lazzarini and Poccetti, 2001: 190–4) are both written in an archaic form of Oscan or a proto-Samnite language, to follow Prosdocimi's most recent terminology (Prosdocimi, 2000: 202–3). Equally relevant is the distribution of Italic votive bronze figurines (Colonna, 1970). These testimonia may not indicate a stable Sabellian presence or a great migration in the sixth century, but they are important documents of the early stages of such phenomena. Thus, it would be appropriate to sum up the aftermath of the alleged Lucanian "take-over" of Poseidonia with the observation that "In some ways this change resembles more a fusion of peoples and ideas than a conquest" (Pedley, 1990: 97). As a matter of fact, there are few, if any, changes that can be detected in the city plan or in the architectural record at large in the course of the fourth century, while the use of most civic buildings is well documented by archaeological finds. There is also convincing evidence that the majority of religious buildings were being used without any noticeable break. Even the major sanctuary at Foce del Sele, the emblem of Poseidonia's cultural and political identity, does not reflect substantial transformation except for the addition, in the course of the fourth century, of the so-called "edificio quadrato" (De La Genière and Greco, 1996). The major assembly hall of the community, the ekklesiasterion of the Greek city, continued in use with only a minor addition, namely the altar with dedicatory inscription in Oscan, a precious documentation of the language of the new inhabitants (see below).

More visible changes resulting from the Lucanian "take-over" are found in the rural landscape of the surrounding territory. In spite of the serious limitations posed by a fairly fragmentary archaeological record (modern, intensive land-use makes a proper systematic surface survey an impossible task), the distribution map of all chance finds of the fourth and third centuries from the territory (Figure 24.2) provides a sufficiently illustrative picture of a densely settled countryside for the later fourth and early third centuries, implying new forms of land exploitation and a phenomenon of agricultural intensification. This is in sharp contrast with a very sparsely settled fifth-century countryside (Pedley, 1990: 98, fig. 60; Greco and Theodorescu, 1996: 192) and points to a process of dense occupation of the countryside set in motion by the Lucanian take-over, well paralleled by comparable changes detected in some districts of the Campanian plain, particularly the area around Samnitic Capua (D'Agostino, 1992). A few systematic analyses of Italic fortified settlements in ancient Lucania and nearby areas indicate similar settlement patterns and related forms of exploitation of the agrarian space controlled by the oppidum (see esp. Roccagloriosa in Gualtieri, 1987, 1993: 245–9, 333–5; Faustoferri and Lloyd, 1998: 5).

Evidence of socio-cultural changes resulting from the "Lucanization" is further provided by an exceptional group of wall paintings in several clusters of chamber tombs from the urban cemeteries. After a period of experimentation with mere decorative/vegetal motifs which betray influences (in both iconography and style) from the Tarentine milieu (Corrigan, 1979: iii), a new, distinctive imagery appearing around the mid-fourth century reflects the values and the ideology of the Lucanian elites settled in the city (Pontrandolfo, 1987a; Rouveret, 1992; contra, Musti, 2005: 295–6; and a note of caution in Crawford, 2006a). True enough, only in one instance can we safely link the painted figures with an Oscan name, significantly in Greek form (the inscribed name in Greek letters *plasos* is the Greek form of the Oscan *plasis*) (Cristofani, 1996: 203, no. 96). It is, however, the sophisticated analysis of Lucanian (and Campanian) culture through a close reading of the iconography which warrants such conclusions (Curti, Dench and Patterson, 1996: 183, n. 132). This implies looking at both the sequences of paintings on the walls of the individual burials and at overall changes in the ideological language of self-representation, an approach rooted in pioneer studies of the 1960s on Campanian culture (Nicolet, 1962; Frederiksen, 1968). In particular, Nicolet concentrated on the representations of combat and return of the victorious warrior bearing spoils, met by a woman offering libations (1962: 474–5). In his conclusions, the combat and return formed part of an indigenous epic cycle, not indebted to Greek contacts, and signifying by its heroic mode the importance (particularly in Campania) of the Italic cavalry nobility (Nicolet, 1962: 497–8). The painted representations of Tomb 61 in the Andriuolo urban cemetery are emblematic in this respect. Its paintings exalt the role of the inhumed person (a woman in this case, as indicated by the grave goods) through a series of images extending over the four slabs, which represent the male members of her family in a sort of procession. Meaningfully, they are organized by age group: a youth, an adult armed individual, and a white-haired man draped and seated on a cart drawn by mules. The adjoining slabs, on the other hand, represent a confrontation between two armed men, then a frieze with weapons followed by the depiction of a horseman carrying a trophy (the tunic seized from the enemy, Figure 24.3): the overall intent of the imagery is to underline the importance of the military role (as well as the ethos of horsemanship and the tradition of individual prowess) of the male family members of the buried person, a necessary precondition of their elite status within the community (Oakley, 1985; Tagliamonte, 2004: 108).

A later group of chamber tombs (noteworthy among them is the so-called Tomb of the Magistrate), dated to the early third century, show the use of a new language (Gualtieri, 1993: 342). Yet it should be noted that Musti (2005: 371, fig. 35) has

Figure 24.3 Short wall of Tomb 61 of the Andriolo urban cemetery: woman facing a mounted warrior. Source: Photo by M. Gualtieri, used by permission of the Museo Archeologico Nazionale di Paestum, Salerno.

proposed to date Tomb 1 from the Spinazzo cemetery at Paestum (the so-called Tomb of the Magistrate) as late as the foundation of the Roman colony, on the basis of the iconography. The popular scenes of the return of the warrior and the hanging weapons are absent and the paintings center on the encounter and salutation of two men depicted on the wall opposite the entrance door. What is more, the elderly family member has the appearance of a magistrate wearing a toga and a golden ring, while allusions to military success refer to the young adults, but in a much more allusive and indirect way (Figure 24.4). Representations such as these are quite explicit about the changing role of the Lucanian elites that took control of the city and are illustrative of key transformations in the socio-political organization of the community at the eve of its direct encounter with Rome. Overall, the kind of social formation emerging from these representations is comparable to what is implied in the Roman concept of the gens (Gualtieri, 1993: 158, 342–4; Smith, 2006: 144–63). It is quite likely that this type of social organization of the Italic communities might have facilitated early contacts between Rome and the local elites, encouraging local aristocrats to forge stronger ties with members of the Roman aristocracy in order to secure their own political standing (Fronda, 2010: 5). It is, indeed, such an environment that underlies Roman expansion in the Italian Peninsula and might, at least in part, explain its success, as convincingly argued for the well-documented and

Figure 24.4 Detail of the Tomb of the Magistrate (Tomb 1), Spinazzo cemetery: seated magistrate. Source: Photo by M. Gualtieri, used by permission of the Museo Archeologico Nazionale di Paestum, Salerno.

recently studied case of Messapia in the later Hellenistic period (Gallone, 2007; also Yntema, 2006; van Dommelen and Terrenato, eds, 2007: 13–22).

5 Aristoxenus on the "Barbarization" of Poseidonia

Closely related to the cultural changes discussed above is a notorious (and somewhat puzzling) passage on the history of early Italy by Aristoxenus (frag. 124 Wehrli, preserved in Athenaeus *Deipnosophistae* 14.632a) referring to Poseidonia/Paestum (Fraschetti, 1981; Mele, 1991; Crawford, 2006a: 59). Aristoxenus complains that "although they were Greeks to start with, they have been barbarized by becoming Tyrrhenians or Roman" mentioning, in support of his statement, the fact that they had changed both customs (*epitedeumata*) and speech (*phonē*), so that in his time they are known to celebrate only one Greek Festival. With the necessary caution required by the evident rhetorical tone of Aristoxenus's discourse, this text is a precious testimony of what might have happened to the "Greek" inhabitants of the fifth-century city in the new situation created by the Lucanian "take-over." It can be excluded right away that the passage is dealing with an alleged economic decline of the Lucanian city as, indeed, some scholars had thought (Zancani Montuoro,

1973). The discourse of Aristoxenus, poet/musician from Tarentum and a pupil of Aristotle, active in the second half of the fourth century (the biographical data are analyzed in Muller *FHG* II, 69–71), is rather on the decline in musical taste and musical education, essential aspects of the *paideia* of Greek *polites* (Fraschetti, 1981; Visconti, 1999: 144–51). The "barbarization" he is emphatically postulating (according to his Greek point of view) obviously has to do with the new situation created in the time of his writing, in the latter part of the fourth century, by an influx of Italic peoples into Paestum and its territory. The figurative repertoire of the painted tombs, starting from the mid-fourth century, discussed in the previous section, and the use of the Oscan language documented also for public inscriptions (although scarce) are new features of the fourth century-city indicated by the archaeological record which might underlie Aristoxenus's concerns. Of particular relevance is a dedicatory inscription to Jupiter set up by the magistrate Statis Statilies on the above-mentioned stele found on the lower tiers of the *ekklesiasterion*, suggesting Italic membership and authority among the ruling elite of fourth-century Paestum (Cristofani, 1996: 203). It is also true, on the other hand, that there is good evidence for continuity of Greek names and language at Paestum throughout the fourth and early third century (Pugliese Carratelli, 1972; Pedley, 1990; 97; Torelli, 2003: 103–4).

What makes the interpretation of such a passage rather puzzling, however, is the mention of the Romans as responsible for the "barbarization" of the city, since Aristoxenus is writing in the second half of the fourth century and certainly long before the foundation of the Latin colony. There have been different ways to explain this evident paradox. Some scholars have gone so far as to eliminate the mention of the Romans, as a spurious later addition to the text. In a different perspective, an authoritative overview of Roman Paestum has interpreted the picture of decline evoked by Aristoxenus as a possible reflection of the social and cultural changes caused by the impact of the Latin colony (Torelli, 1992: 96–7). Obviously, the latter interpretation would necessarily imply a much lower date for the text than our testimonia on the life of Aristoxenus would allow. The obligatory compromise would then be to consider the work to which this fragment pertains as apocryphal, as Torelli suggested.

Yet both of the attempts to solve the puzzle would then imply major revisions to the wording or date and context of the text. A lectio facilior might be to allow for a justifiably incorrect use of the ethnic labels on the part of the Tarentine Aristoxenus, since "this fragment of Aristoxenus reflects a lack of concern in distinguishing accurately between different varieties of Italian barbarians, an attitude which is a familiar feature of modern western ethnocentrism as well" (Dench, 1995: 52). And, indeed, it is quite possible to think that Aristoxenus could have alluded to the Lucanians as Tyrrhenians or Romans, since "the principal concern of Aristoxenus was to find a striking simile for a decline in musical taste, not to analyze the ethnicity of Italy" (Crawford, 2006: 60). Thus, in spite of some inconsistencies in its reference to the cultural geography of contemporary Magna Graecia, this passage provides us with a first-hand description of the socio-cultural situation at Poseidonia/Paestum in the later fourth century, implying the presence and coexistence of both Greek and Oscan elements as a characteristic aspect of the social and cultural framework of fourth-century Paestum. Its greater value, when compared to the few generic (and often highly colored) comments in Strabo's not unproblematic narrative, lies in the very fact that it provides us with a unique, contemporary account of the Lucanian city, including a direct reference to discernible traits of "multiculturalism" (Pedley, 1990: 97; Wonder, 2002; Crawford, 2006). Such a variegated picture is supported by the epigraphic record from a few systematically explored Italic sites on the Tyrrhenian seaboard of ancient Lucania indicating the coexistence of Etruscan, Greek, Oscan and Phoenician/Punic elements (Colonna, 1976; Poccetti, 1990: 147–9; Campanile, 1993).

6 The Latin Colony in 273

The simultaneous foundation in 273 of the Latin colony at the Greek/Lucanian site of Paestum and of the Latin colony of Cosa in the countryside of Etruria marks a decisive step in Rome's hegemony on the Italian Peninsula (Gualtieri, 2003: 18–19). The foundation of the two colonies no doubt was meant to ensure a stable Roman presence in two regions of ancient culture already in contact, in many ways, with the emerging Roman Republic, Magna Graecia and Etruria (Torelli, 1999: 43; Fronda, 2010: 24, n. 54). Quite different, however, was the context of the two colonial foundations (Bradley, 2006: 172–3). Cosa was founded *in vacuo*, on an unoccupied plateau commanding a long stretch of the central Tyrrhenian coast. This might well explain the particular emphasis placed on its Romanitas, as indicated by the large-scale explorations started in the 1950s which have recovered the most extensive archaeological documentation so far available on a Latin colony. The regular city plan, as interpreted by the excavator, shows striking similarities with Republican Rome, although a recent reassessment of the evidence partially deconstructs the excavator's reconstruction of the urban layout (Brown, 1980: 12, 17; Fentress, 2000; Gualtieri, 2002), on the grounds that Brown's interpretation might have been largely inspired by Aulus Gellius's model colony (*NA* 16.13.8–9) as "cuius istae coloniae quasi effigies parvae simulacraque esse quaedam videntur" – "colonies seemed to be miniatures, as it were, and in a way copies [of Rome itself]" (trans. J. Rolfe, 1961, Loeb; see Bispham, 2007a: 75–6, 80–1). As for the immediate impact of the colonial settlement on the surrounding landscape, the systematic archaeological survey of the ager Cosanus and the reconstruction of settlement patterns in the region have provided independent support to literary accounts for hostilities between colonists and local populations (Brunt, 1971: 539; Scott, 1988: 75; Carandini and Cambi, 2002: 62). In the territory where the Latin colony of Cosa was founded, few fourth-century sites survived the Roman conquest. In a quite different context, the establishment of the colony at Paestum implied the dispatch of colonists to the pre-existing Greek and Lucanian settlement. The installation of the colony on the site of Lucanian Paestum must have had a direct impact on the physical organization of the city as well as adding new inhabited areas (Greco, 1988: 82). A number of peculiar aspects of the colonial city are largely due to the many elements of continuity with the previous settlement, not least the continued presence of a substantial Lucanian element and also Greek elements within the population of the Roman city (Torelli, 1992: 92–4, 1999: 45).

Clearer "destructuring" effects of the colonial foundation, on the other hand, have been detected in the occupation of the countryside. Here the gradual disappearance of rural sanctuaries is matched by new evidence for land measurement and division to the north of the city, and have been related, with good arguments, to the Latin colony (Gasparri, 1994: 154). A date corresponding to the foundation of the Latin colony is asserted by Curti, Dench and Patterson (1996: 175, n. 54), although the broad chronological range of the evidence so far recovered might demand caution in establishing a direct connection with possible land assignments to the colonists (Crawford, 2006:

65). An unquestionable element of discontinuity, which also testifies to at least one major institutional change resulting from the deductio of 273, is provided by the third-century numismatic record. In spite of some disagreement concerning its precise dating, the new bronze coinage bearing the Latin legend PAISTANO (a clear reference to the Italic name of the city) has been assigned to the third century. What is more, a careful analysis of coin types issued at Paestum between the Lucanian occupation and the deductio of the colony has highlighted the quintessentially Roman nature of the new issue (Taliercio Mensitieri, 1996: 212). Not only is the legend PAISTANO (the archaic Latin form of a genitive plural) a template of the formula adopted by the early Roman coinage and reproduced in the coins issued by the cities gravitating towards Rome, but the very structure and weight standard of the PAISTANO issues are also Roman (Crawford, 1973; contra, Pontrandolfo, 1983; see also Horsnaes, 2004). The use of the Latin alphabet, by itself, signals a break from the previous coinage while at the same time implying that the issue cannot pertain chronologically to the Lucanian city.

A counterpoint to the issue of Roman coinage just discussed, and the related institutional changes implied, is the continued presence of substantial Greek and especially Lucanian elements which are well attested by the epigraphic record of the urban elites of the third through first centuries as well as precious indications of the phenomena of social mobility within the new socio-cultural framework of the Latin colony (Lomas, 1991; Torelli, 1993: 92–4, 2003; De Bonis, 2007).

As for the urban layout, uncertainties remain in dating the structure, plan and organization of the massive city wall. The evidence so far indicates third-century works of repair and maintenance on a pre-existing late fourth-century fortification which may date to the establishment of the colony. This, perhaps, might also be the case with the creation of the new Forum. It certainly implies reworking of the central public space of the city along Roman lines. But, at the same time, its construction does not substantially alter the century-old layout of the town's civic center, since it is located at the south end of the main public area of the Greek city, the fifth-century agora (Figure 24.1). The impression is much more of a new civic space along Roman lines adapted into the Greek (and Lucanian) city rather than a deliberate attempt at obliterating the pre-existing agora. In any case the picture that emerges from the architectural and archaeological evidence is, generally speaking, not one of marked convulsion. Still a matter of debate are the chronology and circumstances in which the comitium-curia complex came to replace the Greek *ekklesiasterion*, kept in use almost unchanged during the Lucanian period. Here, too, the conventional view that the obliteration of the *ekklesiasterion* "gives the impression of a violent action, definitively cancelling out any trace of the previous political identity of the community" (Curti, Dench and Patterson, 1996: 186) demands caution, since the chronology and hence the speed of the change remain highly controversial and conclusive archaeological evidence about the timing of this change is still lacking (Greco, 1988; Torelli, 1992; Crawford, 2006: 65–6). Overall, it would be fair to conclude that third- and second-century Paestum, after the establishment of the Latin colony, shows several aspects of a reshaping of the city-plan along Roman lines without, however, necessarily implying those "traumatic effects" which have been often postulated in the interpretation of the archaeological evidence (Greco, 1988: 85; Pedley, 1990: 115–18). The reordering of the community along Roman lines was slower and less profound than has previously been presented (Lomas, 2004: 199). To this we should add the undisputed fact that the major Greek temples of the city largely continued in use during the Lucanian and Roman periods even if new forms of religious ritual appear (Zevi, 2005: 805; de Cazanove, 2007). An analogous picture, with more detailed, conclusive evidence about renovations and additions is provided by the recently excavated extra-mural sanctuary of Aphrodite at Santa Venere, to the south of the city, immediately outside of the walls (Pedley and Torelli, 1993). This sanctuary shows an important Lucanian phase of the cult and dedications to Venus through the colonial phase (Pedley, 1990: 157–61) as well as a revival of a traditional Greek priesthood in the later first century (Torelli, 1993: 206–7). It thus provides first-hand evidence about "the pliability ... of local cults within the framework of Roman hegemony and colonization" (Bispham, 2006: 94; see also de Cazanove, 2007).

All told, in the light of the complex and variegated picture emerging from the archaeological record of Roman Paestum the obvious question is: how Roman was the colonial city? Strictly related is also the more general and debated question of "how Roman was Roman colonization during the middle Republic?" (Bispham, 2006). Answers to such questions entail more profound considerations about the very nature of the expanding hegemony of Rome in the Italian Peninsula (and more specifically in Magna Graecia) during the third century. A number of recent studies incorporating ample discussion and reassessment of the most recent archaeological evidence have addressed this problem, often with the declared aim of deconstructing the conventional interpretations (rooted in nineteenth-century historiography) of Roman colonization as a well-structured "statist" (i.e. state-organized) activity designed to meet public aims (strategic, economic, and often settlement of veterans), already by the Mid-Republican period (Bispham, 2006; Fronda, 2010). Recent critical reassessment has resulted in a welcome acknowledgement of a much higher degree of variability of the individual cases (Lomas, 1991, 1993: 174–6), while placing greater emphasis on local factors (Bispham, 2006: 75; van Dommelen and Terrenato, 2007; see also Woolf, 2000: 122–3). This is certainly true with regard to our specific case study, when we take into account the complex multicultural aspects of the successive Greek, Lucanian and Roman phases. We must not forget, however, that while focusing on the analysis of individual case studies of the archaeology of the Mid- and Later Republican period, there has been a growing tendency in very recent scholarship to downplay or even overlook the realities of Roman imperialism in favor of a sort of neutral relativism with respect to the obvious fact of Roman military power and conquest in Italy. A balanced overview of the problem cannot and should not erase from the overall picture the nature of Roman aggression and conquest, while privileging an alleged key role of negotiations, whatever meaning is to be assigned to the fairly ambiguous term of "negotiations" or "dialogue" in the specific historical circumstances of Rome's expanding hegemony in the Italian Peninsula. Curti very sensibly reminds us that "it would be interesting to see if modern colonized peoples would accept the term "debate" as a description of their relationship with colonial powers" (2001: 24–5).

7 Concluding Remarks

A more balanced picture emerging from recent reassessments of the archaeological evidence is, then, one of gradual change of the urban fabric at Poseidonia/Paestum through the Greek, Lucanian and Roman periods. The conventional view, which would allow for a quiet transition of the city through the Lucanian phase, as mentioned in a previous section, also postulates major disruptions in the urban fabric caused by the establishment of the colony. It is very likely, instead, that there was little causal relationship between the institutional and socio-cultural changes which the site underwent in the five centuries of the Republican period and specific developments in the urban framework.

To be sure, the variegated picture of the Latin colony at Paestum which emerges from this survey gradually lost its individuality as a result of the concession of Roman citizenship to the Italian allies after the Social War and of the rapid and widespread phenomena of "municipalization" of the Italian Peninsula in the course of the first century, even if we have to allow for marked disparities of the pace of municipalization across the Italian Peninsula (Bispham, 2006: 409). While there is evidence for the continued use of Oscan as a spoken language and as a mark of identity in public inscriptions of the later third and second centuries, it disappears from the epigraphic record during the first century, probably as a result of the more general phenomenon of cultural homogenization of Italy in the first century (Lomas, 1996: 110; Cristofani, 1996: 202; Crawford, 1996). Between the Late Republic and the principate of Augustus, Paestum would gradually lose the complexity of the socio-political and cultural fabric of the Latin colony, to become one of the many municipia of Roman Italy. The obvious question, arising from this overview of the cultural and institutional changes which the city underwent in the five centuries of the ascendancy of the Roman Republic, is whether Roman Paestum ever completely forgot its Greek roots.

It is very likely that it never did and it is possible to infer that part of the city's Greek character remained, to judge from a number of features indicated by both the archaeological record and the available textual information. It has been duly pointed out that, along with Greek Velia, with which Paestum formed a single circulation area, and Greek Heraklea, Paestum (unlike other Latin colonies) continued to mint coinage through the Late Republic, and even beyond (Crawford, 2006). Furthermore, during the Second Punic War the numismatic record from Paestum shows hoards of victoriati (equivalent to the Greek drachma) rather than Roman denarii (Crawford, 1985). In the same period, along with Greek Neapolis, Paestum is attested as holding some of its communal wealth in the form of plate in temples, a characteristic Greek practice (Livy XXII.32.4–9; XXXVI.9). And, more specifically, in another passage Livy (XXII.36.9) mentions ambassadors in 216 from Paestum presenting the Roman Senate with wealth in the form of golden bowls, just as an embassy from Greek Neapolis had done the year before (Crawford, 1985: 71–2).

One final point: the majority of the traditional accounts about the integration of the Italic peoples into the Augustan tota Italia bring up a good deal of material and linguistic evidence pointing to a speedy and complete Romanization during the Late Republican and, especially, the Augustan periods (Torelli, 1995: 12–14). Certain aspects of the most recent evidence and new approaches to the problem would rather underline the persistence and "reinvention" of ethnic identities (Lomas, 1991, 1993: 184–7, 1996: 108). Not surprisingly, a very recent contribution on the topic does not hesitate to reassess the conventional view of the Augustan period as the end point of the peoples of Italy, privileging rather a likely scenario of "a more … complex situation, in which many older identities continue to be vital and reworked, even when 'overwritten' by the idea of being Roman" (Bradley, 2007: 296; see also De Bonis, 2007).

FURTHER READING

Good recent syntheses on the problems of identity are Jones (1996, 1997). A very recent overview on ethnicity, identity and ancient societies is Ruby (2006). More specifically on Greek/Roman identities, see Wallace-Hadrill (2007). Detailed discussions of Greek Poseidonia and its multicultural aspects can be found in Colonna (1983), Pontrandolfo (1987a), Pontrandolfo and D'Agostino (1990) and Pedley (1990). For Lucanian Paestum, the literary sources are compiled in Mele (1991). An updated, detailed historical picture can be found in Musti (2005). For the painted tombs and self-representation of the populace, see Pontrandolfo and Rouveret (1992) and Rouveret (1992). The best synthesis in English concerning the Latin colony of Paestum is in Torelli (1999). Critical discussions of the conventional view are found in Crawford (2006), Bispham (2006) and Bradley (2006). For discussions on the Romanization and integration of the Italic peoples into the Republic, the general picture and specific case studies (including Paestum) are examined in Lomas (1993, 1996). Discussion of case studies from central and southern Italy can be found in Torelli (1995, 1999). An overall historical picture is in Gruen (1992) and, most recently, in Fronda (2010).

CHAPTER 25

Central Apennine Italy: The Case of Samnium

Marlene Suano and Rafael Scopacasa

1 Introduction

In ancient accounts of the Roman conquest of Italy, the communities of the central Apennines occupy a noteworthy place. Classical authors speak of various peoples inhabiting the region between the fifth and first century, many of which stand out in view of their key roles in the development of the Roman Republic. The Sabines, for example, are linked to the very origins of Rome, while the Marsi, Paeligni, Marrucini and Vestini are counted among the bravest to have fought the expanding Roman Republic (Strabo V.4.2). But of all the central Apennine peoples, the Samnites take the lead as Rome's most challenging adversaries in Italy, whose defeat was crucial for the development of the Republic (Livy VII.29; Dion. Hal. *Ant. Rom.* XVII–XVIII). Classical authors provide slightly different definitions of the Samnite homeland ("Samnium"), but in general the territory in question roughly corresponds to modern-day south Abruzzo, Molise and north/east Campania (see Musti, 1980).

The Samnites enter the historical record in 354 when they concluded a treaty with Rome (Livy VII.19.1–4). In the decades that followed, the Romans subdued both the Samnites and their central Apennine neighbors through a long series of conflicts known as the Samnite Wars, from 343 to 290 (Cornell, 1989). Classical authors assert that the Romans entered these wars to defend themselves and their allies from central Apennine peoples, a claim used to legitimize Roman imperialism in general (Livy VII.29; Dion. Hal. *Ant. Rom.* XVII–XVIII). However, it is likely that the conflicts resulted from increased competition for resources between the expanding Roman Republic and the polities that were developing in the central Apennines. By 304 most of the central Apennine peoples were already subdued and bound to Rome as allies, including the Marsi,

A Companion to the Archaeology of the Roman Republic, First Edition. Edited by Jane DeRose Evans.

Figure 25.1 Map of Samnium, showing sites mentioned in text. Source: Drawing by M. Suano and R. Scopacasa.

Paeligni and Marrucini. A few decades later the Sabines began to be admitted into the Roman citizen body. Also in 304, one of the five Samnite "tribes," the Frentani, obtained a separate treaty with Rome (Livy IX.45.18). The rest of the Samnites fought on, but after being defeated a third time in 293 they lost large tracts of their territory, which became Roman praefecturae and Latin colonies. In the 290s, much of the upper Volturno and Sangro Valleys became Roman praefecturae. In 268 and 263, Latin colonies were established in Beneventum and Aesernia, respectively. Both were strategically situated at crucial nodes of natural communication routes linking Samnium to Rome (Figure 25.1).

Samnite resistance is nevertheless reported to have remained strong until the Second Punic War, at which point Samnite elites started to cooperate with Rome as allies (Polyb.

2.24; Dion. Hal. *Ant. Rom.* XIX.12.1–4). Nevertheless, the fact that the Samnites were among the last of the Italian insurgents to lay down arms in the Social War is noted by classical authors as an indication that the old antagonism had never quite ceased – to the extent that Augustus named his pacifying program "tota Italia," after the central Apennine peoples' name for their idealized state (Vitalia).

The surviving accounts about the Samnites and their neighbors were all written by Greek and Roman authors centuries after the events that they describe. Yet these texts are important because they reveal how the central Apennine communities were perceived by Romans, Greeks and other groups with whom they came into contact. There is a tendency in classical texts to associate the central Apennines in general with rustic simplicity and a lack of cultural sophistication, with either positive or negative overtones. Such imagery probably originated among Roman ideologues of the Middle Republic, who drew on the polarity of Greeks versus barbarians to portray Rome as culturally more developed than its Italian competitors (Dench, 1995: 98–103). The Samnites are a good example. In Livy's narrative they appear as hostile and uncouth barbarians, either in the form of backward mountain folk or decadent in their gold and silver armor (Livy IX.13.6–8, IX.40.1–14). Other literary traditions throw a more positive light on the Samnites as hardy warrior-peasants of Spartan-like austerity. Such a portrayal was also not without its political underpinnings. The Greek geographer Strabo saw the attribution of Spartan origins to the Samnites as a strategy through which Tarentum (a Spartan colony) sought to extend its influence in Italy by appealing to shared origins with their Samnite neighbors (Strabo V.4.12).

The ambiguous case of the Samnites exemplifies the broader attitude of classical writers towards the central Apennines. At one extreme are the Sabines, whose special relationship with Rome meant that they could embody the values of austerity and piety cherished by the Roman Republican aristocracy (Dench, 1995: 80–94). On the other hand, those communities that proved more intractable – such as the Marsi, Paeligni and Marrucini – tended to take on a more menacing aspect, with their harsh rustic lifestyle seen as accounting for their warlike disposition.

Knowing how the Samnites and other central Apennine communities saw themselves is more difficult. No traces of central Apennine literature have been recovered. Perhaps the most reliable traces of their self-identification, as transmitted in the classical sources, are the origin myths known as the "sacred spring" narratives (Strabo V.4.12, VI.1.6; Varro *Ling.* 7.29; Gell. *NA* 11.1.5; Festus p.436L). Salmon (1967: 35) and Dench (1995: 206) note that these narratives are probably authentic as they do not follow the usual pattern of Greco-Roman origin myths, which usually involve founding eponymous heroes. These narratives relate how the Sabines sought to come out of crises by dedicating to Mars all that was born in a given spring, including children. Upon reaching adulthood, the dedicated individuals had to seek out a new home, usually with the guidance of a sacred animal such as a bull, wolf or woodpecker. In so doing, these Sabine migrants are said to have originated many of the central Apennine peoples, such as the Picentes and the Samnites themselves, who in turn are reported to have sent out sacred springs which generated the Lucanians (Strabo V.3.1) and Mamertines (Festus p.150L). The sacred spring myths could have been a means through which central Apennine communities conceptualized their place in the world and their relationship with each other, the memory of which could have been transmitted to the classical writers (Dench, 1995: 185–206). These narratives may also preserve a very faint memory of early Iron Age migrations caused by demographic pressure, although the archaeological evidence for this is still inconclusive.

Epigraphic evidence suggests that the central Apennine communities may indeed have shared a broader identity. The root word *safin-* occurs in a handful of inscriptions in the Oscan language (which was dominant in the region) dating from the fifth to first centuries. In these inscriptions, the earliest of which date to the fifth century, the root word *safin-* occurs both as a noun (*safinim*) and adjective (*safinum, safinas*), in phrases that can be translated as "the community of the Safin-" and "the leaders of the Safin- people" (La Regina, 1981). Such contexts, together with linguistic data, have led scholars to argue that *safin-* refers to the original, self-attributed ethnic of central Apennine peoples, which gave rise not only to the Latin Samnites and Greek *Saunitai*, but also to the Latin name for the Sabines (Sabini) and its derivation Sabelli ("little Sabines" or "children of the Sabines"). The ethnic *safin-* may therefore indicate an overarching identity that was originally shared by people who, in the historical record, are identified separately as Samnites and Sabines (La Regina, 1981: 133).

The case of the *safin-* inscriptions illustrates the disparity between the central Apennine communities' identities and how they are portrayed in classical texts. However, there is still no certainty as to the geographical scope of the ethnic *safin-* or how it changed in meaning during the four centuries in which it was used (Dench, 1995: 200ff.). Nevertheless, in the decades before the Social War, central Apennine elites were clearly employing the ethnic *safinim* to construct a sense of national Italic identity which was instrumental in the subsequent conflict with Rome. This can be deduced from the occurrence of *safinim* on a late second-century inscription at the major sanctuary of Pietrabbondante and later on the coin issues of the Italic insurgents (La Regina, 1991).

It is therefore not impossible to understand how central Apennine communities identified themselves. A much more robust picture of the identity of these communities is revealed through the material remains of their own cultural practices and habits, their religious activities and their economic strategies. By the fifth century, just before their relationship with Rome intensified, these communities were undergoing a crucial transition in their history, from the chiefdoms that had characterized the Iron Age (*c.* 1000–500) to a more complex socio-political organization (Barker and Suano, 1995; Bispham, 2007b).

The sheer amount of archaeological data that has been generated in the past three decades is too large to be dealt with in detail. Therefore, our discussion will focus on the area that is attributed to the Samnites in classical texts. Samnium accounts for a vast portion of the central Apennines and has yielded much of the available archaeological evidence. Also, the ancient communities of Samnium stand out in the historical record. But what were the main characteristics of these communities, and can we really see them as forming a cohesive ethnic or cultural group at any point in their history? It is with such questions in mind that we will survey the archaeological and epigraphic evidence from Samnium.

2 The Region

In light of the classical texts, scholars have conventionally subdivided Samnium into smaller territories, each corresponding to one of the five Samnite "tribes": the Pentri, Caudini, Hirpini, Carricini and Frentani (Tagliamonte, 1996: 6). Although the five tribes are to a great extent a modern construct, the construct does have some grounding in ancient literary sources as well as in the epigraphic evidence. But this conventional framework is problematic because it relies on the ethnic categorizations of classical authors, which do not necessarily represent how the communities of Samnium perceived themselves and their territory (Tagliamonte, 1996: 11–13; Scopacasa, 2009: 29–41). However, given its widespread use, the conventional subdivision of Samnium will be maintained in this chapter, without supporting preconceptions about the true identity of its inhabitants.

Although the archaeology of Samnium originated in the nineteenth century, it was only in the 1960s that archaeologists began systematically to explore the region's culture, society, economy and contacts (La Regina, 1975; *Sannio*, 1980). Regarding the economy, it is clear that the ancient inhabitants of Samnium were agriculturalists, a way of life which was easier in the fertile soil of coastal Abruzzo and Molise than in the rocky terrain of the mountainous Apennine hinterland (Di Niro, 1991d). The need to complement agriculture with an alternative means of livelihood was particularly acute in the Apennine hinterland, which explains the local practice of pastoralism as well as the timber trade (Veal, 2008). The ancient communities of Samnium relied on major natural communications routes to interact with each other and neighboring regions, either through the Adriatic coastal plain, the Trigno and Sangro River Valleys, or the mountain passes east of Venafro. Recent discoveries suggest that connections between Samnium and other regions of Italy were more intense than previously thought. This is suggested by the recovery in coastal Molise of a fragment of an impasto hut urn, which is most likely Villanovan and probably came from the Tyrrhenian coast (pers. comm., A. Di Niro's unpublished excavations at Contrada Arcora).

3 Cemeteries, Sanctuaries and Settlements

Funerary sites provide most of the information on Samnium for the whole of the Iron Age. After the fifth century, funerary sites gradually decline while sanctuaries and settlements become more visible in the archaeological record. Settlements are very difficult to identify archaeologically before the fourth century, as it is only then that masonry structures begin to appear in Samnium. Therefore, the necessary understanding of the relation between settlements and cemeteries in Samnium is still tentative.

The predominance of funerary sites in Samnium poses difficult interpretative challenges. Being the result of ritual practice, mortuary remains are not properly representative of everyday life. Because it is often unclear whether grave goods were personal possessions of the deceased or mortuary offerings, it is difficult to determine whether such artifacts are indicative of the personal identity of the deceased or of community identity in a broader sense. In the case of armor and weaponry there is even a third possibility that such grave goods were spoils taken from vanquished foes. Despite these difficulties, it is possible to approach the mortuary record as a source about social and cultural patterns.

The funerary sites of Samnium vary in size and the burials tend to be individual supine inhumations. Most graves are simple pits, although limestone cists are frequent in inland Abruzzo and Molise (Alfedena, Opi-Val Fondillo, Troccola). Cremation is scarcely attested, being largely confined to Hellenistic-period sites in coastal Molise (Larino-Carpineto and Montorio nei Frentani). In north and east Campania (Montesarchio, Sant'Agata dei Goti, Cairano and Oliveto Citra) simple pit graves predominate, which after the fifth century start to be lined with terracotta tiles.

It is possible to infer from the funerary material that the Iron Age communities of Samnium possessed some level of socio-economic stratification, especially with groups of better furnished burials, with artifacts considered valuable for their material, craftsmanship and provenance. In inland Abruzzo and Molise these artifacts are elaborate pieces of bronze armor and personal ornaments of bronze and amber. Lavish sets of banqueting and drinking vessels are especially common in the funerary sites of north Campania, the area most exposed to Greek and Tyrrhenian influences (Figure 25.2). In contrast, the Iron Age communities of east Campania and coastal Abruzzo/Molise seem more impervious to outside influences, given the marked continuity in the local repertory of pottery forms and metal artifacts throughout the Iron Age (Bailo Modesti, 1980: 203–13; Cuozzo, 2007: 240ff.).

More information can be extracted from the spatial arrangement of burials in funerary sites. Iron Age burials tend to be grouped into large clusters, some of which are placed in circular or concentric patterns, particularly in inland Abruzzo and Molise. Similar circular burial clusters are also found further north in the central Apennines (D'Ercole, Faustoferri and Ruggieri, 2003). There is evidence that each of these burial clusters was used by specific familial groups, suggesting an emphasis on the extended family ties (Bondioli, Corruccini and Macchiarelli, 1986). The pattern suggests a social organization where kinship determines an individual's social role in the absence of a developed state.

One aspect of social organization about which the funerary evidence is especially informative is gender. Scholars have tended to see the occurrence of iron weapons and bronze armor in certain burials as evidence that males were primarily warriors in Iron Age Samnium. The women, on the other hand, are thought to have been relegated to domestic tasks (Tagliamonte, 1996: 116–24). This reasoning overlooks the fact that the men in Iron Age Samnium, including those in elite graves, were usually buried without weapons. Sets of drinking vessels occur frequently with men, which suggest a strong association between men and banqueting (Scopacasa, 2010). This emphasis should not downplay the fact that Samnium was among the regions of Italy that supplied mercenaries to Magna Graecia and Sicily between the sixth and third centuries (Tagliamonte, 1994).

In recent years the analysis of skeletal remains from three funerary sites (Alfedena, San Vincenzo al Volturno and Gildone) has added a crucial dimension to our knowledge of society in Samnium. The data suggest great variability regarding daily activities and lifestyles of individuals. The skeletal evidence from Alfedena indicates a community that did not rely heavily on carbohydrates (and hence on agricultural subsistence) and was free from heavy manual toil, in view of the low occurrence of activity-related stress marks (Macchiarelli and Salvadei, 1986; Capasso, 1989). The individuals at Alfedena were nevertheless exposed to intense physical violence, as a recent study of cranial trauma has demonstrated (Paine *et al.*, 2007). In contrast, at San Vincenzo al Volturno and Gildone the evidence suggests a greater reliance on agricultural subsistence on the part of individuals who were mostly malnourished. Also, the high incidence of activity-related stress

Figure 25.2 Female grave from the necropolis of Opi, at the upper Sangro River Valley, sixth century, with banqueting and drinking vessels plentifully supplied for the dead. Source: Used by permission of the Ministera per i Beni e le Attività Culturali – Soprintendenza per i Beni Archeologici dell'Abruzzo-Chieti.

in young adults shows that agricultural labor was probably more pervasive than warfare (Bowden *et al.*, 2006; Di Niro and Petrone, 1993).

One of the earliest and major sanctuaries known to date is that of Valle d'Ansanto in east Campania, which was dedicated to the Italic goddess Mefitis and was already in use in the sixth century. The sanctuary was situated around a sulfurous pool which Virgil describes as an entrance to the underworld (*Aen.* 7.563–71). Its votive deposits include wooden statues of female figures, whose geometric style is directly evocative of the monumental stone sculpture from further north in the central Apennines, such as the famous "Capestrano warrior" (Rainini, 1985; Fracchia and Gualtieri, 1989).

The rise of sanctuaries from the fifth century onward indicates substantial transformations marking the close of the Iron Age in Samnium, when communities started to undergo a crucial transition from pre-state chiefdoms to a more complex socio-political organization. The old funerary sites decline as focal points for social and cultic activity, while large sanctuaries begin to flourish. This shift in emphasis from the funerary to the cultic sphere probably indicates state formation, inasmuch as it signals the re-channeling of community resources away from the more private or familial context of burials into the more public sphere of communal cult places (La Regina, 1980; Bispham, 2007a: 194). Inscriptions name this rising polity as the *touta*, which was headed by a supreme magistrate called the *meddis tuvtiks.* Such representatives of the *touta* used some of the sanctuaries for centralized administration and decision making, as suggested and confirmed regarding Pietrabbondante (La Regina, 1976; La Regina *et al.*, 2009–10). It is therefore not surprising that these sanctuaries were the focus of monumental building activity from the third century onward. At Pietrabbondante, an Ionic temple from the mid-third century was replaced after the Hannibalic War with a grander temple-theater complex in marble, which drew directly on Hellenistic models filtered through Campania and Lazio (La Regina, 1984: 22–5). It is possible that these buildings were funded by Samnite businessmen at Delos, including slave traders (La Regina, 1976: 229). One of the terracotta roof tiles from the later temple at Pietrabbondante (Temple B) contains a bilingual inscription made by two slave girls in their respective languages, Latin and Oscan. This inscription affords unparalleled insight into cultural change in Samnium, as it reveals the interaction between cultures in the everyday context of personal friendship and raises the key question of whether the Romanization of Samnium involved the wholesale adoption of Roman culture (as is normally assumed) or was instead a more complex two-way process of cultural coexistence. The inscription reads:

In Oscan:	hn.sattiieís.deftri seganatted. plavtad	In Latin:	Herenneis.Amica signavit.qando ponebamus. tegila
Translation:	Deftri, slave of Herennio Sattio signed with shoes		Amica, slave of Herennio signed when we put the roof-tiles (to dry)

A distinctive feature of key Samnite sanctuaries, mainly Pietrabbondante and Campochiaro, is the presence of metal armor and weaponry among the votive offerings. These might have constituted the Samnite victors' dedication of their personal weaponry to the deity before the assembled community (Tagliamonte, 1996: 201), or spoils which the Samnites took from their vanquished enemies, including the Romans. Most of the pieces in question display marked Greek characteristics, and most likely come from Magna Graecia. The possible role of Pietrabbondante as a "sanctuary of a people in arms" has led to its identification as the place where the famous Linen Legion was formed during the Third Samnite War (Figure 25.3; Livy X.38; La Regina, 1976: 226).

In recent years, scholarly attention has focused on the location of the sanctuaries. It has been observed that sanctuaries are aligned with important communication routes such as the drovers' trails, which date back to the Bronze Age and were avenues of transhumance and trade (Gabba and Pasquinucci, 1979: 154–6). The connection between sanctuaries, drovers' trails, pastoralism and trade suggests that sanctuaries also functioned as marketplaces (Torelli, 1996). Scholars note that Hercules, a notorious protector of herdsmen and merchants, was widely worshipped in Samnium (Van Wonterghem, 1999); it has also been suggested that large-scale transhumance funded the building activity in the sanctuaries (Lloyd, 1991: 184–5). On the other hand, Stek (2009: 57) observes that certain sanctuaries are a two-hour walk away from the nearest drovers' trails and therefore would not fit the proposed model. However, we still know little about perceived distances in antiquity and it may be complicated to equate modern capabilities with ancient ones. The importance of pastoralism has been further stressed by De Benedittis (2010), who notes that temples were built so as to directly face the sun at the beginning of the winter solstice, when the flocks used to leave Samnium.

Another recent suggestion is that sanctuaries were territorial markers, since many of them are situated in what might have been boundary zones between communities (D'Ercole, 2000). Such boundary sanctuaries could have been used exclusively by the community to which they belonged, or instead functioned as places of interaction and exchange between communities that bordered on each other (Stek, 2009: 64–5). Nevertheless, such a boundary-marking role should not be assumed for those small and medium-sized sanctuaries situated at the heart of settlements, which seem to have been the focal points of inward-looking communal activity (Stek, 2009: 104–6).

After the Social War, when the communities of Samnium received Roman citizenship, most of the major sanctuaries in the region were abandoned, probably as a result of the great administrative changes following the establishment of the Roman municipal system. Activity nevertheless persisted at the site of some sanctuaries, such as Campochiaro, Vastogirardi and San Giovanni in Galdo (Stek, 2009: 74–6). It is clear that the sanctuaries of Samnium offer great potential for future research. Among the more recently excavated sanctuaries, Monte Pallano (middle Sangro Valley) has already surprised scholars with its elaborate architectural terracottas. Also, important new discoveries at Pietrabbondante of a domus publica and a metal forge seem to corroborate the sanctuary's status as a political and military hub (La Regina *et al.*, 2009–10).

One of the defining features of pre-Roman Samnium and the central Apennines in general is the absence of urban settlements. From the dawn of the Iron Age, Samnium is characterized by dispersed settlements throughout the countryside. Archaeological survey has revealed that, over the Iron Age, there formed an increasingly structured settlement hierarchy in the region consisting of a number of larger, permanent settlements surrounded by villages, hamlets and farmsteads (Barker and Suano, 1995).

The basic settlement pattern detected in the Iron Age continues into the Hellenistic period, the fourth through second century, with certain sites becoming increasingly specialized. The main example is the appearance of hillforts of large polygonal masonry walls (La Regina, 1975; Oakley, 1995). They occupy naturally defensible positions at the summit of hills and mountains, overlooking important crossways. Many hillforts have been identified in the hinterland of Samnium but few have been systematically excavated. The best known are Terravecchia (Sepino), Monte Vairano (upper Biferno Valley), and Monte Pallano (upper Sangro Valley). Other hillforts have been detected in key locations but await excavation, such as Pietrabbondante, Agnone (upper Trigno Valley, a natural passageway across the Apennines) and Cercemaggiore, situated on a natural route to Campania and Rome. Also awaiting excavation is the hillfort on the slopes of Curino, next to the major necropolis at Alfedena. Other networks of hillforts have been identified

Figure 25.3 Plan of the sanctuary of Pietrabbondante: (1) Temple A; (2) Temple B; (3) Theater; (4) Domus publica. Source: Drawing reproduced, with permission from the authors, from Report 2010, www.inasa-roma.it.

in the mountain chains of north Campania, around Trebula, Cubulteria and Telesia (Conta Haller, 1978; Oakley, 1995; Carfora, 2001).

It is unclear when the hillforts started to be built. The largest ones were already in use by the late fourth century, such as Monte Vairano, and it is plausible that they grew in importance as the fighting with Rome intensified. The pressure that Samnium suffered from the advance of the Romans probably led to an improvement of the hillforts, which must be fully taken into account if the Romanization of Samnium is to be understood. Another aspect of the hillforts is their inter-visibility in the landscape, which would have allowed long-distance communication through the passing of fire or light signals (Coarelli and La Regina, 1984: 208). Hillforts were also important as statements of power in the landscape. This has been convincingly argued with respect to Monte Pallano, whose walls are visible from considerable distances. The conspicuous placement of this hillfort would have been a clear statement of power to neighboring Samnite communities, with whom the inhabitants of Monte Pallano need not always have been on friendly terms (Bispham *et al.*, 2000).

In addition to their strategic and military roles, many hillforts also contained stable settlements that were important economic nodes (La Regina, 1975; De Benedittis, 1991). This is clearly the case for the largest centers, which include structures that indicate substantial settled communities, such as monumental buildings and temples (such as the one in Monte Pallano: Lloyd and Faustoferri, 1998; Kane, 2008), organized street plans, pottery kilns and elite housing (as those found in Monte Vairano: De Benedittis, 1991; or Terravecchia (Sepino): Gaggiotti, 1991). Some of the main fortified centers display evidence of large-scale consumption of prestige goods, as suggested by the substantial quantities of imported wine amphoras at Monte Vairano. On the other hand, smaller hillforts might have been only temporarily inhabited as places of refuge for inhabitants of small countryside dwellings, such as the stone farmhouse at Gildone in inland Molise (Di Niro, 1991a).

In contrast with inland Abruzzo/Molise and north Campania, few fortified settlements have until now been found in east Campania and coastal Abruzzo/Molise. In coastal Abruzzo/Molise proto-urban centers developed after the fourth century, such as Larino. In addition to organized streets and monumental buildings, the prosperity of Larino is seen in the occurrence of imported pottery Apulian *kraters* and Gnathian ware. Such economic prosperity probably resulted from Larino's proximity to important communication routes. As for east Campania, an extensive settlement network has been identified in the territory between Avellino and Beneventum (Benevento) (Carfora, 2001). The major Samnite settlement in north-east Campania was converted into the Latin colony of Beneventum in 268. Excavations at pre-colonial Beneventum revealed an entire district of ceramic workshops that produced black-gloss (and perhaps red-figure) pottery. The importance of Beneventum as a central place is further corroborated by the numerous small and medium-sized settlements in its vicinity (Giampaola, 2000).

There is still much debate as to whether the settlements of Samnium were politically autonomous, especially during the period of Roman colonization in the third to first centuries. Coinage is considered a good indicator of political autonomy. However, the earliest coinage in Samnium, from the early third century, comes from settlements in north Campania (Alife and Fistelia), after these settlements had already been conquered by Rome. A few decades later, settlements which were probably still autonomous, such as Aquilonia (Monte Vairano), begin to issue their own coins, but their imagery (helmeted Athena and warrior with patera offering sacrifice) is no different from that of coins from Latin colonies and allied centers (Cantilena, 1991; Rutter, 2001: 58–61). The numismatic evidence is therefore too fragmentary at present to support weighty conclusions about the political status of Samnite settlements.

4 Material Culture

Artifacts found in the cemeteries, settlements and sanctuaries of Samnium can be classified into three main groups: metal artifacts for personal use, pottery vessels and sculpture. Given their crucial role as mediators of social relations (Schlereth, 1988), artifacts are a source of ongoing scholarly debate about Samnite society and culture, especially as regards the complex questions surrounding production, circulation and consumption of goods.

4.1 Metalwork

In spite of the difficulty in obtaining raw materials and in manufacturing technology, metal objects fulfilled important roles in ancient societies, quickly becoming status symbols or prestige goods (Frankenstein and Rowlands, 1978). Bronze and iron artifacts are found mainly in ritual contexts such as burials, sanctuaries and underwater votive deposits, but very rarely in settlements. Since many of these objects were worn or carried by individuals, they have the potential to inform about the fashioning of personal identity in connection with age, gender and social status. The artifacts are also important chronological types, owing to their recurrence in funerary contexts.

Many of the metal weapons, armor, tools and personal ornaments found in Samnium occur throughout Italy, and during the Iron Age they present a fairly homogenous picture with few local variations. Common types (spear heads, greaves, swords, helmets, cheek pieces, necklaces, bracelets, pendants) are classified into broad regional types, such as "Latial" and "Picene," although such regional pigeonholing of archaeological material in Italy has recently been challenged by Bradley (2000b: 111–13). On the other hand, the breastplates, cuirasses and belts found in Samnium tend to be more specific to the region. Circular bronze breastplates, which were strapped on by means of leather bands, are common in seventh- and sixth-century burials in Abruzzo, particularly at Alfedena, and are also depicted as part of the warrior attire in stone sculptures such as the "Capestrano warrior" and the Guardiagrele stele. The breastplates could have been attached to garments as well as wooden shields, and are decorated with geometric patterns and occasionally with images of fantastic animals in orientalizing style.

A later type of bronze breastplate is the triple-disc cuirass, which is found in fifth- and fourth-century burials in the upper Sangro Valley (Alfedena and Troccola-Pietrabbondante) as well as further south in Poseidonia. Smaller and richly decorated bronze discs (sometimes with pierced decoration) are now known to be elements of female attire. Such discs are occasionally suspended at the end of long chains of bronze pendants called *chatelaines* worn over the chest (Colonna, 2007a).

The question of where these bronze artifacts were produced remains unanswered. Until now, no certain evidence of any bronze workshops has been found in Samnium or in the central Apennines. It would be unsafe to assume that breastplates, cuirasses and *chatelaines* were locally produced only because of their considerable presence in Samnium, since they could have been imported from the great bronze-working centers in Campania and Etruria. One way to solve this problem would be through more detailed art-historical analysis of the bronze pieces from Samnium and comparing them with material elsewhere in Italy. Such an analysis could then be cross-checked with a study of the composition of the bronze. The work would involve collaboration between archaeologists and art historians, who have tended to work separately from each other since the rise of anthropological archaeology in the 1960s. Archaeologists have also tended to ignore the potential offered by material culture theory to resolve crucial archaeological questions (Wobst, 1999). Both disciplines can provide solid arguments about the provenance of artifacts based on detailed study of their stylistic and aesthetic properties, which is very helpful given that direct material evidence of workshops is rarely available. Such an approach is already being undertaken in central Italy (see Papi, 1990, whose claims remain to be confirmed by pertinent archaeological data, especially on workshops).

Bronze belts are the only metal artifacts which can be regarded as characteristic of Samnium. Although bronze belts are a constant presence in Etruscan and Greek outfits, those from Samnium present a very particular typology. The Samnite bronze belt consists of a rectangular, thin bronze sheet sewn onto a band of leather or fabric through small holes at its edges. Sets of two to five clasps are normally riveted to the belt, although they were occasionally fused as part of the bronze sheet. The skillfully-made belt clasps have bodies shaped as elaborate palmettes or cicadas, and hooks representing heads of horses or wolves. The bronze belts date from the sixth to first century, and have been used as dating tools for other contexts.

Bronze belts were long considered part of the Samnite warrior outfit. This interpretation has been challenged in recent studies, which show that they can occur in weaponless graves, and even in graves of infants and children, in sets of two and three or as miniature pieces. Such instances suggest that the bronze belt was part of the male outfit, an identity signifier more than a functional piece of armor (Suano, 1991a, 2000). The fact that many belts were restored in antiquity underlines their importance.

Bronze belts with clasps shaped as cicadas support particularly interesting interpretations. Cicadas incubate underground in their nymph stage for as long as 17 years before emerging as adults and flying away. Such a life-cycle is temptingly evocative of the sacred spring narratives, in that the dedicated youths had to leave their homes upon reaching adulthood. The Samnites may have regarded the cicada as a symbol of passage into adulthood, and what that passage meant in their own mythical frame of reference. The association between the cicada clasps and the sacred spring seems even stronger if one considers that the hook-ends of these clasps are shaped as heads of wolves, which figure in the sacred spring narratives as guiding animals (Strabo V.2.12; Festus p.93L). The cicada-type belts may therefore have been linked with rites of passage into adulthood (Suano, 1991a, 2000). However, it is not entirely clear whether the clasps do depict cicadas or simply reproduce the widespread palmette design. It would be useful to compare the cicada-type clasps from Samnium to depictions of cicadas in Greek iconography and material culture, given Thucydides' comment (I.6.3) that the men of Athens used to wear gold cicadas in their hair.

4.2 Pottery

Both the coarse and fine pottery of Samnium tends to be locally made in the Iron Age. The most common type of pottery is impasto, which occurs throughout Italy in the Bronze and Iron Ages. It covers a broad range of colors (light and dark brown, reddish and black) and fabrics, from coarser wares with large inclusions to finer and more lustrous fabrics such as the so-called *impasto buccheroide*. Locally made fine wares (*argilla depurata*) cover the same range of forms as impasto wares, namely large storage and cooking vessels (ollae, amphoras, *pithoi*, *kraters*), serving vessels (dippers, jugs) and vessels for the consumption of drink (cups) and food (platters, plates).

Imported fine wares, on the other hand, are most numerous in northern Campania, which has yielded Etrusco-Corinthian and Attic pottery (black and red figure) as well as Etruscan and Campanian bucchero. In inland Abruzzo and Molise, pottery imports are limited to bucchero wine jugs from Campania. Imports appear to be rarer in coastal Abruzzo, Molise and northern Campania. These areas may have been removed from the main exchange networks with the Greek and Etruscan settlements, but there might also have been conscious conservatism (Suano, 1991b; Cuozzo, 2007). However, a distinctive type of pottery found in the coastal zone of Samnium displays matt-painted decoration that closely resembles the geometric style found in the neighboring region of Daunia (see below; Di Niro, 1991c; Faustoferri, 1991).

After the fifth century, bucchero practically disappears while Apulian red-figure vessels and Gnathian ware become frequent in coastal Molise and Abruzzo (Di Niro, 1991b). The fourth through second centuries witness the dissemination of black-gloss pottery in Samnium, which was imported from Campania as well as locally made at Aquilonia (Monte Vairano) and Beneventum. From the third century, black-gloss vessels gradually outnumber impasto vessels, both coarse and relatively fine. Lastly, in the first century Samian ware spread into Samnium with the establishment of new trade networks resulting from the Roman hegemony in Italy and the Mediterranean.

Scholars normally conceptualize Samnium as divided with regard to the proximity to Greek and Etruscan culture (Tagliamonte, 1996: 51ff.). The portion of Samnium attributed to the Caudini, or northern Campania, is seen as more open to Greek and Etruscan influences, especially in regards to the adoption of the symposium. The Caudini are, therefore, seen as culturally sophisticated while the other Samnite communities are regarded as clinging to simpler forms of social organization. It is generally accepted, for instance, that in these more "backward" Samnite communities power relations were based on the monopoly of violence (Parise Badoni and Ruggieri Giove, 1980: xxxvi; Torelli, 1980: 33; Tagliamonte, 1996: 125–6; Bispham, 2007b: 189). At the apex of society would have stood warrior aristocracies that maintained their supremacy by force, through actual warfare or by legitimizing their authority through the symbolism of warfare.

However, a re-examination of the pottery shows that, even though there are notable contrasts between the Caudini and the other communities of Samnium, the latter were far more complex than usually envisaged. Although weapons did carry some symbolic importance, it is also clear that the ritual consumption of food and drink was also emphasized in most of Samnium. This is indicated by the overwhelming presence of drinking and eating vessels in cemeteries, and, after the fifth century, in major settlements and important sanctuaries (Scopacasa, 2009: 121–9). The Caudini were thus not the only part of Samnium where social prestige and power could be achieved through less violent means, namely through hospitality (Scopacasa, 2010).

In the sixth and fifth centuries, burials in the Adriatic coastal plain are characterized by pottery sets that emphasize the collective preparation and consumption of food and drink (cooking pots, dippers, bowls). The apparent lack of known fine imported wares suggests that this type of consumption was not geared towards asserting differences in social status through the use of prestige items. On the other hand, hinterland areas display a greater contrast between prestigious pottery vessels and common everyday ones. Campanian bucchero occurs in significant numbers, almost always in the form of wine jugs. This suggests that imported fine ware was geared towards the serving and consumption of wine, while the other activities pertaining to consumption were carried out with local ware. The bucchero wine jugs indicate a type of diacritical consumption, where the status of those who take part in the drinking is advertised (Scopacasa, 2010). However, much of the evidence for ritual drinking and funerary feasting consists of large vases that are only partly interred, and are frequently found already broken and displaced when buried. For many generations, such evidence was attributed to graves damaged by plowing. New excavations are revealing pottery remains of funerary feasting at cemeteries such as Fossa, Scurcola and Comino in south Abruzzo (Ruggeri, 2003: 462; Faustoferri, 2003: 476). In addition, results of excavations at the small funerary site at Tornareccio (middle Sangro Valley) suggest not only that ritual drinking took place by the graves, but also that the vessels were ritually broken and discarded in a pit (Suano, in press).

It is still unclear whether this important facet of Samnite society changed after the fifth century, which, as noted above, was a time of great transformation. It is nevertheless striking that although the great cemeteries decline, the ritual drinking that was carried out in them appears to continue in the burgeoning settlements. The clearest evidence is the substantial presence of Rhodian wine amphoras at the key settlement of Monte Vairano (De Benedittis and Bevilacqua, 1980a, 1980b). This might suggest that the new political elites were appropriating an ancient means of structuring power, which casts a new light on the social and political developments of Hellenistic Samnium. It is only through the publication of more data from settlements and sanctuaries that it will be possible to pursue this hypothesis.

4.3 Sculpture

There are four main types of sculpted figural representations from Samnium: stone sculptures, architectural terracottas, terracotta votive offerings and small bronze figurines. There are also isolated examples of large bronze and terracotta sculptures.

Stone sculptures are some of the most distinguishing elements of central Apennine cultures, despite their small numbers, possibly due to the friable nature of the available stone (Colonna, 2007b: 28). Because these sculptures were originally burial markers, they can inform about the statements of power that were made through the placement and signaling of the dead in the landscape. Many of the sculptures are also important as epigraphic sources, as they display key Oscan inscriptions. The tradition of stone sculptures representing either the buried or their ancestors flourished in the central Apennines from the late seventh to the late sixth century. Some of the statues depict local figures of authority. This is the case of the "Capestrano warrior," whose inscription reads: "*ma kupri koram opsut ani[ni]s rakinelis pomp[une]i*," or "Aninis made me, a beautiful image, for king Nevio Populedio" (La Regina, 1989). This inscription reveals that such statues may also have symbolized the continued presence of the king after death, as protector of the realm and its political borders.

Many of the stone sculptures found in the hinterland of Samnium bear striking resemblance to statues from further north in the central Apennines. For example, the so-called Atessa torso from the vicinity of Monte Pallano (eighth century) shows just how much this area of Samnium was linked to the territory of the Marrucini immediately to the north, which produced the "Capestrano warrior" and his fragmentary female counterpart. A series of large circular stone sculptures recovered in Abruzzo (Comino) are very similar to the broad-brimmed hat worn by the "Capestrano warrior," and are probably all that remain of similar statues (Faustoferri, 2007).

Many architectural terracottas were originally antefixes in temples, later to be discarded in refuse pits after renovation work in the sanctuaries. The imagery on these terracottas follows the common trends from Magna Graecia, especially as regards the Archaic motif of the *potnia theron*. Other imported motifs are also present, such as the dolphins on the architectural terracottas from Monte Pallano (Kane, 2008). Refuse pits of sanctuaries have also yielded large numbers of terracotta figurines of Hercules, Mars, Aphrodite, Cybele and Diana. Other figurines depict the worshippers themselves or anthropomorphic figures with undistinguished features. There is also a wide range of anatomical votives. It is difficult to determine whether these artifacts were locally produced, since they are very similar in style and craftsmanship to figurines found in Campania and Magna Graecia between the fourth and first centuries.

Votive figurines (for which there is still no proper typology or chronology) are so frequent in Hellenistic Samnium that Morelli (1997) sees them as ranging from "popular devotion to therapeutic folly." These figurines reveal a connection to Greek and Roman cults, and also suggest the presence of an ancient cult of the earth mother, the "Mother Goddess dispenser of all fruits" (Figure 25.4; Campanelli, 1997: 133). The sheer volume of votive figurines, plus their relatively crude workmanship, has prompted scholars to see them as the product of local workshops integrated into sanctuaries (Tagliamonte, 1996). There are few examples of large cultic statues from Samnium. The most relevant is the Athena of Roccaspromonte (upper Biferno Valley), which displays clear signs of Etruscan craftsmanship (La Regina, 1989; see Tagliamonte, 1996: tab. XXVIII).

The bronze statuary from Samnium follows the same pattern of dependency on the craftsmanship of the Tyrrhenian coast. This statuary ranges from masterpieces such as the head of S. Giovanni Lipioni, which has been compared to the bronze "Brutus" in the Capitoline Museum (Strazzulla, 1997), to numerous votive or cultic figurines representing Mars, Hercules, standing worshippers and priests. Although most of the bronze statuary has been studied with regard to dating and cultural meaning, knowledge of where the pieces were produced (and the important implications that follow) remains tentative.

Figure 25.4 "Dea di Rapino", sixth-century bronze statuette identified as the goddess Cerfia, linked to sacred prostitution at the sanctuary of Rapino, southern Abruzzo. Source: Used by permission of the Ministera per i Beni e le Attività Culturali – Soprintendenza per i Beni Archeologici dell' Abruzzo-Chieti.

5 Recent Trends and Open Questions: Landscape Archaeology and Daunian-Style Pottery

Owing to the rise of landscape archaeology, sites are no longer regarded simply as the repository of material data. Indeed, they are data themselves, informing us about settlement patterns; social, political and economic management of an area; connectivity; and the ways in which people used the landscape to perceive each other and to communicate with each other. The approach to sanctuaries as boundary markers originated in Greek studies (Schachter, 1992), and has recently inspired new perspectives in Italian archaeology (D'Ercole, 2000; Stek, 2009). However, since sanctuaries only become prominent in the central Apennines after the fifth century, it is plausible that, in the Iron Age, cemeteries functioned as landmarks legitimizing land ownership or community boundaries. That funerary sites could fulfill such a role in the ancient world is suggested in classical texts such as Euripides' *Alcestis*, where tombs are regarded as conspicuous elements in the landscape directed at travelers, that is, people within the community as well as outsiders (Eur. *Alc.* 995–1005). Although studies of funerary sites as boundary markers abound in

Figure 25.5 Pottery of the so-called Daunian type, from Grave 7, Guglionesi. Source: used by permission of the Ministera per I Beni e le Attività Culturali – Soprintendenza per I Beni Archeologici dell'Abruzzo-Chieti.

Neolithic and medieval/Anglo-Saxon archaeology (Lucy, 1992, 2002), only very recently have there been similar approaches in Italian archaeology. D'Ercole (2000) has done pioneering work in seeking to reconstruct the visibility of funerary markers and their political significance in south Abruzzo. However, he does not make allowances for stone markers which may have been dispersed or destroyed, or markers made of perishable materials such as wood. Such difficulties are related to the problem that Italian funerary archaeology has overlooked evidence of ritual activity above and around the graves, as mentioned above.

From the late ninth to the fourth century, there circulated in the Adriatic seaboard a distinctive type of fine-ware pottery, which displays a characteristic matt-painted, geometric decoration (Figure 25.5). It is known as Daunian pottery because its production has been pinpointed in Daunia (north Puglia), although it is also found elsewhere in Adriatic Italy and even across the Adriatic (De Juliis, 1977).

Samnium has yielded a comparably decorated type of pottery considered to be local production styled on Daunian ware. This Daunian-type pottery is found in large quantity at Termoli, Larino and Guglionesi (Di Niro, 1981) and as far inland as Tornareccio near Monte Pallano (Faustoferri, 1991). The view that this pottery was locally produced is based on subjective criteria, such as the perceived quality of the craftsmanship and decoration of the vessels, the density of the clay and the color of the slip. Only with the aid of archaeometry will it be possible to determine whether the Daunian-style pottery of Samnium is indeed a local production or a series of imports specifically made for that market. Such a discovery would have considerable implications. If the Daunian-style pots were indeed custom-made imports, they would change the current image of coastal Samnium as a backwater area where external links intensified only with the rise of Larino in the Hellenistic period. Also, if the Daunian-type vessels were imports, they may act as diacritical status markers, much like the bucchero wine jugs in inland sites. It could be that ritual eating and drinking in coastal Samnium was just as diacritical as in the hinterland sites. But if they were locally produced, the Daunian-type vessels might suggest that the inhabitants of coastal Samnium sought to emulate a type of pottery which they may not have been able to afford, or which they could not access for want of proper connections.

6 Conclusion: Conceptualizing Samnium

From the overview presented in this chapter, it is clear that Samnium as a region exhibits distinctive socio-cultural traits in the broader context of Italy. However, it is also true that Samnium had much in common with neighboring regions in terms of material culture, cultural practices and socio-political organization, and that different areas of Samnium tended to gravitate towards their nearest or most accessible neighbors. We have also seen that, at certain points in their history, the inhabitants of Samnium may have subscribed to a much broader *safin-* identity in common with many other central Apennine communities. All of this shows that ancient Samnium is too complex and dynamic to be defined in view of the clear-cut ethnic boundaries taken from classical sources. It is preferable to conceptualize Samnium as a cultural region which possessed great internal diversity and fluid border zones, and where intense cultural interaction with the rest of the peninsula took place. Bradley's approach to cultural regions in Italy as overlapping entities, as opposed to rigidly defined territories, should prove very fruitful in future research on Samnium and the central Apennines in general (Bradley, 2000b: 111–13).

Acknowledgments

We wish to thank colleagues who provided the illustrations for this chapter (Susan Kane, Luigi Scaroina, Amalia Faustoferri and Angela Di Niro) and the Archaeological Superintendencies of Abruzzo and Molise for the publication permits.

FURTHER READING

Salmon (1967) is the pioneering monograph on Samnium, but since its publication an enormous amount of archaeological data has been brought to light: *Sannio* (1980), Barker (1995a, 1995b), Capini and Di Niro (1991) and *Studi sull'Italia dei Sanniti* (2000). Tagliamonte (1996) and Bispham (2007b) provide updated overviews of the current knowledge on Samnium from the Iron Age to the Roman conquest, as does www.sanniti.info, with a good bibliography. Cornell (1989) offers a comprehensive narrative of the Roman conquest of Samnium, and Dench (1995) is essential for understanding the images of Samnites in classical texts. Oakley (1995) remains a fundamental contribution on Samnite hillforts, and Stek (2009) provides an updated and critical assessment of sanctuaries and their importance in Samnite society. There are also more specific works about the economy, such as Gabba and Pasquinucci (1979), Barker (1981) and Lloyd (1991).

CHAPTER 26

Early Rome and the Making of "Roman" Identity through Architecture and City Planning

Ingrid Edlund-Berry

Archaeological finds provide the only primary data we have for the early history of Rome; there are no contemporary documents other than inscriptions, which, though important, are brief and few in number, and are themselves the product of excavations.

(Cornell, 1995: 26)

1 The Foundation of Rome

The traditional history of early Rome consists of a string of stories that suggest a deceptively uniform progression of events from the arrival of the Trojan hero Aeneas in Latium as the last stop on his long journey after the fall of Troy. Once settled there he traveled to the site on the Tiber River that was to become Rome, and was graciously received by the Arcadian king Evander. Destined to become the ancestor of the Romans, Aeneas lived at Lavinium in Latium, and, after his death, his son Ascanius founded the city Alba Longa in Latium; it was from here that, many generations later, the twin brothers Romulus and Remus set out to found the city of Rome. Beginning with Romulus, the city was ruled by kings (Fraschetti, 2002), who each in his own way contributed to the development of the city, until the dramatic expulsion of the last king, Tarquinius Superbus, and the introduction of a new system of government, the Roman Republic, res publica (Cornell, 1995).

These stories are told by both Greek and Roman authors, in prose and poetry, and, even allowing for some inevitable inconsistencies, the overall pattern is that of a city whose history was closely linked to the other cultures of Italy and also to the Greek world. Because of the way stories were told and remembered from early times, both the

A Companion to the Archaeology of the Roman Republic, First Edition. Edited by Jane DeRose Evans.

oral and written traditions saw Rome as destined from the very beginning to take a lead in the political and economic interactions, first with its neighbors in Etruria and Latium, and eventually throughout all of Italy and the rest of the Mediterranean.

The two Roman authors who give a fairly complete description of the first centuries of early Rome are the poet Virgil and the historian Livy, both writing at the time of Augustus. While Virgil's focus was on the adventures of Aeneas after the fall of Troy, traditionally dated to 1184, Livy takes it upon himself to describe the events of each year of Rome's history, with its founder Romulus as a starting point in 753. By mixing stories of war and peace with domestic events, Livy portrays the reign of each king as representing different facets of Rome's growth, from the founder Romulus (753–717), the lawgiver Numa (716–674), the warrior Tullus Hostilius (673–642), the religious reformer Ancus Marcius (641–617) to the Etruscan kings and their dynasty, Tarquinius Priscus (616–578), Servius Tullius (578–534) and Tarquinius Superbus (534–509). Tied to the reign of each king are further references to the creation of buildings in Rome as the city grew and demands on its public and religious functions increased. It is thus from Livy (and other authors) that we learn of important urban changes such as the draining of the Forum valley and the building of the Capitoline Temple.

Even a quick glance at the three traditional dates given, from the fall of Troy, to the founding of Rome, and finally the introduction of the Roman Republic, makes us realize that we are dealing with a time period of close to 700 years. This is a considerable time in a cultural framework with very little, if any, contemporary writing; there are another 500 years to the lifetime of Virgil and Livy, and, of course, three millennia to the world of today. This simple chronological exercise is critical for our understanding of the events of early Rome and explains why this time period is one of the most difficult to document and evaluate in all of Roman history. Influenced by our own surroundings we may think that we can interpret written sources and material remains from 3,000 years ago without prejudice or accumulated preconceptions (Cornell, 2005). Using Virgil and Livy as stepping stones, we may tend to believe that since they were at least closer to the actual events than we are they must be more or less correct in what they describe, or we may discredit their accounts as a mixture of legends and stories, with a kernel of truth, but not much more (Forsythe, 2005; Wiseman, 2008). On the other hand, if we base our view of early Rome on the material remains alone, we are left with a rather bleak picture of primarily fragments of pottery and other small objects, scanty hut foundations, sections of walls and scattered burials (Scott, 2005). To identify and interpret these disparate remains and to place them in a framework of relative or absolute chronology is extremely difficult, and may even seem unsatisfactory to anyone attempting to gain knowledge of human endeavors in this important part of Italy and the Mediterranean. Although it may seem logical to combine the texts and material remains (and such an exercise can be very attractive), this method can create misleading results if it leads to assumptions of a reality that cannot be verified in actual facts, including the historical existence of an Aeneas, a Romulus, or perhaps even the royal family of Tarquins in Rome (Carandini, 1997; Grandazzi, 1997; Poucet, 2008).

Regardless of how successful we may try to be in associating the material remains with names and events in the ancient texts, there are two main questions for which there may be no absolute answer, but for which all types of sources are needed for a clearer understanding of the past. The first point is the extent to which Rome from the very beginning was different from other cities within central Italy, and the second the extent to which it was destined to become the main city of ancient Italy and also the Mediterranean world. Since, however, the focus of this volume is the archaeology of Rome and ancient Italy, the following discussion will include primarily examples of material remains, while at the same time not ignoring the historical background.

2 The Location and Geology of Rome

Roman tradition always argued that the city of Rome had a privileged position, due to its seven main hills and the Tiber River (Figure 26.1). Because of the configuration of the hills in the landscape, Rome's territory exceeded that of other communities in central Italy (Cornell, 2000b: 215). While the Tiber provided means of transport from Rome further inland as well as access to the sea (Coarelli and Patterson, 2009), it also caused continuous problems through flooding of the surrounding areas, including the Campus Martius and the valleys between the hills, including the Forum (Aldrete, 2007) and the Velabrum (Ammerman, 1998; see Chapter 11). An important task for any individual or group in power consisted of flood control to protect the harbor facilities in the area of

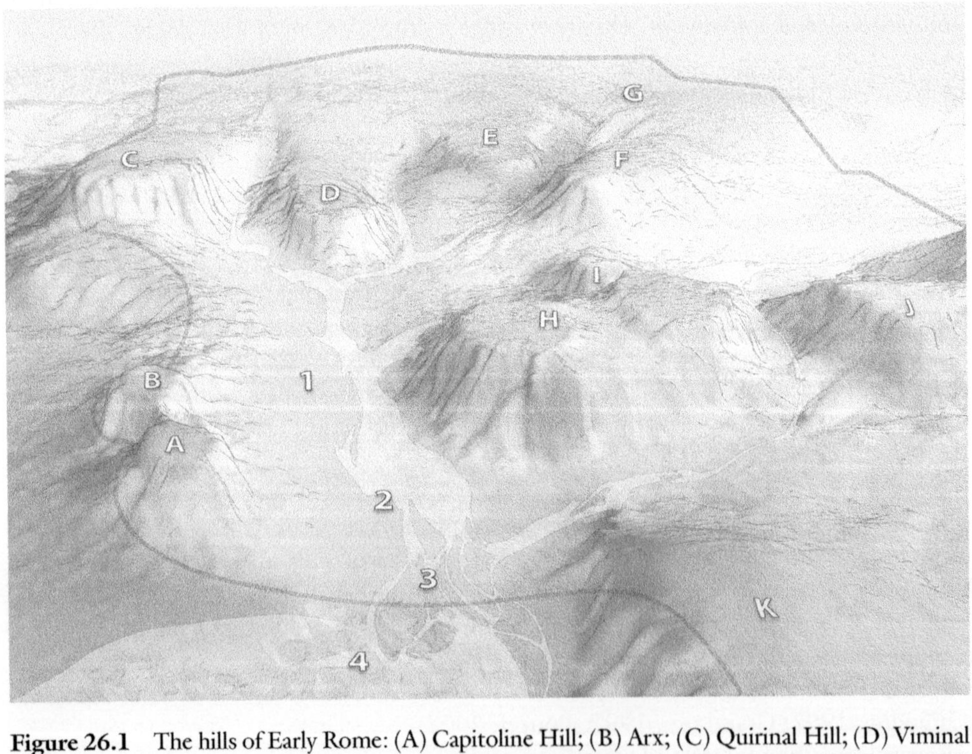

Figure 26.1 The hills of Early Rome: (A) Capitoline Hill; (B) Arx; (C) Quirinal Hill; (D) Viminal Hill; (E) Cispian Hill; (F) Oppian Hill; (G) Esquiline Hill; (H) Palatine Hill; (I) Velia; (J) Caelian Hill; (K) Aventine Hill; (1) Forum Valley; (2) Velabrum; (3) Forum Boarium; (4) Tiber River. The line of the Servian wall is intended to suggest an outline of the fortifications rather than topographically verifiable remains. Source: Image by Jonathan Westin, University of Gothenburg.

the Forum Boarium and Forum Holitorium, and the land reclamation of the Roman Forum became a key issue in the birth and growth of Rome as a city (Ammerman, 1990a; Hopkins, 2010).

Crucial for the development of settlements on the hills of Rome was access to good building materials, and a close analysis of the use of local volcanic tuff (commonly referred to as cappellaccio), as well as tuff from nearby areas, is a useful tool in understanding the configuration of each hill and in dating building remains (Jackson and Marra, 2006; Chapter 17). An important aspect of the geology of the city is also the presence of clay sources in the Velabrum valley between the Capitoline and Palatine hills which allowed for a local production of roof tiles essential for temples as well as public and private buildings from as early as the third quarter of the seventh century (Ammerman *et al.*, 2008; Winter, Iliopoulos and Ammerman, 2009; Chapter 11).

3 Streams, Springs and Water Supply

Due to the geological configuration of the hills and valleys of Rome, the main sources of water came from the Tiber River and its tributaries, and a number of springs (Corazza, Giordano and De Rita, 2006). Many of the springs were well known in ancient myths (such as the spring by the Lupercal cave, *NDTAR*; *LTUR* 3: 198–9 [Coarelli]); some have now disappeared, whereas others are still flowing.

That water conservation was an important issue throughout the history of Rome is shown by the remains of cisterns and wells, useful for identifying both water conservation and engineering efforts. The best preserved round cisterns, cut into the tuff and lined with cut blocks, come from the Palatine, near the so-called House of Livia, the Scalae Caci, and the Domus Augustana (Cifani, 2008: 154–62). Without accompanying small finds, such structures are difficult to date, but the Palatine examples have been dated to the Archaic period. Remains of six wells are located on the Capitoline Hill, of which at least two have been dated to the Archaic period because of the pottery found in them (Cifani, 2008: 109), and as many as 15 have been uncovered on the Velia Hill (Cifani, 2008: 153–4). In interpreting the archaeological records, it is important to distinguish between what is referred to as a "pozzo" for the drawing of water, and a lined pit used as a repository for votive offerings. Depending on the location and construction of the "pozzo," its primary function may have been as a well, only to be re-used for votives.

4 Archaeological Chronology

As can be expected in a city with a long continuous history and habitation, most remains from earlier periods are fragmentary and scattered (Table 26.1). At the same time, new discoveries continue to be made, including re-evaluation of already known and displayed finds, and the events of early Rome are constantly subject to scrutiny and new interpretation (Smith, 2000, 2005). For the earliest archaeological history of Rome, the finds of habitations and burials, including pottery, are usually studied in the context of the area of Latium, to the south of the city. While the scholarly evaluation of the relative and

Table 26.1 Table contrasting historical events, archaeological terminology and archaeological dates with events in early Rome

Historical dates and events	*Archaeological terminology*	*Archaeological dates*	*Events in Rome*
1184 BCE fall of Troy			
Aeneas to Italy	Bronze Age	*c.* 1600–900 BCE	Pottery at Sant'Omobono
	Iron Age	Latial period IIA 900–830 BCE	
		Latial period IIB 830–770 BCE	900–750 BCE "Hut of Romulus" on Cermalus (Palatine)
Romulus 753–717 BCE		Latial period III 770–730/720 BCE	
Numa Pompilius 716–674 BCE		Latial period IVA 730/720–630/620 BCE	*c.* 650 BCE creation of Forum landfill
Tullus Hostilius 673–642 BCE Ancus Marcius 641–617 BCE Lucius Tarquinius Priscus 616–578 BCE	Orientalizing period	Latial period IVB 630/620–580 BCE	*c.* 650–625 BCE earliest roofed building in Rome
Servius Tullius 578–534 BCE			*c.* 580 BCE first temple at Sant'Omobono
Lucius Tarquinius Superbus 534–509 BCE	Archaic period	580–480 BCE	*c.* 509 BCE completion of Capitoline temple
	Classical period	480–400 BCE	Temples built in Rome
396 BCE fall of Veii			
390 BCE Gauls attack Rome			*c.* 378 BCE building of Servian wall

absolute chronology may vary, the basic classification of material remains and cultural development is categorized as the Bronze Age, ending around 900 at the latest, followed by the Iron Age starting at 1020 at the earliest, and ending at 730/720 at the latest (Nijboer *et al.*, 1999–2000; Bartoloni and Delpino, 2005). Depending on the terminology, the following time periods may be referred to as Orientalizing (or Latial period IVA–B) ranging from 730/720 to 580, and the Archaic period from 580 to 480, followed by the Classical period, 480 to 400.

5 Stray Finds

Scattered finds of pottery and tools of flint and stone suggest the earliest presence of individuals or groups in Rome, but it is debatable whether they indicate actual settlements (Gjerstad, 1973: 11–14; Cornell, 1995: 48–53). Especially interesting are Bronze Age sherds found at the site of Sant'Omobono in the Velabrum below the Capitoline Hill, since they may suggest that the material originated on the Capitoline Hill and was moved down the hill through erosion. Regardless of the find context, however, the typology of finds suggests that the hills of the future city of Rome were part of the general pattern of activity in central Italy, both north and south of the Tiber (*Civiltà*, 1976).

Presence of scattered finds can further suggest a continuation of settlements to be verified by excavations, as in the case of the Capitoline Hill. As shown by recent discoveries, there is evidence of terracing, as well as habitation from the Bronze Age, and tombs from the Iron Age (Mura Sommella *et al.*, 2001: 265–8 [A. Cazzella], 269–80 [A. de Santis], 281–90 [F. Lugli, C. Rosa], 291–8 [I. Baroni], 299–306 [P. Boccuccia], 307–20 [S. Brincatt]).

6 Walls and Gates

Like most cities in central Italy, Rome's location in a landscape with hills and valleys early on determined the configuration of settlements perched on hilltops, burials outside the areas for the living, points of crossing the streams that flowed through the valleys into the Tiber, and, ultimately, the need, real or perceived, for marking the boundaries of the hill settlements and for defending the inhabitants from outside attacks.

A few sections of walls consisting of squared blocks in the local grey cappellaccio tuff have been found on the Capitoline hill and in other areas of the city (*Roma medio repubblicana*, 1973: 7–31; Cifani, 2008: 45–73). It has been proposed that these walls are Archaic (seventh to sixth century) and that they formed part of a continuous fortification wall which included the main hills and was laid out in the same line as the subsequent city wall, the so-called "Servian" Wall, built of blocks of a yellow tuff, Grotta Oscura, and dated to after 400. The debate concerning the existence of an early continuous wall is based on the use and date of the building materials, in that the local grey cappellaccio tuff is generally dated to the Archaic period, whereas the yellow Grotta Oscura tuff was quarried primarily in the territory of Veii, a city to which Rome did not gain access until after 396 (Harris, 2001: 543; Chapter 17). The massive sections of the line of wall that is still visible today are usually referred to as the Servian Wall, whether or not any phase of it corresponds with the references in the historical sources to a city wall constructed during the reign of king Servius Tullius (Livy I.44; Cifani, 2008: 255–64). An even more controversial association to a figure in Roman history is found in a stretch of wall located on the northern slope of the Palatine Hill. It has been linked to the first king of Rome, Romulus, based on the historical accounts which describe the founding of the city and a wall erected by Romulus and the corresponding gate, the Porta Mugonia (Livy I.7; Dion. Hal. *Ant. Rom.* 2.50.3; Carandini and Carafa, 1995 [2000]; Carandini, 2004).

7 Roads and Bridges

As in other parts of ancient Italy, paths and roads were created as the need arose to traverse the hills and valleys. Although the purpose of travel varied in different parts of the countryside, ranging from transhumance to commerce and transport, the hills of Rome were undoubtedly connected to each other and also linked to the surrounding settlements. While the later roads of Rome were well documented in the historical sources and named (for example, the Vicus Tuscus or Via Sacra; s.v. Via Sacra, *NDTAR; LTUR* 4: 233–8 [Coarelli]), the actual evidence for the roads is scanty. Some traces of early paved roads have been identified in the Colosseum valley, near the Regia, and by the Temple of Castor (Cifani, 2008: 305–7; Poulsen, 2008: 361–71), and suggest a conscious planning of public spaces and thoroughfares (Chapter 19, and for controversies on the paths of several, see Chapter 7).

Rome's bridges were both important and famous. As suggested by Livy's story of the brave Horatius Cocles, Rome's history would undoubtedly have taken a different course if he had not defended the Pons Sublicius during the siege by the Etruscan king Lars Porsenna (Livy II.10). Although mentioned extensively in the ancient sources (*NDTAR* 299; *LTUR* 4: 112–13 [Coarelli]), there are no remains of this bridge or any other until the Pons Aemilius was built in 179 (*NDTAR* 296–7; *LTUR* 4: 106–7 [Coarelli]; Chapter 19). In the absence of remains of wooden bridges, we can only postulate that the Romans as well as visitors to the city took advantage of the Tiber Island as an easy place to cross the river (unless they selected other points to be crossed by ferries), and to proceed along either shore to the coast as far as the harbor, Ostia, and the saltbeds there, still remembered today through the street name of Via Salaria (*NDTAR* 418–19; *LTUR* 5: 144–5 [Patterson]).

8 Burials

The location of burials is critical for our understanding of the settlement pattern of early Rome since they were usually placed outside the areas of habitation for the living, except in the case of child burials. In addition, they provide useful information about burial types, grave goods, and the social status of the individuals buried. Depending on the state of preservation and excavation history, isolated tombs and groups may be used to suggest the history of different areas of the city.

Although burial sites no doubt were respected, the reason that traces of them have been preserved is usually that the area was later reused for other purposes. Thus, underneath the Forum of Caesar, cremation and inhumation tombs with abundant grave goods have been found, suggesting that a settlement existed nearby, dating from the transition from the Bronze to the Iron Age. Later, but still within the Iron Age, huts were built in the area and the burial ground passed out of use (Conde, 2010).

Likewise connected with the presence of huts are remains of burials on the Cermalus section of the Palatine Hill, near the so-called Scalae Caci. Both the hut foundations (including that of the so-called Hut of Romulus) and the burials suggest continued use, but the relative sequence and chronology from the Iron Age on are subject to much debate (Angelelli and Falzone, 1999; Brocato, 2000).

The so-called Forum necropolis, found at the foot of the Temple of Antoninus and Faustina, serves as one of the key elements for the discussions of early Rome (Gjerstad, 1956; *Civiltà*, 1976: 103–25; Holloway, 1994: 20–50; Barbera *et al.*, 2005). Both its location, in the Forum valley, and its composition (cremation and inhumation burials) have been used to highlight Rome's early habitation (Iron Age phase IIA, ninth century), diversity in burial customs, and a change in the settlement patterns when this cemetery was discontinued, whether replaced by huts or due to the reclamation project which once and for all led to the creation of the Roman Forum as a center for Rome (Ammerman, 1990a; Smith, 1996: 49–54; Chapter 11).

Although often referred to as the Esquiline cemetery, the burial ground on this hill is not confined to a limited area but rather scattered widely, and includes burials from the

Iron Age period phase IIB (or around 800) through the Roman monarchy and Republic (Gjerstad, 1956: 162–266; *Civiltà*, 1976: 125–45; Holloway, 1994: 21–3; Barbera *et al.*, 2005). Many important finds come from this cemetery, including an inscription discussed below, and the area as a whole displays a continuity not known from other parts of Rome.

9 Huts and Houses

Because of the perishable materials used in constructing dwellings and storage areas, there are few actual remains of structures from the Iron Age and later in early Rome. Visually, the best examples of what the dwellings would have looked like come from the funerary urns in the shape of huts found in the Forum necropolis (Gjerstad, 1956). Although executed in varying detail, these models represent oval huts, with a doorway at one short end, and sometimes with windows on the long sides. The modeling of the roofs suggests that they were made of straw and wooden logs, and there are occasional vent holes allowing for the circulation of air from the hearth inside.

These models, in conjunction with the actual remains of hut foundations from the Palatine Hill, provide the best evidence for the huts and hut villages that must have existed in early Rome. The remains on the Cermalus Hill of the Palatine preserve the outline of cuttings in volcanic tuff of a large oval hut measuring *c.* 12 by 8 m (*NDTAR* 74; Cifani, 2008: 265). As suggested by the cuttings of postholes in the tuff, the center of the hut was supported by a main wooden pole, with others forming the outline of the exterior wall, made of clay and reeds.

Because of its location on the Palatine, and references in later historical texts to a hut of Romulus (for example, Dion. Hal. *Ant. Rom.* 1.79.11) that was maintained and continuously repaired as a monument of Rome's past, this hut plays a major role in the discussion of the historical tradition in relation to the archaeological remains (Angelelli and Falzone, 1999; Carandini and Cappelli, 2000: 241–2; Smith, 2005). That there was a settlement on the Palatine in the early Iron Age seems clear, and traces of later huts indicate that the settlement continued even into the Archaic period in the sixth century or even later (Brocato, 2000; Smith, 2005: 96–7).

Other indications of huts in Rome are difficult to verify and have caused a great deal of scholarly debate (Ammerman, 1990a; Hopkins, 2010). While fragments of hut construction, including pieces of wattle-and-daub, may suggest the presence of huts, more evidence is needed to postulate actual dwellings. Especially for the history of the Roman Forum and the huts thought to have existed there, current scholarship suggests instead that before the land reclamation and the creation of the Forum proper around 650, the valley would have been too prone to flooding to allow for any kind of settlement (Ammerman, 1990a; Rathje and van Kampen, 2001; van Kampen and Rathje *et al.*, 2005; Hopkins, 2010; Chapter 11).

As in other parts of central Italy, the form of dwellings changed from oval or rounded Iron Age huts with thatched roofs to Archaic-period rectangular houses, usually with tiled roofs. Evidence for such houses comes from the northern slope of the Palatine Hill, in the vicinity of the so-called wall of Romulus and the Porta Mugonia. Although very poorly preserved, the remains of four houses have been interpreted as having an atrium plan (Carandini and Carafa, 1995 [2000]; Cifani, 2008: 139–45), and as representing the dwellings of very wealthy Romans. As a result of references in the ancient literature that this area was the location of the house of Tarquinius Priscus (for example, Solinus 1.24), the house closest to the Porta Mugonia has been associated with the first Etruscan king of Rome, but the interpretation is much debated (Carandini and Carafa, 1995 [2000]; Wiseman, 2008: 271–92).

If, then, the evidence for houses in early Rome is slim and difficult to interpret, additional information is provided by the presence of architectural elements such as roof tiles and revetments (Winter, 2009). Terracotta objects are quite indestructible, and even small fragments of roof tiles can be studied and thus provide information on missing buildings. While it used to be assumed that architectural decoration such as acroterial statues and revetment plaques were used exclusively on sacred buildings such as temples, this distinction is no longer valid; roof elements are therefore a useful tool in documenting buildings of any kind (Winter, 2009: 2).

For Rome, the earliest roofing tiles (dated to *c.* 650–625) come from a house in the area known as the domus regia in the sanctuary of Vesta (also referred to as the house of king Ancus Marcius; Filippi, 2004; Cifani, 2008: 124–5; Winter, 2009: 8–9). A second building (dated to *c.* 620), in the area of the later Regia, included roof tiles but also horn-shaped acroteria of a type familiar from the Iron Age hut urns (Winter, 2009: 9–11). The first decorated roofs (after *c.* 590) come from areas in the Forum (the area of Comitium and underneath the later Regia) and the Capitoline Hill, and later (*c.* 530) in the same areas but also on the Palatine (Winter, 2009: 554–6). Since the distribution of these fragments seems to be concentrated in important and public areas of the city, it is likely that they belonged to buildings with a sacred or at least a civic function.

While it seems methodologically most correct to describe finds of walls and small objects primarily in terms of find context and topography, it is also important to try to associate the remains with known named buildings in Rome. The ones that are most obvious because of their plan and decoration are the temples, whereas other public buildings are usually identified because of their location and historical context. Many other landmarks in Rome that are known only from the texts must unfortunately remain in a vacuum.

10 Temples and Sacred Places

10.1 Temples

Roman temples were the pride of each generation, and we can follow their creation through the long lists of names of patrons and dedication dates (*NTDAR* 445–6). As can be expected some temples are better known through the historical and literary texts, including the most famous of all temples in Rome, the Capitoline Temple dedicated to Jupiter, Juno and Minerva (Albertoni and Damiani, 2008; Cifani, 2008: 80–109; Hopkins, 2010: ch. 4; see Chapter 11). According to the texts, including Livy, it was built during the reign of the Etruscan kings, the Tarquins, and finally dedicated in 509, just after the fall of the monarchy and at the creation of the Roman Republic (Cifani, 2008: 80–1). Although known from descriptions in the ancient texts and from coins, the preserved remains of the temple consist mainly of sections of the podium and fragments of the roofing system, all of which lead to suggest that it was a monumental complex, larger than any other temple from central Italy.

Perhaps because this structure is so well known in the literature, the real significance of its origin and its importance are easily overlooked. Although in many respects the plan and decoration follow the pattern of Archaic temples in central Italy, the extraordinary size of the foundations alone (*c.* 74 by 54 m, according to Mura Sommella *et al.*, 2001) suggest a unique administrative setup providing the materials and workforce. Likewise, the chronology of the history of the construction straddles the complex question of Rome's identity in the Archaic period, whether the city was perceived as "Etruscan" because of the presence of the Tarquin family, including the rulers, Tarquinius Priscus and Tarquinius Superbus, or had already achieved its own "Roman" identity, granted with inspiration from its Etruscan neighbors (Cornell, 1995: 151–72) and perhaps even a broader international perspective (P. Davies, 2006; Chapter 22).

As indicated by the historical records, the creation of temples was an important act already for the first kings, beginning with Romulus and into the first century of the Republic (*NDTAR*, 445–6). While some were dedicated to rather obscure deities such as Semo Sancus or Fors Fortuna, others played an important role throughout the history of Rome, not just Vesta (the hearth of the city), but Saturn, Diana, and the triad of Ceres, Liber and Libera. Since most temples tended to be repaired and rebuilt as needed, there are few remains for the earliest period of construction. Fragments of revetment plaques dated to *c.* 590–580 have been attributed to the Temple of Jupiter Feretrius (originally founded by Romulus), although there are no remains of foundations (Winter, 2009: 148). In the temple of Castor (dedicated in 484) cappellaccio blocks with an Etruscan round molding of the original podium have been preserved in the much later reconstruction of the building (Nielsen and Poulsen, 1992: 82; Cifani, 2008: 119–23).

As can be expected, find circumstances may be attributed to the fact that some temples are better documented than others and therefore stand out in the study of early temples in Rome. Thus, the area of Sant'Omobono in the Forum Boarium, significantly situated on the banks of the Tiber River (Figure 26.2), has produced evidence for the podium of a temple characterized by its small Etruscan round molding (Edlund-Berry, 2008). To this temple belonged a set of architectural terracottas, including a closed pediment with heraldic felines and a Gorgon, dated to *c.* 580 (Winter, 2009: 349–50). Although the general concept of a temple structure raised on a podium and the use of architectural terracottas fits into the pattern used in central Italy, the early date and some features of the terracotta revetments may suggest that this temple is part of a pioneering effort in establishing an architectural tradition, perhaps to be attributed to king Tarquinius Priscus.

A second phase of the same temple at Sant'Omobono, dated to *c.* 530, likewise preserves a podium, now shown with a larger double Etruscan round; a set of architectural terracottas with frieze plaques of types known also from Etruscan Veii and Latin Velletri; and ridgepole statues of standing figures including Herakles and Athena (Winter, 2009: 316–18). Again, this temple may suggest a link to a ruler in Rome, the king Tarquinius Superbus, and illustrates Rome's leadership in creating temples and temple decorations (Winter, 2009: 581).

The plan and decoration of the temples in early Rome suggest that the Romans, that is the inhabitants of Rome, probably represent a mixture of cultures and traditions, and

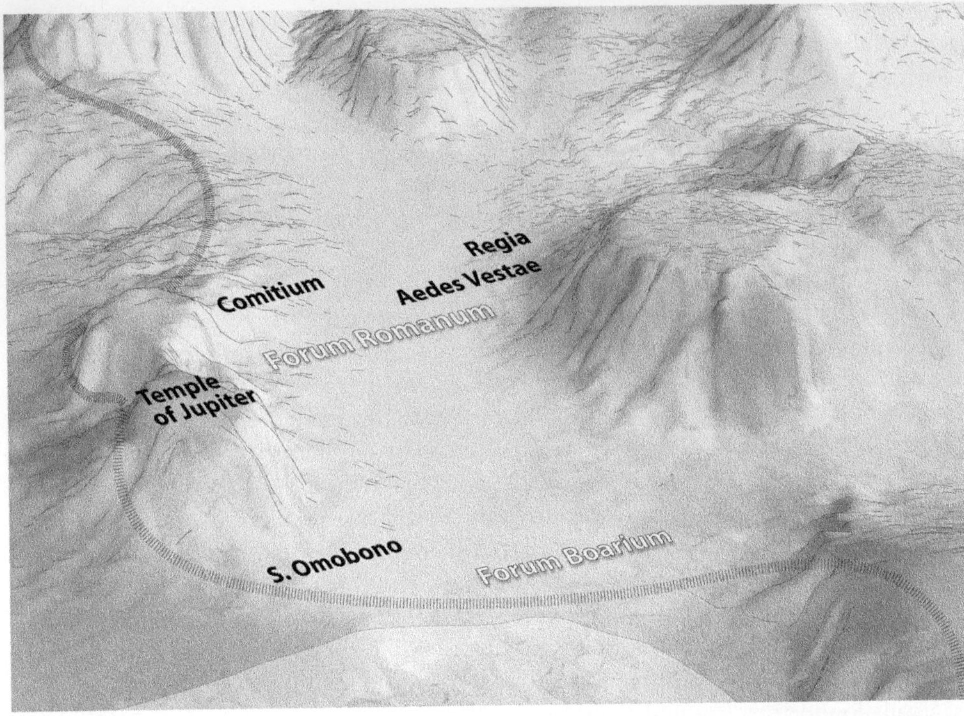

Figure 26.2 Rome around 500 BCE. Source: Image by Jonathan Westin, University of Gothenburg.

may have had a leading role in developing new architectural forms, soon to be shared by their neighbors to the north and south. While unfortunately no names of individuals can be securely attached to these innovative features, we can postulate a strong political system such as that of the ruling kings and their families that had the resources to initiate and complete temples and other types of public buildings. In addition to the architectural innovations, the temples in Rome are also important because of their locations, carefully chosen to correspond with the religious perceptions of where deities were to be venerated, whether on hilltops, near water, or at road crossings.

10.2 Sacred Places

Temple buildings represented only one form of identifying sacred places. Equally common, or probably even more common, were the places marked by natural features such as springs or trees, in addition to altars or votive deposits which contain miniature pottery, votive figurines and other small objects. While many such places are known from the texts referring to an altar (s.v. ara, *NDTAR*; *LTUR* 1), lakes (s.v. lacus, *NDTAR*; *LTUR* 3), springs (s.v. fons, *NDTAR*; *LTUR* 2), specific trees (such as ficus or fig, *NDTAR*; *LTUR* 2) or groves (s.v. lucus, *NDTAR*; *LTUR* 3), actual remains are scarce.

The sanctuary of Vesta, with its sacred hearth and grove, represented the lifeline of Rome, and even today its round temple is a landmark in the Roman Forum. Yet the early history of the area is difficult to interpret (s.v. Vesta, Aedes, *NDTAR*; *LTUR* 5: 125–8 [R. Scott]). Three different contexts, known as the Temple of Vesta (aedes Vestae), the House of the Vestals (Atrium Vestae) and the house of the pontifex maximus (the domus publica or domus regia), are known from the literary sources and have been associated with remains from the mid-eighth century (Filippi, 2004; Cifani, 2008: 124–5; Scott, 2009; Arvanitis, 2010).

11 Votive Deposits

While many votive deposits lack a context, others can be attributed to identifiable places (Bartoloni, 1989–90). Examples of the former category come from the Capitoline Hill, a place with sacred connections (Cristofani, 1990: 63–8, 73–4; Cifani, 2008: 79, 112–13), two locations on the Quirinal, at S. Maria della Vittoria and Villino Hüffer (Bartoloni, 1989–90; Cifani, 2008: 73), and on the Oppian Hill (Cifani, 2008: 74). Others are part of complex features such as the so-called Lapis Niger (black stone) area in the Comitium. Here, in addition to the inscribed stone, described below, and a U-shaped altar of Grotta Oscura tuff, votive objects ranging from pottery (impasto, bucchero, Etrusco-Corinthian, Attic, including miniature pottery) to bronze and terracotta figurines, objects of ivory and bone, and animal bones illustrate the variety of finds common to votive deposits particularly in Rome and Latium (Cristofani, 1990: 54–9). While the area has been variously interpreted as the Tomb of Romulus (although there is no evidence of a burial) or a sanctuary of Vulcan, the continuous use of the site for placing votives, from the sixth to the first century, suggests its importance for the Romans during the monarchy as well as the Republic.

Other votive deposits from distinct sacred contexts come from the sanctuary at Sant'Omobono, already mentioned (Cristofani, 1990: 129–30) and from the area of the Temple of Vesta (Bartoloni, 1989–90). The latter site also contains a pozzo with finds of pottery and other objects, including roof tiles; in a different context, the material could be viewed as domestic rather than sacred (Bartoloni, 1989–90: 749; Cristofani, 1990: 62; Carafa, 2004).

12 Public Spaces

As suggested by areas such as the Sanctuary of Vesta in the Roman Forum, it is difficult to identify the function of structures based on the archaeological remains, and the attribution of names to buildings is often based on their location in relation to other buildings or to textual references. One such building is the so-called Regia in the Roman Forum, known from the texts to have served as the office of the pontifex maximus, and as a storage area for sacred objects connected with the cult of Mars (s.v. Regia, *NDTAR*; *LTUR* 4: 189–92 [R. Scott]). The remains of early structures indicate five phases (*c.* 620 to the early fifth century) with different plans and roofing elements (Cifani, 2008: 126–30; Winter, 2009: 554–5; Hopkins, 2010). Whether these structures are to be interpreted as predecessors of the Republican Regia is uncertain, but throughout the history of the building, the plan included two or three small rooms and an open courtyard of a type known also in private houses. Remains of huts below the area of the later Regia and in the area of Vesta probably indicate habitation that was replaced in connection with the landfill and creation of the Forum as a public open paved space (Ammerman, 2009; Hopkins, 2010; see Chapter 11).

Like the area of the Regia an area known as the Comitium, located below the Capitoline Hill, has a long and complicated history. As a place of assembly throughout the history of Rome and also linked to the sacred area of the Lapis Niger (see above; *NDTAR*; *LTUR* 1: 309–14 and 5: 211 [Coarelli] and s.v. Comitium and Niger Lapis), its main feature was a pavement which probably dates from *c.* 625–600, subsequently replaced by two other pavements during the sixth century, and later expanded with a speaker's platform, the Rostra (Cifani, 2008: 113–18). Fragments of decorated architectural terracottas, including a Gorgon antefix, dated to *c.* 590–580, may belong to a roofed building such as the meeting hall for the senate, the Curia (Winter, 2009: 144).

13 Art Objects

In addition to the many buildings and other structures that gave Rome its public profile during the monarchy and Early Republic, both public and private areas of the city contained works of art that helped glorify the events of Roman history and traditions. As indicated in the texts, these monuments could take the form of figures on foot or on horseback, and, when placed in temples, of cult statues (s.v. statua, *NDTAR*; *LTUR* 4). Since the material used was primarily bronze or terracotta (only later imported Greek marble), most of these statues have disappeared, or remain as tantalizing fragments.

Of the preserved sculptures in Rome, none represents the history and sentiments of Roman tradition better than the bronze statue of a she-wolf with the infant Romulus and Remus. According to tradition this statue has never been buried and remained an outdoor monument until it became one of the key pieces in the Capitoline Museums. While it has long been recognized that the statues of the infants are not ancient, the wolf herself has usually been classified as a masterpiece of Etruscan Late Archaic or Early Classical art. Recently, however, the statue has been analyzed and it has been suggested that she actually dates from the Middle Ages (Carruba and De Masi, 2006; Parisi Presicce, 2000; Bartoloni, 2010; Chapter 23). Whatever the final outcome of this debate may be, there are many artworks (mirrors, reliefs, paintings, etc.) that illustrate the she-wolf as the nurturer of the founders of Rome, Romulus and Remus, and even if this particular statue is not of ancient origin, there most likely were other representations that were displayed in Rome and reminded the Roman public of this important tradition in their early history (Evans, 1992: ch. 4, app. 1).

As in Etruria and Latium, the prime material for works of art as well as architectural ornaments used in Rome was terracotta, usually painted in vivid colors. Fragments of statues have been found in surprising numbers in Rome, mostly connected with buildings, which illustrate the quality of the local terracotta production, especially during the sixth century. The two best preserved terracotta statues, representing Athena and Herakles, come from the sanctuary at Sant'Omobono in the Forum Boarium where they, together with others, decorated the ridgepole of the Archaic temple (see above). Both the location and the themes illustrated have parallels in Etruria and in Latium, and suggest the close connection between Rome and the artistic traditions in central Italy (Winter, 2009: 377–81).

The significance of the terracotta production in Rome is far-reaching. First, the creation of large-scale terracottas requires a well-developed production including presence of clay beds and kilns, technological and artistic expertise, and, ultimately, the requisite economic and cultural milieu. Second, the themes chosen may illustrate myths and stories that supplement the literary texts and reveal Rome's awareness of a world beyond its immediate regional boundaries, expressed in local art works but also in imported ones such as the terracotta acroterial statue of a fallen Amazon, found on the Esquiline Hill (*c.* 500; Lulof, 2007). Unlike later forms of Roman sculpture, which focused on the portrayal of statesmen and historical figures, the artistic tradition of Archaic Rome, with the exception of the she-wolf (if indeed she is ancient) is global rather than local. When this attitude may have changed is a question for historians of the Roman Republic.

14 Inscriptions

As can be expected for early Rome, there is only scant evidence of writing, consisting of inscriptions for a variety of contexts. The texts range from individual words to names and short phrases and an example of a religious law, but together they provide examples of all three languages represented in Rome, that is, Latin, Etruscan and Greek.

The Latin inscriptions include a bucchero sherd from the context of the fourth phase of the Regia complex in the Roman Forum, dated to the last quarter of the sixth century. It preserves the Latin word "rex" (king), which may refer to the king as the ruler in the monarchy, or the priest known as "rex sacrorum" (king of sacrifices) (Cristofani, 1990: 22–3). A form of the same word occurs also on the much longer Latin inscription from the so-called Lapis Niger area in the Roman Forum (see above). The text, carved on four sides of a square block in Grotta Oscura tuff, is fragmentary but seems to contain a religious law protecting the sanctity of the area (Cristofani, 1990: 58–9; Cornell, 1995: 94–5). The block is commonly dated to the sixth century, and would suggest an early use of the Grotta Oscura tuff, otherwise best known from monuments (including the Servian Wall) built after the fall of Veii in 396 (see Chapter 17).

Also written in Latin is a text inscribed on the so-called Duenos vase of buccheroid impasto, consisting of three miniature bowls joined together (Cristofani, 1990: 20–1). The vase was part of a votive deposit found on the Quirinal Hill (Bartoloni, 1989–90; see above) and the inscription, dated to the mid-sixth century, seems to contain a dedication (Colonna, 1979b: 163–72).

Examples of Etruscan inscriptions include names or parts of names and come from areas such as the sanctuary at Sant'Omobono in the Forum Boarium (impasto sherd, late seventh century; Cristofani, 1990: 21), the foot of the Capitoline Hill, near the Temple of Saturn (bucchero cup, *c.* 550–500; Cristofani, 1990: 21–2), and the Esquiline cemetery (Italo-geometric sherd, *c.* 700–650; Colonna, 1987: 58; Bagnasco Gianni, 1996: 305).

Also from the sanctuary at S. Omonbono is an inscription with three names, found in a context dated *c.* 540–530 and inscribed on an ivory plaque in the shape of a recumbent lion, interpreted as a "tessera hospitalis" (guest token) (Cristofani, 1990: 21). At times, the fragmentary state of the inscription makes it difficult to verify whether the wording is in Etruscan or in Latin, as seen in a bucchero sherd from the Capitoline Hill, near the Tabularium (late sixth century; Cristofani, 1990: 22).

Evidence for the presence of Greek speakers comes from an inscription found in the only preserved chamber tomb in the Esquiline cemetery (tomb 125), dated to the late seventh century. A fragmentary name is preserved on the body of a transitional Corinthian olpe, and it has been speculated that it represents a Greek who died and was buried in Rome (Colonna, 1987: 57–8; Mura Sommella, 2000; Hopkins, 2010). The value of these inscriptions for evaluating the composition of the population of early Rome is subject to much discussion (see, for example, Cornell, 1995: 156–9). Regardless of their fragmentary nature, the distribution in findspots and contents suggests that writing was indeed used, and in all likelihood a tool for documenting information that could be accessed by many (Cornell, 1991, 1995: 103–5).

15 Questions of Chronology and Development of Society

Most discussions about the history and culture of early Rome are based on the accounts in the historical and literary sources which include references to the layout of the city as a whole and to its individual buildings. The interpretation of the archaeological remains, on the other hand, is closely tied to the complex documentation of finds from the neighboring areas of Etruria and Latium, and the relative as well as the absolute chronology continues to be a topic of much controversy (Bettelli, 1997). For the purpose of a general understanding of the formation and development of early Rome, it may be necessary to keep the two views of early Rome separate. In other words, the one created from the texts is rightly tied to historical names and events, and as such serves to support the view of a city rich in monuments created by its rulers and inhabitants (Carafa, 1996). The archaeological Rome is much more sketchy and anonymous, yet the presence of walls, foundations, architectural terracottas, graves, pottery and metals suggests very clearly that Rome existed and flourished and that the lack of archaeological evidence should not be viewed as a failure of Rome but rather as a challenge to us to respect the evidence for what it is. We should also take into account the physical setting and, to the extent possible, the natural resources that allowed Rome to make full use of its location and its proximity to its neighbors.

Regardless of which chronological system we follow, the birth and growth of Rome seems to follow a path from the creation of individual settlements (primarily on the Capitoline and Palatine hills), and corresponding burial grounds outside the boundaries for the living. In the next phase, a uniform action is taken to create the Forum valley as a central gathering place, accessible from all the neighboring hills. As a result, sacred and public buildings begin to be erected in the Forum, and this area is accessed through streets leading from the river and from the adjoining hills (Chapter 11). While the archaeological remains do not provide information on how the planning and execution of building projects were accomplished, or on the bureaucratic mechanisms put in place to provide the workforce or necessary equipment, the historical sources suggest that it was the rulers and ruling families that each contributed to the expansion and embellishment of the city.

Seen in the context of archaeological chronology, the Bronze Age represents the beginning of settlements on the hills of Rome; the Iron Age witnesses expansion of the settlements and the unification with the creation of the Roman Forum, and by the Archaic period, the city of Rome is ready to plan for new buildings in the Forum with a focus towards politics through public buildings and towards divine protection through the erection of temples, further expanded into other areas of the city.

16 The Early City

No matter how much the later texts exalt the appearance of the city of early Rome, the archaeological reality suggests that it consisted of modest settlements, spread out over the seven hills and surrounding areas, and that there was little or nothing to suggest that Rome was in any way different from the settlements of its neighbors, except for the unquestionable benefit of the location on the Tiber (Figure 26.2). The layout of the settlements was determined by the hills, and burial grounds were placed away from the habitations of the living. The difficult question to address for this time period is whether there was a substantial change in the mid-eighth century that can be attributed to the figure known from the texts as the founding father, Romulus (Carafa, 1996).

In terms of the physical layout of the city, the first major change in Rome's perception of itself was due to the creation of the Forum as a civic center. The land reclamation of the Forum valley was a massive undertaking, which required organization, engineering skills, a workforce, and someone with a vision to what this space would mean for the city of Rome as a whole. Only two other urban projects could rival this: one was the planning and execution of temples including the Capitoline Temple in the sixth century; and the other was the building of the so-called Servian Wall in the early fourth century.

17 Creation and Growth of the City of Early Rome

17.1 The Forum

The plan of Rome differed from that of its neighbors in one important aspect in that the public center of the city, the Forum, was located in a valley between the major hills. It is through the creation of this space that we can evaluate the force of Rome's birth and development as a city, based on features of geology and engineering as well as politics and economy (Cornell, 2000b).

As indicated by the geologic map of Rome (Figure 26.1), each valley between the hills featured a stream that conveyed water to the main water source, the Tiber. From a practical point such valleys were prone to flooding, and the crossing of streams could provide a challenge as well as a means to prevent movement between individuals and groups.

To facilitate the creation of the Forum, several actions had to be taken: in order to eliminate the recurring flooding of the valley, enormous amounts of soil had to be transported to the area and further be contained to assure stability (Ammerman and Terrenato, 1996; Ammerman and Filippi, 2000, 2004; Hopkins, 2010). Studies of the water and soil levels in the Velabrum and the Forum areas indicate that the reclamation of the Forum valley with soil from the Velabrum took place during the middle of the seventh century

(Ammerman *et al.*, 2008) and that it resulted in access to the clay beds of the Velabrum used for the production of roof tiles but not decorated terracottas. The scope of this effort was huge and involved raising the ground level to make it safe from flooding while at the same time allowing for drainage from the streams in the valleys between the hills. Once the Forum area had been created through the landfill, and had also become paved, public and sacred buildings could be built here, including the area of Vesta and the area of the later Regia (see above) (Colonna, 1979a; Chapter 11). A further technical achievement was made through the construction of the Cloaca Maxima to allow for drainage of the whole area into the Tiber River (Cifani, 2008: 308–13).

17.2 Temples

After the engineering achievement of creating the Roman Forum, the next creative effort is seen in the building of roofed structures, whether intended as temples or as public or private buildings. In addition to the many buildings documented primarily by the presence of roof tiles or architectural terracottas, the earliest temples are the first and second temples at Sant'Omobono (see above; Winter, 2009: 554–5). None of these building efforts, however, rival the scale and engineering program displayed in the creation of the largest structure in early Rome, the Capitoline Temple dedicated to Jupiter, Juno and Minerva (see above, section 10.2; Hopkins, 2010).

Following the completion of the Capitoline Temple, many other temples were constructed, primarily in the Roman Forum but also in other parts of the city. According to the historical records, there were many temples built in Rome between the years 509 and the early fourth century, but only a few have been preserved (*NDTAR* 445–6; Ziolkowski, 1992; Orlin, 1997). Important additions in the Roman Forum were the Temple of Saturn and the Temple of Castor, whereas the Temple of Ceres, Liber and Libera was built on the Aventine, as was the Temple of Juno Regina. The Temple of Semo Sancus was built on the Quirinal, and the Temple of Apollo Medicus in the Campus Martius, by the Tiber River. In the Forum Boarium, the Archaic Temple at S. Ombono may have been replaced by twin temples to Fortuna and Mater Matuta in the early fifth century (Pisani Sartorio, 1982; Hopkins, 2010). For many of the temples known from the later Republic and Roman Imperial period, original features such as isolated blocks may indicate the earliest phase, such as elements of the podium of the Temple of Castor. But for the most part these temples are known only in their later phases.

The plan and decoration of the temples in early Rome suggest that the Romans may have been a leading factor in developing new architectural forms, soon to be shared by their neighbors to the north and south. While unfortunately no names of individuals can be attached to these innovative features, we can postulate a strong political system that had the resources to initiate and complete temples and other types of public buildings.

17.3 The Servian Wall

In addition to the creation of temples such as the Capitoline Temple, and the space of the Roman Forum, the third major achievement of Rome was the building of the so-called "Servian" Wall as the main city wall, constructed after the fall of Veii and the attack by the Gauls in the early fourth century. Although Rome in all likelihood had fortification walls already in the Archaic period and perhaps even earlier (see above), the Servian Wall represents a unified effort to include a major part of the city within a protective wall, built of Grotta Oscura tuff, quarried from the newly defeated Etruscan city of Veii (Chapter 17). The wall measures over 11 km in length, and consists of large rectangular blocks built up in alternating rows of headers and stretchers. It ran from the Capitoline Hill across to the Quirinal, the Viminal and Esquiline, and from there over to the Aventine and the Caelian towards the Forum Boarium and Forum Holitorium by the Tiber River (*NDTAR* 262–3). Regardless of the function of the wall for defense purposes, its construction at a time when Rome had faced two very difficult military endeavors – the siege of Veii and the attack by the Gauls – indicates Rome's ability to produce resources, manpower, and organizational and engineering skills.

18 Rome of the Romans

There are many ways to view and appreciate the city of early Rome. One is to visualize the many buildings and monuments recorded in the ancient texts, but which have disappeared without any trace today. Whether mentioned in historical narratives or isolated inscriptions, the names of sanctuaries, houses, statuary and other urban features help us build up the city from its very beginning and down through the first century of the Roman Republic. In comparison, the archaeological remains are anonymous, and, for the most part, fragmentary and difficult both to interpret and to place in a chronological context. All the same, the earliest remains of dwellings and burials from the Iron Age suggest habitation on the hills of Rome, and by the time of the seventh century the creation of the public gathering place, the Forum Romanum, changes the configuration of Rome from a cluster of hills to a unified whole with a civic center. As this space begins to be filled with buildings and monuments, at least this part of Rome becomes a monumental space that spans chronologically from the Archaic to the Classical period, with no perceptible change as a result of the introduction of the Republic in 509. This monumentality of the Forum Romanum is paralleled with the main temple of the Capitoline Hill, the Temple of Jupiter Optimus Maximus, and temples built on the surrounding hills.

From the point of view of the archaeological remains, it may seem peculiar that the next major building project did not happen for over a century – that is, the creation of the so-called Servian Wall, built in the early fourth century, as a result of the Gallic attack on Rome in 390. Although we have lists of buildings, in particular temples, built between 509 and 390, there is only some evidence of grandeur in the Temple of Saturn, the Temple of Castor and architectural terracottas, yet the overall picture is rather a quantity of smaller structures. It is, of course, difficult to say whether this observation is due to the chance preservation of buildings, or the manner of record-keeping, or whether the political events during the first years of the Republic occupied the attention of the Romans at the expense of construction and embellishment (Cornell, 2005).

A third way of viewing early Rome is to blend the historical and literary records with the physical archaeological remains and to attribute names of famous buildings to the fragmentary walls and other features uncovered. While this approach certainly makes the archaeology of early Rome come alive, it opens up the question of the historical record, and the extent to which we can verify the elements of Roman traditional history, whether based on fact, fiction or a mixture of both (Smith, 2005: 92).

Even if we attempt to document and evaluate the creation and development of early Rome with the help of generally accepted historical facts, the question still remains as to how Rome fits into the culture of central Italy. To create meaningful comparisons of material remains, their manufacture and their distribution, it is easy and often necessary to look south to Latium and north to Etruria. Especially for the time period of the seventh and sixth centuries, the historical tradition highlights the close connections between Rome and the Etruscans, in particular through the arrival of L. Tarquinius Priscus and his wife Tanaquil from Tarquinia. According to the historical accounts, it was during the reign of members of the Tarquin family that Rome grew as a city, and buildings such as the Capitoline Temple are generally perceived as examples of Etruscan architecture. But, since even each Etruscan city was known for its individual traditions in terms of both architecture and culture (see Chapter 23), it should not surprise us if Rome developed its own designs and variations on familiar themes. Therefore we should specify local Roman characteristics of a general central Italic cultural tradition, composed of Etruscans and Latins alike (Ridgway, 2010). While clearly "Rome was not Etruria" (Rasmussen, 1997: 29), it would seem that the presence of Etruscans, speaking their language and familiar with the traditions of the Etruscan cities, played a significant role in the composition of the Roman society in the same way as the Latins, Greeks and other inhabitants of the Italian Peninsula all contributed to the creation of Rome, the city of Romans (Chapter 22).

19 Conclusions

Whether we choose to focus on the historical and literary texts or the material remains, there is little doubt that the Romans were passionately aware of their own identity. As they viewed their family history, the lists of consuls and other officials, the traditions of priesthoods and military exploits, they acted as Romans, citizens of the city of Rome. Unlike their neighbors to the north and south, the Etruscans and the Latins, who were identified by their language, the Romans were recognized as dwellers of the community of Rome. Their government and society was centered on the city, and it was not until they had serious conflicts with their immediate neighbors that they realized that there were other cities and communities that considered themselves equally worthy of power and status. It would take the Romans a long time – with many military defeats and political setbacks – to realize that while their identity provided their best defense, it also was a difficult concept to introduce to other cultures with the goal of superimposing Roman values on others (Chapter 27).

For the understanding of the appearance of early Rome we are to a great extent limited to very fragmentary archaeological evidence that is difficult to interpret. Unlike later periods in Roman history, where building inscriptions and other primary sources define the location of buildings and their history, isolated remains of walls and small finds tend to provide a limited picture. It is therefore tempting to apply the events described by literary sources such as Livy or Virgil to the actual remains, and to equate the monuments mentioned with the ones referred to in these texts. A clear example is the Iron Age hut on the Palatine that traditionally is referred to as the Hut of Romulus, a monument described and cherished throughout Roman history. Archaeologically, however, there is no evidence that the foundations of this hut have anything to do with a presumed Romulus, but it certainly makes the history of postholes and cuttings in the tuff more alive. And, as much as we would like to be able to present the phases of the growing city tied to the familiar names of kings and places, the Rome we thus create must remain an intellectual exercise that needs to be separated from the archaeological evidence. Rome's history was indeed long and glorious, and is represented thus in the later texts. That the physical remains are scattered, fragmentary, and at times few and enigmatic in no way detracts from the glory of Rome, but rather invites us to look very carefully at the evidence and to keep an open mind as to how the city of early Rome appeared, albeit a separate entity from the "Rome of the Romans" and our dreams. While the early city of Rome may not appear to have been destined to the global importance it later achieved, the overall location and the ambitions of the early inhabitants certainly set the city apart from its neighbors and counterparts in Italy, and perhaps even the Mediterranean.

FURTHER READING

The sources for the archaeology of early Rome include ancient historical and literary texts in the original languages or in translation. Due to the often fragmentary nature of the archaeological evidence, it is difficult to access, and, unfortunately, many books and articles refer to monuments and objects without giving information about find context, condition and present location with photographs or drawings. While there are many comprehensive monographs and textbooks on the history of early Rome, the amount of archaeological information is determined by the authors' interest and expertise in the available material. Many of the specialized topographical or chronological studies of early Rome are written for specialists who are well acquainted with the evidence and issues of interpretation. Topographical dictionaries provide useful information on sources and excavated remains, including those pertaining to early Rome (see *NDTAR*, *LTUR*). The publications particularly significant for a general overview are Gjerstad (1953, 1956, 1973), which provides descriptions and illustrations of monuments and tomb groups, but is not up-to-date; Cristofani (1990); Cornell (1995); and Smith (2005).

For archaeological evidence, consult: Anzidei, Bietti Sestieri and De Santis (1985), Bettelli (1997), Bietti Sestieri and De Santis (2000), Colonna *et al.* (1976), Dondero and Pensabene (1982), *Enea nel Lazio* (1981) and Fabbri (2010).

For the history of early Rome, consult: Canali De Rossi (2005), Carandini (2006), Cornell (1997), De Simone (1981), Fraschetti (2002), Menichetti (1994), Smith (1996), Walt (1998) and Wiseman (1995).

For specialized studies and methodology, see Ammerman (2006, 2007), Bietti Sestieri (2000), Bruno (2006), Capanna and Amoroso (2006), Carandini (2000), Carandini and Bruno (2008), Carandini and Greco (2004, 2005, 2007, 2009), Dumser (2005), Farney (2009), Gallone (2000), Ridley (2005), Terrenato (1992) and Wiseman (2000, 2001, 2007a, 2009).

CHAPTER 27

Material Culture and Identity in the Late Roman Republic (*c.* 200–*c.* 20)

Miguel John Versluys

Like all genuine questions, the question about identity will never die.

(Pesic, 2003: 147)

1 Introduction: Rome as Part of a Hellenistic *Koine*

In the final quarter of the second century the city of Praeneste – 32 miles east of Rome – was embellished on a magnificent scale by the building of a new public center with, rising on terraces above and culminating in a porticus in summa cavea, a sanctuary to Fortuna Primigenia (Figure 22.2; see Coarelli, 1987b). With its scenographical orientation on the Via Latina, it was an impressive statement of civic independence and modernity. The (architectural and visual) language the Praenestinians used to express themselves is Hellenistic in every respect (see Lauter, 1986). This is also the case with the famous Nile Mosaic from the town hall of Praeneste: an almost ethnographic picture of Nubia, Egypt and the inundation of the Nile of supreme artistic quality which one would sooner expect in centers of Hellenistic art like Pergamon, Alexandria or Rome (Meyboom, 1995). All this shows that Praeneste was, at least in cultural terms, part of a very large world, a world that stretched from the Pillars of Hercules to the Sudan and Iran. From a contemporary perspective this was, in fact, almost the entire *oikumene* (Nicolet, 1991). Apparently the Praenestinians selected from a repertoire available in that world whatever they considered appropriate for their local concerns. Scholars often designate this "repertoire of material culture at hand" as Hellenistic *koine*. The process of its application could be characterized by the terms universalization and particularization (see Witcher, 2000). First there had been a universalization of the particular: (aspects of) Greek architecture or Egyptian topography had been transferred into Hellenistic *koine*. Subsequently

A Companion to the Archaeology of the Roman Republic, First Edition. Edited by Jane DeRose Evans.

we see a particularization of the universal: selected parts of this Hellenistic *koine* are used for local purposes. When we shift our attention from the monumental to the more mundane, the same picture emerges. Second Style Pompeian wall paintings, for instance, show the existence of a common Hellenistic repertoire found all over the Mediterranean and Near East, together with unique local applications using elements of this *koine* (see Figure 3.4) (Tybout, 1989). Pottery styles and decorations also fit this framework (see Poblome and Waelkens, 2003; see Chapter 5).

Stylistically the repertoire was immense and it included elements from the Etruscan and Italic traditions, the East and Egypt, although it was clearly dominated by styles usually called Greek or Hellenic. The question to what extent "western" cultural traditions like those of the "Celts" added to the *koine* is debated, but the system as such clearly functioned in the Iron Age worlds. A recent interpretation of visual material culture from later Iron Age society (i.e. the last two centuries BCE) characterizes its functioning as: "thinking global, acting local" (Hunter, 2006: 103). A special characteristic of the application of the repertoire in this period is the desire for a combination of a variety of "culture-styles." As many different elements as possible were combined in the most innovative ways that could be devised. Zanker (2007: 38) has characterized and explained this tendency as "stilistische Vielfalt als Bereicherung der Ausdrucksmöglichkeiten" ("expressive possibilities enriched by stylistic diversity").

From this perspective, the combination seen in the reliefs from a monumental base from Rome known as the Ahenobarbus Altar (100–80) in a "Hellenistic baroque" and a "Roman veristic" style do not surprise (see Figure 8.1) (Stilp, 2001). Nor does the Memmius monument (50–25) nor the so-called Octagon (around 40, probably the tomb of a Ptolemaic princess) in Ephesus, which both display a variety of different cultural references in bewildering combinations (see Berns, 2003). All these examples belong to the same cultural *ambiente* that stretched from the Iulii monument in Glanum in the west (Gaul, around 50, combining socle, arch, temple and pyramid form in 18 vertical meters; see Rolland, 1969) to Nemrud Dağ in the East (Commagene, around 50, with Iranian elements also being part of the juxtaposition; see Versluys, in press). Almost nowhere does this *koine* look exactly the same. This is only logical when we recall that, besides being original, its application is all about contextual particularization. Iranian elements that claim a connection with Persian kingship make sense in a dynastic context like Nemrud Dağ; they had no significance in Glanum. In Alexandria, a very important consumer and producer of Hellenistic *koine*, the repertoire therefore looked different. Through a series of tombs and their decorations, it is possible to follow its development in this domain from the early Hellenistic period onwards, as well as follow it in great detail (Venit, 2002). These differences, of course, also depended on the specific history and location of a city or area: in this respect geography matters. The universalization and particularization of Egyptian elements played a much bigger role in Alexandria than Glanum or Nemrud Dağ. But the point is that they potentially *could* have been applied at Nemrud Dağ or in Glanum and, consequently, that also in Egypt itself Egyptian elements are different after they had become part of the Hellenistic repertoire (Versluys, 2010). As we have seen in Praeneste, the outcome was also dependent on the possibilities (financial, social, cultural and practical) of the commissioners and users: the availability of opus incertum, among other things, is what made the sanctuary possible in this particular "Latial" form (see Zanker, 1976). In borrowing from the repertoire the Praenestinians simultaneously added to it and developed it, as did all participants. Still, the locality of all these monuments and places can only be understood against the "global" background that was Hellenistic *koine*. If this "putting together of elements" was more the norm than the exception, then why would we characterize this kind of material culture as "hybrid"?

2 Identity

Sociology has convincingly demonstrated that identity is primarily contextual and that it must be understood as a fluid signifier operating on different levels, often simultaneously. Our personal identity is made up of many differing, and sometimes conflicting, perceptions. In fact, one should rather speak of multiple identities – always in the plural – that are applied in different contexts. Identities are often realized, defined and played out in confrontation with "the Other." Such is the case with collective identities, or images that a group constructs of itself so that its members can identify with them. Nations are probably the clearest examples of these collective self-definitions. If people identify themselves with such an image we speak of cultural identity. Cultural identities are thus a construction and in order for people to believe in them, they need to be made explicit through forms of symbolic expression. Material culture plays an important role in this creation of shared symbolism, a process which is described as cultural formation or ethnogenesis (see Derks and Roymans, 2009). In the guise of cultural memory, the past is often a defining element in processes of ethnogenesis. This is not the past as "historical fact," but the past created by a society at certain locations and occasions in order to make meaningful statements about the present. In this way cultural formation can result in traditions that, when formalized, constitute a canon. Cultural traditions, therefore, are not given or stable, but form the (successfully functioning) imagined collective identity of a group. As such they tend to become a normative instrument, to be used as a survival strategy for cultural identity in times of crisis.

The concept of identity thus includes many related notions relevant to almost every aspect of society, from the level of the individual to that of groups and, ultimately, that of society itself (Halpern, 2004). This is the reason why expressions of identity should be studied in relation to one another and on these different levels, as they are, for instance, in Assmann's concept of "das kulturelle Gedächtnis," or cultural memory (Assmann, 1992). It is clear, therefore, that we cannot understand ancient societies without asking questions about identity. In answering these questions, however, contextual understanding is crucial in every respect. Moreover, a defining element of any understanding of identity expression must be that identities are always a construction put into practice through processes of identification.

3 Thinking about Material Culture and Identity in the Late Roman Republic

The problem with many interpretations regarding identity within Roman archaeology is that all of the above is often not taken sufficiently into account. In general (Roman) archaeologists tend to think in terms of "ethnos" and not "ethnogenesis"; in terms of "cultures" and not "cultural formation"; and in terms of "traditions" and not "the invention of traditions" (for the latter concept, see Hobsbawm and Ranger, 1983). Our research tradition, in other words, has so far "tended to mistake a particular configuration of apparent stabilities for permanent associations between space, territory, and cultural organization" (Appadurai, 2001: 8); identity could be added to that list.

I will highlight here three reasons in particular that might explain this situation. The first one has to do with the Roman period itself. Questions of "national" identity and what it meant to "be Roman" were heavily debated by the Romans themselves. A large part of Roman literature can therefore (also) be read as a discourse on identity and Empire, especially in the Late Republic (Moatti, 1997; Steel, 2002; Dench, 2005). All this is very relevant for our understanding of identity in the Roman world, as long as we realize that material culture has its own discourse that is not necessarily congruent with what we see debated, mediated and suggested in the texts (Cordier, 2005; see also below). Horace's famous statement "Graecia capta ferum victorem cepit et artis intulit agresti Latio" (*Epist.* 2.1.156–7: "Captive Greece captured the barbarous victor and brought the arts into rustic Latium") is important in order to study the presence of Roman identity in relation to Greek material culture in the Late Republican period, but only when it is realized that Horace presents a myth to serve his own ideological agenda. Latium was not rustic before the conquest of Greece; nor was Rome a passive victim of the Greek arts.

The second reason has to do with how, since the nineteenth century, the Roman world and its material culture have been studied. As within the historical disciplines, the framework of understanding has largely been one of "methodological nationalism" (see Terrenato, 1998a; Pitts and Versluys, in press, provide a critique and an alternative). This implies that – although often implicitly – the main unit of analysis used is the nation state, with the inherent notions of imperialism, colonialism, cultural superiority and national identity. As a consequence the study of the Roman world and its archaeology developed into Area Studies. Such studies see the Roman world as a mosaic of many different, distinct "cultures" that interacted with Rome through processes of acculturation, and hence these cultures became more or less "Romanized" in that process. Each of these "cultural-containers" would have its distinct ("national") identity that would be expressed, moreover, in the style of its material culture. Although local identities certainly play a very important role in the Roman Republic and Empire, the Roman–native dichotomy evoked by thinking in "cultural-containers" is particularly unhelpful in understanding material culture (see van Dommelen, 2001 with a case study on Sardinia in the Late Republic). Before this will be further illustrated below it is important first to discuss the third reason why much of Roman archaeology has remained rather static in interpreting identity. This has to do, as we shall see, with its general (mis)understanding of the important concept of "style."

4 Style

For good reasons, style is often considered to be a crucial element – perhaps one could even say *the* crucial element – for the expression of identity by means of material culture. Within archaeology this approach has come into current usage through post-processualism. Prior to this, style was often seen as a kind of DNA of a specific culture, in terms of material characteristics (see Meskell, 2001). Each culture would be characterized by its specific style and hence a map of the "Roman" world consisted of a mosaic of these culture-styles. Post-processualism enlarged this framework by seeing style as having mainly to do with communication between or within groups. Stylistic choices would serve ideologies and strategies of legitimation and the use of a specific style would be a sign of status and identity or serve as a vehicle of meaning within or between cultures.

In this vein, the stylistic choices made for the monumentalization of Praeneste have been interpreted as a symbolic resistance against Rome under the heading of "Gegenarchitektur" (Ley and Strauss, 1982). Theories about the "religious Romanization" of the Italian Peninsula in the Late Republic through the distribution of distinctly Roman material culture and religious practices also fit this framework as well (Torelli, 1999). An example of such an approach for contexts outside Italy is the interpretation of "Roman-style" veristic portraiture erected by Greeks in the first century as an expression of their support for the new ruling power. By using this specific style they would present themselves as *philorhomaioi* or friends of the Romans (Smith, 1988) Although such approaches (rightly) focus on the social functions of style, they still (wrongly) regard it as a kind of cultural index and therefore they often end up interpreting styles of material culture in terms of resistance versus adherence. Where style was first generally used as an indicator of ethnic identity, it is now, as with the three examples mentioned above, often used as an indicator of cultural identity. In practice many archaeological interpretations seem to include aspects of both approaches without serious reflection on the concept itself (see Versluys, 2008).

On theoretical grounds this state of affairs has been severely criticized from both within and outside the field (see, respectively, Pitts, 2007, and Neer, 2005). It can also be refuted on practical grounds, as in the Late Republican and Imperial period the use of styles has been convincingly reconstructed as being a "semantic system" (Hölscher, 2004; originally Hölscher, 1987). Within this system a specific style or visual language was related to a specific *theme* and *association*. For example, within the Ahenobarbus Monument (the Paris-Munich reliefs), a Hellenistic baroque style was used for the mythological subject of a marine *thiasos* while the "historical" subject of the census called for a very different veristic style. Styles were thus not used as an index of ethnic or cultural identity but they were applied to evoke certain associations. The system, of course, was not one of mathematical precision: Zanker (2007: 31) has rightly characterized its application as "'Schlagbild'-Repertoire" (repertoire of impactful visual images). Not only were the semantic elements flexibly used, they were also combined or mixed with newly designed figures – as has been illustrated by the examples from Gaul, the Italian Peninsula, Asia Minor, Commagene and Alexandria; these examples were discussed in the first section of this chapter. The reason for their application was content-driven and the historical background and original meaning of the styles and elements that were used played no real role with their new application (Hölscher, 2004, 2006). This is logical when one realizes that they are styles and elements functioning within a repertoire of Hellenistic *koine*. In the process of the universalization of the particular, these styles and elements will already have lost part of their original meaning and understanding. Through their subsequent particularization, they were changed again to suit their new context. In time, they also acquired new meanings and understandings through this process. What we would need, therefore, to properly understand the contextual meaning of a certain stylistic element in the Late Republican period is its "cultural biography" (see Kopytoff, 1986), from its original associations to its uses, applications and transformations within the Hellenistic *koine*. What we call "Hellenization" and "Romanization" therefore cannot, in regard to material culture and style, be understood as acculturation processes between distinct "cultural-containers," as Veyne (1979) realized long ago. They are not so much about cultural transformation, but about identity and social experience (Cordier, 2005; Wallace-Hadrill, 2007).

It can be concluded, therefore, that style *is* crucial with the expression of identity, not as a kind of ethnic or cultural index, but as something actively significant in itself. Stylistic elements are not only a reflection of their social contexts, they are at the same time constitutive of it (see Tanner, 2000; Tanner and Osborne, 2007; Elsner, 2010). This is the reason why the concept of *bricolage* is so well suited to the study of material culture and style in the Late Roman Republic. *Bricolage* has been defined and applied to the Roman Republic by Terrenato (1998a: 23) as "a process in which new cultural items are obtained by means of attributing new functions to previously existing ones" resulting in "a complex patchwork made of elements of various age and provenance: some of them are new, but many other are old objects, refunctionalized in new forms and made to serve new purposes within a new context." This patchwork has been illustrated by the various examples adduced in the first paragraph of this chapter, where it was argued that processes of universalization and particularization explain its functioning. In terms of material culture, the entire Hellenistic world – of which the Late Roman Republic is part (see below) – is such a collage. This is an important reason why using the term "hybrid" for this context is inappropriate, as everything in the period, in one way or another, is a hybrid. What makes *bricolage* a particularly solid and useful concept is that it suggests agency. It is important to understand this agency as an entanglement of humans and material culture. There is agency on part of *bricoleurs* who appropriate cultural items to create a new kind of patchwork. At the same time, however, there is agency on the part of (some) styles of material culture through their cultural biography.

5 Identity, Material Culture and Style in the Late Roman Republic

Having briefly discussed the three basic concepts (identity, material culture and style), we can now look at how all of this worked in social action and behavior. How do we understand the expression of identity through material culture in the Late Republic? And how do we do that in such a way that: (i) the texts are deconstructed and integrated with our understanding of material culture instead of being unthinkingly followed; (ii) the perspective of "methodological nationalism" and its related concepts are evaded; and (iii) styles are properly understood as part of a constitutive semantic system and not simply as an ethnic or cultural index (alone)?

To answer these questions the practice of code switching is of great importance. A concept developed in the social sciences – Bastide (1955) famously called it "le principe de coupure" (the principle of cutting) – it has been taken up by Wallace-Hadrill to understand relations among identity, material culture and style in the Roman world (a series of important articles is now summarized in Wallace-Hadrill, 2008). With code switching, identity is described as functioning as a persona or theatrical mask: in the one context the wearer takes on identity "x," while a different context asks for the display of identity "y."

This principle has been well established within the language of Late Republican authors (Adams, 2003). Cicero's public speeches, for instance, are markedly Latin in their language while his private letters show an abundant use of Greek. This also applies to the contextual usage of material culture. Wallace-Hadrill (2008) advances the example of the toga and the pallium to illustrate this. The toga was a marker for Roman and hence when going to the forum one was meant to wear a toga. A pallium, on the other hand, was suggestive of "Greekness" (but not "Greek") and as such was not an indication of a Greek ethnic or cultural identity but a dress to be worn for leisure at home. This is underlined by the fact that the concept – and the Latin word – pallium is, in fact, a Roman construction of Greekness, as the *himation* was not considered to be an ethnic marker of Greekness by Greeks themselves. The practice of code switching is in line with what has been said above about styles of material culture functioning as "Schlagbild-Repertoire." Such a "semantic" model, however, seems to be even more useful for understanding material culture than the bilingualism metaphor for three reasons. Firstly there were, within the repertoire of material culture, more theatrical masks available than just "Greek" and "Roman." To properly understand those personae, secondly, we should not think about original meanings but rather about the distinct cultural biographies of certain types, elements or styles as they developed through time. Thirdly, using linguistic models easily leads us back to the understanding of material culture through texts, while an archaeological interpretation should begin with establishing how types, elements or styles constitute their context as cultural operators (see further below). The term "cultural scenario" is therefore perhaps an appropriate way of naming these personae (Versluys, 2008).

In his seminal essay, Hölscher (2004) adduces many examples of cultural scenarios. For instance, he traces the development of a specific representation of the goddess Aphrodite (the so-called Capua Type) and shows how each subsequent phase can only be understood as the contextual appropriation of what is used as a strictly formal, visual example. Because of this continuing process of particularization, original meanings were soon forgotten. The understanding of cultural scenarios thus lies at the intersection of their cultural biography on the one hand and their contextual application on the other. As such, their meaning is a process (see Tanner, 2000; Witcher, 2000). Elements from the Egyptian visual language might serve as another illustration. From Alexandrian funerary culture, for instance, we know that through time, elements of the Egyptian cultural repertoire were to dominate the funerary domain (Venit, 2002). In the first century, Greek, Roman, Egyptian or Syrian inhabitants of Alexandria could all be buried "in the Egyptian manner." The application of this culture-style had nothing to do with ethnic identity or with a desire to become Egyptian. In the cultural scenario Egypt simply "owned the afterlife" and would continue to do so in the Roman Empire – and not in Alexandria alone (Riggs, 2005). In the same period, however, Cleopatra used the cultural scenario "Egypt" for very different reasons – it is of dynastic legitimation – in other contexts. For both the citizen of Alexandria and the Ptolemaic queen the cultural scenario "Egypt" was only one of the masks at their disposal. The Alexandrian might well have carried a Greek name and have played a largely Hellenic role in public life, while in Mediterranean contexts Cleopatra most often behaved in a manner distinctly Greek (see Versluys, 2010). Code switching was thus all about cultural competence, to know what cultural scenario to use in what kind of context. As such it most often has to do with *social* identity (Jenkins, 2008; Hales and Hodos, 2009).

Does this imply that every style we find used in the Late Republic was a conscious appropriation of a specific stage in the biography of a cultural scenario on the part of the user or consumer? Yes and no. No, because there clearly were "waves of fashion" (see Wallace-Hadrill, 2008) as well as (very practical) standardized applications between a specific theme or even type of material culture and the style used for it. Consumers will have followed these unwritten rules unconsciously. For a representation of a god in free-standing sculpture, for instance, a style *a la manière de* Phidias was often used, as the theme "god" was associated with concepts like "majesty" for which the style of Phidias was considered most appropriate (Hölscher, 2004). And yes, precisely because of the fact that "fashion" or popularity should be studied in terms of particularization and universalization to make visible what otherwise remains implicit in the practice of populations. The "blindingly obvious" (pottery types, house decoration, etc.) still needs attention (see Miller and Woodward, 2007, and Kopytoff, 1986, for commodization). However, so far most interpretations of Hellenistic *koine* and cultural scenarios have focused on monumental building projects like those in Praeneste, Rome or Nemrud Dağ, where we know quite a lot about the *bricoleurs* and their motivations. But there clearly is a need to study a wider range of different types of material culture in these terms as well, as becomes clear from the Mahdia shipwreck (sunk probably around 80–60) and its stylistically extremely heterogeneous cargo (Hellenkemper Salies, 1994).

6 Roman Identity and Romanitas

The previous section has illustrated how, in the Late Republic, material culture served to express social identities on the level of the individual and the group. By making the concept of cultural scenarios central to my analysis I have privileged social identity over other forms of identity building. That does not imply that those other forms of identity did not exist or did not play a role in identity building. Styles of material culture certainly *could* be used as an indicator of ethnic identity or evoke associations with a specific cultural identity. It has been illustrated above that "things Greek" could mean very different things in very different contexts – often having nothing whatever to do with the form of Greek identity anymore – but at the same time it is also clear that "things Greek" in the Late Roman Republic were sometimes very well and generally understood as ethnically and culturally Greek. Roman versus Greek was a well-known and often evoked antithesis; molded into a *bon mot* already by Cato the Elder (234–150) who, in 191, explained to the Athenians, in their own city no less, that "the words of the Greeks issue from their lips; those of the Romans come from the heart" (Plut. *Cat. mai.* 12.5–7). The Late Roman Republic certainly saw several instances when reality lived up to this ethnic stereotyping, including the expulsion of Greek philosophers from Rome (see Gruen, 1984; Ferrary, 1988). What then about these identities, which seem to play a role in a more "national" discourse? What about feeling, being and becoming Roman in the Late Republic and what about Roman identity and Romanitas? (see Seel, 1964; Dench, 2005).

Sociology has distinguished several different approaches towards national identity; I will briefly introduce the four most relevant categories (see Smith, 1991). The first one could be called territorial or spatial. It starts from the assumption of the existence of a "homeland" in which a certain people and a certain geographical area would belong historically together as fact. The second one is ethnic, in the sense that it focuses on the people as a kind of super-family and not so much on a specific territory belonging to them. The national identity of these people is seen as inalienable. A third category is cultural and sees national identity as a set of shared values that define "who we are": in this view cultural identities are something like national forms of life. A fourth category has national identity as being political and juridical; here citizenship is the defining characteristic. To understand how these identity-constructions function in social interaction and behavior, it is useful to focus in particular on how people identify with them. In studying identity, identification certainly is a key word. For more recent periods these processes of identification have been well studied and analyzed (see Halpern, 2004). In general sociologists distinguish among functional, normative and emotional identification. Functional identification is the pragmatic adherence to the identity of a group or nation. Normative identification calls for the internalization of specific values by the members of that group or nation: these are their frame of reference. Emotional identification is again stronger and sees normative identification developed to the level of an unconditional loyalty and pride towards a certain national identity.

What is it that we see in the Late Roman Republic? Not surprisingly, we see all these different constructions functioning interchangeably through one another for different contexts and different players and different moments (see Roman and Roman, 2007). The problem with many interpretations using the word "Roman" to indicate identity, perhaps, is that the interpretations do not specify what they mean in terms of specific forms of national identity and identification. But it is clear what Roman national identity actually *was* in the Late Republic. Roman national identity was about a political and juridical status that one could obtain through citizenship. Functional identification with this identity seems to have been the norm, but it could certainly develop into normative and emotional identification. The tension between these categories can be clearly felt from a passage in Cicero (*Leg.* 2.5), where he maintains that: "all men from the *municipia* have two homelands, one by birth and one by citizenship … But that in which we were born is no less sweet to us than that into which we were adopted. And so I will never deny this [i.e. Arpinum] is my homeland, although the other [i.e. Rome] is greater and includes this within it."

Notwithstanding the degree of identification, Roman national identity was a legal status unaffected by cultural choices. The very moment that L. Cornelius Balbus from Gades obtained Roman citizenship from Pompey for military excellence in service of the Republic, he ceased to be a citizen of his Iberian home town. But on returning to Gades he could resume local citizenship – and stop being "Roman" (Wallace-Hadrill, 2007). The first man of non-Italian origin to become consul, Balbus had coins issued in 41 that show the thunderbolt of Melqart/Hercules as the symbol of his native Gades (Steel, 2002). This was no show of resistance. In the same period, but on the other side of the Mediterranean, Antiochos I, the builder of Nemrud Dağ, used the cultural scenarios "Greek" and "Persian" for dynastic self-presentation, notwithstanding the fact that he possessed Roman citizenship and had to thank Pompey for his kingdom (see below). In cultural respects, there was no need for these men to show their Roman identity. This makes sense when we realize that their Roman identity is something political and juridical, in the first instance appearing as a functional identification alone. Moreover, in cultural terms something like a Roman identity did not exist as a fixed element or package to resist or adhere to. What was Roman and what constituted Romanitas was a much-discussed topic in the Late Republic (Seel, 1964; Ferrary, 1988; Moatti, 1997). Wallace-Hadrill (2007: 371; see also Wallace-Hadrill, 2008) has concluded about the "Romanization" of Italy that "the Italians became Roman on the condition that the Romans became Italian." This also seems to apply to the *oikumene* at large: the world became Roman on the condition that Rome became the world. With this there seems to have been a remarkable difference between the west, mainly in the political sphere, and the east that dominated the cultural sphere. But it worked; and one of the main conditions for its success seems to have been a cultural system characterized by an almost indefinite openness. This, of course, is not to underplay what has been characterized as the "ugly side" of Roman imperialism (Mattingly, 2010); but it is to say that to understand (material) culture, identity and style such a colonial model does not work in many contexts.

Be that as it may, it is important to underline that culture and material culture often have nothing to do with constructions of national identity and related processes of identification, as the coins of Balbus and the temple-tomb of Antiochos I have already illustrated. Roman Italy could be constructed in literary texts of the Late Republic as having a distinct territorial and ethnic national identity demanding normative and emotional identification, while at the same time everybody in Neapolis spoke Greek. We should not see this as an inconsistency (see Versnel, 1990/4). Asking about the Hellenization of Rome is therefore asking the wrong question, as in the Late Republican period we cannot speak about a cultural dichotomy between Eastern-Hellenistic on the one side and Italic-Roman on the other. But of course such dichotomies were passionately evoked when the context required them to be. This is the aporia so well analyzed by Lévi-Strauss (2001, originally 1952/71): cultural innovation in societies depends on the intensive collaboration with and the integration of other cultures, while simultaneously these same societies are in need of "the Other" in terms of negative self-definition for their identity and cultural formation.

It can be concluded that most of the national identities we see discussed (and claimed and invented) in the Late Republic are, in fact, cultural scenarios that have to do with social identity and practices of affiliation. We have seen that Cato the Elder constructed Greeks as "Other" to define what was "Roman" at the very moment that Rome saw a sharp intensification of the presence of "things Greek." The correlation between the two developments is, of course, no coincidence and could also work the other way round. The same Cato was probably actively involved in changing constructions of the image of the Sabines from "luxurious and decadent" into "rustic and quintessentially Roman" (Farney, 2007; see Chapters 22 and 25). Consequently, the habit of promoting a Latial family identity became widespread in the Late Republican period to claim aristocratic status. While in the literary discourse the Sabines were made into real Romans and the Greeks into non-Romans, simultaneously the Hellenistic *koine* was functioning and developing with its own, very different cultural logic. It often used categories that looked similar or even were alike, but which had very different meanings.

7 History and Archaeology

The Late Republic is characterized by a series of historical developments that would have a major impact on the ancient world: in all aspects and domains of life things were changing on an unprecedented scale. "Intensification" seems to be the key word that characterizes all these changes. Shipwreck data, for instance, display a threefold rise in number from the third century to the first century (see Hellenkemper Salies, 1994). I purposefully did not begin this article by outlining a historical narrative in order to try and understand material culture only as a reflection or illustration of that historical narrative. Making historical narratives about a small city-state that conquers the world central to our analyses of material culture could easily lead to a conceptualization of "Rome" in nation-state terms of national identity and cultural superiority that would be expressed through a distinctly Roman culture-style (see the three points of critique formulated above). In such a framework (visual) material culture is routinely understood as directly expressive of the social realities of its time. This is only true to a certain extent, however, because material culture is as constitutive as it is reflexive (see Elsner, 2010, and the remarks made on the concept of *bricolage* above). An archaeology of the Late Roman Republic should therefore first focus on concepts such as appropriation, agency, intention, causation, result and transformation.

"Rome" is not a culture but a cultural tradition in development, especially in the Late Roman Republic. The dialogue between the local and the "global" as it takes place within the Hellenistic *koine* is particularly illustrative of this. Above we have seen that large parts of the literary discourse mainly seem to be tied up with the creation of something like Romanitas. However, the literary discourse in fact also testifies to this dialogue. The poem by Catullus on Attis might serve as an example (Nauta and Harder, 2005). Catullus's protagonist is a figure associated with the legendary place of origin of the Roman people, although in the poem Catullus has him challenge almost all the conventional norms of Roman behavior. In terms of literary composition the text testifies to a quite unprecedented merging of genres from the Hellenistic literary and poetic *koine*. Nauta has argued that in terms of interpretation it could be read as a discourse on national identity and Romanitas, as for Romans Phrygia was a component of Roman (constructed) ethnic identity (Nauta and Harder, 2005). Catullus's poem on Attis therefore illustrates the aporia described above in the definition of Lévi-Strauss: Attis is made into the exotic Other on the one hand and simultaneously included as Roman on the other. Being Roman was about doing both things at the same time, in terms of identity. This is the reason why Romans could combine an active use of negative stereotypes about foreigners, as Greeks and other Orientals, while embracing the legends that traced their origins to those same Greeks and Trojans (Gruen, 2006). This is not inconsistent: it is being Roman.

8 Questions of Scale

A recurring problem throughout this chapter is how to define the Late Roman Republic geographically. Late Hellenistic Commagene (see above) is an illustrative example. Antiochos I had to thank Pompey for the stability of the region, which enabled him to build Nemrud Dağ. The Roman general was even directly responsible for the enlargement and enrichment of the kingdom by granting Antiochos I the important city of Seleukeia (Zeugma) (see Versluys, in press). In that respect Commagene can be argued to be part of the Late Roman Republic from the Amisos conference (65/64) onwards. The same reasoning applies to large parts of the Iberian Peninsula from the second century onwards and to Egypt after the Popilius Laenas incident in 156 when it in fact became a Roman protectorate. In this respect large areas of the Hellenistic world were part of the Roman Republican system. The (material) culture of the Roman Late Republic, on the other hand, has been shown in this chapter to be wholly part of Hellenistic developments. Also here "Roman" and "Hellenistic" can only be seen in relation to each another. A Porticus Italica for people from Italy, such as the one preserved on Delos, is as Late Hellenistic as it is Late Republican (Trümper, 2008). Late Republican Rome can only be properly understood as a Hellenistic state.

Chronological definitions are notoriously difficult, but if we take the "intensification" described above as a defining characteristic, a new phase in the development of the Hellenistic Roman Republic seems to begin after the Second Punic War. Visual material culture, at least, can be argued to be largely "realistic" before that period (i.e. being a direct expression of the social and political realities of its time) and largely "semantic" after (as suggested by Hölscher, 2004). Perhaps the Foucauldian transition from "things" to "subjects" might help to describe this development and help us to understand it better (Foucault, 1969). But if not, to understand material culture as an expression of identity it is crucial to realize that the period around 200 is a turning point. In terms of (material) culture the Late Roman Republic continues until the Augustan age is in full swing, around the last decennia BCE. The Mausoleum of Augustus (Hesberg and Panciera, 1994), conceived and designed around 30, is still a full-fledged part of the culture of *bricolage* that has been described above as characteristic of the Late Republic. It was the "cultural explosion" of the first century (Beard and Crawford, 1999) that culminated, with Augustus, in what has been called a truly "cultural revolution" (Wallace-Hadrill, 2008). In identity terms, the Late Roman Republic was a deep and defining process of cultural formation that resulted in the Augustan canon that we call Roman.

FURTHER READING

Literature on identity is abundant: Assmann (1992), Halpern (2004) and Jenkins (2008) can be used with the outline of the discussion, and to find further general literature. Baumann (2004) is an illuminating essay which might help (identity) scholars to overcome "the fashionable but helpless reduction of all social facts to contextual contingency" (xiv). Meskell (2001) and Insoll (2006) are most useful in relation to archaeology and have a large bibliography; for Roman archaeology in particular Pitts (2007) is a fine (theoretical) starting point. An excellent recent overview of discussion, terminology and bibliography regarding material culture and identity is the introduction to Hodos and Hales (2009). General books on the Roman Republic with attention to the problem of identity include Beard and Crawford (1999), Nicolet (1997), Roman and Roman (2007) and Wallace-Hadrill (2008). Rosenstein and Morstein-Marx (2006) has a section on "The creation of a Roman identity" in which Welch offers a useful (interpretative) overview of the art and architecture of the period. Hölkeskamp (2010) is an important historiography, also in this respect, as is, in a different way, Flower (2010). The work of Coarelli on the archaeology of the Late Roman Republic and questions of identity, now conveniently brought together in Coarelli (1996), remains fundamental.

PART V

The Archaeology of Empire during the Republic

CHAPTER 28

The Archaeology of Mid-Republican Rome: The Emergence of a Mediterranean Capital

Penelope J.E. Davies

1 Introduction

Taken in isolation, archaeology reveals a patchy picture of Republican times in Rome. Successive emperors later rebuilt the city so intensively as to render many early structures all but imperceptible in the ground, and what remains is scattered enough to make comparative assessments (for dating purposes, for instance) extremely tentative. Still, Imperial restorations of earlier monuments seem often (though not always) to have been relatively conservative; this means that the extant remains of later buildings, and the Severan Forma Urbis of Rome, can be of some use in envisioning the city several centuries earlier. In order to flesh out a fuller image of Mid-Republican Rome, archaeologists set material evidence against literary texts, principally Vitruvius's treatise on architecture and Livy's history of Rome, both Augustan in date. As problematic as the latter may be for earliest Rome (see Chapter 26), scholars grant it tolerable credulity after about 300 (e.g. Wiseman, 2008). Even if the historian shows less interest in the appearance of monuments than would be useful and does not always record the motivations behind their construction, he was concerned to record which individuals commissioned which buildings and when. Archaeology and literature yield information primarily about public buildings in Rome, which are therefore the chief focus of this chapter; for private and non-elite enterprises the record remains relatively reticent.

The fifth and early fourth centuries appear as a slow time for public building in Rome (in general: La Rocca, 1990; Cornell, 2000a; Torelli, 2006, 2007; Welch, 2006a). A deliberate austerity may have served the Republican ideal of *isonomia*, or resources may have been stretched thin and the social problems been overwhelming for large building projects (Cornell, 2000a: 43; Flower, 2010: 37; Smith, 2006: 316). Archaeological and

A Companion to the Archaeology of the Roman Republic, First Edition. Edited by Jane DeRose Evans.

literary evidence suggests that construction ratcheted up suddenly in both quantity and monumental quality around the last third of the fourth century as Rome reached tentacles of power through the Italian Peninsula, and kept up a lively pace until a slump during the Second Punic War (see Chapter 17). The economic depression induced by the war stalled building enterprises into the early years of the second century, despite the ravages of fires and floods (see, for instance, Livy XXX.26.5, XXXVIII.10). In subsequent decades, as Rome extended its hegemonic reach into Macedonia and mainland Greece, the treasury was replenished; from then until the return of the armies from Corinth and Carthage in 146 the city experienced its greatest building boom before the Caesarian and Imperial periods, and there is little doubt that if Livy's narrative survived after 167 the record would be even fuller (Coarelli, 1977; La Rocca, 1990: 385). During this second period the archaeological and literary record suggests three broad initiatives: refreshing and upgrading the existing infrastructure; embarking on new construction for the city's beautification, conceptually and formally inspired by Greek monumentalism; and expending spoils in the public interest.

Literary and epigraphic sources indicate that in Rome (in contrast to other Italian cities) only elected and appointed magistrates – principally aediles, censors and victorious generals – could engage in public building. Paradoxically, the brevity of their terms of office both kept their construction ambitions within certain bounds and raised the ante on what and how they chose to build. It also accounts for the piecemeal evolution of the city despite evident architectural expertise and financial wherewithal (Davies, forthcoming).

2 The Rise of Individualism and a Regional Capital in the Making: The Onset

The first half of the fourth century witnessed sporadic temple building; but construction appears to have been dominated at this time by a single massive civic project. In response to an incursion of Gauls in *c.* 386, existing fortifications were augmented with a massive ring of ditch-and-rampart fortifications reaching 11 km from the Capitoline to the Aventine, enclosing the Quirinal, the Viminal, the Esquiline, the Oppian, part of the Caelian, and the Palatine (Barbera and Cianetti, 2008). Gates served the principal roads into the suburbia. Tracts of the wall survive in diverse parts of the modern city; the most extensive, outside Termini Station, extends nearly 100 m in length, about 4 m in thickness, and approximately 10 m in height, rising in 17 courses of headers and stretchers of cut tuff from Veii and Fidenae (see Chapter 17). The use of material from recently subordinated territories mitigated the inevitable acknowledgement of vulnerability with a message of conquest. According to Livy (V.55), oversight of the project fell to the censors of 378, Sp. Servilius Priscus and Q. Cloelius Siculus, whose mandate – beyond administering the census – included letting contracts for public works financed by the state treasury; construction was apparently still incomplete in 353. Greek quarry marks incised into some of the blocks suggest that the censors hired architects from south Italy or Sicily (*LTUR* 3: 319–24 [Andreussi]; Barbera and Cianetti, 2008). Thereafter, though Livy asserts that the Romans rebuilt their city quickly after the Gallic fire (V.55.2–5), there is little firm evidence for public building for several decades. Literary sources imply that magistrates went about state business in temples and atrium houses built in the Early Republic (La Rocca, 1990: 322; Gros, 1996–2001: 238; Torelli, 2007: 115). Archaeological evidence and literary sources place examples of the latter, constructed of opus quadratum of tuff, on the lower north slope of the Palatine and around the Forum (Carandini and Papi, 1999).

A change in attitude to public building is evident in the second half of the fourth century. It belongs against the backdrop of a changed political landscape, when competition for office had escalated fiercely. As Rome's territorial reach extended over the Italian Peninsula, from 326 the practice of prorogation (Lintott, 1999: 113) had led to a concentration of authority in a handful of charismatic commanders, for whom the recently deceased Alexander the Great served as a model and an inspiration (Bastien, 2007). Meanwhile, between the Licinian–Sextian law of 367 and the Ogulnian Law of 300 plebeians had progressively gained access to upper magistracies and the pontifical colleges (Lintott, 1999: 37), and for the many (primarily ambitious plebeians, new men without a client base or patricians with poor pedigree) wanting to break the stranglehold of the few, architecture was a powerful means of advancement through visibility.

Excavations reveal that around this time the speaker's platform at the Comitium, the meeting place of the people's assembly on the north-west edge of the Forum, was raised by approximately a meter and lined with four steps (Amici, 2004–5; see also Coarelli, 1985; Carafa, 1998). These modifications may have coincided with the erection of equestrian statues of the consuls C. Maenius and L. Furius Camillus at the Comitium at the behest of the people and the display of rams (*rostra*) from enemy ships on the speaker's platform in connection with the battle of Antium in 338 (Livy VIII.13.9, XIV.12; Eutr. 2.7; Plin. *HN* 34.20). Rome's landmark victory there concluded the First Samnite War and dissolved the Latin League, allowing Rome to emerge as the leading city of Latium (Oakley, 2004: 25–6; Potter, 2004: 70), with an expanded civic body better accommodated by an enlarged Comitium (Torelli, 2007: 121–4; *LTUR* 1: 309–14 [Coarelli]). The victory also filled Rome's coffers: it released the city from an obligation to yield a third of any plunder to the League (Plin. *HN* 34.20) and, by extending the ager publicus, increased revenue from rent (La Rocca, 1990: 323). Literary sources record that around this time, perhaps in Maenius's subsequent censorship in 318, bankers took over the butchers' shops on the north-east side of the Forum and the shops were endowed with wooden balconies named maeniana after him, from which crowds could view spectacles in the center of the Forum (Festus 120L); the butchers relocated further north-east, leaving the reconceived piazza to better express Rome's expanding role as regional capital and economic power, achieved under Maenius's aegis (Varro, frag. 72).

Modest as these changes may have been, the drive to dignify the city by upgrading its amenities, and with it the close association of state-funded development with an individual politician, were quick to take root, as evidenced by the groundbreaking building initiatives of the censorship of 312. Appius Claudius let contracts for the Via Appia, Rome's first major road, and the Aqua Appia, its first aqueduct. The Via Appia ran 212 km from Porta Capena to Capua (see Figure 19.2 and Chapter 19). Surviving sections are paved with volcanic stone, but under Appius Claudius's watch it probably had a gravel surface set on a chipped stone foundation (Staccioli, 2003). Brief tracts of a channel constructed out of cappellaccio blocks have been identified as sections of the Aqua Appia, and Frontinus recorded its course and capacity when serving as curator of the water supply under Nerva and Trajan (*Aq.* 5): it conveyed up to 75,000 m^3 of water a day in underground pipes from a source over 16 km away on Via Praenestina to a distribution point in the commercial zone near Porta Trigemina and the Forum Boarium, on the north-west side of the Aventine (see Chapter 18). Where it entered the city, near Porta Capena, substructures and arches may have raised it above ground for about 0.09 km (Hodge, 2002; *LTUR* 1: 61–2 [Mucci]; Evans, 1994). If so, Appius Claudius's projects – road and aqueduct – came visibly together in the bustling commercial zone located at this gate.

These building projects served a growing community, to be sure; yet in doing so they also served the magistrate who commissioned them. There is no evidence that Rome's existing water supply – from wells, cisterns and the Tiber – was inadequate for drinking and general domestic uses; in fact, the aqueduct's principal purpose seems to have been to benefit plebeians and merchants in the commercial region outside Porta Trigemina (Torelli, 2007: 127–8). Similarly the road, albeit a fast route for military deployment to the south, benefitted newly won territories that were valuable centers of voting power. Thwarted in his bid for a consulship, the censor used both projects to gain a place among the ruling elite. Expending state funds usually controlled by the senate (though he actually neglected to get its approval), he nevertheless named the monuments after himself, using his praenomen, Appius, in imitation of the practices of Philip of Macedon and Alexander (Staveley, 1959; MacBain, 1980; Humm, 1996).

In the afterglow of Appius's exploitation of his censorship, literary texts record a succession of interventions by aediles, whose charge – besides arranging entertainment for religious festivals and keeping order – included maintaining the physical fabric of the city using funds raised through fines (Cic. *Leg.* 3.3.7; Bauman, 1996; Lintott, 1999). They began to expend this income on works of art and architecture that drew political support through symbolically celebrating plebeian legislative successes. As curule aedile in 304, Cn. Flavius dedicated a bronze shrine to Concordia in the Comitium area, ostensibly commemorating the reconciliation between plebeians and patricians, but inevitably rejoicing in the headway made by plebeians like himself (Livy IX.46.6; Plin. *HN* 33.19) (*LTUR* 1: 320–1 [Ferroni]; Torelli, 2007: 122; Levick, 1978). Other aediles embellished the Temple of Jupiter Optimus Maximus and paved sections of the Via Appia and the Clivus Publicius on the Aventine (Livy X.23.13; Varro *Ling.* 5.158; Ov. *Fast.* V.275; Festus 276 L; Staccioli, 2003: 17–19). The grandest such initiative was that of L. Postumius Megellus, aedile in 307 or 306, who raised funds for a Temple to Victoria which was dedicated in 294 (Livy X.33).

Where there is enough evidence to deduce their form, Early Republican temples seem to have had a relatively consistent design, usually termed "Italic." A frontal staircase led to a high podium, framed and reinforced with opus quadratum of tuff and filled with rubble. A columned porch preceded a triple or single cella, occasionally flanked by alae. Lateral and rear walls were usually constructed of tuff, inner walls of tuff, sun-dried brick or opus craticium; the architrave and the armature of the gabled roof were wooden. Wide eaves shielded the walls from rain and terracotta plaques revetted the timbers (Nielsen and Poulsen, 1992). The tensile strength of the wooden architrave permitted wide intercolumniations, and Vitruvius (citing the Hellenistic architect Hermogenes) described these temples as aeraeostyle or diastyle, with a "top heavy, low, and wide" appearance (*De arch.* III.3.4–5).

By contrast, excavations of a set of foundations on the south-west corner of the Palatine, identified as those of Postumius Megellus's Temple of Victoria (Castagnoli, 1964; Wiseman, 1981; Pensabene, 1988; contra, Cecamore, 2002: 122–6) unearthed evidence for a temple with a substantially modified design. Its foundations provide for the first hexastyle façade (and the first peripteros sine postico plan) in Rome since the Temple of Jupiter Optimus Maximus, with much narrower intercolumniations than any temple to date, answering Vitruvius's description of a systyle or eustyle design (*De arch.* III.3.1, III.3.5, III.3.6). This swing from broad, squat proportions to a slender, vertically accented appearance probably indicates the introduction of a stone entablature, in imitation of Hellenic architecture in south Italy, Sicily and beyond. This in turn meant greater durability and a new level of monumentality and implied power; it transported the elegance and refinement of the Greek world to Rome (Davies, in press a).

The Temple of Victoria also marks a shift in temple locations: whereas early Republican temples seem to have been sited to accommodate the nature of the gods they honored and their attendant worshipers, Postumius Megellus's temple was sited for spectacular visibility: with its prospect over the Circus Maximus and the Forum Boarium it was the Palatine mirror of the Capitoline Temple of Jupiter (see Chapter 26). Moreover, according to ancient authors, beginning in these years a new form of decoration appeared in temples. In 304, Fabius Pictor embellished the walls of C. Iunius Bubulcus Brutus's Temple of Salus on the Quirinal with paintings (Plin. *HN* 35.7.19). When Festus mentions paintings in L. Papirius Cursor's Temple of Consus around 272 and Fulvius Flaccus's Temple of Vortumnus of 264, he explicitly indicates that both sets of paintings tied the building to the commissioning general by depicting him in the full glory of his triumph (Festus 228; see Holliday, 1980).

No physical evidence of these paintings remains, but fragments of paintings do survive from the privately funded (and thus less controlled) realm of funerary architecture, where there are commensurate signs of heightened monumentality and individualization in these years. For the most part, burial customs of the fifth and first half of the fourth century appear modest, though inadequate excavation techniques and conservation efforts complicate the record; tuff-lined pits preserved inhumations with few grave-goods on the Esquiline, and small hypogea were cut into tuff embankments nearby (Lanciani, 1875; Taloni, 1973; Scrinari, 1968 [1972]; Albertoni, 1983; Bartoloni, 1987; Colonna, 1996; Barbera *et al.*, 2005). From the middle of the fourth century inscriptions on sarcophagi record the names of the deceased (as in the Tomb of the Cornelii: Pisani Sartorio and Quilici Gigli, 1987/8); by the end of the century or the beginning of the next the politically frustrated L. Cornelius Scipio Barbatus had commissioned a grand chamber tomb near the Via Appia, complete with a long tuff façade decorated with paintings that probably represented military feats (Coarelli, 1972, 1988c; Pisani Sartorio and Quilici Gigli, 1987/8; Talamo, 2008b); other members of the elite had begun to construct free-standing tombs on the Esquiline, at least one of which (associated with a Fabius or a Fannius) had labeled biographical paintings as decoration (Coarelli, 1976; La Rocca, 1983/4; Talamo, 2008c; in general, Talamo, 2008a; Valeri, 2010).

3 The Rise of Individualism and a Regional Capital in the Making: Normalizing

For the bulk of the third century archaeological evidence is patchy, and between 292 and 219, Livy's text is lost. Still, miscellaneous literary notices indicate that war – with other Italic peoples, with Pyrrhus of Epirus and with the Carthaginians – occasioned a great number of temple vows (Orlin, 1997). Surviving remains suggest that at least some of these buildings followed the new aesthetic (Temple A, Largo Argentina). Like the Temple of Victoria, they were sited for visibility, principally along the triumphal route from the Campus Martius to the Capitoline. Though notoriously elusive to scholars in its particulars (Coarelli, 1968b; Bastien, 2007; Beard, 2007), this route presents obvious viewing nodes, where temples clustered: at its starting point (Temples A and C in Largo Argentina), just outside the Porta Triumphalis where the procession slowed to enter the city (Bellona Victrix, Janus, Spes), and the Circus Maximus, where the procession fanned out and spectator seating was abundant (probably Consus, Vortumnus and Jupiter Libertas). This strategic siting seems to have accompanied a change in function. All temples sheltered cult statues; but whereas (at least by the Mid- and Late Republic) most fifth- and earlier fourth-century temples housed diverse forms of state business such as meetings of the senate, rituals of post-war purification and the business of magistrates (Stambaugh, 1978), most temples commissioned thereafter appear to have been more iconic; their principal role was self-promotion on the part of the vower.

Sporadic data also suggest further initiatives to expand the city's amenities and to keep its buildings in step with its civic and foreign policy needs. Frontinus records that as censor in 272 M'. Curius Dentatus commissioned a second aqueduct (*Aq.* 6.1–4), which flowed mostly underground from Tivoli. Longer and with greater capacity (up to 180,000 m^3) than the Aqua Appia, the Anio Vetus tripled Rome's water supply, though it lacked the height to supply the entire city (see Chapter 18). Archaeological evidence for the aqueduct dates mostly to later restorations (Hodge, 2002; *LTUR* 1: 44–5 [Mari]; Evans, 1994). For its construction Curius Dentatus used income from his substantial manubiae earned in the Pyrrhic War, which makes it the first known civic project financed through spoils (Bastien, 2007: 328). Had he not died leaving the unfinished project in the hands of M. Fulvius Flaccus, it might have borne his name.

Some of the civic enterprises of this time may reflect the expansion of Rome's sphere of influence; at the least they indicate a desire to cater to new diplomatic needs. Livy (XXVI.27.1–4, XXVII.11.16) mentions an Atrium Regium on the north-eastern side of the Forum that was destroyed by fire in 210, which, judging by its name, may have been a hybrid between an atrium house and a type of royal hall Greek dynasts built for entertaining embassies and for other business. Inspired by Rome's diplomatic dealings with Hellenistic kingdoms during the war with Pyrrhus of Epirus in South Italy in 280–275, the Roman hall may have served for negotiation with foreigners, for housing and entertaining foreign dignitaries, and as a covered extension of the Forum (Welch, 2003).

Still, local politics remained a driving force. Several texts refer to the articulation of a Circus Flaminius on the southern Campus Martius (Livy *Per.* 20; also Festus; Plut. *Quaest. Rom.* 66; Wiseman, 1974; Develin, 1979; Torelli, 2007: 130); some attribute it to C. Flaminius as censor in 220/218. An enclosure wall may have defined this long open space, but archaeological evidence indicates none of the racetrack appurtenances of the Circus Maximus. In fact, diverse uses are attested – markets, banks, funeral orations and *contiones* (*LTUR* 1: 269–72 [Viscogliosi]) – and the Romans seem to have associated it primarily with plebeian interests, suggesting that it was a sort of people's forum, set apart from the senatorially controlled Forum proper. Two sets of games took place there, the ludi Taurei and the ludi plebeii, the latter established early on in antithesis to the patrician ludi Romani held in the Circus Maximus. The Circus Flaminius was also where armies gathered in preparation for triumphs, so that as well as extolling the family name (Varro believed the circus took its name from the prata Flaminia, so Flaminius may have exploited a false etymological link to his gens: *LTUR* 1: 269–72 [Viscogliosi]) the commission commemorated Flaminius's triumph against the Gauls in 223, which the people granted him in face of the senate's refusal (Zonar. 8.20; Livy XXI.63.2; Yavetz, 1962; Bastien, 2007: 197–8, 241); indeed the Circus bears all the signs of a reward for the people's favor (Develin, 1979).

4 A Mediterranean Capital in the Making: Renewal of the Urban Infrastructure

During the economic depression brought on by the Second Punic War (Crawford, 1974), building slowed dramatically. A few commissions were desperate measures: for instance, the reinforcing of the city walls in 217 (Livy XXII.8.6–7), the restoration of the walls and temples damaged in conflagrations in 213 (Livy XXV.7.5–6), and the construction of a Temple of Magna Mater on the south-west corner of the Palatine in response to oracular directives (Livy XXIX.37.2, 36.36). For the walls of elite houses on the north slope of the Palatine, restored at this time after fire damage, for apparently the first time in Rome builders employed concrete, a low-cost and visually lackluster building material from Campania that would come to revolutionize Roman architecture (Coarelli, 1977; Carandini and Papi, 1999; Lancaster, 2005; see Chapter 17).

According to the picture painted by Livy, as Rome picked itself up after the war the first building interventions aimed primarily at assuaging the ailing economy. In 194, the censors G. Cornelius Cethegus and S. Aelius Paetus Catus reconstructed and enlarged the Villa Publica on the Campus Martius and the Atrium Libertatis near the Forum (Livy XXXIV.54.5), first built in *c*. 435 and some time before 212 respectively (Livy XXV.7.12, XXX.21.12; Varro *Rust*. 3.2.4; *LTUR* 5: 203– 5 [Agache]; *LTUR* 1: 133–5 [Coarelli]; *LTUR* 1: 284 [Coarelli]; Coarelli, 1997). These were the loci of the census and censorial archives, and thus they were critical for processing taxes. Aediles then used revenue from fines to fund projects related to finance and trade, which also secured the grain supply. Thus in 194 M. Iunius Brutus and L. Oppius Salinator restored the bankers' shops on the north-east side of the Forum (Festus 258L), and the following year M. Aemilius Lepidus and L. Aemilius Paullus commissioned a portico leading from the Forum to the census ground. They extended their reach to installations for trade and commerce, erecting a portico in the commercial district outside Porta Trigemina as well as a wharf (Livy XXXV.10.11–12). In 192, M. Tuccius and P. Iunius Brutus added a second portico in the same zone (Livy XXXV.41.9–10). Once the treasury had recovered adequately these initiatives shifted to the purview of the censors. M. Fulvius Nobilior let contracts in 179 for a port and the piles for a bridge, as well as a portico outside Porta Trigemina, "another behind the dockyards (navalia), and near the shrine of Hercules, and behind the Temple of Spes by the Tiber, and near the Temple of Apollo Medicus" (Livy XL.51.3–9). In 174 Q. Fulvius Flaccus and A. Postumius Albinus constructed yet another portico from the Temple of Saturn (which contained the treasury) to the assembly room of the senators on the Capitoline and to the Curia, "they paved the emporium outside Porta Trigemina with stone and enclosed it with posts, and they contracted for restorations to the portico of Aemilius, and built a staircase from the Tiber to the emporium" (Livy XLI.27.7). By 142 the wooden superstructure of Fulvius Nobilior's bridge had probably collapsed and L. Mummius and P. Cornelius Scipio Aemilianus let a contract for arches on the existing piers, whence, probably, it derived its name (Livy XL.51.3–9; Julius Obsequens 16; Pisani Sartorio, Colini and Buzzetti, 1986).

Meshing this literary testimony with archaeological evidence has long been complicated by the identification of the Porticus Aemilia of 193/179 with a concrete structure between the Aventine and the Tiber, upon which the identification of several other sets of remains hinged (Gatti, 1934; see below); that connection is now compellingly dismissed. In fact it is likely that the porticoes of which Livy speaks – probably colonnaded enclosures flanking public spaces or streets – were wooden and have long perished. On the left bank of the Tiber, between the Temple of Spes on the north (a restored version of which survives embedded in the walls of San Nicola in Carcere) and Porta Trigemina on the south, archaeological remains do, however, suggest a systematic, unified and aggressive upgrading of Rome's principal trade port, dating to the first half of the second century (see Chapter 22). Embankment walls of opus quadratum of tuff over a stretch of about 30 m came to light downstream from the Pons Fabricius in 1886, at the one-time end of the ancient bridge known as the Ponte Rotto (Figure 28.1), recognized to be the Pons Aemilius, its earliest phase fashioned in Grotta Oscura blocks (Le Gall, 1953: 87). Embankment walls of Grotta Oscura opus quadratum also came to light in the late nineteenth century around the outlets of the Cloaca Maxima and the Cloaca Circi Maximi, and on the Transtiberim bank near the Pons Cestius (Le Gall, 1953: 109).

Furthermore, in 1936–7, construction work on the Anagrafe (between the Temple of Portunus and temples incorporated into San Nicola in Carcere) unearthed a tuff structure beneath remains of a Trajanic warehouse. It defined a vast trapezoidal space, approximately 120 m in length and just over 20 m in width (giving an area of *c.* 2,600 m^2), walled on three sides and open to the river on the fourth. It may have been part of Fulvius Nobilior's port (Le Gall, 1953; Rodriguez Almeida, 1984; Conticello de'Spagnolis, 1986). Finally, in 1919–20 a section of a wharf built of opus quadratum of tuff (probably Grotta Oscura) was uncovered upstream from the Ponte Sublicio, with two sets of stairs (perhaps peperino) rising from the river to converge at an upper platform. Set beneath each staircase was a travertine mooring ring sculpted in the form of a boar's head. The combination of materials suggests a date in the second century, and the similarity of the wharf's paving to street paving may argue for its attribution to the censors of 174 (Gatti, 1936; Le Gall, 1953: 119–20; see below). In front of the temple of the port god, Portunus, which guarded the bank from at least the fourth century, a bridge was built at around this time to link the area with the ramp to the Pons Aemilia (Ruggiero, 1991/2: 266; Adam, 1994; *LTUR* 4: 153–4 [Buzzetti]).

Figure 28.1 Pons Aemilius (now the Ponte Rotto), first constructed in the second century BCE. Source: Photo by P.E. Davies.

In these years renovations also focused on the city's water supply and cleanliness (Davies, in press b). A massive system of voussoir barrel vaults that closed off the Cloaca Maxima (of Regal or Early Republican times) and concealed its increasingly unpleasant effluence probably dates to the first half of the second century (Figure 28.2), and may belong to M. Porcius Cato's censorship of 184:

> [Cato and Valerius Flaccus] reclaimed all public water flowing into a private dwelling or lot ... Then they let contracts for works to be undertaken from funds designated for that purpose, paving fountain basins with stone, cleaning out the sewers wherever it was necessary, and constructing new sewers on the Aventine and elsewhere where they did not yet exist.
>
> (Livy XXXIX.44.5–5, all translations by author; also Plut. *Cat. Mai.* 19.1)

This would mesh with Dionysius of Halicarnassos's account of a censorial expenditure of 1,000 talents on it at some point (*Ant. Rom.* 3.67.5; Astin, 1978: 84). Road paving, too, was integral to the city's appearance and cleanliness and to easy flow of traffic, and in 174 the censors supplemented the scattered initiatives of past years with a systematic scheme:

> First of all the censors [Q. Fulvius Flaccus and A. Postumius Albinus] let contracts for paving the streets in the city with flint and for laying substrata of gravel for roads outside the city and edging them with foot paths and for building bridges in many places ... And they oversaw the pavement of the Clivus Capitolinus with flint ... And outside Porta Trigemina they paved the emporium with stone ... And inside the same gate they paved the portico to the Aventine with flint.
>
> (Livy XLI.27.7; Wiseman, 1993; Staccioli, 2003: 19)

Figure 28.2 Cloaca Maxima, as vaulted in the first half of the second century BCE. Source: Photo by P.E. Davies.

The same censors installed "stalls in the Circus, and egg-shaped markers to designate the chariots' laps [...] and turning points [...] and iron cages ..." (Livy XLI.27.7).

While these ventures were underway, politicians took care to attend to the maintenance of sacred spaces and the temples of the gods, whose tutelage, often doubted during the course of the Second Punic War, had at length allowed them to prevail and brought them continued success in ensuing years. In 179 Aemilius Lepidus and Marcus Fulvius "caused several shrines and public places, occupied by private persons, to be public and sacred and open to the people" (Livy XL.51.8); the Capitoline Temple was cleaned, its columns whitened, and Mummius contracted to have its ceiling gilded in 142 (Plin. *HN* 33.57). In the first half of the century generous platforms were raised in front of Temples A and C in Largo Argentina; restorations there included monumental new altars, one of them inscribed with the name of a duumvir, A. Postumius Albinus (Coarelli, 1997). In the Temple of Apollo Medicus, a section of white mosaic pavement preserves the end of an inscription within a rectilinear border in black tesserae on a white background, recording a restoration financed from fines (*ILLRP* 45; *CIL* 1^2.2675; Coarelli, 1997: 387). It ran through the center of the cella from the door to the feet of the cult statue. The first part of the inscription – visible in 1953 but now lost – attributed the restoration to aediles, whom, judging by fragments of a similar mosaic under S. Giovanni Calibita on Tiber island, it once named. Letter forms place it in the first half of the second century (*LTUR* 1: 49–54 [Viscogliosi]; Ciancio Rossetto, 1997/8: 192).

This steady stream of urban renewal projects facilitated commerce and movement through the crowded city, and revitalized existing spaces and buildings. The projects also had a powerfully self-serving quality; utilitarian benefactions, they often appealed to particular constituencies of the city, such as the merchants or the urban plebs, and often took the name of the commissioning magistrate. The projects took form against a backdrop of heightened political competition, and many were the work of ex-commanders who could claim a personal stake in the brimming treasury, flush with gains from their own victories (Davies, forthcoming). In the privately funded sphere of funerary architecture, too, patrons enlarged their tombs for greater impact on viewers. Archaeologists recognize a phase of renovation at the Tomb of the Scipios in the mid-second century that they attribute to Scipio Aemilianus: a gallery was added to the existing hypogeum to accommodate additional burials and a columnar façade was built using the existing façade as a podium (Coarelli, 1972, 1988c). Statues of family members (and the poet Ennius) embellished the tomb (Asc. *Pis.* 44; Livy XXXVIII.56.3–5).

5 A Mediterranean Capital in the Making: Monumentalism and the Conquest of Greece

The porticoes that helped to inaugurate Rome's rebirth, which defined and organized the jumbled spaces of the city, drew conceptually on a Greek practice of using stoas to emphasize and articulate public areas (Coulton, 1976; Winter, 2006: 50–70). As income from Greek conquests allowed urban renewal to gain speed, ex-generals infused several of their commissions for new buildings with Greek designs that had made an impression during their campaigns. This trend resulted in a more diverse architectural vocabulary in Rome, a wider range of construction materials, increasingly unified public spaces, and heightened monumentality in urban forms (see, in general, *LTUR* 3: 226–9 [Zevi]; D'Alessio, 2010; Davies, in press c, forthcoming).

After his triumph over Ambracia in northern Greece in 187, Fulvius Nobilior commissioned a Temple of Hercules Musarum on the Circus Flaminius (Eumenius *Panegyric* 9.7.3). As represented on fragments 31eeff of the Severan Forma Urbis, the temple appears to have diverged from the norm by combining a rectilinear *pronaos* with a cylindrical cella (Gianfrotta, 1985; *LTUR* 3: 17–19 [Viscogliosi]; Coarelli, 1997: 452–84). The design evoked – and Romanized – the round *tholoi* that commemorated heroized Greek monarchs, such as the Philippeion at Olympia, built soon after the battle of Chaironeia in 338 (Miller, 1973). Greek sculptural styles also percolated into the traditional terracotta decoration of temples: in the terracotta pediment sculptures discovered in 1878 on Via di San Gregorio, which probably date to the middle decades of the century and depict a suovetaurilia (a sacrifice of a pig, a sheep and a bull) in the presence of divinities, details of drapery and anatomy betray the clear influence of classicism (Figure 28.3) (Ferrea, 2002; Strazzulla, 2006). Fragments of acrolithic sculptures in Greek marble found close to the Temple of Apollo and the *area Capitolina* suggest that some patrons also commissioned cult statues from Greek artists (Martin, 1987); Pliny's writings place this practice as early as 179, when Timarchides sculpted a cult image of Apollo, perhaps for Fulvius Nobilior and Aemilius Lepidus (*HN* 36.35).

By the second half of the century Greek marble had made its way into Rome as a building material, along with experts to work it. There had been earlier efforts: according to Livy, in 173 the senate cried sacrilege when Q. Fulvius Flaccus stripped marble

Figure 28.3 Terracotta figure of Mars from a pediment found on Via San di Gregorio, probably *c.* mid-second century BCE. Source: Photo by P.E. Davies.

tiles from the Temple of Juno Lacinia in Croton to roof his Temple of Fortuna Equestris, "in his extreme eagerness that that there should be no larger or more magnificent temple in Rome," and forced their return (Livy XLII.3). It was Q. Caecilius Metellus after his triumph over Macedonia in 146 who commissioned the city's first temple of marble, dedicated to Jupiter Stator, and he hired Greek architect Hermodorus of Salamis (on Cyprus) for its design (Vell. Pat. 1.11.5; Gros, 1973, 1976; *LTUR* 3: 157–9 [Viscogliosi]). Hermodorus was steeped in the Hellenistic traditions of the Greek east, where he was probably an authority on the refined Ionic style advocated by Pytheos, one of the architects of the famed Mausoleum of Halicarnassos. As described by Vitruvius (*De arch.* III.2.5), Hermodorus's Temple of Jupiter Stator – peripteral and hexastyle with 11 lateral columns – was a canonical example of the Hellenistic Ionic style, characterized by an uncompromising use of proportional relationships between the component elements of a building (Castagnoli, 1955; Gros, 1973, 1976; Pollitt, 1986: 242–7; Winter, 2006: 7).

The Greek architect's presence in Rome ushered in a new phase of temple design that would endure over two decades. Of the original Temple of Jupiter Stator there are no known remains, but the archaeological record does preserve traces of Hermodorus's work elsewhere. Beneath a stable building next to the church of San Salvatore in Campo between Via di San Salvatore and Via de' Specchi, nineteenth-century archaeologists discovered sections of a concrete foundation that once had steps on all sides, forming a

Figure 28.4 Marble temple by the Tiber, probably commissioned by L. Mummius, *c.* 146 BCE, with Temple of Portunus in the background as rebuilt in the early first century. Source: Photo by P.E. Davies.

continuous crepidoma in the Greek style. Fragments of fluted columns indicate a hexastyle façade in Pentelic marble and probably nine lateral columns. Scholars identify the structure as the Temple of Mars that D. Iunius Brutus Callaicus commissioned from Hermodorus after his triumph over Callaecia and Lusitania in 136–135 (Cic. *Arch.* 11.27 schol. Bob.; Val. Max. 3.14.2; Zevi, 1976; *LTUR* 3: 226–9 [Zevi]). It is also likely that the round Temple by the Tiber is Hermodorus's work or that of his school, commissioned by Mummius after his conquest of Corinth, in honor of Hercules Victor (Figure 28.4) (Gros, 1973: 158–9; Ziolkowski, 1988 with bibliography; contra, Coarelli, 1988b: 92–103, 180–204). One of the most radically Hellenizing buildings of its time, it boasted a Greek crepidoma, a circular plan, and what may be the first use of the Corinthian style on the exterior of a building in Rome. The peristyle ceiling was ornamented with coffers with decorative moldings and rosettes in the Greek manner, and the columns and the revetment of the cella were crafted of luminous Pentelic marble. Windows flanking the doorway infused the cella with light, to contrast with the inner gloom of earlier cult buildings. A systemic play of numerical ratios unifies its plan and elevation.

These temples were the hallmark of a lavish triumph. The borrowings of Greek design and material signified conquest and, for many Romans, cultural advance; they served the dual ideological goals of promoting a more cosmopolitan Rome and identifying the patron with his signature conquest (Gros, 1973: 393–7). Little wonder that the acme of such expropriations fell around 146, when Scipio Aemilianus (destroyer of Carthage and commissioner of a round Temple to Hercules), Caecilius Metellus and Mummius were vying with one another for glory and preeminence.

Greek forms also inspired modifications to civic buildings commissioned by censors who had served as generals in Greece. In 184 Cato erected the Basilica Porcia in the north-west corner of the Forum, probably as a sheltered tribunal close to the Comitium and the Curia (Livy XXXIX.44.7; Plut. *Cat. Mai.* 19.3; Astin, 1978: 84; David, 1983: 227; Wallace-Hadrill, 2008: 133). Remains date mostly to a later restoration, but archaeologists reconstruct it as a broad rectangular building, open to the Forum through a colonnade on the south side; inner columns separated a central nave from side aisles and supported an upper ambulatory, along the lines of the Atrium Regium (Coarelli, 1985: 59–62; Gruen, 1992: 68–9; *LTUR* 1: 187 [Steinby]; Welch, 2003). Again Fulvius Nobilior instigated change. Livy XL.51.3–9 records that as censor he let a contract for a second basilica, the Basilica Fulvia, probably as a shelter for Forum business (David, 1983), and archaeologists reconstruct it with a Greek quality of a different order from Cato's atrium–royal hall hybrid. Remains suggest a colonnaded portico placed transverse to the Forum's long north side, with a row of shops and a columned hall behind (*LTUR* 1: 173–5 [Bauer]). In 169, Ti. Sempronius and C. Claudius Pulcher contracted for a Basilica Sempronia on the opposite side of the Forum (Livy XLIV.16.9–11), and meager traces are consistent with a form mirroring the Basilica Fulvia (Carettoni and Fabbrini, 1961; *LTUR* 1: 187–8 [Iacopi]). Wide intercolumniations suggest that these basilicas had wooden entablatures, but their elongated form with the portico on one side evoked the grand stoas of Hellenistic cities, such as the South Stoa (*c.* 250) and the Stoa of Philip V of Macedonia (*c.* 200) in the agora on Delos (Winter, 2006: 46). Though erected 10 years apart, the Roman basilicas transformed the older Italic configuration of the Forum, framed by atrium houses, into the kind of grand public square articulated by majestic colonnades that was rapidly becoming a Mediterranean *koine.*

In new structures of a more utilitarian nature, Greek design concepts combined with existing Roman building practice give a dramatic and innovative effect. Downstream from the commercial port, at the foot of the Aventine, are the remains of a vast building running parallel to the riverbank. Within its walls a forest of pilasters supported 200 barrel vaults set perpendicular to its long sides, covering a series of long narrow rooms that sloped towards the Tiber. In its entirety the structure measured 487 by 60 m and covered about 30,000 m^2. It corresponds with a building indicated on fragments 23 and 24a–c of the Marble Plan and was long considered to be the Porticus Aemilia (Gatti, 1934; Gerkan, 1958). A new identification based on comparanda in Greek lands names the structure as Rome's navalia or shipshed (Cozza and Tucci, 2006; see Chapter 22). Its unparalleled size and adventurous arcuated design relied upon the substitution of concrete for the wooden or stone post-and-lintel construction of Greek *neosoikoi.* The building probably dates to the middle of the second century, and constitutes the first known use of concrete for a large-scale superstructure in Rome. What the material lacked in beauty it more than made up for in its capacity to span the large areas needed in a utilitarian building of this kind; it was also fireproof. Furthermore, economy and speed of construction afforded an astounding new level of monumentality – both real and symbolic – for Rome's extra-urban defenses. An intriguing fragment of Cicero connects a navalia of some sort with Hermodorus (*De or.* 1.14.62), who was probably in Rome at the time of its construction and could conceivably have been responsible (Cozza and

Tucci, 2006). A patron might be sought, hypothetically, in Aemilius Paullus during his censorship of 164 or his natural son P. Cornelius Scipio Aemilianus, censor in 142. The gens Aemilia had a long-established connection with the navy and had only recently contributed to the transformation of the port area further upstream (Davies, in press b).

Among the contracts they initiated in 179, Fulvius Nobilior and Aemilius Lepidus took steps toward the construction of a third aqueduct, which M. Licinius Crassus thwarted by refusing to allow it to cross his property (Livy XL.51.3–9). It was not until 144, after a sudden population increase with the return of armies from Greece and Carthage, that the Senate charged the urban praetor Q. Marcius Rex with the repair of existing water channels and the construction of a new aqueduct (Frontin. *Aq.* 7; *LTUR* 1: 67–9 [Cattalini]; Evans, 1994; Hodge, 2002; De Kleijn, 2001; see Chapter 18). With its origin 91.33 km away on Via Valeria, the Aqua Marcia entered the city at Porta Maggiore before heading towards Porta Tiburtina and a terminal piscina in the zone of Porta Viminalis; again, most identifiable remains of the aqueduct belong to later restorations. It fed a vast swath of the city with Rome's most prized water, which it even introduced to elite houses (Frontin. *Aq.* 89, 91–3; Prop. 3.2.14; Vitr. *De arch.* VIII.3.1; Plin. *HN* 31.24). Though aqueducts were already a proud Roman phenomenon, the scale of the enterprise and its design were unprecedented in the city. It was the first aqueduct with a deliberately monumental appearance: of 61,710 *passi* (= 9,1330 m), most were underground, but 7,463 were raised on arches, 6,472 of them within the seventh milestone (Frontin. *Aq.* 1.7; Robinson, 1980: 49). Moreover, after prolonged discussion it was the first aqueduct to broach the Capitoline Hill, and it is likely that it gained the necessary height through siphons. The impetus for both aspects of the design may have been the triumph of engineering represented by the Madradag aqueduct at Pergamon, probably built by Eumenes II (197–159): for over 3 km a huge siphon propelled its waters along a gradient four times that of the steepest Roman aqueduct (Garbrecht, 1979, 1987; Fahlbusch, 1977: 758–62; Hodge, 1992: 42–5; Davies, in press b).

6 A Mediterranean Capital in the Making: Manubial Structures

In the early Republic, most booty captured from adversaries – weapons, bronze ingots, livestock – was easily recycled. Beginning with the conquest of Syracuse in 211, hordes of Greek sculpture flowed into the city as spoils (Miles, 2008; Östenberg, 2009) and presented a new problem that ongoing Greek conquests only exacerbated: how to dispose of plundered artworks in possession of the state. Archaeological and literary evidence suggest that the problem resulted not only in legal debate but in an age of avid architectural experimentation, as commanders – who had a say in the disposal of spoils seized under their auspices as long as it was in the public interest – vied to display their plunder (Bradford Churchill, 1999; Davies, forthcoming).

Though temporary, the distinctive multi-story aedicular design of the Roman scaenae frons known from literary sources and from later stone versions may have developed in the first decades of the second century as a framework within which to exhibit plundered statues during votive games (Klar, 2006). Other monuments for display were more lasting, among them, probably, Rome's first arches. On returning from Hispania Ulterior in 196, L. Stertinius channeled proceeds from his spoils into the construction of two arches in front of the Temples of Fortuna and Mater Matuta in the Forum Boarium, and another in the Circus Maximus (Livy XXXIII.27.4; Coarelli, 1968b: 82, 89–92, 1995; Calabi Limentani, 1982; De Maria, 1988: 262; Aberson, 1994: 139). On top of the arches he placed gilded statues, probably either plunder or images made or purchased from the proceeds of spoils. Seven years later Cornelius Scipio Africanus erected a similar sculpture-laden monument alongside the Clivus Capitolinus. It probably stood at the point where triumphal processions emerged onto the *area sacra* in front of the Temple of Jupiter Optimus Maximus for the ceremonies' climax, and may even have constituted a grand new entrance to the precinct (Livy XXXVII.3.7; Coarelli, 1968b: 91, 1995; Calabi Limentani, 1982; De Maria, 1988: 263). None of these arches was strictly triumphal; Stertinius had recognized that he stood no chance of a triumph (Pelikan Pittenger, 2008: 116–18) and Scipio's arch preceded his departure to Asia as legate in 189. Arches seem to have originated more specifically for manubial display (Davies, forthcoming).

The most conspicuous form of manubial showcase was a grand style of portico that developed over the course of the second and third quarters of the second century. According to Velleius Paterculus (2.1.1–2), the earliest of them, the Porticus of Octavius on the Circus Flaminius, was also the most resplendent. Admiral of Aemilius Paullus's fleet in 168–167, Cn. Octavius had earned a triumph for his role in the successful outcome of the Third Macedonian War. The considerable spoils he expropriated enriched the triumph of Aemilius Paullus (Livy XLV.42.2), who must have allowed him use of them for the construction of the Porticus. No archaeological trace is known, unless a short line on fragment 31hh of the Marble Plan represents the edge of the complex (Coarelli, 1997: 515–528; contra, Olinder, 1974). It was probably three- or four-sided, and Pliny (*HN* 34.13) calls it a *duplex portico*, a type of building that Vitruvius (*De arch.* V.9.2, V.11.1–2) describes as having two naves separated by columns. Within the building, likely on the interior colonnade, Octavius exhibited bronze Corinthian capitals, perhaps the first Corinthian capitals in Rome. He probably brought them from Greece as spoils (Festus 188 L; Plin. *HN* 34.13; Gros, 1976: 391–2; *LTUR* 4: 130–2 [Viscogliosi]).

Once established, this type of porticus gained currency. P. Cornelius Scipio Nasica built one on the Capitoline, probably in connection with his triumph over Dalmatia in 155 (Vell. Pat. 2.1.2; Gros, 1976), and L. Licinius Lucullus's votive temple to Felicitas in the Velabrum for conquests in Lusitania in 151 stood within a portico embellished with plundered artwork borrowed from Mummius (Cass. Dio 43.21.1; Strabo VIII.6.23; App. *Hisp.* 52; Cic. *Verr.*^{2} 4.126, 2.4.4; Plin. *HN* 34.69, 36.39).

The one porticus for which any substantial archaeological evidence remains is the Porticus Metelli on the northern edge of the Circus Flaminius, which Caecilius Metellus commissioned after he returned to Rome in 146 laden with Macedonian spoils. Traces of tuff foundations indicate four wings arranged around a court; the lateral wings had double aisles, and the south-west wing presented colonnades on the interior and the exterior (Morgan, 1971; Lauter, 1980–1: 37–46; *LTUR* 3: 139–41 [Viscogliosi]). Walls and columns were probably stuccoed tuff. The court enclosed the restored Temple of Juno Regina and the marble Temple of Jupiter Stator (Vell. Pat. 1.11.3–5), and an extraordinary group of bronze sculptures by the Greek sculptor Lysippos representing Alexander and his 24 companions at the battle of Granikos, which Caecilius Metellus had expropriated from Dion (Vell. Pat. 1.11.3; Plin. *HN* 34.64, 36.24; Arr. *Anab.* I.16.4; Cic. *Verr.* 4.126). A magnificent display of manubial income expended in the public interest, the porticus evoked the colonnaded courtyards that characterized Macedonian palaces (Nielsen, 1994: 81–101), as well as the sacred *temenoi* and gymnasium courts that were gaining favor in the Hellenistic regions over which Rome was establishing dominion (Winter, 2006; McKenzie, 2007: 53–8; Nielsen, 1994: 36; Sauron, 1987; Wallace-Hadrill, 2008: 175). Cumulatively, with their rows of stuccoed columns these porticoes resembled the opulent, increasingly regularized spaces of the Hellenistic east and imposed a rectilinear order on the meandering roads and divergent axes of earlier public spaces.

Instigated by commanding generals, these manubial monuments were acts of public benefaction that served the dual purpose of symbolizing, commemorating and justifying imperial expansion on the one hand, and reifying elite authority and promoting the gens of the patron on the other.

7 The Critical Transformation

From available archaeological evidence combined with literary testimony, a vision emerges of a city formed and buffeted by the winds of conflict, economics and political expediency. In the last third of the second century, with Greek conquests drawing to a close, new challenges emerged within the city, on the Italian Peninsula, and further afield in North Africa and Gaul. Politicians addressed these challenges with a changed architectural vocabulary: the hold of Hermodorus's pure Greek forms weakened; travertine superseded marble as a building material, and new types of building emerged to answer popular needs. By the mid-first century, the constitution was so fragile that Pompey and Caesar could build with few of the constraints that applied to earlier politicians, and the stage was set for a new freedom and scale of construction in the empire. Yet as monumental as the city would become in the centuries ahead, it was in the years of the mid-Republic, fuelled by gains in war and the aggressive ambitions of politicians, that it made the critical transformation from a large Italic city graced with a handful of grand temples and rudimentary civic buildings and trade facilities into a capital for the Mediterranean region, which participated in the architectural *koine* of the Greek world and outshone other *metropoleis* with its state-of-the-art amenities (Dio. Hal. *Ant. Rom.* 3.67.5).

FURTHER READING

A critical point of departure for research on the archeology of Rome in the mid-Republic as in any period is the *Lexicon Topographicum Urbis Romae* (*LTUR*, Rome 1993–). In five volumes, the work contains entries by a large team of international scholars on the monuments of Rome, with archaeological and literary data and bibliographies. No single recent text deals with the archaeology of Republican Rome. Introductory books on architecture throughout the Roman period often include an opening chapter on the Republic, emphasizing form and typology; an exception is Gros (1996–2001), which deals with the Republic in greater detail. La Rocca's (1990) excellent book-length essay covers the archaeology of the Republic as a whole; Torelli (2006, 2007),

Cornell (2000a) and Welch (2006a) provide a shorter, overall assessment of the period. Coarelli's books (1985, 1988b, 1997) on the different topographical zones of the city from the beginning to the end of the Republic provide extensive political and cultural context for the city's buildings, and assemble relevant literary sources. Gros (1973, 1976), Gruen (1992) and Zevi (2003) focus on the issue of Greek influence in Rome. Flower (2004) and Rosenstein and Morstein-Marx (2006) contain excellent essays on the political and social context for monuments discussed here; Hölkeskamp (2010) explores the interplay and tension between competition and consensus in Republican Rome especially well.

CHAPTER 29

The Late Republican City of Rome

Jane DeRose Evans

1 Introduction

Writing about the Late Republican city is a daunting task in the few pages allowed in the *Companion*. Each problem in the topography of the first-century city is a complex construction built on references in ancient sources, scraps of walls, inscriptions, maps dating to after the Republican period, rare coin depictions and other such tidbits of evidence. One advantage is that buildings of the Late Republic are described, or at least mentioned, by contemporary authors. Yet often when a scholar accepts or rejects one piece of this evidence, the consequences of the argument affect the placement of other buildings, roads or access stairs. Or a new excavation provides a tantalizing clue that allows us to re-examine our arguments. Excavation within the city of Rome continues at a pace that seems daunting to those of us trying to stay abreast of the work while living outside of Rome. Most recently much new information is available to archaeologists due to the construction of the Metro C line just west of the Forum; inside the Forum in the area of the Lapis Niger; and on the slopes of the Palatine. In order to capture developments since 2000, I will assume that the reader has a basic familiarity with the building projects of the Late Republic in Rome and thus will not repeat entries in *LTUR*.

Since the first century was a period of cascading civil wars and accompanying proscriptions, one would not expect Romans to have the time or money to invest in building projects. Yet there were important building programs occurring in Rome, usually due to powerful generals with imperial ambitions and deep pockets. Marius begins this century by building much in the style and scale of Mid-Republican generals. He was responsible for the Temple to Honos and Virtus (the second one built in Rome) near the Forum, usually thought to be near his Mariana monument ("Marius's trophy") and his house.

A Companion to the Archaeology of the Roman Republic, First Edition. Edited by Jane DeRose Evans.

None of these structures has yet been identified, but the collocation of temple, house and personal victory monument became quite appealing to later builders.

Sulla, too, built in the manner of Mid-Republican generals, as can be seen in the list of temples he built or restored in Rome. Yet Sulla also had the time, money and political control to plan on a broader scale. One of his areas of concentration was the rebuilding of old and significant buildings on the Capitoline. His building projects included erasing a small valley between the Arx and the Asylum on the Capitoline Hill, and he built a *via tecta* connecting the Temple of Saturn and the Aerarium. He began the restoration of the Temple of Jupiter Optimus Maximus, which had completely burned in 83. Sulla added to the city's infrastructure nearby in the Tabularium and thus extended the Capitoline Hill towards the Forum. The Tabularium was finished after Sulla's death and his work on the side of the Capitoline Hill is only known in traces due to Augustan rebuilding. Sulla's other area of interest was in the Forum, as we know that he rebuilt the Curia (and placed an equestrian statue of himself in front) and repaved the Forum, which probably included rebuilding the Comitium (this was only completed after his death). Sulla's programs are heralds of the large building projects later undertaken to provide for the infrastructure of Rome – with buildings named after the patron – by Pompey and Caesar. His large tomb, somewhere in the Campus Martius, foreshadows the substantial projects in this area undertaken by Pompey, Caesar and Augustus. It is to be expected that the building materials reflect the patron's aspirations; in the first century, we see the builders move from opus incertum to opus reticulatum, from tuff and travertine to a more extensive use of marble (see Chapter 17).

With Pompey we shift our focus to a relatively open area of the city, the Campus Martius, where excavations have uncovered small portions of the Theater of Pompey. In the portico attached to his theater, Pompey replicated the old practice of stuffing such buildings with looted art; in a novel gesture, he included a curia, where the Senate could meet outside the pomerium to discuss going to war. The Senate apparently used this curia regularly in the 40s, for this is the building where Caesar was assassinated. Pompey is recorded as having restored part of the Circus Maximus and with that work, restored or rebuilt a Temple to Hercules (now simply called "Hercules Pompeianus") as an old-fashioned Tuscan-type temple. Reflecting new attention to domestic architecture, Pompey had two houses in Rome, one on the Carinae near the Temple of Tellus and a second that looked small compared to the enormous Theater behind which it "floated."

Caesar followed Sulla's lead in concentrating on buildings important to the political life of Rome. Hard by Pompey's entertainment building he began the Saepta, a large building used for voting. He regularized the north and south sides of the Forum to make the area look more like a Hellenistic *agora*, but he also intended the spaces to be used for newly expanded courts (the Basilica Iulia) and, perhaps, merchants (the Basilica Aemilii/Paulli, which Caesar helped pay for). He was given permission to rebuild the Curia Cornelia (then in its second iteration, having been rebuilt after a fire in 52), which would thereafter be called by the Julian name, as well as the area in front of the Curia, including the Comitium and the Rostra. To the east, he may have renovated the Atrium Vestae. Caesar began a new forum, linked to the Forum Romanum, but planned to look like an eastern temple and portico complex. He contributed to the entertainment buildings of the city by planning or building a theater, stadium and naumachia; his most ambitious project, which was quickly abandoned after his death, was the reshaping of the Tiber River to add a great deal of area to the right bank.

2 The First Half of the First Century

As in the Middle Republic, building projects in the city were scattered, being placed wherever the triumphator could obtain land or could restore a building associated with his family; planning of large spaces only occurred after Sulla was named dictator. The Temple of Fortuna Huiusce Diei, vowed by Q. Lutatius Catulus (cos. 102) in 101, has long been recognized in the remains called Temple B in the Largo Argentina. A new study emphasizes the Greek nature of the plan and the neoattic elevation placing it firmly within its Late Hellenistic context. The temple speaks in the language of Hellenism in its decoration, while being clad in local materials and supported by a high podium not found outside of Italy (Caprioli, 2011). It is part of the evidence that Roman architects had assimilated Greek models and forms, and were capable of producing an Italic version of a building type. Yet another temple, which cannot be as easily dated, again displays the innovation occurring in this period: the Temple of Castor and Pollux, long known from a fragment of the pre-Severan Forma Urbis to belong in the Circus Flaminius. This temple was mentioned as having an unusual transverse cella by Vitruvius (*De arch.* IV.8.4; cf. Plut. *Pomp.* 2.8) and its mention in the *Fasti Amiternini* should date the temple prior to 46. Although the only Republican trace of this building is a cocciopesto floor (the remainder of the structure still *in situ* dates to the end of the first or second century CE), temples with transverse cellas were only popular in the first century BCE to the first century CE, and are Italian in origin. It was possibly dedicated in 71 by Q. Caecilius Metellus Pius (cos. 80, following Coarelli, 1997: 504–7) or P. Servilius Isauricus, who triumphed in 74 (Vitti, 2011, who offers both suggestions).

But not all evidence of architectural models that are derived from outside of Italy is as clear as this. The sacred area of Isis, which is thought to have been established on the Arx in the first half of the first century, is known only through literary sources and some Egyptianizing artifacts found on or near the Capitoline. Although Versluys argues that foreign deities were welcomed into Rome long before the establishment of the Iseum and that priests of Isis were seen in the city in the first century, he is reluctant to place a temple on the Arx; he suggests that we should look instead for an altar or shrine (Versluys, 2004).

As can be expected, the shape of Sullan building projects is a major concern for archaeologists. In the Forum, the old Curia Hostilia was restored and possibly greatly enlarged by Sulla in 80 during his reforms of the Senate. This building was itself burnt to the ground in 52 (Plin. *HN* 34.26; Cic. *Mil.* 90; Figure 29.1). No traces of the Sullan phase have yet been found (nor of its replacement in 52), but Amici suggested that the Curia Iulia was built directly on top of the foundations of that building (see below, for the problem of locating the Curia Iulia). Amici also suggested that the Sullan construction thoroughly cleaned the area where the Curia was placed, leaving few Middle Republican remains – perhaps even erasing any traces of the earlier Cura Hostilia (Amici, 1991; Amici *et al.*, 2007).

The shape and size of the Comitium and the function of the pits near the Comitium have come under scholarly scrutiny. Due to a fragment of a wall excavated by Boni in 1899, the Comitium is generally reconstructed as a circular auditorium; the Comitium in Rome then provides the model for the comitia at Cosa and Fregellae (see Chapter 30). Coarelli strongly defends the traditional circular shape (Coarelli, 2005), though recently Mouritsen issued a call for re-evaluating the evidence (Mouritsen, 2004). Within the Comitium are 21 pits of

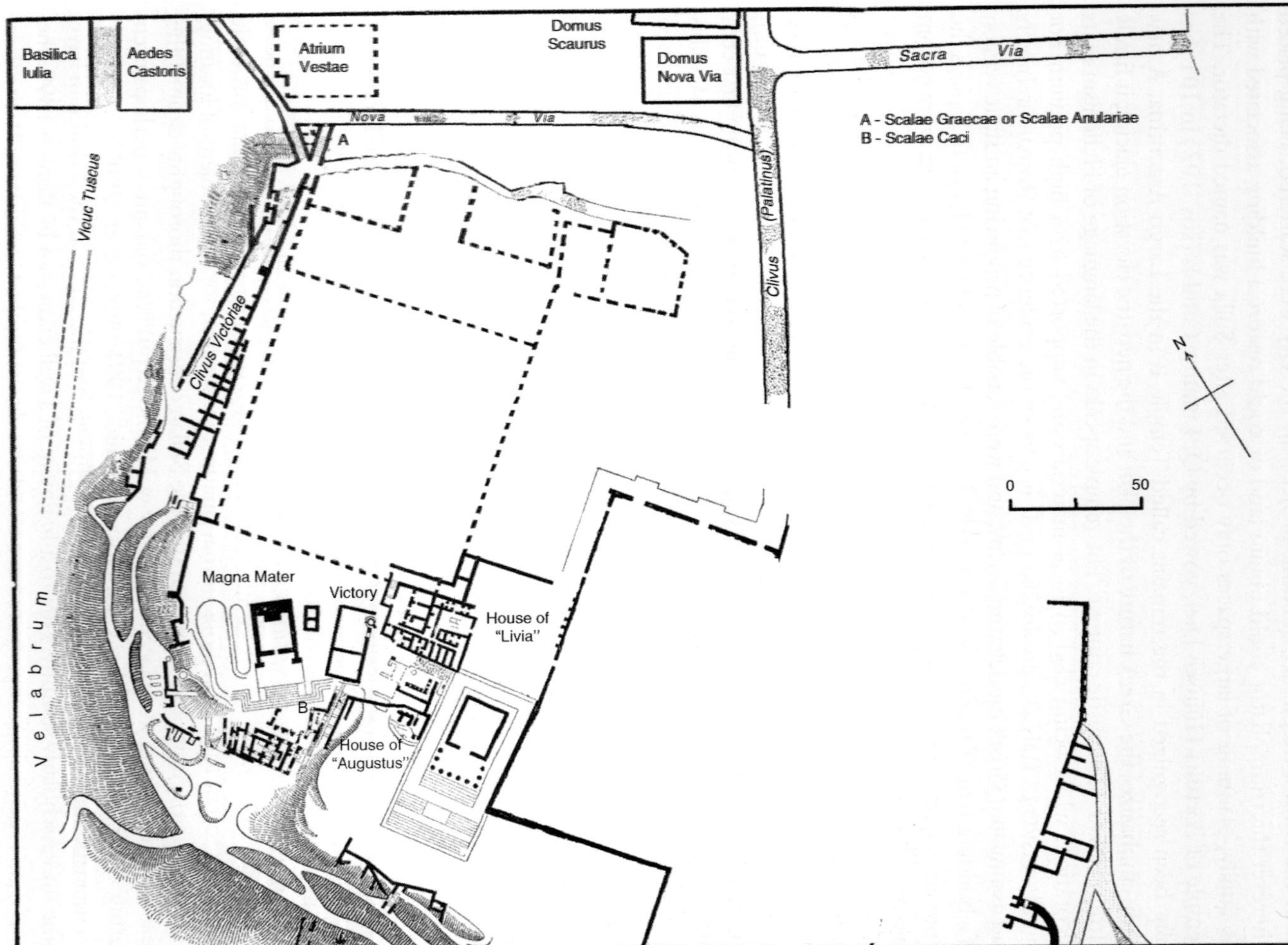

Figure 29.1 Plan of the Palatine in the first century, showing access routes to the hill. Source: Drawing by J.D. Evans.

varying sizes and depths, which Mouritsen and Coarelli agree were used for different purposes, whether for posts to mark religious boundaries, to provide places for audiences of games, to mark out the spot for voting assemblies, or possibly for plants. Coarelli prefers to see the majority of these as belonging to temporary posts erected so that voters could line up to cast ballots; the Saepta followed this tradition, on a much larger scale. If the pits were indeed used to corral voters, the size of the space allotted for this is surprisingly small, given the number of eligible voters in Rome in the first century (cf. Chapter 10).

Visible remains can be seen of the building once confidently labeled the "Tabularium" in the foundations of Michelangelo's Palazzo del Senatore on the side of the Capitoline facing the Forum, including arches between travertine Doric engaged columns (see other suggestions as to the name of the building in Tucci, 2005: 9). Based on a now-lost inscription, the builder is argued to have been Q. Lutatius Catulus (cos. 78) who also finished the restoration of the Temple of Jupiter Optimus Maximus after Sulla's death. Tucci has recently agreed with Coarelli in placing the Late Republican mint in the "Tabularium," a row of rooms that may have been reconstructed by Sulla after the fire of 83 in this area of the city. Yet Coarelli believes that most of what has survived is a *via tecta*, a necessary new route of access linking the Capitolium to the Arx, since the "Tabularium" was built over the original entrance. He notes that a Late Republican building stood where the Temple of Dei Consentes is now; the foundations are under this temple. This building functioned as the annex of the Aerarium Saturni; it stored the metal for the manufacture of coins as well as the finished coins from the mint (Tucci, 2005; Coarelli, 2010). Tucci places the Temple of Juno Moneta on top of the structure, facing the Forum (moving it out of its traditional place in the gardens of Santa Maria in Aracoeli on the Arx). This is a peripteros sine postico temple, in form much like the Temple of Hercules Victor at Tibur (Tucci, 2005). Coarelli does agree that Tucci has identified the remains of a Late Republican temple on top of the "Tabularium," but he does not agree that it is the Temple of Juno Moneta, which he prefers to leave on the Arx. Instead, he suggests that there were three temples here, all of which are only mentioned in the *Fasti*, which note the date of sacrifices offered to Venus Victrix, the Genius populi Romani and Felicitas (Coarelli, 2010). Sulla appears to have been particularly unfortunate – in spite of his nickname – in having almost all of his building projects rebuilt so as to leave very few traces of his vision for the rebuilding of Rome.

The Temple of Jupiter Optimus Maximus was completely ruined in the fire of 83 (e.g. Plut. *Sull.* 27.6). Sulla undertook the task of rebuilding the temple, and if we can trust Pliny (*HN* 36.45 – not all scholars do), he Hellenized the façade by importing columns from Athens while retaining the original footprint of the temple. He did not live to see the end of this task, but the Senate appointed Q. Lutatius Catulus to finish it. Although the temple remains a focal point for studies, nothing has survived of the Late Republican phase (La Rocca, 2011).

The history of the Basilica Aemilia/Paulli, found in the north-east corner of the Late Republican Forum, continues to attract scholarly attention, in part because of the embarrassing abundance of literary sources which mention the building. The basilica, which appears to have undergone several changes of name, is said to have been restored by L. Aemilius Paullus in 55, with money obtained from Caesar (Plut. *Caes.* 29; App. *B Civ.* 2.26; Cic. *Att.* 4.16.14); how much of the earlier structure he retained is a source of great debate (e.g. see Liverani, 2008; contra, Freyberger, Ertel and Bitterer, 2007).

Lipps (2011) argues that all of the remains known today date from the Augustan period or later. Thus, our information on the buildings of the first half of the first century is largely missing, due to later reconstructions, demolitions, or lack of excavation.

3 Approaches to the Palatine Hill

In order to find Republican remains, archaeologists must search at the edges of the Palatine, due to the Imperial buildings found at the hill's summit (see a caveat in section 6 below). It appears that the area saw major renovations in the second to first centuries (Hostetter and Brandt's (2009) Phase II), though the excavators admit that this phase is difficult to date due to the Augustan and Domitianic projects on the hill (Figure 29.2).

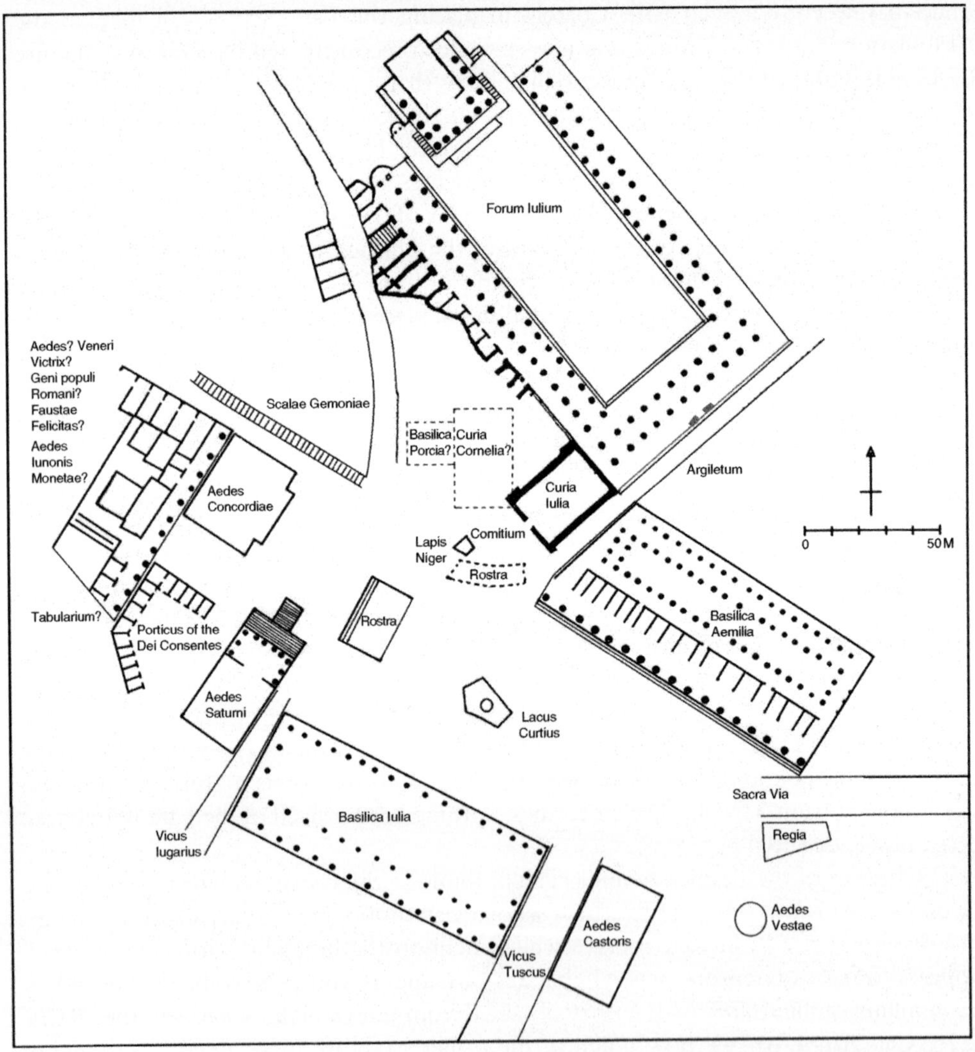

Figure 29.2 Plan of the Forum Romanum in the first century. Source: Drawing by J.D. Evans.

These renovations included repaving of roads of access, restoration of sewers, and rebuilding or renewing of terraces that supported large aristocratic homes. The position of the Nova Via, which was thought by Lanciani to have followed the same route as the imperial version of that street (south of the Atrium Vestae, it runs between the "Scalae Graecae" and the "clivus Palatinus") is a perennial problem, as so much of the area was changed both after the fire of 64 CE and during the Domitianic rebuilding of the Palatine. Besides the route suggested by Lanciani, others have proposed a "high" route (south of Lanciani's proposed route, in a more meandering path) and a "low" route, connecting the Vicus Tuscus south of the Temple of the Castors and the Via Sacra (see Figure 3.1, Carandini's restoration of the route). Hurst and Cirone's (2003) recent excavations have exposed Republican remains in the area next to the "Scalae Graecae" (and thus confirming Lanciani's route in the western section), although Wiseman (2004) suggests that the Republican version of the street was swept away in the rebuilding of the northern flank of the Palatine (contra, Cecamore, 2002, and Hurst, 2007).

Recent excavations on the Palatine have revealed Republican buildings especially along the northern edge of the hill (Figure 29.2). Access routes to the hill in the Late Republican period are now clearer, as it appears that three stairs or ramps provided access to the aristocratic homes found on the sides and top of the hill. The western approach, called for convenience the "clivus Palatinus," rises from the Sacra Via. Its Republican phase is known along the stretch closest to the Forum by its path beside two Late Republican aristocratic houses (the House of M. Aemilius Scaurus and the "domus Nova Via"). The paving and the entire upper portion of this road belong to the Imperial period (Cecamore, 2002). A second entrance, a newly cleaned stair, rises directly from a small paved street that wanders along the northernmost contour of the hill, south of the Atrium Vestae (A in Figure 29.1). The name of this stair has been debated, with Steinby (*LTUR* 4: 241–2) supporting its identification as the Scalae Graecae (see also Krause, 2004); it has as well been called the Scalae Anulariae, due to the remains of a silversmith's workshop found in the vaults which support the stair and an inscription found nearby naming jewelers (*LTUR* 4: 238–9 [Papi]). The excavator notes that the (mostly robbed out) vaults which support the stairs and ramp ascent are made of opus incertum, and probably date to the second century; he remains decidedly neutral on the name of the approach (Hurst, 2006).

The south-west corner of the Palatine, dominated by the prostyle, hexastyle (and perhaps pseudodipteral) Temple of Magna Mater had to be refurbished due to a fire very late in the second century; scholars assign the work to either Caecilius Metellus Caprarius (cos. 113) or Q. Caecilius Metellus Numidicus (cos. 109), the latter favored by D'Alessio (Pensabene and D'Alessio, 2006). In reconstructing the temple, the lower steps were removed to depress the area between this temple and the adjoining Temple of Victory (which was also in the process of being restored, although the restoration was not completed until the Augustan period) in order to make a large open space for public gatherings on the south side of the Palatine. To extend this space, the Clivus Victoriae was lowered, and the ramp to the Palatine was turned into a *via tecta*. The vaults of this *via tecta* supported structures to accommodate the spectators gathered in this area (Favro, 2008: 32–4). The stair here, on which D'Alessio posits an almost-monumental entrance to the temples, continues down the west side of the Palatine. Farther south, the Scalae Caci provided another stepped entrance to the large space in front of the temples (B in Figure 29.1). With this rebuilding, D'Alessio notes that this side of the Palatine becomes an acropolis, a pendant to the Arx, with new building techniques (here, opus reticulatum) confidently used to provide large sacred structures on top of vaulted structures that allowed for performances, much like the temple-theater structures in Latium or in the Hellenistic east (Pensabene and D'Alessio, 2006).

4 The Theater Complex of Pompey

Due to Filippi's and Packer's excavations in the foundations of the cavea of the Theater of Pompey, new attention has been paid to the Theater Complex – both its form and its decoration (Gagliardo and Packer, 2006; Packer, Burge and Gagliardo, 2007; Monterroso, 2010 for the excavations of Filippi). Packer's and Filippi's teams have been able to map more accurately the diameter of the cavea as well as reconstruct the Tuscan order engaged columns on the lowest level of the façade. Packer and his collaborators suggest a placement of stairs leading to the Temple of Venus Victrix which we are told was attached to the cavea (Aul. Gell. 10.1.7); this temple has been thought to be shown on a fragment of the Severan Forma Urbis. The fragment, it should be noted, "joins" the curve of the cavea, although the fragment containing the plan of the cavea no longer exists; it is only known from drawings. Monterroso would now detach the fragment that contains the letter M and one wall of the Temple of Venus from the Theater and add it to the Stadium Domitiana. He then places the shrine to Venus (with the associated shrines of Felicitas, Honos, Virtus and V[ictoria]) in the cavea of the theater, substantially reducing the number of seats available to spectators (Monterroso, 2007, 2010). The same author's exhaustive exploration of each known fragment of the Theater and Portico exposed some traces of pilasters on the façade of the cavea and allowed him clearly to delineate the development of theater architecture from the Mid-Republic to the Early Imperial period. No excavations have been possible in the area of the scaenae frons, leaving the question of a Late Republican building there unsolved (Monterroso, 2010). Recent reinterpretations of the sculptures, paintings, and the Portico ad Nationes (which was attached to or part of the portico behind the Theater) continue to explore the purposes of the patron Pompey in his largest and most public present to the people of Rome (Denard, 2005; Temelini, 2006; Schröter, 2008; Evans, 2009).

5 The Building Projects of Caesar

Caesar focused his efforts on the Forum Romanum, his extension of the Forum in the space he named after himself, and his (today lesser-known) projects in the Campus Martius (for an overview of these works, see Liverani, 2008). His work on the Forum eventually swept up all of the primary political spaces in Rome, from the Curia to the Rostra, the Tribunal to the courts. It was an act of urban redefinition that had not yet been seen on such a scale in Rome. An even more ambitious project, to straighten the course of the Tiber River, was apparently only just begun – and quickly abandoned – after the death of Caesar. Even though the early excavations of the Forum remain largely unpublished (and thus the chronology of the building phases remains problematic), recent work has begun to pull back the curtain on Caesar's intentions.

The recent excavations in the Forum of Caesar have suggested solutions to the problems of the entrance to the Forum and its connection to the Curia Iulia (Figure 29.2). The excavators report traces of an initial building phase along the south-east end of Caesar's Forum, found in a double-aisled portico paved in travertine, with steps down to the interior space of the new Forum. On the wall in the back of this portico was a niche, probably originally to house a monumental statue; the south-west portico was backed by a row of tabernae, which Delfino suggests were used for judicial archives. He argues that the portico was two stories tall and served as a retaining wall for the Capitoline. Thus, the Forum was an enclosed structure which did not open directly onto the Forum Romanum, but was entered through the south-east side, from the Argiletum. It was only after the Senate approved Caesar's plan to rebuild the Curia Cornelia that the south-west side of the Forum was rebuilt to attach that building to the corner of the Forum Iulium, according to the excavator. This is also when the south-east portico was reformed, with the monumental niche replaced by a line of columns and three steps down to the Argiletum along the new face of the portico, allowing for a much more open feeling to the Forum (Delfino, 2008; see also Amici *et al.*, 2007). The present-day remains of the Forum date largely to the Flavian period and both the north-west and north-east sides of the Forum Iulium were likely rebuilt and reshaped with the construction of the Forum Augustum and the Forum Traiani.

Since it is likely that the remains of the Curia Iulia (and the Curia Hostilia?) are located under the Curia built by Diocletian (and restored in modern times), excavations have revealed little about this building. This is unfortunate; the placement of the Curia would help reveal Caesar's plan to rebuild this area of the Forum and his ideas about the relationship between the Forum Romanum and his Forum. If Caesar did rebuild the Curia on top of the Curia Cornelia, he may have been trying to erase all memory of Sulla and his reforms of the Senate. He may have intended to have the Curia Iulia integrate the Forum Iulium with the Forum Romanum and use his new Forum as the entrance to the Forum Romanum. If so, he may have made these plans in order to rewrite history and have the Senate meet in a building that was closely connected with a building project named after him – as Pompey had done by building a curia in the Theater Complex (thus Amici *et al.*, 2007, following Coarelli and most other scholars). But another reading of the placement of the Curia Iulia was a return to the old order, retaining its historical position in the Forum Romanum and making it the cornerstone of his new Forum. The restored Curia thus becomes the "mediator" between the Comitium and the Forum Iulium, and Caesar can be seen as upholding traditional political ideals and authority (Liverani, 2008). As ever, Caesar's intentions remain opaque and controversial.

The work on the Curia – which can be considered an Augustan building project – incorporated the relaying of the paving of the Forum in the area around the Comitium. The Lapis Niger was preserved in this repaving, perhaps because it was thought to be the grave of Romulus or Faustulus (as Festus 184L). The Rostra was rebuilt, which we only know from Cassius Dio (43.49.1), as its remains are encased in Augustan, Severan and Diocletianic rebuildings. Nonetheless, Caesar's ambitions to rework the buildings central to Roman political life and stamp them with his own name are very clear.

The Campus Martius, which had most recently been redefined in the urban life of Rome by Pompey, became part of the Caesarian vision for the rebuilt city. Again mostly known in its later rebuildings, the Saepta was planned by Caesar as a huge voting enclosure; this could only have been built in the relatively open spaces of the Campus Martius. Caesar planned a theater on the edge of the Campus Martius; it may have been built close to the site of the temporary theater built by M. Amelius Lepidus in 179. Aemilius's theater, Monterroso argues, was surely associated with Aemilius's Temple to Diana (a manubial temple also dedicated in 179), linking a theater and temple in the manner so beloved of second- and first-century Latium, as at Praeneste, Terracina and Gabii. Although Caesar did not live to see the construction of his planned theater, Monterroso argues that he did raze both the Temple of Diana and the nearby Temple to Pietas in order to clear a site for the building (Plin. *HN* 7.121; Cass. Dio 43.49.3). The theater was completed by Augustus, who named it after his late nephew Marcellus. Augustus continued to use the cavea of the theater to function as a place for audiences for triumphal processions. He maintained a link to sacred architecture by buying land around the site of the Temple of Apollo Medicus (Sosianus) to associate both this temple and the adjoining Temple to Bellona with the Theater (Monterroso, 2009).

6 Late Republican Houses of the Elite

Late Republican houses have increasingly been found in various sites throughout the city. No house has been excavated in its entirety, and all only survive as foundations, with the exception of the House of the Griffins, the Aula Isiaca, the "House of Augustus" and the "House of Livia" (for the last two, see below). Most of the remains uncovered by archaeologists show how the houses of the aristocrats began to sprawl over the slopes and summits of the hills of Rome, as they accumulated porticoes, ambulatories and private bath-suites (see Chapters 1 and 3). The introduction of these amenities is usually seen as more evidence of Greek influence on Rome as aristocrats conquered the old Hellenistic capitals and brought Greek architects to Rome. Indeed, Pliny tells us that it is in the first century that marble columns were built into private houses, though only those of the richest men (temporarily, M. Aemilius Scaurus, aed. 58; Plin. *HN* 17.5–6, 36.6; permanently, L. Crassus, cos. 95; Plin. *HN* 37.3–6; the house of Q. Lutatius Catulus, cos.78, is called magnificent, but no other details are given in Plin. *HN* 17.2). On the other hand, some aristocrats proudly lived in old-fashioned houses that looked quaint even to their peers (as did T. Pomponius Atticus, Nep. *Att.* 13, though admittedly, he was of equestrian rank).

The stubs of the walls of these houses are identified by the use of opus incertum or opus reticulatum. One house on the Velian has a portico/gymnasium attached to a row of rooms with cocciopesto floors; the middle room is round, and the whole is underlain by a large cistern. The excavator (who finds an identity for the Late Republican owner) suggests this is part of a luxurious residence, and it backs onto another, larger portico of a second domus (Fraioli, 2009). For another house, this time on the Arx, the excavator draws parallels to the layout of the Villa of the Papyri, and through those parallels finds an identity for the owner of the house in Rome. Although only traces of the house survived the Imperial building projects in the area, the house seems to have been richly decorated with mosaics and paintings (Tucci, 2005).

The House of M. Aemilius Scaurus (pr. 56) was said to be at the meeting of the Sacra Via and the Clivus Palatinus (Asc. *Scaur.* 23); the identity of the owner, Carandini argues, is secure, given the location of the house and the huge dimensions of its atrium (Carandini, 2004). Scholars continue to debate the function of the 30 windowless rooms found in the basement of this house. Most texts continue to call this basement an ergastulum, or secure slave quarters, though the small rooms with stone beds have also been called guest rooms, prison cells, or the working quarters of a lupinar. Like the other aristocratic houses mentioned here, it has a bath-suite and mosaic floors. Less is known about a house on the Palatine, though it included a two-story "portico-like" building made of travertine blocks (Panella and Zeggio, 2004). Hostetter and Brandt (2009) propose that this house must at least be near the house of C. Octavius, Augustus's father, and it helped to monumentalize the north-east slope of the Palatine.

The location of Cicero's house continues to draw the attention of scholars, in part due to the many hints we have as to its location on the Palatine, and in part due to its history of having been acquired by Cicero's enemy, partly rebuilt as a shrine to Libertas, and then restored to Cicero. If scholars can locate it, then several other houses of Late Republican Rome can also be located. Although Krause (2004) suggests it is near the top of the Palatine hill, by the recently discovered "clivus tectus," Wiseman (2012) cautions that we should not look for the house at this height and he would expect to find it on the side of the hill, closer to the house of Caesar.

The "House of Augustus" and the "House of Livia" continue to be re-evaluated by archaeologists (Iacopi and Tedone, 2006; Carandini and Bruno, 2008). Some scholars have questioned why the "House of Augustus" was buried, but the "House of Livia" remained accessible throughout antiquity (Wiseman, 2009, 2012; Zink, 2012). Although the houses have now been dated to *c.* 36, a full discussion of their plans, decorations and building phases is better placed within the context of Augustan Rome.

7 Rome Poised to Become an Imperial City

That the face of Rome changed dramatically during the first century is not in question. What is still being debated are the motives of the builders and the extent of the changes. For instance, the beginnings of the Imperial system of horti ringing the city with a "green belt" is thought to be seen in the gardens of Lucullus, who bought land in the 60s on the Pincian Hill. Here, we are told, he could relax as if he were in his country estate in Campania, but still be close enough to the city to be able to participate fully in its political life (Plut. *Luc.* 39). After Lucullus, other aristocrats began to amass "country estates" within an easy ride to the Forum, until the city was ringed with something like parks filled with Greek-inspired art, fountains, ponds, aviaries and luxury houses containing libraries, baths and porticoes, at least by the Early Imperial period (Purcell, 2007b; Wood, 2010). It was Pompey who first opened his gardens to the plebs; Caesar's gardens across the Tiber were used to stage public events, but only opened to the public after his death, when he willed them to the people of Rome. By such means was the larger public educated in aristocratic taste in art and the art of living, especially as the neighborhoods around the edges of these horti were of mixed use – domestic, industrial and infrastructure (Coates-Stephens, 2004).

Perennially interesting to scholars is the route of the triumph: Marcattili (2009) continues to uphold the traditional interpretation of the parade passing through the

Circus Maximus from the Forum Boarium (he relies on a rereading of texts and visual representations of the Circus; there have been no new excavations). Beard (2007) has argued that the route was not fixed, but was uniquely arranged for each triumph, a view that has much to recommend it, especially given the growing length and complexity of triumphal parades in the first century (see also Wiseman, 2007b).

The builders of first-century Rome had their counterparts in cities within Italy (Lomas, 2003, cautions against a broad brushstroke here; also Pobjoy, 2000b). The impact of the individual builder constructing large programs that would alter the city can be best seen in the Roman colonies, where perhaps it is the Roman colonists who stimulated this kind of development. In Pompeii, the duumvir M. Porcius helped provide funds for a new altar in the (relatively) new Temple of Apollo and joined with Quinctius Valgus to build the Odeon and Amphitheater. Both men were veterans of Sulla's army and likely had amassed their wealth due to the Sullan proscriptions. Valgus's efforts were not confined to Pompeii, as he gave money for walls and gates for at least two towns (again, joining with other patrons), and in one of those he built porticoes around a forum and a curia; for the third, he is listed as patron, but his building project is unclear (*ILLRP* 523, 565, 598, 645, 646).

The rise of the powerful individual ruler who uses building programs to shape his identity and to present himself to the ruled has been clearly linked to the increasing contact Rome had with the cities of the east, where such building programs had been common since the time of Alexander the Great (see Chapter 27). And it is the will to reshape entire portions of the city that is new to first-century Rome. While triumphators used their manubial coin to pay for isolated temples, horrea or porticoes (a pattern repeated with Marius), Sulla formed a vision for rebuilding the most revered temple in Rome and extended that project down the sides of the Capitolium to embrace the fundamental areas of Roman political life in the Forum. By the time of his death, the Forum was changed in the permanent buildings given over to economic functions, in the basilicas that ringed the north and south sides of the Forum. And the fundamental political buildings had been completely restored or rebuilt. Pompey's ambitions were lesser, in a sense, as he combined in his Theater Complex the perceived needs of the city's infrastructure with a temple and a curia. Perhaps he lacked the far-reaching vision of Caesar who appeared ready to leave no area of the city untouched as he swept away old buildings, restored and renewed, and gave (or intended to give) to the city a new forum, law courts, theater, senate house and entertainment buildings. With Caesar Rome made a great leap towards what would become the Imperial city.

FURTHER READING

Given the riches of ancient strata in Rome, it is no wonder that archaeologists stagger under the weight of excavating necessary current building projects. Full publications of excavations must take second place to the necessity of excavating, but the scholar who is not living in Rome can get glimpses of what is going on in several ways. Many excavations – especially if the archaeologist has attached a famous ancient name to the building – do appear in Italian newspapers; scholars can access www.patrimoniosos.it , under "Rassegna Stampa" for brief mentions, especially of spectacular finds. Excavations will appear in one or two paragraphs under "Latium" in www.fastionline.org, which replaces the periodical *Fasti Archaeologici*. Summaries of excavation areas all over Rome are provided annually (since 2008) by R. Coates-Stephens in the *PBSR*; for the activities of the German School the *Archäologische Anzeiger* provides annual summaries. Reviews of literature occasionally appear in *BullComm*, under "Notiziario bibliografico di Roma e suburbio." Preliminary results of the Metro C line excavations are found in Egidi, Filippi and Martone (2011). Notices of various excavations dating from 1980 to 2006 can be found in Tomei (2006). More synthetic articles regularly appear in the new series *Workshop di archeologia classica* (e.g. Carandini and Greco, 2004, 2005, 2006, 2007, 2009).

Two English-language guidebooks to Rome itself, Claridge (1998) and Coarelli (2007b) (which also encapsulates his earlier work in Coarelli, 1985, 1988b, 1997), provide good overviews of the Late Republican building projects as well as more detailed notices of particular buildings. For specific buildings, the entries in *LTUR* are essential. Platner-Ashby, *NTDAR* and Haselberger (2008) are other good places to start. Torelli, Menichetti and Grassigli (2008) has a chapter dedicated to the Late Republic; Gros (1996–2001) looks at various building types, from the Republican through the Imperial period. Gros and Torelli (2007) offers a larger picture of Rome, with a section devoted to the Late Republic. Patterson (2010) covers the Mid-Republic to the Mid-Empire in a discussion of new excavations. Less archaeological in nature is de Chaisemartin (2003). Holz (2009) provides an overview of Late Republican building programs in light of imperial conquests.

CHAPTER 30

Cosa

Stephen L. Dyson

those colonies seem to be miniatures, as it were, and in a way, copies [of Rome] ...
(Aulus Gellius *NA* 16.13.9)

1 Introduction

The Latin colony of Cosa was founded in 273 on a limestone plateau overlooking the Tyrrhenian Sea 138 km north-west of Rome. It was the fifteenth Latin colony created by Rome after the Latin League was dissolved in 338 (Brown, 1951). The settlement's role was to watch over the territory of nearby Vulci after the major Etruscan revolt of 280. Pliny (*HN* 3.52) mentioned that there was already an Etruscan settlement in the area, which the Romans removed from the territory of Vulci. Wherever that "Etruscan Cosa" was, it was not at the site of the Roman colony. The new settlement probably also functioned as a coastal watch-station at a time when Rome did not control the Mediterranean. The twin role was likely reflected in the colony's first issue of bronze coins, which had a head of Mars on the obverse and on the reverse, symbols of Poseidon, a horse head over a dolphin.

The exact size of the original foundation population is not known, but it certainly involved several thousand families. Frank Brown, the founding excavator at Cosa, estimated the original settlement at 2,500 families (Brown, 1980: 15–16). As a Latin colony, Cosa was largely autonomous, for by becoming citizens of the city, they lost voting rights at Rome. That isolation was enhanced by the fact that Cosa was not located close

A Companion to the Archaeology of the Roman Republic, First Edition. Edited by Jane DeRose Evans.

to any major road connecting it to Rome. That changed with the construction of the Via Aurelia in the middle of the third century (see Chapter 19). The roads out of two of the gates then linked the city to the Via Aurelia. It was in the area of those roads that the most important parts of the necropolis of Cosa were located.

Three series of events must have shaped the early history of Cosa. The First Punic War started less than a decade after the colony was founded and lasted until 241. Cosa was a coastal colony and must have been on high alert, until Rome secured dominance at sea. Then in 225, the last major invasion of the Gauls was stopped at Telemon, only 10 miles from Cosa. Finally, the Second Punic War placed serious demands on the remaining resources of the colony.

Livy (XXXIX.55) noted that the colony was reinforced in 197, as a result of losses stemming from the Second Punic War. Otherwise, Cosa is little recorded in the ancient literary sources. The last ancient reference that we have on Cosa comes from the fifth-century writer Rutilius Namatianus, who, in 416 CE, sailed by the site and commented on its desolation (*De reditu* 1.285ff.). Antiquarians noted its picturesque remains in the desolate Maremma. George Dennis, who visited Cosa in the 1840s, gives in his *Cities and Cemeteries of Etruria* an evocative account of the site as it appeared in the mid-nineteenth century (Dennis, 1848).

2 Frank E. Brown and the Early Excavations of Cosa

In the later 1940s Cosa attracted the attention of Frank Brown of the American Academy in Rome. He saw it as an excavation site that could provide insight into the Etruscan–Roman transition, into the processes of Roman colonization and into the formative phases of Roman Republican architecture. He took advantage of the new Italian openness to foreign excavations to launch an American Academy research program there. American excavations started in 1948 and continued, with inactive interludes, down into the 1980s. While only a small part of the total site has been excavated, representative sections have been cleared and sampled. Occupation and construction activity have been documented from the foundation in 273 to the last, limited settlements of the Middle Ages.

The first stage in the American Academy operations involved the clearing and mapping of the site. The colony was founded on a limestone plateau that overlooked the Tyrrhenian Sea. On the south-west was the acropolis or Arx, clearly the religious center of the colony (Figure 30.1). On the east was another high point, designated the Eastern Height. Both contained structural remains. The basic grid plan of the colony could be reconstructed even without excavation. Rectangular blocks with remains of buildings covered much of the site. The remains of a number of cisterns were identified (one large example has been excavated); these provided the only supply of water throughout the long history of the colony, for Cosa never had an aqueduct. Remains of a small bath structure that has never been fully excavated were identified. Finally, the site of the forum was located, but not excavated in this phase.

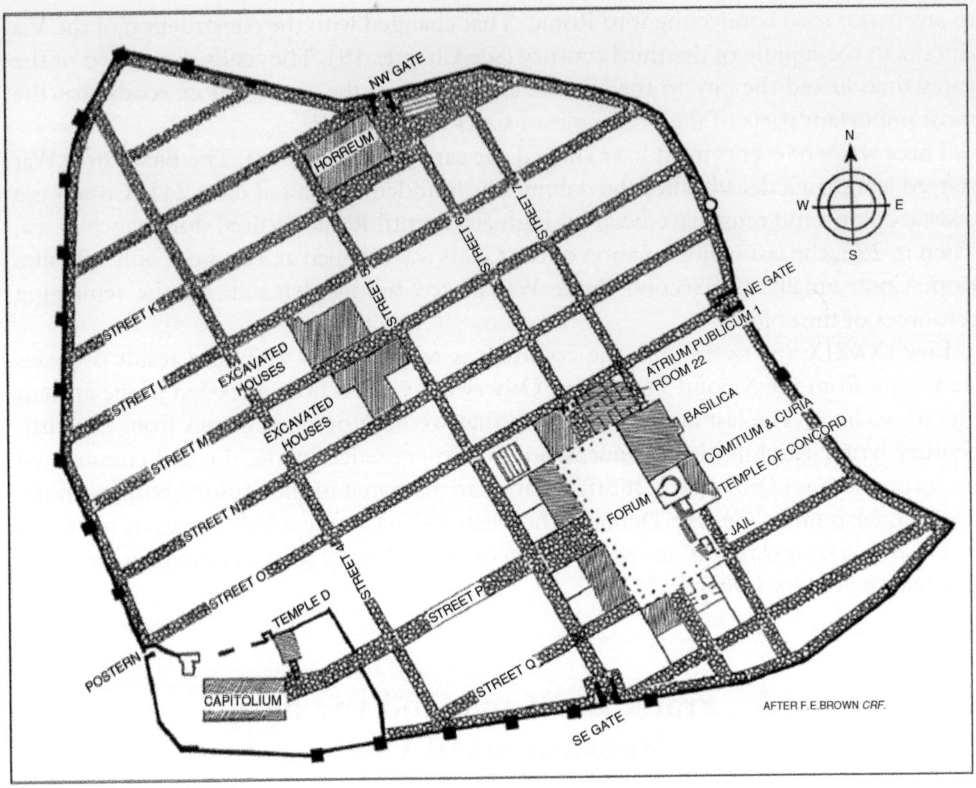

Figure 30.1 Plan of Cosa, second phase of colonization. Source: Map by Clementsalvi, after Brown (1980), http://commons.wikimedia.org/wiki/File:Plan_cosa.png. GNU Free Documentation Licence 1.2.

3 The Walls of Cosa

The most striking feature at the site of Cosa today is the impressive wall circuit formed of massive polygonal blocks of limestone (Figure 30.2). It is smoothed on the outside with the spaces chinked with smaller pieces of limestone. The style and construction techniques recall those of the towns of Latium from which many of the colonists would have come. However, the some 20 towers, placed at strategic points along the perimeter, reflect acquaintance with the most advanced, early third-century fortification techniques. Recent cleaning and study have provided new information on the details of construction (Poggesi, 2004). Three gates, designed with interior courts, led out from the settlement. The walls and gates represented the only major feature at the site which can be dated to the settlement period of the colony.

4 The Arx

After the walls, the most impressive remains at Cosa are those of the temple complex on the Arx. The ritual nature of this southern height was emphasized by the construction

Figure 30.2 Polygonal masonry of the northwest gate, Cosa. Source: Photo by Mac9, http://it.wikipedia.org/wiki/File:Cosa02.jpg. Creative Commons Attribution-Share Alike 2.5 Generic License.

of a perimeter wall that set the sacred space apart from the inhabited city. Cosa's version of the Via Sacra linked the temples to the Forum. The first American excavations (1948–50) were concentrated there, and Academy archaeologists returned at later dates to that part of the site (Brown, Richardson and Richardson, 1960, 1993; Fentress, 2003; Dyson, 2005).

Brown claimed to have found there evidence of rituals associated with the foundation of the colony (Brown, Richardson and Richardson, 1960; Brown, 1980: 51–4). Under the remains of the cellae of the Capitolium was discovered a square platform area cut into the rock but oriented differently from the later temple. There was also a crevasse/pit which had within it vegetative remains which suggested ritual activity associated with the religious foundation of Cosa. Brown provided a reconstruction of that earliest ceremonial center with an auguraculum, where the augurs read the heavens, and a mundus, where the foundation offerings were deposited, all part of the ritual site of Cosa Quadrata. The colony in its foundation rituals supposedly recalled the early world of Roma Quadrata. Yet those supposed testimonies to the colonial foundation rituals have remained a source of controversy and skepticism among students of Roman ritual and Roman colonization (Frederiksen, 1962; Fentress, 2000; Dyson, 2005).

Today the Arx at Cosa is dominated by the remains of the Capitolium (Brown, Richardson and Richardson, 1960). With its dominant frontal porch and three interior cellae the temple was obviously modeled on the great sixth-century temple of Jupiter, Juno and Minerva at Rome. That "archaic" quality was enhanced by the walls of the supporting temple platform, which were constructed of large irregular blocks, in the same construction technique as the city walls, and by the heavy foundations topped with moldings that support the temple, which recall the building traditions of Etruscan and early Roman architecture. In front was an altar, not oriented to the main temple, but to the walls of the auguraculum which were now buried under the new construction.

While the Capitolium was initially thought to have been founded in the initial building phase of the colony – that is, as another expression of Cosa as a "mini-Rome" – a detailed study of both the construction details and of its sculptural and decorative terracottas demonstrate that it was built in the early to mid-second century, probably as an affirmation of Roman loyalty and identity in the aftermath of the Second Punic War (R. Taylor, 2002). It was, as far as we know, the only Capitolium constructed in a Latin colony; it thus says something about Cosa's special relation to Rome.

Also located on the Arx, just to the right of the Capitolium are the foundations of a smaller temple (Temple D), probably dedicated to the distinctly Roman-Italic goddess Mater Matuta (Brown, Richardson and Richardson, 1960; Brown, 1980: 47–9). It, too, appears to have been built after the reinforcement of the colony in the 190s, probably slightly earlier than the Capitolium. It was remodeled in the early first century, not long before disaster hit the colony.

The excavations produced a large number of architectural terracottas associated with various phases of the temple decoration and redecorations. They included pedimental sculptures and revetment plaques that lined the edges and ridges of the roofs. These colorful decorations in baked clay, mostly dating to the late third to early first century, displayed common qualities with finds from Etruscan and Latin sites in Hellenistic Italy. Their evolving styles reflected the influences of the larger Hellenic Mediterranean world that Rome was coming to dominate.

Two sets were clearly associated with the Capitolium and the Mater Matuta temples. However, a third group of relatively early date did not fit either sacred structure. They seemed to have been part of an earlier temple located on the Arx. The excavators assigned this somewhat hypothetical temple to Jupiter (even though the imagery of the decorations related more to Minerva and Hercules) and dated it to 240–220, just after the First Punic War. Locating that temple has proved a major problem. Initially, the archaeologists placed it to the left of the Capitolium in an area then covered by modern constructions (Brown, Richardson and Richardson, 1960). Later excavations found no traces of a temple there, but some exiguous remains that suggested the Jupiter Temple was located behind the later Capitolium (Taylor, 2002). While the Arx was mainly a religious center, some evidence of Republican housing has been found there as well.

5 The Forum

After the excavation of the Arx, in 1950 the American Academy turned to the excavation of the forum. The two areas were the foci of life in the colony and linked by a wide processional street. The civic importance of the forum was highlighted by the construction of a three-aisled monumental entrance arch (fornix) in faced concrete. Its ruins are still among the most impressive remains at the site. Cosa's fornix is the oldest surviving example of that form in Roman architecture and it reinforced the vision of Cosa as a mini-reflection of Mid-Republican Rome. While difficult to date precisely, the Forum Arch seems to have been added not long after the reinforcement of the colony (Brown, Richardson and Richardson, 1993).

Detailed survey and limited soundings provided evidence for the basic layout of the forum as well as the location of major public buildings. Three major phases of American excavation in the forum have provided evidence for a complex history of occupation, construction and reconstruction (Brown, Richardson and Richardson, 1993; Fentress, 2003). As with the temples on the Arx, the forum saw the gradual development of its civic and religious buildings. Early, and most imposing architecturally, were the two political centers, the Curia, where the local order of the decuriones met, and the Comitium, where the Latin citizens of Cosa assembled. The Curia was a circular theatrical structure, the meeting-place of the citizens, which was backed by a raised, rectangular room that was the local senate house. The design of the Comitium was similar to that found at several other colonies. The two buildings were then thought to have been very similar to the contemporary Comitium and Curia at Rome, another instance of how the civic forms in the colony reflected those in the Mid-Republican capital. Recent research suggests that the Comitium-Curia at Rome itself had not been so regular in plan and was the product of long historical evolution, not one specific moment of foundation (see Chapter 26).

By the mid- to late second century, the forum had other buildings. One of the most important was the basilica, a large structure erected toward the middle of the century (Brown, 1980: 56–8; Brown, Richardson and Richardson, 1993). It had a four-sided colonnade that supported a raised roof and a tribunal that projected back from the wall opposite the forum. Again, it imitated a form that was becoming popular in the capital. It presumably served a variety of political, legal and commercial functions. The basilica at Cosa had a long and colorful history, but civic decline meant it had fewer public functions. In the era of Claudius and Nero it was damaged by an earthquake and then turned into a mini-theater and a Julio-Claudian shrine (Collins-Clinton, 2000). Nero came from the family of the Domitii Ahenobarbi, who were major property owners in the Cosa area during the Later Republic.

Other public buildings were located on the same side of the forum. Of early date was a structure that has been identified as a prison (Brown, Richardson and Richardson, 1960). Just before the middle of the second century a temple (Temple B) was inserted into the north side of the forum. Temple B appears to have replaced a simple sacellum erected earlier on the same spot. The temple itself was raised on a platform faced with irregular stone blocks like those of the walls and the Capitolium foundation. It had a single cella with four columns in front. Concordia, and more recently Ceres, have been proposed as the deity honored in that structure (Brown, 1980: 38–9; Bruno and Scott, 1993). Remarkably, Cosa was not cluttered with religious buildings. That forum temple and the poorly preserved remains of a temple, probably of the second century, on the so-called Eastern Height are the only religious structures so far identified outside of the Arx.

In the original plans drawn up by the Academy archaeologists, the forum was shown lined on the west and south side by buildings, which on superficial examination recalled the forms of elite houses (Figure 30.1) (Brown, Richardson and Richardson, 1993). That was not surprising, since the houses of senators would have been found in the forum area of contemporary Rome (see Chapter 3). One such structure (Atrium Building I), located on one side of the monumental arched entrance to the Cosan forum, was excavated. Scanty evidence suggested a construction in the early second century. It had the classic atrium plan but without sleeping or service rooms. The structure was interpreted as some type of public building. The so-called Atrium Libertatis in Rome provided possible parallels and again identified the colony with the mother city. Less clear was the role of the other "atrium-type" buildings that lined the Cosa forum.

More recent excavations have demonstrated that at least some of those were indeed elite residences. Most important is the early second-century atrium house (Atrium Building V) that faced onto the forum on the south (Fentress and Rabinowitz, 1996; Fentress, 2003). Rather similar in form to the House of Sallust at Pompeii, it had an impluviate atrium surrounded by rooms. At one end was the tablinum, decorated with a mosaic floor, and a room that originally served as a triclinium. The domus also had a kitchen garden and a small bath house. Two "shops" on its front faced onto the forum and related the residence to the commercial life of the community. The dating evidence suggests an original construction in the early second century with much subsequent reordering and redecoration, especially in the Augustan period. It is possible that other elite houses were located on the forum, much as they were in earlier Republican Rome. These elite residences might well have been the dwellings of the decuriones of the community. That possibility is reinforced by the fact that the Atrium Building house-plot size was considerably larger than those in other parts of the city.

6 Houses

Detailed information on housing is not limited to the forum. Excavations for a museum/laboratory complex in blocks VD and VID during the late 1960s and early 1970s provided the first carefully researched samples of domestic architecture at Cosa (Bruno and Scott, 1993). Initially the blocks were divided into modest residences of more or less uniform size. Those excavated seem to date to after the reinforcement of the colony in 197. While there is evidence of basic amenities such as cisterns and cesspits, decoration was modest. Remains of material culture were limited. However, graffiti on black-gloss pots suggest that one may have belonged to the Salvii family.

By the early first century very different structural types had emerged. The largest (the so-called House of the Skeleton) was an elegant atrium house with tessellated flooring, a tablinum, a small bath, First Style wall painting, and a small elegant hortus. The house appears to have assumed its final form around 80, not long before its destruction, which occurred at the same time as the destruction of much of urban Cosa. The second block was composed of much more modest dwellings of traditional plan with little sign of elegance or luxury.

The diversity of housing within one block by the early first century shows that the overall development of residential architecture within the walls of Cosa from the third to the first centuries was complex. Study of the evidence from the detailed survey within the walls suggests that some 300 block units could have been created in the town center and that probably less than 100 houses were actually built. Whatever uniformity may have initially existed probably changed over time. Block VD would not have been unique in having dwellings of varied degrees of elegance.

The houses revealed dramatic evidence of violent events that led to the end of Republican Cosa. The elegant domus bears the designation of the House of the Skeleton (Bruno and Scott, 1993). That derives from the discovery of a human skeleton, a 30- to 35-year-old male, in one of the water cisterns, a clear sign of a violent end for the house and its family. In the more modest adjoining structure, which on the basis of graffiti has been assigned to a certain Q. Fulvius, was found a modest clay pot. It contained a massive hoard of 2,004 silver denarii, wealth that contrasted greatly with the modest dwelling place. The terminal date was 73–71 (Buttrey, 1980). The fact that such a fortune was never recovered again argues for a dramatic event that impacted the city. Given the coastal vulnerability of Cosa, a piratical attack in the years before Pompey swept them from the seas seems most likely. These blocks, as with so much of Cosa, remained abandoned until the Augustan era.

While port and countryside recovered and flourished in the Later Republic, the town itself seems to have languished. Surface survey within the walls of the colony shows extensive areas of abandonment or limited occupation in the years after the disaster of the late 70s (Fentress, 1994). The Capitolium seems to have been restored, since a number of the terracotta decorations were dated to the middle of the first century (Taylor, 2002). Since the temple was the major religious center for the area, and life continued and even prospered outside of the town, such a restoration was to be expected.

Only with the Augustan period was any significant effort made to revive the town of Cosa. That was best reflected in the group of houses that were built on one side of the forum. Most important was the so-called House of Diana, an elite residence of the Augustan and Julio-Claudian periods (Figure 3.3). The effort at revival was not a great success. The Augustan revival sputtered into the Late Flavian period with some Severan revival. Some community persisted at Cosa until the Early Middle Ages (Fentress, 2003).

7 The Port

Significant research has been done at the port and in the countryside of Cosa, so we have a full and complex history of the Roman community in all of its significant aspects. Below the hillside community, the Romans developed a small port in a little protected bay (Figure 21.1; see Chapter 21; McCann *et al.*, 1977; McCann, 2002). The site has the only major fresh water spring in the area. Two projects were undertaken by Roman engineers to complete the anchorage. The first was the construction of breakwaters; then came the building of piers, some using the waterproof cement that the Romans had developed by the second century (see Chapter 17). Cosa also appears to have had a small lighthouse.

The second project involved the improvement of a series of natural clefts in parts of the small, northern headlands and especially that of the cutting known as the Tagliata. Detailed study has revealed a complex series of artificial cuttings and insets for sluice gates. Initially explained as a means of preventing the sanding-up of the harbor, those improvements are now explained more as helping control the fresh/sea water mix in the fish tanks that developed along the shore.

While never a major port of transit, the Cosa harbor was still a useful roadstead for Roman ships heading to the expanding Roman spheres of interest in south Gaul and Iberia. The harbor developed its own small community, complete with a temple dedicated to either Portunus or Neptune. The temple resembled the Temple on the Arx in form and decoration and probably dates similarly to 170–160. While the town itself declined in the first century, the port assumed a new importance with the growth of the villa economy in the countryside of Cosa. Indeed, one of the major villas in the area was built on the beach itself. Remains of fish tanks show investment in aquiculture and the production of garum. The natural spring at the site was improved by the construction of a springhouse during the early first century.

The beach area and the waters just offshore have yielded numerous fragments of transport amphoras, testimony to the flourishing trade in wine based in the Cosa area. The earliest amphoras date to the late third century. Many were stamped with the sign of the Sestius family (see Chapter 6). The Sestii were a major exporter of wine whose trade network extended well into Gaul. Based on the concentration of their stamps, it seems likely that the family enterprise was based in part in the Ager Cosanus. Their earliest stamped amphoras at Cosa date to 175–150, and they continue into the first century (Manacorda, 1978, 1979).

8 The Hinterland of Cosa

The colony of Cosa controlled a hinterland of some 500 square kilometers, whose rough boundaries can be reconstructed through post-classical land records. It was of uneven quality with much of its wild upland good only for pasturage and forage. However, the coastal plane and several interior valleys provided an abundance of quality farm land. Early in the research program at Cosa, Castagnoli reconstructed some areas of centuriation in the Ager Cosanus, through the use of Second World War vintage aerial photographs (Castagnoli, 1957). They represented the land divisions associated either with the settlement of the original colony or with its reinforcement in the 190s. Recent research has reconstructed more widespread evidence for centuriation in the landscape of Cosa (A. Wilson, 2004).

The initial programs of archaeological research at Cosa showed relatively little interest in the rural ager, where most of the colonists would have lived and made their livelihood. In the late 1960s and the 1970s that changed. Since then, American, Italian and British teams have applied landscape archaeology, survey and excavation to both the territory of ancient Cosa and the Albegna River Valley that marked its northern border. The Ager Cosanus is now one of the best studied of Roman rural territories (see Chapter 13).

Some of the most important research has illuminated the settlement history in the area of Cosa before the Romans came on the scene. When excavations started in the town just after the Second World War it was assumed that they would uncover an Etruscan Cosa that preceded the Latin colony. Several generations of excavation on the hill have yielded almost no evidence for an Etruscan occupation. Finds of Etruscan material have also been sparse in the rural surveys near the colony and near the shore.

The relative absence of Etruscan settlement evidence was considered strange. The written sources clearly indicate the existence of an Etruscan Cosa. The Roman settlement was carved out of the territory of Vulci, one of the most important of Etruscan centers (Carandini, 1985b). Rich Etruscan tombs of the seventh and early sixth century had been found at Marsigliano at the north-west border of the ager. As the survey projects extended in range and their methodology became more sophisticated and flexible, much more evidence for an Etruscan presence emerged. Excavations at the medieval hilltop settlement of Capalbiaccio, some 15 km south-east of Cosa, revealed traces of an Iron Age–Etruscan occupation. Survey archaeologists working in the Albegna Valley have discovered a number of Etruscan settlements, including the remains of a major Etruscan center at a place called Doganella, just north of the river (A. Wilson, 2004). It was possibly ancient Kalousion. The center was fortified and provided evidence of artisan production and a flourishing amphora and ceramic industry. Clearly the Etruscans lived in the territory that later became the Ager Cosanus. However, they had a system of settlement very different from that of the Roman colonists, with more focus toward the river valley and the interior.

The first stages of the Roman colonial occupation in the Ager Cosanus have proved hard to document. The Etruscan center at Doganella seems to have been destroyed at the time of the suppression of the uprising at Vulci. Survey evidence suggests a general drop in population in the Albegna Valley, probably also the result of the Roman conquest. At least some of the ager of Cosa was then centuriated and the bounded plots given to colonists. While plot size may have varied according to status, most of the first farmsteads were probably modest affairs. They left little in the way of archaeological traces.

The reinforcement of the colony in 197 led to the creation of more farmsteads. The rural settlements of the second century are archaeologically more visible. As a result of scattered literary references and the area's history of intensive survey, the Ager Cosanus has been one of the most important foci in debate about the fate of the Italian countryside in the decades after the Hannibalic Wars. Tiberius Gracchus passed through the territory of Cosa on his way to and from military service in Spain. It has often been claimed that the sight of the colony's depopulated countryside helped spur him to push reforms that would restore the Roman citizen-farmers. The countryside of Cosa figured several times in the troubled events of the first century. In 87, Marius arrived at nearby Telemon from North Africa and rallied forces in the area. In 77, Aemilius Lepidus gathered forces at Cosa for transport to Sardinia. In 49, L. Domitius Ahenobarbus gathered slaves, freedmen and peasants from his estates around Cosa. They were then transported to Massilia. All of these references suggest a rather complex countryside with more than just chained estate slaves.

Until the 1970s the best-known rural sites around Cosa were the large ruined villas at Settefinestre and Le Colonne, whose terraces and towered facades suggested the opulent residences of major Roman landowners, who would have controlled the latifundia that dominated the landscape of Late Republican Cosa (Figure 30.3) (Carandini, 1985a; Dyson, 2002). The work at Settefinestre, directed by Andrea Carandini, was designed both to illuminate the occupation history of the villa and to provide a training program that would introduce Italian students to the sophisticated methods of British classical archaeology. The excavations showed a structural and occupation history that extended

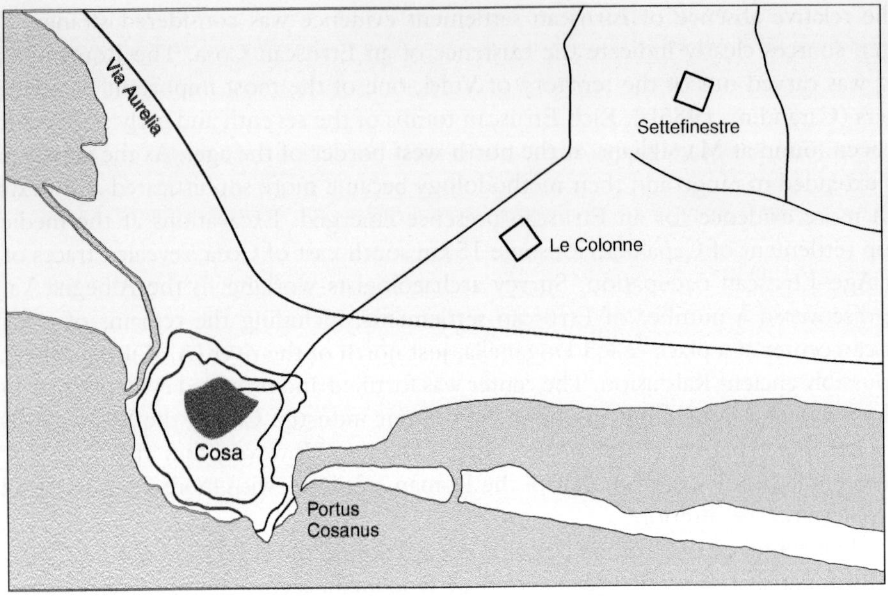

Figure 30.3 Plan showing the site of Cosa, Le Colonne and Settefinestre. Source: Illustration by Amy Yandek, adapted from Dyson (2002).

from the second century BCE to the second century CE. They recovered significant evidence for the elegant lifestyle of the villa owners, as well as facilities related to agrarian productive activities. Carandini was at that time much influenced by Marxist interpretations of Roman history. The Settefinestre villa was presented as an example of the "slave mode of production" in the later Republic. Its agrarian activities were tied into those of the family of the Sestii at the port. Indeed the Sestii were even suggested as possible owners of the villa (Carandini, 1985a). Another possible owner might have been the family of the Domitii Ahenobarbi, ancestors of the emperor Nero, who (as has been mentioned) also had estates in the area.

The American excavations at Le Colonne showed a similar developmental history with a foundation in the later second century BCE and a serious decline in the second century CE. Like Settefinestre, the villa had a number of elegant features including mosaics, wall paintings, and a garden whose form imitated a stadium.

Excavations at the villa were accompanied by several major field surveys that intensively explored much of the Ager Cosanus and the Albegna Valley (Dyson, 1978; Rathbone, 1981; Perkins and Walker, 1990; Perkins, 1999; Carandini and Cambi, 2002). A large number of Roman sites were identified and dated, mainly on the basis of ceramic finds. Relatively few sites from the first decades of the colony were discovered. Much more abundant were those of the second century. More controversial was the reconstruction of what happened to many of the rural sites in the later second and early first century. Advocates of Marxist interpretations argued for a period of abandonment of smaller sites and the flourishing of great villas; such as reconstruction would suit models based on the dominance of the slave mode of production (Carandini and Settis,

1979). Others have downplayed the extent and significance of the decline of smaller landholders (Dyson, 1978, 2002; Rathbone, 1981). There is considerable evidence of significant occupation in the Early Empire, probably reflective of revival of the Augustan colony. It appears that a significant crisis occurred in the second century CE, one that impacted rural settlements both big and small. Such evidence is paralleled in the decline in the city center itself.

9 The Impact of the Work at Cosa

The collection and documentation of material culture remains such as ceramics, glass, and lamps during the early Cosa excavations was, for its time, of high quality (Fitch and Goldman, 1994). Much potential information was recovered. The decision was taken to publish the artifacts as discrete collections, rather than as elements in the consideration of the excavated units. The publication record of the artifacts has been uneven. While several ceramic types (such as black-gloss and utilitarian wares) have been fully published, other important categories of material culture, such as amphoras, glass, sculpture and inscriptions, have no full, published study. While Cosa was not a major site during the Empire, it was important during the Middle Republic and studies of archaeological materials like black-gloss pottery and the utilitarian pottery have made important contributions to our understanding of the material culture of the Republic (see Chapter 5).

It is not the task of this chapter to take the history of the colony into the Empire and beyond. More than 50 years of quality excavation and field survey have made it one of the best-studied Roman sites in Italy. Excavation goals have naturally changed since Cosa was first excavated. Frank Brown's desire to find at the colonial site a mirror of Mid-Republican Rome has been only partly realized. The historical limits of the settlement have been increasingly understood. Like many of the Mid-Republican Latin colonies, it was founded for a specific purpose – that is, to be a fortress that watched over recently restless Etruscan territory and the vulnerable coast. The small harbor and the passage of the Via Aurelia through its territory linked it partially to the wider Roman world. When disaster struck in the late 70s there was no immediate impulse to restore the colony. The Augustan efforts at revival were only partly successful. Long before the general crisis of the Empire, Cosa was a marginal, rather pathetic place, whose massive walls alone remained a reminder of former glories.

FURTHER READING

While the final publications of the site must first be addressed, Cosa remains a much-discussed example of Roman colonization and urban planning. For final publications of the buildings on the site, see Brown (1951), Brown, Richardson and Richardson (1960, 1993), Bruno and Scott (1993), McCann (1987, 2002) and McCann *et al.* (1987). For final publications on the material culture from the site, see Buttrey (1980) (coins), Fitch and Goldman (1994) (lamps), Goldman (2001) (various); for the pottery, see Taylor (1957), Marabini Moevs (1973, 1980a, 1980b, 2006), Scott (2008) and Jolivet (2009). For the settlement patterns and agricultural activity around Cosa, see Castagnoli (1957), Manacorda (1978), Dyson (1981) and Rathbone (1981). For the excavations at Le Colonne, see Dyson (2002); for Settefinestre, see Carandini and Ricci (1985). For recent rethinking of the chronology and function of Cosa, see Fentress (1994, 2000, 2003), Fentress and Rabinowitz (1996), Gualtieri (2002), Taylor (2002) and Gazda (2008).

CHAPTER 31

Becoming Roman Overseas? Sicily and Sardinia in the Later Roman Republic

R.J.A. Wilson

1 Introduction

These two islands, of very different character, became Rome's first provincial acquisitions as a direct result of its clash with the other great Mediterranean superpower of the third century, Carthage. For both places, however, the term "Roman Republican archaeology" sits uneasily. Sicily had been settled by Greeks between the late eighth and early sixth centuries. As part of a gradual process of cultural assimilation and interaction, many of the indigenous hilltop settlements of the interior had taken on the trappings of "Greek" culture by the end of the fourth century. Only in the far west, in the Phoenician coastal settlements of Palermo, Motya and Soluntum, that had by then became part of the Carthaginian power-base in Sicily, was a different, Punic, cultural milieu dominant. Much of Sicily, however, retained its overwhelmingly Hellenistic Greek character throughout the period of the Roman Republic (Wilson, forthcoming a, forthcoming b). The term "Roman Republic" must be seen principally, therefore, as a chronological, not a cultural, label. In Sardinia, an island that saw no interest from the Greek colonization movement, Phoenicio-Punic culture remained dominant down into the first century BCE and beyond (Wilson, 1980–1, 1996: 442–8). What impact, if any, therefore, did "Rome" make on these two provinces from the third quarter of the third century, when they first came under Roman suzerainty, down to the end of the Republic?

A Companion to the Archaeology of the Roman Republic, First Edition. Edited by Jane DeRose Evans.

2 Sicily

2.1 The First and Second Punic Wars

The First Punic War resulted in enormous havoc and destruction. Rome and its allies lost 100,000 men at their own reckoning; thousands more were sold into slavery after brutal sieges and the subsequent storming of enemy cities such as Camarina, Agrigentum and Palermo. Some, such as Selinus, were never to revive afterwards. Even if the casualty figures are inflated, the toll of war was huge. Lipari, for example, in the Aeolian Isles off Sicily's north coast, a key strategic base in Carthaginian hands, resisted a Roman siege in 258, but it was attacked again in a renewed Roman offensive in 252/1 (Polyb. I.39). Stone blocks from the Greek walls were ripped out by the defenders to hurl down on their attackers; catapult balls in non-local stone were perhaps Roman projectiles, and huge quantities of iron arrow-tips and lead sling-shot had also been stockpiled (Bernabò Brea and Cavalier, 1998: 182–3, 191–2).

The unsettled conditions at the time of the First Punic War are reflected in a remarkable collection of eight bronze documents, in Greek, all but one of them municipal decrees set up by Entella, a hill-town in the Belice Valley in western Sicily. The remaining one was a decree of the so-far-unidentified town of Nakone. All were intended to be displayed to promulgate their contents publicly: Nakone's was placed in its Temple of Zeus Olympios, while four of Entella's tablets were put up in the city's *bouleuterion*, and the other three in the Temple of Hestia. They most likely date between 254 and the close of the war in 241. The mention in one of them of a probable Roman official (perhaps a praefectus), Tiberius Claudius of Anzio, seems inconceivable before the start of the First Punic War (Corsaro, 1982), and the mention of a Mamertine, Minatos Corvios, in another decree, as being among the people of Petra (a further hill-town in western Sicily), must post-date the Mamertine capture of Messina in 288. The explicit mention of synoecism in four decrees is also in keeping with the unsettled political and military climate of the First Punic War. One, for example (the Ninth Decree), speaks of the council and assembly of Entella's expression of good will (*eunoia*) "for all time" to the people of Segesta, and the agreement of reciprocal citizen rights (*isopoliteia*) between the two communities. This was in return for favors offered by the Segestans, which had included working for the release and safe return of Entellans captured during the hostilities (Ampolo, 2001).

On March 10, 241, the First Punic War ended off the Egadi Islands with a major naval battle, its location now pinpointed through underwater finds off the north-west coast of Levanzo. These include the discovery (up to 2012) of no less than 10 ships' rams of bronze, each with a triple beak (Famà, 2007; Tusa, 2010: 219–28). Some with Latin inscriptions, indicating official approval from a quaestor, must come from Roman galleys; only one to date (with the same type of ram) has a Punic inscription. The finds raise questions about the context of a Punic ship excavated in the 1970s in shallow waters off Motya. If this really was a mid-third-century warship, it was presumably limping away from the battle when it ran aground (Frost, 1981), but it may have been an approximately contemporary merchant ship, although one emptied of its cargo (Parker, 1992: 262–4).

It seems unlikely that Sicily's recovery from the war was fast. There is little building activity in the towns of central and western parts of the island that can be ascribed to the second half of the third century, and the evidence for coin circulation points to a general impoverishment in western Sicily before a resurgence of monetary supply in the second century (Frey-Kupper, 2006: 44). All this is hardly surprising; and in any case, before long, the storm-clouds of military conflict were gathering once again.

Throughout this period one part of Sicily was initially untouched by the military mayhem. Hieron II, in power at Syracuse from *c.* 270, had first allied himself with Carthage at the start of the First Punic War, but then had smartly changed sides. As a result, he was able, with Rome's support, to keep his independent kingdom, which embraced much of south-eastern Sicily, as a separate entity untouched by war. The resulting period of economic prosperity and cultural renaissance made a stark contrast with what was happening elsewhere in Sicily (Lehmler, 2005). On Hieron's death in 215 his son Hieronymus nominally switched allegiance to Carthage. His assassination a year later and Rome's prompt military intervention led not only to the siege and sack of Syracuse and the destruction of other cities nearby, but also to the inclusion of the whole of the kingdom into the fledgling Roman provincia. In 210 M. Valerius Laevinus, after a fact-finding tour, reported to the Senate, no doubt with more than a touch of exaggeration, that not a single Carthaginian remained in the island, and that Sicilian agriculture was in full swing again. Provincial administration had begun in earnest – Sicily became, as Cicero was later to put it, the "first such ornament of empire, the first to show our ancestors how glorious a thing it is to rule over foreign nations" (*Verr.* 2.1.2).

2.2 Roman Changes in Sicily

Until the arrival of the Roman administration, Sicily was a collection of independent Greek-style *poleis*, each with their own government, their own citizenship, and their own *territoria*. During the First Punic War, in either 252 or 248, the consul C. Aurelius Cotta saw to the upgrading of the road between Palermo and Agrigentum, commemorating it with a rare surviving milestone of Republican date (*CIL* I^2.2877; Prag, 2006b). There must of course have been a road system of sorts in Sicily from the Archaic period, but improving long-distance communication was no doubt a priority in any attempt to unify the island. The process of road improvement doubtless continued *c.* 210 with M. Valerius Laevinus, who gave his name to the Via Valeria, the north-coast road linking Messina with Palermo, while the Via Pompeia along the east coast may have been created by Pompey the Great, propraetor in Sicily in 82 (Uggeri, 2004: 23, 2007; Prag, 2007a: 252; Sirena, 2011). The upgrading process must have included bridge-building, and a Latin inscription *c.* 90/85 by another governor, C. Norbanus, explicitly refers to road- and bridge-repair near Syracuse (*CIL* I^2.2951). The scanty remains of Roman bridges in Sicily, however, are all of Imperial date, replacements no doubt for earlier structures washed away in winter floods.

Rome's economic and political interest in Sicily was above all based on its potential as a major supplier of grain (Soraci, 2011). Rome introduced a new uniform tax-system, based closely on the Lex Hieronica that had operated so successfully in Hieron's kingdom; it was probably applied province-wide by M. Valerius Laevinus at the close of the

Second Punic War. The details need not concern us here, but communities and individuals had to pay a tithe of 10% of their wheat and barley crop; a tithe (probably of the same amount) of wine, olives, fruit and vegetables; and a pasture tax paid in cash. Rome also imposed the right to take a second tithe annually if needed, in return for cash, at a price fixed by the senate. Any remaining surplus Sicilians could sell on the open market. One might have thought that the new taxation burden (in addition to municipal taxation) might have crippled the Sicilian communities, and 30 years ago that indeed was the orthodox view of Sicily during the Republic – that the province was depressed, weighed down by exploitative Roman overlords (Coarelli, 1980: 373–80; cf. Campagna, 2003a; Prag, 2009a). Yet, as we shall see, the Sicilian cities enjoyed a building boom in the second and early first centuries, funded by the euergetism of the Sicilian urban elite (Campagna, 2007, 2011). That elite gained its wealth from its landed estates and their agricultural productivity. Perhaps the economic stimulus was provided by the chance to sell their surplus overseas to generate extra cash, resulting in more land coming under cultivation than ever before. So far from being oppressed by excessive taxation, the Sicilian communities rose to the fresh challenge by shifting their economic gear and broadening their horizons.

Another major change under Rome was the introduction of a new coinage (Crawford, 1987). At the end of the Second Punic War in 211, Rome withdrew from circulation all previous silver issues from city mints and replaced them with Roman denarii; so Sicilians had to get used to Roman imagery and Latin legends on their coinage for the first time (see Chapter 7). Bronze issues from Rome also circulated in Sicily, but alongside it some local coinages were active as well – a small issue of so-called "Romano-Sicilian" coins, which gave names of Roman officials in Latin; and bronze issues from a number of continuing city mints. The latter generally followed traditional Sicilian Greek coinage in their designs, and legends remained almost exclusively Greek until the period of Sextus Pompey's control (42–36). This civic coinage was still based, at least into the second century, on Greek weights; how two separate denominations circulating side by side worked in practice is unknown. Yet that duality had already disappeared by the late 70s, because Cicero alludes to a unitary monetary system, with everyone using "a single kind of coinage" (*Verr.* 2.3.181). Local practice, however, died hard. Greek financial accounts on stone from Taormina of the mid-first century, for example, still reckoned in *nomoi*, *tetralitrai* and *litrai*. By then these units must have been compatible with the Roman system, with perhaps the *tetralitron* being treated as the equivalent of the Roman sestertius, and the *litra* of an *as* (Crawford, 1985: 113–15, 344–5).

2.3 Urban Settlement and Society

Once change had been absorbed and war damage repaired, Sicily settled down to an era of relative peace and prosperity. The blessings of the pax Romana were not inconsiderable: after centuries of periodic unrest, Sicilians could enjoy the fruits of the countryside without the constant worry of military upheaval. That in turn had an effect on the demography of settlement patterns. Many urban settlements had been founded on waterless hilltops for sound defensive reasons, understandable when the threat of violent invasion was never far away; its removal considerably diminished the attractions of life there.

Although the gradual abandonment of hill-towns had been going on since the sixth century (Congiu, Miccichè and Modeo, 2009; for western Sicily, Vassallo, 2000), it continued steadily during the period of the Roman Republic (Wilson, 1985). Places such as Camarina, Morgantina and Lentini (the last according to Cicero a "wretched and poverty-stricken community"; *Verr.* 2.2.160) were shadows of their former selves during the last two centuries BCE, and petered out completely in the Early Empire (Tsakirgis, 1995; Di Stefano, 2006: 157–60; Frasca, 2009: 147–52).

Most of those communities that did survive the upheavals of the Punic Wars, however, flourished during the second century, when they undertook major rebuilding programs that substantially reshaped both their civic centers and their private houses. Places such as Segesta, Monte Iato, Palermo, Soluntum, Agrigentum, Licata, Halaesa and Tindari all show major building activity at this time. At Segesta the work embraced a new agora with impressive two-story stoa, and a *bouleuterion* and theater close by, all constructed during the second half of the second century (Ampolo, forthcoming). At Monte Iato, the theater, two *bouleuteria* and two-story peristyle houses with lavish interior decor all probably also belong to the second century (Isler, 2000; Wilson, 1990b). Halaesa received a new agora and stoa *c.* 125/100, and Agrigentum a temple *c.* 100/75 (Scibona and Tigano, 2009; De Miro and Fiorentini, 2011: 34–43). Palermo, Licata and Tindari all show energy rebuilding their domestic quarters, with some new houses boasting elegant and sophisticated interiors. Nor is there any likelihood that Sicily went into reverse in the 70s as a result of Verres' alleged depredations: the *Verrines* are a rhetorical flourish, and not to be treated as "history" (Wilson, 2000b; see Chapter 37).

There was, in short, an extraordinary building boom that transformed the appearance of many Sicilian towns by the early first century. Did ideas coming from the Italian Peninsula help shape urban development in this Roman province? Or did Sicily look elsewhere for its architectural inspiration? Space precludes detailed analysis here (Campagna, 2006, 2011; Wilson, forthcoming a), but the short answer is that the predominant aspect of these Sicilian towns at the time of the later Republic is their essentially "Greek" character: their physical appearance had more in common with the Hellenistic Greek world of the eastern Mediterranean than it did with Italy (Torelli, 2011: 213–14). Sicilian architects were not slavish copyists, however, but created their own distinctive brand of Sicilian Hellenism, and here the debt to the creativity and originality of Hieronian Syracuse in the second half of the third century is not to be underestimated. Sicilian theaters of the second century, for example, at Monte Iato, Segesta and Soluntum, use decorative sculpture in their scene buildings which take their lead from the Hieronian theater at Syracuse, and their fondness for semi-circular or near semicircular auditoria (rather than the two-thirds circle preferred by Greek theaters in the eastern Mediterranean) probably also derives from the Syracusan prototype (Figure 31.1) (Wilson, 1990b; Campagna, 2006). West Sicilian theaters also have small projecting wings on either side of the stage building which may have come from the same source. Both these last features recur in Campanian theaters such as at Pompeii and Sarno: far from being the passive receptor of influence from the Italian Peninsula, Sicily served as the propagator of new ideas in the opposite direction (see Chapter 2).

Hieronian Syracuse may have been influential in other ways. The impressive two-story stoas at Segesta and also Soluntum were probably inspired by similar structures in Hieronian Syracuse, rather than on more distant models in the Hellenistic east (Wilson,

Figure 31.1 View of the remains of the second-century theater at Soluntum (Solunto), Sicily, looking south. The original Punic settlement of *c.* 700 lay on the small bay in the middle distance, left of center. Source: Photo by R. Wilson, by permission.

forthcoming a, forthcoming c). New Sicilian forms of Ionic and Corinthian capitals were introduced in the third century – the Corinthian taller than the standard, with spiky acanthus leaves, the Ionic with volutes on all four faces and an additional pair of palmettes set obliquely. The type spread in the second century not only to other parts of Sicily but also to Italy and North Africa (Wilson, 1990b; Campagna, 2003b, 2003c). Some of the grander private houses have been seen as mirroring Macedonian precedents (Isler, 1996), and certainly the superimposition of orders (Doric below and Ionic for the upper story), as in Peristyle House 1 at Ietas (Monte Iato: Dalcher, 1994), ultimately derives from there. Once again, however, Hieronian Syracuse may have played a role as intermediary in the transmission: the almost total loss of domestic architecture there hinders assessment of its true significance.

Another aspect of Sicily's Hellenism during the period of the Republic is the continuing vitality of a quintessentially Greek institution, the gymnasium, of which over 20 examples are attested. Places of relaxation and exercise for the local elite, they also served as recruiting grounds for citizen soldiery, a function also performed by gymnasia in the Greek east (Prag, 2007b). Such locally trained militias also helped to maintain law and order, obviating the need for central government to send Roman garrisons to the island except in exceptional circumstances, such as to quell two major slave revolts in the later second century (Bradley, 1989: 46–82).

So did Italy make any contribution at all to shaping the visual culture of Sicily during the time of the later Roman Republic? One type of building that does make a limited

Figure 31.2 The Italic podium temple (the "Oratory of Phalaris"), Agrigentum, Sicily, *c.* 100/75; its columnar façade and frontal steps are now missing. Part of its Doric frieze is visible on the ground. The square altar and curved seat in the foreground are contemporary to the temple. The piazza was originally paved in white limestone, which is visible in the center foreground. This paving originally covered the semi-circular seating of the *ekklesiasterion*, which is now visible. Source: R. Wilson, by permission.

appearance is the Italic podium temple, with examples at Monte Iato, Agrigentum (Figure 31.2) and probably Eryx (the last on the testimony of a coin of *c.* 57). The podia at both Monte Iato and Agrigentum, however, use rubble fill rather than concrete, and the latter employs the same large-block building technique that Agrigentine architects had been using for centuries. Furthermore, the mixture of orders displayed at the latter, with Ionic capitals supporting a Doric frieze, also shows an irreverence for the standard architectural rules that is another aspect of Sicilian distinctiveness at this period (Wilson, 1990b).

Of the scores of Sicilian houses at this date, only a couple at Marsala and Agrigentum have the atrium and peristyle combination so familiar from Campania. Instead the type modeled around a central peristyle dominated the house-plans of the local Sicilian elites. In decorative detail, such as mosaic borders of rich festoons of fruit and flowers with theatrical masks (such as that framing the Hunt of Alexander mosaic in a wealthy house at Palermo of *c.* 100: Wootton, 2002; Spatafora and Montali, 2006), one might think that the hand of Campania, where close parallels occur, can be detected. But Sicily was part of the wider Hellenistic mosaic *koine* (such borders are known in Hellenistic Delos, for example), and the island very probably played a role in influencing the Italian mainland rather than the other way round (Dunbabin, 1999: 38). Certainly some building types that took root in Republican Italy, like the basilica, are completely absent in contemporary Sicily, where stoas almost certainly served a juridical function when required, as they did in the Greek east.

Culturally, therefore, the impact of Italy on Sicily during the Late Republic was extremely limited, nor is any sense of a distinctive, emerging Sicilian "identity" detectable from among the Sicilians themselves (Perkins, 2007; Prag, 2009b, 2011b: 89–92). There were groups of Italians with business interests present in all the major towns (Pinzone, 1999b), and Cicero refers to the existence of "conventus civium Romanorum" for them at Syracuse, Agrigentum, Palermo and elsewhere. But they made little impact on the overwhelmingly Hellenistic Greek culture of Republican Sicily. Latin was still a foreign language spoken and understood by a small minority. Cicero has to remind his audience in the *Verrines*, when talking about the curia at Syracuse, that the Syracusans themselves called it the *bouleuterion*; he needed to elucidate, too, the workings of the Greek calendar system in use throughout Sicily; and it is clear that Verres while governor had conducted all his business with Sicilians through interpreters (*Verr.* 2.2.50, 2.2.129, 2.3.84). Urban magistrates bore Greek titles such as *tamiai* (the equivalent of quaestors), *agoranomoi* (aediles) and *strategoi* (at Taormina), and decrees were made on the authority of the "council and people" (*boule kai demos*) in the Greek manner. Latin inscriptions were few and far between until the Augustan period (Prag, 2007a: 257). They were either set up by Roman government, like those mentioned above attesting road-building activity by Cotta and Norbanus, or by Italian immigrants, such as the Italici who erected a commemorative stone at Halaesa (*CIL* I².612). Latin inscriptions of Republican date (there are only a handful) are less surprising at the Temple of Venus-Astarte at Eryx (Erice) in western Sicily, known to have been popular with Roman visitors (Diod. 4.83.6).

Rome did, however, have political interest in certain Sicilian communities. Eryx was historically a settlement of the Elymians, whose origins Rome traced back to the Trojans (Sammartano, 2003). Most authors made Aeneas stop there en route from Troy to his Italian destiny. Segesta, which had responsibility for the Eryx temple, was one of the five Sicilian towns privileged by Rome as a "free and tax-free community" (civitas libera ac immunis); the Trojan connection may have helped (Kunz, 2006: 151–3, 304–11). In the same category of town was Centuripe, which also boasted Trojan connections. Lanaios, founder of Lanuvium near Rome was allegedly a Centuripan follower of Aeneas, and the links between the two communities were formally recognized in a treaty in the first century recording Lanuvium as a "foundation" (*apoikia*) of Centuripe (*SEG* 42.837; *AE* 1990.437; Prag, 2011), although another Sicilian inscription, painted on plaster in the gymnasium at Taormina, implies the opposite, that Centuripe was founded by Lanuvium (Patané, 2002: 135–7). Cicero alludes more than once to the close bond of "kinship" that both Centuripe and Segesta enjoyed with the Roman people (*Verr.* 2.5.125, cf. 2.2.163, 2.4.72, 2.5.83). Enna in central Sicily, though not one of the favored communities, gained increasing fame during the Republic as the center of the worship of Demeter/Ceres, with its much-admired cult statues (Cic. *Verr.* 2.4.109–10). Enna was in the heart of the grain-lands, and, in another piece of convenient mythmaking that first surfaces in the third century, her daughter's abduction by Pluto was set on the shores of nearby Lake Pergusa. Such was the cult's popularity that it was introduced to Rome itself; so too was that of Venus Erycina (Kunz, 2006: 65–6, 312–13).

Apart from the five "civitates liberae ac immunes" (the others were Halesa, Halicyae and Palermo), three Sicilian towns, Taormina, Noto and Messina, were "civitates foederatae," enjoying a "treaty" with Rome. The first two were probably so rewarded for siding with M. Claudius Marcellus during the Second Punic War (Wilson, 1990a: 353, n. 58). Messina was a special case. In 288 it was seized by the Campanian mercenaries who called themselves Mamertines. It was their appeal to Rome for help in 264 that sparked the outbreak of the First Punic War (Pinzone, 1999a: 121–72). This Oscan population rarely shows up in the archaeological record. As well as brick stamps and graffiti, there are a handful of Oscan inscriptions on stone, including one to Apollo recording two *meddices*, the Oscan name for the chief magistracy (Crawford, 2006b, 2007). The coinage, apart from two issues using the Oscan form "Mamertinoum," continued to have its ethnic inscribed in Greek and the gods were still worshipped under their Greek names. Apart from a couple of chamber tombs that appear to be of Italic type, and the name of what sounds like an Oscan lady, Pakia Pomptia, written (in Greek) on a pot (*SEG* 44.773), the material from Messina does not look very different from that of the rest of eastern Sicily in the second and first centuries. By *c.* 100 all traces of Oscan elements fade completely from the archaeological record (Mastelloni, 2005).

2.4 Western Sicily: The Carthaginian Heritage

Palermo, Soluntum and Motya's successor Marsala were part of the Carthaginian jurisdiction (*epikrateia*) in Sicily until 241, and thus had a significant Punic legacy when they passed under Roman jurisdiction. We know most about Soluntum, another place essentially rebuilt during the second-century building-boom, with agora, theater, *bouleuterion* and peristyle houses, many sharing the same characteristics as their counterparts elsewhere in Sicily (Campagna, 2011: 168–75). But there are differences too. Soluntum contains a striking example of a second-century Punic sanctuary, consisting of elongated cells (four in two pairs and one on its own) set in their own precincts. Each cell has a rear area, sometimes a raised shelf, for various cult statues, but the sanctuary as a whole is set entirely on ground level. Elsewhere at Soluntum, some of the houses are equipped with Punic-type cisterns, long and thin with parallel sides and rounded ends, while several dwellings show a Punic love of privacy by having the principal room tucked away out of sight from the street. Walls are sometimes in opus africanum, a technique of Punic origin – the use of orthostats (large, dressed stones set vertically) at intervals, separated by coursed rubble walling in between (Wilson, 2005). In addition, there is the unsurprising evidence that the town was laid out using the Punic foot as the basic unit of measurement, also detected in the town-plan of Marsala (Greco, 2005: 18; Caruso, 2005: 777–82). Perhaps the most important Punic legacy was the crushed brick and mortar flooring known as opus signinum. Widely used not only in western Sicily but also all over the island from early in the third century, it was inherited from North Africa where it was invented towards the end of the fourth century (if not earlier). Later, opus signinum also began to be used on the Italian mainland, a further example of the island serving as a transmitter of new building fashions. Several "Punic" features, including opus signinum floors and elongated rounded-ended cisterns, have been found also on Pantelleria, another place which enjoyed a spurt of new building in the second century (Osanna, 2006; Schäfer, 2006).

A further Punic legacy in western Sicily was language. Punic inscriptions continued there into the first century BCE, and Punic was spoken, in the countryside at least, into the second century CE (Apul. *Met.* 11.5), but Punic was not as strong nor as long-lasting in Sicily as it was in Sardinia (see below). In Marsala, at least, Greek had replaced Punic by the end of the second century. Its inscriptions give the impression that by then the population was essentially a mixed Helleno-Punic one with a range of Latin and Italic names as well (Wilson, forthcoming a). As for the standard of their spoken Greek, Cicero implies it was not of the purest strain (*Div. Caec.* 12.39).

2.5 *Rural Sicily*

The richness of archaeological evidence from the towns is not yet matched by that from the countryside. Diodorus gives the impression that the growth of latifundia ruined Sicily in the second century; but it is equally clear from Cicero's *Verrines* that farmers with small and medium-sized estates did not disappear from the scene (Wilson, 1990a: 21). No doubt a myriad of different settlement patterns existed in the countryside and the sweeping generalizations of the classical sources give only a very incomplete picture. Several field survey projects in Sicily suggest a decline in the number of rural sites occupied in both the west and the south of the island after the First Punic War, possibly the result of the amalgamation of smallholdings into large estates at this time. Although some surveys are now fully published, most notably those in the hinterlands of Himera, Halaesa and Gela, much remains to be done (Bergemann, 2010, I: 13–18, and Spanò Giammellaro, Spatafora and van Dommelen, 2008: 134–41, assess all Sicilian survey results).

What is also needed is far more excavation of individual rural sites of the last two centuries BCE before detailed archaeological evidence can be used to qualify the literary sources (Bejor, 2007). A farm partially excavated at Aguglia near Acrae in south-east Sicily, with rooms around at least two sides of a central court, produced agricultural implements, grain measures and a range of amphoras, including Rhodian, Greco-Italic and Dressel 1 types. The period of occupation probably extends from the mid-third to the end of the second century (Wilson, 1990a: 21–2, with different dating; cf. now Walthall, 2011, for the grain measures). A farm at Gazzi south of Messina produced Rhodian, Coan, Greco-Italic and Dressel 1 amphoras as well as Punic types; occupation was between the mid-third and the mid-first centuries (Bonanno, 2001: 195–7). A different sort of settlement, a small agricultural village of *c.* 3 hectares at Campanaio near Montallegro (25 km west of Agrigentum) yielded the plan of an L-shaped building with nine rooms (part of it is lost to the plow) of the second and first centuries, and two tile kilns with an associated workshop close by (Wilson, 2000a). But the number of such rural sites known in Sicily that were active between the mid-third and first centuries is still small (Wilson, 2000b: 207, n. 58; for western Sicily, see Spanò Giammellaro, Spatafora and van Dommelen, 2008: 146–8).

We do not really know when the urban elites started building villas in the countryside that matched the chic sophistication of their town houses, but Cicero mentions them in the 70s in the context of wealthy men such as Sthenius of Termini Imerese and Gn. Pompeius Philo of Tindari (*Verr.* 2.2.92, 2.4.48). One of the earliest is that at

Timpone Rasta near Mazara in western Sicily, where excavation has revealed part of a main villa block about 40 m^2 set around a central peristyle; it may be as early as the second century, but was certainly in use in the first century. There were some fragments of opus signinum and limestone-chip floors; a piece of opus scutulatum and a stuccoed cornice also give a hint of its former elegance. Other villas of first-century date have been explored at San Luca in the deep interior of western Sicily, and on the sea at Marina di Avola south of Syracuse; but in both cases the plans are very incompletely known (Vassallo, 2009; Wilson, 1990a: 197–8).

Much work needs to be done also on the material culture, and in particular on improving our understanding of just what was imported and what was locally made. For the fine wares we know that Syracuse and Morgantina were making versions of black-gloss "Campana" wares in the second and first centuries, because wasters and kiln sites are known, but many more such production centers must lie undetected (Malfitana, 2006, 2011; cf. also Patané, 2006, for a Centuripan perspective; see Chapter 5). The same is also true of amphora production. One direct result of the Roman conquest of Sicily was the beginning of a massive import of wine amphoras from Italy; but even after the Second Punic War Carthage continued to supply especially western and southern Sicily with significant quantities of Punic cylindrical amphoras (perhaps carrying oil), down to its demise in 146 (Bechtold, 2007). Increasingly, the Sicilian towns were producing their own wine: half a dozen put the emblem of grapes on their coinage, and vinum Murgentinum was good enough to rate mention by Cato (*Agr.* 6.4). The containers that carried this wine both within Sicily and further afield are largely unknown: Sicilian versions of later versions of the Greco-Italic and Dressel 1 amphoras, as well as Punic types, await identification (Wilson, 1990a: 22–3; Malfitana, 2008: 150; Bechtold, 2011, for "Punic" amphora production near Marsala of 150/50). Amphora stamps also require further study: "Naxios" and "Onasou" ("[produce] of Onasus") are two examples on Sicilian amphoras of this date. The latter perhaps refers to the homo nobilis of that name mentioned by Cicero at Segesta, where the amphora stamp is attested (*Verr.* 2.5.120; Garozzo, 2006: 728–30).

3 Sardinia

3.1 Introduction

Sardinia had largely escaped the conflict that embroiled Sicily during the First Punic War, but three years after its conclusion, Rome seized the island in an act of naked military aggression. Further campaigns were necessary before the island was completely wrested from Carthaginian control, and it was only in 227 that it was formally to become, with Corsica, Rome's second provincia. Even then the mountainous interior remained untamed, and continued forays by Roman military units were needed to suppress a series of revolts until the end of the second century. No less than 25 triumphs were celebrated by victorious Roman generals down to the end of the second century – a telling indication of the fierce and proud resistance of the islanders to a foreign aggressor (Dyson, 1985: 239–63; Meloni, 1990; Mastino, 2009: 91–105); indeed, the Ilienses was still not fully pacified in Augustan times (Livy XL.34.12). No surprise, then, that

Sardinia continued to retain a reputation for brigandage down into the Late Republic (Varro *Rust.* 1.16.2).

Sardinia's economic prosperity came not from the hills but from the rolling agricultural lands on the south and west sides of the islands. A series of natural harbors here had attracted Phoenician settlement from the eighth century, and by the third century this had been consolidated through the foundation of more towns by the Carthaginians. They also established at Olbia in the fourth century, at the only decent harbor on the north-east coast, a Punic outpost to keep an eye on the growing ambitions of Rome. The major urban settlements of Sardinia at the time of the Roman conquest were thoroughly Punic in character. How much and how quickly did that change with the advent of the Roman administration?

Sardinia does not have the wealth of archaeological evidence for the later Republic that characterizes Sicily at this period (Portale, Angiolillo and Vismara, 2005: 189–315; Vismara *et al.*, 2011). The two Phoenicio-Punic towns where extensive excavation has occurred, Nora on the south coast and Tharros in the mid-west, continued to flourish throughout the Roman Empire, and much of the evidence for the third, second and first centuries lies buried beneath later levels (Meloni, 1988; Mastino, 2009: 205–332). Other major urban centers such as Cagliari and Sant'Antioco remain vibrant cities to this day, hindering opportunities for extensive excavation (Colavitti, 2003; Zucca, 2003: 203–74). Nevertheless, taken together with the evidence from the countryside, the broad outlines of Sardinia's development during the last two centuries of the Republic are clear (Rowland, 2001: 105–25; Dyson and Rowland, 2007: 127–43; Mastino, 2009: 91–123).

3.2 *Punic Conservatism*

The Roman interventions discussed above in relation to Sicily – establishment of roads, and introduction of a tax regime and the denarius-based monetary system – equally applied to Sardinia; but what emerges most emphatically is that, as with Sicily, Rome's takeover of Sardinia had no immediate impact on the island from a cultural point of view. The strength of centuries-long Phoenician and then Punic culture in the towns was initially undiminished. Nomenclature, inscriptions, burial customs, the built environment – all point to the same conclusion. The chief magistrates in the towns, for example, retained the Carthaginian title of *sufetes* (for which we have evidence at Tharros and Cagliari), held by men with straightforward Punic names such as Himilkat, Abdeshmun and Mutumbal. It is a sign of the continuing strength of Punic conservatism in Sardinia that *sufetes* were still current as late as the later second century CE, as at Bithia (Zucca, 2004: 84–101).

Very few public buildings erected in the Sardinian towns during the last two centuries of the Republic survive. For example, there are no theaters of Late Republican date, and amphitheaters and bath buildings did not reach Sardinia until the first century CE. "Roman" building materials such as rubble-aggregate "concrete" do not make an appearance before Augustan times. Most public buildings of Republican date that do survive are temples, and they are of two principal types. One, raised on a podium and approached by a frontal staircase, will be discussed below; the other consists of a simple

Figure 31.3 Nora, Sardinia, temple thought to be of Asclepius, apsed hall of the inner shrine (*c.* 300 CE). A second-century BCE shrine is known to lie beneath, probably dedicated to Eshmun, the Punic equivalent of Asclepius. Source: R. Wilson, by permission.

shrine of the "Punic" type described above, with or without an adjacent open space or courtyard, all built at the same level. The chronology is often imprecise, but examples of Republican date include the Capo San Marco temple at Tharros, the first phase of the Temple of Sardus Pater (Sardinia's national god) at Antas, and the two-celled example at Monte Sirai (Ghiotto, 2004; Tomei, 2008).

This temple design continues into the Imperial period, and some of those undoubtedly replace earlier buildings of the same type that lie below. One example is that at Nora. Strikingly situated on the promontory of Su Punta de su Colòru, it comprises in its visible form (fourth century CE) an open space from which lead steps to a large apsed hall, which is the shrine itself. Excavations have located elements of a second-century BCE predecessor below (Figure 31.3). From a votive deposit of *c.* 100 came some terracotta statuettes of offerands, their hands outstretched, and two larger statues, also of terracotta. The better preserved is of a sleeping naked youth, his eyes closed, entwined by a snake – the god of healing was supposed to repair the wounds of sick devotees through the restorative lick of his sacred serpents. Given the overwhelmingly Punic character of second-century Nora, the god was likely worshipped here under his Punic name, Eshmun, but epigraphic confirmation is lacking, and indeed the identity of the divinity has recently been questioned (Ghiotto, 2004: 47; Tomei, 2008: 180–98).

Other temples of Republican date are known only from inscriptions. An intriguing base at Sant'Antioco, for example, with a bilingual inscription in Latin and Punic, records the dedication of a temple to the obscure Near Eastern goddess Elat, "the goddess"

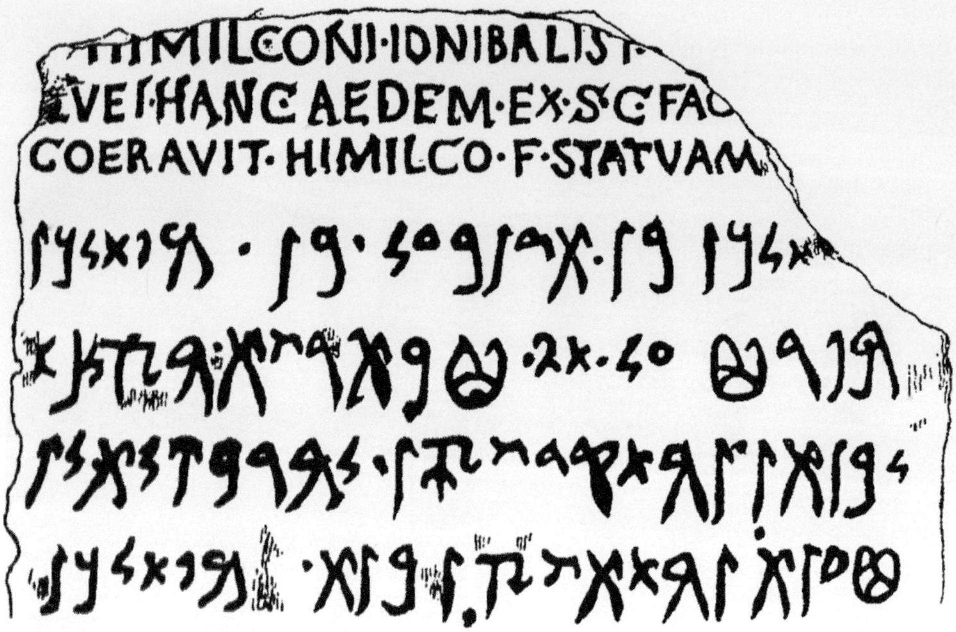

Figure 31.4 Statue base from Sulcis (Sant'Antioco), Sardinia, with bilingual inscription in Latin and neo-Punic in honor of Elat, "the goddess", from the middle of the first century. The Latin reads: "To Himilco son of Idnibal, who saw to the making of this temple by decree of the Senate. Himilco his son [set up] the statue". The Punic version makes clear that the Senate is the Senate of Sulcis. Source: *CIL* X.7513.

(Figure 31.4; *CIL* X.7513; Guzzo Amadasi, 1967: Sard. Neopunic 5; Adams, 2003: 212–13). Himilco, its dedicator, set up a statue of his father, also named Himilco, who had seen to the construction of the sanctuary. The statue may well have stood in the temple precinct. The date is *c*. 50 or a little later; and although the provision of a Latin version of the text is a nod in the direction of the language of the governing power, the content of this inscription, recording the establishment of a Semitic cult and dedicators with purely Punic names, shows how little had changed in Sardinia after nearly two centuries of Roman political control.

In private housing there is again little that has escaped later redevelopment. Some of the modest housing at Tharros and Nora seems to belong to the third/second century, and is of a type found elsewhere in the Punic world. Each unit tends to be small, with the entrance from the street set off-center in relation to the house-plan, entailing that the main room is not visible from outside the house. Another legacy of Punic housing to Roman Sardinia was the fondness for the type of walling known as opus africanum. It is found all over the island, especially in private housing but in public structures too, into the Late Empire (Ghiotto, 2004: 13–14, 158–61).

Of the more up-market houses, only the "House of the Punic Emblems" at Cagliari definitely falls within the period of the Roman Republic, but its plan is only very incompletely known – a small central court, with a reception room opening off it (Ghiotto, 2004: 163). Both the corridor around the court and the main chamber have pavements of opus signinum with white tessera inlay, of standard second/first century type; the corridor includes in its design a depiction of a Tanit symbol (a triangle surmounted by a horizontal bar with a circular disc above: Angiolillo, 1981: no. 110) – further evidence of the persistence of Punic religious symbolism well after Sardinia had become a Roman province. To call the court in this house an atrium and the reception room a tablinum, as some have done (Tronchetti, 1990: 13), is to borrow unnecessarily a "Roman" terminology that the owner of the house would surely not have recognized. In fact, despite the mislabeling by modern scholarship of some Sardinian houses of later date as having "atria," when the features in question are nothing more than small central courts, no atrium/peristyle houses of the type so familiar from Campania are known in Sardinia.

Other aspects of Sardinian culture during the last two centuries of the Republic show the continuing dominance of Punic traditions. The *tophet*, places of child sacrifice in the Carthaginian religion according to Roman sources, continued to be frequented at Tharros, Monte Sirai, Bithia and elsewhere down into the second century, and at Sant'Antioco even until *c.* 50 (Zucca, 2003: 216; Mastino, 2009: 406). Whether human sacrifice was still practiced at that time or animals had been substituted for children is not clear. Stelai, of which there is a particularly long series from Sant'Antioco, continue down into the mid-first century and display increasing elements of Hellenization, such as architectural aediculae with columns and pediments within which the dedicant stands. But the Punic link is never far away, as with those that are shown grasping the Tanit symbol in one hand (Bartoloni, 1986; Moscati, 1986). Burial customs essentially show conservatism. Although cremation in Sardinia became more widely used by the urban middle classes during the second century, possibly under Roman influence, inhumation still predominated, a predominance even more striking in rural cemeteries (Mastino, 2009: 437; van Dommelen, 2007: 63). Many burial practices (like inhumations inside large amphoras or *pithoi*, the so-called *enchytrismos* type) show little sign of change between pre-Roman times and the first century (Dyson and Rowland, 2007: 124).

The survival of Punic names down to the end of the Republic, with little trace of Roman overlay, is one striking aspect of Punic continuity in the first two centuries of Sardinia's existence as a Roman province. The same might be said of the continuing strength of the Punic language and of the overwhelming predominance of Punic inscriptions at the same period. The number of attested Latin inscriptions during the Roman Republic is very small. Only 13 inscriptions in Latin on stone or bronze belong to the 200 years after 227, and of those, one is bilingual (Punic and Latin) and one trilingual (Zucca, 1996).

3.3 Italian Influence

Rome's attitude towards Sardinia was wholly negative. The interior of the island was of no account, being dismissed as "Barbaria" – the ancient world's label for anywhere completely uncivilized. The agriculturally fertile part of the island, by contrast, had been under the control of the Carthaginians, for whom the Romans wore their hatred on their sleeve. The Carthaginians were descendants of the Phoenicians, "a most treacherous race" according to Cicero; and the Sardinians, whom he claims were of Carthaginian descent "with an admixture of African blood," were no better (*Scaur.* 42). Sardinians were widely thought to be untrustworthy (Festus *Glos. Lat.* 428 L). The climate was unhealthy, with malaria an endemic problem (Sallares, 2002: 90–3). Martial was later laconically to comment that Sardinia was synonymous with death (4.60). No wonder that Cicero, on learning that his brother was to be posted to Olbia, told him to look after his health (*Q Fr.* 2.3.6); no wonder, either, in view of Rome's poor opinion of the island and its inhabitants, that no Sardinian town during the Republic was given the status of "free and tax-free" (libera ac immunis) or given treaty status with Rome (civitas foederata): all Sardinian civitates were stipendiariae (tax-paying). Cicero drives home that point in his defense of M. Aemilius Scaurus, provincial governor in 55, against Sardinian charges of extortion: "for what province is there besides Sardinia which has not a single city in it on friendly terms with the Roman people, not a single free city?" (*Scaur.* 44).

Rome's principal economic interest in Sardinia was its grain: Cicero, for example, names the island along with Sicily and Africa as one of the "reserves" (subsidia) of the grain supply (*Leg. Man.* 12.34). Also important to Rome was the island's mineral wealth, mainly in lead but also in iron. The principal mining area lay in the south-west, with the shrine of Sardus Pater at Antas at its heart. The mines were probably exploited by Rome soon after the conquest, but archaeological evidence for extraction during the Roman Republic is minimal. Only a single lead ingot with a Punic inscription which might be of this date and from these mines is known, from Monastir, north-east of Cagliari; the earliest dated ingot from the mining district itself is not before Augustan times (Le Bohec, 1992: 258).

Roman official involvement in Sardinia would have brought in its wake Italian and other immigrants keen on exploiting economic possibilities in the new province, whether as businessmen or tax collectors, but as in Sicily their numbers and the significance of their impact at any one time during the Roman Republic is hard to assess (Colavitti, 1999). No doubt they were especially active at major harbor cities with good commercial contacts, such as Olbia and Cagliari. Some of the few inscriptions of Republican date in Latin might refer to such individuals, such as C. Apsena Pollio from a tomb monument at Cagliari, of *c.* 50 (Angiolillo, 1985). Apsena is a nomen of Etruscan origin, and he might therefore be an immigrant Italian, but we know nothing else about him (*AE* 1986.721; Zucca, 1996: no. 18). Also Italian immigrants were "the Faliscans who are in Sardinia" who dedicated a bronze tablet to the Capitoline Triad at Falerii Novi, a town in southern Etruria, in the second century (*CIL* I^2.364a; *ILS* 3083), but again we have no idea what they actually did in Sardinia or why they were there.

Little in the archaeological evidence suggests that Italy made much of a contribution to shaping the visual culture and the built environment of Later Republican Sardinia. The most tangible witness, as in Sicily, is in temple building. In addition to the Punic-style type noted above, some temples in Sardinia were built in the Italian manner, with porch and cella raised on a podium and approached by frontal steps. At Tharros, the Distyle Temple of the second half of the second century may be the earliest, but although the cella is raised and there are four frontal steps, the two porch "columns" are chunky square piers, and the whole shrine is built of large ashlar construction in the Punic fashion; this tiny temple is a "Punico-Roman" hybrid rather than a straightforward borrowing from Italy. Larger tetrastyle temples at Tharros and Nora, the latter built *c.* 50 on the north side of the forum (and perhaps its Capitolium), are more traditionally planned in the Italian manner, but both are now in very ruinous condition (Ghiotto, 2004: 38–43).

Of greater interest is the tetrastyle temple excavated over 70 years ago in Via Malta at Cagliari. It lay within an imposing precinct, some 120 m by 43 m, and included a theater on the slope in front of the temple. The sanctuary therefore belongs to a discrete group of Late Republican theater-temple complexes in central Italy (see Chapters 3 and 22), and the Sardinian example presumably represents a direct borrowing from there (Angiolillo, 1986–7; Ghiotto, 2004: 36–7). Clay terracottas found in a well in the Cagliari precinct, which include standing female figures and a plaque of Victory holding a trophy, are likely to have been influenced by Republican models from sanctuaries in central and southern Italy, but the presence of molds at Cagliari shows that they were being made close by (Comella, 1992). The same can be said of other Sardinian ex-votos. Demeter figurines from Santa Margherita di Pula, for example, also of the second century, show parallels with terracotta votives from Fregellae, Veii and elsewhere (Angiolillo, 1987: 203–5), while the rich series offered at the shrine of Bes at Bithia between the later third and first centuries, including clay representations of specific body parts for which pilgrims were seeking a cure, mirrors the popularity of similar medical votives in central Italy at the same time (Dyson and Rowland, 2007: 138).

3.4 *The Sardinian Countryside*

Most of the evidence for Sardinia during the Roman Republic discussed above refers to the towns; we are much less well informed about the contemporary countryside. We do, however, have the results of intensive field survey, especially those conducted in the area around the Gulf of Oristano in western Sardinia and also in the hinterland of Nora (van Dommelen, 1998a, 1998b, 2006: 8–14). The conclusions reinforce the impression from urban settlements that rural sites throughout the later third, second and early first century have a thoroughly Punic character, to judge from the surface pottery which includes locally made wares and Punic amphoras, but much less in the way of imported goods from Italy. There is a good deal of black-gloss "Campana" pottery, especially that with a gray fabric, but this was likely being manufactured in Sardinia as well as elsewhere, although detailed evidence, as in Sicily, is as yet forthcoming. Greco-Italic amphoras from southern Italy and Sicily also reached Sardinia in the third and second centuries, and later Dressel 1 wine amphoras from central Italy. But the overwhelming impression is a strong persistence of Punic material culture on rural sites down into the Late Republic. Van Dommelen has suggested (e.g. 1998b: 40–5, 2001, 2007: 58–60) that this apparent aversion to imported goods from Italy was part of a conscious Sardinian "resistance" to all things Italian, out of local distaste for their Roman overlords, but it might be questioned whether pottery can be used in this way as a barometer of cultural attitudes. More likely the absence of imported goods is a reflection of deep-seated Sardinian conservatism and aversion to change, and in any case the presence of such imports in the countryside presupposes the existence of appropriate economic supply mechanisms, as well as the capability of local populations to pay for them. Neither may have been in place at some rural settlements, especially the further one moves away from the coastal towns, with their easy access to sea-borne supplies.

To date there has been little excavation of these rural sites. Whether the urban elite built comfortable villas on their country estates during the second or first centuries is at present unknown, although there are hints from surface survey of some appearing in the first century (van Dommelen, 2006: 14). Farms have equally seen little attention from the archaeologist's spade (Sanciu, 1998). S'Imbalconadu, 5 km south of Olbia, stands out by itself for the quality and quantity of the information it has yielded (Sanciu, 1997). It consists of an enclosure at least 33 m by 30 m, with a narrow entrance on its south side. A number of rooms, perhaps up to six, were built up against the inside of this southern perimeter wall; one room had pressing facilities and two collection basins, probably for wine-production, while millstones in another suggest the preparation of flour for bread-making. In addition there was a freestanding residential block, 9 m square, near the center of the enclosure. It had four small rooms of equal size and a central access corridor. Basalt blocks were used throughout the farm for the lowermost courses of walling; mud-brick composed the superstructure. Floors were mainly of beaten earth, but fragments of plain opus signinum show that more substantial pavements were present too. Roof tiles suggest a sloping roof of conventional type. The farm was built in the third quarter of the second century and abandoned *c.* 75. Black-gloss ware was mainly from Campania and central Italy, but a few scraps have been identified as belonging to a Sardinian production. The plain ware was mostly local but some Punic types were imported from North Africa, as were Punic oil amphoras from what is now Tunisia and a single example of another cylindrical container, from Tripolitania. In addition Dressel 1 amphoras carried wine from central Italy, but one fragment of this type was identified scientifically as local production, raising the likelihood of a Sardinian version or versions, presumably to hold local wine, of this widely copied amphora shape. Some of the Punic amphoras may also be local. Sheep/goat predominated among the animal bones (26%), followed by deer, cattle and pig; there was surprisingly high percentage (14%) of dog bones. The discovery of two granite blocks inscribed with the sign of Tanit is a further reminder of the continuing strength of the Punic tradition in Republican Sardinia. The excavations at S'Imbalconadu have, in short, given us a rare insight into the workings of a Sardinian farm and its agricultural production during the second and first centuries (for others, see van Dommelen and Finocchi, 2008: 181–8).

Dominating the landscape in Sardinia to this day are the extraordinary Bronze Age round tower-fortresses, the *nuraghi*, together with their ancillary buildings: found in nearly every corner of the island, there are some 7,000 of them in all. Not surprisingly these symbols of former power and prestige were not entirely neglected in later times, even though most of the structures had fallen out of use at least by the sixth century. Yet a few show evidence of renewed activity in the Roman Republican period and later. Sometimes this takes the form of reusing parts of nuraghic complexes for dwellings, material culture alone betokening habitation; in other cases new, simple domestic structures were erected, either in the central courtyards of *nuraghi* (one wonders whether rural insecurity was a factor) or else alongside the nuraghic ruins (e.g. at Sa Tanca 'e sa Mura, in north-west Sardinia; van Dommelen and Finocchi, 2008: 184); in other cases again, the former strongholds were revisited for cultic purposes. At the *nuraghe* of Villanovafranca, for example, 50 km north of Cagliari, incense burners of Punic type and other objects suggestive of a Punic-style Demeter cult were found inside a nuraghic Iron Age shrine – the memory of a former sacred place still resonating centuries later.

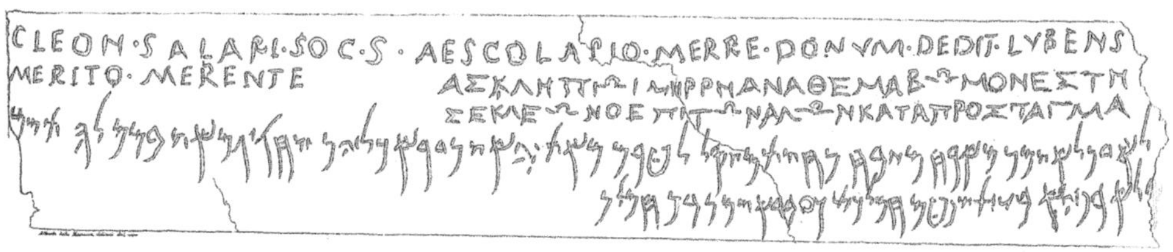

Figure 31.5 Trilingual inscription in Latin, Greek and neo-Punic from San Nicolò Gerrei, Sardinia, early first century BCE (?). The Latin probably reads: "Cleon slave of the organization of salt-workers willingly and deservedly gave [this] as a gift to the deserving Aescolapius Merre". The final word is in the ablative case (not the normal dative), suggesting hesitancy in the use of Latin. The Punic version adds more detail: the altar dedication was made of bronze; it weighed 100 pounds; the *sufetes* (at nearby Cagliari?) were Himilkat and Abdeshmun. Source: *CIL* X.7586.

Another type of ritual structure contemporary with the *nuraghi*, the well or water-shrine, was also sometimes chosen as a renewed focus of worship. One at Cuccuru s'Arriu, on the north side of the Gulf of Oristano, was the site of a Punic fertility cult active between the late third and early first century (van Dommelen, 2001: 75–8). Another at San Nicolò Gerrei, in the mountains north-east of Cagliari, produced a bronze altar probably erected in the early first century (Figure 31.5). The dedicator was a certain Cleon, a slave belonging to a societas of saltworkers (*CIL* I^2.2226, X.7856; Zucca, 1996: no. 24; Culasso Gastaldi, 2000; Adams, 2003: 210–13). The sanctuary was dedicated to Eshmun-Merre-Asclepius, three different names for the same deity; and the date was recorded by reference to two *sufetes*, Himilkat and Abdeshmun, mentioned above. The inscription, in Latin, Greek and Punic, records therefore a dedication by a Greek-speaking slave (with a Greek name) working for a Roman type of organization, a societas, citing the Punic names of chief magistrates with a Punic title; and all this took place at a nuraghic water shrine (Merre might have been its indigenous deity). This is an unusually complex example of multiculturalism in Late Republican Sardinia; but of the various cultural strands that made up its society it was, as we have seen, the Punic above all that had the strongest and most visible identity.

Sardinia and Sicily, by the end of the Roman Republic, both display in their very different ways one shared characteristic: that while politically and administratively they had been under Roman hegemony for more than 200 years, the cultural impact made by "Rome" on these islands by the later first century was still extremely limited.

FURTHER READING

There are no dedicated accounts of the Republican period alone for either Sicily or Sardinia in monograph form. An introductory volume that covers both provinces over the entire Roman period is Portale, Angiolillo and Vismara (2005). On Sicily, the historical background is available in Finley (1979, chs. 9–11), and, for the early period, in Zambon (2008). Collections of essays that also have coverage of the Republican period include Smith and Serrati (2000, chs. 9–12), Miccichè, Modeo and Santagati (2007) and Engels, Geis and Kleu (2010). There are also two essays on

Republican Sicily in Colivicchi (ed., 2011). Other assessments of the relevant archaeological material can be found in Wilson (1990a: 17–32, forthcoming a, forthcoming b). On architecture, Osanna and Torelli (2006) is fundamental.

On Sardinia, the standard accounts are Meloni (1990) and Mastino (2009); there are also relevant chapters in Rowland (2001) and Dyson and Rowland (2007). Vismara *et al.* (2011) is splendidly illustrated, as is Angiolillo (1987). On architecture, Ghiotto (2004) is fundamental; on temples the reappraisal by Tomei (2008) is thoughtful if at times controversial. On the countryside, a good starting point for both provinces is van Dommelen and Gómez Bellard (2008, chs. 6–7).

CHAPTER 32

The Archaeology of Africa in the Roman Republic

David L. Stone

1 Early Intervention and the Conquest of the Province

What are the origins of the province of Africa? According to Pliny (*HN* 5.25), the Roman general Scipio Africanus divided the territory of Carthage from that of the Numidian kings in 202 by digging a line in the earth leading from Thabraca in the north to Thaenae in the south. The line was called the fossa regia, "royal ditch." Studies of North Africa have traditionally claimed that land on the Carthaginian side of this frontier became Rome's first province in North Africa after the final defeat of Carthage in 146 (Gsell, 1928; Broughton, 1929; Bullo, 2002). The province of Africa was extended in 46 when land belonging to the Numidian kings was added as a new province, Africa Nova, and the original was renamed Africa Vetus. Not long afterward, probably in 35, both of these provinces were amalgamated into Africa Proconsularis, which stretched from Hippo Regius in the north-west through the Tripolitanian city of Lepcis Magna in the south-east (Figure 32.1). In 33, at the very end of the Republic, a separate province named Numidia was annexed. The remainder of North Africa was administered by client kings, but did not formally become Roman territory until 40 CE. Today, North Africa is commonly defined in studies of the ancient world to include the modern countries of Libya, Tunisia, Algeria and Morocco – and, as in ancient times, Egypt is considered separately from North Africa.

It is clear that Rome's relationship with North Africa is more complicated than the conventional presentation of annexation dates suggests. The Roman presence in Africa goes back as far as the expedition of Marcus Atilius Regulus's army to Cap Bon in 256. In the early second century Rome had asymmetrical relationships with Carthage and the Numidian kingdoms; neither kingdom acted independently and both made voluntary

A Companion to the Archaeology of the Roman Republic, First Edition. Edited by Jane DeRose Evans.

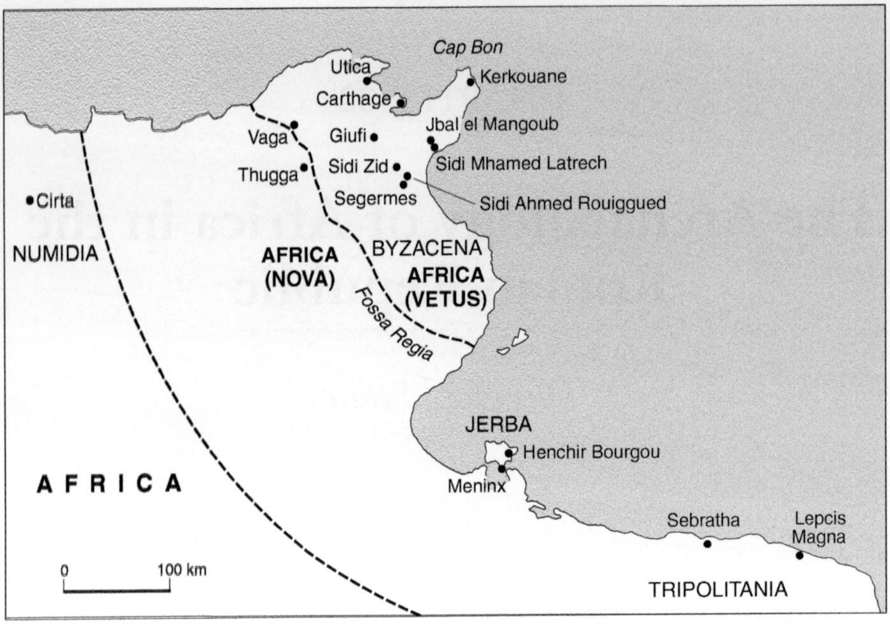

Figure 32.1 North Africa in the Republican period, with provincial boundaries and sites mentioned in the text. Source: Illustration by D. Stone.

contributions of grain and mercenaries to the Roman army during the Second and Third Macedonian wars (Livy XXXI.19.2–4, XXXII.27.2, XXXVI.4.9, XLII.29.8 and 62.2, XLIII.6.11–13; Polyb. XV.18.6). Further, there is the question of what "provincia" meant in the second century. Recent discussions have emphasized that the meaning of provincia was not well established at the time Carthage was conquered (Kallet-Marx, 1995; Quinn, 2004). It underwent evolution from a term that signified the task handed to a Roman magistrate to a designation closer to the modern usage – that of a territorial unit subject to taxation, the use of force, and jurisdiction. Thus, it may be better to recognize the "provincialization" of Africa as a progressive development of the second and first centuries, rather than a sudden rupture of the political situation occurring in the mid-second century, as it has traditionally been viewed.

What implications does such a reassessment have for an archaeological evaluation of North Africa during the Roman Republic? One is that a focus on the "conjonctures" or the "longue durée," rather than the "histoire événementielle" – to speak in Braudellian terms – appears best suited to examining the gradual imposition of Roman control. Indeed, too much effort has been spent on assigning remains a specific date after the conquest in 146, and not enough on assessing the broader changes that took place during the Republican period. We may never know whether the tower mausoleum at Thugga was erected for Massinissa after his death in 148, what happened to the land near Carthage on which Gaius Gracchus attempted to established a colony of veteran soldiers in 122, or how soon after the Lex agraria of 111 the centuriation of the province began, but we can evaluate tombs, colonies and field boundaries as important aspects of patterns affecting the region over the long term. This chapter will therefore assess medium- and long-term changes in the archaeological record of the last two centuries in North Africa.

It will focus on the part of North Africa containing the Republican provinces of Africa Vetus and Africa Nova (later Africa Proconsularis), which for simplicity's sake I will refer to as Africa, despite the alternative designations given to it in this period.

2 People

The inhabitants of Africa comprised Libyan, Punic and Roman peoples. "Libyan" refers to indigenous inhabitants who were a diverse group of sedentarized port-, hillfort-, lowland- and oasis-dwellers, as well as semi-pastoralists. Greek and Roman sources refer to them by more than 100 names, including Berbers, Numidians, Gaetulians, Musulamii and Afri. Around 800, colonists from the Levant arrived in Carthage to settle. They are generally known as "Phoenician," but "Punic" is more often the term applied to their descendants remaining in Carthage after it was no longer dependent on Phoenicia and had become a major power, ruling the North African coastline and parts of Spain, Sicily and Sardinia. We know rather less about the ethnic identities of Africa's inhabitants than we have often assumed. The archaeological and textual records demonstrate that these names do not have straightforward definitions or material correlates. There was clearly a good deal of cultural fusion, resulting in combinations known – even in antiquity – as "Libyphoenician." What this, and "Punic," meant is just starting to be addressed in a rigorous fashion (Prag, 2006a; van Dommelen, 2006; Ben Younes, 2007). In addition, literary sources confirm that a small number of Italians settled in Africa during the Republic. Sallust (*Iug.* 26 and 47) mentions Italian negotiatores living at Vaga and Cirta. Gaius Gracchus's attempt to found a colony with 6,000 veterans at Carthage in 122 failed, but some soldiers were assigned property, and those who chose to remain rather than sell it were granted ownership (*FIRA* 1, 102–21). There must also have been a settlement of Romans, including the governor and his staff, at Utica.

3 Approaches to Africa in the Republican Period

Earlier approaches to the Republican period in North Africa have often reflected wider trends in the study of imperialism, which we may define as the process by which political, military and economic sovereignty was established and maintained over subject peoples. The subject of Roman imperialism was shaped in the nineteenth and early twentieth century, when the dominant view held that the expansion of Rome's political control over dependent subjects was primarily defensive in nature. As developed by Mommsen (1885), the theory of defensive imperialism argued that, under the Republic, Rome was not interested in acquiring foreign territory, but participated in wars to defend itself and its allies, and when it emerged victorious, it had no imperial strategy or interests. Thus Mommsen regarded Rome as uninterested in Africa after the Punic Wars:

> Rome strongly held the territory which Carthage had possessed at the time of its defeat, less to develop to its own advantage than to prevent others from so doing, not to resurrect the region but to guard its corpse. Fear and jealousy, not a desire for power and possessions, gave rise to the province of Africa which had no history under the Republic.
>
> (Mommsen, 1885: 623)

Mommsen's opinions influenced subsequent studies of Africa and many other Roman provinces (see Jullian, 1908; Haverfield, 1912; Frank, 1914; Holleaux, 1921; Freeman, 1997). Stéphane Gsell, a prominent historian of the early twentieth century, wrote eight volumes on North Africa, one of which he devoted to the Roman Republic. Gsell went further than Mommsen in recognizing Rome's active involvement in the province through confiscation of Carthaginian land, which it sold to members of the senatorial and equestrian orders, but fundamentally agreed with the thesis of defensive imperialism. His view summarized the opinions of scholars about the way Africa remained undisturbed by the conquest throughout the Republican period:

> Care was not taken to spread Latin civilization there; it was not for devoting itself to this task that the Republic had annexed the territory of Carthage, and perhaps with the exception of Gaius Gracchus, no one saw a moral obligation, nor even a means to affirm conquest. On the contrary, in general the nobility thought that it was dangerous for the citizens of Rome to create equals – that is to say eventual rivals – outside Italy. The locals were left to live their lives.
>
> (Gsell, 1928: 118)

Many scholars of North Africa have rejected a number of the positions held by Gsell, but have remained in agreement with his views on the Republic. Even Bénabou (1976), whose ground-breaking study of "resistance" has been credited with overturning the positive assessment of Rome's experience in North Africa and inaugurating post-colonial scholarship within the Roman world more generally, did not contradict Gsell on this point. Only in the last decade has a different position emerged. In three articles, Quinn (2003, 2004, 2009) has argued that Africa was influenced by Rome as well as societies of the eastern Mediterranean under the Republic: "the social and cultural history of North Africa in this period is better understood in the larger context of the Hellenistic Mediterranean as a whole, than in terms of its relationship to one or another imperial power" (Quinn, 2003: 9). Her analysis offers an important new reading, yet the conclusions of both Mommsen and Gsell have not been entirely rejected and remain influential (Février, 1989: 93–8; Le Bohec, 2005: 29).

The period from the fall of Carthage to the end of the civil war between Pompey and Caesar offers the best example of this tendency to regard Rome as uninvolved in Africa. Bénabou considered the first century after the conquest a time of "bold and ambiguous initiatives, half-failures, and renunciations … when no definitive policy was established" (Bénabou, 1976: 26). Likewise, Lassère argued that "nothing had changed in the lives of the Libyans, who passed from subjects of Carthage to subjects of Rome, without their condition improving much" (Lassère, 1977: 142). For most Africanists including Gsell, Bénabou and Février, the "tempo of Roman imperialism" quickened after Caesar's victory in 46, and progressed even more rapidly with the advent of the principate. This shift is symbolized by the settlement of veteran soldiers in newly founded colonies and by well-known monuments such as the Altar of Publius Perelius Hedulus in Carthage and the Theater of Annobal Tapapius Rufus in Lepcis Magna, both of which were erected under Augustus and expressed connections to his ideological program. It is equally possible to point to the emergence of new building types, artistic styles, epigraphy and rituals of worship (e.g. Rives, 1995: 51–63; MacMullen, 2000: 30–49). In the historiography of North Africa, this has often been connected with the "advantages" of Roman rule, and since much of the research in North Africa has been carried out in a colonial framework by scholars affiliated with the French or Italian governments, which colonized Algeria, Tunisia, Morocco and Libya, the aims of ancient and modern colonial enterprises have been closely intertwined (Février, 1989: 23–66; Mattingly, 1996b).

Developments in North Africa have also been seen in comparison to those in other areas of the Roman Empire. For instance, a thesis classifying the Republic as "static" and the Augustan age as "fast-paced" has been endorsed by scholars who have examined the spread of Roman culture. Note that the spread of Roman culture is a process many refer to as "Romanization" but in my view this term insufficiently explains the complexity of developments in provincial contexts. I therefore avoid it except when discussing how other scholars employ the word (see Chapters 22 and 27; Keay and Terrenato, 2001; Fentress, 2006; contra, Hingley, 2005: 14–48; Mattingly, 2002). Woolf has argued that the adoption of Roman lifestyles in the provinces was limited before 31, but that during reign of Augustus the formation of Roman provincial cultures appears to have taken place over a vast area and to have united the Empire (Woolf, 1995, 1998: 91–8, 2001). As examples of unity, Woolf has identified the widespread appearance not only of tangible material culture such as red-slipped pottery, Latin inscriptions, amphitheaters and temples, but also of intangible items such as the Latin language, gestures, behaviors and rituals, which proliferated at the same time. Woolf's thesis draws on the plethora of surviving material from the Augustan period to argue that changes in provincial cultures primarily occurred with the distribution of Roman material culture, language and practices. It holds that the Roman conquest had a limited effect prior to the diffusion of those identifiably "Roman" things at the start of the Imperial era and does not consider how imperialism may have affected provincial cultures prior to the widespread diffusion of material goods. In effect, then, the position that provincial cultures formed under Augustus is a more complex restatement of Mommsen's and Gsell's conclusions that Africa "had no history under the Republic" and "the locals were left to live their lives."

It is now possible to sketch an alternative approach to the formation of provincial culture in Africa more deeply grounded in the – admittedly limited – evidence from the Republican period. First, we must recognize that the extraordinary spread of Roman culture in the Augustan period took place as the result of strategies carefully crafted in Rome to support the political leadership victorious in the Civil War (see Zanker, 1988; Wallace-Hadrill, 2008). Although the actions of provincial elites were essential to promote this proliferation, it did not originate with changes in provincial cultures (see Terrenato, 2001c: 59). Therefore, it is not appropriate to characterize the Republic as a period of slow uptake of Roman culture on the basis of new features appearing in North Africa, Spain, Gaul and elsewhere more than a century after the initial conquest. Rather, it seems more fitting to recognize that these new features occurred when they did because of changes to the political system in Rome, and corresponding changes in the expression of ideology within the Roman Empire. While members of indigenous society may make ideological statements, they are often following the lead of imperial

Table 32.1 Phases of Roman imperialism in Africa

Phase	*Date*	*Activities*
1. Conquest	Late third–first century BCE	Punic Wars, destruction of Carthage, Jugurthine war, "client" kingdoms, enslavement, military campaigns
2. Tribute	Second century BCE–fifth century CE	Lex agraria, land confiscation, tax and tribute collection
3. Expansion of urban settlements	First century BCE–second century CE	Establishment of colonies, system of laws, municipal promotion, public building and euergetism, private domestic building
4. Infrastructure	First century BCE–fifth century CE	Port construction, road network, centuriation
5. Economic exploitation	First century BCE–fifth century CE	Senatorial and imperial estates, Lex Manciana, agricultural production, "villas," fishing, pottery production, stock-raising, wild animal capture, mining, slaves
6. Maintenance of order	First–fifth century CE	Quelling of indigenous revolts (Tacfarinas, Gaetulians), peace agreements, legionary and auxiliary bases, border construction, military recruitment
7. Ideological promotion	First–fifth century CE	Artistic and architectural propaganda, emperor worship, religious syncretism, language, education, dress, burial rituals
8. Collapse	Third–fifth century CE	Wars, withdrawal of troops, reduction in tax and tribute collection, decreased trade

ruling elites, and that appears to be the catalyst here. Several studies have highlighted the important contributions of local elites to Roman imperialism in the last 25 years, but they have rightly called attention to initiatives directed from the center too (e.g. Millett, 1990a; Woolf, 1998; Terrenato, 1998a; Creighton, 2000; Revell, 2009). This was certainly the case in Africa. What we must also recognize is that imperialism was not a static process but a dynamic one, and it went through several phases during the time in which Rome dominated Africa (in my view from the late third century BCE to the early fifth century CE).

Table 32.1 documents the progressive evolution of Roman imperialism (by which I mean much more than is traditionally associated with the word Romanization) throughout the history of the province of Africa. The table is necessarily schematic and is incapable of recording minor phases or activities, but it makes clear that we should expect to find in the third to first centuries only what is present in the first five phases: conquest, tribute, expansion of urban settlements, infrastructure and economic exploitation. Not every activity was present for the duration of each phase, and in particular many activities started partway through phases. This is significant, because several activities in the urban settlements, economic exploitation and infrastructure phases were only characteristic of the later Imperial era, including municipal advancement, port construction, pottery production, domestic building and imperial estates. Activities traditionally associated with the spread of Roman culture, such as emperor worship, artistic propaganda and education, also began after the end of the Republic. Since these activities did not take place anywhere until the Imperial era, it would seem unwise to consider the Republican period in Africa a static one on the basis of their absence.

To judge from the archaeological and epigraphic evidence available, Roman imperialism appears to have developed at an uneven pace. For the most part, evidence from the Republic is thin on the ground, by comparison to both earlier and later periods. One reason for its infrequent presence may be that the main period of African ceramic production did not occur until the second century CE, and earlier ceramic materials are less easily datable. Another may be the lack of a well-excavated major type site from the Republican period. The chief explanation, however, must be that the number of settlements and the population size were much smaller. Carthage and Kerkouane offer a picture of developed urban centers in the third and second centuries; both went out of existence during the Punic Wars. The devastation of these wars would have taken a toll on many other towns, and the extraordinary growth of Africa was not manifest until the second century CE (though actions taken in the Republican period would certainly have set the region on course for this development). An additional cause, however, is that the main interest of Roman archaeologists has long been the well-preserved towns of the second to fifth centuries CE. Temples, baths, amphitheaters, houses and Christian basilicas of this period in many cases remain standing, and have painted a picture of North Africa in the Roman period as a "lost Golden Age." These monuments not only obscure earlier phases of urban history, but also make it impossible for archaeologists to investigate further without dismantling extant structures which provide impressive attractions for modern site visitors and museums. Their preservation has also been due to use as symbols legitimizing modern colonial regimes which identified themselves as a "second coming of the Roman Empire" (Dondin-Payre, 1994; Mattingly, 1996b; Oulebsir, 2004).

In the rest of this chapter I use archaeological evidence to focus on medium- and long-term history through a study of three "landscapes." The urban, rural and mortuary landscapes that I will explore are spaces actively constructed by individuals that allow us to examine how people conceived of their surroundings, and made changes to them through time.

4 Urban Landscapes

Settlements of the second and first centuries took a variety of forms (Figure 32.1). Before its destruction in 146, Carthage deserved the label of "city," with its temples, military and commercial harbors, and organized domestic quarters. Otherwise, we are dealing with incipient urban centers, many of which could be categorized as port towns, agricultural towns or hillforts. The following diachronic evaluation will show how they developed significantly over the long term.

Port towns include those "Libyphoenician" settlements founded along the coastal strip to serve as way-stations for long-distance trade along the southern coast of the Mediterranean. Carthage had pressed heavily on them, extracting tribute and exploiting the agricultural resources of their hinterlands. The largest burdens must have fallen on the ports of Cap Bon, but we hear of Carthaginian imperialism affecting Byzacena and Tripolitania too. The decision of a number of these ports to side with Rome in the Third Punic War, which earned them the status of "civitates liberae et immunes" in its aftermath, must be regarded at least in part as a rejection of Carthaginian oppression.

In the port towns of Tripolitania, both Lepcis Magna and Sabratha expanded upon earlier foundations in the second and first centuries. At Lepcis Magna terrace walls were erected to create a public square close to where the Wadi Lebda enters the sea. Temples to Shadrapa and Milk'Ashtart were built on the square (later known as the Forum Vetus), and the two adjacent insulae were also developed. These additions date to approximately 100, as do city walls and parts of the port structure, though as the remains lie beneath later buildings, further precision on the basis of current evidence is impossible. At Sabratha, during the early second century there appears to have been a public square surrounded by domestic quarters. Mausolea A and B (see below) were constructed to the south of the settlement at this time. The area of the settlement expanded so much that further housing had encroached on the mausolea by the middle of the first century. The successive alignments, at least two of which predate Roman incorporation, exemplify the early growth of the town (Bullo, 2002: 167–9, 189; Mattingly, 1995: 118, 125–6).

On the nearby island of Jerba, two towns also indicate substantial occupation in the last two centuries. Excavations at the port of Meninx have identified traces of buildings beneath the later forum, and documented murex dye production, for which the town would become famous, as early as the second century. One estimate suggests that Meninx spread over at least 19 hectares during the second and first centuries. Another town, Bourgou, located 3 km from the coast, appears to have been primarily agricultural, but was undoubtedly connected with one of the ports on the island. Surface survey has shown an extensive distribution of second- and first-century pottery (Greco-Italic amphoras, black-gloss fine ware) over nearly the entire 20-hectare site. As the center of Bourgou has not been excavated, it is impossible to discuss the presence of contemporary buildings within the town, but a second-century mausoleum on its outskirts (see below) also attests to the likelihood of their presence (Fentress, Drine and Holod, 2009: 75–80).

Utica allied itself with Rome during the Third Punic War and was pronounced a "civitas libera et immunis." It was one of perhaps six ports (Acholla, Hadrumetum, Leptiminus, Thapsus and possibly Usala) and one town in the interior (Theudalis) that did so (cf. Gsell, 1928: 38–46). All presumably received favorable treatment and did not suffer extensive confiscation of land. Utica had a significant pre-Roman settlement and was made the capital of the first African province. Unlike Lepcis Magna or Sabratha, it has yielded little evidence for development in the second or early first century. Near the middle of the first century, however, it may rapidly have been embellished with public buildings. An orthogonal grid plan covering an extended area was laid out. Temple A was constructed overlooking a public square on the spot where an earlier temple had stood. This new structure had a blend of Italic and African elements; it sat on a podium of 4 m in elevation with six columns in front of a cella, while its walls were constructed in opus africanum (cobbles held together by larger uprights stationed at regular intervals) and the unit of its measurement was the local foot. A theater built into the slope of a hill, with a semi-circular cavea spanning more than 100 m in diameter, appears to date to the same building phase. Since the theater and defensive walls constructed adjacent to it were both mentioned in the account of the Civil War between Caesar and Pompey, it has a secure terminus ante quem (Caes. *B Civ.* 2.25). The use of the natural topography of this hill to accommodate seating for the theater and to enhance the defensibility of the town indicates well-organized aspects of town planning (Lézine, 1968, 1970).

Settlements away from the coastal strip consist of hillforts (oppida) and agricultural towns. They have generally been regarded as inhabited by indigenous Libyans. Ferchiou has examined small fortified settlements, which are the most frequent type of the third to first centuries in the Tell region of Africa (Ferchiou, 1990, 1994). These hillforts, generally not much larger than two hectares in size, are characterized by location in a defensible position, surrounding enceinte walls, remains of interior walls and the occasional presence of cisterns. Most settlements were located on an elevation, were not regularly plowed, and did not produce abundant ceramic evidence. Fragments of black-glaze pottery or Punic amphoras and some cases of stylistically well-defined architectural elements allowed Ferchiou to date some of the sites between the late fourth and first century. An example of such a fortified hilltop is Sidi Zid, some 60 km south of Carthage and 12 km north of Segermes. Wadis run along two sides of the hilltop which overlooks a plain to the south. Large limestone blocks surround a defensible plateau covering 16,000 m^2. Two gates permitting entry seemed to be identifiable. Near one, a block was carved with a throat cornice, a motif derived from Egyptian architecture but commonly used at Carthage and neighboring sites. Black-gloss pottery and amphoras of the third and second centuries were abundant, and a stela with a Punic design was also found (Ferchiou, 1990: 73–5). Several architectural elements, including two Doric column capitals and an Ionic column base and shaft, suggest a possible Doric distyle temple (Ferchiou, 1994: 16). Not far from the settlement, approximately 26 *haouanet*, or one-room rock-cut funerary chambers, have been located (see below). If Sidi Zid is comparable to other hilltop sites, they are almost certainly small agricultural villages inhabited by indigenous farmers. Their economy no doubt involved some combination of sedentary and pastoral elements, but in general they reflect the transition to settled agriculture.

To summarize, over the course of the second and first centuries, we find evidence that does not conform to the picture of a "static" period. Development of urban centers proceeded at a slow pace, but aspects of urbanization are noticeable. Construction of public buildings occurred at Sabratha, Lepcis Magna, Utica, and other towns including Thugga and Simitthus which I have not considered. Hillforts and agricultural towns show signs of increased habitation and construction of walls: these features are signs of incipient urbanization. The embellishment of many of these urban centers with buildings expressing "Roman" architectural and artistic styles took place in the Imperial period rather than the Republic, but the progressive advances during the Republic are unmistakable.

5 Rural Landscapes

I focus on two areas in the discussion of rural landscapes: surface surveys and centuriation (see Chapters 12 and 16). The results of archaeological surface surveys conducted in Tunisia and Libya during the last 25 years provide the most reliable means of evaluating the countryside of Africa in the last two centuries. I have selected for consideration here the Lepcis Magna, Jerba and Segermes surveys because they employed the most rigorous methodologies, spacing fieldwalkers at 10 to 25 m apart, collecting artifacts, and analyzing them in detail. These techniques have been proven to identify the largest number of sites, including those with small and unobtrusive surface remains, and to offer the most reliable means of dating. The three projects have also published detailed analyses of their results. They indicate that the countryside was characterized by considerable variation; in some regions small farmsteads grew increasingly prominent in the rural landscape, while in others small fortified villages and tombs remained the most visible elements throughout the period.

In Tripolitania, rural settlements first appear in the immediate hinterland of Lepcis Magna during the second century (see Munzi *et al.*, 2004, 2004–5, 2010). Forty-one sites in a 6 by 8 km area between Ras el-Mergheb and Ras el-Hammam to the south-west of the city can be dated to this century by the presence of Campanian black-gloss fine wares. A further 14 sites have been found in coastal locations to the east (Wadi Caam) and west (Silin) of the city. During the subsequent century, the number of sites to the south-west increased to 64, while 18 were occupied in the coastal zones. A single site may be identified as a hilltop village, but the majority possess the features of small farms, initially settled in this period and expanded in subsequent centuries. On the island of Jerba, archaeologists identified 4 urban centers, 16 villages, 3 cemeteries, 19 villas and 76 farms in the period 250–50 (the numbers are drawn from Fentress, Drine and Holod, 2009: 86, fig. 7.1). The presence of a more elaborate settlement hierarchy than at Lepcis Magna is clear even if, as the surveyors themselves admit, some categories may overlap. Villages (or hamlets) were identified by less concentrated remains or several distinct concentrations over a large area. Villas were characterized by evidence of luxury or deliberate planning including pieces of plaster, querns and architectural fragments. Most of the farms were small and yielded little pottery. The earliest sites could be dated to the late third century on the basis of fine wares, but the second century saw the spread of Greco-Italic amphoras and Campanian A and B wares to many small sites, making clear that these goods circulated beyond a narrow elite (see Chapter 5). Jerba's rural landscape in the second and first centuries can be linked to the production of wine, and perhaps also olive oil. Kilns manufacturing amphoras in the Punic tradition were found at eight locations, and were mainly connected to agricultural settlements (Fentress, Drine and Holod, 2009: 87–95).

By contrast, the region of Segermes looks to have been very different from territories near Lepcis Magna and on Jerba in the second and first centuries. In the Segermes Survey, there were two hilltop villages (Sidi Zid and Sidi Ahmed Rouiggued) dating to the second and first centuries. They yielded Campanian black-gloss fine ware, Maña C and D amphoras, Punic coins, and a number of architectural elements. At a small number of other sites, generally located in the northern part of the Segermes Valley, similar materials – with the exception of architectural elements – were also found. This evidence suggests that a few dispersed farms or tombs began to occupy the landscape in this period, though since many of these sites demonstrated evidence for habitation in subsequent periods, it was not possible to determine how the sites were exploited in the last two centuries BCE. Nonetheless, given the intensive fieldwalking techniques the Segermes Survey employed, the scarcity of material implies that dispersed farms were not prevalent. All indications suggest that this occupation and exploitation was much less intense than near Lepcis Magna or on Jerba.

Several other archaeological surveys of the rural landscape of the northern portion of Africa have also been conducted. These projects covered much larger areas and examined the landscape in a less detailed fashion. The picture they have produced is therefore less reliable, but important nonetheless as nearly all found patterns similar to that near Segermes: hilltop villages and nearby tombs shared the countryside with very few farmsteads (on wider examinations of northern Tunisia, see Ben Baaziz, 1992; Ferchiou, 1990, 1994). The sole exception was the Carthage Survey. Here dispersed rural settlement appeared by the third century, possibly earlier than elsewhere in Africa, as Carthaginian exploitation of the countryside grew (see Greene, 1983, 1992; van Dommelen and Gómez Bellard, 2008).

Remains of centuriation (individual parcels of farmland demarcated as part of a large land measurement scheme) offer a second means of assessing rural activity within the province under the Republic. In Africa, three areas of centuriation were laid out by surveyors or gromatici for the Roman state (Caillemer and Chevallier, 1959; Trousset, 1977; Peyras, 1994). They cover nearly 15,000 square kilometers, more territory than any other such pattern in the Roman world (Figure 32.2). A "northern" zone, oriented 32 degrees from true north, stretched from Hippo Diarrhytus to Uppena and from the Cap Bon Peninsula to Vaga (100 by 100 km). The "east-central" group, oriented 42 degrees from true north, extended from Horrea Caelia to Caput Vada (100 by 30 km). The "south-eastern" group, oriented 32 degrees from true north, was placed in the countryside near Acholla (30 by 30 km).

The dates of the three centuriation schemes are not known, but there are several reasons to conclude that they were conducted during the Late Republic (Chevallier, 1958; Peyras, 1994: 224–5). First, the Lex agraria of 111 established the regulations for Romans to take advantage of property gained in Rome's conquests. Some of this conquered land may have previously belonged to individuals, but much is likely to have been held communally, by towns, villages or tribal groups. The law stipulated how the land was to be sold, leased, allotted to individuals or communities, and taxed. The section of the law concerning Africa mentioned "fields within centuries" ("ager locus in ea centuria" and "limitesque inter centuria[s]"; *FIRA* 66 and 90), implying that the measurements had begun in the second century. Second, from the speeches of Cicero we know of several prominent Italians who acquired land in Africa: one was the equestrian M. Caelius Rufus; another, the senator L. Aelius Lamia (Cic. *Cael.* 73; *Fam.* 12.29). A "saltus Lamianus" appears in inscriptions of the second century CE and may relate to the property formerly owned by this senator. Third, the land surveyed was for the most part restricted to the eastern side of the fossa regia, within the borders of the initial

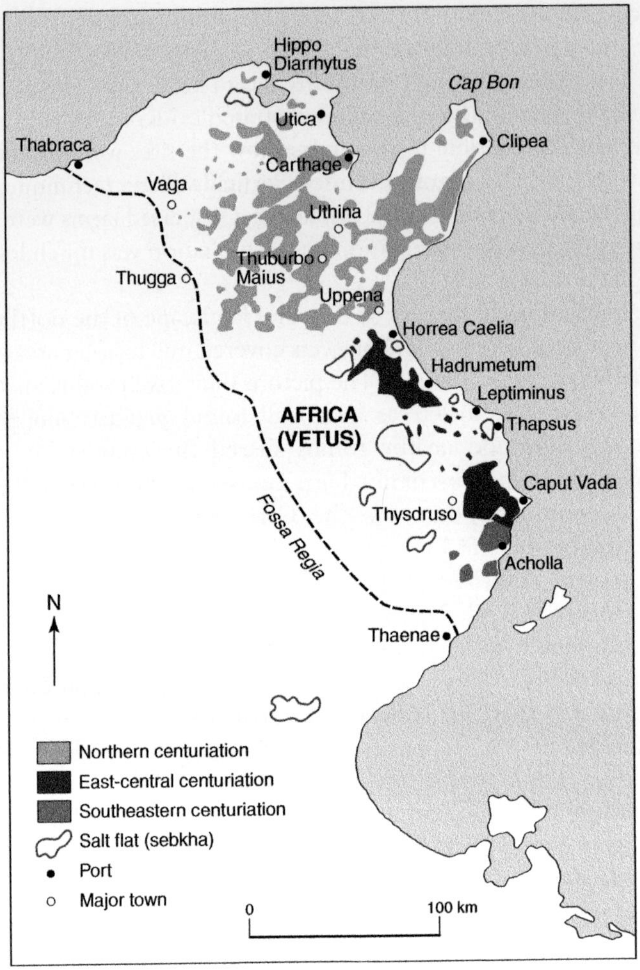

Figure 32.2 Centuriation in North Africa, with sites mentioned in the text. Source: Illustration by D. Stone.

province. Although boundary stones for a large area of centuriation have been found between Ammaedara and Capsa, this is clearly part of a different surveying operation (Trousset, 1978). Fourth, veteran soldiers settled by Roman generals in colonies will have received allotments of land. During the Republic phases of colonization these have been attributed to Gaius Gracchus, Marius, Julius Caesar and Octavian. Almost all commentators agree that there were colonies at Carthage, Uthina and Thuburbo Minus but the problems involved in determining how many additional colonies there were, and the total numbers of soldiers in each, are irresolvable (Lassère, 1977: 103–233; Février, 1989: 101–4; Bullo, 2002: 19–45). There are at least three colonies within the province, and probably 11 more in other regions of North Africa. Estimates vary widely; MacMullen, for example, cited 38 colonies founded by Caesar and Augustus in all of North Africa (MacMullen, 2000: 32–3). This number is surely too high. Still, it is reasonable to conclude that in this period many veterans received land allotments within centuriated plots. Finally, an argument can be made that "civitates liberae" carried out their own centuriations under the Republic. There are large gaps in the "northern" and "east-central" centuriations. Portions of this territory may have been excluded from these centuriation schemes because the land was uncultivated or marginal, but it is striking that territory in the vicinity of the "civitates liberae et immunes" of Utica, Hadrumetum, Leptiminus and Thapsus comprises much of the excluded land. Another suggestion that the land remained under the control of these towns is the presence of different orientations and units of measurement. Between Leptiminus and Thapsus there are five orientations close to 20 degrees (clearly different from the east-central orientation of 42 degrees). The Punic cubit (51.5 cm) also appears to be the basis for measuring the land, instead of the Roman foot (29.6 cm) (Ouni, 1999; Ouni and Peyras, 2002). Although the evidence is not conclusive, it may be reasonable to think that these towns retained the authority to organize their territoria during the Republic, but such local control probably did not persist far into the Imperial period. For instance, Leptiminus controlled aspects of municipal affairs such as the minting of coins in the reigns of Augustus and Tiberius, but it seems that local control in general ceased around this time (Ben Romdhane, 2001).

The evidence of both archaeological surface surveys and centuriation shows the development of the rural landscape of Africa during the last two centuries. Substantial restructuring of land-holdings in Africa took place by the middle of the first century. This was a time when the rural landscape began to be occupied by permanent settlements, when properties came into the hands of wealthy Italians, and when the seeds were sown for later exploitation on a massive scale. Both the actions of indigenous Libyphoenicians who chose to occupy the territory of Lepcis Magna and Jerba, as well as of the Roman state, which confiscated and sold, leased or assigned land, are recognizable Although only certain aspects of Roman imperialism were present at this time, the state was hardly uninvolved in the governance of this province.

6 Mortuary Landscapes

As we have seen in our examination of rural landscapes, tombs featured prominently in the countryside of Africa throughout the third to first centuries. There were major cemeteries of shaft graves surrounding Punic settlements, clusters of dolmens and ***haouanet*** near hillforts, and individual monumental tombs occupying conspicuous positions in the countryside. The diversity of funerary monuments has spawned attempts to associate Libyan, Punic and Libyphoenician peoples with specific tomb types, but the evidence does not support such blanket correlations of a culture-historical nature. What I aim to show in this section is that, like the urban and rural landscapes we have already considered, the mortuary landscape experienced considerable change due to the profound shifts in power and cultural influence during this time. Given the difficulties of dating poorly preserved and incompletely recorded tombs, and the perils of assigning specific tombs to known individuals or to cultural groups, this landscape can be presented only at a more general level, but nonetheless its development can be assessed.

The mortuary landscape, by which I refer to the articulation of burials within cemeteries and the wider environment, may have been where novel ideas were tested and explored during this period. One new type of burial monument is the so-called "Numidian royal architecture," a group of large monuments including eight mausolea and two altars built at (or near) major settlements of the third and second centuries (Camps, 1961b, 1995; Rakob, 1983; Coarelli and Thébert, 1988; Stone, 2007b; Quinn, forthcoming). Four of these mausolea lie within Africa (two at Sabratha, and single examples at Thugga and Bourgou) and were probably constructed by the middle of the second century. The mausolea are either in the form of towers two or three stories in height, or in large stepped monuments on circular bases. With ashlar masonry, false doors, Egyptian cornices and Ionic and Doric colonnades, their architecture contained references to other Mediterranean cultures. They also made use of indigenous elements; they were placed next to towns or in contested locations, dominated other tombs within surrounding cemeteries, and utilized images of cavalry as coins and other Numidian iconography did (see Quinn, forthcoming). Some of these mausolea have been interpreted as the resting places of Numidian dynasts, but the location of others, such as those at Sabratha and Bourgou, makes this sort of connection improbable, as does the inscription on the Thugga tomb. At Bourgou, the mausoleum was situated on the southern edge of the town (cf. above). It was a hexagonal monument with alternating curved and straight sides *c.* 5 m in length. Above ground, it rose in a stepped fashion; below ground lay a burial chamber. Pottery from the foundations dated the tomb to 175–150. A single Doric capital and the bust of a male figure have been discovered, but the remains are insufficient to afford a precise arrangement of the superstructure (Akkari-Weriemmi, 1985; Ferchiou, 2009). The tomb at Bourgou and the other examples of Numidian royal architecture are novel features in the landscape of this period. What functions did they serve? Fentress has suggested that the tombs are the "heroa of local nobles," an attractive idea which finds support in the evidence for ancestor worship in North Africa (Fentress, 2006: 8; cf. Camps, 1986). On the other hand, even if the mausolea housed the remains of local nobles, we must also explain why they were all constructed in (or close to) the first half of the second century. This was a period when towns were expanding, dispersed farmsteads were appearing in the countryside, and contacts with external powers were increasing as Rome, Carthage and the Numidian kingdoms were struggling to assert supremacy within Africa. If times were turbulent, as the evidence suggests, then local elites may have sought to erect mausolea as a new means of asserting power and legitimacy within their own communities. Such structures connote power by dominating a surrounding town or cemetery or asserting ownership of land; status by employing expensive items or utilizing costly labor; tradition via worship of ancestors; and knowledge through inscriptions and references to symbols in the *koine* of Mediterranean elites. They sent these messages in terms which both local and foreign audiences could understand.

Numidian royal architecture is seen as marking a fundamental shift in elite mortuary practices, but beyond wealthy tombs the funerary world of North Africa has been regarded as "in essence conservative" (Ghaki, 1999: 234). We ought to dismiss this notion, which is similar to those we have rejected concerning the static nature of Africa under the Republic. A study of the tombs labeled *haouanet* provides evidence for the changes in the mortuary landscape of non-elite members of society. More than 400 *haouanet* ("rooms" or "shops") are known today from Morocco to Tunisia, but cluster mainly within the confines of Africa. These tombs were excavated into escarpments; they contained a narrow opening leading to a single room between 1 and 3 m on a side. The opening was sealed with rocks, but normally remained visible and above ground. Thus, very few *haouanet* have survived intact to modern times, and artifacts found within them are scarce, though they often contain sculpted, painted or incised decoration. Where chronological information can be ascertained, usually through iconographic parallels, the *haouanet* date broadly to the fourth through second centuries, but perhaps most belong near the end of this period. They are commonly associated with Libyan or Libyphoenician peoples, but not with the elites of those groups (cf. Longerstay, 1995; Ghaki, 1999; Stone, 2007a).

At Sidi Zid, one of the two hilltop villages in the Segermes Survey area occupied from the third to first centuries, a mortuary landscape surrounds the settlement. No fewer than 26 *haouanet* have been identified along prominent limestone escarpments within 3 km of the village, most grouped in two *necropoleis* (Ben Younes, 2000). These tombs have common features, such as the funerary beds carved into the rock within 23 of the tomb chambers. Some tombs contain one bed, others as many as three or four. Some beds have sculpted pillows which represent the gender of the deceased, as Ben Younes has argued. He has also suggested that the beds reflect a borrowing from the Etruscan world, specifically the tumuli of Caere of the late seventh to fifth centuries (Ben Younes, 2000: 213–14). Most of the tombs include additional sculpted images in their decoration. Bovine heads, human heads, snakes, sphinxes and an oval disk guard the entranceways and interiors (Ben Younes, 2000: 209–11). The insignia of Tanit appears once above a tomb entrance at Sidi Zid. Niches on interior walls may be associated with funerary rituals or with offerings.

At Sidi Zid, and other sites such as Jbel el Mangoub and Sidi Mhamed Latrech (Ghaki, 1999), the appearance among Libyan and Libyphoenician peoples of tombs demonstrating affiliations with local as well as external cultures is a striking characteristic of the third and second centuries. These *haouanet* represent greater integration with the Mediterranean world, but also complex displays of status in local contexts. As I have argued in this chapter, Africa was gradually coming under greater external control at this time, the landscape was being divided up, and individual ownership (as opposed to communal organization) was becoming more prominent. It is not surprising then to find that non-elite members of society would be affected, and would adopt strategies similar to those of elites in demonstrating identities and negotiating the new social environments of changing times.

7 Conclusion

Beginning with Theodor Mommsen, scholars have regarded Africa as "without history" after the Fall of Carthage. Woolf has argued that little development took place in Roman provinces in the Republican period, prior to the formation of provincial cultures in the Augustan era. Consideration of Africa during the last two centuries, however, shows that much was occurring. The important work of Quinn has shown that in the Republican period Africans built monuments with references drawn from Libyan, Punic, Roman, Egyptian and Greek contexts and pursued strategies of emulation. For Quinn, the choice of Africans to emulate many cultures lies not in "resistance" to Rome, but in the integration of Africa within a wider world: "the social and cultural history of North Africa in this period is better understood in the larger context of the Hellenistic Mediterranean" (Quinn, 2003: 9). By contrast in the Augustan era there was a much greater (although not exclusive) emulation of Rome.

My re-evaluation of Africa over the longer term has indicated that instead of being a static environment, its towns, countrysides and cemeteries were full of change. Since its occupants did not make use of the more visible material culture of the Augustan and Early Imperial period, this change has been harder to detect, but nonetheless provides an understanding of this increasingly complex region during the Late Republic. As I have argued, Roman imperialism appears to have developed at an uneven tempo; it may appear to advance rapidly at certain times, such as when the province of Africa was formally annexed, when many veteran colonies were settled by Caesar and Augustus, and when initiatives spread from Rome in the reign of Augustus. At other times, it seems to progress more slowly, but is not absent. Such differences in speed are easier to detect over periods of a century or more. It is important to evaluate Roman imperialism by means of both short-term and long-term perspectives, since understanding the varied tempo is essential to a fuller understanding of the history of provinces like Africa.

Acknowledgments

I thank Stephen Collins-Elliott and the students in my graduate seminar on Roman imperialism at Florida State University in spring 2011 for valuable comments on a draft of this chapter.

FURTHER READING

The first detailed coverage of the Republican era in North Africa appeared in Gsell (1928). It was followed shortly afterward by Broughton (1929). Teutsch's (1962) analysis of the municipal organization of Republican Africa also covered the period in depth. While no monographs focused specifically on the period have effectively surpassed these works, numerous other studies shed light on particular areas. On the history of Carthage in North Africa, see Lancel (1995), Fantar (1984–6) and Hoyos (2003). On the Numidian kingdoms, one can consult Camps (1961a), Berthier (1981) and Horn and Rüger (1979). Lassère (1977) has written an excellent analysis of population growth in North Africa beginning in the Roman Republic. A number of studies have made important contributions to the understanding of Roman imperialism in North Africa: Bénabou (1976), Fentress (2006), Mattingly (1995, 1997).

The long history of landscape archaeology in North Africa has contributed to the understanding of urban, rural and mortuary landscapes. The recent and comprehensive treatment of urban centers in Bullo (2002) contains references to previous work on urban landscapes. Ferchiou has written several important articles on the landscape of the Tunisian *tell* in the second and first centuries (Ferchiou, 1978, 1987, 1990, 1994). For rural landscapes, Fentress and Docter (2008) present an overview of the Punic period. The surface survey of Jerba produced important material concerning the agricultural development of that island during the second and first centuries (Fentress, Drine and Holod, 2009). Comparisons can be drawn to the results of the Segermes Survey (Dietz, Ladjimi Sebaï and Ben Hassen, 1995; Ørsted *et al.*, 2000), the Lepcis Magna Survey (Munzi *et al.*, 2004, 2004–5, 2010) and the Tunisian national archaeological maps (Ben Baaziz, 1992). The culture of the mortuary landscape of North Africa in this period is discussed in two recent articles (Ben Younes, 2007; Quinn, forthcoming). My own views on aspects of the rural and mortuary landscapes of this period can be found in three articles (Stone, 2004, 2007a, 2007b).

CHAPTER 33

Hispania: From the Roman Republic to the Reign of Augustus

Isabel Rodá

1 Introduction

We commonly talk about "Hispania" even though we are conscious of the fact that within this place name there are diverse plurals. It would be more correct to speak of "Hispaniae," and by using that term, cover all of the people-groups on the peninsular territory that Rome made its own after having set foot on its Mediterranean coasts only 200 years before. On the other hand, the singular name Hispania corresponds with the name the Greeks gave to the peninsula. But Iberia was the name that Strabo used around the turn of the first century (IV.4.19). Iberia had previously maintained intense commercial relations with large zones of the Mediterranean including the Roman world, before Rome's conquest and continuous presence in the territory (Anello and Martínez-Pinna, 2008). There is no doubt that the Iberian Peninsula constituted the authentic learning grounds of Roman expansion beyond Italy. We do not believe we exaggerate by saying that Rome learned how to govern an overseas province in Iberia.

The gradual conquest of the Iberian Peninsula, although not initially intentional, began in 218 and ended in 19. This is a long period and, as we would expect, varying degrees of assimilation and resistance were found throughout the peninsula, providing an authentic mosaic of very diverse realities that the Romans encountered. In this chapter, we will summarize the recent contributions of archaeology to our knowledge of some of the major phases from the end of the third century until the total conquest of Hispania during Augustus's time. We will situate this framework within the process of Romanization in the western Mediterranean.

A Companion to the Archaeology of the Roman Republic, First Edition. Edited by Jane DeRose Evans.

2 The Second Punic War

Although direct or indirect relationships, contacts and exchanges preceded the continuous presence of the Romans in Hispania, the first Roman military contingent landed in the allied city of Emporion (Ampurias) in August of 218 (Figure 33.1). This army, under the command of Cn. Cornelius Scipio, intended to cut off the rearguard of Hannibal's army, which was threatening to head to Rome. This Roman landing was in response to Hannibal's siege of Saguntum (Sagunto), the primary trigger of open conflict with the Romans. It is in Emporion, in the area of Neapolis, that we find the archaeological footprints of these historical movements at the end of the third century, in a powerful wall some 2 m wide, in parapet style (*proteichisma*), that was constructed in front of the gates of the old Greek city in order to prevent the attack by contemporary siege machines (Mar and Ruiz de Arbulo, 1993; see also Olesti, 2000).

From Emporion the Romans went to Kesse-Tarraco (Tarragona). We believe that the Celtiberian city named Kesse was situated on a small rise in the ground in the southern part of modern Tarragona, in the area which would later expand to become the forum of the Roman city of Tarraco (Otiña and Ruiz de Arbulo, 2000; cf. Arrayás, 2005; Fiz and Macías, 2007; Panosa, 2009). Next to this indigenous city, the Romans established their base, from which they successfully attacked the Punic camps in the Ebro River Valley, as Livy relates (XXI.60). The archaeology in this area corroborates the validity of Livy's words, with the discovery of the Roman camps at La Palma (L'Aldea) and Castellet de Banyoles (Tivissa), which provide evidence of military actions at the mouth of the Ebro in this point in time (Noguera, 2008, 2009; Noguera and Tarradell-Font, 2009). Other fortifications in the interior zone of Catalonia reveal the troops' movements due to the military conflict. To control the pass through the Montseny massif, the Ausetani Iberians constructed a prominent fort in the enclave of El Brull (Montgròs) in the last quarter of the fourth century; it was destroyed around the year 200, during the clash of the Ausetani with Rome (Figure 33.2) (López Mullor, 2002; Riera and López Mullor, 2004).

Recent archaeological work increases our knowledge of the diverse attitudes caused by the first stable presence of Rome among the inhabitants of what is today Catalonia at the end of the third century (Nolla and Sanmartí, 1984). In the areas of prior Phocaean presence on the coast, there was a receptive attitude to the Romans, but greater resistance is found in the interior, where the Ilergetes, Ausetani and Lacetani did not submit easily; the uprising of the Indibil and Mandonio peoples perhaps constitutes the best-known episode of resistance (Garcés and Rovira, 1996/7; Moret, 2002/3).

But let us return to Tarraco. On the acropolis, Roman engineers, using local labor, planned and began to construct the impressive walls almost the day after they set foot in the area. We know that the first phase of the cyclopean walls and their towers can be dated to the end of the third century, that is, to the first phase of Roman occupation (Figure 33.2) (Lamboglia, 1974; Hauschild, 1988, 2006; Aquilué *et al.*, 1991). As evidence of the Roman presence, we possess a relief and an inscription naming Minerva that constitutes the most ancient testimony of its genre in the Iberian Peninsula (Grünhagen, 1976; Alföldy, 1981; Pina Polo, 2003; contra, Sordi, 2006). There are only a few parallels to these cyclopean walls in Italy itself, such as the spectacular walls

Figure 33.1 Map of Republican Hispania, with sites mentioned in text. Source: Drawing by A. Yandek.

Figure 33.2 Tower of El Brull (Turó de Montgròs), dating to the mid-third century. Source: Photo by I. Rodà.

of Cosa (Figure 30.2; see Chapter 30) and other colonies such as Cremona and neighboring Placentia (Piacenza), founded in 218 in Cisalpine Gaul.

From Emporion and Tarraco, the Romans began to expand south, in pursuit of the Carthaginian army, creating allies with the indigenous elite whenever possible (Aquilué *et al.*, 1991; Keay, 2006). At the end of 218, Punic dominion came to an end north of the Ebro. In 209, Cartagena (renamed Carthago Nova) was taken, only 20 years after being founded by Hasdrubal (Ruiz, 2009). In 206, the Second Punic War was considered finished in the peninsular territory after the victory at Ilipa (Millán León, 1989); the signing of a foedus with Gades (Cádiz) (Remesal, 2011), which was surely expanded to include other cities and communities; and the establishment of Scipio Africanus's veterans in the new colony of Italica (Marín Díaz, 1988; Caballos and León, eds, 1997; Corzo, 2002; Gil de los Reyes and Pérez Paz, 2005; Caballos, 2010). Roman expansion runs parallel to the history of minting in Hispania, which prior to the Second Punic War had been confined to the emissions of Emporion and Rhode and coins from Punic mints; those coins were precisely the amount of money necessary to pay the army (see Chapter 7 for a slightly different view; Burnett, Amandry and Ripollès, 1992; Chaves, 2009; García-Bellido, 2000/1, 2006). The Romans were on the brink of their first experience of establishing provinces in Hispania.

3 The Second Century

If at the end of the third century the overriding interest of the Romans in the Iberian Peninsula was to defeat the Carthaginians, at the beginning of the second century, Rome was to consolidate and administratively organize its territories in the peninsula from the Pyrenees to Gades. For this initial stage we have the valuable testimony of the bronze plaque of Lascuta (near Gades) engraved with the edict of 189, by which the slaves who were inhabitants of Turris Lascutana were given their freedom; their land and oppidum was also returned to them (*CIL* II.5041; González, 1982: no. 520; García Moreno, 2001). The Romans made the most of the pre-existing physical and social structures. The cities of Cerro de los Santos and Torreparedones, in the east and south, respectively, are emblematic examples of this continuity (Morena, 1989; Noguera, 1994; Bendala, 2009). It is logical that during this era, Emporion continued to be an essential platform for the stability of the region; as we have already noticed, the first Roman installation coincided with the blossoming of the old Greek city. The excavations of the area of Neapolis confirm this prosperity; the statue of a Greek male god (either Asclepius or Serapis or a syncretic image) is dated at the end of the second century (Ruiz de Arbulo, 1995, 2009b; Rodà, 2004; *L'Esculapi: El retorn de déu*, 2008).

In other pacified provinces, Rome founded new urban nuclei near pre-existing indigenous populations or, better yet, settled these populations in cities, fostering in them and conferring on them a Roman appearance. The urban phenomenon in the south was a long tradition, beginning with the Bronze Age through the Carthaginian occupation; it was further ingrained through Hellenistic culture (Bonsor, 1997; Bendala and Blánquez, 2002/3; Niveau de Villedary, 2008; Prados, 2008). It is not surprising, therefore, that the density of cities in the Baetis valleys of the Guadalquivir River is truly spectacular, and otherwise unheard-of in the west (Roldán Hervás, 1996; Keay, 1998). On the other hand, we cannot forget that changes in social structures occurred simultaneously, since Rome always tried to develop the support, whenever possible, of local leaders and aristocrats. This support was well-developed in the southern part of the peninsula. Rome also had the ability to channel the diverse customs of those different peoples to its benefit, as in the case of hospitality and fidelity agreements.

In 197, the Mediterranean coasts and the Guadalquivir Valley, respectively, constituted the provinces of Hispania Citerior (the closest one to Rome, which extended into Catalonia and south to Carthago Nova) and Hispania Ulterior (the furthest one from Rome and corresponding to a part of modern Andalusia). They were the territories most accustomed and open to external contacts, with a long urban tradition and with laws that controlled the outlying areas. Rome discovered how to attract these previously urbanized territories habitually to its cause.

Yet the process of integration was not completely without friction. In 195, Rome had to send Cato the Censor to the north-east to repress an uprising of some of the Iberian peoples (Knapp, 1980; Martínez Gázquez, 1992; Ñaco Del Hoyo, 2003; Cadiou, 2008). The action was decisive; Cato's camp in the garrison of Emporion was the area in which afterwards the public zone of the Roman city was developed. For example, Cato destroyed the walled city of Ullastret; he also reduced the Hellenistic city of Rhode (Roses) (Martin *et al.*, 1999; Martin and Plana, 2001; Nolla, Palahí and Vivó, 2010; Puig and Martin, 2006). He likewise crushed the Lacetani, who were living on the northern slopes of the coastal mountain range (García, Martin and Cela, 2000). Cato's intervention was definitive and took place over the entire peninsula, from the Guadalquivir to the interior, from the Ebro Valley to the north. With this campaign, Emporion became especially prominent. In the extreme north-east of the peninsula (in modern Girona), we find an authentic hodge-podge in which the Hellenistic and indigenous models united in an interesting syncretic process over which Rome imposed a new order. An example of this process can be seen in the extraordinary site of Mas Castellar de Pontós, where a magnificent Pentelic marble altar and some refined stucco architectural decorations show the strong Hellenistic influence that was later replaced by Roman cultural artifacts (Pons, 2002; Ruiz de Arbulo, 2002/3; Asensio, Pons and Fuertes, 2007; Nolla, Palahí and Vivó, 2010).

At this moment, Rome still had two dangerous neighbors in the peninsula. In Hispania Citerior, the Celtiberian threat loomed; in Hispania Ulterior, the Lusitanians remained unpacified (Alarcão, 1988). Between 180 and 178, the governor of Hispania Citerior, Ti. Sempronius Gracchus, tried to establish a stable border with the Celtiberians in the Ebro Valley, after many confrontations and treaties (Ñaco Del Hoyo, 2003; Cadiou, 2008). In the Rioja region, at present-day Alfaro, he founded a city to which he would give his name – Gracchurris (Marín Díaz, 1988; Hernández Vera, 2002; Gómez-Pantoja, 2008) – initiating in the north-east of the peninsula a tradition that other victorious generals such as Pompey and Lepidus would continue in the first century (Bandelli, 2002a; Maselli Scotti, Mandruzzato and Tiussi, 2007).

An intense effort to build urban centers followed in the second century, with the foundation of new cities, the strengthening of those already established, and the progressive abandonment of the fortified dwellings that had previously been used by the people of the peninsula (Moret, 2003; see the special case of Ilerda: Pérez Almoguera, 1994). The exceptions to this rule are the centers of Turdetania (in modern Andalusia) and the province of Hispania Ulterior, which remained continuously inhabited (Keay, 1998). The presence of Italic populations continues to be very powerful and we can prove this in the building of new sacred architecture and in commercial exchanges (Ramallo, 1997; Barreda, 1998; Ruiz de Arbulo, 2009b; González Román, 2010; Gorgues, 2010).

A well-known case is that of Carteia; in 171, it became a Latin colony that contained an important contingent of Italian freedmen, as we know from a curious passage in Livy (XLIII.3.1–4). The status of the city was revealed in its name "Colonia Libertinorum Carteia" (Pena, 1988; García Fernández, 2009). The excavation of the city permits us to see the reality of this strategic city from the first, Carthaginian, phase (as seen in its very name) to the enclave which the Romans valued as a colony (Marín Díaz, 1988; Roldán *et al.*, 1998; Bendala, Roldán and Blánquez, 2002; Roldán *et al.*, 2003; Bendala and Roldán, 2005; Ruiz de Arbulo, 2009b). Additionally, the sources of wealth of the area of the Strait were developing, especially the salt industry in the area of Gades (Lagóstena, 2001; Etienne and Mayet, 2002; Arévalo, Bernal and Torremocha, 2004; Lagóstena, Bernal and Arévalo, 2007). The industry developed in the second half of the second century in Carteia and Baelo (Bolonia). These cities controlled tuna fishing, and the export of fish and fish products was one of the pillars of their economies. The archaeological investigations beyond the limits of the city have highlighted the relationship, and even the dependence, of Baelo on Carteia (Sillières, 1997; for recent excavations, see Arévalo and Bernal, 2007).

Another city that presents continuity throughout its history is Carmo (Carmona), with an urban occupation that dates to the eighth century; it became a key center of Punic power, and remained an important city through the first century (Moret, 1996). The investigations at the Seville Gate have highlighted the different phases of remodeling. Recently, the very Roman character of the gate has been vindicated, though it was possibly erected upon the location of a primitive Punic entrance. The bastion of the town was not only built for defensive purposes, but with an eye to propaganda and spectacle, crowned as it was by a large Roman temple (Jiménez, 1989; contra, Schattner, 2005; also Bendala, 2001).

Not long after the founding of Carteia came the Roman phase of the city that would be the capital of Hispania Ulterior, Corduba (Cordoba). The Roman settlement was placed across the valley from the primitive indigenous settlement of Colina de los Quemados (Murillo, 1994; Murillo and Vaquerizo, 1996; Vaquerizo, 2006). Because of Strabo (III.2.1) we know that it was Marcellus's foundation, the consul who was in Hispania twice, the first time in 169–168 and the second time in 152–151. We are not exactly sure when the foundation of Roman Corduba took place, though it was perhaps in his first term (Knapp, 1983; Marín Díaz, 1988; Stylow, 1996; Dupré, 2004; Murillo and Jiménez, 2002; Panzram, 2002). This city was later said to be one of the Empire's most splendid cities, due to its commercial activities, thanks to the navigability of the Guadalquivir River, which permitted easy exit to the sea. The excavations of the last few years, which are still continuing, have permitted us to draw more precisely the urban layout and its evolution; they even show evidence of a stretch of the era's foundational wall, built in a technique that finds its parallels in Rome (León, 1996; Ventura, León and Márquez, 1998; Murillo and Jiménez, 2002; Molina and Valdivieso, 2007).

At the same time, remodeling of the walls was undertaken in Tarraco, the other Hispanic capital; these grew spectacularly in width, though without the construction of towers. Thanks to archaeological investigations, the wall has been studied and dated to after 150, and thus placed in relation to the growing urban environment of the end of the second century. This is when other singularly great monuments were built into the urban fabric, like the Fountain of the Lions, placed by the harbor (Arrayás, 2005; Hauschild, 2006; Fiz and Macias, 2007; cf. Remolà and Pociña, 2005; Ruiz de Arbulo, 2007).

Between the second and first centuries, the Republican walls of Roman Emporion were built, with a polygonal base and opus caementicium faces that constitute a purely Italic construction technique. This is the earliest known example found outside of Italy (Aquilué *et al.*, 1991; Nolla, Palahí and Vivó, 2010). Hispania continued to be, then, the laboratory for the first great Roman experience of expansion.

To Pliny (*HN* 3.21) we owe the famous assertion that "Tarraco Scipionum opus sicut Carthago Poenorum" (Tarraco was built by the Scipios, just as Carthago [Nova] was by the Carthaginians), placing in parallel the two great cities, the first built by the Romans, the second captured by the Romans at the end of the third century. There are many similarities between both and these similarities were so overwhelming that scholars even thought that Carthago Nova, not Tarraco, was the first capital of Hispania Citerior (Ruiz de Arbulo, 1992, 2006; cf. Hauschild, 1995/6, and Panzram, 2007). Although we have neither literary nor archaeological evidence for those initial moments of creation of the Iberian province, it appears that Tarraco was the seat of government, and exercised capital status.

In Carthago Nova there was clear continuity of occupation, even in the name, which is a literal translation from the Punic "Qart Hadasht." Logically, the continuity to the Roman period is due to the suitable characteristics of the location and the port; from the beginning, the city was walled, and these Punic walls were later restructured for continued service (Ramallo and Ruiz, 1994, 2002; Ramallo, 1997, 2006a, 2006b; Ramallo *et al.*, 2008; Díaz Ariño, 2008a; Ramallo *et al.*, 2009). The current interventions of the Cerro del Molinete, directed by J.M. Noguera, are uncovering new sections of the wall. But the glory of Roman Carthago Nova would come later, with a cosmopolitan character and a solid economic base, thanks to the richness of lead and silver deposits from the area.

But it was not only the urban areas that were privileged by building projects. In the area of Carthago Nova's influence, the two Italic temples comprising the sanctuary of La Encarnation, in Caravaca, were built in the first half of the second century. Temple "A" is very small and situated directly on bedrock; it is the older of the two. Temple "B" is placed on a stone platform. The latter could date between 176 and 160, based mainly on the date of the terracotta revetments that adorned the temple (Ramallo, 1992, 1993; Brotons and Ramallo, 1994; Ruiz de Arbulo, 2009b).

The taking of Corinth and Carthage marks a high point for Rome. After 146, Rome felt more secure and as a result, grew into its dominant position in the Mediterranean and also in the Iberian Peninsula. In Hispania in the second half of the second century, the panorama is very diverse in the different zones, whether under Roman administration or on the borders. For instance, we see that in Hispania Citerior, new provincial centers surged and ones already in existence became more powerful.

Valentia was founded in 138, and the city was built in a decidedly Italic manner. Thanks to intense archaeological investigations, the physical reality of the city and its materials are very eloquent, although legal aspects of the colonial settlements (deductio) continue to provoke controversy (Ribera, 1998, 2006, 2008; Jiménez and Ribera, 2002; Cadiou, forthcoming). Saguntum, Rome's allied city, where the Second Punic War began when the city was attacked, was not forgotten. The Romans conferred the status of "civitas foederata" or "ally" on the city; excavations have uncovered diverse structures from throughout the second century, most specifically on the northern face of the Castillo Mountain and in the forum area.

We can confirm the continuity of the main urban centers, like Ilici (La Alcudia de Elche) (Abad, 2006; Ramos and Uroz, 1992), although we still know little about Saetabis (Játiva), where the best evidence of the city comes from the coins minted here at the end of the third and the second century (Ripollès, 2007). We have more evidence for Roman Lucentum (Alicante), which replaced the old Iberian city on this site (Olcina, 1998, 2002, 2006a, 2006b). And in 123, when Q. Caecilius Metellus conquered the Balearic Islands, he founded Palma and Pollentia on the island of Mallorca (*Symposium de Arqueología*, 1983; Marín Díaz, 1988; Orfila, 2005; cf. Orfila, 2000; Orfila *et al.*, 2006; Orfila, Chávez and Cau, 2006; Orfila, Cau and Chávez, 2008).

On the modern-day Catalan coast between Tarraco and Emporion, there were no other Roman urban centers in the second century, but there were important Iberian centers (Moret, 1996). Olèrdola, for example, was refortified at the end of the second or beginning of the first century (Molist, 2000; Rodà, 2001). We should highlight the great Layetani village of Burriac (Cabrera de Mar) that experienced an important Roman-style urban reform in the second half of the second century (Olesti, 2000; Zamora, 2006/7). In the middle of the second century in the Cabrera de Mar Valley, an extraordinary Italic settlement was created, with settlers possibly coming from Campania, with a probable aim of making the town the administrative center of the region. The settlers built their own hot baths, the oldest known on the peninsula (Olesti, 1995, 2000; Martin, 2000; Garcia, Martin and Cela, 2000; Martin and Garcia, 2002), even earlier than those in Valentia (Ribera, 1998; Tsiolis, 2008b). It has been suggested that the mint that produced Iberian coins labeled "Ilturo" was located in this Roman establishment

(Villaronga, 1982). What is certain is that this minting preceded the foundation of the municipality of Iluro (Mataró). The Italians and their descendants were settled in this city, along with the inhabitants of the Iberian towns in the area, including those from the oppidum of Burriac; we will return to this issue later (Rodà, 2002).

A parallel case is seen in Girona (Sant Julià de Ramis), which had a system of defense walls dating to the beginning of the second century. Later, some terrace walls were built, which permitted the construction of a platform which acts as a foundation for an Italic-style temple. This temple lacks a podium, lessening its Italic flavor; it was constructed around 130–110. These contacts with the Roman world made possible the founding of the municipium of Gerunda around 70; the city absorbed many of the inhabitants of Girona's oppidum, which was peacefully permanently abandoned around 50 (Burch, 1995; Burch *et al.*, 2001; Rodà, 2002; Ruiz de Arbulo, 2009b).

A topic that has caused much controversy in the last few years has been the chronology of the watchtowers in the Late Republican era. Since the towers were in use through the Arabic era (the eighth to tenth centuries CE), the Roman character of the construction has been called into question. It is not outlandish to suppose their continued use, or even repair in the era of the emirates, but it does not seem prudent at this time to move all of the phases of construction to that era (Pera, 2008; Nolla, Palahí and Vivó, 2010). While the Romanization of Hispania Citerior and Hispania Ulterior, which were controlled by Rome and in which the development of urban structures advanced without interruption, in the Hispania Lusitania and Celtiberian zones to the north-west the panorama was very different.

Almost simultaneously around 155, the Celtiberians rebelled and the Lusitanians entered the Guadalquivir Valley, triggering the decisive point of the Lusitanian Wars. The best Lusitanian general was Viriathus; he was betrayed to the Romans and assassinated in 139 (Alarcão, 1988; García Moreno, 2001; Pastor, 2006; Gómez-Pantoja, 2008). The defeat of these tribes allowed Rome to move the border of Hispania Ulterior to the Tagus River, as far as Olisipo (Lisbon). The proconsul D. Junius Brutus moved his army north of the Douro River, into the territory of the Callaeci, an expedition which, according to Velleius Paterculus (2.5), gave him the nickname of Callaicus. He also gave his name to the city of Brutobriga (Schulten, 1937; Tranoy, 1981; Marín Díaz, 1988; Bandelli, 2002b). Control of these territories was ephemeral, but it made way for the first contacts with Rome and corresponding repercussions on the Castro culture. And in Celtiberia in 154, the circumstance of the inhabitants of Segeda rebuilding their wall to increase its size was the detonator of 20 years of bloody Celtiberian wars, as narrated by Appian (*Hisp.* 44–5) (Ñaço del Hoyo, 2003; Burillo, 2006, 2008; Gómez-Pantoja, 2008; Burillo *et al.*, 2009). In 133, Scipio Aemilianus ended this war with the taking and destruction of Numantia, after strong resistance (Figure 33.3) (Jimeno, 2002, 2005; Chaín and Torre, 2005; Morales, 2005; Boullosa, 2010).

Two Hispanic cities were destroyed in the same dramatic way but for different reasons: on the Mediterranean coast, Saguntum was devastated for its loyalty to Rome; Numantia was destroyed for ferociously opposing Rome. Numantia resisted the Romans futilely in the second century. Today it is again endangered due to the threat of massive construction projects that would destroy the beauty of its surroundings by invading the area and surrounding the monuments and sites in such a way that would dishonor the site, especially one as emblematic as Numantia. The efforts of various institutions are praiseworthy in attempting to prevent the new destruction of the city; among these institutions the actions of the Universidad Nacional de Educación a Distancia stand out.

Figure 33.3 Wall of Tarragona, late third century to first decades of the second century, with a cyclopean masonry foundation; a relief of Minerva can be seen on the tower. Source: Photo by J.M. Macias, with permission.

Rome joyously celebrated Scipio Aemilianus's triumph and began a massive phase of restructuring of the conquered territories and the rear areas, in spite of the fact that at the end of the second century there were episodic clashes with the Celtiberians and the Lusitanians, although none with the magnitude of the previous wars. These conflicts were taken up by various governors who, as a result of these battles, celebrated official triumphs in Rome (Pina Polo, 2009). The instability permitted the raid of the Cimbri in 104, who descended from the Pyrenees. They were pushed back by the Celtiberians. The end of the second century was also of great importance for the structuring of southern Gaul. Narbo Martius (Narbonne) was founded in 118; this is when Cn. Domitius Ahenobarbus paved the old trade route between Hispania and southern France which ran through the Pyrenean pass of Panissars (El Pertús-La Jonquera). This road is known as the Via Domitia in France and the Via Augusta in Spain (Castellví *et al.*, eds, 1997; Soto, 2010).

Hispania's reorganization at the end of the second century is witnessed by epigraphic sources and archaeology from the Atlantic to the Mediterranean (Pina, 1997; Díaz Ariño, 2008c). We have already discussed the cities from the second half of the century; we can now mention that agricultural exploitation of the rural world was intensified, with the installation of numerous towns that would, for the most part, be continuously occupied in the succeeding periods. On the other hand, public works that were military in flavor continued, as is the case of the systemization of the Republican roads in the care of the proconsuls, indicated to us by a milestone in Lleida that names Q. Fabius Labius; it dates to around 118–114. The three milestones in the province of Barcelona that name M'. Sergius are dated between 120 and 110 (Fabre, Mayer and Rodà, 1984:

Figure 33.4 Funerary monument from Malla, late second century, depicting Hercules, Nessus and Deianira, and on the opposite side, the retinue of a magistrate. Source: Photo by R. Manent, with permission.

I, nos. 175, 176 and 181, II, no. 89; Díaz Ariño, 2008c: no. C3–C6). The latter marked a road through the modern-day regions of Vallés and Osona; next to it, in the municipality of Tona, it has been possible to excavate a short-term military-type establishment dating from 120–80, with metallurgical workshops for the elaboration of bronze and lead objects (Durán and Mestres, 2008; Durán, Mestres and Principal, 2008). Another singular monument is found next to the road that M'. Sergius built. Next to the church in modern Malla were found the remains of an exceptional funerary monument with reliefs depicting scenes from the legend of Hercules, specifically Hercules, Nessus and Deianira, and the introduction of the hero at Olympus. It is clear to us that it is a Republican monument dating to the end of the second century (Figure 33.4) (Rodà, 1993, 1998; Vivó, 2007).

If in northern Hispania Citerior there are examples of Republican sculpture in the second century, it is logical that in the south-east and south, examples of such sculpture grow annually. The total corpus of statuary is already very considerable from the end of the third century to the first century, especially demonstrating in their frequency the continuity of Iberian sanctuaries (Rodríguez Oliva, 1996; Noguera, 2005; Noguera and Rodríguez Oliva, 2008; *CSIR España*, forthcoming).

4 The First Century

In 93, the last deadly blows in the wars against the Celtiberians were dealt by Valerius Flaccus. Soon thereafter, in 87, the very same governor, who gave his name to the city of Valeria, was the authority commemorated in a bronze plaque in Contrebia Belaisca

(Botorrita) (Díaz Ariño, 2008c: no. C9). Another singular document, a bronze plaque found in Ascoli, testifies to the concession of Roman citizenship in 89 to native constituents from the "turma Salluitana" (*CIL* VI.8, fasc. 3.37045; Criniti, 1970; Mattei, 2001: no. 27). But the events that strongly marked the history of the peninsula in the first half of the first century were not the clashes with non-Roman people, but the military conflicts of Roman against Roman. A repercussion of the war between Marius and Sulla was their respective supporters taking advantage of the existing Lusitanian and Celtiberian enmity in Iberia, though these conflicts were never an attempt to free Hispania from its Roman yoke.

The protagonist was Q. Sertorius, named governor of Hispania Citerior in 83. He was suddenly and devastatingly removed from office by Sulla the following year. Considering himself the legitimate proconsul, in 80 he put himself at the forefront of a new Lusitanian rebellion against Rome. He defeated both the governor of Hispania Ulterior, Q. Caecilius Metellus Pius, and the new governor of Hispania Citerior, M. Domitius Calvinus (Alarcão, 1988). Sertorius left his quaestor, Hirtuleius, in Lusitania and, allying himself with the Celtiberian tribes, marched towards the Ebro Valley. Sertorius received reinforcements from the soldiers sent by Perperna Vento, the governor of Narbonne, also in the anti-Sullan party. He devastated opposing cities, such as Bursau (Borja), Cascantum (Cascante) and Gracchurris (Alfaro). Archaeological investigations have revealed the harshness of the Sertorian War in Ebro, which provoked destruction by the two competing armies; some of this destruction led to the end of very important enclaves, such as Azaila (Ruiz de Arbulo, 2009b). Many cities of the eastern Mediterranean were added to the allies of Sertorius, including Dianium (Denia) which became the main Sertorian naval base, as seen in the abundant ceramic material from these years on the site (Gisbert, 1998; Grau, 2002). Sertorius appeared to have every advantage, but in 76, the Senate of Rome decided to send Cn. Pompey, a staunch supporter of Sulla, to Hispania with an extraordinary imperium.

At first, Pompey suffered a defeat in Lauro, on the eastern coast, after crossing the Ebro. But in the following year, in 75, the tables began to turn, after Metellus defeated Hirtuleius in Lusitania. Metellus and Pompey joined forces and this was the beginning of the end of Sertorius. He was assassinated in 73 in "his city" of Osca (Huesca), where he had tried to reproduce Rome's system of government for Hispania (Juste, 2000). Pompey stayed in the area in the following months to put out the embers of the rebellion; he also consolidated clients in the area, who were later acquired by Julius Caesar (*B Civ.* II.18.7), although more work is needed to clarify the picture (Pina Polo, 2008). The Roman name of Pamplona, Pompelo, may be due to Pompey himself. In Tarraco an inscription was dedicated to him (Alföldy, 1975: no. 1; Díaz Ariño, 2008c: no. C59), and in Emporion we have inscriptions naming people of the gens Pompeia (Fabre, Mayer and Rodà, 1991: 79, 80), a name frequently found on the north side of the Pyrenees (Fabre, Mayer and Rodà, 1991: 188; *CIL* XIII.20, 65, 66, 70, 236, 237, 315). The foundation of some cities in the area of Catalonia happened due to the Pompeian program, as we will see for Gerunda and Iluro (Olesti, 1994).

Pompey marched in triumph in Rome in 71, after erecting trophies in the Pyrenees; Latin authors refer to them in a very laconic manner. Fortunately in the 1980s and 1990s, they could be identified and excavated; they are in the natural pass of the Pyrenees, Panissars, flanking the Via Augusta-Domitiana. They are impressive architectural

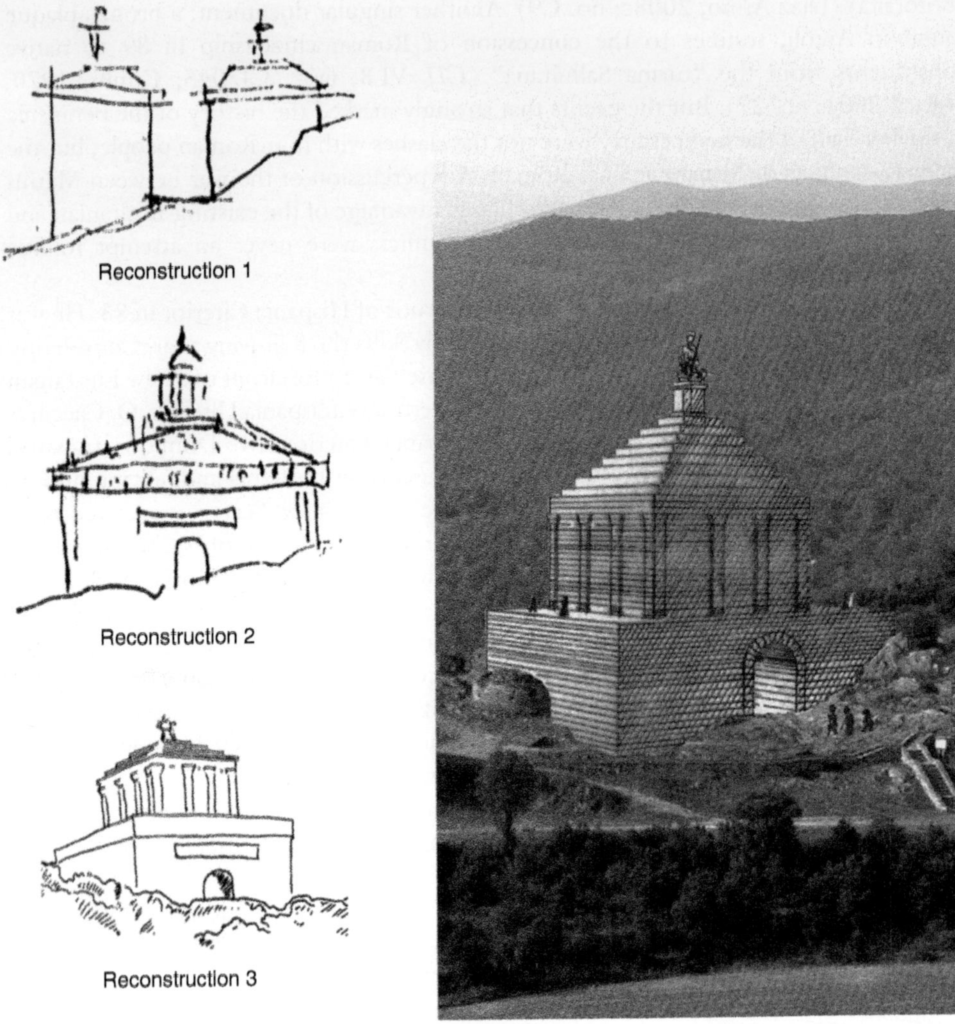

Figure 33.5 Suggested reconstructions of the Trophy of Pompey in the Pyrenees (Panissars), according to D. Mar (1); and J.-L. Paillet (2); Rodà after R. Mar (3) and J.-L. Paillet (right). Source: *Gallia* Suppl. 58 (2008). Used by permission.

monuments, the first of their kind; previously, the Romans celebrated victories by mounting captured armor, mannequin-style, on artificial supports. The remains have been ravaged by time but a possible reconstruction is a monument in the form of a turret, towering above the Via Augusta (Figure 33.5).

Halfway between the second and first centuries, the task of municipalization continued in Hispania. We have already mentioned above the city of Emporion and its walls. Newly founded cities during this stage, such as Iesso (Guissona) and Aeso (Isona), helped to integrate their indigenous inhabitants in the Roman way of life (Moret, 1996; Rodà, 2001). Land registry seems to correspond to retaliation for support of Sertorius's party, as it could have happened in Osca and Calgurris (Ariño, 1991). In Iesso a first-century stele in the Iberian language allows us to meet "Neitin(ke), son of Suba(ke)," but in a form already distinctly Roman – a display of clear cultural integration (Guitart *et al.*, 1996; Pera, 2005). On the other hand, in Aeso, the epigraphy of the Imperial era reveals members of a family with cognomina such as "Celtiber" or "Numantina," which highlights a clear geographic origin within an aristocratic and closed society; this was a society in which women played an important role in a matriarchal system (*IRC* II 26, 27, 28, 29, 38; Castellví, Nolla and Rodà, 2008).

Throughout the first century, we see a continuous progression in the assimilation of indigenous inhabitants into the Roman system. The epigraphic documents of Botorrita's third bronze plaque and in the mosaics of Andelos (Navarra) and Caminreal (Teruel) are very significant examples. Different versions in Iberian script of the name Licinus or Licinius are also found in the latter. Licinus could be derived from an indigenous name, Licnos, of Celtic origin. Signers of the two mosaic pavements could be, perhaps, the same person (F. Beltrán Lloris, 1996; Untermann, 2000; Wodko, 2000). These testimonies show us how it was no longer possible to reverse the process of integration that would end up converting all of the inhabitants of the Iberian Peninsula into Hispanic Romans or, better yet, into Romans from Hispania.

The end of the Sertorian War also instigated the formation of some urban centers, as in the case of Gerunda (Gerona) founded around 70 (Nolla, 1987, 2001; Rodà, 2001; Burch *et al.*, 2000). The Romans became aware of the territory thanks to their presence in the Iberian town of Girona (whose Italic temple without podium we noted above). Around the same time, other cities in the Catalan region were also founded, such as Iluro (Mataró) and Baetulo (Badalona), although their dates range between the second quarter and the middle of the first century. We have compared Iluro to Gerunda, due to the primitive settlement dating to the second century at the foot of an Iberian town, as in the case of Burriac, that precedes the founding of Roman Iluro and the Roman presence in the town of Gerunda that prefigures the future municipality of Girona. It is probable that Iluro corresponds to the urban program of Pompey, and thus post-dates the victory over Sertorius (Cerdà *et al.*, 1994; Olesti, 1994, 1995; García, Martin and Cela, 2000). The neighboring city of Baetulo presents a curious characteristic, uncovered in recent excavations at the site of Illa Fradera, where a settlement of clear Italic character functioned between 100 and 50. These new discoveries can be explained as an ephemeral phase that could have been the primitive installation or the headquarters from which the building of a new city could have been supervised, as materials began to be more abundant in the second quarter of the first century, even if it has been established that residual materials dating from around 100 were present (Antequera *et al.*, 2010; cf. Guitart and Padrós, 1990; Comas *et al.*, 1994; Guitart, Padrós and Fonollà, 1994; Guitart, 2008).

The coastal zone of Gerunda in this era became solidly Roman in organization, due to the fact that certain points of strategic control were abandoned. For instance, the recently excavated sites of Montmeló and Can Tacó seem to be military-style settlements that could perhaps correspond to the road-house of Praetorium, due to the Vicarello-type vases found in all three sites, although the distances between Praetorium and the other sites do not seem to make sense, if Praetorium functioned as a stop. The excavations suggest that the enclave was occupied between, approximately, 150 and 50 (Choren, Mercado and Rodrigo, 2007; Guitart *et al.*, 2006a, 2006b; Mercado *et al.*, 2008).

Towards the middle of the first century, civil war erupted once again among the Romans, and was carried into Hispania. Both Pompey and Caesar were very familiar with the terrain. Pompey had first set foot on Hispania during his fight against Sertorius; according to the terms of the Conference at Luca, in 56, the two provinces of Hispania belonged to him. For his part, Caesar had been named governor of Hispania Ulterior in 61. This is when he consolidated the Roman position between the Tagus and Douro Rivers, opposite the Lusitanians, with an important military establishment in Santarem (Scallabis Praesidium Iulium, which could have also been founded during the time of the civil war) (Beltrán Fortes, 2008).

In 49, Caesar crossed the Rubicon River and initiated the hostilities. Pompey, surprisingly, left Italy for Greece, but Caesar took the war to Hispania. There he vanquished Pompey's troops sent by Petreius from Ilerda, dismantling the Pompeian army on the Iberian Peninsula. Between 48 and 46, Caesar had entrusted Hispania Ulterior's government to Q. Cassius Longinus. The appointment provoked an uprising that allowed Pompey's son, Cnaeus, to move to Hispania in the hopes of opposing Caesar. Many cities aligned themselves with the Pompeian faction. Caesar quickly returned to Hispania from Egypt and, in 45, defeated both of Pompey's sons in Munda. Caesar's reprisals were swift and Corduba, because it embraced the Pompeian cause, was set on fire and destroyed. Loyal cities were liberally rewarded, as in the case of Los Balbos de Cádiz (*B Hisp.* 34.1).

Tarraco also was privileged, since it seems to have received the status of colony in 49 when Caesar received the delegations of the various tribes (*B Civ.* II.21.4–5). The leader of the colonists was P. Mucius Scaevola, in whose honor an inscription was engraved from a spoliated plaque that had previously carried a dedication to Pompey; this was, in a manner, a damnatio memoriae (Alföldy, 1975: no. 2, 2004; Díaz Ariño, 2008c: nos. C59, 60; Ruiz de Arbulo, 2002, 2009a; Arrayás, 2005). This colonization has been related to the centuriation of the territory around the city, with some plots organized according to an auguraculum (place to observe the omens), which has been identified. A little after this colonization, Tarraco named Cn. Domitius Calvinus its patron, as Alföldy has been able to demonstrate, thanks to her restoration of a partial inscription (Alföldy, 2000; Salom, 2006; Palet and Orengo, 2010). Another privileged city was Emporion. Livy (XXXIV.9) comments that "coloni ab divo Caesare post devictos Pompei liberos adiecti" (colonists were added by the divine Caesar after the defeat of Pompey's sons). And after Caesar's death, Emporion chose as its patrons the closest collaborators and supporters of Caesar and Augustus, beginning with Cn. Domitius Calvinus and Agrippa (Fabre, Mayer and Rodà, 1984: 24–9; Rodà, 1986–9).

5 Augustus and the End of the Conquest of Hispania

Under Augustus, Rome became the center of the Mediterranean world. This was a world whose cities and provincial communities wished to be seen as Roman, with the integration of the Italian and Hispanic elites, and a capacity for personal promotion (see Chapter 27). It is important to understand the power and the well-being entailed by being part of the Roman state, as much at the collective level as at the private level (Le Roux, 2006). Social mobility, the possibility of citizenship and positions of public office constituted powerful incentives, even for the dependent classes that did not aspire to any of these, as in the case of the freedmen, powerful contributors in many Hispanic cities (Schulze-Oben, 1989).

The beginning of the Empire meant very important changes for Hispania. Traditionally the major change is said to have come about in 27, when Hispania was newly divided into three provinces: Citerior, Baetica and Lusitania. In fact, the process seems to have been more complicated, but a discussion of the division of the province would entail the writing of a separate chapter.

In 25, it was erroneously believed that the submission of the Cantabrians and Asturians had been finalized; this same year, the veterans of the war founded the new city of Augusta Emerita (Mérida), to be the capital of the Lusitanian province (Álvarez and Nogales, 2003; Mateos, 2006). But the hostilities had not ended, and a second phase of the Cantabrian Wars had to be undertaken. Agrippa undertook this, subduing the tribes in 19 (Roddaz, 1994; Rodà, 2005; Morillo, 2007). After that, it was possible to proceed with a global plan of structuring the northern third of the peninsula. The restructuring was likely a trial and error affair, and to make it happen, components of the same legions that had been instrumental in the final victory were employed. In this way, after being an instrument of conquest, they were the means of integration of the territories, the stimulus of the execution of great public works and the path to personal promotion, creating the necessary structures for the incorporation of recently conquered towns into the Roman sphere.

In the rear zones, the restructuring was less complex, given the long period of Roman presence. Roads and cities were the main objectives; the urban settlements in the mountain regions, like Labitolosa's center, in the Aragonese Pyrenees, were monumentalized later (Chasseigne *et al.*, 2006; Magallón and Navarro, 2010). During the time of Augustus the road network grew, making the system that we know of as the Via Augusta, which stretches from Cádiz to Rome – it was the only road from Hispania to Rome – with other branch roads to connect interior zones, as is seen in Aragon (Lostal, 1992). New cities were born, such as Colonia Iulia Augusta Faventia Paterna Barcino (Barcelona) and Caesar Augusta (Zaragoza). Both came from the same political program in 19 and both received propagandistic names. The founding of Caesar Augusta led to the decline of Celsa (Velilla de Ebro); founded in 44, it was suddenly depopulated. It is likely that the city paid a high price for calling itself Lepida (before changing its name to Celsa) in the competition with a new well-situated metropolis, given great power by Augustus, due to his enmity with Lepidus (M. Beltrán Lloris, 1985, 1990, 1997; Mostalac and Beltrán Lloris, 1994; Amela, 2002b; M. Beltrán Lloris and Mostalac, 2008). Other capitals, like Corduba and Carthago Nova, experienced a strong wave of urbanization, as seen in their theaters and other urban city projects conceived during the Augustan era (Ramallo and Ruiz, 1998; Ventura, León and Márquez, 1998; Ventura *et al.*, eds, 2002; Monterroso, 2004; Ramallo *et al.*, 2009).

In the north-western quadrant, the peoples living in the Galician zone did not participate in the wars; it was in the Cantabrian north where the Asturians and Cantabrians formed the focal point of the resistance. Excavations have offered a substantially new vision of the main cities, such as Lugo, Astorga, Braga and León, and the camps of Legions IV, VI and X (see also Chapter 14). The Augustan renovation of the northern part of the peninsula was spectacular and profound. It began after the end of the second phase of the Cantabrian Wars in 19, and was only possible because of the work of Augustus's great collaborator, Agrippa, who was on the peninsula from 19 to 18. In order to inspect these projects, Augustus returned to Hispania in 16/15; Cassius Dio (54.23.7) says that during that time, he founded many cities in Hispania.

We close this chapter with the reforms undertaken during Augustus's era in order to complete the historic cycle of the Roman conquest of Hispania (218–19), to better understand the decisions that were made and the actions that were taken that would situate the entire Iberian Peninsula within the Roman sphere, especially between the last years of the Roman Republic and the final decades of the first century, when two figures were key and complementary: Caesar and Augustus, with the crucial work carried out by Agrippa. Therefore, we propose to leave our discussion here, as the reforms that affected the political-administrative framework that were created in the early Augustan era began to show multiple aspects of the economic, social and religious life that deeply marked the historic evolution of Hispania.

Acknowledgments

This chapter has been translated by Kellye Hawkins (Temple University) and J.D. Evans.

FURTHER READING

For an overall view of the history of the period from the Second Punic War until the age of Augustus, see Curchin (1996), Ñaco del Hoyo (2003: 127–93), Bravo (2007) and Cadiou (2008). For more specialized discussions of the third through first centuries, see: Knapp (1977a, 1979), Keay (1988), Marín Díaz (1988), González Román (1990), Rodríguez Neila (1992, 2006), Pena (1994), Richardson (1998), Roddaz (2009), Bandelli (2002b), Laffi (2002), Melchor, Mellado and Rodríguez Neila (2005), Pina Polo (2007), Bravo (2008), Gómez-Pantoja (2008), Díaz Ariño (2008b), Le Roux (2010) and Barrandon (2011). For the Augustan period, see Gómez-Pantoja (2008) and Roddaz (2002).

A summary of the diversity of the Iberian Peninsula is presented in Almagro-Gorbea, Palomero and Osuna (1997). A discussion about the Hellenization and Romanization of the province can be found in the important volume François, Moret and Péré-Noguès (2006). Fundamental publications after 2000 about the relations between Hispania, Gaul and Italy, with studies of the stage that interests us here, are Keay and Terrenato (2001), Luik (2005), Sartori and Valvo (2006), Uroz, Noguera and Coarelli (2008), Gorgues (2010: 53–68), Navarro (ed., 2010), Pons Pujol (2010) and Rodríguez Gutiérrez (2011).

For a discussion of Hannibal's itinerary, cf. F. Beltrán Lloris (1984). The Scipios in Hispania are discussed by Roddaz (1998) and Zecchini (2002). Recent investigations in Saguntum have contributed interesting archaeological facts about this phase of the city (Aranegui, 2002, 2004, 2006). For Iberians in Hannibal's army, see Brizzi (2006) and a brief synthesis in Noguera (2011). For an overall discussion of Roman camps, see Chapter 14 and Morillo, Cadiou and Hourcade (2003: 25–156). For the Republican colonies, see the papers presented by Caballos (forthcoming) and Cadiou (forthcoming).

For scholars concentrating on Sertorius in Hispania, see García Morá (1991), Scardigli (2002), Roddaz (2006) and Meister (2007). A good summary of the war against Sertorius in the northeastern section of the peninsula can be found in Olesti (1995: 56–63). A large bibliography about the Sertorian episode in the Ebro exists: F. Beltrán Lloris (1990), Pina (2000, 2007), Ferreruela and Mínguez (2002) and Pina and Pérez Casas (1998).

Concerning the administration of the Hispanic provinces, see Salinas de Frías (1995), Roldán Hervás and Wulff Alonso (2001) and F. Beltrán Lloris (2008). Hispania Citerior's first monuments, the relationship of the indigenous people with Rome and the striking of coin are discussed in Ripollès (2002a, 2002b), García-Bellido (2006), Chaves (2009) and Gorgues (2010: 253–68). For Pompeius's patronage activity, see Amela (2002a), Arrayás (2005: 64–71) and Badian (1997: 278–84, 303–5). For Iesso, see Guitart and Pera (1994, 2004), Guitart (2006; 2008: 18–24, 28) and Rodrigo (2004). For Aeso, see Payà, Puig and Reyes (1994) and Reyes, González Villaescusa and García (1998). An analysis of the elite in Hispanic cities has constituted an interesting line of study in recent years: see Rodríguez Neila and Navarro (1999), Navarro and Demougin (2001), Panzram (2007) and Rodríguez Neila and Melchor (2006).

The archaeology of Augustan Spain has advanced greatly due to the Congress in León in 1968 (*Legio VII Gemina*, 1970), the exhibition in Astures (Fernández Miranda and Fernández Ochoa, 1995), a Congress in Lugo (Rodríguez Colmenero, 1998) and an exhibition in Cantabrós (Iglesias and Muñiz, 1999). For recent overall views, see Rodà (1998), Morillo and Aurrecoechea (2006) and Morillo (2007). For excavations in Bracara Augusta (Braga), see Martins (2006) and Garrido, Mar and Martins (2008). For legionary camps, see also Rodríguez Colmenero and Ferrer (2006).

CHAPTER 34

The Archaeology of Palestine in the Republican Period

J. Andrew Overman

1 Introduction

The terms for the chronological period of the archaeology of the Levant and ancient Palestine during the Republican period differ somewhat from the rest of the Mediterranean. The period between the third and first centuries is divided into the Hellenistic, Hasmonean and Herodian eras, the names of which reflect the shifting political control of the region. There was significant growth and development in the region precisely during this period. While other areas of the Mediterranean demonstrate heavy Roman influence and material presence within this time span, this is not so in Palestine. In the region of Palestine the era reveals virtually no evidence of Roman material presence until very late in the first century. Indeed, Roman presence can be thought to be a product of the building projects of Herod the Great. Yet few other Mediterranean cultures boast a material culture that demonstrates clearly how a region could straddle two worlds, west and east – Graeco-Roman on the one hand and Persian-Parthian on the other (Rast, 1992). This cultural diversity and dynamic placed Palestine between two powerful and surging economic and cultural forces and this informs much of the history, development and archaeology of Palestine. This was the case in Palestine from at least the Late Bronze Age and is most certainly an important feature of Palestine in the Late Hellenistic through the Early Imperial periods.

Much of the material and archaeological development in the Hellenistic period in Palestine, in light of recent excavations, appears in the most northern part of the region (see map, Figure 34.1). In this chapter I will discuss some of the more significant Hellenistic and Hasmonean period excavations and discoveries over the last generation.

A Companion to the Archaeology of the Roman Republic, First Edition. Edited by Jane DeRose Evans.

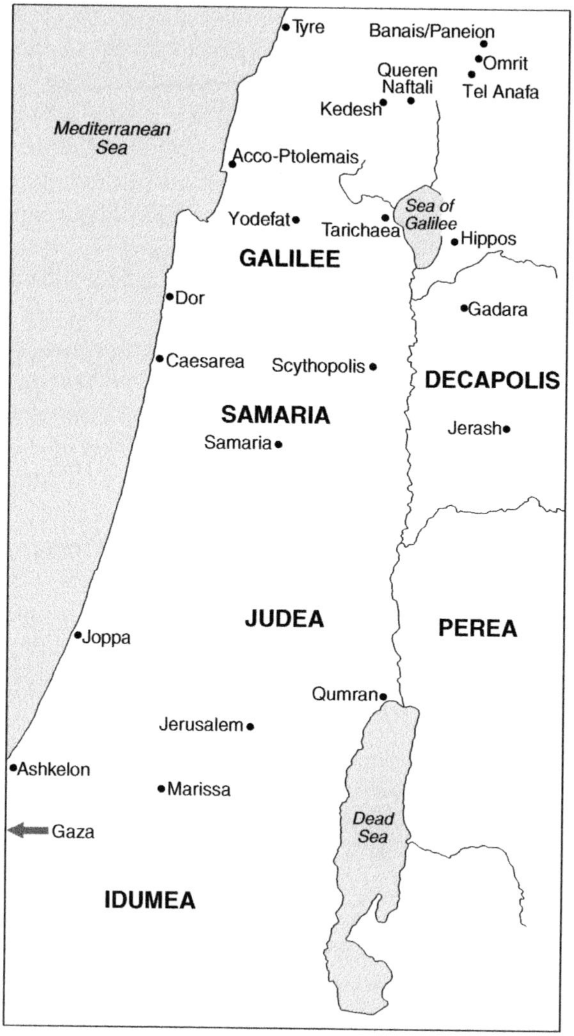

Figure 34.1 Map of Palestine, with sites mentioned in text. Source: Drawing by A. Yandek.

I will conclude by discussing the building and development that occurred under Herod, whose accomplishments bridge the Late Hellenistic and Early Imperial periods in Palestine.

2 The Late Hellenistic Period, *c*. 200–*c*. 140, in the North

In the year 198 the most accomplished of Seleucid leaders, Antiochus III, recaptured Syria and Palestine for the Seleucid Kingdom, driving their nemeses, the Ptolemies, back across Raphia and Gaza into Egypt. The beginning of our era is the decisive Battle of Paneion, which took place on the slopes of Mount Hermon, just south-west of the Cave of Paneion and a mere 50 km south-west of Damascus. The Battle of Paneion, settled by a famous clash of elephants and heavy-armed cavalry (Polyb. 16.18–19, quoting Zeno) with the superior Seleucid elephants taken from India carrying the day, settled the matter between the Ptolemies and the Seleucids, and it signaled a change in development and trade, and also rising tension in the region (J. Wilson, 2004). Some scholars argue that the Seleucids denied people and cities their freedom and adherence to their own customs, as there are mentions of revolts and rebellions throughout Antiochus's reign; in Palestine, it is said, Antiochus's victory led to a division between pro-Seleucid and pro-Ptolemaic factions (see Jerome's commentary on *Daniel* 11.14; contra, Gera, 1998). According to Josephus, Antiochus III was beneficent toward the local Jewish population. Compared to Ptolemaic policies, he provided some tax breaks. He declared freedom for the Jews and protection of their native traditions. For instance, they were allowed *isonomia*, or some degree of self-governance, according to their own traditions (Joseph. *AJ* 12.138–42). This rising tension is most likely a reflection of Seleucid rule in Palestine closer to the time of Polybius, Antiochus IV Epiphanes in particular, and does not resonate with recent discoveries pertaining to Seleucid management of Palestine in our period, as we shall see below.

The Seleucids, and especially Antiochus III, were builders and developers. Hellenistic foundations are a characteristic of their realm whether one looks at Bactria or Palestine from this period (Bilde, 1990; Walbank, 1992; Holt, 1999). We can see this feature of the Seleucid period emerging in the historical and archaeological record of our region. Yet archaeological finds in Palestine relative to the Seleucids have been relatively rare. Recent archaeological excavations go a long way toward filling out what heretofore has been a minimal amount of material remains from the Seleucid presence in Palestine.

The northern part of the region, from the port city of Acco-Ptolemais on the Mediterranean coast to Mount Hermon in the east and from the Bet Natopha Valley and the Sea of Galilee north to the Beqaa Valley, experienced a concentration of growth precisely during this period, as recent excavations have demonstrated. Acco-Ptolemais is a port city with a long history. Acco has suffered or enjoyed a crucial, strategic location throughout history. During the Hellenistic period, Ptolemaic and Seleucid powers vied for control of this city as they did in Tyre and Sidon to the north. This strategic site grew in importance as Hellenistic culture sank its roots in the city and environs (Freyne, 1980). The Acco mint reached its apex of production in the transition from Ptolemaic to Seleucid occupation, around 200. Extensive remains of Hellenistic temple and sacral artifacts have been uncovered in excavations in recent years. These include figurines, unguentaria, lamps, stamped amphora handles (found in the entire region), and votives of women, a dove and others. These finds – all in a Hellenistic context from Acco – clearly reflect temple and sacrificial activity from the Seleucid period in the city. These date from the third century through the second century (Dothan, 1976; Aviam, 2004). Many statue fragments have been found in this context, as well as fresco wall fragments, fluted pillars and faux architectural plaster fragments, all suggesting the growth of Hellenistic art, design and culture in the Acco region. Excavations continue today at Acco-Ptolemais under the direction of Pennsylvania State and Haifa Universities. Acco has also recently been declared a UNESCO preservation site.

A new excavation along the Tyre–Damascus corridor at the Hellenistic tel of Kedesh has revealed a large administrative center. The site sits atop a tel on the modern Israel–Lebanon border and is one of a string of settlements from the coastal city of Tyre to Damascus. Tel Kedesh features a Hellenistic period building measuring 56 by 40 m, which the excavators refer to as the Hellenistic Administrative Building. This was built on an earlier Persian-period structure, but was abandoned in the middle of the second century. The excavations, under the auspices of the Universities of Michigan and Minnesota, also uncovered an extraordinary, large hoard of over 2,000 stamped clay seal impressions in this Administrative Building. The images and inscriptions from these seal impressions provide rich new data on the religious and political milieu of the Upper Galilean and Mediterranean coastal region from the Hellenistic period (Herbert and Berlin, 2003; Ariel and Naveh, 2003). Not surprisingly, the seals demonstrate a close connection to the influential city of Tyre. Tyre is referred to often and dates are usually given according to the Tyrian calendar. Ariel and Naveh report that a Tanit symbol (a triangle forming the body or torso of a figure), typical of Phoenician culture, is found on at least nine seal impressions. One of those also contains a Phoenician inscription referring to "he who is over the land" – that is, a Phoenician official in the town. Another seal impression represents the city of Kedesh itself. It depicts grapes hanging from a vine, with the inscription – in Greek – "Kedesh." Most of the seal impressions reflect Greek culture, such as the sealing that depicts the head of the Greek goddess Tyche; it also bears a date, in Greek, of 167.

Buildings on the site reveal very well preserved plastered walls and floors, bossed, faux architectural styles characteristic of later Hellenistic structures in the region. The finds from Kedesh are beginning to shed new light on inter-regional relations, contacts between Phoenicians and Seleucids, and the interaction of both with the Jewish minority sandwiched in between these two substantial eastern Mediterranean powers.

A Hellenistic fortress east of Kedesh and still situated along the west–east line from Tyre to Damascus is Qeren Naphtali. A survey and small excavation has occurred there which also reflects the expansion and activities of the Hellenistic diadochs in the northern region of Palestine. Qeren Naphtali rises above the western slope of the Hulah Valley and holds a commanding view of the pass up Mount Hermon, and the road that went to Damascus. First surveyed in the 1880s by Guérin, at the beginning of the last century, Masterman conducted a survey from which he published two inscriptions from the site (Masterman, 1908). The longer of the two, carved on a lintel, was dedicated to Athena. The other, on a table, was carved into a table dedicated to Zeus Heliopolitanus.

Aviam surveyed more of the site in 1995. He found parallels at Qeren Naphtali to other known excavated sites in the region, such as Dor and Yodefat. The pottery collected was primarily from the Hellenistic period. This included so-called Galilean Coarse Ware (GCW) indigenous to the region, and Eastern Terra Sigillata (Aviam, 2004). From these diagnostic wares, it appears that the fortress at Qeren Naphtali was built in the early part of the second century. As in the case of Kedesh, just west of Qeren Naphtali, this site also became an important battle site between the Seleucids and Hasmonean forces as Seleucid influence and strength began to wane in the middle part of the second century (I *Macc.* 11:63–74; Joseph. *BJ* 4.105, *AJ* 13.154–62).

A recently excavated site along the same west–east corridor in Upper Galilee which dates from our period is Tel Anafa. Tel Anafa is situated in the Hulah Valley, along its eastern edge. It is positioned at the northern extent of the ancient north–south road which runs along the Jordan rift toward the Dead Sea, passes Jericho and then turns west and heads up to Jerusalem. This is the so-called Way of the Kings. According to its first

Figure 34.2 Aerial view of the excavations of the temple at Omrit. Source: Macalaster College Excavations, with permission of J. Andrew Overman.

excavator, Tel Anafa was a Seleucid village (Weinberg, 1971). Excavations by the University of Missouri began in 1968 for five seasons. A second series of excavations at Tel Anafa for five summer seasons between 1978 and 1986 were co-directed by the Kelsey Museum of the University of Michigan and the Museum of Art and Archaeology of the University of Missouri.

The most substantial find for our period from Tel Anafa is the Late Hellenistic Stucco Building (LHSB). The building has suites of rooms on three sides opening onto a central courtyard. The date for the LHSB is the last quarter of the second century. This was likely Tel Anafa's zenith as a Hellenistic settlement. The fresco and stucco from the site is impressive, as are the imported fine wares associated with the building. The majority of the coins are Seleucid, and they date from the last quarter of the third century to the middle of the second century (Herbert and Ariel, 1994). There are later remains from Tel Anafa, but by the end of the first century Tel Anafa was in substantial decline.

Around the time Tel Anafa was in decline another Late Hellenistic site, on the eastern edge of the Hulah Valley in the south-west foothills of Mount Hermon, began to develop. This site is Horvat Omrit, which probably began as a rural sacral and temple site that lasted through the early Byzantine period (Figure 34.2). The site has been excavated since 1999 by Macalester College and was joined in 2006 by Carthage College (Overman and Schowalter, 2011).

Figure 34.3 Detail of the excavations of the temple at Omrit, showing the small sacred site overlaid by the Imperial temple foundations. Source: Macalaster College Excavations, with permission of J. Andrew Overman.

In the last quarter of the second century a small, rural structure was built on the hill near the site of the Battle of Paneion, on a spot which appears to have been a cult site (Figure 34.3). This structure was subsequently buried by larger and later phases of a temple complex. But excavators were able to uncover a plaster surface in association with this earliest Hellenistic phase, along with parts of a wall that may have been part of a courtyard or rural shrine wall. Figurines and votive material was found in the makeup of the plaster in the floor and foundation of the wall. This second-century structure was superseded by a small, ornate mausoleum or heroon, which buried the earlier structure but preserved portions of the building.

What excavators at Omrit refer to as the Early Shrine measures approximately 8 by 6 m and is 12 m in height. The Shrine has ornate and well-preserved stucco and fresco. It rests on a podium 2 m in height. The podium is stuccoed all around and has an upper and lower course molding. The Shrine also had a finely frescoed temenos wall surrounding the entire structure. A second phase of the building features the addition of plastered steps to enter the building and the imposition of a door to enter the complex and Shrine itself. This second phase marks the transition of the Shrine to a proper temple architecturally speaking. This marks the establishment of a Hellenistic cult site in Upper Galilee during the waning years of Seleucid presence in northern Palestine.

From about the last quarter of the first century, the Early Shrine site transitioned into a far larger, more substantial Roman temple, coinciding with the establishment of the reign of Herod the Great in the region (Figure 34.4). From this period to the Flavian period, a temple in the form of a traditional Augusteum, or imperial cult building, stood at Omrit. The temple was prostyle tetrastyle in form and measured approximately 13 by 20 m and was 20 m high. At the end of the first century CE this Herodian-period temple was transformed into a peristyle temple. The steps were widened. Additional statuary was added, and the building was lengthened and widened by 3 m each direction (Overman, Olive and Nelson, 2007).

Another rural cult site was also established in this period by the headwaters of the Jordan River, at Paneion (Banias). In the Hellenistic period Paneion was a rural cult site dedicated to the god Pan (Berlin, 1999). This site developed into a significant polis only in the Imperial period.

Several striking features emerge in this review of recently excavated sites along the Tyre–Damascus corridor. A distinct line of Hellenistic sites can be traced along a trade route which led from the coast to Damascus and, presumably, beyond. These sites reflect the emergence of Hellenistic art and architecture in the region. The introduction of fresco and stucco walls, both domestic and public, is part of these developments. The excavation of these sites reveals also the introduction of Hellenistic and Greco-Roman style architectural elements. These architectural innovations are reflected also in the faux-architectural designs in the buildings and larger, more impressive houses and installations. This includes both Ionic and Corinthian capitals, ashlar construction, stucco imitating drafted blocks, and egg and dart molding, to name but a few. At Tel Anafa and

Figure 34.4 Reconstruction of the Early Imperial temple at Omrit. Source: Drawing by Michael Nelson, with permission.

Omrit, the Hellenistic buildings also contain gilded blocks and cornices, developments dating from the Ptolemaic–Seleucid transition in the region. The coins of Antiochus III are prominent at several of these sites. In all, material innovations date roughly from 200 and reach an apex around the last quarter of the first century. The end of the Hellenistic period in the region coincides with the beginning of the building projects of Herod (Netzer, 2006). He transformed much of the region. But we will address his work separately.

The sites along this important northern corridor of Palestine, skirting the northern edge of the Hulah Valley and the southern extent of the Beqaa Valley, provide insights into the political and administrative workings of the region, as in the case of Tel Kedesh. But we also see religious features of life in the area in several of the seal impressions from Kedesh, which feature Hellenistic deities and cults. Omrit and Banias/Paneion are also Hellenistic cult sites which develop into significant cities with long histories. Finds from these sites reflect a combination of influences and styles. Parallels for some of the Hellenistic buildings from these sites are difficult to find, as Herbert commented in her final report on Tel Anafa. The LHSB at Tel Anafa and the so-called Mausoleum or Early Shrine at Omrit both have a synthesis of elements that may be best explained, as Herbert suggested, as examples of broader Syrian–Seleucid architecture and style (Herbert and Ariel, 1994).

Finally, these sites introduce a level of style and finish that is impressive. The Hellenistic buildings mentioned here are ornate and striking in their design and beauty. An economic and artistic stratum is here introduced to the region that is rightfully considered a new element. How this reflects cultural and economic changes and currents is beyond the scope of this chapter. We do see herein, however, new emphasis and resources being devoted to a region that is fast becoming a strategic realm for competing powers.

3 The Hellenistic Period, *c.* 200–*c.* 140, in the South

The southern Hellenistic sites can be differentiated by their size and intensity of Hellenization. South from Omrit and Tel Anafa, a short distance down the Jordan River fault to the east coast of the Sea of Galilee, is the highly significant, recent excavation of a Decapolis city of the Hellenistic period, Hippos (Sussita), excavated by Haifa University since 1998 (Figure 34.5). Hippos appears on several ancient lists of the Decapolis League. It is close to the only Decapolis city west of the Jordan River, Nyssa-Scythopolis, as well as the large Decapolis centers of Gadara and Jerash (in modern-day Jordan). Hippos was inhabited in the Late Hellenistic through the Umayyad period. A modest (13 by 25 m) forum was uncovered early on in the excavations. It is thought to date to the first century CE, but it may reach back to the Late Hellenistic period. Adjacent to the forum, on the western or cliff edge, is a bathhouse and a monumental building, which was originally thought by the excavators to be a building associated with the imperial cult or at least have something to do with imperial honors.

What the excavators have termed a Hellenistic compound has now been fully exposed. The excavators suggest that there was a third century (Late Ptolemaic period) occupation at Hippos but that it was when the area was under Seleucid control that the city began to flourish. The so-called Hellenistic Compound, situated in the center of Hippos, began to be built in this later Hellenistic period. It consists of a flat paved area with a

Figure 34.5 Aerial view of the compound of Hippos, probably dating to the Late Hellenistic period. Source: Michael Eisenberg, with permission.

stepped podium in the center, likely for a temple building; this later formed part of a Roman Imperial temple. This prominent and impressive central architectural feature of Hippos dates from the last quarter of the second century. A cardo and decumanus maximus form the main outline of the Hellenistic Compound. Hellenistic Hippos appears to have had two main gates and a civic arch (Stern, 2008: 1782–7 [Epstein]).

The architectural remains and elements from the Hippos excavations are truly impressive and well-preserved. A large and seemingly affluent, well-funded Hellenistic city on the shores of the Sea of Galilee – now substantially excavated – must inform our assessment of the material culture of northern Hellenistic Palestine. Hellenistic cults such as that of Tyche are present at the site. And the Hellenistic Compound featured a major temple structure at its center. No deity has been associated with the temple to date (Segal *et al.*, 2004: 28–9). While Hippos evolved into a Christian center after the fourth century CE, boasting several churches with impressive mosaics, the Late Hellenistic period is precisely the period in which Hippos bursts onto the scene in the eastern Mediterranean world.

Crossing back across the Sea of Galilee the Decapolis city, Nysa-Scythopolis (Bet She'an), is one of the largest Greco-Roman cities excavated in the ancient world over the last 25 years. Excavated by the Israel Antiquities Authority and the Hebrew University, the excavations of the Hellenistic stratum at Scythopolis have revealed an extensive, well-planned city. The Hellenistic city is surrounded by a fortification wall. It had paved streets of 2–3 m wide, residential insulae and public buildings and structures (Stern, 2008: 1623–8 [Mazor]). A private archive of 120 bullae was also excavated. Rhodian stamped amphora handles are in abundance, along with large amounts of

Hellenistic pottery. The Hellenistic homes exhibit fresco walls and often imitate an ashlar masonry style. The excavators suggest that Scythopolis was first founded as a military post by Ptolemaic forces, *c.* 260. Other Decapolis sites close by were founded or expanded by the Ptolemies as part of their strategy in the region in precisely this period. But the city was multicultural, as Jews and Greeks lived in the city. Evidence for a cult of Zeus, Dionysus, Serapis and Tyche-Nysa have been found and these cults reflect the broad cultural and religious milieu of the Hellenistic city.

One of the most dramatic recent finds in Palestinian archaeology in recent years has been the excavation of Hellenistic Magdala (argued to be ancient Taricheae) on the north-west shore of the Sea of Galilee. These salvage excavations done over a few short seasons by the Israel Antiquities Authority and an Italian team headed by Stefano De Lucca have revealed a large port and massive pylons sunk into the lake (De Lucca, 2009). The dock was made of huge cut stone masonry and the main road through the small city led directly to the port. Underwater surveys have revealed large submerged pillars, ashlars and Corinthian capitals at the port of Taricheae. Numerous other impressive and finely constructed structures were exposed during these short seasons. Liebner is correct in observing that this port site on the Sea of Galilee was undoubtedly a significant Hellenistic and Early Imperial city (Leibner, 2009).

Taricheae is known to us through Roman sources that depict one of the seminal catastrophes of the Late Republican period – the loss of the Roman standards to the Parthians by Crassus at the Battle of Carrhae in 53. Crassus was killed, along with most of his troops; the rest were taken captive. Those who survived, including Cassius, took refuge in the city of Carrhae and later withdrew to Taricheae, subduing it and resting his troops (Joseph. *BJ* 1.180; *AJ* 14.120). Taricheae, since it was a hellenized city, would have felt more familiar to Roman officials and administrators than surrounding cities. The region east in the Hauran plateau and north of Mount Hermon would have been more pro-Parthian in sentiment and not secure enough for the already depleted spirits of Cassius's troops. A few years after Cassius's departure, Herod would undertake the task of expunging the pro-Parthian groups from the region.

The city of Jerusalem experienced a significant transformation into a true polis in the Hellenistic period. The Seleucids did construct an *akra* or defensive tower in the city from the time of Antiochus IV. Tsafrir has suggested a location for that *akra* as a result of excavations in a part of the Old City a generation ago (Tsafrir, 1975). It is also stated by Josephus (*AJ* 12.141) that the Seleucids provided funding for the rebuilding of the Temple in Jerusalem during the reign of Antiochus III.

One of the most important Hellenistic sites in Palestine excavated over the last generation is the site of Marissa (Maresha). South of modern Ashkelon in what is considered the eastern Shephelah, Marissa was a substantial Hellenistic site where excavations have been ongoing for about 25 years. The site has an upper and lower city. It possesses a Hippodamian plan and a large temple complex dating from the early second century (Arav, 1989). It was known for pigeons and for the production of olive oil, as any visit to Marissa reveals. Large, numerous dovecotes are on the site and in caves in the very soft limestone characteristic of the region (Kloner, 2003). Some of these working subterranean chambers were converted from cisterns. The Hellenistic paintings and tombs of Marissa are spectacular and one of the most famous features of the site (Jacobson, 2005). The paintings feature faux architectural styles, gabled burial façades, and ornate, very colorful representational designs.

The site was doubtless involved in the protracted Syrian Wars between the Ptolemies and Seleucids. As a result of Marissa's proximity to the Mediterranean and the so-called Phoenician plain, it was involved in a string of battles, and changed hands often. Marissa is mentioned in the Zenon Papyri as an important trading site as early as *c.* 259. Marissa is close to the main route between Gaza and other strategic sites along the coast. The troop movements of especially the Fourth Syrian War were focused on this part of ancient Idumea. Antiochus III concentrated on controlling this part of the Phoenician plain against his Ptolemaic rivals (Gera, 2009; Grainger, 2010). The Seleucids did secure the region briefly, though they lost to the Ptolemies further south, in Raphia.

Of particular interest to the Hellenistic period in Palestine has been the discovery of an inscription from Marissa, which provides much-needed information about Seleucid rule and administration of the area. The inscription, written in Greek, is well preserved but missing the last portion, at the bottom of the stele (Cotton and Wörrle, 2007). It is a letter from Seleukos to one Heliodoros, with two other cover letters, and can be dated to *c.* 178. The main portion of the inscription reads:

> King Seleukos to Heliodoros his brother greetings. Taking the utmost consideration for the safety of our subjects, and thinking it to be of the greatest good for the affairs in our realm when those living in our kingdom manage their lives without fear, and at the same time realising that nothing can enjoy its fitting prosperity without the good will of the gods, from the outset we have made it our concern to ensure that the sanctuaries founded in the other satrapies receive the traditional honours with the care befitting them. But since the affairs in Koilē Syria and Phoinikē stand in need of appointing someone to take care of these (i.e. sanctuaries) … Olympiodoros …
>
> (trans. Cotton and Wörrle, 2007: 193)

Amazingly, the remaining portion of this important inscription, in three pieces, was uncovered in excavations recently at Marissa, in subterranean complex 57. The pieces have been confirmed through laboratory tests to belong to the original stele published by Cotton and Wörrle; they have recently been published by Gera (2009). Some corrections and emendations to Gera's reconstruction of the inscription have been further supplied by Jones (2009). The inscription continues:

> … Olympiodoros will preside prudently over the orderly conduct of associations, [he, who] has demonstrated his loyalty to us because of his attitude, from times gone by, as he had been raised <with> us and had gained for himself the best disposition in all things, he was, on the one hand, appointed chamberlain with reason, because he has proven himself worthy due to his long-standing loyalty, while on the other, he was justifiably introduced into (the ranks of the) First Friends because of his love for us, having made the most assiduous demonstrations (of loyalty) while in such a [rank], and following the way that we are inclined to [increase] the honour of the gods in accordance with the [attitude] of our ancestors and the care of the temples in which we believed from [times gone by] …
>
> (trans. Gera, 2009: 126)

This important inscription provides an insight into Seleucid administration of sacred precincts in the Seleucid realm. We see a concern in the Seleucid kingdom at the highest level that the sanctuaries in the entire realm be properly maintained. This will help foster the good will of the gods and ensure order in the kingdom so that subjects may manage their lives "without fear." The inscription is explicit that Seleucid policy concerning temples has not been sufficiently managed in at least part of Palestine–Coele Syria and Phoinike. Seleukos's informing Heliodorus of the appointment of Olympiodoros to the role of overseer of these affairs in broader Syria, including Palestine, will remedy this problem. The Heliodoros mentioned in the first line may very well be the Heliodoros mentioned in II *Maccabees* 3 at an earlier stage of his career. Olympiodoros's qualifications for this role are listed in the royal letter. His loyalty to Seleukos himself is stressed. He is well-known to Seleukos, even perhaps having been raised with him. He was appointed chamberlain, and awarded the illustrious rank of "First Friend" within the Seleucid court.

The relationship between Seleucid royal ideology and administration and the Temple in Jerusalem has been a subject of long-standing fascination. These two relatively recent epigraphical discoveries have helped to fuel that enduring interest (Bikerman, 1946–8; Habicht, 1976; Bencivenni, 2011). The story from II *Maccabees* also has to do with the maintenance of the sanctity of temples – in this case the Temple in Jerusalem. In this piece of Hasmonean propaganda Heliodoros emerges as a pious authority who, in the end, gladly honors the god of the Jerusalem Temple (cf. II *Macc.* 3.35ff.).

This recent discovery complements the uncovering of another rare Seleucid inscription from Palestine found 7 km from Scythopolis (Bet She'an) at Hefzibah in the mid-1960s (Landau, 1966). This inscription, actually a dossier containing eight documents from an exchange between Antiochus III and Ptolemaios, while not concerned with temple and sacral maintenance expressly, is concerned with order and propriety in Seleucid Palestine. The inscription is earlier than the Marissa inscription and dates to the beginning of the Fifth Syrian War (*c.* 201). An authoritative edition of the inscription, *SEG* 29, 1613, is based on a revised text by Fischer (Fischer, 1979; Cotton and Wörrle, 2007: 194). What is significant here is that the Ptolemaios in the inscription, and the person who established contact with Antiochus III, earlier fought in the Fourth Syrian War for Egypt but came over to the Seleucid side. He stayed in Palestine and maintained lands in the Jezreel Valley, not far from Scythopolis. He has the titles of "Strategos and Chief Priest." In the ensuing years his villages and land had suffered violence due to the presence of Seleucid troops. This affected the raising of crops and trade among neighboring villages and it brought unlawful billeting of troops on his land. In the royal dossier Antiochus orders that a list of saleable and lawful items to be traded be made public and the *oikonomos* of the region be in charge of this and regulate fair commerce. Also, should unjust billeting or damage to his property continue, damages must be repaid tenfold to Ptolemaios. We find in the inscription that this is Ptolemaios's "hereditary land." Thus, according to the inscription, Antiochus follows the requests of Ptolemaios himself. Antiochus honored Ptolemiaos's right to the land and its produce, and drafted regulations in response to Ptolemaios's request of him.

These two finds over roughly the last generation, and the scholarly discussion around them, have shed additional light on the Seleucid presence and administration of Palestine in the Hellenistic period. This is a subject where our historical knowledge has been wanting. What emerges from these inscriptions is an outline of Seleucid policy and their general approach to maintaining order in parts of the realm, including Palestine. Seleucid rulers and their local administrators did this in part by honoring local gods and sacred precincts or temples. The risk inherent in failing to do this is something taken quite seriously at the highest levels of Seleucid imperial management. Thanks to Ma's recent work, we see that these policies that were imposed on Palestine are common to all other Seleucid holdings in Coele Syria and Asia Minor (Ma, 2002). This new information does suggest how aberrant the presentation of what is perhaps the most famous Seleucid event in Palestine, the Maccabean Revolt of 165, truly was. The portrayal of a Seleucid ruler failing to place the highest priority on civil order and respect for temples in particular does not match well with this new archaeological data from Hellenistic Palestine.

4 The Hasmonean Period, *c.* 165–*c.* 40

By the last third of the second century the expansionistic Hasmonean leader John Hyrcanus sought to expand his Jewish rule by engaging the now-weakening Seleucid forces around Marissa, as well as other locales. Following the establishment of the rule of the Hasmonean house in 165 and the beginning of relative political independence in the wake of the Maccabean Revolt, a new phase in the archaeology of Palestine begins. Hasmonean expansion and building emerges with some distinction by the middle part of the second century. This stands in contrast to Seleucid foundations which left few obvious archaeological footprints.

Hyrcanus had captured Marissa; he forced the Idumean inhabitants of the city to be circumcised and to "observe the laws of the Jews" (Joseph. *AJ* 13.257–8). But Antiochus VII Sidetes forced Hyrcanus to sue for peace and take a less than optimal settlement, with the result that Marissa – and other captured Idumean cities – were freed of Hasmonean control in *c.* 132. Hyrcanus then, as promised, supported Antiochus VII in his campaign against the Parthians (Joseph. *AJ* 13.236ff.). Following the complete collapse of the Seleucid force in Palestine, Marissa was retaken again by the Hasmoneans, in 112. Pompey granted Marissa independence following his annexation of the country in 63.

We are told that Marissa was rebuilt by Gabinius (the Roman proconsul of Syria, called in by Hyrcanus) in *c.* 55, following the cessation of the Hasmonean wars and civil unrest (Joseph. *AJ* 14.88). Gabinius was a major builder and architectural force in Palestine in the middle of the first century. Had he not been followed so closely by perhaps the greatest ever builder in the region, Herod, much more would be known about Gabinius's building prowess. He reclaimed and rebuilt many cities following the destruction that occurred during Hasmonean decline and civil war, including many of the Decapolis cities (cf. Joseph. *AJ* 14.75). Marissa did come under Parthian attack once more, in 40, because the Parthians suspected Herod of having hidden most of his treasures in that city (Joseph. *AJ* 14.365). After this, Marissa was replaced and renamed Bet Guvrim and became, during the Imperial period, the major city of this region.

Hasmonean royalty assumed the traditional oasis of Jericho as a locale for a palace and a winter residence, like other rulers and elites before them. The Hasmonean Palace at Jericho was excavated by the Hebrew University. It was founded most likely during the long reign of Hyrcanus (134–104). It is situated north of the defining topographical feature of the site, the Wadi Qelt, and is comprised of a central building and two adjacent small swimming pools. The central building was decorated with fresco walls in a style common in the later Hellenistic period (Netzer, 1975, 2001; Rosenberg, 2008). The excavators suggest a development and expansion of the site during the subsequent reign of the Hasmonean leader Alexander Jannaeus (103–76). This expansion included two larger swimming pools east of the central building, and a Doric pavilion surrounded by paved areas and gardens.

While Jannaeus did build in Jericho, it should not be overlooked that he was also characterized as a "destroyer of Greek cities" during his long but tumultuous reign, according to Josephus (*AJ* 13.395ff.). His apparent opposition to Hellenistic influences is well recorded. Josephus details his reducing of Gaza, Raphia and others (*BJ* 1.87). The Roman legate Gabinius rebuilt many of the cities razed or partially destroyed by Jannaeus.

Toward the end of his reign Jannaeus constructed a fortified palace atop a mound in Jericho. This may certainly be an indication of the civil struggles associated with the end of his career. Jannaeus's widow, Alexandra, ruled following his death, from 76 to 67. She added two "mansions" or adjacent twin palaces, and two more pools and other gardens south of the central complex. The buildings from this era are well preserved and have also been excavated by Netzer.

Hasmonean fortunes and strength declined greatly through the middle part of the first century. Those that were left of the Hasmonean line were now mere puppets of Roman administrators. By the mid-50s Pompey, and more directly Gabinius, directed Hasmonean activities. They did allow Jericho and its environs to remain under Hasmonean control. This began to change in late 40s and early 30s when a struggle between the Hasmoneans and the young Herod began to consume the area. There are mosaics and wall paintings from this period from a bathhouse but the structure is precisely from this transitional time and may be either Hasmonean or Herodian. Fine and dramatic fresco decorations are emblematic of buildings from both periods. Jericho may have remained in Hasmonean hands as late as 31. Shortly after 31 some of the structures at Jericho were damaged by an earthquake, which affected many sites across the region (Deg, 1990).

One of the best-known and most dramatic archaeological discoveries from the second and first centuries in Palestine is the Dead Sea Scrolls. Wadi Qumran, where the scrolls were discovered in 1947 by a Bedouin shepherd boy, lies along the Dead Sea, the lowest point on earth (400 m below sea level), south of Jericho and south-east of Jerusalem. Excavations at the site began in the early 1950s under the direction of Roland De Vaux of the École Biblique in Jerusalem and G.L. Harding, the Chief Inspector of Antiquities in Jordan. Subsequent to these excavations, surveys of the area have located 11 caves containing over 900 fragmentary scrolls. Most are in Hebrew; roughly 20% are in Aramaic and a few are in Greek. They contain biblical material (such as a complete scroll of the Prophet Isaiah), extra-biblical material or retelling of biblical stories. There are hymns, messianic materials, and a large number of laws, codes and rules for the community's life together. Some of the rules include the details of a long apprentice period before being allowed to join the community; harsh penalties (including expulsion) for breaking the community rules; and detailed purity and halakic commands and procedures (Schiffman, 1975).

The group associated with the scrolls, possibly the group that authored them, emerges as an apocalyptic group. They believe they live in the end times. They are a persecuted minority. They are completely alienated from the main cult and the Temple in Jerusalem; they view the priests in Jerusalem who control the Temple as corrupt and perverse. God will act soon to judge those wicked priests. The Qumran community will be vindicated and the true sons of Zadok will be restored. The language of dualism, light and darkness, truth and falsehood, sinful and righteous, punctuate many of the documents. The community has detailed regulations and strict halakic or legal interpretations. The Hebrew Bible is important to the group but usually only in so far as texts can be related specifically to the Qumran community. This mode of interpretation within the scribal community has been referred to as *pesher*. By copying and preserving the texts at Qumran, the group sees its own history and articulates its divinely guided future. The scrolls' dates cover a period from *c.* 250 to 25 (Collins, 1992; Magness, 2002).

Archaeologically, the ruins on the site of Qumran date from the last third of the second century to its destruction by the Romans during the First Jewish Revolt in 68 CE. It is likely that the founding of the Qumran community dates to the time of John Hyrcanus in the period of Hasmonean expansion. These Hasmonean rulers and administrators, who had consolidated both political and religious power in one ruler, may well have been the object of the vitriol evident in the scrolls (Eshel, 2008).

The archaeological remains from Qumran in the Hellenistic period now boast a kitchen, a so-called scriptorium, cisterns, an assembly room, *miqvaot* (ritual baths) and more. Khirbet Qumran developed into a substantial Hasmonean-period complex over the second and first centuries. But the interpretation of the archaeological finds at Qumran remains greatly contested. The association of the site with the scrolls themselves also remains a complicating and debated feature of work at Qumran and itself a contested thesis (Davies, 1983; Magness, 2002).

Hasmonean expansion, spurred no doubt in part by Hellenistic influences recorded with mixed emotions in the Maccabean literature, developed Jerusalem along the lines of a traditional polis. The old upper and lower parts of Jerusalem were joined during the Hasmonean period by a massive bridge spanning the Tyropoeon Valley. Remains of that bridge have been discovered along the Western Wall (the "Wailing Wall") of the Temple. An arch from that bridge is intact and still *in situ*. The Hasmoneans added a defensive structure of their own along the north side of the Temple Mount which was later taken over and expanded by the Romans during their occupation of the city (Avigad, 1983). Massive Hasmonean period ashlars have been exposed at the very foundation of the Temple Mount beneath the present-day Western Wall. The size and sophistication of these late second-century cut stones hint at the magnitude of some of the Hasmonean building projects which, on the whole, were dismantled by Herod in his subsequent pervasive building projects.

Other Hasmonean period building projects point to an increasing sense of nationalism. Perhaps the most unusual find at Taricheae has been the discovery of an early synagogue dating to the middle of the first century, in excavations conducted by Dina Avshalom-Gorni and Arfan Najar of the Israel Antiquities Authority (Zangenberg, 2010: 476–7). This synagogue, located in the center of Taricheae, is ornate, prominent and one of the earliest synagogues found to date in the region. It is a square building with benches around the walls. A path around the benches has a mosaic floor and its walls and pillars are covered in fresco. The use of the Masonry Style walls and ornate houses, temples and public buildings is beginning to emerge as a characteristic of the Late Hellenistic, Hasmonean and Herodian sites throughout the region. For instance, we also see this Hellenistic architectural convention at Tel Anafa, Kedesh and Omrit.

The synagogue had a rectangular stone table, with four legs, decorated on all four sides and on top. The table was found in the center of the floor of the synagogue. One of the sides of the table depicts an eight-branch menorah flanked by two vases and framed by two pillars and an arch, in low relief. It has been suggested that the panel actually depicts the façade of the Temple in Jerusalem. Other synagogues from a similar period have been excavated at Gamla, in the Golan Heights, and Masada, further south near the Dead Sea. Both of these small and relatively simple synagogues date to the Herodian period. At Taricheae we have an ornate, refined synagogue that pre-dates both of these structures. Netzer has suggested that there is a synagogue at Jericho that belongs to the Hasmonean period (Netzer, 2001). If so, this would be the earliest synagogue excavated in ancient Palestine. But we have in Taricheae a significant Jewish civic space in the middle of a flourishing Hellenistic center. This is a phenomenon that just a few years ago would have been associated only with the diaspora communities.

Scythopolis figured in the Hasmonean wars of expansion in the later second century (I *Macc.* 12:40ff.). According to Josephus, the city surrendered to the Hasmoneans at the end of that century (*AJ* 12.280). A new period of revival and growth at Scythopolis occurred at the very end of the Hellenistic period when Pompey and Gabinius put an end to Hasmonean rule and restructured Syria and the east for Rome. Gabinius built up the city and made Scythopolis one of the five *Synedria* of the country (Mazor and Najjar, 2007).

Qeren Naphtali became a Hasmonean site following the defeat of the Seleucid forces. A *miqva* (ritual bath), excavated in the 2000 season, dates to no later than mid-first century, since the pottery from the fill dated to the third and second centuries. The *miqva* has parallels to those *miqvaot* excavated in Jericho which date to the Late Hellenistic period (Aviam, 2004). It is likely that Qeren Naphtali was one of the fortresses conquered by Herod when he first took over Upper Galilee prior to the battle of Actium (Joseph. *BJ* 1.238–9).

5 Herod, *c.* 40–4

The Republican period in Palestine draws to a close with one of history's most accomplished and productive builders. Herod, the great builder, transformed the landscape of Palestine in ways that are still easily understandable to scholars and visitors to the land. Herod's own career begins at the close of the Hasmonean Period. He emerges as a political and military personality to be reckoned with during the Late Republic, especially during the civil war that closed the period. Herod's political, military and diplomatic career is spectacular in its own right, but cannot be rehearsed in detail here.

He came from a family with deep roots in regional politics and intrigue. His father, Antipater, was well-known and used by Roman, Parthian and Nabatean leaders as a regional diplomat. Herod himself, through his father but also through relations with the Hasmonean family, was known to leading Romans at an early age. In 47, when he was only 26 years old, he was appointed governor of Galilee; the next year he became governor of Coele Syria. He was reappointed to the same post by Cassius in 42. During this period he raised troops, put down unrest, and guarded the region from foreign influences, especially Parthian. He supported Julius Caesar in 45 by sending troops to Apamea. Herod aligned himself with Mark Antony after they met face-to-face in Bithynia in 41, and continued to support him throughout the civil war. The following year, Herod helped resist a Parthian invasion of Palestine and simultaneously faced a revolt against his governorship. He fled the country, making his way eventually to Rome. While in Rome he was named "King of the Jews and Friend of the Romans" by the Senate; he was also appointed King in Galilee and Perea.

In the year 39, Herod returned to Palestine and fought his way from Acco-Ptolemais to Joppa and then Jerusalem. He resettled in Galilee and finally took Jerusalem from his rival Antigonos in 37. But Herod lost his patron, Mark Antony, after the Battle of Actium. In one of the Hellenistic world's greatest moments of political brinksmanship, Herod sailed to Rhodes and confronted Octavian directly (Joseph. *BJ* 1.388ff.; *AJ* 15.188ff.). He knew Octavian well by this point. The three men had been colleagues and friends in Rome during Herod's visit to the city earlier in the decade. Herod should have surrendered or committed suicide. Instead he made no apologies for his devotion to Antony. But he made the point that he would be every bit as loyal and ferocious a client for Octavian as he was for Antony. In a famous phrase he said to Octavian, "ask not *whose* friend I was, but *what kind of friend* I was" (Joseph. *BJ* 1.388–92). Herod retained his status as Rome's chief client lord in the region and remained "King of the Jews" until his death in 4 (Richardson, 1996).

During his time as governor of Galilee and then governor of Coele Syria (in the mid-40s) and extending to the time when he was named "King of Jews" by the Roman Senate, Herod was busy building cities, structures and sites that forever changed the landscape of Palestine. These projects announced both his building and engineering prowess and his intimate *philia* with Rome and the emperor Augustus in particular. Roughly 15 years of control, development and building activity can be attributed to Herod before he aligned himself with the most powerful patron in the world following the Battle of Actium. Herod's building projects changed dramatically in scope and size, and in style, after he became Augustus's chief client in the east. These later undertakings by Herod are today easily accessible in light of recent archaeological research in Palestine. The resources at Herod's disposal to pursue these projects are difficult to calculate (see Evans, 2006). But it is also the case that Augustus himself had a keen watch over the east and the enduring Parthian threat. It was in Augustus's best interest to maintain a strong and impressive presence in the east. That is where Herod comes in; many of these impressive Herodian building projects played a role in Augustus's foreign policy concerning the east (Millar, 1993; Overman, 2009).

Herod's earliest building projects and monuments may have actually been outside of Palestine, for on Rhodes, he restored the Pythian Temple of Apollo beginning *c.* 40. He already provided funds for buildings and cities in the diaspora at this time as well. By the early to late 30s, Herod was busy making his mark across his own kingdom. In Jerusalem he built the Antonia Fortress (named after his then patron), together with porticoes connecting it to the Temple platform. In Jerusalem and its environs he built many new water systems, reservoirs, aqueducts and pools. We are told he built a theater, amphitheater and hippodrome in Jerusalem all prior to Actium; while it is almost certainly true, these projects have not been identified as yet in surveys or excavations (Joseph. *AJ* 15.8.1, 17.10.2; *BJ* 2.3.1). Herod also claimed and rebuilt the Hasmonean Palace in Jerusalem around this time. In Jericho he reconstructed and greatly expanded the Hasmonean winter palace, which served as the first of Herod's own winter palaces (Netzer, 2006). In Alexandreion, Jericho, Cypros (north of Jericho) and Hycania he built, rebuilt and fortified Hasmonean foundations and palaces in the early and mid-30s (Netzer, 2001). Horvat Omrit, in the very north of Herod's kingdom, most likely has a pre-Actium Herodian stage to the many-phased temple complex at that site. The so-called Early Shrine most likely belongs to a pre-Actium stage and may owe its founding to the period when Herod was governor of the region in the mid-40s of the first century (Overman and Schowalter, 2011). There he built another cult building in Palestine honoring his patron Augustus. He expanded the complex, buried (and thus preserved) much of the earlier Shrine, and emulated Roman imperial temples from other regions of the Empire (Overman, Olive and Nelson, 2007).

Of particular interest to archaeologists of the Herodian period has been the desert fortress of Masada. Excavations began there under Yigael Yadin in the early 1960s. This fortress also was a Hasmonean site which was fortified by Herod in the mid-30s. For the next decade he most likely added the synagogue, fortifications and casement walls, and the impressive western Palace, other palaces and the northern villa. Masada is one of the first locations where we begin to see Herod's mastery of new Roman building techniques, and the art and decoration associated with the Late Hellenistic and Herodian phases in Palestine (Foerster, 1995). Although Masada was besieged in 70 CE by the Roman Tenth Legion and mostly destroyed, the Herodian buildings, art and architecture are still on display atop the site.

After Actium and the cementing of his relationship with Augustus, Herod's building projects gained in number and in sheer size and increasingly reflected his Hellenized taste. Most of these memorable sites were intended to secure Herod's position in Palestine in light of his competition, but they also demonstrated the close relationship between this eastern client and his lord, the emperor. Samaria (which he renamed Sebaste in honor of Augustus) housed the first imperial cult building in Palestine. Sebaste had a forum, a colonnaded street and a stadium. Excavations were conducted on the site toward the end of the British Mandate (Crowfoot, Kenyon and Sukenik, 1942), although they have not continued there since the 1960s.

Perhaps the most famous of the Herodian projects is Caesarea (which modern scholars have dubbed Maritima to differentiate from all the other Caesareas in the east). Here Herod developed a new harbor to attract trade and funds to his kingdom; he hoped it would rival the historic and dominant port to the north, Tyre. The engineering, excavation and building of the new harbor remains a technological feat of extraordinary proportion (Hohlfelder, 1983). Here too Herod built an imperial cult building to honor the emperor; this was the Temple to Roma and Augustus. Today, one can see there a restored theater, and remains of the hippodrome (Porath, 1995; Patrich, 2001, contra, Porath, 2003), Promontory Palace (Burrell, 1993) and water systems all commissioned by Herod. Excavations recently have been carried on by the University of Maryland and Haifa University (Raban and Holum, 1996; Holum, Raban and Patrich, 1999). Caesarea Maritima was probably constructed in the mid- to late 20s and thus was his second regal city in Palestine. Herod constructed another desert palace outside of modern-day Bethlehem called Herodium.

Today, after years of excavation by Netzer and the Hebrew University, one can see among Herod's projects a cylindrical tower-fortress, synagogue and small forum, built with beautiful masonry and decorated with frescoes, all on top of the artificial mound built by Herod (Netzer, 1981). At the bottom of the hill are a large swimming pool, stoa-colonnade, and baths with mosaic floors. In 2007, a small, beautiful and well-preserved theater or odeon was uncovered by the team in the massive fill placed against the fortress walls. Next to it is what has now been widely, though by no means unanimously, accepted as Herod's own mausoleum (Netzer *et al.*, 2010). His burial place originally consisted of a small, ornate structure on top of a Roman-style podium, with finely cut ashlars, using no mortar, with pristine and stunning faux-architectural stucco – a characteristic of Herod's work.

Further north in Galilee, the city of Sepphoris (Zippori) has been excavated since the mid-1980s by the University of South Florida, Duke University and the Hebrew University. Previously, one long season of excavations had been carried out in the mid-1930s by L. Waterman of the University of Michigan. While Sepphoris clearly has Herodian foundations, excavators are not sure which prominent architectural features and buildings specifically belong to this phase, or to subsequent members of the Herodian family (Strange, 1992; Meyers, 1999). A theater has been excavated at Sepphoris which may belong to Herod, his son Antipas, or most likely both, since it appears to have two distinct phases – a common feature of theaters in the region. The earlier Herodian phases, such as they are, have been obviated by later larger buildings, spectacular mosaics from the Imperial period, and later Byzantine structures.

Herod perhaps made his biggest mark in Jerusalem itself. His greatest project there was the restoration and expansion of the Temple in Jerusalem. To this day Herod's work and the magnitude of this undertaking are easily apparent. The so-called Western Wall, the holiest site in modern Judaism, is the only remaining portion of Herod's Temple. In the mid-20s, Herod undertook the massive project of expanding the Temple Mount to *c.* 200 by 150 m. He then rebuilt the sacred precinct, and added a temenos and courtyards. He remodeled porticoes, built a huge bridge to the Upper City, added several new gates, and adorned the entire temple sanctuary with ashlars with drafted margins, stucco, and ornate frieze, cornice and other architectural elements. Excavations have continued in and around the Temple Mount over the last generation, directed largely by the Israel Antiquities Authority, the Hebrew University and others. Herod's Temple is certainly exceptional among the great temples and monumental buildings of Hellenistic and Roman periods (Jacobson, 2007).

Herod's building activities continued in many areas around Palestine and he continued his building and largesse outside of Palestine until his death in 4. His sons Antipas and Philip attempted to continue the development initiated by their father, but with far less success and acclaim. Herod's building across Palestine has provided, and will continue to provide, challenges for archaeologists in the region for a very long time.

FURTHER READING

For overviews of the historical period, besides the sources mentioned above, see Applebaum (1989); and for an overview of the archaeology of the area, see Chancey and Porter (2009). For questions concerning particular periods, see Habicht (1990), Sievers (1990) and Berlin (1997). For Roman temples in Palestine, see Ovadiah and Turnheim (2011); for the development of city centers, see Jones (1931), Arav (1989) and Regev (2004). For specific sites, see Holum *et al.* (1988) (Caesarea), Vann (1992) (Caesarea), Bahat (1990) (Jerusalem), Weiss and Netzer (1996) and Eshel (2009) (Masada). Among many new studies of Herod's building program, see Fittschen (1996), Roller (1998) and Japp (2000). Specialist studies can be found in Adan-Bayewitz (1993) (pottery), Rosenberger (1975) and Evans (2006) (coins).

CHAPTER 35

Greece and the Roman Republic: Athens and Corinth from the Late Third Century to the Augustan Era

Michael C. Hoff

1 Introduction

During the timeframe of the Republic, as Rome began the climb towards world domination, the cities of ancient Greece were falling precipitously from the pinnacle of political and military hegemony. The free city-states of Achaia, whose military once felled the armies of the Persian Empire, could no longer field forces to defend even their own territories. Over the centuries Greece had become weakened. The Macedonians, from Philip II to Philip V, capitalized on Greece's enfeebled state to keep the once free cities in political check. What causes brought about the decline of Greece during the post-Classical period? There are in fact many factors, at local and regional levels, one can point to as contributory. Prominent among all is loss of population, brought about by warfare, disease and shifting populations.

But is it a fair assessment that a weakened and depopulated Greece resulted in irrelevance in the larger Hellenistic world, as is often suggested? Certainly power and wealth within the Greek world had shifted from the ancient city-states of Greece proper to the new and vibrant urban landscapes of Asia, Macedonia and Egypt. One is, in fact, hard pressed to find a city in mainland Greece whose political situation within the overall Greek world could be considered as significant. In a list of major Greek cities, Athens, Corinth and Sparta are no longer mentioned in the same breath as Alexandria, Antioch and Pergamon – to name a few. Yet even as political and military significance shifted away from Greece towards the east, "Old" Greece retained its prestige and to some extent influence over the nascent Hellenistic powers. One only has to note how transformed the urban landscape of Athens became under the patronage of Hellenistic monarchs to recognize the rewards prestige brought to the city. Greece continued in its role as an

A Companion to the Archaeology of the Roman Republic, First Edition. Edited by Jane DeRose Evans.

intellectual center of the Greek world; the philosophical schools at Athens so attest almost to the very end of antiquity. And in terms of art and culture, the cities of Greece extolled their own past and rich traditions to maintain their significance within the broad spectrum of Hellenism.

In the course of events Rome "inherited" a politically weakened, but still culturally vibrant Greece. Although "old" Greece – and in particular Athens – benefited during the Hellenistic period from patronage by the "new" Hellenistic cities, Rome's overall effect on Greek culture has been described as negligible (Jones, 1963: 3; cf. Alcock, 1993: 1–3), an attitude that adequately sums up scholarly opinion on the subject. But new research over the last 20 years indicates that Roman presence in the Greek east had a more significant impact than had been previously considered. In addition, scholarship has generally considered at face value the paradox referred to by the oft-quoted line of Horace: "Graecia capta ferum victorem cepit et artis intulit agresti Latio" ("Captive Greece took her savage victor captive and brought arts into rustic Latium"; *Ep.* 2.1.156). Yet a recent trend towards more balanced and nuanced attitudes by scholars is bringing into greater clarity the complex relationship between Rome and Greece (Alcock, 1993).

Cities and sanctuaries have always received greater attention than the countryside. Traditional archaeology concerns excavated material remains and naturally is predisposed towards urban and cultic sites that are rich in artifacts. Yet increasingly regional surveys are employed to examine non-urban landscapes (see Chapter 12). Greece has been the focus of intense archaeological interest over the past half-century, particularly as its landscapes serve as among the earliest laboratories for regional field surveys. Because of its wealth of sites, material assemblages, and relatively open spaces available for survey, Greece is often held as a model for the benefits that regional survey can bring to archaeological research (Alcock, 1993). The strength of regional survey lies in its ability to gauge diachronic transformation in the long term. Although this "macro-archaeological" approach might not be greatly fruitful in the search for nuanced change, which is the strength of traditional "micro-archaeological" excavations, its benefits in analyzing data within a broad landscape make it an ideal barometer for revealing diachronic trends in agriculture and demographics.

Survey projects in Greece have tested a wide variety of landscapes and the published reports largely verify the loss in population from the Hellenistic period through the early Roman era. But contrary to ancient literary laments of a depressed Greece, survey reports indicate that in the Late Classical and Hellenistic periods, as a result of population decline, agricultural land that was once held by small family units falls into the hands of elites who amass these holdings into larger estates. The quantity of agricultural product remains virtually unchanged. These same population trends continue into the Roman period. What cannot yet be understood is which came first: demographic decline or larger elite-held estates.

As elites possess increasingly amounts of land, rural communities that once depended on small family units begin to disappear as the rural population shifts towards the greater urban centers. In fact Greece becomes one of the most urbanized provinces of the Roman Empire. Many of the "old" cities of Greece benefit from these population shifts, but others do not fare so well. The number of cities in Aitolia, for example, shows a decline during the Hellenistic and Roman periods. It appears that at least in the case of

Aitolia, the decline in communities may be a result of *synoecism*, in which communities are forced to give up populations to other cities, often newly established colonies. These population realignments begin to appear in the archaeological record in the regions bordering Nikopolis and Patras, both Augustan foundations (Alcock, 1993: 136–7). These colonies do not appear to have exerted a "romanizing" influence on the Greeks; instead these cities adopt or reflect a more "hellenic" demeanor. This appears to be the case despite the Italian character of their populations and, in particular for Nikopolis, Patras and Corinth, the fact they all share a general westerly orientation, towards the nexus of their foundation, Rome.

As Rome becomes increasingly entrenched in the affairs of Greece, the most visible traces of Roman authority are not found in the rural landscape, but in the city. Here, within the confines of city walls that no longer are viewed as important as long as Rome is able to defend the province, interaction between city population and Rome is shown by the embellishment of the urban landscape in a manner that generally reflects a sense of the "new order." We, as observers of these new architectural embellishments, can often trace the development of civic architecture during the Roman period, but often fail to discern the motives and agents behind such architectural construction. As an example, one may note the Roman Agora of Athens, whose building is known to have been funded first by Julius Caesar and later by Augustus. What we are generally unable to discern are the basic questions involved in public benefactions such as this: Whose idea was it to build the new market? Was it an Athenian-inspired project or was it Julius Caesar himself who ordered it built? How autonomous were the builders from the Roman benefactors? Although paid for by Caesar and Augustus, how "Roman" was the so-called Roman Agora? The answers to questions as simple as these are difficult to ferret out from available evidence. But it is precisely these aspects of civic patronage and benefactions that go to the heart of questions concerning "Romanization," acculturation and cultural entrenchment.

Unfortunately, few cities in Greece have been archaeologically probed enough to offer tangible answers to such questions about the insinuation of Rome into the Greek urban fabric. Cities such as Athens and Corinth, however, through long-term excavation and a treasury of surviving epigraphical and literary sources, offer the best opportunities to view their embellishments and understand any possible civic transformations through the media of public architecture and sculpture.

2 Roman Athens

2.1 Athens from 211 to 88

> And she is of such authority that the name of Greece, now enfeebled and almost broken, rests upon the glory of this city.
>
> (Cic. *Flac.* 26.62)

In 211, during the First Macedonian War, the Roman proconsul Sulpicius Galba captured the island of Aigina, a scant 30 km from Athens and, once plundered, handed the former Macedonian possession over to the Aitolians. The Athenians had professed neutrality in the war but these operations occurring within view of their walls convinced the Athenians to help secure a peace between the warring parties. This first instance of Roman forces operating in proximity to Athens provided the Athenians an opportunity to observe first-hand the destructive power of the Romans, an awareness that would be reinforced a few years later by Galba's lightning raids on Corinth and Sicyon which were also under Macedonian control.

In 200 Athens turned aside its neutrality and declared war on Philip V after Macedonian forces and their allies raided the Attic countryside, inflicting heavy damage. Livy's description of the devastation (XXXI.23–6) provides a colorful account of one of the most destructive events in Athenian history. Philip's animosity towards the Athenians resulted in widespread ruin, vented particularly upon rural shrines of Attica. Although it is difficult to note clear evidence of Macedonian destructions in the Attic countryside archaeologically, it may be possible to discern more indirect repercussions in the shrines themselves. The excavations of the Athenian Agora have uncovered *membra disiecta* of several structures of primarily Classical date that were removed from their original positions elsewhere in Attica and brought into the Agora during the Augustan and Julio-Claudian periods. Most notable among these transplanted structures is a fifth-century temple which once was thought to have been originally located in Acharnai, but foundations of a temple have been recently uncovered in Pallene which more likely pertain to this temple. The excavators of the Agora propose that this temple, thought to have been originally dedicated to Athena Pallenis, was rededicated to Ares under Augustus (Camp, 2001: 116–17). The building apparently was not intact when it was moved, as it was repaired by blocks cannibalized from the fifth-century temple of Poseidon at Sounion. The implication of these repairs is that both temples were in a ruinous condition at their original locations and that parts of both temples were removed to complete a whole temple erected within the new topographical fabric of the Early Imperial Agora. Also brought to the Agora from Sounion at around the same time were elements from the Temple of Athena to be incorporated into the Southeast Temple; a Classical structure from Thorikos in eastern Attica had parts removed and transported to the Agora for reuse in the so-called Southwest Temple. Damage to other Attic shrines is noted in the Augustan period repair to the Temple of Nemesis at Rhamnous as well as the sanctuary of Artemis at Brauron (Camp, 2001: 169). Philip is regarded as the likely culprit responsible for the devastation to these sacred sites, although it is nevertheless possible that damage occurred at a later date or simply as a result of neglect. An inscription most recently dated to the later Julio-Claudian period (*IG* II^2, 1035) describes a program for the restoration of Attic shrines that had been damaged in the Hellenistic wars or through natural disrepair (see Shear, 1981: 365). There is also the possibility that damage to the Attic countryside was caused by Julius Caesar's lieutenant, Calenus, whom Cassius Dio reports (XLII.14.1–2) ravaged the countryside during the civil war against Pompey in 48.

The Athenians at the beginning of the second century were no longer capable of waging any kind of effective offensive action on their own. When war was declared the Athenians joined Attalus of Pergamon in his offensive against Macedon. But even with Pergamon's assistance more help was needed because, even in defense, Athens was poorly prepared. As a result the Athenians sent a delegation to Rome to appeal for aid. For their part, Rome had no prior obligation towards Athens, but used this appeal to re-enter hostilities against Philip. To aid Athens in their quarrel, and to further their own strategic interests in Achaia, the Roman consul Sulpicius Galba, who had already gained extensive experience in fights close to Athens in 211, dispatched 20 naval vessels and 1,000 soldiers to Athens, where they would be garrisoned in Piraeus to serve as a protective force should Philip attempt to capture the strategic harbor. This is a significant event in Athens's relationship with Rome as it marks the first known occasion in which Rome's military first set foot in the city. It is also important to note that the Romans were welcomed to the city as key allies, and not as invaders.

On several occasions during the course of the Second Macedonian War the Roman garrison engaged Philip's forces who attempted to capture the city. At one time the Macedonians were nearly successful in breaking through the city walls near the Dipylon Gate, but were forced to retreat when the Romans arrived. Philip also failed to capture the fortress at Eleusis, again because of Roman and Pergamene reinforcements. Ultimately Philip abandoned his efforts to gain control of Athens, and withdrew his forces from Attica.

In the ensuing years Athens would remain in good relations with Rome and it is likely the military force occupying Piraeus stayed at least through the mid-180s. There is no report that Roman forces were withdrawn from Athens; indeed following the victory at Cynoscephalae the Romans maintained legions in several of the former Macedonian strongholds, such as Corinth and Chalchis. The harbor at Piraeus would have served a critical strategic need for Roman occupying forces. Roman presence in Piraeus was augmented from 191 to 188 by additional Roman warships that used the harbor against Antiochus and the Aitolians (Habicht, 1997a: 209).

These early ties between Athens and Rome were likely formalized by treaty, perhaps shortly after 191 during the campaign against Antiochus, when Roman forces first arrived in Piraeus. No archaeological evidence survives to denote Rome's military activities in Attica. Nevertheless it may be possible to detect their presence in a new type of coinage issued by the Athenians. Bronze coinage dated to the 190s, depicting a helmeted Athena, appears to be borrowed from the Roman denarius (Kroll, 1993: 51). This coinage was likely meant at least in part as military pay, although it should also be seen as commemorative to Rome honoring the city as a key ally in this time of strife.

During the Third Macedonian War (171–68) Athens was bound to Rome as a result of the treaty it had signed. Although the Athenians offered Rome their manpower towards the war effort against Perseus, the Roman Senate instead requisitioned the city's grain supplies to feed the soldiers (Livy XL.6.2–3). Athens proved to be a compliant ally in the war and was rewarded with the islands of Imbros, Skyros, Lemnos and Delos. The acquisition of Delos would prove to be extremely lucrative for Athens in the ensuing years, as much of the eastern Mediterranean trade was funneled through its harbor and sent on towards the west and Italy (Rauh, 1993).

By and large the Athenians suffered little throughout the three Macedonian wars and, with the exception of the devastation of the Attic countryside in 200, the city witnessed no fighting within its borders. For over a century, from the signing of the peace treaty in 197 ending the war against Philip until 88, Athens enjoyed a long period of peace. It could also be claimed that the city regained a modicum of prosperity during the second century as a result of the re-acquisition of the commercial territories of Imbros and Delos.

The second century brought a tremendous energy in new construction; a spate of building occurred that had not been witnessed for over two centuries. But these new buildings were not constructed by the Athenians. In spite of an upturn in the economy, the city was no longer capable of – or perhaps no longer interested in – monumental building projects. Instead, foreign potentates took it upon themselves to enhance the topographical flavor of the city with often large and sumptuous structures. It has already been noted that in spite of Athenian loss of power – political, military and economic – since the end of the fifth century, the city nevertheless still wielded enormous prestige throughout the Greek-speaking Hellenistic east. Athens still was considered unsurpassed in culture, education, and often diplomacy. And in no small way did the city's illustrious past add to its reputation. The status of the city was such that the newer Hellenic powers afforded it pre-eminence within the Greek world. The kings and dynasts of these new power centers of the Hellenistic Greek world paid homage to Athens in the form of architectural benefactions, each one of which would rival the previous donation. Prominent among these Hellenistic royal houses were the Attalids of Pergamon.

Among the most significant architectural features of the Athenian cityscape is the Stoa of Attalos II, constructed in the Agora between 159 and 138 to serve the market needs of the city. The large stoa was a gift to the city from an alumnus, the Pergamene king who studied in the philosophical schools in his student days. The building, excavated by the American School of Classical Studies in the 1930s, has been reconstructed to serve as the excavation's storerooms and offices, and also houses the museum of the Agora. Attalos's gift was not the first Pergamene benefaction to the city. Attalos II's father, Attalos I, erected on the Acropolis a monumental sculptural group commemorating the Pergamene battles against the barbarian Gauls (on which, see Stewart, 2004). Eumenes II, brother of Attalos II, had constructed a stoa just west of the Theater of Dionysos to be used as a shelter for the theater-goers. In all three examples Pergamene architects were employed as opposed to local builders; in the case of the Eumenes Stoa some of the materials of the building were pre-carved in Pergamon and then shipped to Athens for erection onto the building. One should not necessarily suppose that the state of Athenian architecture had reached its nadir because architects, workmen and even materials were being imported into the once proud architecture-rich city. It may simply be a statement of preference on the part of the dynast who was footing the bill. However, one is hard pressed to find a major construction in Athens that was not funded principally by foreign interests. The most recent work on the Tower of the Winds, another structure paid for by a wealthy foreigner, attempts to settle the controversy of its dating; Kienast (1997) comes down on the side of its dating in the second-century. Building by foreign interest groups would also occur as Rome became more involved in the affairs of Athens.

Not to be outdone by the Pergamene Attalids, Antiochus IV Epiphanes of Syria (175–164) also became a benefactor of Athens. Antiochus saw the unfinished giant temple of Olympieion Zeus south-east of the Acropolis as an opportunity for a contribution to the architectural panoply of the city. Construction of this temple had begun during the tyranny of Peisistratos near the end of the sixth century. Construction stopped at the death of the tyrant and, not coincidently, at the inauguration of democracy. The abandoned Peisistratid temple was perhaps viewed as a symbol of tyranny and its incomplete state for over 350 years, during the time in which Athens had gained and then lost its political and military power, was meant to be understood as testimony of civic antipathy towards the monarchial state. But by the time of Antiochus, the Athenians no longer had the will or desire to oppose his benefaction.

Vitruvius informs us about some of the details of Antiochus's benefaction, particularly the identity of Antiochus's architect for the temple, Cossutius, whom he purposely refers to as "cives romanus" (VII. praef. 15). This appears to be a particularly intriguing choice on the part of Antiochus, to give a commission of a Greek temple – and one directly in the nexus of Classical architecture – to a Roman architect; even Vitruvius hints that it was an unusual commission as he specifically labels him as Roman. But Cossutius was an architect endowed, as Vitruvius adds, "with great skill and scientific attainments." For Cossutius to have been awarded such a high-profile commission, one must assume that Antiochus already had formed a positive opinion of Cossutius's architectural skill. It has been suggested that Antiochus may have become acquainted with Cossutius during the former's long sojourn in Rome; it is also significant that the name of Cossutius was found scratched into the mortar of a second-century aqueduct in Antioch (Downey, 1963: 58; cf. Thompson, 1987: 2–3; Rawson, 1975: 37). That Cossutius should be considered as Antiochus's primary court architect is therefore likely and that other Antiochan structures in Antioch such as another temple to Olympieion Zeus may be ascribed to him.

The nomen Cossutius is not unknown in Athens within epigraphical sources. A well-known inscribed base (now lost) names Δέκμος κοσσούτιος Ποπλίου Ρωμαιος (Dec(i)mus Cossutius, of the Roman people; *IG* II^2, 4099). This was found near the Olympieion and is usually referred to as a statue base honoring the Olympieion's architect (see most recently Thompson, 1987: 3; Rawson, 1975: 37; Abramson, 1974: 3). Recently the inscribed base has been shown, however, to be a grave marker that dates after the second century, and therefore does not belong to the architect (Byrne, 2003: 4, n. 7). The same formula in the nominative case is seen on another grave columella to a Marcus Cossutius in the Kerameikos (*IG* II^2, 10154). Although neither individual is the Olympieion's architect, they are likely to be related to him. A study shows a number of Cossutii operating in the marble-bearing islands of the Aegean, Asia Minor, the Greek mainland and Rome from the late second century and into the first century CE. These Cossutii seemed to have been part of a dynasty, operating in the east as mainly quarry operators or owners (although some were indeed sculptors), eventually becoming apparently quite wealthy in the marble trade (Rawson, 1975: 36–47).

The building that Cossutius envisioned was revolutionary in that it was the first Corinthian dipteral temple, constructed wholly with Pentelic marble, complete with octostyle facades. Clearly Cossutius (and Antiochus) was looking into the past at the original intent of the Peisistratid architects who modeled the building on the great giant Ionian temples of the sixth century, such as those at Samos and Ephesus. In that respect Cossutius's building was archaizing, although it is true that the temple was simply utilizing the existing footprint laid out by the sixth-century architects. It was also designed with a contemporary flavor, in the sense that it had incorporated Corinthian columns, as opposed to the Doric, which the Peisistratid original was meant to have included. The building conformed to the standards of Hellenistic construction; if one did not have the historical evidence of Cossutius's role, one would have assumed the architect to be Greek, which makes the fact that a Roman designed it more surprising. If one searches for any Roman contribution to the building, one would be disappointed, except for the fact that the proportional system imposed on the building was the Roman septunx proportional system (Tolle-Kastenbein, 1994: 57, 169–70). Most students of Hellenistic architecture consider the Olympieion to be a *tour de force* and as such it compels a reconsideration of certain biases concerning the qualifications of Romans as architects and helps to blur the line between Greek and Roman architecture.

It is still unclear how much of the Athenian temple was completed when construction operations ceased at the death of Antiochus in 164. It appears that the building may have been complete up to the level of the architraves at its abandonment (Tolle-Kastenbein, 1994; Camp, 2001: 173–6, 266–8). The temple would again lie unfinished for over 275 years until ultimately completed by Hadrian, although some work was apparently undertaken under Augustus.

The only other evidence of a structure of second-century Athens that appears to have Roman associations comes from an inscription that refers to the festival of the Theseia, in which there is a single enigmatic reference to setting military equipment ἐν τη στοα του Ρωμαίου (*IG* II², 958, line 29). The ambiguity of the name causes confusion – Stoa of the Roman or Stoa of Romaios; both could be correct. In either case a Roman is involved in either the building's construction or its outfitting (cf. Ferguson, 1911: 367; Miller, 1995: 230). It may be possible that the only Roman citizen in Athens attested as serving in a public function *c.* 150, a Spurius Romaios, is connected with the construction of this stoa (*IG* II², 1938, line 40; on Spurius Romaios, see Habicht, 1997b: 13). Hesberg (1980: 28) suggests the "Romaios" stoa may be an Attalid honorific hall for Romans.

The Olympieion was not the only the structure in Athens built, or rather reconstructed, by a Roman architect. The Odeion of Pericles, built into the south-eastern slope of the Acropolis directly east of the Theater of Dionysos, was originally constructed during Pericles' reign, *c.* 445, and according to Plutarch (*Perikles* 13) and other sources it would have been used for musical events, philosophical discussions and theater rehearsals, and even as law courts (Camp, 2001: 100–1, 256–7). The Odeion was a large, square hypostyle hall, measuring approximately 60 m on each side; in many respects it resembled the Periclean Telesterion at Eleusis, complete with an internal forest of columns. A recent reconstruction based on excavations and previous studies suggests that the interior contained four sets of banked seating (Izenour, 1992: 30–5). The Odeion suffered grievous damage during the Sullan siege of the Acropolis (see below), having purposely been put to the torch by the Athenian defenders (App. *Mith.* 38), in order to deny the Roman besiegers the extensive wood beams of the ceiling rafters to build siege engines (Hoff, 1997: 37).

Clearly the loss of the Odeion had a major impact on the Athenian psyche, understandable since tradition held that Xerxes' tent, which he used before the pivotal battle at Plataea in 479, was preserved inside the building. Just as Antiochus funded the completion of the Olympieion, so too would another Hellenistic monarch seize the opportunity to demonstrate his euergetism towards Athens. Vitruvius records that Ariobarzanes II Philopator of Cappadocia, a former student in the philosophical schools of Athens, paid for the reconstruction of the Odeion (V.9.1). Ariobarzanes, who ruled *c.* 63 to 51, was honored by the demos with a statue erected near the Odeion and Theater (*IG* II², 3427). A second statue base honoring Ariobarzanes (*IG* II², 3427) lists the dedicators as the architects commissioned by Ariobarzanes, "ἐπι τήν του ᾠιδείου κατασκευήν" (upon the equipment of the Odeion). Three names are given in the nominative: the Roman brothers, Gaius and Marius Stallius, and a Greek, Menalippos (for the last, most commentators of the inscription follow Day (1942: 129) in reading "Melanippos" instead of the correct

Menalippos). The commission of non-Athenians to serve as architects for the rebuilding of the Odeion has caused at least one recent commentator to suggest Athens was suffering from a "dearth of native talent" (Palagia, 1997: 81). This may well be the case, but it is more likely that Hellenistic monarchs, who are the ones providing the funding, prefer the services of their own architects whom they have likely employed on previous commissions. As noted previously this appears to be the case for the Pergamene construction of the Stoas of Eumenes and Attalos II. So the Periclean Odeion contained ties to the age of glory in Athenian history; its reconstruction therefore demonstrated a desire on the part of the Athenians for a continued connection to their past and its rich traditions. This is a very different scenario than the one described for the Olympieion.

The architecture of Late Hellenistic Athens generally conforms to the styles already present in the cities of the Greek east. It is therefore not surprising to observe that extremely little in the architecture of Hellenistic Athens demonstrates the stamp of Roman presence. There is one relatively obscure structure discovered in the Agora, however, that is unmistakably Roman: the Bema. No other type of building so clearly reflects the power of Rome than the speaker's platform, or rostra, in a city's forum from which Roman *strategoi* would address the people. From Poseidonius we have an historical record of its existence in Athens (*FGrH* 87, F 36.50) dating at the latest to 88. Plutarch records that while visiting Athens on his way towards Cilicia in 67 Pompey addressed the demos (*Pomp.* 27); one should assume that this address was delivered from the Bema (see Hoff, 2005: 328). The Agora excavators note its placement in front of the retaining wall and directly at the middle of the Attalos Stoa (Thompson and Wycherley, 1972: 50–1; Stephanidou-Tiveriou, 2008: 14–15), an axis-favoring position that is usually associated more with Roman topography than Greek.

The Athens Bema was not the only example of its kind in Roman Achaea. At Corinth, the Bema was built also in a central location in front of the South Stoa within the forum, and at Philippi the Bema can be found also in the forum. Both Corinth and Philippi are colonies with a distinct Italian character; a rostra included in a forum of clearly Roman type should not be considered unusual. But certainly for Athens its presence marks a definite uniqueness. The Bema appears to be the first structure in the city that is directly tied to Rome's hegemony. In the early first century the Agora square was empty of buildings, purposely left so for generations, as the mechanics of independent Athenian democracy would occur within its confines. The placement of a Roman rostra within the nearly open Agora therefore should be seen symbolically as a shift away from independence towards acceptance of Roman authority – or to demonstrate at least a desire to hear Romans speak.

2.2 Athens, from 88 to 31

Athens's relationship with Rome in the years following the Third Macedonian War is generally described as cordial. Rome acts as a benefactor to Athens and there is nothing in the ancient sources to indicate issues regarding strained relations. Nevertheless in the early 80s Athens voluntarily joined with Mithridates VI Eupator, ruler of Pontus, in a war against Rome and would suffer grievous harm as a result (Habicht, 1997a: 297–314; Hoff, 1997). Most historians of this episode note that Mithridates came to be viewed as "champion" of the Greek people against the Romans and that a wave of anti-Roman sentiment, that likely had been festering in the background, suddenly exploded, particularly after Mithridates' early successes against the Roman military. In Asia Minor, Mithridates ordered the surprise slaying of thousands of Roman citizens. Greeks celebrated this massacre everywhere, including Athens, as an expression of panhellenic zeal against the foreign "invader." In Athens our sources indicate that factional politics arose in which a sudden and irresistible anti-Roman faction came to power and literally drove out pro-Roman citizens who were forced to flee the city. More significantly, Mithridates' army descended on Athenian possessions in the Aegean, including Delos, where allegedly 20,000 mostly Italian traders were massacred (App. *Mith.* 28). The Pontic general Archelaus then established his Aegean headquarters at Piraeus, from which he could lead raids against Rome's interests in the Mediterranean. Much of Greece quickly fell under his control.

Rome's response to the revolts and massacres is not surprising. In the spring or summer of 87, L. Cornelius Sulla arrived in Attica with five legions with the intent of dislodging the entrenched Archelaus from Greece (see Hoff, 1997). Archelaus's control of fortified Piraeus allowed his fleet to sail in and out of the harbor unmolested, as Sulla had no naval force strong enough to defeat the Pontic navy. Sulla also had to deal with the rebellious Athenians, ensconced within their hastily repaired walls. Sulla ultimately decided to wait out both strongly defended forces. While Archelaus's forces were in a more advantageous position, his Athenian allies on the other hand were besieged during the winter of 87/86 within their walls and suffered severe deprivations (Plut. *Sull.* 13.3; App. *Mith.* 38).

On March 1 Sulla's troops were successful in breaching the walls of Athens. The carnage that ensued was, according to all historical sources, horrific. Pausanias wrote, albeit 250 years later, that Sulla's actions were "more brutal than one would expect from a Roman" (I.20.7). Plutarch graphically recounts – perhaps with some hyperbole – the deaths of Athenians at the hands of rampaging Roman soldiers: "the blood that was shed in the Agora extended all over the Kerameikos inside the Dipylon and, as many have said, it flowed into the suburbs beyond the Gate" (*Sull.* 14. 3–4).

The physical damage to the city caused by the rampaging Roman troops was enormous. Although few of our sources specifically indicate which buildings were damaged or destroyed in 86 (the Odeion of Pericles noted above is an exception), the archaeological record is much more revealing. A recent study has brought together evidence that demonstrates how pervasive was the destruction within the city, with damage so severe that it would take decades to recover (Hoff, 1997). Many of the buildings within the Agora, including most of the Middle Stoa and South Square, the Tholos (the headquarters of the Athenian Boule), and the buildings usually associated with the military, the Arsenal and Strategeion, were targeted by Sulla. Religious buildings were apparently spared, but culturally significant structures were affected. Even the Acropolis was not spared from harm. Evidence points to the possibility that the Erechtheion may have suffered accidental damage during the Roman siege, and was left unrepaired until the time of Augustus.

Perhaps even more grievous than damage to buildings was the damage caused to the city's economic infrastructure. We have evidence that Athens's pottery industry was affected, a blow to the local economy that took decades to ameliorate (Rotroff, 1997: 100–6). But more importantly, the docks and wharves at Piraeus must have been heavily damaged when Sulla's troops ravaged the city. Sulla destroyed the harbor city – including the famous Arsenal of Philo – and thereby crippled Athens's maritime commerce. It would appear that the destruction by Sulla to Athens's primary means of commerce was wholly deliberate, a punishment to the city that was calculated to limit Athens's quick recuperation.

Athens was indeed slow to recover. As with much of Greece, Athens was in a wretched state. Although some scholars have tried to downplay the ruinous condition of the Greek countryside in the post-Sullan years as painted by the ancient sources, ultimately Athens and the rest of Greece were clearly at a low point (Alcock, 1993: 14; Shear, 1981: 356). Rebuilding the economy would likely have been the first order of business, but if so, the ancient sources render this renovation sketchily. Apart from the reconstruction of the Odeion mentioned above, the earliest notice of any kind of recovery occurred in 62 when Pompey came to the city following the Mithridatic War. While in Athens, according to Plutarch, he donated to the city 50 talents to be used for the restoration of its monuments (*Pomp.* 15.35). Plutarch fails to mention, however, which monuments were restored. An Augustan inscription (*IG* II^2, 1035) implies that at least part of the funds Pompey donated was meant to restore the *Deigma*, or bazaar, in Piraeus. This would suggest that recovery efforts were still going on long after Sulla departed and at least some efforts were earmarked for repairs to the city's economic infrastructure (Hoff, 1989b: 2, n. 8; 2005: 332–3).

Continued focus on repairing the city's commerce can still be seen in mid-century. The dedicatory inscription preserved on the epistyle of the West Gate of the Roman Agora (*IG* II^2, 3175) expressly states that funds for the market's construction were provided by Julius Caesar and Augustus (see Hoff, 1989b: 1–8). The date of Julius Caesar's donation can be narrowed down to 51, implied in a letter from Cicero to Atticus in which he writes that Caesar paid 50 talents for some unnamed purpose. Cicero further records that Pompey became angered at the donation; he believed that his own benefaction 11 years earlier, surely referring to the Piraeus repairs, was "squandered by the Athenians" (*Att.* 6.1.25). This is a significant passage. It suggests that the initiative behind these donations came from the Athenians; they requested a construction and the Roman benefactor responded in coin. Pompey likely considered Caesar's gift a greater benefit to the Athenians, hence his anger because Caesar's funds were earmarked for use in the heart of the city, rather than his own on the wharves of Piraeus where it enjoyed less visibility. Caesar's benefaction to the city was designed to emulate the euergetism of previous Hellenistic kings, like the Attalids and Antiochus, and Caesar is so honored on two statues dedicated in Athens (Raubitschek, 1954). Whether he purposely meant to trump Pompey's donation is unknown, but the fact that his donation exactly matched Pompey's is at least suggestive.

It is unlikely that much of the market was constructed in the few years from 51 to the outbreak of the civil war in 48. It is now considered that part of the Augustan building program includes the visible remains of the quadriporticus, which wholly dates to the Augustan era. Nevertheless it is possible that the site was at least prepared for the building's construction before Pharsalus. Substantial alteration of the ground, involving excavation of the Acropolis slope on one side and terracing on the opposite side, had to have occurred before any building could have begun. In the civil war Athens joined the side of Pompey. The war's engagements largely took place in Greece. Similar to the events of

Figure 35.1 Fragmentary base with inscription mentioning Brutus (I3366). Source: Photo by permission of the American School of Classical Studies at Athens, Agora Excavations.

the Second Macedonian War, Caesar's lieutenant Calenus devastated Attica and, although prevented from storming Athens, captured Piraeus. After the war, the Athenians came to Caesar as suppliants who, according to Dio, forgave the city but admonished the citizens for their bad decision-making (XLII.49.1–4). Along with other pro-Pompeian cities, Athens likely was compelled to pay massive war indemnities. The war must have interrupted the market's construction; it is difficult to imagine that Caesar's construction funds would have survived untapped.

Compared to Pompey, whom once the Athenians showered with nearly divine honors (Hoff, 2005), Caesar was particularly reviled. He was judged responsible for devastating the countryside during the war and exacting war reparations that must have stressed an already fragile economic recovery. More significantly, in the eyes of the Athenians, Caesar had become dictator, an office that was repugnant to the democratic traditions of Athens. After Caesar was assassinated in 44 the Athenians dedicated honorific bronze statues to Brutus and Cassius in the Classical Agora next to the famous statue group of Harmodius and Aristogeiton, another pair of tyrannicides (Cass. Dio XLVII.20.4). A fragment of the Brutus base survives (I 3366), which was recovered in 1936 (Raubitschek, 1959), and it serves as a remarkable testament to the fiercely held notions of independence even while the city was in the firm grip of Rome (Figure 35.1).

Yet in addition to expressions of patriotism, the Athenians believed Rome and its officials acted disingenuously towards the city. The ancient sources are replete with episodes of Romans arriving in the city and departing with some of the city's treasures (see Hoff, 1989a). Some notable examples include Sulla, who removed the expected gold and silver treasures. He also took away capitals from the Olympieion and the contents of an entire library. From the testimony of Cicero we hear that Verres was accused of stripping the Parthenon of some of its gold (Cic. *Verr.* II.i.17–45), and that *c.* 58 L. Calpurnius Piso was responsible for despoiling the city (Cic. *Pis.* 40.96). Clearly Cicero viewed these acts by his compatriots as unethical; in a letter to Atticus he proudly claims that when he passed through Athens in 51 his visit was not a burden upon the Athenians (Cic. *Att.* 5.10.2, 11.5; see Chapter 37, and Miles, 2008: 130 on Verres stripping gold from the Parthenon). Not all Roman officials took advantage of their superior status to rob the city. Some, like Pompey, attempted to better Athens's condition by sumptuous benefactions. Although Caesar initially played the role of benefactor, his memory was vilified by the Athenians for his "barbarous," Persian-like behavior in the devastation of Attica.

Some of the same attitudes that fueled the Hellenistic building program, in which a wealthy individual provides an architectural embellishment to a cultural icon or religious sanctuary, can also be found at Eleusis. In this case, Appius Claudius Pulcher paid for the construction of the Lesser Propylaea of the Sanctuary of Demeter in fulfillment of a vow he made while consul of Rome in 54. A recent study suggests that the vow occurred after the Tiber flooded low-lying regions of Rome in that year, resulting in loss of life and severe damage, and in which much of the city's grain stores were destroyed (Clinton, 1997: 164–5; 2008, 2: 298–9, no. 290). While the form of the structure is not extraordinary in itself, its embellishment strikes an unusual tone in a form of Neo-Attic "baroque" (Sauron, 2001: 267–83). Prominent in the decoration were the caryatids in the porch that echo the figures from the South Porch of the Erechtheion, but their style appears to be more of a hybrid between fifth-century forms and Hellenistic proportions. In other words, the sculptures present a completely new style, but one that also represents, as one commentator suggests, the final breath of Athenian sculpture (Palagia, 1997: 83–91).

2.3 Athens during the Augustan Period, 31 BCE to 14 CE

Immediately following his victory at Actium in 31 – before he set out to follow Antony and Cleopatra to Alexandria – Octavian visited Athens. His short stay there was meant to reconcile Rome with those Greek cities, Athens included, that participated in the civil war on the side of Antony. The prestige of venerable Athens is what drew Octavian to the city, an echo of the city's role in years past. Octavian tarried in Athens long enough to participate in the first of two initiations of the Eleusinian Mysteries. Whether he was sincerely interested in the religious aspects of the Mysteries is unknown, but clearly his participation in the city's most significant religious institution was designed to foster goodwill with the city. Octavian surely recognized the now long-standing animosity the Athenians held towards Rome, given how Roman–Athenian relations had soured from Sulla on, and reconciliation with Athens must have been a primary concern (Hoff, 1992). But as events unfolded, his early efforts were apparently for naught.

Nine years after Actium Augustus returned to Athens. In spite of Augustus's efforts to establish good relations with the Athenians, the historical sources indicate that demonstrations of anti-Roman sentiment occurred while the emperor was in the city, indicating that at least some of Athens's citizenry still felt deep rancor towards Rome (Hoff, 1989a).

Figure 35.2 Athena Gate to the Roman Forum, Athens. Source: Image AT 68, American School of Classical Studies at Athens, Archives, Alison Franz Photographic Collection.

Augustus's reaction to these anti-Roman expressions in the winter of 22/1, according to Dio, was carefully measured, as he primarily enacted economic sanctions against the city. It must be assumed that Augustus's modest response must have been successful as he returned to the city in 19, only two years later. At that time it should be assumed that there was reconciliation, since Dio reports that Augustus participated in his second initiation in the Eleusinian Mysteries (LIV.9.7–10), an event that would have been unlikely if the emperor's ire towards the city had not softened. Once the situation had been resolved, Augustus may have used the opportunity to reward the city with new benefactions. It has been argued that the circumstances surrounding Augustus's visit in 19 resulted in the donation of funds to complete the market that Caesar had earlier financed (Hoff, 1989b).

The Roman Agora, often referred to as the Market of Caesar and Augustus, was constructed approximately 200 m east of the Classical Agora and was built along an earlier road that led from the Agora along the contour of the north slope of the Acropolis; the west and east propylaia of the Roman Agora correspond to the positioning of this road (Figure 35.2). Once completed, the building, 111 by 104 m, was the largest structure ever constructed in the city, and would serve the civic market needs for the rest of antiquity (Hoff, 1988; Sourlas, 2008). The building was constructed as a quadriporticus, with a curious employment of Doric capitals on the inside colonnades and Ionic on the exterior, which is the reverse of the usual classical colonnade form. Elements of the building also demonstrate aspects of incompleteness and, even more significant, the colonnades and the interior were not originally paved in stone, but left purposely unpaved; the courtyard would only be paved during the Antonine period. It would appear that the building, while large and evocative of grandeur, was nonetheless constructed within a finite budget with funds directly provided by Augustus, as recorded in its dedicatory inscription (*IG* II², 3175). In spite of its Roman funding, the building's antecedents belong to the Hellenistic world, again indicative of looking towards the east, rather than the west, for inspiration and modeling. Recent excavations within the northern half of the market demonstrate that the building was constructed within one phase, thereby confirming that the construction belonged wholly to the Augustan age rather than the Caesarian (Hoff, 1988; Sourlas, 2008).

As many of the commercial activities of the Classical Agora were transferred to the new market, the open space of the older Agora became filled by new or refurbished religious and cultural buildings: the construction of a new Odeion, altars and transplanted temples (referred to above). The old Agora became more of an architectural showcase than a functioning market and the new market became the "forum" of Augustan Athens.

The Athenians surely would have responded to the benefaction of the new market, as well as to the re-affirmed reconciliation. It appears that the response came in the form of a monopteral structure on the Acropolis that most scholars date to the time of Augustus's visit in 19. Traditionally the structure is referred to as the Temple of Roma and Augustus, although its identification as a temple has been called into question (Camp, 2001: 187–8). The completely preserved dedicatory inscription (*IG* II², 3173) in fact does not refer at all to the building's identification, leaving its purpose open to interpretation, although it is clear that it was the demos who initiated the dedication. The *monopteros* was constructed on the precise east–west axis of the Parthenon, a position that is generally suggestive of a more Roman topographical interest. But the *monopteros* was constructed at the usual position of a temple's altar. In the case of the Parthenon there was no built altar, leading some scholars to suggest the *monopteros*'s role was more of an elaborate baldacchino for altars and statues of Augustus and Roma that served the nascent imperial cult (Camp, 2001: 188; Stephanidou-Tiveriou, 2008: 21–3). Its diminutive size compared to the massive Parthenon is indicative of the inability of the Augustan era demos to construct large-scale public buildings.

The institution of emperor worship is often viewed as one of the principal markers of Romanization. Athens was no different than other Greek cities with respect to the manifest displays of statuary, altars and structures dedicated to the imperial cult. The so-called Temple of Augustus and Roma is one such example. Inscriptional evidence indicates that there may have been at least one other venue for a cult to Augustus. Included among the candidates is a small annex constructed immediately behind the Stoa of Zeus Eleutherios in the Agora (Thompson and Wycherley, 1972: 102–3). Although some cult association with emperor worship is possible, the structure, unobtrusively hidden as it is behind the massive stoa, appears unsuited to serve as a focus for the imperial cult, particularly as its unobtrusiveness is the opposite of the high-profile Acropolis *monopteros* (contra, Spawforth, 1997: 193). Instead one should consider a more likely placement for the cult to be situated near the newly built Augustan Market. To the east of the market was constructed a building that may have functioned as a *sebasteion* (building to the imperial cult) (Hoff, 1996). Although the building in its present form was likely constructed in the late

Julio-Claudian period, it may supersede an earlier Augustan-period cult locale associated with the Augustan Market. A building that served as a focus for the imperial cult is a frequent neighbor to forums or agoras in the Greek east, a logical placement where multitudes of people congregate on a daily basis. The Augustan Market offers a more logical location – and more visible – than the Classical Agora. A similar combination of forum and imperial cult may be seen in Corinth.

3 Corinth, from 146 BCE to 14 CE

> On my return voyage from Asia, as I was sailing from Aigina towards Megara, I began to survey the surrounding localities. Behind me was Aigina, in front Megara, on my right Piraeus, on my left Corinth, cities which at one time were most flourishing, but now lay before my sight in ruin and decay.
>
> (Cic. *Fam.* 4.5.4)

The letter Servius Sulpicius wrote to Cicero dates to 45, just the year before Julius Caesar refounded Corinth as the Colonia Laus Julia Corinthiensis. There is little doubt that Sulpicius's letter is imbued with literary hyperbole, but nevertheless the chord he strikes of a devastated Corinth, similar in circumstances to the other victims of Roman aggression such as Megara and Piraeus, rings true. Corinth was indeed in a wretched state. But there are indications that the city was not so ruined by the events of 146 nor was it so desolate in the years prior to its refoundation as the sources would have us believe.

Corinth was destroyed in 146 by the consul L. Mummius after the annihilation of the Achaean League's forces at Leucopetra on the Isthmus. The members of the Achaean League had sought complete freedom from Roman authority; Rome had offered a grant of libertas, or conditional freedom, but to the democratic-minded Greeks, the offer was insufficient and thus refused. The event that precipitated the disaster at Leucopetra and the subsequent destruction of Corinth was the declaration of war against Sparta, which was at the time Rome's close ally. Corinth was singled out for destruction because of its leadership role within the Achaean League and it was also closest to the battleground on the Isthmus. Mummius's forces sacked the city and the proverbial wealth of Corinth was carted off to adorn both public and private buildings in Rome. The nearby city of Sikyon was later allowed to farm the land of Corinth, now under the ownership of Rome (ager Romanus), and the panhellenic festival of Isthmia was also given to the Sikyonians.

Since 1896 ancient Corinth has been undergoing excavation by the American School of Classical Studies. These ongoing excavations have been able to document not only the destruction of the city under Mummius, but also that the damage was not thorough enough to inhibit some post-146 life in the old city (Wiseman, 1979: 494–5). After all, though many of the pre-destruction inhabitants of the city (with an estimated population of 85,000) were either killed or sold into slavery by the Romans, it is logical to assume that some survived and continued to live nearby or within the confines of the now ruined city. There is documented evidence that after the destruction of the city, modest clean-up operations occurred and the city's cults continued to be cared for (Wiseman, 1979: 494–5;

Romano, 1994: 62). Studies also indicate that during the interim period from 146 to the city's refounding in 44 – while one should not postulate any type of civil administration – there was a population existing within the city that conducted some commercial activity and engaged in international trade (Romano, 1994: 62).

Just before his death in 44, Julius Caesar ordered the colonization of Corinth and Carthage. Caesar's motives behind these foundations were to provide land grants for his veterans and Rome's urban plebs. In so doing, Caesar was able to benefit several classes of Roman society with new land, not to mention appeasing the wealthy landowners of Italy whose land holdings were under pressure from these same groups (Engels, 1990: 17). With their removal from Italy, the pressure was now off the agricultural elite. Caesar must have also envisioned the new Corinth becoming a major commercial center, as it had been in its previous incarnation, because of its strategic location.

The colony of Corinth was laid out by Roman surveyors in a grid system at the time of Caesar's foundation, although there is substantial evidence that an earlier centuriation occurred near the end of the second century (see Chapter 16; Romano, 2003: 280–3). Nevertheless it appears that Corinth is not considered by scholars to have been primarily an agricultural center but, as mentioned above, was focused on commercial trade (Engels, 1990; Williams, 1993: 31). Although Caesar was responsible for Corinth's colonization and, one assumes, its centuriation, the construction of roads and other elements of infrastructure, public structures and temples apparently date to the Augustan period. It is also clear that many Classical era buildings were rehabilitated during the Roman period (Wiseman, 1979: 494).

The heart of the new colony was its forum, perhaps overlaying the pre-Mummius agora, encompassing an open space of over 15,300 m^2, and bounded on all sides by religious and secular structures (Figure 35.3). The prevailing view has been that the Greek agora at Corinth had been located somewhere other than the area of the Roman period forum, perhaps under the *plateia* of the current village (cf. Wiseman, 1979: 489). Recent research on state-owned objects found in association with Greek levels under the Roman forum now suggests the placement of the Greek agora is to be found under the later Roman forum (Donati, 2010).

Many of the non-religious buildings were commercial but their functions apparently served more than the quotidian market needs of the colony. The defining single structure of the south side, the South Stoa, was originally built in the fourth century and apparently survived the city's destruction to be rehabilitated by the Roman colonists. Constructed in two stories, the building housed 26 shops originally, and likely continued its commercial functions into Late Antiquity. Newly constructed within the forum and occupying its higher south half is a row of shops identified as the Central Shops. Their position on the central strip of the forum and attached to the Bema suggests a significance beyond mere retail. One commentator suggests that at least part of the Central Shops dealt with precious metals or served as bankers' offices, a necessary inclusion in the fabric of a robust commercial city (Williams, 1993: 37). The space defined within the refurbished South Stoa was largely open, the row of Central Shops notwithstanding. The construction of the Corinthian Forum was likely undertaken at about the same time as the building of the Roman Agora in Athens, another large space that was open within the interior. While details in construction differ between the two building programs, the inherent Roman predilection towards enveloping space through architecture is clearly similar.

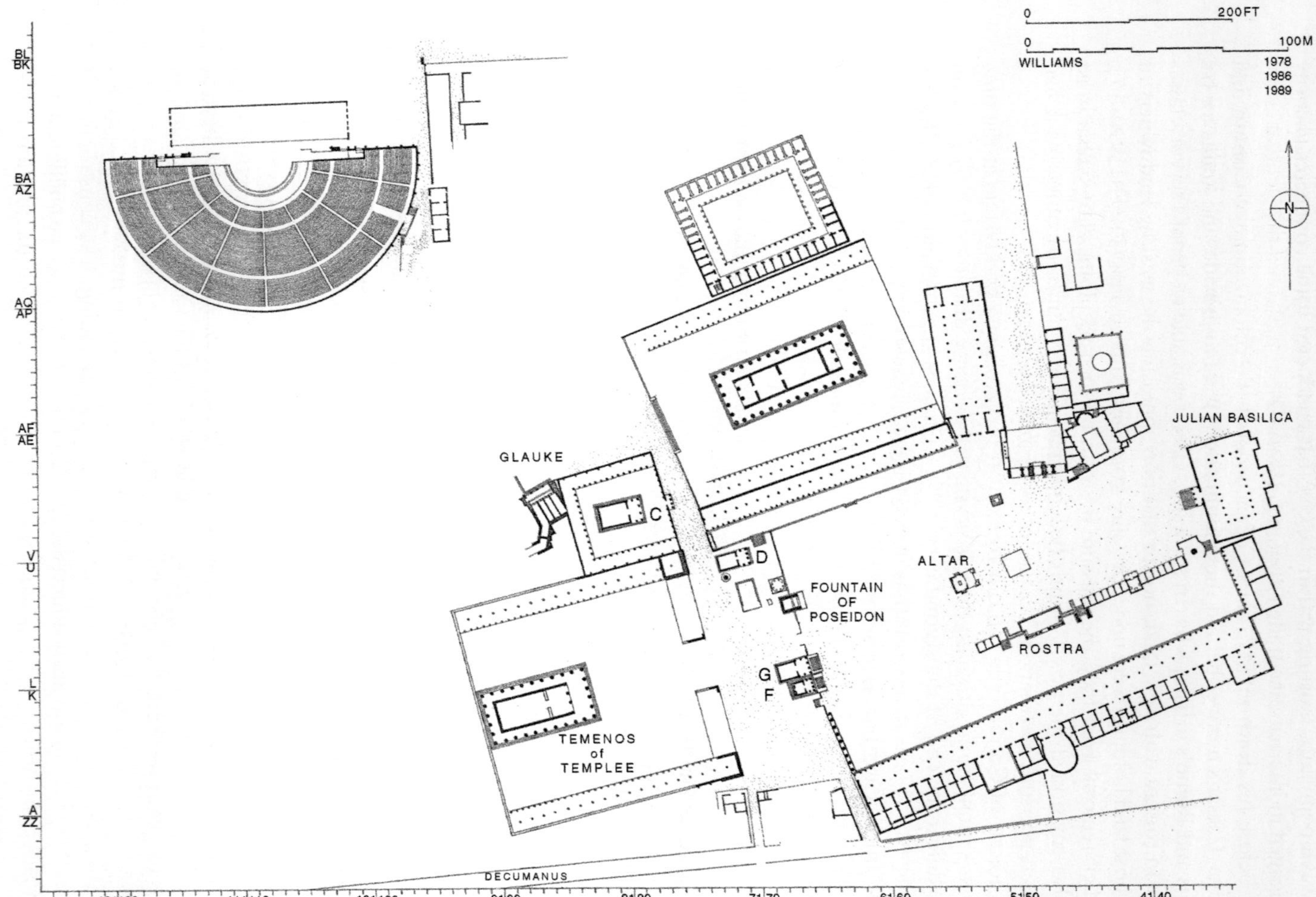

Figure 35.3 Plan of the center of Roman Corinth, early 1st century CE. Source: Corinth Plan 004–009a, from *The Corinthia in the Roman Period*, *JRA* Suppl. 8, courtesy of the American School of Classical Studies, Corinth Excavations.

As with Athens, there are also imperial cult activities occurring within the Corinthian Forum. Located at the eastern flank of the forum in Corinth is the so-called Julian Basilica, which has been most recently dated to the late Augustan period (Scoton, 2005). The structure apparently served as a tribunal, but recent research points to the likelihood that the basilica also served an important role as a venue of the imperial cult (Scoton, 2005). Together with Athens then, the focal points for the cult to the emperor in the two cities seem to be localized in highly visible areas in, or in the vicinity of, the new commercial centers. Cult and commerce among the Romans are seemingly associated.

4 Conclusion

Just outside the entrance of the Apollo sanctuary at Delphi was the Roman-period agora; Pausanias ignores it prior to his description of the sanctuary but everyone who entered the sanctuary would have passed through it. It is not difficult to imagine the kinds of items available for sale at this agora in the Roman period – probably items associated with a tourist's or pilgrim's visit to Delphi. The sanctuary had long since passed its prime in offering oracular advice to the Classical Greek city-states; during the Roman period Delphi had become a stop on the tour of Greece. By the time of the end of the Roman Republic the political stage of the ancient world had completed its shift from old Greece to Rome. Greece was now the refuge of the intellectually and culturally curious, and those interested in the past. As Bowersock famously said in describing Greece during the Augustan period, "[Greece] was a country learning how to be a museum; cultivated Romans admired Greece romantically for what she had been" (Bowersock, 1965: 90–1).

Acknowledgments

I want to thank my colleague Rhys Townsend for reading this chapter and for offering suggestions and corrections.

FURTHER READING

Finlay's early study, *Greece under the Romans* (1844) was ahead of its time, as intensive scholarly interest in Roman Greece is a recent phenomenon. The most important synthesis for Roman Greece on a whole is Alcock's masterful 1993 work. For Athens, the most important early scholar was the Belgian Paul Graindor, who produced from 1921 to 1934 a series of monographs and articles all dealing with the Imperial period. This was followed by the first major synthesis of the twentieth century, Day (1942). An important Italian contribution was Accame (1946). In more recent times, one can trace scholarly wakening of Roman Greece to a seminal synthesis by Geagan (1979). This was followed by a critical examination by Shear (1981). In the last few decades, scholarship on Roman Athens has been particularly disseminated in conference proceedings: Walker and Cameron (1989), Torelli (ed., 1995), Hoff and Rotroff (1997) and, most recently, Vlizos (2008).

CHAPTER 36

Computer Technologies and Republican Archaeology at Pompeii

Michael Anderson

When the first pick-strike landed at Pompeii in 1748, the disciplines of archaeology and art history can scarcely be said to have existed (Trigger, 1989: 36–9). Indeed, from the start, the long and still on-going process of clearing Pompeii from the ashes and volcanic debris that buried the city, along with Herculaneum, Stabiae and other nearby sites, has played a major role in the development of the very disciplines that are at the core of every chapter of this volume. Even today, research at Pompeii continues to push the practice of archaeological science and art history forward, acting as a laboratory for the integration of new computerized techniques into archaeological practice and permitting an assessment of how digital technology is changing the ways that the archaeological problems of the site are studied. This is particularly true for current projects of subsurface excavation, which more than ever emphasize the study of the city's pre-Roman and Republican past and at the same time make deliberate and extensive use of technologically sophisticated methods in achieving this aim.

1 Questions of Urban Development in Pre-Roman and Republican Pompeii

Study of the pre-Roman and Republican periods of Pompeii has long been hampered by an imperfect understanding of the underlying archaeological remains. The very preservation of walls and floors for which Pompeii is so famous has naturally tended to discourage deeper excavation to reveal the full chronology (Ling, 2005: 27). As a result, the earliest narratives of urbanization and cultural interaction at Pompeii were almost entirely dependent on the historical framework derived from occasional references in the ancient

A Companion to the Archaeology of the Roman Republic, First Edition. Edited by Jane DeRose Evans.

accounts, conceived of as a series of conquests, first by Etruscans and Greeks, then by Oscan speaking Samnites, and finally by the Romans (Strabo V.4.8; Plin. *HN* 3.60–2; Livy IX.38.2–3; Vell. Pat. 2.16.2; App. *B Civ.* 1.39 and 50; and Oros. 5.22). Specifics of the urban environment and its development were then situated within this narrative by reference to relative chronologies of building techniques and wall decoration styles (Fiorelli, 1873: 78 ff.; Mau, 1882, 1908: 37; Carrington, 1933; contra, Nissen, 1877), largely without controls from ceramic or numismatic dating. This has tended to isolate Pompeian studies from the broader discussion of cultural change and interaction within peninsular Italy from the mid-sixth century to the first century BCE (cf. Keay and Terrenato 2001), and at the same time to limit the contributions by Pompeianists to the on-going debate about the processes of "Romanization" (Webster, 2001; Mattingly, 2004). I use the word "Romanization" here in the "weak" sense (after Keay and Terrenato, 2001: ix), referring to the creation of a political integration and devoid of connotations of either nationalist or colonial value-laden terminology; nor do I imply a planned or deliberate program of change initiated from Rome.

This situation began to change in the middle of the twentieth century, when the great pioneer of sub-surface research, Amadeo Maiuri (1973), expanding upon the preliminary work of Antonio Sogliano (1937), laid much of the groundwork for current archaeological excavation at the site. Present degradation of many floors in Pompeii also means that it is now possible to undertake deeper excavation without damage to intact 79 CE materials, a fact reflected by the large number of international projects of excavation active within the city (cf. Guzzo and Guidobaldi, 2005, 2008, for an overview). For this reason, though considerable research continues to be undertaken on preserved final phase remains, the past 30 years have witnessed a particular emphasis on clarifying issues of urban development and architectural chronology. This work has produced precise archaeological data for a number of locations throughout the city, serving to refine, detail, and sometimes challenge the pre-existing developmental sequence.

The earliest history of Pompeii yet remains the most incompletely understood phase of the site, a situation that has only partially been addressed by the increasing discovery of Neolithic (Varone, 2008), Bronze Age (Wynia, 1982: 331, 337; Albore Livadie *et al.*, 1986: 60; Brunsting and Wynia, 1993: 6, 15; Ranieri, 1998: 137 f.; Nilsson, 2008) and early Iron Age remains (Carafa, 1997: 16, 25–6; D'Alessio, 2008). It is, however, clear that settlement began in the area somewhat earlier than had long been supposed and that it did not correspond precisely with the area occupied by the later city (Robinson, 2008: 133). To this may be added the considerable debate surrounding the earliest phases of the circuit wall and development of the urban layout. The oldest foundations of the wall, built in the so-called "pappamonte" stone (De Caro, 1985; Maiuri, 1930, 1943), a granular local tuff, and following roughly the same course as the final-phase wall, have been found to be contemporary with the earliest traces of building activity from throughout the site as a whole (*c.* mid-sixth century) (Maiuri, 1973: 161–82, 212–16; Pesando, 1990: 195–9; Carafa, 1997: 22, 1999: 27 ff., 31; Varone, 2005; Berg, 2005). The physical evidence for the earliest wall, however, survives in only a few courses of elevation and seems never to have been very tall (Chiaramonte, 2007: 140). This implies that it might never have been intended to play a defensive role. Indeed, it might simply have been one component of a system of terracing (Sakai, 2000–1: 92; Esposito, 2008: 78) intended to equalize the internal topography and enclose the major routes of travel in the area (Chiaramonte, 2007: 145).

Whatever the original function of the circuit wall, the revelation of its early date has undermined an older theory that Pompeii's street layout was the product of periodic additions to an original urban core, long identified in the south-eastern corner and named the *altstadt* (Haverfield, 1913: 63–8; Gerkan, 1940; Laurence, 1994: 12–19; Geertman, 2007). Subsequent efforts to reconcile this idea with the sixth-century date for the city wall have considered the *altstadt* zone as a sort of arx within the larger circuit wall, in the manner of an oppidum. However, sufficient traces of mid-sixth-century structures between the *altstadt* and outer wall have now been recovered to be certain that this area was not composed entirely of vineyards or agricultural lands (Esposito, 2008: 76; Coarelli, 2008: 174). At the same time, it is clear that no great concentration of inhabitation was present in the area of the *altstadt* at this time. In fact, excavations in Insula VI 5 have hypothesized the presence of a sanctuary centering on the Tuscan order column that was later incorporated into the so-called Casa della Colonna Etrusca (Bonghi Jovino, 1988). This has been interpreted as a sacred beech grove or lucus (Varro *Ling.* 152; Coarelli, 2002; De Caro, 2007: 74) possibly dedicated to Jupiter, on analogy with the Temple of Jupiter Fagutalis at Rome (Coarelli, 2008: 174; *NTDAR* 148). To this may be added the contemporary original foundations of the Temple of Apollo on the western side of the forum (De Caro, 1986) and the so-called Doric Temple situated at the southern edge of the *altstadt* (De Waele, D'Agostino and Lulof, 2001), as well as numerous sporadic finds of ashlar foundations in pappamonte stone throughout the city (Maiuri, 1973: 161–82, 212–16; Pesando, 1990: 195–9; Carafa, 1997: 22, 1999: 27ff., 31).

The overall impression is of a low-density but consistent occupation throughout the area between the walls. This is reinforced by the fact that the major arterial routes also seem to have aligned with the earliest mid-sixth-century remains and the city gates, suggesting that some form of city layout may have been in place at this time. This planning process probably incorporated earlier pathways, such as the Via Stabiana and Via Consolare, which had developed following the natural topography (Carocci *et al.*, 1990: 208). While some of these details remain uncertain, it is clear that from the start Pompeii experienced significant influence from surrounding cultural groups. The sanctuary around the Casa della Colonna Etrusca, the Temple of Apollo and the so-called Doric Temple all display Italic features (De Caro, 2007: 73ff.), while excavations in these areas have produced sherds of imported Corinthian and Attic pottery, as well as bucchero vessels, several with Etruscan inscriptions (De Caro, 1986; Cristofani, Pandolfini Angeletti and Coppola, 1996: 59–64; Fulford and Wallace-Hadrill, 1999: 82–4, 110–12). Sufficient evidence therefore has been found to suggest interconnections with nearby Greek centers at Neapolis and Cuma as well as Etruscan sites such as Capua and Pontecagnano, which, while confirming the general outline of Pompeii's history presented by Strabo (V.4.8), provide little evidence for the political domination of local peoples assumed by ancient writers.

These earliest traces of activity at Pompeii were followed by a notoriously problematic "empty" period from about 470 onwards (Ling, 2005: 35). This may relate to the defeat of Etruscan forces by Hieron I of Syracuse in a sea battle off Cumae in 474 (Descœudres, 2007: 14). A second phase of the outer circuit wall, possibly a short-lived response to the initial movement of Oscan-Samnites or similar migrations in the early fifth century (Coarelli, 2008: 175; Esposito, 2008: 76), was followed by almost complete silence from the archaeological record. Excavations in Insula VI.10 produced further evidence of abandonment overlying earlier remains (Coarelli and Pesando, 2008: 24), and a lack of activity in Insula VI.1 (Jones and Robinson, 2005; Anderson, Jones and Robinson, forthcoming) and in the southern area of Insula VII.6, all of which also supports the identification of a reduced occupation (Anderson *et al.*, 2012). The greatest evidence of a break in the history of inhabitation, however, rests upon the cessation of temple dedications and votive offerings identified in the Temple of Apollo during this period (De Caro, 1986; Descœudres, 2007: 15).

Current excavations have provided considerable evidence that many of the features of Pompeii's urban layout were established within a second major phase of occupation, probably toward the end of the fifth century. Though no source specifically mentions the conquest or reoccupation of Pompeii, it was likely included in the general migration of Oscan-Samnites indicated by the conquest of Capua in 423 (Livy IV.37.1; Cornell, 1995: 155), and the Lucanian occupation of Paestum (Cornell, 1995: 347). Certainly the presence of inscriptions in Oscan (Vetter, 1953: 46–7) suggests that speakers of this language had arrived in Pompeii by this time, if they had not always been present. The first stage of settlement seems to have been rather small scale, possibly simply a continuation of reduced circumstances during the fifth century, concentrated into the area of the *altstadt* in the south-east corner of the ridge and directly overlooking the Sarno Valley (Coarelli, 2008: 175). Certainly Livy's (IX.38.2–3) description of Roman military actions in 310 implies that at the end of the fourth century no great settlement was present at the mouth of the Sarno River that might have resisted them (Descœudres, 2007: 15; Coarelli, 2008: 175). The zone long identified as the *altstadt* therefore may represent not the earliest settlement, but rather a secondary or reduced second stage, marked by the infiltration of this new cultural group. A possible fossa identified via subsurface geophysical examination and through traditional excavation combined with the recovery of traces interpreted as a gate in the Casa dei Postumii (Ribera, Olcina and Balester, 2007: 89; Dickmann and Pirson, 2005; Bustamente *et al.*, 2010) reveal that a wall probably encircled the *altstadt* at the end of the fifth or the beginning of the fourth century (Coarelli, 2008: 175) and not before.

The re-expansion from this limited occupation into the area occupied by the later city appears to have been a development from the end of the fourth to the early third century, when the outer city walls were rebuilt in Sarno-stone blocks, roughly following the course of the original circuit (Chiaramonte, 2007: 142; De Caro, 1985; Maiuri, 1930, 1943). The urban street grid was also finalized at this time, building upon or reusing what had already been achieved in the sixth century and coordinating with the axes of the major streets (Befani, 2005; Geertman, 2007; contra, Ling, 2005: 30–4). The insulae themselves, laid out according to a loosely "Hippodamian" orthogonal grid (Arist. *Pol.* 2.8) measured in rough Oscan feet (*c.* 27.534 cm) (Schoonhoven, 2006: 14, 37), imply that the internal layout of the city was planned at the same moment, though plausibly in a series of separate phases over an as yet undetermined period of time (Geertman, 2007: 86–8). The resulting plot divisions within each city block do not generally seem to have been fossilized in stone-built house construction until the end of the third century, but some earlier houses both in stone (De Haan *et al.*, 2005) and in more ephemeral materials (Jones and Robinson, 2005) demonstrate occupation had occurred earlier in some areas.

Coarelli (2008: 175) connects the process of urban planning at Pompeii strongly with the beginning of its participation in the Nucerian league and a resulting early

"Romanization" of the city, inspired by Latin colonies. Certainly the timing of this change is persuasive, since many Roman colonies such as Paestum, Cosa (273) and Alba Fucens (303) also exhibited urban planning in their foundations (Cornell, 1995: 383), but the phenomenon of urban planning was not restricted either to the third century or to Roman foundations (cf. fifth-century Marzabotto; Haynes, 2000: 188). However, perhaps it is best not to interpret these developments as an active process of alignment with the new ally and leader of the federation, but rather a sympathetic and shared response to a wider Mediterranean *koine* brought about through socio-political alliance and a broader world view as the result of a form of peer-polity interaction (Renfrew and Cherry, 1986: 8). In a way, this would suggest a process of acculturation not dissimilar from that which had occurred in the previous century (see Chapter 22). In fact, recent discoveries of Greek style public banqueting facilities (*hestiatorion*) under the Casa delle Forme di Creta (D'Ambrosio and De Caro, 1989), paralleled by a similar private *andron* (Brun, 2008) in Insula I.5, complete with traces of painted decoration that precede First Style decoration, are more suggestive of continuing Greek and Etruscan influences than acculturation to Roman practices.

By the late third century and continuing throughout the second century, Pompeii experienced a boom in prosperity and population reflected in extensive domestic construction in stone. The unifying characteristic of this work was building either in ashlar blocks (opus quadratum) or framework construction (opus africanum/*opera a telaio*), at first without lime-based mortar, but soon to be augmented by its addition (Peterse, 1999). These techniques, long taken as a sign of early construction at Pompeii (Fiorelli, 1873: 78 ff.; Nissen, 1877; Mau, 1879, 1908: 37; Overbeck and Mau, 1884: 30 ff.; Peterse, 1999), have been confirmed by excavation as the earliest methods, but have been dated to the end of the third century (Jones and Robinson, 2007: 391 ff.), perhaps with a few late fourth-century exceptions (De Haan *et al.*, 2005). In some areas, notably in Regio I and II, this took the form of blocks of nearly identical row houses, generally known as *casette a schiera* (Hoffman, 1979; Nappo, 1997), in a style remarkably similar in plan to houses built at Cosa between the end of the third and first quarter of the second century (see Chapters 3 and 30; Bruno and Scott, 1993: 29). Elsewhere, houses were also created according to more spacious layouts, corresponding at least in part to the concept of the so-called "atrium house," such as the Casa del Chirurgo, and the Casa di Sallustio (Jones and Robinson, 2005; Stella and Laidlaw, 2008; Pesando, 2008). One plausible explanation for the sudden rise in the population of Pompeii at this time might be a movement of refugees from the sacking of Capua and Nuceria that occurred during the Second Punic War (Varone, 1988; Nappo, 1997). However, since some of the new constructions, such as the Casa del Chirurgo, clearly replaced earlier buildings (Jones and Robinson, 2005, 2007: 391 ff.), population growth cannot be the only factor in this phenomenon.

Such major changes in the urban fabric of the city must have occurred not simply because of war-time destruction, but also because of rising prosperity and the after-effects of Roman military successes in the Second Punic War (Morel, 1989: 493). In fact, it is clear that by the beginning of the second century, the fortunes of Rome had begun to have significant economic impact at Pompeii. It might naturally follow that the city was also heavily within the Roman sphere of influence. However, even if the political and economic reality may have begun to focus on Rome at this time, acculturation was expressed in terms of the incorporation of aspects of eastern Hellenistic culture, especially reflecting royal luxuries (Zanker, 1999; Ling, 2005: 43–6). Indeed, it appears that the most significant cultural connection between Rome and the Campanian region at this time remained a continued interest in the incorporation of elite cultural practices from the eastern Mediterranean. Major additions to the urban layout at this time included a theater built largely in a Greek fashion (Mau, 1906; Richardson, 1988: 75–80), at least one palaestra (Ling, 2005: 46–7) and public bathing facilities (see Chapter 1; Eschebach, 1979), plus the addition of a portico on the southern end of the forum. Such might easily be seen as the natural result of close involvement with Roman affairs at the elite level, which was itself experiencing a similar transformation, so famously decried by traditionalists like Cato the Elder (Plut. *Cat. Mai.* 23). The classic example of this phenomenon at Pompeii is the Casa del Fauno, with its large collection of mosaic copies of Hellenistic paintings, First Style wall painting, and two grand peristyles. It surely can be seen as an expression of Hellenistic luxuria, transmitted either directly from Greece, through the cities of Magna Graecia, or through the elites in Rome, and realized by a wealthy Pompeian family (see Chapter 3; Zanker, 1999: 39–41).

At the same time, the continued use of Oscan political titles in Pompeian politics (Zanker, 1999: 52–6) is witnessed by the *tholos* of the *meddix* Numerius Trebius, a new opus sectile floor for the Temple of Apollo (Vetter, 1953: 52; L. Richardson, 1988: 93), and the inscription on the portico of V. Popidius, q(uaestor?) (*CIL* X.784), at the south end of the forum (Descœudres, 2007: 15) suggests continuity of local traditions. Recent excavations in the Temple of Venus that have produced evidence of earlier phases of the structure imply that, already in the second century, a strong connection was being forged between the Oscan Samnite deity Mefitis Fisica and the Roman goddess Venus (Curti, 2008). This means that the rebuilding of this temple after 80 can no longer be interpreted as an explicit example of Roman cultural imposition, but rather must be seen as a gradual and potentially locally-driven shift in focus tending to incorporate aspects of Roman culture within a pre-existing religious framework. Changes focused on the "Capitolium" at the northern end of the forum (Maiuri, 1973: 124), and in the floor of the Temple of Apollo, both also underway prior to the Roman colony, might therefore be interpreted in a similar light. Significantly, the motivating force of such "Romanization" would therefore be reversed, involving the subordination of nominally dominant Roman cultural features within the pre-existing cultural norms of locals for their own benefit, perhaps explaining some of the motives behind the acculturation (Keay and Terrenato, 2001: 2).

The process of "Romanization" at Pompeii would logically have received the greatest impetus in the years following the fateful involvement of the Campanian sites in the Social War (91–89) against Rome. Though Pompeii was not destroyed like Stabia (Plin. *HN* 3.70), the area near to the Porta Ercolano was clearly devastated (Strocka, 1991: 61; Jones and Robinson, 2005, 2007). In the years following, Pompeii was punitively converted into a Roman colonia, the Colonia Veneria Cornelia Pompeianorum, and hosted an influx of new inhabitants: retired soldiers from Sulla's armies. These Roman citizens presumably represented a major force promoting the adoption of Roman cultural characteristics. They do seem to have brought a taste for Second Style wall painting with them (see Chapter 3; Stroka, 1991: 68–9; Ling, 2005: 58), which came to characterize the suburbs to the north of the city where this significant addition to the Pompeian population may well have settled (Zevi, 1991, 1996). However, not even this period witnessed anything but superficial and largely cosmetic changes to bring the site into line with its new colonial role.

Inscriptions in Latin began to dominate the public sphere of the new colonia, now with renamed Latinized political offices. One of these records the euergetism of Quinctius Valgus and Marcus Porcius in the donation of a stone amphitheater (spectacula) in the eastern area of the city (*CIL* X.852; see Chapter 2). The fact that this action required the removal of houses to make space (De Franciscis, 1975) may plausibly have caused friction between colonists and locals. The same officials were responsible for a covered theater (theatrum tectum; *CIL* X.844–5) that may have served primarily as a meeting house for the colonists (Zanker, 1999: 65–8). At this time, the temple on the northern side of the forum was converted into a Capitolium appropriate for a Roman colony, though the works in progress prior to the Social War (Maiuri, 1973: 101–24) suggest that these may simply have completed a process that was already underway, such as in the Temple of Venus (cf. above). While the donation of a new altar to the Temple of Apollo (*CIL* X.800) could have been intended to co-opt this important religious sanctuary, it equally could be seen as a respectful integration into local traditions. Electoral evidence also suggests that the colonia brought about few changes to the socio-political makeup of the city, as the old ruling families seem to have regained control shortly after a temporary period of suppression (Moritsen, 1988), despite Cicero's remarks regarding political and social tensions between colonists and locals (*Sull.* 60–2).

Indeed, just as the building of urban villas cascading over the now obsolete city walls seems to have continued processes of urban expansion that were already underway before the Sullan siege (Jones and Robinson, 2005: 259), so the provision of the Forum Baths, the donations by Valgus and Porcius, and alterations to the major temples may simply have been the continuation of the program of city embellishment that had begun during the Samnite period. Shops in the Forum Baths and those in Insula VII.6, which were likely built as a single commercialization of the area, both use Oscan feet in their divisions, whether from habit, convenience, or persistence (Anderson, Jones and Robinson, forthcoming). The longest-lasting social effect of Pompeii's revolt may have been the considerable period of abandonment of the area just within the Porta Ercolano, which, following the destruction wrought by missile fire from siege in 89, seems to have been left in ruins for at least a generation. This would have been a forceful reminder of the military defeat suffered by the Pompeians, evident to anyone using the Via Consolare, until the Herculaneum gate was monumentalized in the Augustan period (Chiaramonte, 2007: 193), and the surrounding structures rebuilt.

It was during the Augustan to early Flavian period that Pompeii received for the first time structures of unambiguously Roman derivation. These included new constructions on the eastern side of the forum with particular emphasis on the celebration of the emperors (cf. Dobbins, 1994) as well as the Eumachia building, a structure plausibly inspired by the new fora in Rome given by Julius Caesar and Augustus. Excavations in the Casa del Chirurgo also suggest that the adoption of "Roman" features may not have been limited to the public sphere. Modifications undertaken at this time provided the house with the full axial vista and alae of a "Roman" atrium house, small details that it previously seems to have lacked (Anderson, Jones and Robinson, forthcoming). However, it seems unlikely that this apparently new emphasis on "Roman" features reflects a fundamental change in the processes of acculturation at Pompeii that had prevailed during the previous centuries. Rather, it reveals a change in vocabulary, suited to the new political situation presented by the early principate. At Rome, Augustus sought to recreate and rebuild the idea of "romanitas" itself as a major component of his system of propaganda (Zanker, 1988: 100–2). Perhaps the reproduction of characteristically "Roman" forms became an expression of allegiance to the new political system and simply resulted from precisely the same processes of acculturation that earlier had led to the adoption of Etruscan, Greek and Hellenizing features.

The absence of Pompeii in general from discussion about the processes and meanings of Romanization (Keay and Terrenato, 2001) has been the result of a traditional isolationism on the part of Pompeian studies themselves and a lack of precise chronological understanding at the site. The results of recent and on-going archaeological excavations have now revealed a complex process of acculturation at Pompeii, suggesting both continuities in motivation and differences in execution. This reveals the general inadequacy of traditional explanations of Romanization (as noted already by Webster, 2001; Millett, 1990b; Mattingly, 2004, inter alia), and suggests that a focus upon the motivations of local peoples may prove to be the most profitable future avenue of research. While it is unlikely that the study of pre-Roman and Republican Pompeii will be able to provide a complete explanation of this complicated phenomenon, nevertheless it will permit the particular history of Pompeii at last to participate in and contribute to the broader archaeological study of the Republican period in Italy.

2 Digital Technologies in the Archaeology of Pre-Roman and Republican Pompeii

The same projects of archaeological excavation that have served to produce new data on the Republican period of Pompeii have also witnessed an increasing integration of "cutting-edge" technology into their methodologies. For this reason, while the information they have produced has played a key role in clarifying the history of the site, the manner in which it was produced illuminates the evolution of archaeological methods. However, although the integration of digital technologies in archaeological projects at Pompeii has been very widespread, it has not experienced equal application across all aspects of the discipline. Projects generally have sought to augment the process of data collection by increasing the quantity and quality of information recorded, to speed the process of data entry through "born-digital" recording, and to facilitate the process of field research by sharing and copying data among team members. Much more rarely have digital technologies been used to produce new means of analysis or interpretation. Nor have the advantages of digital data been fully exploited to open new avenues for scholarly collaboration and data sharing. It is primarily the traditional and substantially unchanged method of the published site report that has continued to move the understanding of Pompeii forward, documenting the presence or absence of features in Pompeii's urban history and detailing the local chronological sequence. Even the movement towards rapid online publication in FastiOnline through FOLD&R, while greatly assisting timely dissemination of data, has not facilitated the release of digital data. In the end, though the application of digital technology at Pompeii has done much to advance archaeological field methodology, it has yet to reach its full potential.

The particular focus of digital technologies at Pompeii derives from the unique nature of the site and the history of its study. The depth of detail preserved by the site is at once a tremendous opportunity and a considerable challenge to recording. Modern researchers are painfully aware that a significant amount of data went unrecorded by previous generations of archaeologists (e.g. Laidlaw, 2007), and that considerable information has been lost subsequently to exposure, decay and war-time bombing (Wallace-Hadrill, 1994: 67). At the same time, such is the depth of information still preserved and length of research history that any archaeological project planning to investigate the urban development of an area of the city must not only come to terms with a wealth of archaeological detail, but an extensive pre-existing scholarly record. This might include architectural features of many thousands of walls surfaces, their related interconnections, fabric, mortars and mutual developmental history; often poorly recorded information on the original condition of the area, including finds recovered during clearance from eruption debris; the results of any previous sub-surface excavations; plus all the scientifically excavated stratified data produced by the project itself. These factors have combined to produce a near obsession with comprehensive recording in ever-increasing detail that seems to be a major motivating factor behind the particular form that digital technologies have taken at Pompeii. The simplest expression of this is the explosion of digital photography in all aspects of archaeological research (Wallrodt, 2011), producing a volume of photography that dwarfs the previous record. Other uses of advanced technology may be organized into the categories of geophysics, 3D data acquisition and modeling, excavation and data entry, and data analysis.

Geophysical examination (see Chapter 12) of subsurface remains is a technology that has only gradually come to be widely used in archaeological research at Pompeii. Initially, difficulties posed by standing architecture and small, room-sized spaces, extremely dry soils, and extensive surviving solid flooring, seem to have discouraged the use of this technology, which also may have been considered unnecessary given that the primary original purpose of geophysics was to locate new sites or to define the overall layout of known locations as an adjunct to excavation. Furthermore, earth electrical resistance (resistivity) and gradiometry (magnetometry) techniques, when applied at first produced disappointing results. The campaigns undertaken by the University of Bradford (AAPP), for instance, produced reasonable data in open areas such as the Praedia di Giulia Felice (II.4 2–12), and Palaestra Grande (II.7 1–10) but did not function well within the tight areas of the Casa del Chirurgo (VI.1, IX.10.23) (Frankel and Cheetham, 1995). The subsequent development of more powerful techniques of geophysical analysis, combined with convincingly useful implementation in a number of Roman sites outside of Campania, such as at Portus and Falerii Novi has helped to promote the use of these technologies at Pompeii. Recently many projects have undertaken successful geophysical study. Ground penetrating radar was employed by Albert Ribera, in efforts to locate traces of the fossa of the *altstadt* which were recovered on the southern side of the Casa Arianna. The use of magnetometry and electro-resistive pseudo-sections within the heavily bombed Insula VII.6 produced not only signs of buried features and preserved stratigraphy (Figure 36.1), but also confirmed the absence of unexploded bombs (Anderson, Jones and Robinson, forthcoming). Further geophysical prospection undertaken in the area of the Villa delle Colonne a Mosaico permitted the preliminary testing of theories of urban development within the shops

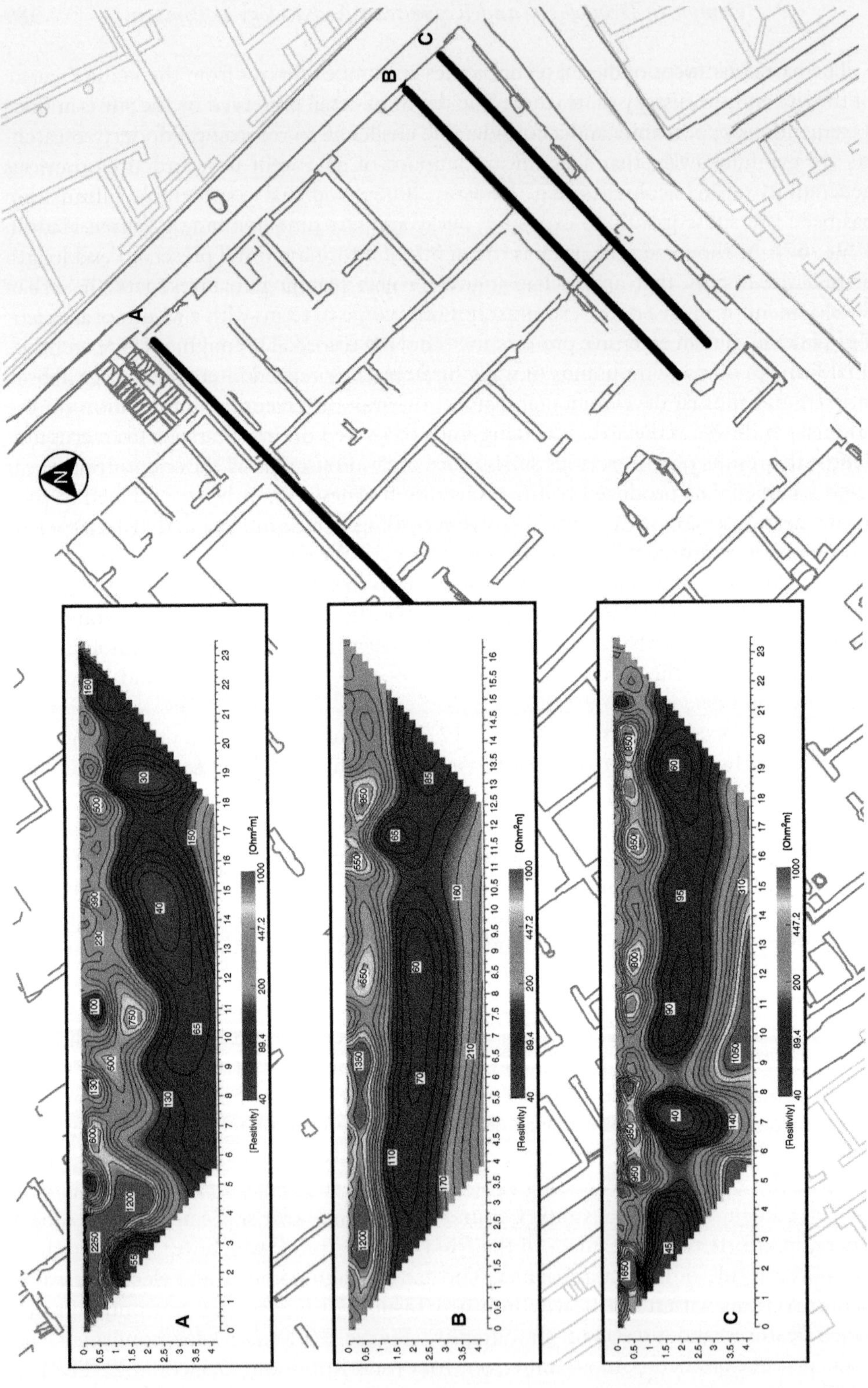

Figure 36.1 Three resistivity sections undertaken in the south-east corner of Pompeii Insula VII 6 by Geostudi Astier s.r.l. for the Via Consolare Project. Source: Used by permission of the Via Consolare Project.

on the Via Sepolchri and the planning of targeted excavation. While the most recent work directed towards sub-surface study on the site, utilizing satellite imaging similar to work successfully undertaken in Egypt (Parcak, 2009), focused deliberately on the discovery of new eruption-phase sites, future advances in this type may yet produce meaningful results for the study of Pompeian Republican and pre-Roman subsurface remains.

One of the most unusual characteristics of Pompeii and the sites buried by Vesuvius is the three-dimensionality of the remains, which retain information about their construction phases, decoration and function to a greater height and quantity than most ancient sites. This has served generally to promote the integration of technologies that ease or improve the process of recording. Electronic distance measuring devices and "total stations" (theodolites with electronic distance meters to read a point in three dimensions) were used relatively early for survey in Pompeii (Eiteljorg, 1995; Dobbins and Ball, 2005). Such devices are now common in archaeological work, but the early phase in Pompeii is an expression of this phenomenon. Furthermore, the provision of an accurate plan of surrounding structures has always been a primary requirement of work on the site, since available plans, carried out with older surveying methods (e.g. Eschebach, 1970; Van der Poel, 1986), are often inaccurate in a number of significant points of detail. While survey can produce considerably more accurate ground plans complete with topographic relief, as well as 3D wire-frame models of lines tracing the edges of standing architecture, these cannot record every tiny surface detail of the walls or the colors, textures and fabrics of features. For this reason, work at the site has more recently seen the integration of methods intended to capture all of these details with millions of data points, producing dense "point clouds" of 3D measurements. Several distinct methodologies, including so-called "time-of-flight" (the time it takes for a laser beam to bounce off the object in order to measure distance), that is, laser scanning, LiDAR ("Light Detection And Ranging"), and photogrammetry are all used to record standing structures in three dimensions. Work undertaken to scan Pompeii's forum (Balzani *et al.*, 2004; Guidi *et al.*, 2009), the Casa di Giulio Polybio (Varone, 2005) and the Casa della Diana (VII.6, 3) (cf. Luzón Nogué *et al.*, 2011) have all proven the utility of large-scale scanning in the city, serving to record wall surfaces, their locations, surface details and coloration with sub-centimetric accuracy. From a single setup of the device, all surfaces, structures and archaeological details in view of the device may be obtained in a single measurement sweep, which are then combined via "point registration" in order to provide a comprehensive set of readings that trace all surfaces of the study area. The end product can be a highly accurate 3D model of standing architecture, its surfaces, details and features, preserving the state of the standing remains in nearly perfect detail, and permitting its examination and measurement away from the site (cf. Stanco, Battiato and Gallo, 2011, for an overview). However, since laser scanners and LiDAR systems are expensive and bulky, and require significant power sources, "structure-from-motion" algorithms that are currently open-source and freely available can be employed in order to achieve similar high-resolution results with low-cost consumer grade point-and-shoot cameras. Computerized matching of points between groups of photographs taken from slightly different angles can be used to produce dense point clouds of accurate 3D measurements, albeit after several-to-many hours of processing time (Ducke, 2011; Anderson, Jones and Robinson, forthcoming).

The natural end product of point cloud measurements is a full 3D model of the subject suitable for virtual reality reconstructions and computer simulations. To date, with only a few exceptions (Curti, 2008; Murgatroyd, 2008), research directed towards the pre-Roman and Republican periods at Pompeii has tended not to make use of this type of technology, despite its advantages for theory testing, particularly in environments that are no longer preserved. For instance, recent computerized 3D reconstruction of the physical environment of the Casa del Fauno focused on its most famous mosaic treasure, the so-called Alexander Mosaic, examined the seasonal changes in light and shadows in this room, concluding that architectural changes had been made in response to these conditions (Dobbins and Grueber, in press). Since the current state of the remains does not replicate their original lighting conditions, observations such as these would be impossible without reconstruction. Nevertheless, virtual reality reconstruction at Pompeii and surrounding sites has tended to focus on 79 CE materials and the reconstruction of ancient society within them at the moment of the eruption (e.g. M. Anderson, 2004, 2008; Frederick, 2011). Furthermore, many of these initiatives appear to have a greater web occurrence than print presence (e.g. Institute for Advanced Technology in the Humanities, Museo archeologico virtuale di Ercolano) and sometimes are the result of private individuals and professional companies (e.g. Procedural, SketchUp Warehouse objects, etc.), which can render them difficult to assess for quality or accuracy, and seem to miss the potential research advantages presented by accurate 3D models (cf. Frischer and Dakouri-Hild, 2008).

However, for all of the power presented by 3D point clouds and virtual reality reconstructions, frequently a simple section or plan drawing is more appropriate for publication or to aid the process of interpretation. For this reason, many projects also employ rectified digital photography as a method of field recording (De Haan and Wallat, 2008; Boman, 2008: fig. 4; Rispoli *et al.*, 2008; Di Maio, Giugliano and Rispolli, 2008; Anderson, 2009). As a straightforward substitution for traditional section drawings, rectified photography produces flattened perspective images of wall surfaces, recording minute details of the surface (mortar details, brushstrokes and fine painted details in wall paintings) at a resolution that might not be practical with 3D point-acquisition methods and would even be difficult with hand-drawn techniques. In one version of this method, many photographs are layered in image matching software and automatically stitched together, while perspective and lens-distortion effects are removed in order to produce a flattened image of each wall surface (Figure 36.2) (Anderson, 2009). Similar, hand-produced methods have been employed by the Via dell'Abbondanza Project on a massive scale, to create photographic sections along whole streets, matched for accuracy with total station surveyed frontages and merged with traditional image-editing software (Stevens and Stevens, 2007).

Technology has also been employed directly to facilitate and streamline the process of excavation. In some instances, this has simply involved the application of similar methods of 3D data acquisition and virtual reality reconstruction to the recording of subsurface deposits (e.g. Anderson, Jones and Robinson, forthcoming). The replacement of traditional pencil and permatrace methods of excavation planning with total station survey has occurred rather widely in professional archaeology worldwide but is not yet a universal practice at Pompeii. This permits the recording of deposits in true 3D, rather than in 2D with a scattering of elevation data, but is most useful in situations where traditional

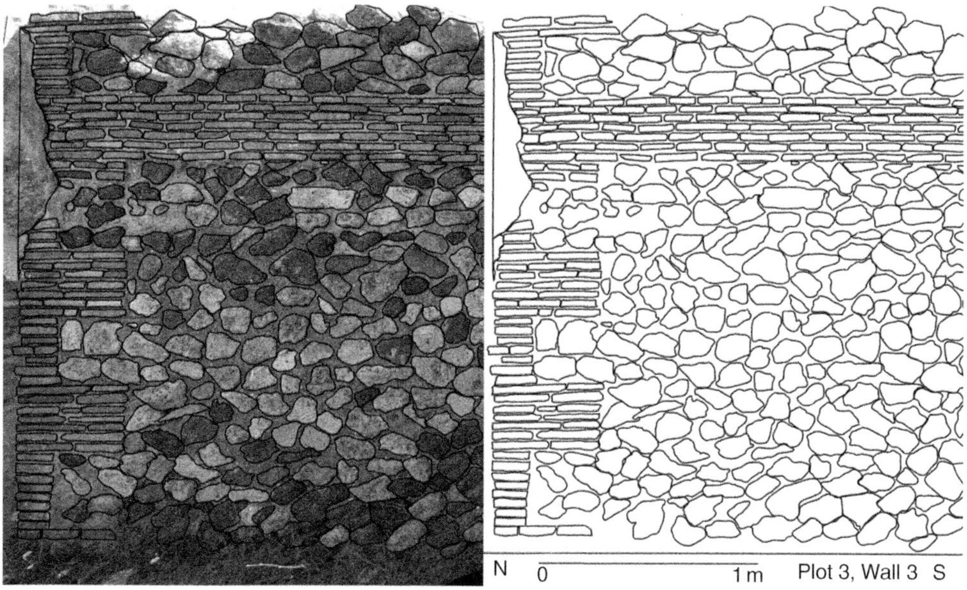

Figure 36.2 Photograph of a wall rectified using Hugin panoramic stitching software and resulting section drawing in Pompeii Insula VII 6. Source: Used by permission of the Via Consolare Project.

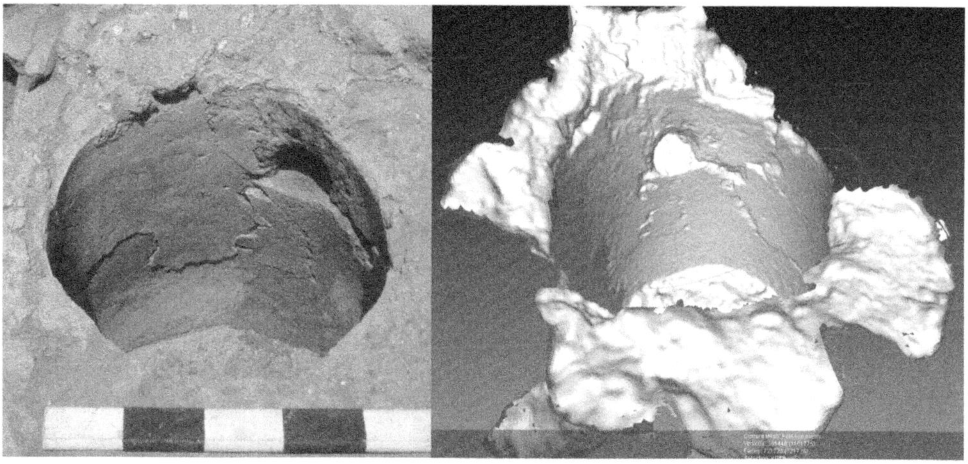

Figure 36.3 Cistern recovered in Pompeii, VII 6, 26: photograph and 3D surface mesh produced using Bundler, PMVS and Meshlab. Source: Used by permission of the Via Consolare Project.

methods would prove to be highly inaccurate due to trench depth or extreme topography. While the use of laser scanners in such undertakings in archaeological research is often limited by the cost involved, lower-cost structure-from-motion methods may serve to change this situation. Indeed, in 2011 the Via Consolare Project captured 3D surfaces of all excavated deposits including the successful "scanning" of the inside of a cistern or well-head recovered in excavation through the narrow opening at the top (Figure 36.3), which would have been nearly impossible with a tripod-mounted scanner because of the oblique viewing angles involved.

Other projects have focused the application of technology towards facilitating the process of data entry and the creation of a record that is "born digital," that is entered directly into the computer without a paper equivalent. The most recent and heavily publicized of these efforts involves the use of Apple iPads by the Pompeii Archaeological Research Project: Porta Stabia (PARP:PS) and Pompeii Quadriporticus Project (PQP) to enter and coordinate access to project data in the field (Ellis *et al.*, 2011). In a manner similar to earlier experiments successfully undertaken at Silchester by the VERA project (Fisher, Terras and Warwick, 2009), the whole system was wirelessly interconnected and updated in real time, reducing or eliminating subsequent data-entry work, and permitting real-time access and collaboration among individual team data sets. Plans of excavated materials were also sketched directly into the iPads, and context sheets and observations about wall phases entered into a digital database at trench-side, significantly reducing the work of data entry in the post-season. The wireless connectivity of this system, like that undertaken at Silchester, was also a new addition to the long-time use of field computers, PDAs, netbooks and similar devices for live data entry in the field and on site.

The most significant effect of technology in archaeology at Pompeii has been upon the recording and acquisition of data itself. In some cases the increased resolution of those data themselves can lead to new observations and conclusions. However, computer technology can also augment the process of analysis and interpretation, providing new avenues for discovery that would be impossible via traditional methods of observation. Digital relational databases are one such tool. These have seen long and profitable use in archaeology all over the world, and certainly have had similar impact when used in Pompeian archaeology. All current projects of archaeological research maintain digital databases of their finds, their excavated context records, and their archaeological excavations. Geographic information systems (GIS) present the natural extension of these abilities towards the coordination, examination and analysis of spatially related data (cf. Chapman, 2006; Conolly and Lake, 2006). An excellent example would be the linking of information on artifacts, resulting dates, soil colors and compositions, to the actual spatial location of the deposit from which they were excavated. Data from wall analysis, research into legacy publications, and the like, may all be integrated in the same way. Beyond simply displaying this data in plans, GIS-based databases can be queried spatially, in order to identify interconnections between the data that are revealed through proximity, groupings, or distribution that would otherwise not be apparent. Though currently limited by the lack of true 3D support in most GIS software, certain projects of excavation and research currently do make use of GIS applications. One of the most complete GIS implementations to date in Pompeian excavation has been carried out by the Università Ca' Foscari di Venezia in Insula VI.7 (Zaccaria Ruggiu and Maratini, 2008). However, despite wide use in archaeological research overall, the most frequent use of GIS to produce new interpretations of data has been witnessed in research focused on the final phase of the city (e.g. Knikman, 2005; Poehler, 2006; Weiss, 2009; Anderson, 2005, 2010, 2011).

3 Potentials and Pitfalls in Digital Archaeology at Pompeii

There can be no doubt that the use of digital technology in Pompeian archaeology has brought a number of distinct benefits to the practice of research and the nature of results. The changes wrought, however, also present new challenges, many of which will require further evolution of the discipline in order to be resolved. Geo-referenced plans in sub-centimetric accuracy, detailed and informative architectural sections, and 3D records of standing and subsurface deposits all significantly add to the preservation and recording of details at the site and have revolutionized the practice of field excavation and primary research. Similarly, these have increased the speed with which data can be recorded. Digital imaging, rectified photography, total station survey, laser scanning, structure-from-motion point capture, and even the use of iPads for born-digital records all reduce the time required for the acquisition of data. Often, such as in the case of digital photography or structure-from-motion data capture, there is also a clear reduction in cost of data acquisition. It should be noted however that for most digital technologies, the initial costs can be much greater than traditional excavation equipment and it remains to be seen whether the potentially numerous issues of maintenance of the equipment in harsh conditions or through rugged use mean that the costs of replacement outweigh the savings produced (cf. Fisher, Terras and Warwick, 2009).

Digital data also hold the potential for greater accessibility and collaboration. Copies of the data may easily be shared between researchers, sent electronically via the Internet, and, once produced, may be put to many subsequent and additive uses, both for local interpretation and for site-wide analysis. To date, these advantages have been experienced largely within individual projects, since rarely are the raw digital data released as a component of publication (e.g. Allison, 2004). In order for the computerized aspect of digital data to be of the greatest benefit, there must be a process by which the primary digital results of individual researchers are combined and coordinated into an ever-growing and accessible data resource. The project-specific use of GIS, for instance, if expanded into a universal and collaborative scale, could transform our understanding of Pompeii and would begin to bring together the considerable volume of data produced by 250 years of archaeological investigation. The global integration of all of this information into a single unified source would provide a tool that would change dramatically the way in which archaeological research is conducted at the site. While this observation is not new (cf. Pareschi *et al.*, 2000; Knikman, 2005: 239), a recent UNESCO review (C829 nr. 6–8) requiring the provision of a GIS for the site of Pompeii may mean that the implementation of such a system is not far away. Certainly data preservation and coordination of this type could and should only be achieved by a centralized coordinating body with the proper authority such as the Soprintendenza speciale per i beni Archeologica di Napoli e Pompei (SANP). The universal release and sharing of digital data for collaborative use generally has been hampered by concerns of intellectual property rights and a lack of unified data standards. Efforts by the SANP to collect digital data from projects and to provide a systematic grid of geodetic points (Morichi *et al.*, 2008) in order to facilitate survey of the city seem to have been aimed directly toward this goal, but in order for digital technology to equal or exceed the distribution potential of traditional print publication of results, data must not only be maintained and coordinated centrally, but also be easily accessible, perhaps through open or broad Internet access.

However, while the nature of digital data generally presents new opportunities, it can also create difficulties. The diversity of data formats from surveyed 3D data files, digital photos, clouds of point data and digital databases can make the process of integration of data not only difficult, but costly. Similarly, the rate at which technology develops means that the imposition of data standards is unlikely ever to be truly effective or useful, and the end result of digital archaeological research, unlike the relatively compatible form of analogue plans, notes and photographs produced by traditional methods, is likely to be a diverse and non-standardized set of data, recalcitrant to integration. Furthermore, the sheer volume of digital information produced by new technologies is not insignificant in size. Though in theory this information is much more easily handled and cheaply stored than traditional forms of data, nevertheless it requires significant and on-going investment to maintain, organize and make available in a manner quite dissimilar to traditional library conditions. The difficulties in upkeep experienced by the Soprintendenza regarding access to and control of previous comprehensive database projects (cf. Allison, 1992: 22–3, on the Fiat-IBM Project; IBM Italia *et al.*, 1992: 118–27) hint that while a digital record may be easier for individual projects to maintain, this advantage does not necessarily extend to controlling bodies at the institutional level. If these problems are not solved, then the positive effects of digital data on the archaeological process will likely continue to be circumscribed.

Pompeian archaeology is currently in the process of an important transformation. While the data produced by new excavation mean the urban history of Pompeii at last may begin to participate in the discussion of cultural change and interaction throughout the Republican period, many key aspects of Pompeii's development remain to be answered. At the same time, the integration of "cutting-edge" technological approaches into archaeological research at the site is a process that is far from complete and has witnessed an uneven application across the discipline. While digital technology means that the remains are examined and recorded in ever greater detail, the ways in which these new forms of computerized data will be shared, published and ultimately interpreted yet remain to be decided. It is, however, certain that the digital transformation of the archaeological process, though far from complete, holds great potential to alter fundamentally the way in which scholars go about the reconstruction of ancient history, culture and society at Pompeii.

FURTHER READING

Though the complete publication of the results of University of Bradford excavations in Insula VI.1 have not yet been published, I have been working toward the publication of the stratigraphic record for the Casa del Chirurgo (VI.1.10.21) and the Casa del Triclinio (VI.1, 1–4) which are now near completion. The details presented throughout this chapter therefore represent an accurate summary of data that shortly will become available in the forthcoming publications. The report on Via Consolare Project Field Seasons 2007–11 is available on FastiOnline (www.fastionline.org). The conclusions concerning the area around the Porta Ercolano are the results of recent stratigraphic and dating analysis by H.E.M. Cool, D. Griffiths and M. Anderson. A publication of this area is currently under production.

General information about geophysical processes can be found in Gaffney and Gater (2003), Witten (2006) and Garrison (2010). For geophysical analysis in projects in Pompeii, see Dickmann and Pirson (2005), Albiachi *et al.* (2008), Beneduce *et al.* (2008), Rispoli *et al.* (2008), Di Maio, Giugliano and Rispoli (2008) and Ellis and Devore (2010). For projects outside of Pompeii, see Ogden *et al.* (2010) and Keay *et al.* (2008) (for Portus), and for Falerii Novi, see Keay *et al.* (2000).

Besides FastiOnline, there are several web sites or web pages devoted to the study of Pompeii, including Virtual Theatre District (2007), "Pompeii Forum Project under way" (1994) and Hanna (1996). Two that are specifically concerned with 3D reconstructions are Martini (1998) and Di Domenico (2011).

PART VI

Republican Archaeology and the Twenty-First Century

CHAPTER 37

Archaeology and Acquisition: The Experience of Republican Rome

Margaret M. Miles

1 Introduction

A distinctive aspect of the archaeology of the Roman Republic is the complementary (and at times disputed) relationship between the material remains of the past and the literary tradition about the period, investigated in many of the chapters in this book. The combination of archaeology and historical study based on literature allows us to pursue such questions as what Romans during the Republic thought about the material world around them, parts of which are preserved. We can only deduce what some Romans thought about some categories of material objects at a given time, but even that glimpse of interiority helps us understand more broadly Roman cultural attitudes toward their material, visual world. Such comments made by ancient authors about material objects illustrate ancient views on the role of artifacts and the material environment in society.

The ongoing ethical problem about acquisition of archaeological artifacts is intertwined both with the written history of Republican Rome and with the history of archaeology in Italy. The debate about the ethics of acquisition is an old one, rooted in antiquity, when issues about the ownership of art and artifacts became significant because art and artifacts were so highly valued that they were taken regularly as spoils of war and at times became disputed or problematic. Booty also included human captives to be sold as slaves, arms and armor, land, bullion and coined money, but these are mentioned matter-of-factly by authors of the period. During the Republic, when Roman power was expanding throughout the peninsula of Italy and into the Mediterranean, a huge quantity of artifacts was taken from defeated Etruscans, Gauls, Spaniards, Carthaginians and Greeks. This included statuary, paintings, gold and silver vessels, tapestries and textiles, furniture, art objects of unusual design (made of precious stones), cameos and gems, and

A Companion to the Archaeology of the Roman Republic, First Edition. Edited by Jane DeRose Evans.

other artifacts such as libraries of books (in the form of scrolls), much of it what we would call today "cultural property." This booty was publicly paraded through the streets of Rome in the triumph, the greatest spectacle Rome offered, and the visible means by which the quality of the victory was judged by the public. So politically potent a spectacle was this that a triumph was awarded to conquering generals only by permission of the Senate during the Republic, and later under the Empire usually only members of the imperial family were allowed to stage them.

In their accounts of warfare, Roman authors often note the captured art among the booty, and because the art typically had been created and used in sanctuaries and dedicated to gods, in some cases they felt apprehension about appropriating for their own use the objects that technically belonged to a god or were used to worship a god. As the foreign artifacts became better known and more widely available, there was a shift in attitude about ownership, from public, sacred use to private enjoyment, which contemporary Romans quickly noticed. Some authors complained that old Roman virtues of austerity were being undermined by the new fashions for a more abundant array of objects and furnishings, and worried about the use of luxury objects and how the new richness and refinement of the captured objects might contribute to ethical decline and physical "softness," thus potentially reducing Roman valor and might. Yet the age-old assumption that "to the conqueror go the spoils" nonetheless was strongly rooted in Roman society. The discourse on what was appropriate ownership of what we would call "cultural property" also had a symbolic value and served as a larger metaphor for discussing the effects of war, both on the conquerors and the conquered.

The commentary by Roman authors about the experiences of the Republic in turn influenced more recent debates in Early Modern Europe, especially during the Napoleonic Wars, when once again conquerors took home artistic spoils of war. Roman experience illustrates the moral and ethical issues that arise when a stronger power (a nation, government, institution, individual or commercial force) appropriates items that belong to the artistic and cultural heritage of another people, and Roman reactions to this were used as cultural references in the late eighteenth and early nineteenth century, when educated people were very well read in the classics. In turn, these issues are of on-going concern for archaeologists, typically at work in "source" countries whose antiquities are exploited in various ways. This chapter explores the experience of Romans during the Republic in dealing with the artistic possessions of a defeated enemy taken in war, and how that experience has had an impact on current thinking about cultural property.

2 Livy's Depiction of Problematic Booty

In his grand account of the expansion of Roman power within Italy and beyond into the Mediterranean, Livy frequently describes the aftermath of Roman victory with imagery of captured art from the fallen city. In the triumph, a public spectacle and elaborate ritual, the captured art, arms, armor and other booty were gathered up together with human prisoners from the defeated city and processed through Rome along with the victorious troops and the celebrating general, in a formalized sequence of events that included sacrifices and offerings on the Capitoline Hill (see Holliday, 2002; Beard, 2007; Pittenger, 2008). The triumph glorified the victors and demonstrated the value of the victory to the people of Rome: spectators saw painted scenes of battles, the defeated enemy leader in person or represented by statuary, models of defeated cities, placards with numerical data about funds acquired for the treasuries (and consequent tax relief), and hundreds of wagons carrying the actual spoils. The statues, furniture, vessels and other precious objects were then dedicated in temples, or set up in special displays elsewhere in the city, or sometimes sold, and the proceeds distributed (Waurick, 1975; Pietilä-Castrén, 1987). Rome acquired a great deal of art and luxurious objects as spoils of war, and because so much of it was taken from a religious context, or represented gods, according to Livy they sometimes felt uneasy about taking ownership of it forcibly (Miles, 2008).

At times the booty became controversial: whether all of it was declared publicly, or how it was distributed, whether some of it was "sacred" and needed to be returned or treated in a special way, and also whether it was "good" for Romans generally to acquire so much ("avaritia" or greediness) are all issues recounted by Livy. He also makes clear in his account that booty and its distribution were almost always potent political issues as well, and he uses booty as a theme to illustrate the events surrounding an early episode of Roman expansion, the aftermath of the capture of Veii in 390.

Livy describes the special procedure ("evocatio" or invitation) taken by the victorious Romans in dealing with the cult image of Juno from Veii (V.22). In their prolonged struggle against the nearby Etruscan city, the Romans eventually prevailed, after a 10-year siege that would have reminded readers of the Trojan War. The leading general, Marcus Furius Camillus, vowed a temple to the chief Veientine deity, Juno, and the victorious Romans wished to take back to Rome her cult image from her temple at Veii. Chosen young men bathed and put on special garments, and asked Juno if she wanted to go: the statue seemed light, thus signifying agreement, and the statue was taken to the Aventine Hill, where eventually a new temple was constructed. Thus the concern about plundering a sacred image is resolved by ritually *inviting* the deity to transfer to Rome.

In later episodes also described by Livy, Roman concern arose about statues taken from Capua in 211, and again in 189 after the fall of Ambrakia in Greece; in both instances, the "concern" over booty had a political subtext as well. Charging that sacred images had been taken was a means to delay or potentially even block the awarding of a triumph to a political opponent. Such conflict was resolved by asking the pontifical college of priests to examine the booty to determine whether it was sacred or profane (Livy XXVI.34, XXXVIII.44). Evidently contemporary Romans assumed that specialists could determine the intrinsic sacredness of the captured statuary in a credible way, and that disputes could be resolved by carrying out this procedure.

His detailed descriptions of booty carried in a long series of triumphs give a fascinating vividness to Livy's account. Since the triumphs occur regularly in his narrative, the triumph itself becomes for the reader an overt demonstration of Roman military might: so much so that in later eras, triumphs were staged in explicit emulation of those described by Livy. Famous ones include the "triumph" staged by Pope Julius II from Bologna into Rome, 1506, and in the Early Modern period, that of Napoleon Bonaparte, who returned from Italy to Paris in 1802, and processed statuary, paintings, books and other spoils taken from Italian principalities and the Vatican (Gould, 1965; Johns, 1998; McGowan, 2000; Miller, 2001; Miles, 2008; Curran, 2011).

3 The Fall of Syracuse and Greek Art

For Livy looking retrospectively at the achievements of the Republic, the capture of Syracuse by Marcus Marcellus in 211 marked the beginning of a new decadence in Rome, due to the flood of Greek art brought to Rome from the fallen city (Livy XXV.40). What made the fall of Syracuse so notable in retrospect was that it was the first major Greek city captured and sacked by Romans. Founded in 734 (not long after Rome's traditional founding date of 753), Syracuse was notable for its wealth and luxury, and its many cultural contributions in literature, rhetoric and science, including those of Archimedes, who was killed during the siege (Finley, 1979; Jaeger, 1997: 124–31; Flower, 2003; Jaeger, 2003, 2008). Livy does not blame Marcellus personally for the aftermath of the battle, and says the spoils were legitimate, but he views that terrible moment as the beginning of general spoliation of both sacred and profane buildings, which would continue to his own day. He notes that the very temples in Rome dedicated by Marcellus that had received spoils from Syracuse are now empty (because they were plundered in recent civil wars). Thus in Livy's view, the statues and paintings gathered up by Marcellus's army fueled a strong appetite for Greek art among Romans which had no end and caused further destruction in his own time.

Livy's negative view of the impact of booty echoed the warnings of his predecessor Polybius, the Greek historian who wrote in the mid-second century about the rise of Roman power and subsequent events in Greece, Italy and Carthage that he witnessed himself. Polybius stated that the artistic spoils of war did no good for Rome ("a city is not adorned by external splendors, but by the virtue of its inhabitants," he begins, 9.10, trans. W. Paton, 1922, Loeb), that Roman conduct was wrong then and was still wrong; bullion and coin were practical and could be used by the victors, but art does not contribute to power, and excites hatred in the defeated and arouses jealousy in others; taking other people's art does not lead to a peaceful settlement. He goes on to urge future rulers of empires not to strip fallen cities as though misfortunes of others were an ornament for their own country. He adds that the artistic booty should have been left in its original place. Polybius makes it very clear that he sees no gain in acquiring the goods of others for what should be the objective of war, the appropriate, just rule in an empire: let conquered peoples keep their art objects, and let dignity and magnanimity be the adornment of Rome. Thus for Polybius the problem is not the art itself, or an appetite for it that results in a decline in vigor, but that once it is removed from its original context, the former owners are so deeply distraught at their loss they feel renewed hatred of the victors and there can be no true peace.

Polybius's strong criticism of Roman behavior at Syracuse was picked up and repeated by other authors, such as Cicero, Sallust and then Livy. The historians' complaints about supposed decadence and decline in their own time thus seem to be a way of shifting blame for the cultural trends they deprecate to a specific time in the past, to the fall of Syracuse, as a sort of repeated, commonplace lament about changing times. The fall of Syracuse was a convenient point for historians to comment generally on the effects of Greek artistic spoils of war on Republican Rome: the art triggered a cultural appetite that would last for two centuries at least, until the reign of Augustus, when Romans confidently commissioned their own styles to suit a new domestic and political environment.

4 The Fall of Carthage and Scipio Aemilianus's Repatriations

As Roman power expanded further into the Mediterranean, there is a shift in the focus on the booty as depicted in ancient authors. Now the victorious Roman generals were remembered for the particular decisions they made about newly won art, plundered from distinguished ancient cities; they are depicted as both more fastidious about what they took, and more abstemious personally (Shatzman, 1972; Churchill, 1999). The fall of Carthage and of Corinth in 146 resulted in an especially large flow of artistic booty into Rome. In their discussions of these two events that occurred in the same year, ancient authors judge the ethical actions and character of the commanding generals by how they dealt with the issue of captured art, what art was taken from what locations and what they did with it. The captured booty also becomes a motif in illustrating differences between Greek and Roman cultural attitudes and accomplishments.

Scipio Aemilianus, for example, was remembered for abstaining from taking any of the captured booty from Carthage for himself, and he is said to have forbidden his household to participate in the plundering, or even to purchase anything captured from the fallen city (e.g. Polyb. XVIII.35.9 and other examples in Astin, 1967: 76, n. 2, 78). Although the bulk of Carthaginian booty was paraded in triumph in Rome, Scipio Aemilianus also, quite remarkably, sent envoys to Sicily, offering to repatriate significant statues taken earlier by the Carthaginians from several towns on the island. Scipio Aemilianus then rededicated them with his own name attached, as Cicero remarks in the *Verrines*; three preserved inscriptions verify this (*IG* XIV.315; *Syll.*^{3} 677; *ILS* 8769).

This idea, the repatriation to their original owners of significant (usually religious) art that had been taken as booty, recalled earlier precedents noted in ancient accounts of Cyrus the Great and Alexander the Great in Persia and Mesopotamia, and how those rulers handled captured art. Scipio also gave away the scrolls from the Punic libraries at Carthage to local chieftains, after having some texts translated into Latin, thus sharing out some of the cultural booty within the Punic ambit. The portrait of Scipio painted by Cicero in the *Verrines* as a just, austere and magnanimous military hero was repeated in Valerius Maximus's collection of exemplary deeds. Much later Scipio was represented as a personification of Magnanimity in medieval Siena (North, 1979: 225). Repatriation of captured art was unusual in antiquity, but associated with "great" men who had the authority and understanding to give back significant (usually religious) items as a gesture of compassion, or, in the eyes of some beholders, out of political astuteness.

5 The Fall of Corinth and L. Mummius's Rededications

Ancient authors give varying descriptions of Lucius Mummius's treatment of the artistic booty from Corinth, taken in the same year as Scipio Aemilianus's destruction of Carthage. Perhaps best known is his supposed ignorance of art; some later authors allege that he could not tell the difference between Zeus and Philip II, referred to statues of youths as Priam and Nestor, and thought one statue or painting could be replaced with another, if it became damaged. They even blame him for allowing soldiers to play dice on famous paintings. As Gruen argued some time ago, these comments cannot withstand scrutiny (Gruen, 1992: 123–9). They seem to reflect ancient rhetorical exercises rather than actual anecdotes about Mummius; Gruen effectively rehabilitates Mummius's intelligence and judgment.

Apart from rhetorical anecdotes, the epigraphical record shows that Mummius distributed artistic war booty in carefully chosen locations over a wide geographic area (Waurick, 1975; Philip and Koenigs, 1979; Knoepfler, 1991; Tzifopoulos, 1993; Arafat, 1996: 92–7; Graverini, 2001; Yarrow, 2006). Outside of Rome, some 17 dedications are known. In Greece, after Mummius punished Corinth with utter destruction for its resistance to Rome, he made a concerted effort to reward compliant cities that had been cooperative with Roman interests. He apparently sought to make Roman presence in Greece respectful of local religious traditions in sanctuaries in the Peloponnesus and in Boeotia. This was done at Olympia first by dedicating 21 gold shields that were hung up on the metopes of the Temple of Zeus (Paus. V.10.4). The dowel cuttings in some of them are still visible. Mummius also dedicated at least two statues of Zeus, seen by Pausanias at Olympia (V.24.4). In turn, the Eleans honored him and his Roman commission of 10 senators with statues set up in the sanctuary.

In central Greece, some cities in Boeotia and along the Euboean Gulf received attention from Mummius, or honored him for some service. He was involved in agonistic festivals; apparently one festival was named after him at Eretria, the Mummieia, a form of unusual special cultic honors. An inscribed Doric frieze block found there commemorates victories in the games named for him, and for Artemis (Knoepfler, 1991). Elsewhere in the area, bases for statues dedicated by him have been found at Oropos, Aulis (probably originally from Tanagra), and two from Thebes that supported dedications to Apollo or simply "to the gods." We may surmise that statuary from Corinth was set up in these cities. In a different context, and with reference to Sulla's activities in Greece some 60 years later, Pausanias makes a caustic remark about "worshipping the gods with other people's incense" (IX.30.1). By that time this kind of dedication clearly became proverbial.

Mummius is said by Vitruvius to have taken bronze vessels used to amplify or improve sound in theaters, by which he brought this technical innovation to Rome (Poulle, 2000). So, too, the appetite for "Corinthian" bronze, a specific alloy much prized by Roman collectors, may date from the aftermath of the destruction of Corinth. Several Italian towns benefited directly from yet more dedications made by Mummius. Inscribed statue bases have been found at Parma, Nursia in Umbria, and four of them at three sites in Latium, Cures, two from Monteleone, one from Fregellae, and one from Pompeii, found in the sanctuary of Apollo. An additional base was found at Italica in Spain, and the list keeps growing as more inscriptions are discovered in excavations.

Naturally a major part of the manubial booty was taken by Mummius from Greece to Rome. There was a highly esteemed painting from Corinth of Dionysos, by the painter Aristides, which he dedicated in the Temple of Ceres. The best known of all the preserved statue bases was found in Rome, a dedication to Hercules Victor. Like the other Italian dedications, this one reminds the reader that Mummius was the conqueror of Achaea and specifically that Corinth was obliterated under his command; no such grim reminder is included on the bases set up in Greece. The inscription (*CIL* I^2.626), now in the Vatican, reads:

> When Achaea had been taken and Corinth wiped out under his leadership, his auspices and his command, Lucius Mummius, son of Lucius, consul, returned to Rome in triumph. For these successes he, the commander, dedicates this shrine and statue of the Victorious Hercules, as he had vowed during the war.

Mummius is also said to have built a round temple to Hercules Victor. Its location and identification are not certain and still in dispute, but possibly the round temple in the Forum Boarium is the one paid for by Mummius ex manubiis (cf. Chapter 28; Ziolkowski, 1988; Orlin, 1997).

What is of special interest about the find spots of the preserved dedications is the focus on two major places – Olympia, and the city of Rome – and after them, the smaller cities and sanctuaries in Achaea, Boeotia and central Italy. The link between Olympia and Rome might have seemed stronger with the choice of a circular structure in Rome, like the Philippeion at Olympia. Some scholars have suggested that the central Italian dedications may reflect Mummius's personal and familial relationships; in this view, the distribution of the booty deliberately fostered these connections. In general, the Latin inscriptions dedicate the booty to a community, whereas the Greek dedications are usually to the gods.

There are far more preserved inscriptions from Mummius than any other Republican general, which seems to support the ancient assertions that Mummius spread his dedications over all of Italy. From an Italian perspective, sharing booty would not have the same connotations as "worshipping the gods with other people's incense," as it might have had in Greece, still under the shock of the total destruction of Corinth, one of Greece's most magnificent and wealthy cities (see Chapter 35). Yet we find that the view of Mummius in Greek authors – both those of his contemporary associate Polybius, and those of Pausanias, who speaks with outrage about the fate of Corinth – is nonetheless surprisingly positive (notwithstanding the much later rhetorical exercises that question his art-historical knowledge). Even though Corinth was "wiped out" by Mummius, the authors condemn Sulla, whose actions against Athenians – starved into submission – were so brutally vicious. The collective ancient view of rededicating statues clearly depends on the context and the overall actions of the dedicator; Mummius seems to have made an effort to respect numerous sanctuaries within Achaea, and reward other Greek cities for their loyalty.

As for Mummius personally, he followed the examples set by his predecessor L. Aemilius Paullus and his sometime rival Scipio Aemilianus, and did not keep manubial booty for himself. In fact, Mummius is said to have left nothing for his household and no dowry for his daughter, which then had to be supplied by the Roman Senate (Cic. *Off.* 2.76; Plin. *HN* 34.36). For Romans during the Republic, it seems that the monetary value and art-historical value were secondary to the symbolic value of a captured item, given by the conqueror. Captured statuary and "old master" paintings were perhaps most valuable as social and cultural capital and by rededicating captured statues at many places within Greece and Italy, Mummius left a widespread record of his achievement at Corinth. He found a public and sometimes "sacred" use for the artistic booty, and escaped personal opprobrium as a conqueror, despite the thorough devastation of Corinth.

6 Stolen Statues in Peacetime: Sicily Endures Verres

A vivid picture of a collector of art and artifacts is given by Cicero in the *Verrines*, a set of speeches he wrote in 70 after successfully prosecuting Gaius Verres for extortion. Verres was sent to Greek Sicily, then a Roman province, as its governor in 73. He was in office three years rather than the normal one year, because other problems at the time kept his potential successors busy: pirates in the Mediterranean were causing an ongoing crisis for shipping; and Spartacus was leading an uprising of gladiators and slaves in southern Italy, which Romans perceived as a serious threat to domestic stability. During his three years in office as governor in Sicily, Verres abused the power of his office with systematic extortion, manipulation of the grain supply, horrific judicial abuses, and even the execution of innocent people. Cicero's speeches of prosecution focus on the main charges, extortion of 40 million sesterces from Sicilians, and in the trial itself he presented a wealth of documentary evidence to support the charges (Vasaly, 1993; Butler, 2002; Miles, 2002, 2008).

Cicero includes in the published speeches accounts of other criminal activity: Verres was an avid collector of Greek art, who would stop at nothing to acquire something he wanted. He swept the island and systematically gathered up statues, paintings, gold and silver vessels, furniture, tapestries, gems and cameos: the sorts of things usually paraded in a triumph, but now taken in peacetime. In fact, Cicero remarks that Verres took items that Marcellus himself did not, such as paintings that were in the Temple of Athena at Syracuse, and its chryselephantine doors with gold and ivory panels that Verres wrenched off (Cic. *Verr.* 2.4.123). Verres took dedications and consecrated cult images from ancient sanctuaries, public statuary set up in agoras, and family heirlooms from private household shrines. He even took for himself some of the statues that Scipio had returned to the island after the fall of Carthage. He hoarded bronze couches, tables made of exotic wood, panel paintings, jewelry and gems of all sorts. Sicilian families who invited him for dinner had their dinner service (typically gold or silver antiques, embellished with mythological scenes) "borrowed" by him but never returned. Even an elaborate bejeweled lamp stand, owned by a young prince from Asia Minor and intended as an offering to Jupiter Optimus Maximus in Rome, was seized by Verres (Cic. *Verr.* 2.4.32).

Although the formal charges were for extortion, the other accusations made the case seem all the more significant to his immediate audience of jurors and the crowd in the Forum in Rome, because Cicero presented many details about Verres' illegal acquisition of art as religious and ethical violations (Leen, 1991). Through the lens of this trial as presented by Cicero, we learn his views on appropriate ownership of art. Cicero is the first preserved author to raise issues about public ownership and use of art, versus private ownership and use of art: "Why do we [Romans] value art? Who should 'own' art? Does art have a fixed location where it belongs? Should the victors in war ever allow the defeated to retain their art?" Cicero plays upon the dichotomy between public uses of art and its private enjoyment and repeats in various ways the rhetorical theme of good uses of art (meaning public, commemorative and religious) and bad uses of art (private, consumptive and decadent); these were not new ideas about art and its purposes, but he formulated them in ways that his readers would have understood and appreciated. These widely held views, as articulated by Cicero, were then taken up and repeated in new contexts by later Roman authors, such as Pliny the Elder.

Cicero's damning portrait of Verres as a greedy, criminal collector was so potent that in the twilight of the Roman Republic the fashions for collecting and displaying Greek art began to shift to new pursuits (Miles, 2008: 218–40). The impact of the speeches may be traced in subsequent Roman authors, and because generations from the Renaissance onward also admired Cicero and his powerful rhetoric, the story of Verres' peacetime plundering of Sicily continued to be read. When northern European travelers began to visit Italy as part of the Grand Tour in the eighteenth century, Cicero's speeches on Verres became part of their "background" reading, just as Pliny the Elder's chapters on art were read eagerly by artists and architects. Thus, Roman views on art and art ownership helped shape more recent views in the early modern period.

7 Archaeology and Acquisition in Peace and War

The history of archaeology in Italy is intertwined with antiquarianism, the activities of collectors, and scholarly, text-based pursuits. During the Renaissance, the eagerness to learn about classical culture through newly discovered texts, monuments and chance finds was pervasive and gave its name to the period, a time of "rebirth" of classical culture. A farmer plowing a field on January 14, 1506 in the area over the Baths of Trajan suddenly came upon a hole that when opened up, revealed the marble statue of Laocoön. It was instantly identified as "Pliny's Laocoön" by a well-read group called to the site that included Michelangelo and the architect Giuliano da Sangallo and his young son, who, as a grown man, wrote about the experience (cf. Plin. *HN* 36.37; Hibbard, 1974: 90–1; Haskell and Penny, 1981: 243). Pope Julius II purchased it after a couple of months of debate over who should have it; it soon inspired poetry and was considered by all to be the finest ancient statue ever found until well into the nineteenth century, when higher regard for the Parthenon marbles began to put it into eclipse (Curran, 2011: 113–14). For artists, architects and scholars of the Italian Renaissance, the texts and physical remains of the Roman past around them were inspiring exemplars, a constant muse for new creations, new commentaries and histories.

In the eighteenth century, the lure of contemporary Rome brought educated visitors who were interested in the past and who had read Livy, Cicero and Pliny in Latin; these wealthy Grand Tourists brought with them or hired in Italy artists and architects to document their travel (Black, 1992; Jenkins, 1992; Chaney, 1998; Jenkins and Sloane, 1996). For those of a scholarly turn of mind, texts were influential in directing early efforts at historical research, but in many instances such early research was stimulated by happenstance, whatever became available to study, rather than an endeavor systematically pursued with an agenda set up to answer specific questions. The serendipity of study sometimes followed the serendipity of preservation: the discovery of Herculaneum (1738) and Pompeii (1748) caused a sensation, and people flocked to see the diggings. The idea of being able to walk on actual ancient Roman streets and step into ancient Roman houses was thrilling, and of course avid interest in the cities overpowered by the volcano continues to this day. Initially, however, the finds were tightly controlled and visitors were not even allowed to sketch. The purpose of those diggings was to acquire marble and sculpture for the palace and villas of the local Bourbon monarchs and the digs were closely guarded as royal property.

Acquisition was central to the motivation for early efforts to uncover the past: while popes and kings were able to command and collect the more monumental items and statuary, antiquarians and collectors such as Sir William Hamilton, British envoy to the Bourbon court at Naples, sought small-scale items that could be sold easily (Jenkins and Sloane, 1996). Traveling architects made drawings and recordings, but they also took home with them pieces of buildings to be used as models, from Rome and elsewhere in Italy, and Greece and Turkey (Lethaby, 1908). In general, early explorers kept or sold what they found.

As an academic discipline, archaeology grew out of antiquarianism from several different starting points. Ideas about ordering the art of the past into a broader history are credited to J.J. Winckelmann, author of *History of Art* (1764), who spent much time in Italy but never went to Sicily or Greece, yet contributed a new way of looking at art that is still fundamental to both art history and archaeology (Potts, 1994). In northern Europe, local histories were pursued with on-site investigations and these efforts contributed new archaeological ideas about taxonomies and typologies of artifacts (Schapp, 1996). But the move from antiquarianism toward scientific archaeology required a new way of thinking about time. Until the early nineteenth century, calculations in historical time by scholars were restricted by reckonings based on the Bible (with creation set in 4004). After the publication of Sir Charles Lyell's *Principles of Geology* (1830–3), layered time could be postulated, and Darwin's *On the Origin of Species* (1859) provided a longer evolutionary view of the past. From this arose the concept of a sequence of physical layers in the earth, which could be systematically excavated, providing a historical context for artifacts.

Once the physical relationship between artifacts and historical context was appreciated and understood, archaeology could become a more rigorous scientific endeavor, directed at historical understanding rather than ownership of objects. The early pioneer Giuseppe Fiorelli put these techniques into use as Director of the excavations at Pompeii (1860–75), and under his leadership stratigraphy was observed carefully and the first scientific excavations at Pompeii were carried out. He also invented new techniques specific to the site, such as using plaster to fill voids that allowed the recovery of "casts" of human bodies, furniture, food items and other finds usually not preserved in a typical excavation. Under Fiorelli, some finds from Pompeii were left *in situ* and the houses restored to protect them, or they were put into the Naples Museum, rather than into private collections. The collections in the Naples Museum became state property after 1860, and Fiorelli used that moment to re-organize them.

Italy suffered extensive loss of both antiquities and later Italian art during the Napoleonic Wars, when the French invasion and occupation resulted in an enormous transfer of Italian-owned art to Paris. Napoleon and his troops systematically plundered Italy during his invasion of 1796, in a continuation of seizures begun under the new French republic in northern Europe during the early 1790s. In Italy the seizures began as a series of unequal treaties that included handovers to the victors of artistic property, but expanded as the French moved south and began gathering up Italian art from churches and private collections to take back to Paris. The Laocoön was taken from the Vatican, still then the most famous statue from antiquity; the Apollo Belvedere, the Dying Gaul, much of the papal library and even the pope's working papers followed that statue. Other booty from Italy included the Medici Venus, geological samples and collections, and hundreds of paintings large and small. Considerable public outcry arose, not only in Britain (which was opposed to the French) but also in France. Some 50 French artists and scholars signed an open letter of protest written by the architectural theorist A. Quatremère de Quincy, who modeled some of his arguments against the French plundering on those of Polybius and Cicero (Quatremère de Quincy, 1989; Lavin, 1992: 150–7). Nonetheless, the seizure of art continued, and the collections of the Louvre (for the time being designated the Musée Napoléon) in Paris grew ever larger (Gould, 1965; Parker, 1965; Johns, 1998).

When Napoleon organized his fleet and equipment for the invasion of Egypt in 1798, he brought along some 167 scientists and scholars, including engineers, draftsmen, naturalists, artists and architects whose job it was to assist with the infrastructure for the military (roads, canals, food supply, etc.) but also to make illustrations of Egypt and its antiquities. This resulted in the *Description de l'Égypte* (published in 1809–21) which became the foundation for the field of modern Egyptology. Napoleon's men also collected many antiquities, including the Rosetta Stone, but these were taken as spoils of war by the British after the final defeat of the French in Egypt in 1801, and are now in the British Museum. An explosion of interest in Egyptian antiquity followed, including "Egyptomania" in fashion and consumables, but also new scholarly efforts which led soon to the decipherment of hieroglyphics, based on the Rosetta Stone.

After Waterloo, the British under the leadership of the Duke of Wellington made the decision that the art looted by Napoleon and the French should be repatriated, to German principalities and to the Netherlands, as they had participated in the allied defeat of Napoleon, but also to Italy. In a public letter written on September 23, 1815, the Duke of Wellington outlined his reasons for repatriation of the artistic spoils: the art would only fuel future animosity in France as it was a reminder of military triumph; and it belonged in its original place. This marked the first modern repatriation of art taken in time of war, by fiat, from the commanding general (Miles, 2011). The fate of the art in Paris was followed avidly in newspapers throughout Europe (repatriation was strenuously resisted by the French, who felt they had "won" the art) and it was a primary focus of reporting for many months after the battle (for views in British newspapers, see Miles, 2008: 336–43; for French views, see Johns, 1998: 183–94; Pommier, 1999: 354–7). The journalists often invoked the discourse about booty from Polybius, Livy and Cicero and noted the many parallels with the Roman Republic, a comparison made easier by the assessments of Napoleon himself. In all countries that had been protagonists in the recent war the discussion was followed closely, and it raised keen public awareness about the disposition and ownership of art. Current tabulations show that about 55% of what was taken was returned to Italy, but the rest of the pieces had been sent out to provincial museums in France where they were more difficult to retrieve.

In the aftermath of the repatriations after Waterloo, art collecting intensified on a national scale as museums and monarchs in Europe competed to acquire new finds. The idea of a "universal" museum, which could provide visitors and local artists with examples of art from many different parts of the globe, was rooted in the artistic and educational ideals of the eighteenth century, and began to become a reality in the nineteenth century with the expansion of the collections in the Louvre and the British Museum (Duncan and Wallach, 1980; McClellan, 1994; Marchand, 2000). By the later nineteenth century, European museums and private organizations sponsored collecting expeditions into the Near East, Turkey and Greece. The explorers were sometimes required to share their finds with the "host" government.

In all the examples of forcible acquisition noted above, there is a plain commonality in artistic appreciation among various cultures: the Roman victors wanted the Etruscan image of Juno from Veii, and the Greek images from Syracuse, Ambrakia and Corinth. The spoils from Carthage were purchased by others, if not Scipio's household. Much later, Napoleon was eager to bring ancient Roman art from the Vatican to Paris, as well as Italian Renaissance painting, Flemish tapestries and paintings, granite columns from Charlemagne's royal church at Aachen and other artistic trophies. The British who defeated the French in Egypt took over much of what had been gathered up there. Collecting and ownership of items from other cultures has a very long history, and the explicitly "universal" ambitions of collectors in the late eighteenth and early nineteenth century had ample precedent in Republican Rome. For example, after Pompey's third triumph in 61, in which the lavish personal effects of the defeated Mithridates VI were paraded, Pompey used some of his manubial funds to build a new theater-temple complex, complete with a garden of trees exotic to Italy (ebony and balsam from Judea), statues of conquered nations, and other reminders that he had been victorious on three continents (Kuttner, 1999; Mayor, 2010: 357–60).

Despite advances in archaeology and art history as academic disciplines, and international legal efforts to limit looting in wartime, the systematic seizure of art by the Nazis in the Second World War, and the more recent looting of the Baghdad Museum during the Gulf War, illustrates how both legal and cultural expectations may be ignored (Nicholas, 1994; Simpson, 1997; Bogdanos, 2005a, 2005b). International agreements about protecting cultural property in wartime, including the Hague Conventions of 1907 and 1954, were based in part on a memorandum drawn up by Francis Lieber for President Abraham Lincoln for Union troops in the American Civil War, *General Orders No. 100* (1863). Lieber was a professor of political science who had emigrated to the United States from Prussia; when he was 16 years old, he served in the Prussian army at Waterloo, and would have been familiar with the debates surrounding the art in the Louvre. He was trained as a classicist and had studied with the great historian of Rome, Barthold Niebuhr. The Lieber Code (as it is known) is based on a long tradition of legal and juridical reflection that came to see "cultural property" as a category of real property that should be treated in an exceptional way in warfare (Merryman and Elsen, 1998; Sax, 1999). In the evolution of modern thinking about this issue, Roman authors made a signal contribution. Although international agreements have not guaranteed protection in war, at least today there is a common ground of agreement that cultural property should be protected.

The existing agreements about cultural property in wartime serve as a model to address another current problem, the looting of antiquities from archaeological sites. Modern excavators are professionally trained and want to see finds properly conserved in local museums. Because private individuals and some public museums are willing to pay high prices for objects to put in their collections, however, scavenging and looting is carried out by thieves hoping to sell their finds, thus posing another type of acute threat to cultural heritage. The destruction of archaeological sites is an on-going problem in "source" countries such as Italy, Greece and Turkey and in many sites in South and

Central America, Africa and China. Orphaned finds without context represent a loss of history, for there is both an interpretive problem for the archaeologist, and an ethical one: there is much less we can say about an undocumented find – assuming it is authentic and not a fake. In actuality, one can comment only on how it might fit into an established typology, or appreciate and describe its appearance. Because publication confers a certain "legitimacy" to stolen objects, we do not publish undocumented finds, and professional journals will not accept papers about them. Further efforts for international agreements controlling the exportation and marketing of antiquities are underway.

Ancient art has been contested in war, treated as booty, and cherished in collections at least since the Roman Republic, when it had an urgency and immediacy because the art was actually changing Roman culture. People take pride in ownership of artifacts, whether privately held or by some larger corporate body such as a community or nation. For the Romans of the Republic, captured art was a military trophy as well as something to be enjoyed and appreciated, perhaps to be recycled in other sanctuaries as Mummius did, and, very rarely, to be given back to its owners. The perceived value of an object from the past may be based on aesthetic considerations, or it may be prized for its singularity or rarity, and it can be a symbol of educated taste, and of leisure, but it is always much more valuable if we know its original find spot and context: then the object is truly a historical artifact.

FURTHER READING

For discussion of tensions and resolutions between Greek and Roman culture with an analysis of individual issues, see the standard account of the era, Gruen (1984), with other cultural studies, Rawson (1985b), Gruen (1992, 1996); on administration, Lintott (1993); on the issues viewed through art, Pollitt (1974), Ridgway (1984), Isager (1991), Holliday (2002), Edwards (2003) and Dillon and Welch (2006). On Livy and his techniques, see Jaeger (1997) and the essays in Chaplin and Kraus (2009); on Cicero and the trial of Verres, see Mitchell (1979), Vasaly (1993), Butler (2002) and Miles (2008). For the impact of Roman texts and art in later periods, see Greenhalgh (1989) on the medieval period, and Haskell and Penny (1981). Bober and Rubenstein (1986), Schapp (1996), Ayres (1997), Rowland (1998) and Barkan (1999) provide various perspectives on how ancient art and texts were used in the Renaissance and later. For histories of collecting and of museums, see Jenkins (1992), Pearce (1992, 1995), Elsner and Cardinal (1994) and Crane (2000). Renfrew (2000) is very judicious and balanced on problems related to museums purchasing illicit antiquities. For investigative journalism, see Watson (2006) and Felch and Frammolino (2011). Muensterberger (1994) is an excellent book on the psychology of collectors. Legal histories of international agreements begin with Hugo Grotius and continue onward: see Ballis (1973), Ruddy (1975), Bull, Kingsbury and Roberts (1990) and Buckle (1991). For a full and detailed account, see Grewe (2000); for a discussion of the impact of ancient Greek and Roman authors on legal thinking, see Bederman (2001); for a biography of Francis Lieber, see Freidel (1968). Sax (1999) gives a very readable account of the crucial legal issues and competing interests surrounding cultural property. For archaeological ethics, see Messenger (1989), Vitelli (1996), Carman (1996) and Meskell (1998). On restitution, see Merryman (2000), Warren (2004), Gill and Chippendale (2007) and Miles (2011). Extensive bibliographies on cultural property and the problem of looting can be found on the web site: http://www.davidgill.co.uk/looting/returning.html.

References

Abad, L. (ed.) (2005) *De Iberia in Hispaniam. La adaptación de las sociedades ibéricas a los modelos romanos.* Alicante.

Abad, L. (2006) The juridical promotion of Oppida of the southeast of the Iberian Peninsula: the cases of Ilici and Ilunum, in Abad, Keay and Ramallo (eds), pp. 118–32.

Abad, L., Keay, S. and Ramallo, S. (eds) (2006) *Early Roman Towns in Hispania Tarraconensis* [*JRA* Suppl. 62]. Portsmouth, RI.

Abadie-Reynal, C. (ed.) (2003) *Les céramiques en Anatolie aux époques hellénistique et romaine: productions et échanges.* Paris.

Aberson, M. (1994) *Temples votifs et butin de guerre dans la Rome républicaine.* Rome.

Abramson, H. (1974) The Olympieion in Athens and its connections with Rome. *CSCA* 7, 1–25.

Actes del XIV Congrés Internacional d'Arqueologia Clàssica. La ciutat en el món romà (Tarragona, 1993) (1994), vol. 2. Tarragona.

Accame, S.S. (1946) *Il Domino romano in Grecia dalla guerra acacia ad Augusto.* Rome.

Adam, J.-P. (1984) *La construction romaine. Matériaux et techniques.* Paris.

Adam, J.-P. (1994) *Le Temple de Portunus au Forum Boarium.* Rome.

Adamesteanu, D. and Lejeune, M. (1971) Il santuario lucano di Macchia di Rossano di Vaglio. *MemLinc* 16, 39–83.

Adams, D. (ed.) (1997) *Festschrift for Eric H. Hamp*, vol. 2. Washington DC.

Adams, J.N. (1993) The generic use of *mula* and the status and employment of female mules in the Roman world. *RhM* 136, 35–61.

Adams, J.N. (2003) *Bilingualism and the Latin Language.* Cambridge.

Adams, J.N. (2007) *The Regional Diversification of Latin 200* BC–AD *600.* Cambridge.

Adams, J.P. (1982) Polybius, Pliny and the Via Egnatia, in Adams and Borza (eds), pp. 269–302.

Adams, W.L. and Borza, E.N. (eds) (1982) *Philip II, Alexander the Great and the Macedonian Heritage.* Washington DC.

Adan-Bayewitz, D. (1993) *Common Pottery of Roman Galilee: A Study of Local Trade.* Ramat Gan.

Agostinelli, M. and Bietti Sestieri, A.M. (eds) (1985) *Roma. Archeologia nel centro*, 2 vols. Rome.

A Companion to the Archaeology of the Roman Republic, First Edition. Edited by Jane DeRose Evans.

Agustí, B., Burch, J. and Merino, J. (eds) (1995) *Excavacions d'urgència a Sant Julià de Ramis, anys 1991–1993.* Girona.

Aicher, P. (1995) *Guide to the Aqueducts of Ancient Rome.* Wauconda, IL.

Aigner Foresti, L., Barzanò, A. and Bearzot, C. (eds) (1994) *Federazioni e federalismo nell'Europa antica. Bergamo, 21–25 settembre 1992. Alle radici della casa comune europea, 1.* Milan.

Aitken, R. (1956) Virgil's plough. *JRS* 46 (1–2), 97–106.

Akkari-Weriemmi, J. (1985) Un témoinage spectaculaire sur la presence libyco-punique dans l'ile de Jerba: le mausolée de Henchir Bourgou. *Reppal* 1, 189–96.

Alarcão, J. (1988) *Roman Portugal.* Warminster.

Albers, J. (ed.) (2008) *Das Marsfeld in Rome.* Bern.

Albertoni, M. (1983) La necropoli Esquilina arcaica e repubblicana 1870–1911, in Pisani Sartorio and Quilici (eds), pp. 140–55.

Albertoni, M. and Damiani, I. (eds) (2008) *Il tempio di Giove e le origini del colle capitolino.* Milan.

Albiachi, R., Ballester, I., Escrivá, E. *et al.* (2008) Estudios estratigráficos y geofísicos entre la casa de Ariadna y el Vicolo Storto (VII, 4), in Guzzo and Guidobaldi (eds), 2008, pp. 249–64.

Albore Livadie, C. (ed.) (1986) *Tremblements de Terre, Éruptions Volcaniques et Vie des Hommes dans la Campanie Antique.* Naples.

Albore Livadie, C., D'Alessio, G., Mastrolorenzo, G. and Rolandi, G. (1986) Le eruzioni del Somma-Vesuvio in epoca protostorica, in Albore Livadie (ed.), pp. 55–66.

Alcock, S.E. (1993) *Graecia Capta. The Landscapes of Roman Greece.* Cambridge.

Alcock, S.E. (1997) Greece: a landscape of resistance?, in Mattingly (ed.), pp. 103–16.

Alcock, S.E. (2002) *Archaeologies of the Greek Past: Landscape, Monuments, and Memories.* Cambridge.

Alcock, S.E. (2007a) The essential countryside. Introduction, in Alcock and Osborne (eds), pp. 118–19.

Alcock, S.E. (2007b) The eastern Mediterranean, in Scheidel, Morris and Saller (eds), pp. 671–97.

Alcock, S. and Cherry, J. (eds) (2004) *Side-by-Side Survey: Comparative Regional Analysis in the Mediterranean Region.* Oxford.

Alcock, S. and Osborne, S. (eds) (2007) *Classical Archaeology.* Oxford.

Aldrete, G.S. (2007) *Floods of the Tiber in Ancient Rome.* Baltimore.

Aldrete, G.S. and Mattingly, D.J. (1999) Feeding the city: the organization, operation, and scale of the supply system for Rome, in Potter and Mattingly (eds), pp. 171–204.

Alföldy, G. (1975) *Die römischen Inschriften von Tarraco.* Berlin.

Alföldy, G. (1981) Die alteste römische Inschrift der iberischen Halbinsel. *ZPE* 43, 1–12.

Alföldy, G. (2000) Wann wurde Tarraco römische Kolonie?, in Paci (ed.), pp. 3–22.

Alföldy, G. (2004) Introducción histórica, in Dupré (ed.), pp. 7–14.

Al-Harthi, A.A., Al-Amri, R.M. and Shehata, W.M. (1999) The porosity and engineering properties of vesicular basalt in Saudi Arabia. *Engineering Geology* 54, 313–20.

Allen, H.L. (1970) Excavations at Morgantina (Serra Orlando) 1967–1969: preliminary report X. *AJA* 74, 359–83.

Allison, P. (1992) The distribution of Pompeiian house contents and its significance. PhD thesis. University of Sydney.

Allison, P. (1993) How do we identify the use of space in Roman housing?, in Moorman (ed.), pp. 1–8.

Allison, P. (2004) *Pompeian Households: An Analysis of the Material Culture* [Cotsen Institute of Archaeology Monograph 42]. Los Angeles.

Almagro-Gorbea, M. (ed.) (1982) *El Santuario de Juno en Gabii. Excavaciones 1956–69.* Madrid.

Almagro-Gorbea, M., Palomero, S. and Osuna, M. (eds) (1997) *Ciudades Romanas en la Provincia de Cuenca. Homenaje a Francisco Suay (Cuenca 1992).* Cuenca.

Álvarez, J. and Nogales, T. (2003) *Forum Coloniae Augustae Emeritae.* Mérida.

Alvarez, W., Ammerman, A.J., Renne, P.R. *et al.* (1996) Quaternary fluvial-volcanic stratigraphy and geochronology of the Capitoline Hill in Rome. *Geology* 24 (8), 751–4.
Alvarez, W., Claeys, P. and Montanari, A. (2009) Time-scale construction and periodizing in big history: from the Eocene-Oligocene boundary to all of the past. *Geological Society of America Special Paper* 452, 1–15.
Amandry, M., Burnett, A. and Ripolles, P.P. (1992) *Roman Provincial Coinage*, vol. 1. Paris.
Amela, L. (2002a) *Las clientelas de Cneo Pompeyo Magno en Hispania*. Barcelona.
Amela, L. (2002b) Lépido en Hispania. *HispAnt* 26, 35–8.
Amendolea, B. (ed.) (1995) *I siti archeologici. Un problema di musealizzazione all'aperto*. Pisa.
Amici, C.M. (1991) *Foro di Cesare*. Florence.
Amici, C.M. (2004–5) Evoluzione architettonica del comizio a Roma. *RendPontAcc* 77, 351–79.
Amici, C.M. *et al.* (2007) *Lo scavo didattico della zona retrostante la Curia (Foro di Cesare): Campagne di scavo 1961–1970.* Rome.
Ammerman, A.J. (1990a) On the origins of the Forum Romanum. *AJA* 94, 627–45.
Ammerman, A.J. (1990b) Morfologia ed ambiente dell'area del Foro Romano. *Archeologia Laziale* 10, 13–16.
Ammerman, A.J. (1998) Environmental archaeology in the Velabrum, Rome: interim report. *JRA* 11, 213–23.
Ammerman, A.J. (2006) Adding time to Rome's *Imago*, in Haselberger and Humphrey (eds), pp. 297–308.
Ammerman, A.J. (2007) Review of G. Heiken, R. Funiciello, and D. De Rita, *The Seven Hills of Rome. A Geological Tour of the Eternal City*. *JRA* 20, 403–4.
Ammerman, A.J. (2009) Environmental setting, in Scott (ed.), pp. 153–63.
Ammerman, A.J. (2010) Forum in Rome, in Gagarin and Fantham (eds), pp. 212–22.
Ammerman, A.J. and Filippi, D. (2000) Nuove osservazioni sull'area a nord del Comizio. *BullCom* 101, 27–38.
Ammerman, A.J. and Filippi, D. (2004) Dal Tevere all'Argileto: nuove osservazioni. *BullCom* 105, 7–28.
Ammerman, A.J., Iliopoulos, I., Bondioli, F. *et al.* (2008) The clay beds in the Velabrum and the earliest tiles in Rome. *JRA* 21, 7–30.
Ammerman, A.J. and Montanari, A. (2000) Palatium e Sacra Via: sourcing the walls. *Bollettino di Archeologia, Ministero Beni Culturali e Ambientali* 31–3, 189–93.
Ammerman, A.J. and Terrenato, N. (1996) Nuove osservazioni sul Colle Capitolino. *BullCom* 97, 35–46.
Ampolo, C. (1992) Onomastica e mobilità sociale: da Strinpon a Stremponius. A proposito dell'iscrizione da Pontecagnano. *AION* 14, 63–8.
Ampolo, C. (2001) *Da un'antica città di Sicilia: i decreti di Entella e Nakone*. Pisa.
Ampolo, C. (ed.) (2009) *Immagine e immagini della Sicilia e di altre isole del Mediterraneo antico*, 2 vols. Pisa.
Ampolo, C. (ed.) (forthcoming) *"Agorai" di Sicilia, "agorai" d'Occidente*. Pisa.
Ampolo, C., Cassola, F., Fabbrini, F. *et al.* (eds) (1990) *Roma e l'Italia: radices imperii*. Milan.
Anderson, G. (ed.) (2004) *Reinventing the Museum. Historical and Contemporary Perspectives on the Paradigm Shift*. Walnut Creek, CA.
Anderson, J.C., jr. (1997) *Roman Architecture and Society*. Baltimore.
Anderson, M. (2004) Computer games and archaeological reconstruction: the low cost VR, in Magistrat der Stadt Wien (ed.), pp. 521–24.
Anderson, M. (2005) Houses, GIS and the micro-topology of Pompeian domestic space, in Bruhn, Croxford and Grigoropoulos (eds), pp. 144–56.
Anderson, M. (2008) Putting the reality in virtual reality: new advances through game engine technology, in Posluschny, Lambers and Herzog (eds), pp. 144ff. (CD-ROM).

Anderson, M. (2009) Precision recording of Pompeian standing remains via stitched rectified photography, in J. Crawford, Frischer and Koller (eds), pp. 22–26.

Anderson, M. (2010) Mapping the domestic landscape: GIS, visibility and the Pompeian house, in Niccolucci and Hermon (eds), pp. 183–9.

Anderson, M. (2011) Disruption or continuity? The spatio-visual evidence of post earthquake Pompeii, in Flohr, Cole and Poehler (eds), pp. 74–87.

Anderson, M., Jones, R. and Robinson, R. (forthcoming) *Excavations in the Casa del Chirurgo*. Oxford.

Anderson, M., Weiss, C.J., Edwards, B.R. *et al.* (2012) Via Consolare Project – 2007–2011 field seasons in Insula VII.6. *Journal of Fasti Online*, FOLD&R, http://www.fastionline.org/docs/FOLDER-it-2012-247.pdf (accessed November 16, 2012).

Andreau, J., Broise, H., Catalli, F. *et al.* (2002) *Musarna I : Les Trésors Monétaires*. Rome.

Andreau, J. and Descat, R. (2006) *Esclave en Grèce et à Rome*. Paris.

Andrén, A. (1939) *Architectural Terracottas from Etrusco-Italic Temples*, 2 vols. Lund.

Andreu, J., Cabrero, J. and Rodà, I. (eds) (2009) *Hispaniae. Las provincias hispanas en el mundo romano* [Documenta 11, ICAC]. Tarragona.

Anello, P. and Martínez-Pinna, J. (eds) (2008) *Relaciones interculturales en el Mediterráneo antiguo: Sicilia e Iberia*. Málaga.

Angel, J.L. (1964) Osteoporosis: thalassemia?, *American Journal of Physical Anthropology* 22, 369–74.

Angelelli, C. and Falzone, S. (1999) Considerazioni sull'occupazione protostorica nell'area sud-occidentale del Palatino. *JRA* 12, 5–32.

Angelelli, C. and Rinaldi, F. (eds) (2008) *Atti del XIII Colloquio dell'Associazione Italiana per lo Studio e la Conservazione del Mosaico. Conosa di Puglia, 21–24 febbraio 2007*. Puglia.

Angiolillo, S. (1981) *Mosaici antichi in Italia. Sardinia*. Rome.

Angiolillo, S. (1985) A proposito di un monumento con fregio dorico rinvenuto a Cagliari. La Sardegna e i suoi rapporti con il mondo italico in epoca tardo-repubblicana, in Sotgiu (ed.), pp. 99–110.

Angiolillo, S. (1986–7) Il teatro-tempio di via Malta a Cagliari: una proposta di lettura. *AnnPerugia* 29: 57–81.

Angiolillo, S. (1987) *L'arte della Sardegna romana*. Milan.

Antequera, F., Padrós, P., Rigo, A. and Vázquez, D. (2010) El suburbium occidental de Baetulo, in Vaquerizo (ed.), pp. 173–210.

Antonaccio, C.M. (2003) Hybridity and the cultures within Greek culture, in Dougherty and Kurke (eds), pp. 57–76.

Antonaccio, C. (2005) Excavating colonization, in Hurst and Owen (eds), pp. 97–113.

Antonaccio, C.M. (2009) (Re)defining ethinicity: culture, material culture, and identity, in Hales and Hodos (eds), pp. 32–53.

Anzidei, A.P., Bietti Sestieri, A.M. and De Santis, A. (eds) (1985) *Roma e il Lazio dall'età della pietra alla formazione della città*. Rome.

Appadurai, A. (ed.) (1986) *The Social Life of Things. Commodities in Cultural Perspective*. Cambridge.

Appadurai, A. (ed.) (2001) *Globalization*. Durham.

Applebaum, S. (1989) *Judea in Hellenistic and Roman Times: Historical and Archaeological Essays*. Leiden.

Aquilué, X., Dupré, X., Massó, J. and Ruiz de Arbulo, J. (1991) La cronologia de les muralles de Tarraco. *RArq* 1, 271–301.

Arafat, K.W. (1996) *Pausanias' Greece. Ancient Artists and Roman Rulers*. Cambridge.

Aranegui, C. (2002) Ob restitutam Saguntum Bello Punico Secundo, in Jiménez and Ribera (eds), pp. 245–54.

Aranegui, C. (2004) *Sagunto. Oppidum, emporio y municipio romano*. Barcelona.

Aranegui, C. (2006) From Arse to Saguntum, in Abad, Keay and Ramallo (eds), pp. 63–74.

Arav, R. (1989) *Settlement Patterns and City Planning, 337–301 BCE [BAR-IS 485]*. Oxford.

Arce, J., Ensoli, S. and La Rocca, E. (eds) (1997) *Hispania romana. Da terra di conquista a provincia dell'Impero*. Rome.

Architecture et société de l'archaïsme grec à la fin de la république romaine (1983) [Coll ÉFR 66]. Rome.

Arévalo, A. and Bernal, D. (2007) *Las cetariae de Baelo Claudia. Avance de las investigaciones arqueológicas en el barrio meridional 2000–2004*. Seville.

Arévalo, A., Bernal, D. and Torremocha, A. (2004) *Garum y salazones en el Círculo del Estrecho*. Algeciras.

Ariel, D. and Naveh, J. (2003) Selected sealings from Kedesh in Upper Galilee. *BASOR* 329, 61–80.

Ariño, E. (1991) *Catastros romanos en el convento jurídico Caesaraugustano: la región aragonesa*. Zaragoza.

Arrayás, I. (2005) *Morfología histórica del territorio de Tarraco (siglos III–I a.C.)*. Barcelona.

Arvanitis, N. (2010) *Il santuario di Vesta. La casa delle vestali e il tempio di Vesta, VIII sec. a.C.–64 d.C. Rapporto preliminare*. Pisa.

Arza, R.Á. and Argente, M.C. (1999) Los *pila* del poblado ibérico de Castellruf, *Gladius* 19, 121–42.

Asensio, D., Pons, E. and Fuertes, M. (2007) La darrera fase d'ocupació al Mas Castellar de Pontòs (Alt Empordà, Girona), in Palahí, Nolla and Vivó (eds), pp. 97–131.

Ashby, T. (1910) The classical topography of the Roman Campagna part III. Section II. *PBSR* 5, 215–425.

Ashby, T. (1935) *The Aqueducts of Ancient Rome*. Oxford.

Ashby, T. and Fell, R.A. (1921) The Via Flaminia. *JRS* 21, 125–90.

Assmann, J. (1992) *Das kulturelle Gedächtnis. Schrift, Erinnerung und politische Identität in frühen Hochkulturen*. Munich.

Astin, A.E. (1967) *Scipio Aemilianus*. Oxford.

Astin, A.E. (1978) *Cato the Censor*. Oxford.

Attema, P. (1993) An archaeological survey in the Pontine region. PhD thesis. University of Groningen.

Attema, P., Burgers, G.-J., van Joolen, E. *et al.* (eds) (2002) *New Developments in Italian Landscape Archaeology: Theory and Methodology of Field Survey, Land Evaluation and Landscape Perception, Pottery Production and Distribution*. Oxford.

Attema, P., Burgers, G.J. and van Leusen, M. (2011) *Regional Pathways to Complexity. Settlement and Land-Use Dynamics in Early Italy from the Bronze Age to the Republican Period*. Amsterdam.

Attema, P., Delvigne, J. and Haagsma, B.-J. (1999) Case studies from the Pontine region in central Italy on settlement and environmental change in the first millennium BC, in Leveau, Walsh, Trément and Barker (eds), pp. 105–21.

Attema, P. and Schoerner, G. (eds) (in press) *Comparative Issues in the Archaeology of the Roman Rural Landscape, Site Classification between Survey, Excavation and Historical Categories* [*JRA* Suppl.]. Portsmouth, RI.

Attolini, I., Cambi, F., Castagna, M. *et al.* (1991) Political geography and productive geography between the valleys of the Albegna and the Fiora in northern Etruria, in Barker and Lloyd (eds), pp. 142–52.

Attolini, I. and Perkins, P. (1992) The excavation of an Etruscan farm at Podere Tartuchino. *PBSR* 60, 1–76.

Aufderheide, A.C. and Rodríguez-Martín, C. (1998) *The Cambridge Encyclopedia of Human Paleopathology*. Cambridge.

Aviam, M. (2004) *Jews, Pagans and Christians in Galilee*. Rochester.

Avigad, N. (1983) *Discovering Jerusalem*. Jerusalem.

Ayres, P. (1997) *Classical Culture and the Idea of Rome in Eighteenth Century England*. Cambridge.

Baatz, D. (1982) Hellenistische Katapulte aus Ephyra (Epirus). *AM* 97, 211–33.

Bacci, G. and Marinelli, M. (eds) (2003) *Studi Classici in onore di Luigi Bernabò Brea*. Palermo.

Bacci, G. and Tigano, G. (eds) (2001) *Da Zancle a Messina: un percorso archeologico attraversi gli scavi II*. Palermo.

Badian, E. (1997) *Foreign Clientelae (264–70 BC)*. Amsterdam.

Badolati, E. (ed.) (2010) *First Meeting on Cultural Astronomy*. Naples.

Bagnall, R. and Frier B. (1994) *The Demography of Roman Egypt*. Cambridge.

Bagnasco Gianni, G. (1996) *Oggetti iscritti di epoca orientalizzante in Etruria*. Florence.

Bahat, D. (1990) *Historical Atlas of Jerusalem*. New York.

Bailey, S. (1960) The Roman nobility in the Second Civil War. *CQ* 10, 253–67.

Bailo Modesti, G. (1980) *Cairano nell'età arcaica: l'abitato e le necropoli*. Naples.

Baker, C. and Baker, R. (1994) *Classical Ingenuity: The Legacy of Greek and Roman Architects, Artists and Inventors*. Peterborough.

Balch, D. and Osiek, C. (eds) (2003) *Early Christian Families in Context*. Grand Rapids, MI.

Balland, A. (1969) *Céramique étrusco-campanienne à vernis noir. Fouilles de l'École Française de Rome à Bolsena (Poggio Moscini) III. 1* [*MÉFRA* Suppl. 6].

Ballis, W. (1973) *The Legal Position of War, Changes in Its Practice and Theory from Plato to Vattel*. London.

Balzani, M., Santopuoli, N., Grieco, A. and Zaltron, N. (2004) Laser scanner 3D survey in archaeological field: the Forum of Pompeii, *International Conference on Remote Sensing Archaeology*, October 18–21, 2004, Beijing, http://www.pompeiana.org/research/22-Balzani_Santopuoli.pdf (accessed November 16, 2012).

Bandelli, G. (2002a) Aquileia colonia latina, in Jiménez and Ribera (eds), pp. 57–69.

Bandelli, G. (2002b) La colonizzazione romana della peninsola Iberica da Scipione Africano a Bruto Callaico, in Urso (ed.), pp. 105–42.

Bandelli, G. (2009) Agricoltura e allevamento nella Cisalpina repubblicana, in Carlsen and Lo Cascio (eds), pp. 369–94.

Bang, B., Ikeguchi, M. and Ziche, H. (eds) (2006) *Ancient Economies, Modern Methodologies: Archaeology, Comparative History, Models and Institutions*. Bari.

Bannon, C.J. (2009) *Gardens and Neighbors: Private Water Rights in Roman Italy*. Ann Arbor.

Banti, L. (1973) *The Etruscan Cities and Their Culture*. Berkeley.

Barat, Y. and Laubenheimer, F. (forthcoming) Importations et consommation du vin dans l'Ouest parisien (Yvelines, Val-d'Oise, Essonne).

Barbera, M. and Cianetti, M.M. (2008) *Archeologia a Roma Termini*. Milan.

Barbera, M. and Paris, R. (eds) (1996) *Antiche stanze di Termini. Un quartiere di Roma antica*. Milan.

Barbera, M., Pentiricci, M., Schingo, G. *et al.* (2005) Ritrovamenti archeologici in piazza Vittorio Emanuele II.2 La Necropoli. *BullCom* 106, 302–37.

Barbieri, G. (ed.) (1987) *L'alimentazione nel mondo antico: gli etruschi*. Rome.

Barkan, L. (1999) *Unearthing the Past: Archaeology and Aesthetics in the Making of Renaissance Culture*. New Haven.

Barker, G. (1977) The archaeology of Samnite settlement in Molise. *Antiquity* 51, 20–4.

Barker, G. (1981) *Landscape and Society. Prehistoric Central Italy*. London.

Barker, G. (1988) Archaeology and the Etruscan countryside. *Antiquity* 62, 772–85.

Barker, G. (ed.) (1995a) *Biferno Valley Survey: The Archaeological and Geomorphological Record*. Leicester.

Barker, G. (ed.) (1995b) *A Mediterranean Valley: Landscape Archaeology and Annales. History in the Biferno Valley*. Leicester.

Barker, G. (ed.) (1999) *Companion Encyclopaedia of Archaeology*. London.

Barker, G. (2004) Agriculture, pastoralism, and Mediterranean landscapes in prehistory, in Blake and Knapp (eds), pp. 46–76.

Barker, G. and Grant, A. (1999) Food and farming, in Barker (ed.), 1999, pp. 546–607.

Barker, G. and Hodges, R. (eds) (1981) *Archaeology and Italian Society. Prehistoric, Roman and Medieval Studies. Papers in Italian Archaeology II* [*BAR-IS* 102]. Oxford.

Barker, G. and Lloyd, J. (eds) (1991) *Roman Landscapes. Archaeological Survey in the Meditrerranean Region* [*BSR Archaeological Monograph 2*]. London.

Barker, G. and Rasmussen, T. (1991) The Tuscania Survey, in Barker and Lloyd (eds), pp. 104–14.

Barker, G. and Rasmussen, T. (1998) *The Etruscans*. Oxford.

Barker, G. and Suano, M. (1995) Iron Age chiefdoms 1000–500 BC, in Barker (ed.), 1995b, pp. 159–80.

Barnes, E. (2003) The dead *do* tell tales, in Williams and Bookidis (eds), pp. 435–43.

Barra Bagnasco, M., De Miro, E. and Pinzone, A. (eds) (1999) *Magna Graecia e Sicilia. Stato degli studi e prospettive di ricerca. Atti dell'Incontro di Studi, Messina 2–4 dicembre 1996* [Pelorias 4]. Messina.

Barrandon, N. (2011) *De la pacification à l'intégration des Hispaniques (133–27a.C.)* [Scripta Antiqua 35]. Bordeaux.

Barreda, A. (1998) Gentes itálicas en Hispania Citerior (218 a.C.–14 d.C.). Los casos de tarraco, Carthago Nova y valentia. PhD thesis. Universitat Autònoma de Barcelona.

Bartoloni, G. (1987) Esibizione di ricchezza a Roma nel VI e V secolo: doni votivi e corredi funerari. *Scienze dell'Antichità* 1, 143–59.

Bartoloni, G. (1989–90) I depositi votivi di Roma arcaica: alcune considerazioni. *Scienze dell'Antichità* 3–4, 747–59.

Bartoloni, G. (ed.) (2010) *Lupa capitolina*. Rome.

Bartoloni, G. and Delpino, F. (eds) (2005) *Oriente e Occidente: metodi e discipline a confronto. Riflessioni sulla cronologia dell'età del ferro in Italia. Atti dell'Inctontro di Studi, Roma, 30–31 ottobre 2003. Mediterranea, I (2004)*. Pisa.

Bartoloni, P. (1986) *Le stele di Sulcis: catalogo*. Rome.

Bastide, R. (1955) Le principe de coupure et le comportement Afro-Brésilien, in *Anais do XXXI Congresso Internacional de Americanistas, S. Paulo, 1954*, pp. 493–503. São Paulo.

Bastien, J.-L. (2007) *Le triomphe romain et son utilisation politique à Rome aux trois derniers siècles de la République*. Rome.

Bats, M. (ed.) (1996) *Les céramiques communes de Campanie et de Narbonnaise: La vaisselle de cuisine et de table*. Naples.

Battaglini, G. (2009) Le terme pubbliche di Fregellae: forma e funzione degli spazi, in Padovano (ed.), pp. 345–63.

Bauman, R.A. (1996) *Crime and Punishment in Ancient Rome*. New York.

Baumann, G. (2004) Grammars of identity/alterity: a structural approach, in Baumann and Gingrich (eds), pp. 18–50.

Baumann, G. and Gingrich, A. (eds) (2004) *Grammars of Identity/Alterity: A Structural Approach*. New York.

Beacham, R.C. (1991) *The Roman Theatre and Its Audience*. London.

Beacham, R.C. (2007) Playing places: the temporary and the permanent, in McDonald and Walton (eds), pp. 202–26.

Beard, M. (2007) *The Roman Triumph*. Cambridge, MA.

Beard, M. and Crawford, M. (1999) *Rome in the Late Republic: Problems and Interpretations* (2nd edn). London.

Beard, M., North, J. and Price, S. (1998) *Religions of Rome*. Cambridge.

Beare, W. (1964) *The Roman Stage: A Short History of Latin Drama in the Time of the Republic* (3rd edn). London.

Beazley, J. (1947) *Etruscan Vase Painting*. Oxford.

Bechtold, B. (2007) Alcune osservazioni sui rapporti commerciali fra Cartagine, la Sicilia occidentale e la Campania (IV–metà II sec. a.C.): nuovi dati basati sulla distribuzione di ceramiche campane e nordafricane/cartaginesi. *BABesch* 82, 51–76.

Bechtold, B. (2011) Amphorae production in Punic Sicily (7th–3rd/2nd centuries BCE): an overview. *FACEM* (version 06.06.2011), http://facem.at/img/pdf/Amphorae_in_Punic_Sicily_06_06_2011.pdf (accessed November 16, 2012).

Becker, H. (2007) Charismatic power and the private armies of Rome and Etruria: *condottieri*, *bella privata*, and mercenaries. Unpublished paper delivered at the 138th meeting of the American Philological Association, San Diego, CA, January 6, 2007.

Becker, J.A. (2003) *An architectural reappraisal of the Roman villa at Grottarossa (vocabolo Monte delle Grotte)*. MA thesis. University of North Carolina at Chapel Hill.

Becker, J. (2007) The building blocks of empire: civic architecture, central Italy, and the Roman Middle Republic. PhD thesis. University of North Carolina at Chapel Hill.

Becker, J.A. and Terrenato, N. (eds) (2012) *Roman Republican Villas: Architecture, Context, and Ideology*. Ann Arbor.

Bederman, D. (2001) *International Law in Antiquity*. Cambridge.

Bedini, A. (1984) Scavi al Torrino. *Archeologia Laziale* 6, 84–90.

Beeser, J. (1979) *Pilum Murale? Fundberichte aus Baden-Württemberg* 4, 133–43.

Befani, V. (2005) Vico prolungato della Fullonica. *RStPomp* 16, 183–6.

Bejor, G. (2007) Gli insediamenti rurali in Sicilia tra Repubblica e Impero, in Miccichè, Modeo and Santagati (eds), pp. 14–26.

Bek, L. (1983) Questiones conviviales: the idea of the triclinium and the staging of convivial ceremony from Rome to Byzantium. *AnalRom* 12, 81–108.

Belayche, N. (ed.) (2006) *Rome, Les Césars et la Ville aux deux premiers siècles de notre ère*. Rennes.

Bell, C. (ed.) (2007) *Teaching Ritual*. Oxford.

Bell, M. (1965) Tactical reform in the Roman Republican Army. *Historia* 14, 404–22.

Bell, M., III (1988) Excavations at Morgantina, 1980–1985: preliminary report XII. *AJA* 92, 313–42.

Bell, S. and Nagy, I. (eds) (2008) *New Perspectives on Etruria and Rome: Papers in Honor of Richard D. De Puma*. Madison, WI.

Bell, T., Wilson, A. and Wickham, A. (2002) Tracking the Samnites: landscape and communications routes in the Sangro Valley, Italy. *AJA* 106 (2), 169–86.

Beloch, J. (1890) *Campanien: Geschichte und Topographie des Antiken Neapel und seiner Umgebung* (2nd edn). Breslau.

Beloch, K.J. (1886) *Die Bevölkerung der griechisch-römischen Welt*. Leipzig.

Beltrán Fortes, J. (2008) La arqueología cesariana de la provincia Hispania Ulterior, in García-Bellido, Mostalac and Jiménez (eds), pp. 41–8.

Beltrán Lloris, F. (1984) El año 218 a.C. Problemas en torno al comienzo de la Segunda Guerra Púnica en la península ibérica, in González and Díez (eds), pp. 147–71.

Beltrán Lloris, F. (1990) La "pietas" de Sertorio. *Gerión* 8, 211–26.

Beltrán Lloris, F. (1996) *El tercer bronce de Botorrita (Contrebia Belaisca)*. Zaragoza.

Beltrán Lloris, F. (2008) Les debuts de l'Hispania Citerior: précédents de la régionalisation de l'administration provinciale, in Piso (ed.), pp. 123–41.

Beltrán Lloris, F., Martín-Bueno, M. and Pina, F. (eds) (2000) *Roma en la cuenca media del Ebro. La Romanización en Aragón*. Zaragoza.

Beltrán Lloris,M. (1985) *Celsa*. Zaragoza.

Beltrán Lloris, M. (1990) El valle medio del Ebro y su monumentalización en época republicana y augustea. Los ejemplos de Celsa y Cesaraugusta, in Trillmich and Zanker (eds), pp. 183–6.

Beltrán Lloris, M. (1997) *Colonia Celsa. Velilla de Ebro*. Madrid.

Beltrán Lloris, M. and Mostalac, A. (2008) La Colonia Lepida/Celsa y Salduie: sus testimonios arqueológicos durante el segundo triunvirato y comienzos del Imperio, in García-Bellido, Mostalac and Jiménez (eds), pp. 107–27.

Bénabou, M. (1976) *La résistance africaine à la romanisation*. Paris.

Benassai, R. (2000) La tomba dipinta nolana Weege 30, in Capelli (ed.), pp. 78–81.

Benassai, R. (2001) Due tombe capuane: contesto architettonico e committenza. *RendNap* 70, 341–55.

Ben Baaziz, S. (1992) La carte nationale des sites archéologiques et des monuments historiques. *Bulletin des Travaux de l'Institut National du Patrimoine* 6, 1–15.

Bencivenga, M., Di Loreto, E. and Liperi, L. (1995) Il regime idrologico del Tevere, con particolare riguardo alle piene nella citta di Roma, in Funiciello (ed.), pp. 125–72.

Bencivenni, A. (2011) "Massima considerazione": forma dell'ordine e immagini del potere nella corrispondenza di Seleuco IV. *ZPE* 176, 139–53.

Bendala, M. (2001) La Carmona bárquida, in Caballos (ed.), pp. 37–51.

Bendala, M. (2009) Continuidad y renovación en los centros sacros de las ciudades hispanorromana, in Mateos *et al.* (eds), pp. 345–70.

Bendala, M. and Blánquez, J. (2002/3) 'Arquitectura militar púnico-helenística en Hispania', *Cuadernos de Prehistoria y Arqueología, Universidad Autónoma de Madrid*, 28–29, 145–59.

Bendala, M. and Roldán, L. (2005) El templo romano republicano de Carteia (Hispania) y su problemática arqueológica e histórica, in Lafon and Sauron (eds), pp. 153–66.

Bendala, M., Roldán, L. and Blánquez, J. (2002) Carteia: de ciudad púnica a colonia latina, in Jiménez and Ribera (eds), pp. 157–72.

Beneduce, P., Gallipoli, M.R., Guarino, P. *et al.* (2008) Il contributo delle geoscienze per l'individuazione dell'area portuale di Pompei: primi risultati, in Guzzo and Guidobaldi (eds), 2008, pp. 551–3.

Bennett, J. (1982) The Great Chesters "Pilum Murale." *Archaeologia Aeliana* 5 (X), 200–4.

Benquet, L. (2004) L'emporion de Toulouse, in Brun, Poux, and Tchernia (eds), p. 158.

Ben Romdhane, K. (2001) Leptiminus: atelier monétaire romain, in Stirling, Mattingly, and Ben Lazreg (eds), pp. 457–60.

Ben Younes, H. (2000) Les tombes rupestres (haouanet) du Jebel Sidi Zid, in Ørsted, Carlsen, Ladjimi Sebaï and Ben Hassen (eds), pp. 191–216.

Ben Younes, H. (2007) Interculturality and the Punic funerary world, in Stone and Stirling (eds), pp. 32–42.

Berg, R. (2005) Saggi archeologici nell'Insula dei Casti Amanti, in Guzzo and Guidobaldi (eds), 2005, pp. 200–15.

Bergemann, J. (ed.) (2010) *Der Gela-Survey. 3000 Jahre Siedlungsgeschichte in Sizilien, I–III* [Göttinger Studien zur Mediterranen Archäologie 1]. Munich.

Berggren, E. and Andrén, A. (1969) Blera (località Selvasecca). Villa rustica etrusco-romana con manifattura di terrecotte architettoniche. *NSc* ser. 8.23, 51–71.

Berghaus, P. (1991) Roman coins from India and their imitations, in Jha (ed.), pp. 108–21.

Bergmann, B. (1995) Visualizing Pliny's villas. *JRA* 8, 406–20.

Berlin, A. (1997) Between large forces: Palestine in the Hellenistic period. *BiblArch* 60, 2–51.

Berlin, A. (1999) The archaeology of ritual: the sanctuary of Pan at Banias/Caesarea Philippi. *BASOR* 315, 27–45.

Berlin, A. (in press) Review of *The Roman Temple Complex at Horvat Omrit. BASOR*.

Bernabò Brea, L. and Cavalier, M. (1998) *Meligunis-Lipára IX*. Palermo.

Bernard, S. (2010) Pentelic marble in architecture at Rome and the republican marble trade. *JRA* 23, 35–54.

Berns, C. (2003) *Untersuchungen zu den Grabbauten der frühen Kaiserzeit in Kleinasien*. Bonn.

Berry, J. (1997) Household artefacts: towards a re-interpretation of Roman domestic space, in Laurence and Wallace-Hadrill (eds), pp. 183–95.

Bersani, P., Canalini, A. and Dragoni, W. (2010) First results of a study of the Etruscan tunnel and other hydraulic works on the Ponte Coperto Stream (Cerveteri, Rome, Italy). *Water Science and Technology* 10 (4), 561–69.

Berthier, A. (1981) *La Numidie: Rome et le Maghreb*. Paris.

Bettelli, M. (1997) *Roma. La città prima della città: i tempi di una nascita*. Rome.

Bettini, M.C. and Poggesi, G. (eds) (2000) *Archeologia 2000. Un progetto per la provincia di Prato*. Montespertoli.

Bianchi, E., Meneghini, R., Jackson, M. *et al.* (2011) Archaeological, structural, and compositional observations of the concrete architecture of the Basilica Ulpia and Trajan's Forum. *Commentationes Humanarum Litterarum (Societas Scientiarum Fennica)* 128, 72–93.

Bianchini, S. (1996) La ceramica a vernice nera di Lucca. *RStLig* 62, 169–210.

Bieber, M. (1961) *The History of the Greek and Roman Theater* (2nd rev. and enlarged edn). Princeton.

Bietti Sestieri, A.M. (2000) The role of archaeological and historical data in the reconstruction of Italian protohistory, in Ridgway *et al.* (eds), pp. 13–31.

Bietti Sestieri, A.M. and De Santis, A. (2000) *Protostoria dei popoli latini. Museo Nazionale Romano, Terme di Diocleziano*. Milan.

Bikerman, E. (1946–8) Une proclamation séleucide relative au temple de Jérusalem. *Syria* 25, 67–85.

Bilde, P. (1990) *Religion and Religious Practice in the Seleucid Kingdom*. Aarhus.

Bintliff, J. (1991) *The Annales School and Archaeology*. Leicester.

Bintliff, J. and Sbonias, K. (2000) Demographic and ceramic analysis in regional survey, in Francovich and Patterson (eds), pp. 244–58.

Bintliff, J. and Sbonias, K. (eds) (2000) *Reconstructing Past Population Trends in Mediterranean Europe (3000 BC–AD 1800)*. Oxford.

Birley, E. (1982) The dating and character of the tract *De munitionibus castrorum*, in Wirth (ed.), pp. 277–81.

Bisel, S. and Angel, J.L. (1985) Health and nutrition in Mycenaean Greece: a study in human skeletal remains, in Wilke and Coulson (eds), pp. 197–209.

Bishop, M. and Coulston, J. (2006) *Roman Military Equipment from the Punic Wars to the Fall of Rome* (2nd edn). Oxford.

Bispham, E. (2000) Mimic? A case study in early Roman colonisation, in Herring and Lomas (eds), pp. 157–86.

Bispham, E. (2006) *Coloniam deducere*: how Roman was Roman colonization during the Middle Republic?, in Bradley and Wilson (eds), pp. 73–160.

Bispham, E. (2007a) *From Asculum to Actium*. Oxford.

Bispham, E. (2007b) The Samnites, in Bradley, Isayev and Riva (eds), pp. 179–223.

Bispham, E. (ed.) (2008) *Roman Europe: 1000 BC–AD 400*. Oxford.

Bispham, E., Bradley, G., Hawthorne, J. and Kane, S. (2000) Towards a phenomenology of Samnite fortified centres. *Antiquity* 74, 23–4.

Bispham, E. and Smith, C. (eds) (2000) *Religion in Archaic and Republican Rome and Italy. Evidence and Experience*. Edinburgh.

Bjur, H. and Santillo Frizell, B. (eds) (2009) *Via Tiburtina. Space, Movement and Artefacts in the Urban Landscape*. Stockholm.

Black, J. (1992) *The British Abroad, The Grand Tour in the Eighteenth Century*. New York.

Blackman, D. (1982) Ancient harbours in the Mediterranean, parts 1 and 2. *IJNA* 11, 79–104, 185–211.

Blackman, D. (2008) 'Roman shipsheds', in Hohlfelder (ed.), pp. 23–36.

Blackman, D. and Hodge, A.T. (eds) (2001) *Frontinus' Legacy: Essays on Frontinus' "De aquis urbis Romae."* Ann Arbor.

Blagg, T. and Millett, M. (eds) (1990) *The Early Roman Empire in the West.* Oxford.

Blake, M.E. (1947) *Ancient Roman Construction in Italy from the Prehistoric Period to Augustus.* Washington DC.

Blake, E. and Knapp, A.B. (eds) (2004) *The Archaeology of Mediterranean Prehistory.* Oxford.

Blois, L. de, Bons Ton Kessels, J. and Schenkeveld, D.M. (eds) (2005) *The Statesman in Plutarch's Works*, vol. 2, *The Statesman in Plutarch's Greek and Roman Lives. Proceedings of the Sixth International Conference of the International Plutarch Society. Nijmegen/Castle Hernen, May 1–5, 2002* [*Mnemosyne* Suppl. 250]. Leiden.

Boatwright, M.T. et al. (1982) *I Volusii Saturnini: una famiglia romana della prima età imperiale.* Bari.

Bober, P.P. and Rubinstein, R.. (1986) *Renaissance Artists and Antique Scupture: A Handbook of Sources.* Oxford.

Bodel, J. (1997) Monumental villas and villa monuments. *JRA* 10, 3–35.

Bodel, J. (2012) Villaculture, in Becker and Terrenato (eds), pp. 45–60.

Böethius, A. and Ward-Perkins, J.P. (1970) *Etruscan and Roman Architecture.* Harmondsworth.

Bogaers, J. and Rüger, C. (1974) *Der niedergermanische Limes.* Bonn.

Bogdanos, M. (2005a) Casualties of war: the truth about the Iraq Museum. *AJA* 109, 477–526.

Bogdanos, M., with Patrick, W. (2005b) *Thieves of Baghdad.* New York.

Bököyni, S., Costantini, L. and Fitt, J. (1993) The farming economy, in Gualtieri (ed.), pp. 281–307.

Boman, H. (2008) The interaction between street and houses in the north-west corner of Insula V 1, in Guzzo and Guidobaldi (eds), 2008, pp. 87–92.

Bon, S. and Jones, R. (eds) (1997) *Sequence and Space in Pompeii.* Oxford.

Bonanno, C. (2001) L'insediamento rurale suburbano nei pressi della via per Catania, in Bacci and Tigano (eds), pp. 195–203.

Bonatti, E. (1963) Stratigrafia pollinica dei sedimenti post-glaciale di Baccano, lago craterico del Lazio. *Atti della Societa' Toscana di Scienze naturali* 70, 40–8.

Bonatti, E. (1970) Pollen sequence in the lake sediments, in Hutchinson (ed.), pp. 26–31.

Bonaudo, R. (2004) *La culla di Hermes. Icongrafia e immaginario delle hydriai ceretane.* Rome.

Bondioli, L., Corruccini, R.S. and Macchiarelli, R. (1986) Familial segregation in the Iron Age community of Alfedena, Abruzzo, Italy, based on osteodental trait analysis. *American Journal of Physical Anthropology* 71 (4), 393–400.

Bonello Lai, M. (ed.) (1992) *Sardinia antiqua. Studi in onore di Piero Meloni in occasione del suo settantesimo compleanno.* Cagliari.

Bonetto, J. (1999) Ercole e le vie della transumanza: il santuario di Tivoli. *Ostraka* 8, 291–307.

Bonfante, L. and Swaddling, J. (2006) *Etruscan Myths.* London and Austin.

Bonghi Jovino, M. (ed.) (1988) *Ricerche a Pompei: l'insula 5 della Regio VI dalle origini al 79 d.C.* Rome.

Bonnemaison, S. and Macy, C. (eds) (2008) *Festival Architecture.* New York.

Bonnet, C., Bonnano, D., Cusumano, N. and Péré-Noguès, S. (eds) (2011) *Alleanze e parentele: Le "affinità elettive" nella storiografia sulla Sicilia antica. Convegno internazionale, Palermo 14–15 aprile 2010.* Caltanissetta.

Bonsor, G. (1997) *Las colonias agrícolas prerromanas del Valle del Guadalquivir.* Seville.

Bordenache Battaglia, G. (1979) *Le ciste prenestine I–II.1.* Rome.

Borsari, L. (1894) Terracina: del Tempio di Giove Anxur. *NSc* 96–111.

Boschung, D. (1987) Die republikanischen und frühkaiserzeitlichen Nekropolen vor den Toren Ostias, in von Hesberg and Zanker (eds), pp. 111–24.

Boullosa, E. (2010) Escipión en Numancia ¿un triunfo de la disciplina? *Revista de Historia Militar* 107, 41–72.

Bound, M. (1991) The Giglio wreck: a wreck of the Archaic Period c.600 BC off the Tuscan island of Giglio. An account of its discovery and excavation: a review of the main finds. *Enalia* 1, 3–36.

Bourbou, C. (2003) Health patterns of Proto-Byzantine populations (6th–7th centuries AD) in South Greece: the cases of Eleutherna (Crete) and Messene (Peloponnese). *International Journal of Osteoarchaeology* 13, 303–13.

Bourbou, C. (2005) Biological status in Hellenistic and Roman elites in western Crete (Greece). *Eres* 13, 87–110.

Bourbou, C. (2010) *Health and Disease in Byzantine Crete (7th–12th centuries AD)*. Farnham.

Bowden, W., Burgess, A., Castellani, A. and Moran, M. (2006) The Samnite cemetery and the origins of the Samnite vicus, in Bowes, Francis and Hodges (eds), pp. 49–92.

Bowersock, G. (1965) *Augustus and the Greek World*. Oxford.

Bowes, K., Francis, K. and Hodges, W. (eds) (2006) *Between Text and Territory: Survey and Excavations in the Terra of San Vincenzo al Volturno*. London.

Bowman, A. and Brady, M. (eds) (2005) *Images and Artefacts of the Ancient World*. New York.

Bowman, A., Champlin, E. and Lintott, A. (eds) (1996) *Cambridge Ancient History, New Edition, Volume X (31 BC–AD 70)*. Cambridge.

Bowman, A. and Wilson, A. (eds) (2009) *Quantifying the Roman Economy. Methods and Problems*. Oxford.

Bowman, A. and Wilson, A. (eds) (2011) *Settlement, Urbanization and Population*. Oxford.

Braccesi, L. and Antonelli, L. (eds) (2003) *Hesperìa 17: Studi sulla Grecità di Occidente*. Rome.

Bradford Churchill, J. (1999) *Ex qua quod vellent facerent*: Roman magistrates' authority over *praeda* and *manubiae*. *TAPA* 129, 85–116.

Bradley, G.J. (1997) Iguvines, Umbrians and Romans. Ethnic identity in central Italy, in Cornell and Lomas (eds), 1997, pp. 53–67.

Bradley, G.J. (2000a) *Ancient Umbria: State, Culture, and Identity in Central Italy from the Iron Age to the Augustan Era*. Oxford.

Bradley, G.J. (2000b) Tribes, states and cities in southern Italy, in Herring and Lomas (eds), pp. 108–29.

Bradley, G.J. (2002) The Romanisation of Italy (Long review of Mouritsen 1998). *JRA* 15, 401–6.

Bradley, G.J. (2006) Colonization and identity in Republican Italy, in Bradley and Wilson (eds), pp. 161–87.

Bradley, G.J. (2007) Romanization: the end of the peoples of Italy?, in Bradley, Isayev and Riva (eds), pp. 295–322.

Bradley, G., Isayev, E. and Riva, C. (eds) (2007) *Ancient Italy. Regions without Boundaries*. Exeter.

Bradley, G. and Wilson, J. (eds) (2006) *Greek and Roman Colonisation: Origins, Ideologies and Interactions*. Swansea.

Bradley, K.R. (1984) *Slaves and Masters in the Roman Empire: A Study in Social Control*. Brussels.

Bradley, K.R. (1989) *Slavery and Rebellion in the Roman World, 140 BC–70 BC*. London.

Bradley, K.R. (1994) *Slavery and Society at Rome*. Cambridge.

Bradley, K.R. and Cartledge, P. (eds) (2011) *The Cambridge World History of Slavery, Volume 1: The Ancient Mediterranean World*. Cambridge.

Bradley, M. (ed.) (2012) *Pollution and Propriety: Dirt, Disease and Hygiene in Rome from Antiquity to Modernity*. Cambridge.

Brandon, C., Hohlfelder, R., Oleson, J.P. and Stern, C. (2005) The Roman Maritime Concrete Study (ROMACONS): the harbour of Chersonisos in Crete and its Italian connection. *Revue géographique des pays méditerranéens* 104, 25–9.

Brandt, R. and Karlsson, L. (eds) (2001) *From Huts to Houses. Transformations of Ancient Societies* (Skrifter utgivna av Svenska Institutet i Rom, 4°, LVI). Stockholm.

Braudel, F. (2001) *Memory and the Mediterranean*. New York.

Bravo, G. (2007) *Hispania. La epopeya de los romanos en la Península*. Madrid.

Bravo, M. (2008) *El largo camino de los hispani hacia la ciudadanía*. Madrid.

Brecciaroli Taborelli, L. (ed.) (2007) *Forme e tempi dell'urbanizzazione nella Cisalpina (II secolo a.C.–I secolo d.C.)*. Firenze.

Bricault, L. (ed.) (2004) *Isis in occident* [Actes du IIème Colloque Internationale sur les Études Isiaques]. Leiden.

Bricault, L. and Versluys, J. (eds) (2010) *Isis on the Nile. Egyptian Gods in Hellenistic and Roman Egypt*. Leiden.

Briquel, D. (2008) L'espace consacré chez les Étrusques: réflexions sur le rituel étrusco-romain de fondation des cités, in Dupré Raventós, Ribichini and Verger (eds), pp. 27–47.

Briscoe, J. (1973) *A Commentary on Livy. Books XXXI–XXXIII*. Oxford.

Brisson, J.-P. (ed.) (1969) *Problèmes de la guerre à Rome*. Paris.

Brizzi, G. (2006) Gli Iberi nell'armata di Annibale: armamento e funzioni, in Sartori and Valvo (eds), pp. 157–66.

Broadhead, W. (2001) Rome's migration policy and the so-called *ius migrandi*, *Cahiers du Centre Gustave-Glotz* 12, 69–89.

Broadhead, W. (2003) The local élites of Italy and the crisis of migration in the second century BC, in Cébeillac-Gervasoni and Lamoine (eds), pp. 131–48.

Broadhead, W. (2004) Rome and the mobility of the Latins: problems of control, in Moatti (ed.), pp. 315–35.

Brocato, P. (2000) Dalle capanne del *Cermalus* alla *Roma quadrata*, in Carandini and Cappelli (eds), pp. 284–7.

Brödner, E. (1992) *Die römischen Thermen und das antike Badewesen*. Darmstadt.

Bronson, R.C. (1965) Chariot racing in Etruria, in Camporeale (ed.), pp. 89–106.

Brookes, A.C. (1974) Minturnae: the Via Appia bridge. *AJA* 78, 41–8.

Brotons, F. and Ramallo, S. (1994) Un santuario suburbano: La Encarnación de Caravaca (Murcia), in *Actes del XIV Congrés*, pp. 74–5.

Broughton, T.R.S. (1929) *The Romanization of Africa Proconsularis*. Baltimore.

Brouquier-Reddé, V. (1997) L'équipement militaire d'Alésia d'après les nouvelles recherches (prospections et fouilles), in Feugère (ed.), pp. 277–88.

Brown, F.E. (1951) Cosa I. History and topography. *MAAR* 20, 5–113.

Brown, F.E. (1980) *Cosa: The Making of a Roman Town*. Ann Arbor.

Brown, F.E., Richardson, E.H. and Richardson, L., jr. (1960) *Cosa II: The Temples of the Arx* [*MAAR* 26].

Brown, F.E., Richardson, E.H. and Richardson, L., jr. (1993) *Cosa III: The Buildings of the Forum* [*MAAR* 37].

Bruckner, E.-C. (1995) Forum Appi, in De Rossi and Mello (eds), pp. 189–221.

Bruhn, J., Croxford, B. and Grigoropoulos, D. (eds) (2005) *TRAC 2004: Proceedings of the Fourteenth Annual Theoretical Roman Archaeology Conference*. Oxford.

Brun, J.-P. (2004) *Archéologie du vin et de l'huile. De la préhistoire à l'epoque hellénistique*. Paris.

Brun, J.-P. (2008) Uno Stile zero? Andron e decorazione pittorica anterior al primo stile nell'Insula I 5, di Pompeii, in Guzzo and Guidobaldi (eds), 2008, pp. 61–70.

Brun, J-P., Poux, M. and Tchernia, A. (eds) (2004) *Le vin nectar des dieux, génie des hommes*. Golio.

Bruneau, P. (1995) Deliaca (X). *BCH* 119, 45–54.

Bruno, D. (2006) La topografia dei culti dell'Aventino ricostruita, in Carandini and Greco (eds), 2006, pp. 113–19.

Bruno, V.J. and Scott, R.T. (1993) *Cosa IV: The Houses* [*MAAR* 38].

Brunsting, H., and Wynia, S.L. (1993) Gli scavi e i mobilia, in Peters (ed.), pp. 3–37.

Brunt, P.A. (1965) Italian aims at the time of the Social War. *JRS* 55, 90–109.
Brunt, P.A. (1971) *Italian Manpower, 225 BC–14 AD*. Oxford.
Brunt, P.A. (1975) Two great Roman landowners. *Latomus* 34 (2), 619–35.
Bruun, C. (1991) *The Water Supply of Ancient Rome: A Study of Roman Imperial Administration [Commentationes Humanarum Litterarum 93]*. Helsinki.
Bryson, N. (1990) *Looking at the Overlooked: Four Essays on Still Life Painting*. London.
Buckle, S. (1991) *Natural Law and the Theory of Property, Grotius to Hume*. Oxford.
Buikstra, J. and Ubelaker, D. (1994) *Standards for Data Collection from Human Skeletal Remains* [Arkansas Archaeological Survey Research Series 44]. Fayetteville, AR.
Bull, H., Kingsbury, B. and Roberts, A. (eds) (1990) *Hugo Grotius and International Relations*. Oxford.
Bulliet, R. (1992) "Annales" and archaeology, in Knapp (ed.), pp. 131–4.
Bullo, S. (2002) *Provincia Africa: Le città e il territorio dalla caduta di Cartagine à Nerone*. Rome.
Buranelli, F. (ed.) (1987) *La tomba François a Vulci*. Rome.
Burch, J. (1995) El poblat ibèric de Sant Julià de Ramis, noves aportacions, in Agustí, Burch and Merino (eds), pp. 85–94.
Burch, J., Nolla, J.M., Palahí, L. *et al.* (2001) *Excavacions arqueològiques a la muntanya de Sant Julià de Ramis 1. El sector de l'antiga església parroquial*. Girona.
Burch, J., Palahí, L. Sagrera, J. and Vivó, D. (2000) La fundació de Gerunda. Dades noves sobre un procés complex de reorganització del territori. *Empúries* 52, 11–29.
Burgers, G.-J. (1998) *Constructing Messapian Landscapes: Settlement Dynamics, Social Organization and Culture Contact on the Margins of Greco-Roman Italy*. Amsterdam.
Burgers, G.-J. (2001) L'archeologia e l'Italia meridionale post-annibalica: una prospettiva regionale e diacronica, in Lo Cascio and Storchi Marino (eds), pp. 249–66.
Burgers, G.-J. and Recchia, G. (eds) (2009) *Ricognizioni archeologiche sull'altopiano delle Murge: La carta archeologica del territorio di Cisternino*. Foggia.
Burillo, F. (ed.) (2006) *Segeda y su contexto histórico. Entre Catón y Nobilior (195–153 a.C.). Homenaje a Antonio Beltrán Martínez*. Zaragoza.
Burillo, F. (2008) Una propuesta sobre los "ciudadanos" de las ciudades Estado celtibéricas de Segeda y Numancia a partir de la batalla de la Vulcanalia, año 153 a.C., in Uroz, Noguera and Coarelli (eds), pp. 509–20.
Burillo, F., Cano, M., López, R. and Saiz, E. (2009) La ciudad celtibérica de Segeda I, in Mateos, Celestino, Pizzo and Tortosa (eds), pp. 195–202.
Burnett, A. (1982) The currency of Italy from the Hannibalic War to the reign of Augustus. *AIIN* 29, 125–37.
Burnett, A. and Crawford, M. (eds) (1987) *The Coinage of the Roman World in the Late Republic* [*BAR-IS* 326]. Oxford.
Burns, M. (2003) The homogenisation of military equipment under the Roman Republic. *"Romanization"? Digressus Supp.* 1, 60–85.
Burns, P.E. (1979) Log-linear analysis of dental caries occurrence in four skeletal series. *American Journal of Physical Anthropology* 51, 637–48.
Burrell, B. (1993) Palace to praetorium. The Romanization of Caesarea, in Raban and Holum (eds), pp. 228–47.
Burri, R., Delacrétaz, A., Monnier, J. and Nobili, M. (eds) (2004) *Ad limina II*. Alexandría.
Bustamente, M., Chover, I.E., Férnandez, A. *et al.* (2010) Pompeya. Alrededor de la via "degli Augustali": el macellum (VII, 9, 25) y la casa del "Forno a riverbero" (VII, 4, 29). Campana de 2009. *Journal of Fasti Online*, FOLD&R http://www.fastionline.org/docs/FOLDER-it-2010-210.pdf (accessed November 17, 2012).
Butler, S. (2002) *The Hand of Cicero*. London.

Buttrey, T.V. (1980) Cosa: the coins. *MAAR* 34, 11–153.

Buttrey, T.V. (1989) The Morgantina excavations and the date of the Roman denarius. *Atti del Congresso Internazionale di Numismatica* 261–7.

Buttrey, T.V. (1994) Calculating ancient coin production II: why it cannot be done. *NC* 154, 341–52.

Buxó, R. and Pons, E. (eds) (2000) *L'hàbitat protohistòric a Catalunya. Rosselló i Llenguadoc Occidental. Actualitat de l'arqueologia de l'Edat del Ferro*. Girona.

Byrne, S. (2003) Early Roman Athenians, in Jordan and Traill (eds), pp. 1–20.

Caballos, A. (ed.) (2001) *Carmona romana. Actas del II Congreso de Historia de Carmona (Carmona, 1999)*. Carmona.

Caballos, A. (2010) *Italica-Santiponce: Municipium y Colonia Aelia Augusta Italicensium*. Rome.

Caballos, A. (forthcoming) Les colonies romanes de Bétique: migrations, integration, vertebration, in *XVe Rencontre franco-italienne d'épigraphie du monde romain (Paris, 2008)*. Paris.

Caballos, A. and León, M. (eds) (1997) *Actas de las Jornadas del 2.200 aniversario de la fundación de Itálica*. Seville.

Cadiou, F. (2001) Les guerres en Hispania et l'émergence de la cohorte légionnaire dans l'armée romaine sous la République: une révision critique. *Gladius* 21, 167–82.

Cadiou, F. (2008) *Hibera in Terra Miles. Les armées romaines et la conquête de l'Hispanie sous la République (218–45 av. J.-C.)*. Madrid.

Cadiou, F. (forthcoming) Les colonies de l'Hispanie républicaine: le dossier de Valence, in *XVe Rencontre franco-italienne d'épigraphie du monde romain (Paris, 2008)*. Paris.

Cadiou, F. *et al.* (eds) (2008) *La guerre et ses traces dans la péninsule Ibérique à l'époque de la conquête romaine: approches méthodologiques. Actes de la table ronde internationale (Madrid, novembre 2007)* [*Saldvie* 8]. Zaragoza.

Cagniart, P. (2007) The Late Republican Army (146–30 BC), in Erdkamp (ed.), pp. 80–95.

Caillemer, A. and Chevallier, R. (1959) *Atlas des centuriations romaines de Tunisie*. Paris.

Calabi Limentani, I. (1982) I fornices di Stertinio e di Scipione nel racconto di Livio (XXXIII, 27, 1–5 e, XXXVII, 3, 7), in Sordi (ed.), pp. 123–35.

Calastri, C. (2007) Albinia. L'Albegna, l'entroterra e il mare : problemi relativi all'assetto del territorio nell'Antichità, in Vitali (ed.), pp. 15–24.

Calbi, A. and Susini, G.C. (eds) (1995) *Pro poplo arimenese. Atti del Convegno internazionale: Rimini antica. Una respublica fra terra e mare. Rimini, ottobre 1993*. Faenza.

Caldelli, M.L. (2001) Gladiatori con armaturae etniche. Il Samnes. *ArchCl* 52, 279–95.

Callataÿ, F. de (1995) Calculating ancient coin production: seeking a balance. *NC* 155 (289), 311.

Callataÿ, F. de (2011) More than it would seem: the use of coinage by the Romans in Late Hellenistic Asia Minor (133–63 BC). *AJN* 23, 55–86.

Callebat, L., Gros, P. and Jacquemard, C. (1999) *Vitruve, de l'Architecture*. Paris.

Camassa, G., de Guio, A. and Veronese, F. (eds) (2000) *Paesaggi di potere. Problemi e prospettive*. Rome.

Cambi, F. and Terrenato, N. (1994) *Introduzione all'archeologia del paesaggi*. Rome.

Cameron, A. (1976) *Circus Factions: Blues and Greens at Rome and Byzantium*. Oxford.

Camp, J.M. (2001) *The Archaeology of Athens*. New Haven.

Campagna, L. (2003a) La Sicilia di età repubblicana nella storiografia degli ultimi cinquant'anni. *Ostraka* 12, 7–31.

Campagna, L. (2003b) Il capitello della cosidetta agorà di Siracusa e la cronologia dei più antichi capitelli corinzio-sicelioti, in Bacci and Marinelli (eds), pp. 149–68.

Campagna, L. (2003c) Un capitello ionico a quattro facce nel Museo Regionale di Messina, in Fiorentini, Caltabiano and Calderone (eds), pp. 167–79.

Campagna, L. (2006) L'architettura di età ellenistica in Sicilia: per una rilettura del quadro generale, in Osanna and Torelli (eds), pp. 15–34.

Campagna, L. (2007) L'architettura pubblica ed evergetismo nella Sicilia di età repubblicana, in Micciché, Modeo and Santagati (eds), pp. 110–34.

Campagna, L. (2011) Exploring social and cultural changes in Provincia Sicilia: reflections on the study of urban landscapes, in Colivicchi (ed.), pp. 161–83.

Campanelli, A. (1997) Gli dei degli Italici, in Campanelli and Faustoferri (eds), pp. 132–41.

Campanelli, A. and Faustoferri, A. (eds) (1997) *I luoghi degli dei: Sacro e natura nell'Abruzzo italico.* Pescara.

Campanile, E. (1993) Un fenicio a Roccagloriosa. *StEtr* 58, 369–71.

Campardou, M.J. (1949) Un milliaire de Cneus Domitius Ahenobarbus Imperator découvert à Treilles (Aude). *Gallia* 7, 195–205.

Campbell, B. (2000) *The Writings of the Roman Land Surveyors.* London.

Campbell, D. (2005) *Siege Warfare in the Roman World 146 BC–AD 378.* Oxford.

Camporeale, G. (ed.) (1965) *Studi in onore di Luisa Banti.* Rome.

Camporeale, G. (ed.) (2004) *The Etruscans Outside Etruria.* Los Angeles.

Camporeale, G. (2008) The deified deceased in Etruscan culture, in Bell and Nagy (eds), pp. 220–50.

Camps, G. (1961a) *Aux origines de la Berbèrie: Massinissa ou les débuts de l'histoire.* Algiers.

Camps, G. (1961b) *Aux origines de la Berbèrie: rites et monuments funéraires.* Paris.

Camps, G. (1986) Funerary monuments with attached chapels from the Northern Sahara. *African Archaeological Review* 4, 151–64.

Camps, G. (1995) Modèle hellénistique ou modèle punique?: les destinées culturelles de la Numidie, in Fantar and Ghaki (eds), pp. 235–48.

Canali De Rossi, F. (2005) Review of G. Forsythe, *A Critical History of Early Rome. BMCR* 2005.08.40.

Canina, L. (1841) *Descrizione dell'antico Tuscolo.* Rome.

Cantilena, R. (1991) Le emissioni monetali di Larino e dei Frentani, in Capini and Di Niro (eds), pp. 141–8.

Capanna, M.C. and Amoroso, A. (2006) Velia, Fagutal, Oppius. Il periodo arcaico e le case di Servio Tullio e Tarquinio Superbo, in Carandini and Greco (eds), 2006, pp. 87–111.

Capasso, L. (1989) Modelli biologici e modelli tecnologici nell'Abruzzo preromano, in *Mostra Museo Archeologico Nazionale*, pp. 3–12. Campli.

Capelli, R. (ed.) (2000) *Studi sull'Italia dei Sanniti: In occasione della mostra "Italia dei Sanniti." Roma, Museo Nazionale Romana, Terme di Diocleziano.* Milan.

Capini, S. and De Benedittis, G. (eds) (1984) *Sannio, Pentri e Frentani dal VI al I sec. a. C.: Atti del convegno 10–11 novembre 1980.* Naples.

Capini, S. and Di Niro, A. (eds) (1991) *Samnium: Archeologia del Molise.* Rome.

Capogrossi Colognesi, L. (2002) *Persistenza e innovazione nelle strutture territoriali dell'Italia romana.* Naples.

Caprioli, F. (2011) Forma architettonica, linguaggio, decorative e committenza della prima fase del Tempio B di Largo Argentina, in La Rocca and D'Alessio (eds), pp. 89–107.

Caputo, P. (2000) Su un frammento di pittura funeraria di Cuma, in Capelli (ed.), pp. 74–7.

Carafa, P. (1996) La "grande Roma dei Tarquini" e la città romuleo-numana. *BullCom* 97, 7–34.

Carafa, P. (1997) What was Pompeii before 200 BC? Excavations in the House of Joseph II in the Triangular Forum and in the House of the Wedding of Hercules, in Bon and Jones (eds), pp. 13–31.

Carafa, P. (1998) *Il Comizio di Roma dalle origini all'età di Augusto.* Rome.

Carafa, P. (1999) Cercando la storia dei monumenti di Pompeii. Le recherché dell'Università di Roma "La Sapienza" nelle Regioni VII e VIII, in Senatore (ed.), pp. 17–33.

Carafa, P. (2004) L'*Aedes* e il *Vicus* di Vesta. I Reperti, in *Workshop di archeologia classica* 1, 135–43.

Carandini, A. (1980) Il vigneto e la villa del fondo di Settefinestre nel Cosano: un caso di produzione agricola per il mercato transmarino. *MAAR* 36, 1–10.
Carandini, A. (1981) Sviluppo e crisi delle manifatture rurali e urbane, in Giardina and Schiavone (eds), vol. II, pp. 249–60.
Carandini, A. (ed.) (1985a) *Settefinestre: una villa schiavistica nell'Etruria romana*. Modena.
Carandini, A. (ed.) (1985b) *La Romanizzazione dell'Etruria: il Territorio di Vulci*. Florence.
Carandini, A. (1986/8) Domus e Insulae sulla pendice settentrioniale del Palatino. *BullCom* 91, 263–78.
Carandini, A. (1988) *Schiavi in Italia*. Rome.
Carandini, A. (1989) La villa romana e la piantaggione schiavistica, in Momigliano and Schiavone (eds), vol. 4, pp. 101–200.
Carandini, A. (1990) Domus aristocratiche sopra le mura e il pomerio del Palatino, in Cristofani (ed.), 1990, pp. 97–9.
Carandini, A. (1994) Rome: excavations in the Forum. *Current Archaeology* 139, 261–5.
Carandini, A. (1997) *La nascita di Roma. Dèi, eroi, e uomini all'alba di una civiltà*. Turin.
Carandini, A. (2000) *Giornale di scavo*. Turin.
Carandini, A. (2004) *Palatino, Velia e Sacra Via. Paesaggi urbani attraverso il tempo* [*Workshop di Archeologia Classica* Quad. I]. Rome.
Carandini, A. (2006) *Remo e Romolo. Dai rioni dei Quiriti alla città dei Romani (775/750–700/675 a. C.)*. Turin.
Carandini, A. and Bruno, D. (eds) (2008) *La casa di Augusto: dai "Lupercalia" al Natale*. Rome.
Carandini, A. and Cambi, F. (eds) (2002) *Paesaggi d'Etruria*. Roma.
Carandini, A. and Carafa, P. (eds) (1995 [2000]) *Palatium e Sacra Via*, 1 [*Bollettino di Archeologia* 31–4]. Rome.
Carandini, A. and Cappelli, R. (eds) (2000) *Roma. Romolo, Remo e la fondazione della città*. Rome.
Carandini, A., D'Alessio, M.T. and Di Giuseppe, H. (eds) (2006) *La fattoria e la villa dell'Auditorium nel quartiere Flaminio di Roma*. Rome.
Carandini, A. and Greco, E. (eds) (2004) *Workshop di archeologia classica*, 1. Pisa.
Carandini, A. and Greco, E. (eds) (2005) *Workshop di archeologia classica*, 2. Pisa.
Carandini, A. and Greco, E. (eds) (2006) *Workshop di archeologia classica*, 3. Pisa.
Carandini, A. and Greco, E. (eds) (2007) *Workshop di archeologia classica*, 4. Pisa.
Carandini, A. and Greco, E. (eds) (2009) *Workshop di archeologia classica*, 6. Pisa.
Carandini, A. and Papi, E. (eds) (1999) *Palatium e Sacra Via*, 2. Rome.
Carandini, A. and Ricci, A. (eds) (1985) *Settefinestre: una villa schiavistica nell'Etruria romana*. Modena.
Carandini, A., Ricci, G., D'Alessio, M.T. *et al.* (1997) La villa dell'Auditorium dall'età arcaica all'età imperiale. *RM* 104, 117–48.
Carandini, A. and Settis, S. (1979) *Schiavi e padroni nell'Etruria romana*. Bari.
Carettoni, G. (1956) Le gallerie ipogee del foro Romano e i ludi gladiatori forensi. *BullCom* 76, 23–44.
Carettoni, G. and Fabbrini, L. (1961) Esplorazioni sotto la Basilica Giulia al Foro Romano. *RendLinc* 16, 53–60.
Carfora, P. (2001) Nuove acquisizioni in area caudina, in Quilici and Quilici Gigli (eds), pp. 171–8.
Carlsen, J. and Lo Cascio, E. (eds) (2009) *Agricoltura e scambi nell'Italia tardo-repubblicana*. Bari.
Carlsen, J., Ørsted, P. and Skydsgaard, J.E. (eds) (1994) *Land Use in the Roman Empire*. Rome.
Carlson, D. (2007) Mast-step coins. *IJNA* 36, 317–24.
Carman, J. (1996) *Valuing Ancient Things. Archaeology and the Law*. London.

Carocci, F., De Albentiis, E., Gargiulo, M. and Pesando, F. (eds) (1990) *Le Insulae 3 e 4 della regio VI di Pompei : un'analisi storico-urbanistica.* Rome.

Carpenter, T. *et al.* (eds) (forthcoming) *Beyond Magna Graecia. New developments in south Italian Archaeology.* Cambridge.

Carrington, R.C. (1931) Studies in the Campanian "villae rusticae." *JRS* 21, 110–30.

Carrington, R. (1933) Notes on the building materials of Pompeii. *JRS* 23, 125–38.

Carrington-Smith, J. (1982) A Roman chamber tomb on the south-east slopes of Monasteriaki Kephala, Knossos. *BSA* 77, 255–93.

Carruba, A.M. and De Masi, L. (2006) *La Lupa capitolina: un bronzo medievale.* Rome.

Carson, R.A.G. (ed.) (1978) *Scripta Nummaria Romana. Essays Presented to Humphrey Sutherland.* London.

Carter, J.C. (2006) *Discovering the Greek Countryside at Metaponto.* Ann Arbor.

Caruso, E. (2005) L'insula di Capo Boeo nell'impianto urbanistico di Lilibeo, in Spanò Giammellaro (ed.), pp. 777–85.

Casey, J. and Reece, R. (1988) *Coins and the Archaeologist* (2nd rev. edn). London.

Castagnoli, F. (1953/5) I piú antiche divisioni agrarie romane. *Bulletino del Museo della Civiltà Romana* 18, 3–10.

Castagnoli, F. (1955) Peripteros sine postico. *RM* 62, 139–43.

Castagnoli, F. (1957) La centuriazione di Cosa. *MAAR* 25, 149–65.

Castagnoli, F. (1964) Note sulla topografia del Palatino e del Foro Boario. *ArchCl* 16, 173–99.

Castagnoli, F. (ed.) (1966) *Studi di Urbanistica Antica.* Rome.

Castagnoli, F. (1977) Communication. Les sanctuaires du Latium Archaïque. *CRAI* 121 (3), 460–76.

Castagnoli, F. (1980a) Installazioni Portuali a Roma. *MAAR* 36, 35–42.

Castagnoli, F. (1980b) Santuari e culti nel Lazio Arcaico. *Archeologia Laziale* 3, 164–7.

Castellví, G., Nolla, J. and Rodà, I. (eds) (2008) *Le trophée de Pompée dans les Pyrénées (71 av.J.-C.)* [*Gallia* 58]. Paris.

Castellví, G., Pézin, A., Kotarba, J. and Comps, J.-P. (eds) (1997) *Voies romaines du Rhône à l'Èbre: via Domitia et via Augusta.* Paris.

Castor, A. (2010) Late Classical representations of jewelry: identifying costume trends in Etrusco-Italic art. *EtrStud* 13, 29–48.

Catalli, F. (1988) Monete straniere in Etruria. *PACT: revue du Groupe européen d'études pour les techniques physiques, chimiques et mathématiques appliquées à l'archéologie* 20, 465–93.

Cavalieri Manasse, G. (2000) Un documento catastale dell'agro centuriato veronese. *Athenaeum* 88, 5–52.

Cébeillac-Gervasoni, M. (ed.) (1996) *Les élites municipales de l'Italie péninsulaire des Gracques à Néron. Actes de la table ronde de Clermont-Ferrand (28–30 novembre 1991)* [*Collection de l'École Française de Rome 215, Collection du Centre Jean Bérard 13*]. Naples.

Cébeillac-Gervasoni, M. and Lamoine, L. (eds) (2003) *Les élites locales dans le monde Hellénistique et Romain. Les élites et leurs facettes* [Actes du colloque international de Clermont-Ferrand]. Rome.

Cecamore, C. (2002) *Palatium. Topografia storica del Palatino tra III sec. a.C. e I sec. d.C.* Rome.

Ceci, F. (2005) *Blera e le sue necropoli.* Rome.

Ceglia, V. (2008) San Martino in Pensilis, Campobasso, Molise, Italy: the "villa" of Contrada Mattonelle, in Lock and Faustoferri (eds), pp. 191–204.

Cerchiai, L. (1995) *I Campani.* Milan.

Cerdà, J. *et al.* (1994) Iluro, oppidum civium Romanorum: estado de la cuestión, in *Actes del XIV Congrés*, pp. 97–9.

Chaín, A. and de la Torre, J. (eds) (2005) *Celtíberos: tras la estela de Numancia.* Soria.

Chancey, M. and Porter, A. (2009) The archaeology of Roman Palestine. *NEA* 64, 164–203.

Chaney, E. (1998) *The Evolution of the Grand Tour: Anglo-Italian Cultural Relations since the Renaissance*. London.

Chaplin, J. and Kraus, C. (eds) (2009) *Livy*. Oxford.

Chapman, H. (2006) *Landscape Archaeology and GIS*. Stroud.

Chasseigne, L., Fincker, M., Magallón Botaya, M.A. *et al.* (2006) Labitolosa and other Roman towns on the south side of the Pyrenees, in Abad, Keay and Ramallo (eds), pp. 213–22.

Chaves, F. (2009) Las amonedaciones hispanas en la Antigüedad, in Andreu, Cabrero and Rodà (eds), pp. 47–97.

Chellini, R. (2002) *Acque Sorgive Salutari e Sacre in Etruria (Italiae Regio VII): Ricerche Archeologiche e di Topografia Antica* [*BAR-IS* 1067].

Chevallier, R. (1958) Essai de chronologie des centuriations romaines de Tunisie. *MÉFRA* 70, 61–126.

Chevallier, R. (1976) *Roman Roads*. London.

Chevallier, R. (ed.) (1992) *Les eaux thermales et les cultes des eaux en Gaule et dans les provinces voisines: Actes du Colloque, 28–30 septembre 1990, Aix-les-Bains*. Tours.

Chiaramonte, C. (2007) The walls and gates, in Dobbins and Foss (eds), pp. 140–9.

Chiarucci, P. (1983) *Lanuvium*. Rome.

Childe, V.G. (1956) *Piecing Together the Past: the Interpretation of Archaeological Data*. London.

Chilton, E.S. (ed.) (1999) *Material Meanings. Critical Approaches to the Interpretation of Material Culture*. Salt Lake City.

Chiranky, G. (1982) Southern Italy before the Romans: Sabellians, Sicilians and Italian Greeks in Lucania and Bruttium (450–367 BC). PhD thesis. University of California, Berkeley.

Choren, J., Mercado, M. and Rodrigo, E. (2007) El jaciment de Can Tacó: un assentament de caràcter excepcional al Vallès Oriental. *Ponències* (Anuari del Centre d'Estudis de Granollers) 2006, 57–76.

Chouquer, G., Clavel-Lévêque, M., Favory, F. and Vallat, J.-P. (1987) *Structures agraires en Italie centro-méridionale. Cadastres et paysages ruraux (CollEFrR 100)*. Rome.

Christian, D. (2010) History and science after the chronometric revolution, in Dick and Lupisella (eds), p. 441.

Christie, N. (ed.) (1995) *Settlement and Economy in Italy 1500 BC to AD 1500*. Oxford.

Christol, M. and Darde, D. (eds) (2009) *L'expression du pouvoir au debut de l'Empire: Autor de la Maison Carrée à Nîmes* [Actes du Colloque]. Nîmes.

Churchill, J.B. (1999) *Ex qua quod vellent facerent*: Roman magistrates' authority over praeda and manubiae. *TAPA* 129, 85–116.

Ciancio Rossetto, P. (1973) *Il sepolcro del fornaio Marco Virgilio Eurisace a Porta Maggiore*. Rome.

Ciancio Rossetto, P. (1995) Indagini e restauri nel Campo Marzio meridionale: Teatro di Marcello, Portico d'Ottavia, Circo Flaminio, Porto Tiberino. *Archeologia Laziale* 12, 93–101.

Ciancio Rossetto, P. (1997/8) Tempio di Apollo: nuove indagini sulla fase repubblicana. *RendPontAcc* 70, 177–95.

Cibecchini, F. and Capelli, C. (forthcoming) Nouvelles données sur les épaves républicaines du IIIe s. avant notre ère en Méditerranée occidentale, in Olmer (ed.).

Cibecchini, F. and Principal, J. (2004) Per chi suona la Campana B?, in De Sena and Dessales (eds), pp. 159–72.

Cicala, V. (1995) Diana ariminense: tracce di religiosità politica, in Calbi and Susini (eds), pp. 355–65.

Cifani, G. (1995) Aspetti dell'edilizia romana arcaica. *StEtr* 60, 185–226.

Cifani, G. (1998) Caratteri degli insediamenti rurali nell'Ager Romanus tra VI e III secolo a.C., in Pearce and Tosi (eds), pp. 53–64.

Cifani, G. (2002) Notes on the rural landscape of central Tyrrhenian Italy in the 6th–5th c. BC and its social significance. *JRA* 15, 247–60.

Cifani, G. (2008) *Architectura romana arcaica*. Rome.
Cima, M. and La Rocca, E. (1986) *Le tranquille dimore degli dei*. Rome.
Cima, M. and La Rocca, E. (eds) (1998) *Horti Romani: atti del convegno internazionale, Roma, 4–6 maggio 1995*. Rome.
Cipriani, M. (1996) Prime presenze italiche organizzate alle porte di Poseidonia, in Cipriani, Longo and Viscione (eds), pp. 119–39.
Cipriani, M., Longo, F. and Viscione, M. (eds) (1996) *Poseidonia e i Lucani*. Naples.
Civiltà del Lazio primitivo. Palazzo delle esposizioni (1976). Rome.
Claridge, A. (1998) *Rome: Oxford Archaeological Guide*. Oxford.
Clarke, J.R. (1991) *The Houses of Roman Italy 100 BC–AD 250*. Berkeley.
Clavel-Lévêque, M., Jouffroy, I. and Vignot, A. (eds) (1994) *De la terre au ciel I. Paysages et cadastres antiques (Besançon, 1993)*. Paris.
Clavel-Léveque, M. and Orejas, A. (eds) (1998, 2002) *Atlas historique des cadastres d'Europe*, 2 vols. Luxembourg.
Clinton, K. (1997) Eleusis and the Romans, in Hoff and Rotroff (eds), pp. 161–81.
Clinton, K. (2008) *Eleusis. The Inscriptions on Stone. Documents of the Sanctuary of the Two Goddesses and Public Documents of the Deme*, 3 vols. Athens.
Coale, A. and Demeny, P. (1966) *Regional Model Life Tables and Stable Populations*. Princeton.
Coarelli, F. (1968a) Navalia, Tarentum e la topografia del Campo Marzio meridionale. *QITA* 5, 27–38.
Coarelli, F. (1968b) La porta trionfale e la via dei trionfi. *DialArch* 2, 55–103.
Coarelli, F. (1971) Discussione. *AttiTaranto* 11, 330–1.
Coarelli, F. (1972) Il Sepolcro degli Scipioni. *Dialoghi di Archeologia* 6, 36–106.
Coarelli, F. (1976) Cinque frammenti di una tomba dipinta dall'Esquilino, in *Affreschi romani dalle raccolte dell'Antiquarium Comunale*, pp. 22–8. Rome.
Coarelli, F. (1977) Public building in Rome between the Second Punic War and Sulla. *PBSR* 45, 1–23.
Coarelli, F. (1980) La cultura figurative in SicIlia: dalla conquista romana a Bisanzio, in Gabba and Vallet (eds), pp. 373–92.
Coarelli, F. (1985) *Il Foro Romano. Periodo repubblicano e augusteo*. Rome.
Coarelli, F. (1987a) Munigua, Praeneste e Tibur. I modelli laziali di un municipio della Baetica. *Lucentum* 6, 91–100.
Coarelli, F. (1987b) *I santuari del Lazio in età repubblicana*. Rome.
Coarelli, F. (1988a) Colonizzazione romana e viabilità. *DialArch* 6, 35–48.
Coarelli, F. (1988b) *Il Foro Boario dalle origini alla fine della Repubblica*. Rome.
Coarelli, F. (1988c) *Il Sepolcro degli Scipioni a Roma*. Rome.
Coarelli, F. (1996) *Revixit ars: Arte e ideologia a Roma dai modelli ellenistici alla tradizione repubblicana*. Rome.
Coarelli, F. (1997) *Il Campo Marzio. Dalle origini alla fine della Repubblica*. Rome.
Coarelli, F. (1998a) The Odyssey frescoes of the Via Graziosa: a proposed context. *PBSR* 66, 21–37.
Coarelli, F. (1998b) Il culto di Mefitis in Campania e a Roma, in *I Culti della Campania Antica. Atti del Convegno Internazionale di Studi in Ricordo di Nazarena Valenza Mele*, pp. 185–90. Rome.
Coarelli, F. (2001) Il rescritto di Spello e il santuario etnico degli Umbri, in *Umbria cristiana. Dalla diffusione del culto al culto dei santi. Secoli IV–X. Atti del XV Congresso internazionale di studi sull'alto medioevo, Spoleto 23–28 ottobre 2000*, pp. 39–51. Spoleto.
Coarelli, F. (2002) Lo sviluppo urbano, in Coarelli, Foglia and Poglia (eds), pp. 27–45.
Coarelli, F. (2004) Le terme di Fregellae, in Ghini (ed.), pp. 73–6.
Coarelli, F. (2005) Pits and fora: a reply to Henrik Mouritsen. *PBSR* 73, 23–30.

Coarelli, F. (2007a) Horrea Cornelia? in Leone, Palombi, Walker (eds), pp. 41–6.

Coarelli, F. (2007b) *Rome and Environs: An Archaeological Guide*. Berkeley and Los Angeles.

Coarelli, F. (2008) Il settore nord-occidentale di Pompei e lo sviluppo urbanistico della città dall'età arcaica al III secolo a.C., in Guzzo and Guidobaldi (eds), 2008, pp. 173–6.

Coarelli, F. (2010) *Substructio* et *tabularium*. *PBSR* 78, 107–32.

Coarelli, F., Batocchioni, G. and Romagnoli, L. (1995) Il parco archeologico di Fregellae, in Amendolea (ed.), pp. 133–44.

Coarelli, F., Foglia, A. and Poglia, P. (eds) (2002) *Pompei: la vita ritrovata*. Udine.

Coarelli, F. and La Regina, A. (1984) *Guide archeologiche Laterza: Abruzzo e Molise*. Bari and Rome.

Coarelli, F. and Monti, P.G. (eds) (1998) *Fregellae Vol. 1. Le Fonti, la Storia, il Territoria*. Rome.

Coarelli, F. and Patterson, H. (eds) (2009) *Mercator Placidissimus: The Tiber Valley in Antiquity: New Research in the Upper and Middle River Valley*. Rome.

Coarelli, F. and Pesando, F. (eds) (2006) *Rileggere Pompei. 1. L'Insula 10 della Regio VI*. Rome.

Coarelli, F. and Thébert, Y. (1988) Architecture funéraire et pouvoir: réflexions sur l'hellenisme numide. *MÉFRA* 100, 761–818.

Coarelli, F. and Torelli, M. (1984) *Sicilia* (2nd edn). Rome.

Coates-Stephens, R. (2004) *Porta Maggiore: Monument and Landscape: Archaeology and Topography of the Southern Esquiline from the Late Republican Period to the Present* [*BullCom* Suppl. 12]. Rome.

Cohen, B. (2010) New light on a master bronze from Etruria. *AJA* Online Museum Review, 114.3.

Colavitti, A.M. (1999) *La presenza dei "negotiatores" italici nella Sardegna di età romana. Oristano*. Colavitti.

Colavitti, A.M. (2003) *Cagliari. Forma e urbanistica* [Città antiche in Italia 6]. Rome.

Colini, A.M. (1967–8) La stipe delle Aque salutari di Vicarello: Notizie sul complesso della Scoperta. *RendPontAcc* 40, 35–56.

Colini, A.M. (1980) Il porto fluviale del Foro Boario a Roma, in D'Arms and Kopff (eds), pp. 43–53.

Colivicchi, F. (2011) The long goodbye: the local elites of Daunia between continuity and change (3rd–1st c. BC), in Colivicchi (ed.), pp. 112–37.

Colivicchi, F. (ed.) (2011) *Local Cultures of South Italy and Sicily in the Late Republican Period: Between Hellenism and Rome* [*JRA* Suppl. 83]. Portsmouth, RI.

Colivicchi, F. (forthcoming) "Native" vase shapes in south-Italian red-figure pottery, in Carpenter *et al.* (eds).

Collart, P. (1976) Les milliaires de la Via Egnatia. *BCH* 100, 177–200.

Collins, J.J. (1992) The Dead Sea Scrolls, in Freedman (ed.), vol. 2, pp. 85–101.

Collins-Clinton, J. (2000) The Neronian Odeum at Cosa and its sculptural program: a new Julio-Claudian dynastic group. *MAAR* 45, 99–130.

Colonna, G. (1970) *Bronzi votivi umbro-sabellici a figura umana*. Florence.

Colonna, G. (1976) Nuovi dati epigrafici sulla protostoria della Campania. *Atti XVII Riunione Istituto italiano di preistoria e protostoria*, pp. 151–69. Florence.

Colonna, G. (1979a) The Later Orientalizing Period in Rome, in D. Ridgway and F. Ridgway (eds), pp. 223–35.

Colonna, G. (1979b) Duenos. *StEtr* 47, 163–72.

Colonna, G. (ed.) (1981) *Gli Etruschi e Roma. Atti dell'incontro di studio in onore di Massimo Pallottino. Roma, 11–13 dicembre 1979*. Rome.

Colonna, G. (1983) Identità come appartenenza nelle iscrizioni di possesso dell'Italia preromana. *Epigraphica* 45, 49–55.

Colonna, G. (ed.) (1985) *Santuari d'Etruria*. Milan.

Colonna, G. (1987) Etruria e Lazio nell'età dei Tarquinii, in Cristofani (ed.), 1987, pp. 55–66.

Colonna, G. (1991) Il posto dell'Arringatore nell'arte etrusca di età ellenistica. *StEtr* 56, 99–119.

Colonna, G. (1996) Roma archaica. I suoi sepolcreti e le vie per i Colli Albani, in Pasqualini (ed.), pp. 335–54.

Colonna, G. (2006) Sacred architecture and the religion of the Etruscans, in de Grummond and Simon (eds), pp. 132–68.

Colonna, G. (2007a) Dischi-corazza e dischi di ornamento femminile: due distinte classi di bronzi centro italici. *ArchCl* 58, 3–30.

Colonna, G. (2007b) La scultura in pietra in Abruzzo, in Ruggeri and Tavano (eds), pp. 27–31.

Colonna, G., Bartoloni, G., Colonna di Paolo, E. and Melis, F. (eds) (1976) *Civiltà del Lazio primitivo*. Rome.

Colonna Di Paolo, E. and Colonna, G. (1978) *Norchia I: Le necropoli rupestri dell'Etruria meridionale*. Rome.

Colucci-Pescatori, G. *et al.* (eds) (1996) *Gli Irpini. Storia di Avellino*, vol. 1. Salerno.

Comas, M., Llobet, C., Padrós, P. *et al.* (1994) Un espai d'ús públic a l'àrea central de Baetulo (Hispania Tarraconensis). Evolució històrica i transformacions urbanístiques, in *Actes del XIV Congrés*, pp. 110–12.

Comella, A. (1981) Tipologia e diffusione dei complessi votivi in Italia in epoca medio- e tardo-repubblicana. Contributo alla storia dell'artigianato antico. *MÉFRA* 93, 717–803.

Comella, A. (1992) Matrici fittili dal santuario di via Malta a Cagliari, in Bonello Lai (ed.), pp. 415–23.

Comella, A.M. and Mele, S. (eds) (2005) *Depositi votivi e culti dell'Italia antica dall'età arcaica a quella tardo-repubblicana*. Bari.

Conde, M. (2010) Rome: the Imperial Fora, http://rometheimperialfora19952010.wordpress.com/2010/08 (accessed November 22, 2012).

Congiu, M., Miccichè, C. and Modeo, S. (eds) (2009) *"ΕΙΣ ΑΚΡΑ," insediamenti d'altura in Sicilia dalla preistoria al III sec. a.C.* [Atti del V Convegno di Studi, Caltanissetta, May 10–11, 2008]. Caltanissetta.

Connolly, P. (1978) *Hannibal and the Enemies of Rome*. London.

Connolly, P. (1981) *Greece and Rome at War*. London.

Connolly, P. (1989) The Roman army in the age of Polybius, in Hackett (ed.), pp. 149–68.

Connolly, P. (1997) Pilum, gladius and pugio in the Late Republic. *Journal of Roman Military Equipment Studies* 8, 41–57.

Connolly, P. (2000) The reconstruction and use of Roman weaponry in the second century BC. *Journal of Roman Military Equipment Studies* 11, 43–6.

Connolly, P. (2002) The pilum from Marius to Nero – a reconsideration of its development and function. *Journal of Roman Military Equipment Studies* 12/13 (2001/2002), 1–8.

Conolly, J. and Lake, M. (2006) *Geographical Information Systems in Archaeology*. Cambridge.

Consentino, R. and Sabbatini Tumolesi, P. (1989) L'edificio termale delle Aquae Caeretanae. *Miscellanea Ceretana* 17 (1), 95–112.

Conta Haller, G. (1978) *Ricerca su alcuni centri fortificati in opera poligonale in area campano-sannitica*. Naples.

Conticello de'Spagnolis, M. (1986) Nuove osservazioni sull'area del tempio dei Dioscuri in circo Flaminio. *BullCom* 91, 91–6.

Conyers, L. and Goodman, D. (1997) *Ground-Penetrating Radar: An Introduction for Archaeologists*. Alta Mira.

Cool, H. and Philo, C. (eds) (1998) *Roman Castleford, I. The Small Finds* [*Yorkshire Archaeology* 4]. Wakefield.

Cooley, A.E. (ed.) (2000) *The Epigraphic Landscape of Roman Italy*. London.

Cooley, A.E. (2006) Beyond Rome and Latium. Roman religion in the age of Augustus, in Schultz and Harvey (eds), pp. 228–52.

Corazza, A., Giordano, G. and de Rita, D. (2006) Hydrogeology of the city of Rome, in Heiken (ed.), pp. 113–18.

Corazza, A., Lombardi, L. and Marra, F. (2004) Geologica, idrogeologia e approvvigionamento idrico del Colle Capitolino (Roma, Italia). *Il Quaternario* 17, 413–41.

Cordier, P. (2005) *Varius in omni genere vitae*: l'acculturation, les identités, le métissage et les faits culturels romains entre histoire et anthropologie. *Mètis. Anthropologie des mondes grecs anciens*, n.s. 3, 295–305.

Cornell, T.J. (1989) The conquest of Italy, in *CAH* VII (2), 351–419. Cambridge.

Cornell, T. (1991) The tyranny of the evidence: a discussion of the possible uses of literacy in Etruria and Latium in the Archaic age, in Humphrey (ed.), 1991, pp. 7–33.

Cornell, T.J. (1995) *The Beginnings of Rome: Italy and Rome from the Bronze Age to the Punic Wars (c. 1000–264 BC)*. London.

Cornell, T.J. (1996) Hannibal's legacy: the effects of the Hannibalic War on Italy, in Cornell, Rankov and Sabin (eds), pp. 97–117.

Cornell, T.J. (1997) Ethnicity as a factor in early Roman history, in Cornell and Lomas (eds), 1997, pp. 9–21.

Cornell, T.J. (2000a) The city of Rome in the Middle Republic (c. 400–100 BC), in Coulston and Dodge (eds), pp. 42–60.

Cornell, T.J. (2000b) The city-states in Latium, in Hansen (ed.), pp. 209–28.

Cornell, T.J. (2004) Deconstructing the Samnite Wars: an essay in historiography, in Jones (ed.), pp. 115–31.

Cornell, T. (2005) The value of the literary tradition concerning Archaic Rome, in Rauflaub (ed.), pp. 47–74.

Cornell, T.J. and Lomas, K. (eds) (1996) *Urban Society in Roman Italy*. London.

Cornell, T.J. and Lomas, K. (eds) (1997) *Gender and Ethnicity in Early Roman Italy*. London.

Cornell, T., Rankov, B. and Sabin, P. (eds) (1996) *The Second Punic War. A Reappraisal*. London.

Corrigan, E.H. (1979) Lucanian tomb paintings excavated at Paestum 1969–1972. PhD thesis. Columbia University.

Corsaro, M. (1982) La presenza romana ad Entella: una nota su Tiberio Claudio di Anzio. *AnnPisa* 3 (12), 993–1032.

Corsi, C. and Polito, E. (eds) (2008) *Dalle sorgenti alla foce. Il bacino del Liri-Garigliano nell'antichita*. Rome.

Corzo, R. (2002) La fundación de Itálica y su desarrollo urbanístico, in Jiménez and Ribera (eds), pp. 123–35.

Costantini, L. and Giorgi, J. (2001) Charred plant remains of the Archaic Period from the Forum and Palatine. *JRA* 14, 239–48.

Cotton, H. and Wörrle, J. (2007) Seleukos IV to Heliodoros. A new dossier of royal correspondence from Israel. *ZPE* 159, 191–205.

Cotton, M.A. (1979) *The Late Republican Villa at Posto Francolise*. London.

Cotton, M.A. and Métraux, G. (1985) *The San Rocco Villa at Francolise*. London.

Coudry, M. and Humm, M. (eds) (2009) *Praeda: Butin de guerre et société dans la Rome républicaine* [*Collegium Beatus Rhenanus* 1]. Stuttgart.

Couissin, P. (1926) *Les Armes romaines*. Paris.

Coulston, J. and Dodge, H. (eds) (2000) *Ancient Rome. The Archaeology of the Eternal City*. Oxford.

Coulton, J.J. (1976) *The Architectural Development of the Greek Stoa*. Oxford.

Cousin, G. and Deschamps, G. (1888) Inscription de Magnésie du Méandre. *BCH* 12, 204–23.

Cowan, R. (2009) *Roman Conquests: Italy*. Barnsley.

Cozza, L. and Tucci, P.L. (2006) Navalia. *ArchCl* 57, 175–201.
Crane, S.A. (ed.) (2000) *Museums and Memory.* Stanford, CA.
Crawford, J.W., Frischer, B.D. and Koller, D. (eds) (2009) *Making History Interactive: Computer Applications and Quantitative Methods in Archaeology (CAA). Proceedings of the 37th International Conference on Computer Applications and Quantitative Methods in Archaeology (CAA) Williamsburg, Virginia, USA, March 22–26* [*BAR* 2079]. Oxford.
Crawford, M. (1968) Plated coins – false coins. *NC* ser.7, 8, 55–9.
Crawford, M. (1969) *Roman Republican Coin Hoards.* London.
Crawford, M. (1973) Paestum and Rome: the form and function of a subsidiary coinage. *La monetazione di bronzo di Poseidonia-Paestum, AIIN* Suppl. 1a: 18–19, 47–103.
Crawford, M. (1974) *Roman Republican Coinage,* 2 vols. Cambridge.
Crawford, M. (1985) *Coinage and Money under the Roman Republic.* London.
Crawford, M. (1987) Sicily, in Burnett and Crawford (eds), pp. 43–51.
Crawford, M. (1995) La storia della colonizzazione romana secondo i romani, in Storchi Marino (ed.), pp. 187–92.
Crawford, M. (1996) The west: Italy and Rome from Sulla to Augustus, in *CAH* X, 414–33, 979–89.
Crawford, M. (2006a) From Poseidonia to Paestum via the Lucanians, in Bradley and Wilson (eds), pp. 59–72.
Crawford, M. (2006b) The Oscan inscriptions of Messina, in Michelini (ed.), pp. 521–5.
Crawford, M. (2007) The Mamertini, Alfius and Festus, in Dubouloz and Pittia (eds), pp. 273–9.
Crawford, M. (2008) States waiting in the wings: population distribution and the end of the Roman Republic, in de Ligt and Northwood (eds), pp. 631–43.
Crawford, M. and Burnett, A. (eds) (1987) *The Coinage of the Roman World in the Late Republic* [*BAR-IS* 326]. Oxford.
Creasy, G.L. and Creasy, L.L. (2009) *Grapes.* Cambridge, MA.
Creighton, J. (2000) *Coins and Power in Late Iron Age Britain.* Cambridge.
Cresci Marrone, G. and Tirelli, M. (eds) (2005) *Terminavit sepulcrum: I recinti funerari nelle necropoli di Altino. Atti del convegno, Venezia, 3–4 dicembre 2003.* Rome.
Criniti, N. (1970) *L'epigrafe di Asculum di Gn. Pompeo Strabone.* Milan.
Cristofani, M. (1996) La scrittura e la lingua, in Cipriani, Longo and Viscione (eds), pp. 201–3.
Cristofani, M. (ed.) (1987) *Etruria e Lazio arcaico* [Quaderni del centro di studio per l'archeologia etrusco-italica 15]. Rome.
Cristofani, M. (ed.) (1990) *La grande Roma dei Tarquini: Palazzo delle espossizioni, Roma 12 giugno–30 settembre 1990.* Rome.
Cristofani, M., Cristofani Martelli, M., Fiumi, E. *et al.* (eds) (1975–7) *Corpus delle urne etrusche di età ellenistica, 1 Urne Volterrane 1.* Florence.
Cristofani, M., Pandolfini Angeletti, M. and Coppola, G. (eds) (1996) *Corpus Inscriptionum Etruscarum 2.2 Inscriptiones et in Latio et in Bampania repertae.* Rome.
Crowfoot, J., Kenyon, K. and Sukenik, E. (1942) *Samaria-Sebaste I: The Buildings.* London.
Crowther, N.B. (1983) Greek games in Republican Rome. *AntCl* 52, 268–73.
Culasso Gastaldi, E. (2000) L'iscrizione trilingue del Museo di Antichità di Torino (dedicante greco, ambito punico, età romana). *Epigraphica* 62, 11–28.
Cullhed, M. and Slej, K. (eds) (2008) *The Temple of Castor and Pollux II.2. The Finds and Trenches.* Rome.
Cunliffe, B. (ed.) (1988) *The Temple of Sulis Minerva at Bath II.* Oxford.
Cuozzo, M. (2007) Ancient Campania: cultural interaction, political borders and geographical boundaries, in Bradley, Isayev and Riva (eds), pp. 224–67.
Curchin, L. (1996) *España romana. Conquista y asimilación.* Madrid.

Curran, B. (2011) Love, triumph, tragedy: Cleopatra and Egypt in High Renaissance Rome, in Miles (ed.), pp. 96–131.

Curti, E. (2000) From Concordia to the Quirinal. Notes on religion and politics in mid-Republican-Hellenistic Rome, in Bispham and Smith (eds), pp. 77–91.

Curti, E. (2001) Toynbee's legacy: discussing aspects of the Romanization of Italy, in Keay and Terrenato (eds), pp. 17–26.

Curti, E. (2008) Il Tempio di Venere Fisica e il porto di Pompei, in Guzzo and Guidobaldi (eds), 2008, pp. 47–60.

Curti, E., Dench, E. and Patterson, J.R. (1996) The archaeology of central and southern Roman Italy. Recent trends and approaches. *JRS* 86, 170–89.

D'Agostino, B. (1992) Greci, Campani e Sanniti: città e campagna nella regione campana, in *La Campania fra VI e III secolo* (Atti XIV Convegno di Studi Etruschi e Italici), pp. 83–95. Florence.

Dalcher, K. (1994) *Studia Ietina VI. Das Peristylhaus 1 von Iaitas: Architektur und Baugeschichte.* Zürich.

D'Alessio, A. (2010) Fascino greco e "attualità" romana: la conquista di una *nuova* architettura, in La Rocca, Presicce and Lo Monaco (eds), pp. 49–64.

D'Alessio, M.T. (2008) La Casa delle Nozze di Ercole (VII,9,47): storia di un isolato presso il Foro alla luce dei nuovi dati ceramici, in Guzzo and Guidobaldi (eds), 2008, pp. 275–82.

D'Amato, C. (1989) Terme e cure termali nell'antica Roma. *Terme Romane e Vita Quotidiana: Supplemento all'Edizione Romana della Mostra*, pp. 11–16. Rome.

D'Amato, R. and Sumner, G. (2009) *Arms and Armour of the Imperial Roman Soldier: From Marius to Commodus, 112 BC–AD 192.* London.

D'Ambrosio, A. and De Caro, S. (1987) La necropoli di Porta Nocera: Campagna di scavo 1983, in von Hesberg and Zanker (eds), pp. 199–228.

D'Ambrosio, A. and De Caro, S. (1989) Un contributo all'architettura e all'urbanistica di Pompei in età ellenistica. I saggi nella casa VII 4, 62. *AION* Sezione di Archeologia e Storia Antica II, 173–215.

Damgaard Andersen, H. (1993) The Etruscan ancestral cult – its origin and development and the importance of anthropomorphization. *AnalRom* 21, 8–66.

D'Andria, R. (1989) Il teatro tardo-ellenistico di foce Sarno. *Romana Gens* 78, 9–14.

D'Arms, J. (1998) Between public and private: the *epulum publicum* and Caesar's *horti trans Tiberim*, in Cima and La Rocca (eds), pp. 33–44.

D'Arms, J.H. and Kopff, E.C. (eds) (1980) *The Seaborne Commerce of Ancient Rome: Studies in Archaeology and History.* Rome.

David, J.-M. (1983) Le tribunal dans la basilique: évolution fonctionnelle et symbolique de la République à l'Empire, in *Architecture et société: de l'archaïsme grec à la fin de la République romaine: Actes du colloque international organisé par le Centre national de la Recherche scientifique et l'École française de Rome (Rome 2–4 décembre 1980)*, pp. 219–45. Rome.

Davies, G. (2006) *Roman Siege Works.* Stroud.

Davies, J. and Jones, R. (2006) *Roman Camps in Wales and the Marches.* Cardiff.

Davies, P.J.E. (2006) Exploring the international arena: the Tarquins' aspirations for the Temple of Jupiter Optimus Maximus, in Mattusch, Donohue, and Brauer (eds), pp. 186–9.

Davies, P.J.E. (2012) Pollution, propriety and urbanism in Republican Rome, in Bradley (ed.), pp. 67–80.

Davies, P.J.E. (in press a) On the introduction of stone architraves in Republican temples in Rome, in Thomas and Meyers (eds).

Davies, P.J.E. (in press b) Rome and her neighbors: Greek building practices in Republican Rome, in Ulrich and Quenomoen (eds).

Davies, P.J.E. (forthcoming) *Architecture, Art and Politics in Republican Rome.* Cambridge.

Davies, Ph. (1983) *Qumran*. Grand Rapids.

Day, J. (1942) *An Economic History of Athens Under Roman Domination*. New York.

De Benedittis, G. (1991) L'abitato di Monte Vairano, in Capini and Di Niro (eds), pp.127–30.

De Benedittis, G. (2010) L'orientamento astronomico dei templi sannitici, in Badolati (ed.), pp. 103–13.

De Benedittis, G. and Bevilacqua, G. (1980a) Le anfore, in *Sannio*, pp. 342–8.

De Benedittis, G. and Bevilacqua, G. (1980b) Bolli rodi, in *Sannio*, pp. 306–8.

Deberge, Y. and Guichard, V. (2000) Nouvelles recherches sur les travaux césariens devant Gergovie (1995–1999). *RACentre* 39, 83–111.

De Bonis, R. (2007) Una nuova iscrizione di A. Vinicius Lucanus da Paestum. *AION* 13/14, 13–18.

De Caro, S. (1985) Nuove indagini sulle fortificazioni di Pompei. *AION* 7, 74–115.

De Caro, S. (1986) *Saggi nell'area del tempio di Apollo a Pompei. Scavi stratigrfici di A. Maiuri nel 1931–32 e 1942–43, Naples 1986*. Naples.

De Caro, S. (2007) The first sanctuaries, in Dobbins and Foss (eds), pp. 73–81.

De Caro, S. and Greco, A. (1981) *Campania* [Guide archeologiche Laterza 10]. Rome.

De Casa, G. and Lombardi, G. (2007) Caratteri fisico-meccanici del Tufo Giallo della Via Tiberina (Roma). *RendLinc* 18, 5–25.

De Casa, G., Lombardi, G., Meucci, C. *et al.* (1999) Il tufo lionato dei monumenti Romani: caratteri petrografici, geomeccanici, e trattamenti conservativi. *Geologia Romana* 35, 1–25.

de Cazanove, O. (2000) Some thoughts on the "religious Romanisation" of Italy before the Social War, in Bispham and Smith (eds), pp. 71–6.

de Cazanove, O. (2005) Le aree interne dal III al I sec. a.C.: Il quadro archeologico. *Atti del quaranta quattresimo convegno di studi sulla Magna Grecia Taranto* 44, 763–99.

de Cazanove, O. (2007) Pre-Roman Italy, before and under the Romans, in Rüpke (ed.), pp. 43–57.

de Cazanove, O. (2008) *Civita di Tricarico I: Le quartier de la maison du monolithe e l'enceinte intermédiaire*. Rome.

de Cazanove, O. and Scheid, J. (eds) (2003) *Sanctuaires et sources dans l'antiquité: Les sources Documentaires et leur limites dans la description des lieux de culte*. Naples.

de Chaisemartin, N. (2003) *Rome: Paysage urbain et idéologie des Scipions à Hadrien*. Paris.

D'Ercole, V. (2000) I "paesaggi di potere" dell'Abruzzo protostorico, in Camassa, de Guio and Veronese (eds), pp. 121–52.

D'Ercole, V., Faustoferri, A. and Ruggieri, M. (2003) L'età del ferro in Abruzzo, in *Atti della XXXVI Riunione Scientifica Preistoria e Protostoria dell'Abruzzo*, pp. 451–85. Florence.

De Franceschini, M. (2005) *Ville dell'Agro Romano*. Rome.

De Franciscis, A. (1975) Attività archeologica. *CronPomp* 1, 245.

De Franciscis, A. (ed.) (1982) *La regione sotterrata dal Vesuvio: studi e prospettive; atti del Convegno Internazionale, 11–15 novembre 1979, Napoli*. Naples.

Deg, M. (1990) A database of historical earthquake activity in the Middle East. *Transactions of the Institute of British Geographers* n.s., 15, 294–307.

Degrassi, A. (1969) Epigraphica IV. *MemLinc* 14, 111–41.

de Grummond, N.T. (2006) *Etruscan Myth, Sacred History, and Legend*. Philadelphia.

de Grummond, N.T. (2009) *The Sanctuary of the Etruscan Artisans at Cetamura del Chianti*. Florence.

de Grummond, N.T. and Edlund-Berry, I. (eds) (2011) *The Archaeology of Sanctuaries and Ritual in Etruria* [*JRA* Suppl. 81]. Portsmouth, RI.

de Grummond, N.T. and Simon, E. (eds) (2006) *The Religion of the Etruscans*. Austin.

De Haan, N., Peterse, K., Piras, S. and Schipper, F. (2005) The Casa degli Scienziati (VI.14.43): elite architecture in fourth-century BC Pompeii, in Guzzo and Guidobaldi (eds), 2005, pp. 240–56.

De Haan, N. and Wallat, K. (2008) Le Terme Centrali a Pompei: ricerche e scavi 2003–2006, in Guzzo and Guidobaldi (eds), 2008, pp. 15–24.

De Juliis, E.M. (1977) *La ceramica geometrica della Daunia*. Florence.

De Kleijn, G. (2001) *The Water Supply of Ancient Rome*. Amsterdam.

de La Genière, J. and Greco, G. (1996) Riflessioni intorno al c.d. *thesauròs* nel santuario di Hera alla Foce del Sele, in Montepaone (ed.), pp. 455–67.

DeLaine, J. (1988) Recent research on Roman baths. *JRA* 1, 11–32.

DeLaine, J. (1989) Some observations on the transition from Greek to Roman baths in Hellenistic Italy. *MeditArch* 2, 111–25.

DeLaine, J. (1990) Structural experimentation: the lintel arch, corbel and tie in western Roman architecture. *WorldArch (Architectural Innovation)* 21, 407–24.

DeLaine, J. (1995) The supply of building materials to the city of Rome: some economic implications, in Christie (ed.), pp. 555–62.

DeLaine, J. and Johnston, D. (eds) (1999) *Roman Baths and Bathing*, 2 vols [*JRA* Suppl. 37]. Portsmouth, RI.

De Lachenal, L. (ed.) (1992) *Da Leukania a Lucania. La Lucania centro-orientale fra Pirro e i Giulio-Claudii*. Rome.

Delfino, A. (2008) Il foro di Cesare nella fase cesariana e augustea, in Gentili (ed.), pp. 52–4.

de Ligt, L. (2006) The economy: agrarian change during the second century, in Rosenstein and Morstein-Marx (eds), pp. 590–605.

de Ligt, L. (2012) *Peasants, Citizens and Soldiers. Studies in the Demographic History of Roman Italy 225 BC–AD 100*. Cambridge.

de Ligt, L. and Northwood, S. (eds) (2008) *People, Land, and Politics: Demographic Developments and the Transformation of Roman Italy 300 BC–AD 14*. Leiden.

Delorme, J. (1960) *Gymnasion*. Paris.

del Tutto Palma, L. (ed.) (1996) *La tavola di Agnone nel contesto italico (Convegno di studio, Agnone 13–15 Aprile 1994)*. Florence.

De Lucca, S. (2009) La cittá ellenistico-romana di Magdala/Taricheae. Gli scavi del Magdala Project 2007–2008. *Liber Annus* 59, 343–562.

De Maria, S. (1988) *Gli archi onorari di Roma e dell'Italia romana*. Rome.

De Marinis, G. (1985) Ceramica a vernice nera, in Vannini (ed.), 1985, pp. 303–7.

De Miro, E. and Fiorentini, G. (2011) *VI. Agrigento romana. Gli edifici pubblici civili*. Pisa and Rome.

Denard, H. (2005) At the foot of Pompey's statue: reconceiving Rome's *Theatrum Lapideum*, in Bowman and Brady (eds), pp. 69–76.

Dench, E. (1995) *From Barbarians to New Men: Greek, Roman and Modern Perceptions of Peoples of the Central Apennines*. Oxford.

Dench, E. (2003) Beyond Greeks and Barbarians: Italy and Sicily in the Hellenistic Age, in Erskine (ed.), 2003, pp. 294–310.

Dench, E. (2005) *Romulus' Asylum. Roman Identities from the Age of Alexander to the Age of Hadrian*. Oxford.

De Neeve, P.W. (1984a) *Colonus: Private Farm-Tenancy in Roman Italy during the Republic and the Early Principate*. Amsterdam.

De Neeve, P.W. (1984b) *Peasants in Peril. Location and Economy in Italy in the Second Century BC*. Amsterdam.

Dennis, G. (1848) *Cities and Cemeteries of Etruria*, 2 vols. London.

De Puma, R.D. and Small, J.P. (eds) (1994) *Murlo and the Etruscans*. Madison, WI.

De Rita, D., Funiciello, R. and Parotto, M. (1988) *Carta geologica del complesso vulcanico dei Colli Albani (scale 1:50,000)*. Rome.

De Rita, D., Funiciello, R., Corda, L. *et al.* (1993) Carta geologica del Complesso Vulcanico Sabatino (1:50,000). *Quaderni de La Ricerca Scientifica* 114, 33–79.

De Rossi, G.M. and Mello, M. (eds) (1995) *Tra Lazio e Campania. Ricerche di Storia e di Topografia antica 16.* Naples.

Deroux, C. (ed.) (1980) *Studies in Latin Literature and Roman History* [Coll. Latomus 168]. Brussels.

De Ruyt, C. (1983) *Macellum: Marché alimentaire des Romains.* Louvain-La-Neuve.

Derks, T. and Roymans, N. (eds) (2009) *Ethnic Constructs in Antiquity. The Role of Power and Tradition.* Amsterdam.

Desbat, A. (2004) Le commerce du vin à *Lugdunum*, in Brun, Poux, and Tchernia (eds), pp. 254–63.

Descœudres, J.-P. (2007) History and historical sources, in Dobbins and Foss (eds), pp. 9–27.

De Sena, E. and Dessales, H. (eds) (2004) *Metodi e approcci archeologici: l'industria e il commercio nell'Italia antica/Archaeological Methods and Approaches: Industry and Commerce in Ancient Italy.* Oxford.

De Simone, C. (1981) Gli Etruschi a Roma: evidenza linguistica e problemi metodologici, in Colonna (ed.), pp. 93–103.

De Simone, C. (1992) Lingue e culture nelle Puglie nel III–II sec. a.C., in Uggeri (ed.), 25–8.

De Tommaso, G. (1987) Ceramica a vernice nera, in Vannini (ed.), 1987, pp. 3–6.

Develin, R. (1979) The political position of C. Flaminius. *RhM* 122, 268–77.

de Vos, A. and de Vos, M. (1982) *Pompei, Ercolano, Stabia [Guide archeologiche Laterza 11].* Rome.

De Waele, J., D'Agostino, B. and Lulof, P. (eds) (2001) *Il tempio dorico del foro triangolare di Pompei.* Rome.

Díaz Ariño, B. (2008a) Las murallas romanas de Cartagena en la segunda mitad del siglo I a.C. *Zephyrus* LXI, 225–34.

Díaz Ariño, B. (2008b) Un quaestor pro praetore republicano en Carthago Nova. *JRA* 21 (1), 255–63.

Díaz Ariño, B. (2008c) *Epigrafía latina republicana de Hispania.* Barcelona.

Di Capua, F. (1939/40) Appunti su l'origine e sviluppo delle terme romane. *RendNap* 20, 81–160.

Dick, S. and Lupisella, M. (eds) (2010) *Cosmos & Culture: Cultural Evolution in a Cosmic Context [NASA SP–4802].* Washington DC.

Dickmann, J.-A. (1997) The peristyle and the transformation of domestic space in Hellenistic Pompeii, in Laurence and Wallace-Hadrill (eds), pp. 121–36.

Dickmann, J.-A. and Pirson, F. (2005) Il progetto "Casa dei Postumii," in Guzzo and Guidobaldi (eds), 2005, pp. 156–69.

Didierjean, F. (2008) Camps militaires romains et archéologique aérienne: méthodologie et données nouvelles, in Cadiou *et al.* (eds), pp. 95–115.

Didierjean, F. and Petit-Aupert, C. (forthcoming) Les nouveautés des prospections aériennes. *La guerre et ses traces. Conflits et société en Hispanie à l'époque de la conquête romaine (IIIe–Ier s. av. J.-C.). Colloque International* [Bordeaux, November 2010].

Di Domenico, S. (2011) Pompei a un anno dal primo crollo: rivive la Schola Armaturarum, ma in 3D. Gli esperti informatici più veloci degli archeologi, torna alla luce la palestra dei gladiatori in macerie. *Corriere del Mezzogiorno*, November 7.

Diebner, S. (1979) *AESERNIA-VENAFRUM: Untersuchungen zu den römischen Steindenkmälern zweier Landstädte Mittelitaliens.* Rome.

Diebner, S. (1982) Frühkaiserzeitliche Urnen aus Picenum. *Röm Mitteil* 89, 81–102.

Diebner, S. (1983) Un gruppo di cinerari romani del Lazio meridionale. *DialArch* (3rd ser.) 1, 65–78.

Diebner, S. (1986) *Reperti funerari in Umbria a sinistra del Tevere I sec. a.C.–I sec. d.C.* Rome.

Diebner, S. (1986–8) Cippi carsulani. *ArchCl* 38/40, 35–66.

Diebner, S. (1987) Sonderformen von Urnen und Grabstelen in den Regionen Mittelitaliens, in von Hesberg and Zanker (eds), pp. 229–37.

Diebner, S. (1991) Testimonianze di arte funeraria: il Sannio nel contesto delle altre regioni dell'Italia centrale, in *La Romanisation*, pp. 231–43.

Diebner, S. (1991–3) La sfera dei sepolcri: Apporti alla conoscenza dell'aspetto socio-economico del territorio. *Le Marche: Archeologia, Storia, Territorio* 4, 83–97.

Diebner, S. (2007) Landstädtische Sepulkraldenkmäler aus Picenum. *Röm Mitteil* 113, 95–145.

Diebner, S. (2009) Frühkaiserzeitliche Grabsteine aus Mittelitalien. *ArchCl* 60, 161–94.

Dietler, M. (2005) The archaeology of colonization and the colonization of archaeology: theoretical reflections on an ancient Mediterranean colonial encounter, in Stein (ed.), pp. 33–68.

Dietler, M. (2010) *Archaeologies of Colonialism: Consumption, Entanglement and Violence in Ancient Mediterranean France*. Berkeley.

Dietz, S., Ladjimi Sebaï, L. and Ben Hassen, H. (eds) (1995) *Africa Proconsularis: Regional Studies in the Segermes Valley of Northern Tunisia*. Copenhagen.

Di Giuseppe, H. (2005) Un confronto tra l'Etruria settentrionale e meridionale dal punto di vista della ceramica a vernice nera. *PBSR* 73, 31–84.

Di Giuseppe, H. (2008) La villa romana di S. Gilio (Oppido Lucano-PZ) tra aristocrazie urbane e locali, in Russo and Di Giuseppe (eds), pp. 305–54.

Di Giuseppe, H. (2011) Hannibal's legacy and black glaze ware in Lucania, in Colivicchi (ed.), pp. 57–76.

Di Lieto, M. (2011) The "North-Lucanian" district in the Republican Period, in Colivicchi (ed.), pp. 44–56.

Dilke. O.A.W. (1971) *The Roman Land Surveyors: An Introduction to the Agrimensores*. Newton Abbott.

Dilke, O.A.W. (1985) *Greek and Roman Maps*. London.

Dilke, O.A.W. (1987a) Maps in the service of the state: Roman cartography to the end of the Augustan era, in Harley and Woodward (eds), pp. 201–11.

Dilke, O.A.W. (1987b) *Mathematics and Measurement*. London.

Dillon, S. and Welch, K.E. (eds) (2006) *Representations of War in Ancient Rome*. Cambridge.

Di Maio, G., Giugliano, I., and Rispoli, P. (2008) Pompei via del Tempio di Iside – collettore delle Terme Stabiane: Indagini preliminare al resanamento idrogeologico, in Guzzo and Guidobaldi (eds), 2008, pp. 529–31.

Di Niro, A. (1981) *Necropoli arcaiche di Termoli e Larino. Campagne di scavo 1977–78*. Campobasso.

Di Niro, A. (1982) *La villa romana di S. Fabiano e il sistema di produzione schiavistico*. Campobasso.

Di Niro, A. (1991a) Cercemaggiore-Gildone: la casa, le tombe e il sacello, in Capini and Di Niro (eds), pp. 121–6.

Di Niro, A. (1991b) La zona frentana tra IV e I sec. a.C., in Capini and Di Niro (eds), pp. 131–4.

Di Niro, A. (1991c) Le Necropoli dell'Area Interna. Le Necropoli della Zona Costiera, in Capini and Di Niro (eds), pp. 61–71.

Di Niro, A. (1991d) Una societa agricolo-pastorale, VI–V a.C.: Introduzione, in Capini and Di Niro (eds), pp. 53–5.

Di Niro, A. and Petrone, P.P. (1993) Gildone: mortalità, stress nutrizionali e da attività lavorativa in un campione di Sanniti del V–IV séc. a.C. *PBSR* 59, 33–49.

Di Stefano, G. (2006) Aspetti urbanistici e topografici per la storia di Camarina, in Pelagatti, Di Stefano and De Lachenal (eds), pp. 157–76.

Dobbins, J. (1994) Problems of chronology, decoration, and urban design on the Forum at Pompeii. *AJA* 97, 629–94.

Dobbins, J.J. and Ball, L.F. (2005) The Pompeii Forum Project, in Guzzo and Guidobaldi (eds), 2005, pp. 60–72.

Dobbins, J.J. and Foss, P.W. (eds) (2007) *The World of Pompeii*. London.

Dobbins, J.J. and Grueber, E. (in press) Modeling hypotheses in Pompeian archaeology: the House of the Faun. *Proceedings of the 38th Conference on Computer Applications and Quantitative Methods in Archaeology Granada, Spain, April 2010*.

Dobson, M. (2008) *The Army of the Roman Republic. The Second Century* BC, *Polybius and the Camps at Numantia, Spain*. Oxford.

Dobson, M. (forthcoming) Tents, huts or houses? Soldiers' accommodation at Numantia, in *La guerre et ses traces. Conflits et société en Hispanie à l'époque de la conquête romaine (IIIe–Ier s. av. J.-C.). Colloque International* (Bordeaux, November 2010).

Dobson, M. and Morales, F. (2008) Monedas inéditas de los campamentos romanos republicanos de Numancia y Renieblas: consideraciones cronológicas. *Archivo Español de Arqueología* 81, 213–28.

Donati, J. (2010) Marks of state ownership and the Greek Agora at Corinth. *AJA* 114, 3–26.

Dondero, I. and Pensabene, P. (eds) (1982) *Roma repubblicana fra il 509 e il 270 a.C.* Rome.

Dondin-Payre, M. (1994) *La commission d'exploration scientifique d'Algérie: une héritière méconnue de la commission d'Égypte*. Paris.

Dothan, M., (1976) Akko: interim excavation report, first season 1973/4. *BASOR* 224, 1–50.

Dougherty, C. and Kurke, L. (eds) (2003) *The Cultures within Ancient Greek Culture*. Cambridge.

Downey, G. (1963) *Ancient Antioch*. Princeton.

Drago Troccoli, L. (ed.) (2009) *Il Lazio dai Colli Albani ai Monti Lepini tra preistoria ed età moderna*. Rome.

Drerup, H. (1959) Bildraum und Realraum in der römischen Architektur. *RM* 66, 147–74.

Driel-Murray, C. van (1990a) New light on old tents. *Journal of Roman Military Equipment Studies* 1, 109–37.

Driel-Murray, C. van (1990b) The Roman army tent. *Exercitus* 2, 137–45.

Driel-Murray, C. van (1991) A Roman tent: Vindolanda I, in Maxfield and Dobson (eds), pp. 367–72.

Driel-Murray, C. van (1998) The leatherwork from the forts, in Cool and Philo (eds), pp. 285–334.

Driel-Murray, C. van (2001) Les restes d'une tente césarienne en cuir, in Reddé and Schnurbein (eds), pp. 362–8.

Dubouloz, J. and Pittia, S. (eds) (2007) *La Sicile de Cicéron. Lectures des* "Verrines." Besançon.

Ducke, B. (2011) Multiview 3D reconstruction of the archaeological site at Weymouth from image series. *Computers and Graphics* 35 (2), 375–82.

Dumser, E.A. (2005) Review of A. Carandini, *Palatino, Velia e Sacra Via. Paesaggi urbani attraverso il tempo. Workshop di archelogia classica, Quaderni*, I. *BMCR* 2005.08.18.

Dunbabin, K. (1989) Baiarum grata voluptas: pleasures and dangers of the baths. *PBSR* 57, 6–49.

Dunbabin, K. (1999) *Mosaics of Greek and Roman World*. Cambridge.

Duncan, C. and Wallach, A. (1980) The Universal Survey Museum. *Art History* 3, 448–69.

Duncan-Jones, R. (1982) *The Economy of the Roman Empire: Quantitative Studies*. Cambridge.

Duncan-Jones, R. (1990) *Structure and Scale in the Roman Economy*. Cambridge.

Duncan-Jones, R. (2003) Roman coin circulation and the cities of Vesuvius, in Lo Cascio (ed.), pp. 161–80.

Dunkeld, M., Campbell, J., Louw, H. *et al.* (eds) (2006) *Proceedings of the Second International Congress of Construction History*, 2 vols. London.

Dupré, X. (2004) *Córdoba. Colonia Patricia Corduba*. Rome.

Dupré, X. (ed.) (2004) *Colonia Iulia Vrbs Triumphalis Tarraco*. Rome.

Dupré Raventós, X., Ribichini, S. and Verger, S. (eds) (2008) *Saturnia Tellus. Definizioni dello spazio consacrato in ambiente etrusco, italico, fenicio-punico, iberico e celtico* [Atti Del Convegno Internazionale Svoltosi a Roma dal 10 al 12 Novembre 2004]. Rome.

Durán, M. and Mestres, I. (2008) El Camp de Les Lloses (Tona, Osona). *Monografies del Montseny* 23, 197–209.

Durán, M., Mestres, I. and Principal, J. (eds) (2008) *Les colleccions de l'exposició permanent del Camp de Les Lloses*. Barcelona.

Duval, P.-M. (1949) A propos du milliaire de Cneus Domitius Ahenobarbus trouvé dans l'Aude en 1949. *Gallia* 7, 207–31.

Dyson, S.L. (1978) Settlement patterns in the Ager Cosanus: the Wesleyan University Survey, 1974–1976. *JFA* 5, 251–68.

Dyson, S.L. (1981) Settlement reconstruction in the Ager Cosanus and the Albegna Valley: Wesleyan University research, 1974–1979, in Barker and Hodges (eds), pp. 269–74.

Dyson, S.L. (1985) *The Creation of the Roman Frontier*. Princeton.

Dyson, S.L. (1992) *Community and Society in Roman Italy*. London.

Dyson, S.L. (2002) The excavations at Le Colonne and the villa culture of the Ager Cosanus. *MAAR* 47, 209–28.

Dyson, S.L. (2003) *The Roman Countryside*. London.

Dyson, S.L. (2005) Success and failures at Cosa (Roman and American). *JRA* 18, 615–20.

Dyson, S.L. and Rowland, R.J., jr. (2007) *Archaeology and History in Sardinia from the Stone Age to the Middle Ages. Shepherds, Sailors, and Conquerors*. Philadelphia.

Eckardt, H. (ed.) (2010) *Roman Diasporas: Archaeological Approaches to Mobility and Diversity in the Roman Empire* [*JRA* Suppl. 78]. Portsmouth, RI.

Eckert, M. (1988) *Capuanische Grabsteine: Untersuchungen zu den Grabsteinen römischer Freigelassener aus Capua* [*BAR-IS* 417]. Oxford.

Edlund-Berry, I. (1994) Ritual destruction of cities and sanctuaries: the "un-founding" of the archaic monumental building at Poggio Civitate, in De Puma and Small (eds), pp. 16–28.

Edlund-Berry, I. (2006a) Ritual space and boundaries in Etruscan religion, in de Grummond and Simon (eds), pp. 116–31.

Edlund-Berry, I. (2006b) Hot, cold or smelly: the power of sacred water in Roman religion, in Schultz and Harvey (eds), pp. 162–80.

Edlund-Berry, I. (2006c) Healing, health and well-being: archaeological evidence for issues of health concerns in ancient Italy. *Archiv für Religionsgeschichte* 8, 81–8.

Edlund-Berry, I. (2008) The language of Etrusco-Italic architecture: new perspectives on Tuscan temples. *AJA* 112, 441–7.

Edwards, C. (1993) *The Politics of Immorality in Ancient Rome*. Cambridge.

Edwards, C. (2003) Incorporating the alien: the art of conquest, in Edwards and Woolf (eds), pp. 44–70.

Edwards, C. and Woolf, G. (eds) (2003) *Rome the Cosmopolis*. Cambridge.

Egidi, R., Filippi, F. and Martone, S. (eds) (2011) *Archeologia e infrastrutture* [*BdA* Volume Speciale]. Rome.

Eichberg, M. (1987) *Scutum: die Entwicklung einer italisch-etruskischen Schildform von den Anfängen bis zur Zeit Caesars*. Frankfurt.

Eigler, U., Gotter, U., Luraghi, N. and Walter, U. (eds) (2003) *Formen römische Geschichtsschreibung von den anfängen bis Livius. Gattungen – Autoren – Kontexte*. Darmstadt.

Eiteljorg, H. (1995) New total station – surveying in Pompeii. *CSA Newsletter* VIII (2), http://csanet.org/newsletter/aug95/nl089506.html (accessed November 17, 2012).

Ekstrom, L. and Hecht, R. (2007) Ritual performance and ritual practice: teaching the multiple forms and dimensions of ritual, in Bell (ed.), pp. 229–49.

Eliakis, C., Eliakis, E. and Iordanidis, P. (1966) Détermination de la taille d'après la mensuration des os longs. *Annales de Médecine Légale* 46, 403–21.

Ellis, S. and Devore, G. (2010) The fifth season of excavations at VIII.7.1–15 and the Porta Stabia at Pompeii: preliminary report. *Fasti On Line Documents & Research* 202, http://www.fastionline.org/docs/FOLDER-it-2010-202.pdf (accessed November 17, 2012).

Ellis, S., Emmerson, A., Pavlick, A. and Dicus, K. (2011) The 2010 field season at I.1.1–10, Pompeii: preliminary report on the excavations. *Fasti On Line Documents & Research* 220, http://www.fastionline.org/docs/FOLDER-it-2011-220.pdf (accessed November 17, 2012).

Elsner, J. (1995) *Art and the Roman Viewer: The Transformation of Art from the Pagan World to Christianity*. Cambridge.

Elsner, J. (2010) Alois Riegl and classical archaeology, in Noever, Rosenauer and Vasold (eds), pp. 45–57.

Elsner, J. and Cardinal, R. (1994) *The Cultures of Collecting*. London.

Enea nel Lazio. Archeologia e mito (1981) Rome.

Engels, D. (1990) *Roman Corinth. An Alternative Model for the Classical City*. Chicago.

Engels, D., Geis, L. and Kleu, M. (eds) (2010) *Zwischen Ideal und Wirklichkeit. Herrschaft auf Sizilien von der Antike bis zum Spätmittelalter*. Stuttgart.

Ennabli, A. (ed.) (1992) *Pour sauver Carthage: Exploration et conservation de la cité punique, romaine, et byzantine*. Paris.

Erdkamp, P. (ed.) (2007) *A Companion to the Roman Army*. Oxford.

Erskine, A. (ed.) (2003) *A Companion to the Hellenistic World*. Malden, MA.

Erskine, A. (ed.) (2009) *A Companion to Ancient History*. Chichester.

Eschebach, H. (1970) *Stadtplan von Pompeji im Massstab 1:1000*. Mainz.

Eschebach, H. (1979) *Die Stabianer Thermen in Pompeji* [*Denkmäler antiker Architektur* 13]. Berlin.

Eshel, H. (2008) *The Dead Sea Scrolls and the Hasmonean State*. Grand Rapids.

Eshel, H. (2009) *Masada: A Carta Field Guide*. Jerusalem.

Esposito, D. (2008) Un contributo allo studio di Pompei arcaica. I saggi nella *Regio V, ins. 5* (Casa dei Gladiatori), in Guzzo and Guidobaldi (eds), 2008, pp. 71–80.

Etienne, R. and Mayet, F. (2000) *Le vin hispanique*. Paris.

Etienne, R. and Mayet, F. (2002) *Salaisons et sauces de poisson hispaniques*. Paris.

Ettlinger, E. (1990) Campana und Schwarze Sigillata, in *Conspectus Formarum Terrae Sigillatae Italico Modo Confectae*, pp. 3–4. Bonn.

Evans, H.B. (1994) *Water Distribution in Ancient Rome: The Evidence of Frontinus*. Ann Arbor.

Evans, J.D. (1992) *The Art of Persuasion*. Ann Arbor.

Evans, J.D. (2006) *The Joint Expedition to Caesarea Maritima: Excavation Reports v.6, The Coins and the Hellenistic, Roman, and Byzantine Economy of Palestine*. Boston.

Evans, J.D. (2009) Prostitutes in the theater of Pompey? A reconsideration. *TAPA* 139, 123–45.

Fabbri, M. (2010) La maison des rois de Rome. *Dossiers d'Archéologie* 339, 56–61.

Fabre, G., Mayer, M. and Rodà, I. (1984, 1985, 1991) *Inscripcions romaines de Catalogne*, 3 vols. Paris.

Fabricius, E. (1932) Some notes on Polybius's description of Roman camps. *JRS* 22, 78–87.

Fabricotti, E. (1976) I bagni nelle prime ville romane. *CronPomp* 2, 29–111.

Fagan, G. (1996) Sergius Orata: inventor of the hypocaust? *Phoenix* 50, 56–66.

Fagan, G. (1999) *Bathing in Public in the Roman World*. Ann Arbor.

Fahlbusch, H. (1977) The development of the Pergamon water supply between 200 BC and 300 AD, in *Proceedings of the 17th Congress of the International Association of Hydraulic Research [IAHR]*, pp. 758–62. Baden-Baden.

Famà, M.L. (2007) *Antiche testimonianze di pace e guerra dal mare di Trapani. Mostra, Trapani, Museo Pepoli 29 settembre–30 novembre 2007*. Trapani.

Fant, J.C. (1988) *Roman Marble Quarrying and Trade*. Oxford.

Fantar, M. (1984–6) *Kerkouane: cité punique de Cap Bon (Tunisie)*, 3 vols. Tunis.

Fantar, M. and Ghaki, M. (eds) (1995) *Actes du IIIe Congrès international des études phéniciennes et puniques*, 3 vols. Tunis.

Farney, G. (2007) *Ethnic Identity and Aristocratic Competition in Republican Rome*. New York.

Farney, G. (2009) Review of T.P. Wiseman, *Unwritten Rome*. *BMCR* 2009.06.11.

Fasold, P., Fischer, T., von Hesberg, H. and Witteyer, M. (eds) (1998) *Bestattungssitte und kulturelle Identität: Grabanlagen und Grabbeigaben der frühen römischen Kaiserzeit in Italien und den Nordwest-Provinzen. Kolloquium in Xanten vom 16. bis 18. Februar 1995 "Römische Gräber des 1. Jhs. n.Chr, in Italien und den Nordwestprovinzen."* Cologne.

Fasolo, M. (2003) *La Via Egnatia I da Apollonia e Dyrrachium ad Herakleia Lynkestidos*. Rome.

Faustoferri, A. (1991) I rapporti con l'Apulia. La ceramica di argilla depurata, in Capini and Di Niro (eds), pp. 72–5.

Faustoferri, A. (2003) L'Abruzzo centro-meridionale dal tardo-orientalizzante all'età tardo-arcaica, in *Atti della XXXVI Riunione Scientifica Preistoria e Protostoria dell'Abruzzo*, pp. 470–86. Chieti-Celano.

Faustoferri, A. (2007) Il Torso di Atessa e i Lucani del Sangro, in Ruggeri and Tavano (eds), pp. 56–63.

Faustoferri, A. and Lloyd, J. (1998) Monte Pallano: a Samnite fortified centre and its hinterland. *JRA* 11, 5–22.

Favro, D. (1996) *The Urban Image of Augustan Rome*. Cambridge.

Favro, D. (2008) The festive experience: Roman processions in the urban context, in Bonnemaison and Macy (eds), pp. 10–42.

Fears, J.R. (1981) The cult of virtues and Roman imperial ideology, in *ANRW* II.17.2, pp. 827–948.

Felch, J. and Frammolino, R. (2011) *Chasing Aphrodite. The Hunt for Looted Antiquities at the World's Richest Museum*. Boston.

Fenelli, M. (1975) Contributo per lo studio del votivo anatomico: i votivi anatomici di Lavinio. *ArchCl* 27, 206–52.

Fentress, E. (1994) Cosa in the Empire: the unmaking of a Roman town. *JRA* 7, 208–22.

Fentress, E. (2000) Frank Brown, Cosa and the idea of a Roman city, in Fentress (ed.), pp. 11–24.

Fentress, E. (ed.) (2000) *Romanization and the City. Creation, Transformation and Failures* [*JRA* Suppl. 38]. Portsmouth, RI.

Fentress, E. (2003) *Cosa V: An Intermittent Town. Excavations 1991–1997*. Ann Arbor.

Fentress, E. (2006) Romanizing the Berbers. *Past and Present* 190, 3–33.

Fentress, E. (2009) Peopling the countryside: Roman demography in the Albegna Valley and Jerba, in Bowman and Wilson (eds), 2009, pp. 127–61.

Fentress, E. and Docter, R. (2008) North Africa: rural settlement and agricultural production, in van Dommelen and Gómez Bellard (eds), pp. 101–28.

Fentress, E., Drine, A. and Holod, R. (eds) (2009) *An Island Through Time: Jerba Studies Volume 1. The Punic and Roman Periods* [*JRA* Suppl. 71]. Portsmouth, RI.

Fentress, E. and Rabinowitz, A.S. (1996) Excavations at Cosa 1995, Atrium Building V and a new Republican temple. *MAAR* 41, 221–36.

Ferchiou, N. (1978) Trois types de monuments funéraires situés dans (ou sur les franges de) l'ancien territoire de la Carthage punique. *Africa* 5–6, 191–214.

Ferchiou, N. (1987) Le paysage funéraire pré-romain dans deux régions céréalières de Tunisie antique. *AntAfr* 23, 13–70.

Ferchiou, N. (1990) Habitats fortifiés pré-impériaux en Tunisie antique. *AntAfr* 26, 43–86.

Ferchiou, N. (1994) Le paysage pré-impérial à l'est et au sud de Zaghouan (Tunisie). *AntAfr* 30, 7–55.

Ferchiou, N. (2009) Recherches sur le mausolée hellénistique d'Hinshir Burgu, in Fentress, Drine and Holod (eds), pp. 107–28.

Ferdière, A. and Villard-le-Tiec, A. (1993) La tombe augustéenne de Fléré-la-Rivière (Indre) et les sépultures aristocratiques de la cité des Bituriges en Berry au début de l'époque gallo-romaine. *Revue Archéologique du Centre de la France* Suppl. 7.

Ferguson, W.S. (1911) *Hellenistic Athens*. London.

Fernández Miranda, M. and Fernández Ochoa, C. (eds) (1995) *Astures, pueblos y culturas en la frontera del Imperio romano*. Gijón.

Fernández Ochoa, C. and García-Entero, V. (eds) (2000) *Termas romanas en el Occidente del Imperio. Actas del II Coloquio Internacional de Arqueología de Gijón*. Gijón.

Ferrandes, A.F. (2006) Produzioni stampigliate e figurate in area etrusco laziale tra fine IV e III secolo a.C. Nuove riflessioni alla luce di vecchi contesti. *ArchCl* 57, n.s. 7, 117–74.

Ferrary, J.-L. (1988) *Philhellénisme et impérialisme: aspects idéologiques de la conquête romaine du monde hellénistique*. Rome.

Ferrea, L. (2002) *Gli dei di terracotta. La ricomposizione del frontone da via di San Gregorio*. Milan.

Ferreruela, A. and Mínguez, J. (2002) La Cabañeta (El Burgo de Ebro, Zaragoza), in Jiménez and Ribera (eds), pp. 205–14.

Feruglio, A.E. (1982) Le tombe dipinte Golini di Settecamini e la tomba degli Hescanas, in *Pittura Etrusca a Orvieto*, pp. 21–64. Rome.

Feugère, M. (1994) L'équipement militaire d'époque républicaine en Gaule. *Journal of Roman Military Equipment Studies* 5, 3–23.

Feugère, M. (ed.) (1997) L'équipement militaire et l'armement de la république (IVe–Ier s. avant J.-C.) (Proceedings of the Tenth International Roman Military Equipment Conference, Montpellier, 1996). *Journal of Roman Military Equipment Studies* 8.

Feugère, M. (2002) *Weapons of the Romans*, trans. D.G. Smith. Charleston, SC.

Février, P.-A. (1989) *Approches du Maghreb romain: pouvoirs, différences, et conflits*, 2 vols. Aix-en-Provence.

Fields, N. (2010) *The Roman Conquests: North Africa*. Barnsley.

Filippi, D. (2004) La domus regia, in Carandini and Greco (eds), 2004, pp. 101–21.

Finlay, G. (1844) *Greece under the Romans*. Edinburgh.

Finley, M. (1973) *The Ancient Economy* (1st edn). Berkeley.

Finley, M.I. (1979) *A History of Sicily. Vol. I. Ancient Sicily to the Arab Conquest* (rev. edn). New York.

Fiorelli, G. (1873) *Relazione sugli scavi di Pompei dal 1861 al 1872*. Naples.

Fiorenti, G., Caltabiano, M. and Calderone, A. (2003) *Archeologia del Mediterraneo. Studi in onore di Ernesto De Miro*. Rome.

Fiorentino, G. and Magri, D. (eds) (2008) *Charcoals from the Past: Cultural and Palaeoenvironmental Implications. Proceedings of the Third International Meeting of Anthracology, Cavallino – Lecce (Italy) June 28th–July 1st 2004* [*BAR-IS* 1807]. Oxford.

Fischer, T. (1979) Zur Seleukideninschrift von Hefzibah. *ZPE* 33, 133–8.

Fischer, W. (1914) *Das Römische Lager insbesondere nach Livius*. Leipzig.

Fischer-Hansen, T. (ed.) (1995) *Ancient Sicily* [*Acta Hyperborea* 6]. Copenhagen.

Fisher, C., Terras, M. and Warwick, C. (2009) Integrating new technologies into established systems: a case study from Roman Silchester, in Crawford, Frischer and Koller (eds), http://www.caa2009.org/articles/Fisher_Contribution191_c%20%281%29.pdf (accessed November 17, 2012).

Fisher, N. and van Wees, H. (eds) (1998) *Archaic Greece: New Approaches and New Evidence*. London.

Fisher, R. and Schmincke, H.-U. (1984) *Pyroclastic Rocks*. Berlin.

Fitch, C.R. and Goldman, N.W. (1994) *Cosa: The Lamps* [*MAAR* 39].

Fittschen, K. (ed.) (1996) *Judaea and the Graeco-Roman World in the Time of Herod*. Göttingen.

Fiz, I. and Macias, J. (2007) Forma Tarraconis: una descoberta en evolució, in Solé *et al.* (eds), pp. 25–40.

Fless, R. and Moede, K. (2007) Music and dance: forms of representation in pictorial and written sources, in Rüpke (ed.), pp. 249–62.

Flohr, M., Cole, K. and Poehler. E. (eds) (2011) *Pompeii: Art, Industry and Infrastructure.* Oxford.

Flourentzos, P. (ed.) (1996) *Excavations in the Kouris Valley*, vol. 3, *The Basilica of Alassa.* Nicosia.

Flower, H.I. (1996) *Ancestor Masks and Aristocratic Power in Roman Culture.* Oxford.

Flower, H. (2003) "Memories" of Marcellus. History and memory in Roman Republican culture, in Eigler *et al.* (eds), pp. 39–52.

Flower, H.I. (ed.) (2004) *The Cambridge Companion to the Roman Republic.* Cambridge.

Flower, H.I. (2010) *Roman Republics.* Princeton.

Foerster, G. (1995) *Masada V: The Yigael Yadin Excavations 1963–1965, Final Reports, Art and Architecture.* Jerusalem.

Forbes, H. (1989) Of grandfathers and grand theories: the hierarchised ordering of responses to hazard in a Greek rural community, in Halstead and O'Shea (eds), pp. 87–97.

Forbes, H. and Foxhall, L. (1995) Ethnoarchaeology and storage in the ancient Mediterranean, in Wilkins, Harvey and Dobson (eds), pp. 69–86.

Forcey, C., Hawthorne, J. and Witcher, R. (eds) (1998) *TRAC 97. Proceedings of the Seventh Annual Theoretical Roman Archaeology Conference.* Oxford.

Forsythe, G. (2005) *A Critical History of Early Rome.* Berkeley.

Forsythe, G. (2007) The army and centuriate organization in early Rome, in Erdkamp (ed.), pp. 24–41.

Foucault, M. (1969) *L'archéologie du savoir.* Paris.

Fox, S.C. (1996) The human skeletal remains from Alassa-*Ayia Mavri*, Cyprus, in Flourentzos (ed.), pp. 39–64.

Fox, S.C. (2003) Human skeletal remains, in S. Manning (ed.), pp. 38–40.

Fox, S.C. (2005) Health in Hellenistic and Roman times: the case studies of Paphos, Cyprus and Corinth, Greece, in King (ed.), pp. 59–82.

Foxhall, L. (1990) The dependent tenant: land leasing and labour in Italy and Greece. *JRS* 80, 97–114.

Foxhall, L. (1998) Snapping up the unconsidered trifles: the use of agricultural residues in ancient Greek and Roman farming. *Environmental Archaeology* 1, 35–40.

Foxhall, L. (2007) *Olive Cultivation in Ancient Greece. Seeking the Ancient Economy.* Oxford.

Foxhall, L., Jones, M.K. and Forbes, H. (2007) Human ecology and the classical landscape, in Alcock and Osborne (eds), pp. 91–117.

Fraccaro, P. (1934) Polibio e l'accampamento Romano. *Athenaeum* n.s. 12, 154–61.

Fracchia, H. (2001a) The Romanization of the Ager Buxentinus (Salerno), in Lo Cascio and Storchi Marino (eds), pp. 55–74.

Fracchia H. (2001b) The period 250–50 BC in the Mingardo/Bussento region, in Gualtieri and Fracchia (eds), 2001, pp. 117–26.

Fracchia, H. and Gualtieri, M. (1989) The social context of cult practices in pre-Roman Lucania. *AJA* 93 (2), 217–32.

Fracchia, H. and Gualtieri, M. (1999) Roman Lucania: the Upper Bradano Valley. *MAAR* 43, 295–344.

Fracchia, H. and Gualtieri, M. (2009) Roccagloriosa (SA): organizzazione insediativa e sviluppi istituzionali (IV–III sec. a.C.), in Osanna (ed.), pp. 119–42.

Fracchia, H. and Gualtieri, M. (2011) The countryside of Regio II and Regio III (300 BC–AD 14), in Colivicchi (ed.), pp. 11–29.

Fraioli, F. (2009) Domus tardo-repubblicane di Roma tra Velia e Carinae. *Workshop di archeologia classica* 6, 123–35.

Franchi de Bellis, A. (1997) *I Cippi prenestini.* Urbino.

Franchi dell'Orto, L. (ed.) (1988) *Pompei. L'informatica al servizio di una città antica.* Rome.

Franciosi, G. (1997) Regime delle acque e paesaggio in età repubblicana, in Quilici Gigli (ed.), pp. 11–20.

François, P., Moret, P. and Péré-Noguès, S. (eds) (2006) *L'hellenisation en Méditérranée occidentale au temps des guerres puniques (260–180 av. J.-C.) [Pallas 70].* Toulouse.

Francovich, R. and Patterson, H. (eds) (2000) *Extracting Meaning from Ploughsoil Assemblages: The Archaeology of Mediterranean Landscapes 5.* Oxford.

Frank, F. (1924) *Roman Buildings of the Republic: An Attempt to Date Them from Their Materials.* Rome.

Frank, T. (1914) *Roman Imperialism.* New York.

Frankel, R. and Cheetham, P. (1995) untitled, in Bon, S., Jones, R. and Robinson, D. (eds) Anglo-American Research in Pompeii, 1994. *Old World Archaeology Newsletter* XVIII (2), 19–21.

Frankenstein, S. and Rowlands, M. (1978) The internal structure and regional context of Early Iron Age society in southwest Germany. *BIALond* 15, 73–112.

Frasca, M. (2009) *Leontinoi. Archeologia di una colonia greca.* Rome.

Fraschetti, A. (1981) Aristosseno i Romani e la "barbarizzazione" di Poseidonia. *AION* 3, 97–115.

Fraschetti, A. (2002) *The Foundation of Rome.* Translation of *Romolo il fondatore.* Edinburgh.

Frayn, J.M. (1979) *Subsistence Farming in Roman Italy.* London.

Frederick, D. (2011) Digital Pompeii Project, http://pompeii.uark.edu/Digital_Pompeii/Welcome.html (accessed November 17, 2012).

Frederiksen, M. (1962) Review of F.E. Brown, E.H. Richardson and L. Richardson *Cosa II: The Temples of the Arx. JRS* 52, 266–7.

Frederiksen, M. (1968) Campanian cavalry: a question of origins. *DialArch* 2, 3–31.

Frederiksen, M. (1970/1) The contribution of archaeology to the agrarian problem in the Gracchan period. *DialArch* 4/5, 330–7.

Frederiksen, M. (1984) *Campania* (ed. with additions by N. Purcell). London.

Freedman, D. (ed.) (1992) *Anchor Bible Dictionary*, 6 vols. New York.

Freeman, P. (1997) Mommsen through to Haverfield: the origins of Romanization studies in late 19th-c. Britain, in Mattingly (ed.), pp. 27–50.

Freeman, P., Bennett, J. and Fiema, Z.T. (eds) (2002) *Limes XVIII. Proceedings of the XVIIIth International Congress of Roman Frontier Studies.* Oxford.

Freidel, F. (1968, reprint of 1947) *Francis Lieber, Nineteenth-Century Liberal.* Gloucester, MA.

French, D. (1988) *Roman Roads and Milestones of Asia Minor* [*BAR-IS* 392]. Oxford.

Freyberger, K., Ertel, C. and Bitterer, T. (2007) Neue Forschungen zur Basilica Aemilia auf dem Forum Romanum. *RM* 113, 493–525.

Frey-Kupper, S. (2006) Aspects de la production et de la circulation monétaires en Sicile (300–180 av. J.-C.): continuités et ruptures. *Pallas* 70, 27–56.

Freyne, S. (1980) *Galilee: From Alexander the Great to Hadrian: 323 BCE to 135 CE.* Wilmington, DE.

Frézouls, E. (1982) Aspects de l'histoire architecturale du théâtre romain, in *ANRW* II.12.1, pp. 343–411. Berlin.

Frier, B.W. (1969) Points on the topography of Minturnae. *Historia* 18, 510–12.

Frier, B. (1982) Roman life expectancy: Ulpian's evidence. *HSCP* 86, 213–51.

Frier, B. (1983) Roman life expectancy: the Pannonian evidence. *Phoenix* 37, 328–44.

Frischer, B. and Dakouri-Hild, A. (eds) (2008) *Beyond Illustration. 2D and 3D Digital Technologies as Tools for Discovery in Archaeology* [*BAR-IS* 1805]. Oxford.

Fronda, M. (2010) *Between Rome and Carthage: Southern Italy during the Second Punic War.* Cambridge.

Frontini, P. (1985) *La ceramica a vernice nera nei contesti tombali della Lombardia.* Como.

Frontini, P. and Grassi, M.T. (eds) (1998) *Indagini archeometriche relativi alla ceramica a vernice nera: nuovi dati sulla provenienza e la diffusione, Milano 22–23 novembre 1996*. Como.

Frost, H. (1981) Lilybaeum (Marsala). The Punic ship. Final excavation report. *NSc* ser. 8.30, 7–311.

Fuentes, N. (1991) The mule of a soldier. *Journal of Roman Military Equipment Studies* 2, 65–99.

Fulford, M. and Wallace-Hadrill, A. (1999) Toward a history of Pre-Roman Pompeii: excavation beneath the House of Amarantus (I 9 11–12) 1995–8. *PBSR* 67, 17–44.

Funiciello, R. (ed.) (1995) *Memorie descrittive della carta geologica d'Italia 50, la geologia di Roma I*. Rome.

Gabba, E. (1954) Le origini della Guerra Sociale e la vita politica romana dopo l'89 a.C. *Athenaeum* n.s. 32, 41–114, 295–345.

Gabba, E. and Pasquinucci, M. (1979) *Strutture agrarie e allevamento transumante nell'Italia romana*. Pisa.

Gabba, E. and Vallet, G. (eds) (1980) *Storia della Sicilia II*. Naples.

Gaffney, C. and Gater, J. (2003) *Revealing the Buried Past: Geophysics for Archaeologists*. Stroud.

Gagarin, M. and Fantham, E. (eds) (2010) *The Oxford Encyclopedia of Ancient Greece and Rome*, vol. 3. Oxford.

Gaggioti, M. (1991) Saepinum, in Capini and Di Niro (eds), pp. 243–6.

Gagliardo, M.C. and Packer, J.E. (2006) A new look at Pompey's theater: history, documentation, and recent excavation. *AJA* 110, 93–122.

Galasso, G. and Romeo, R. (eds) (1991) *Storia del Mezzogiorno*, vol. 1.2. Rome.

Gallant, T. (1991) *Risk and Survival in Ancient Greece. Reconstructing the Rural Domestic Economy*. Cambridge.

Gallini, C. (1973) Che cosa intendere per ellenizzazione. Problemi di metodo. *DialArch* 7, 175–91.

Gallone, A. (2000) Sepolture alle pendici settentrionali del Palatino, in Carandini and Cappelli (eds), pp. 291–2.

Gallone, A. (2007) Agricultural landscapes in Hellenistic Messapia: socio-economic structures and cultural implications, in van Dommelen and Terrenato (eds), pp. 23–32.

Gambogi, P. and Palladino, S. (eds) (1999) *Castiglioncello. La necropoli ritrovata. Cento anni di scavi e scoperte*. Florence.

Gamer, G. and Frías, T. (1969) Neue Beobachtungen am römischen Lager bei Almazán (Prov. Soria). *Madrider Mitteilungen* 10, 172–84.

Garbrecht, G. (1979) L'alimentation en eau de Pergame. *Dossiers d'Archéologie* 38, 26–33.

Garbrecht, G. (1987) Die Wasserversorgung des antiken Pergamon, in Garbrecht (ed.), pp. 2, 13–47.

Garbrecht, G. (ed.) (1987) *Die Wasserversorgung antiker Städte (Geschichte der Wasserversorgung 2)*. Mainz.

Garbrecht, G. (ed.) (1988) *Die Wasserversorgung antiker Städte 3*. Mainz.

Garcés, I. and Rovira, M. (1996/7) *Indíbil i Mandoni. Reis i guerrers*. Lleida.

García, J., Martin, A. and Cela, X. (2000) Nuevas aportaciones sobre la romanización en el territorio de Iluro (Hispania Tarraconensis). *Empúries* 52, 31–53.

García-Bellido, M. (1993) Origen y función del denario ibérico, in *Sprachen und Schriften des antiken Mittelmeerraums. Festschrift für Jürgen Untermann zum 65 Geburtstag*, pp. 97–123. Innsbruck.

García-Bellido, M.P. (2000/1) Roma y los sistemas monetarios provinciales. Monedas romanas acuñadas en Hispania en la Segunda Guerra Púnica. *Zephyrus* 53/4, 551–77.

García-Bellido, M. (ed.) (2006) *Los campamentos romanos en Hispania (27 a.C.–192 d.C.). El abastecimiento de moneda* [Anejos de *Gladius* 9]. Madrid.

García-Bellido, M., Mostalac, A. and Jiménez, A. (eds) (2008) *Del Imperium de Pompeyo a la auctoritas de Augusto. Homenaje a Michael Grant* [Anejos de *ArchEsp* XLVII]. Madrid.

García Fernández, E. (2009) Reflexiones sobre la latinización de Hispania en época republicana, in Andreu, Cabrero and Rodà (eds), pp. 377–90.

García Morá, F. (1991) *Un episodio de la Hispania republicana: la guerra de Sertorio.* Granada.

García Moreno, L. (2001) *De Gerión a César. Estudios históricos y filológicos de la España indígena y Romano-republicana.* Alcalá de Henares.

Gargola, D.J. (1995) *Land, Laws and Gods: Magistrates and Ceremony in the Regulation of Public Lands.* Chapel Hill.

Garmy, P., Michelozzi, A. and Py, M. (1981) Une nouvelle sépulture à Beaucaire (Gard): la tombe du Mas Jallon. *Revue Archéologique de Narbonnaise* 14, 71–87.

Garnsey, P. (1988) *Famine and Food Supply in the Graeco-Roman World.* Cambridge.

Garozzo, B. (2006) Sicilia occidentale e la Campania: bolli anforari, in Michelini (ed.), pp. 719–32.

Garrido, A., Mar, R. and Martins, M. (2008) *A fonte do Idolo. Analise, interpretaçao e reconstituiçao do santuario.* Braga.

Garrison, E.G. (2010) *Techniques in Archaeological Geology.* Berlin.

Gasparri, D. (1994) Nuove acquisizioni sulla divisione agraria di Paestum, in *Le ravitaillement en blé de Rome* (Proceedings of the Conference held at the Centre J. Bérard, CÉFR 196), pp. 149–58. Rome and Naples.

Gasparri, D. (2000) La colonia latina di Paestum: indagini sulla centuriazione, in Greco and Longo (eds), pp. 219–26.

Gasperini, L. (1988) Gli Etruschi e le sorgenti termali, in *Etruria meridionale: conoscenza, conservazione, fruizione: Atti del Convegno, Viterbo, 29/30 novembre–1 dicembre 1985*, pp. 27–35. Viterbo.

Gasser, F. (1999) *Germana patria. Die Geburtsheimat in den Werken römischer Autoren der späten Republik und der frühen Kaiserzeit.* Stuttgart.

Gatti, G. (1934) "Saepta Iulia" e "Porticus Aemilia" nella "Forma Severiana." *BullCom* 62, 135–44.

Gatti, G. (1936) L'arginatura del Tevere a Marmorata (un manoscritto inedito del P. Luigi M. Bruzza). *BullCom* 64, 55–82.

Gazda, E.K. (2008) Cosa's hydraulic concrete: towards a revised chronology, in Hohlfelder (ed.), pp. 265–90.

Geagan, D. (1979) Roman Athens: some aspects of life and culture. I. 86 BC–AD 267, in *ANRW* II.7.1, pp. 371–437. Berlin.

Gébara, C. and Morhange, C. (2010) *Fréjus (Forum Julii): Le Port Antique/The Ancient Harbor* [*JRA* Suppl. 77]. Portsmouth, RI.

Geertman, H. (2007) The urban development of the pre-Roman city, in Dobbins and Foss (eds), pp. 82–97.

Gejvall, N.-G. and Henschen, F. (1968) Two Late Roman skeletons with malformation and close family relationship from ancient Corinth. *OpAth* 8, 179–93.

Gentili, G. (1952) Nuovo esempio di theatron con gradinata rettilinea a Siracusa. *Dionisio* 15, 122–30.

Gentili, G. (ed.) (2008) *Giulio Cesare: L'unomo; le imprese; il mito.* Rome.

Gentili, M.D. (2005) Riflessioni sul fenomeno storico dei depositi votivi di tipo etrusco-laziale-campano, in Comella and Mele (eds), pp. 367–78.

George, M. (1997) *Servus* and *domus*: the slave in the Roman house, in Laurence and Wallace-Hadrill (eds), pp. 15–24.

Gera, D. (1998) *Judaea and Mediterranean Politics, 219 to 161 BCE.* Leiden.

Gera, D. (2009) Olympiodoros, Heliodoros and the Temples of Koilē Syria and Phoinikē. *ZPE* 167, 125–55.

Gerding, H. (2002) The tomb of Caecilia Metella: Tumulus, Tropaeum, and Thymele. Dissertation. Lund University.
Gerkan, A. von (1940) *Der Stadtplan von Pompeji*. Berlin.
Gerkan, A. von (1958) Review of G. Lugli, *La tecnica edilizia romana I*, 1957. *GGA* 212, 178–97.
Gerkan, A. von and Krischen, F. (1928) *Thermen und Palästren von Milet (Milet, I.9)*. Berlin.
Gerkan, A. von, Messerschmidt, F. and Ronczewski, K. (1930) *Nekropolen von Vulci*. Berlin.
Ghaki, M. (1999) *Les haouanet de Sidi Mhamed Latrech*. Tunis.
Ghedini, F., Bonetto, J, Ghiotto, A.R. and Rinaldi, F. (eds) (2005) *Lo Stretto di Messina nell-antichità*. Messina.
Ghedini, F. and Rosada, S. (eds) (1997) *Via per montes excisa. Strade in galleria e passaggi sotterranei nell'Italia romana*. Rome.
Ghini, G. (ed.) (2004) *Lazio e Sabina 2* [Atti del Secondo incoltro di studi sul Lazio e la Sabina, Roma 7–8 maggio 2003]. Rome.
Ghini, G. (2010) Il sepolcro detto degli Orazi e Curiazi ad Albano: Monumento funerario o memoria gentilizia?, in Valenti (ed.), pp. 86–91.
Ghiotto, A.R. (2004) *L'architettura romana nelle città della Sardegna* [Antenor Quaderni 4]. Rome.
Giampaola, D. (2000) Benevento. Dal centro indigeno alla colonia Latina, in La Regina (ed.), pp. 36–46.
Gianfrotta, P.A. (1985) Indagini nell'area della Porticus Philippi, in Agostinelli and Bietti Sestieri (eds), vol. 2, pp. 376–84.
Giardina, A. (1997) *L'Italia romana. Storie di un'identità incompiuta*. Rome.
Giardina, A. and Schiavone, A. (eds) (1981) *Società romana e produzione schiavistica*. Rome.
Giardina, B. (2010) *Navigare necesse est: Lighthouses from Antiquity to the Middle Ages* [*BAR-IS* 2096]. Oxford.
Gibbon, E. (1897) *The History of the Decline and Fall of the Roman Empire*, ed. J.B. Bury, vol. 1. London.
Gil de los Reyes, M.-S. and Pérez Paz, A. (2005) *Itálica. Guía oficial del conjunto arquelógico*. Seville.
Gill, D.W.J. and Chippindale, C. (2007) From Malibu to Rome: further developments on the return of antiquities. *International Journal of Cultural Property* 14, 205–40.
Gilliver, C. (1993) Hedgehogs, caltrops and palisade stakes. *Journal of Roman Military Equipment Studies* 4, 49–54.
Gilliver, C. (2001) *The Roman Art of War*. Stroud.
Ginouvès, R. (1959) *L'établissement thermal de Gortys d'Arcadie*. Paris.
Ginouvès, R. (1962) *Balaneutike*. Paris.
Gioia, P. and Volpe, R. (eds) (2004) *Centocelle I. Roma S.D.O. le indagini archeologiche*. Soveria Mannelli.
Girosi, M. (1936) L'anfiteatro di Pompei. *MemNap* 5, 27–57.
Gisbert, J. (1998) Ànfores i vi al territorium de Dianium (Dènia). Dades per a la sistematizació de la producció anforal al País Valencià, in *Actes del 2on Colloqui Internacional d'Arqueologia romana. El vi a l'Antiguitat. Economia, producció i comerç al Mediterrani occidental (Badalona, 1998)*, pp. 385–8. Badalona.
Giuliani, C.F. (1966) Norchia, in Castagnoli (ed.), pp. 5–12.
Giuliani, C.F. (2004) *Tivoli: il santuario di Ercole Vincitore*. Tivoli.
Giuliani, L. (ed.) (2005) *Meisterwerke der antiken Kunst*. Munich.
Gjerstad, E. (1953) *Early Rome I*. Lund.
Gjerstad, E. (1956) *Early Rome II*. Lund.
Gjerstad, E. (1973) *Early Rome VI*. Lund.

Gleason, K. (1994) Porticus Pompeiana: a new perspective on the first public park of ancient Rome. *JGH* 14,13–27.

Gleba, M. (2008) *Textile Production in Pre-Roman Italy*. Oxford.

Gleba, M. and Becker, H. (eds) (2009) *Votives, Places, Rituals in Etruscan Religion. Studies in Honour of Jean MacIntosh Turfa*. Leiden.

Gleba, M. and Horsnaes, H. (eds) (forthcoming) *Communicating Identity in Italian Iron Age Communities* [Proceedings of the International Conference held at the University of Copenhagen, Nov. 2008]. Oxford.

Glinister, F. (2006) Reconsidering "religious Romanization," in Schultz and Harvey (eds), pp. 10–33.

Gnade, M. (2002) *Satricum in the Post-Archaic period. A Case Study of the Interpretation of Archaeological Remains as Indicators of Ethno-cultural Identity*. Leuven.

Goffredo, R. (2010) Persistence and change in settlement patterns in the Ofanto Valley near Canusium and Cannae (Apulia) (late 4th c. BC–1st c. AD). *JRA* 23, 7–34.

Goldman, N. (ed.) (2001) *New Light from Ancient Cosa* [Hermeneutics of Art 10]. New York.

Golvin, J.-C. (1988) *L'amphithéâtre romain: essai sur la théorisation de sa forme et de ses fonctions*, 2 vols. Paris.

Golvin, J.-C. and Landes, C. (1990) *Amphitheatres et gladiateurs*. Paris.

Gómez-Pantoja, J. (2008) Historia romana: de Escipión a los visigodos, in Sánchez-Moreno (ed.), pp. 283–623.

González, A. and Díez, M. (eds) (1984) *Hannibal Pyrenaeum transgreditur. XXII Centenari del pas d'Anníbal pel Pirineu 218 a.J.C.–1982 d.J.C.* [5th Colloqui Internacional d'Arqueologia de Puigcerdà 1982]. Puigcerdà.

González, A. and Díez, M. (eds) (2002) *I Biennal de la Restauració Monumental: L'Hospitalet del Llobregat*. Barcelona.

González, J. (1982) *Inscripciones de la provincia de Cádiz*. Cádiz.

González Román, C. (1990) *La república tardía: cesarianos y pompeyanos*. Madrid.

González Román, C. (2010) Romanos e itálicos en la Hispania republicana, in Pons Pujol (ed.), pp. 13–32.

Goodchild, H. (2009) The productive potential of the Middle Tiber Valley, in Coarelli and Patterson (eds), pp. 769–96.

Goodchild, H. and Witcher, R. (2009) Modelling the agricultural landscapes of Republican Italy, in Carlsen and Lo Cascio (eds), pp. 187–221.

Goodman, D., Nishimura, Y. and Rogers, J. (1995) GPR time-slices in archaeological prospection. *Archaeological Prospection* 2, 85–9.

Gorgues, A. (2010) *Économie et société dans le nord-est du domaine ibérique (IIIe–Ier s. av. J.-C.)* [Anejos de *ArchEsp* LII]. Madrid.

Gosselain, O.P. (1998) Social and technical identity in a clay crystal ball, in Stark (ed.), pp. 78–106.

Goudineau, C. (ed.) (1989) *Aux origines de Lyon, Documents d'Archéologie en Rhône-Alpes 2* [Série Lyonnaise 1]. Lyon.

Gould, C. (1965) *Trophy of Conquest, The Musée Napoléon and the Creation of the Louvre*. London.

Gowland, R. and Garnsey, P. (2010) Skeletal evidence for health, nutritional status and malaria in Rome and the Empire, in Eckardt (ed.), pp. 131–56.

Grace, V. (1979) *Amphoras and the Ancient Wine Trade*. Princeton.

Graepler, D. (1990) Neue Forschungen zum hellenistischen Tarent, in Schwinzer and Steingräber (eds), pp. 88–95.

Graepler, D. (1997) *Tonfiguren im Grab. Fundkontexte hellenistischer Terrakotten aus der Nekropole von Tarent*. Munich.

Graepler, D. (2002) Le necropoli e la cultura funeraria, in Stazio (ed.), pp. 195–218.

Grahame, M. (1997) Public and private in the Roman house: the spatial order of the Casa del Fauno, in Laurence and Wallace-Hadrill (eds), pp. 137–64.

Grainger, J. (2010) *The Syrian Wars*. Leiden.

Grandazzi, A. (1997) *The Foundation of Rome. Myth and History*. Ithaca.

Grau, I. (2002) *La organización del territorio en el área central de la Contestania ibérica*. Alicante.

Graverini, L. (2001) L. Mummio Acaico. *Maecenas* 1, 105–48.

Graves-Brown, P., Jones, S. and Gamble, C. (1996) *Cultural Identity and Archaeology. The Construction of European Communities*. London.

Gravina, A. (ed.) (2008) *Atti del 26th Convegno sulla Preistoria, Protostoria e Storia della Daunia*. San Severo.

Greco, C. (2005) *Solunto. Ein kleiner Füher*. Palermo.

Greco, E. (1988) Archeologia della colonia latina di Paestum. *DialArch* ser. 3, 6/2, 79–86.

Greco, E. and Longo, F. (eds) (2000) *Paestum: Scavi, studi, recherché. Bilancio di un decennio (1988–1998)*. Paestum.

Greco, E. and Théodorescu, D. (1987) *Poseidonia-Paestum. Forum Nord* [CÉFR 42.3]. Rome.

Greco, E. and Theodorescu, D. (1990) Poseidonia-Paestum: un exemple d'intégration urbaine à l'époque médio-republicaine, in *Akten des XIII Internationalen Kongresses fur Klassische Archaeologie, Berlin 1988*, pp. 85–100. Mainz.

Greco, E. and Theodorescu, D. (1996) La città ed il territorio nel IV secolo a.C., in Cipriani, Longo and Viscione (eds), pp. 192–200.

Greco, G. and Pontrandolfo, A. (eds) (1990) *Fratte. Un insediamento etrusco-campano*. Modena.

Greco Pontrandolfo, A. and Rouveret, A. (1983) La rappresentazione del barbaro in ambiente magno-greco, in *Modes de contacts*, pp. 1051–66.

Green, C.M.C. (1997) Free as a bird: Varro *De Re Rustica* 3. *AJP* 118 (3), 427–48.

Green, C.M.C. (2007) *Roman Religion and the Cult of Diana at Aricia*. Cambridge.

Green, C.M.C. (2012) The shepherd of the people: Varro on herding for the villa publica in *De re rustica* 2, in Becker and Terrenato (eds), pp. 32–44.

Green, R. (1955) *Asclepiades: His Life and Writings*. New Haven.

Greene, J. (1983) Carthage survey, in Keller and Rupp (eds), pp. 197–9.

Greene, J. (1992) Une reconnaissance archéologique dans l'arrière-pays de la Carthage antique, in Ennabli (ed.), pp. 195–7.

Greenhalgh, M. (1989) *The Survival of Roman Antiquities in the Middle Ages*. London.

Gregory, T.E. (ed.) (1993) *The Corinthia in the Roman Period Including the Papers Given at a Symposium Held at the Ohio State University on 7–9 March, 1991* [*JRA* Suppl. 8]. Ann Arbor.

Gregory, T.E. and Gebhard, E. (eds) (forthcoming) *Half a Century on the Isthmus*.

Grelle, F. (2009) L'Agro Pubblico nella Puglia del secondo secolo a.C., in Carlsen and Lo Cascio (eds), pp. 317–40.

Grewe, W.G. (2000) *The Epochs of International Law*. Berlin.

Grimal, P. (ed.) (1983) *Roman Cities*, trans. G.M. Woloch. Madison.

Gros, P. (1973) Hermodoros et Vitruve. *MÉFRA* 85, 137–61.

Gros, P. (1976) Les premiers générations d'architectes hellènistiques à Rome, in *L'Italie préromaine et la Rome républicaine. Mélanges offerts à Jacques Heurgon* (CÉFR 27), vol. 1, pp. 387–410. Rome.

Gros, P. (1996–2001) *L'architecture romaine: du début du IIIe siècle av. J.-C. à la fin du Haut-Empire*. Paris.

Gros, P. and Torelli, M. (eds) (2007) *Storia dell'urbanistica. Il mondo romano* (2nd edn). Rome-Bari.

Gruat, P. (2004) Puits et cavités à offrandes du Sud-ouest, in Brun, Poux and Tchernia (eds), pp. 182–4.

Gruen, E.S. (1984) *The Hellenistic World and the Coming of Rome*, 2 vols. Berkeley.

Gruen, E.S. (1992) *Culture and National Identity in Republican Rome*. Ithaca, NY.

Gruen, E.S. (1996) *Studies in Greek Culture and Roman Policy*. Leiden.

Gruen, E.S. (2006) Romans and others, in Rosenstein and Morstein-Marx (eds), pp. 459–77.

Grünhagen, W. (1959) Die Ausgrabungen des Terrassenheiligtums von Munigua, in *Neue deutsche Ausgrabungen im Mittelmeergebiet und im Vorderen Orient*, pp. 329–43. Berlin.

Grünhagen, W. (1976) Bemerkungen zum Minerva-Relief in der Stadtmauer von Tarragona. *MM* 17, 209–25.

Gsell, S. (1928) *Histoire ancienne de l'Afrique du Nord. Vol. 7. La république romaine et les rois indigènes*. Paris.

Guaitoli, M. (1981) Gabii. *PP* 36, 152–73.

Gualandri, I. and Mazzoli, G. (eds) (2003) *Gli Annei. Una famiglia nella storia e nella cultura di Roma imperiale*. Como.

Gualtieri, M. (1987) Fortifications and settlement organization. An example from Pre-Roman Italy. *WorldArch* 19, 30–46.

Gualtieri, M. (ed.) (1993) *Fourth Century* BC *Magna Graecia: A Case Study*. Jonsered.

Gualtieri, M. (2002) Cosa, the Roman city and "Romanization." *JRA* 15, 397–401.

Gualtieri, M. (2003) *Lucania Romana. Cultura e società nella documentazione archeologica* [*Ostraka*, Quad. 8]. Naples.

Gualtieri, M. (2008a) La Lucania centro-settentrionale in età romana: la nuova documentazione archeologica, in Russo and Di Giuseppe (eds), pp. 205–22.

Gualtieri, M. (2008b) Lucanian landscapes in the age of "Romanization" (third to first centuries BC): two case studies, in de Ligt and Northwood (eds), pp. 387–416.

Gualtieri, M. (2009a) Villae e uso del territorio nell'alto Bradano (Regio III) tra tarda repubblica e primo impero, in Carlsen and Lo Cascio (eds), pp. 341–68.

Gualtieri, M. (2009b) La romanizzazione del territorio: "Grumentum" e l'alta Val d'Agri nel contesto della Lucania romana, in Mastrocinque (ed.), pp. 217–33.

Gualtieri, M. and Fracchia, H. (eds) (1990) *Roccagloriosa I*. Naples.

Gualtieri, M. and Fracchia, H. (eds) (2001) *Roccagloriosa II. L'oppidum Lucano e il territorio*. Naples.

Gualtieri, M., Fracchia, H. and Ferrari, S. (2010) La villa di Osaia (Cortona, AR) ed il territorio cortonese in età romana. *Amoenitas* I, 131–62.

Gudea, N. (ed.) (1999) *Roman Frontier Studies XVII* [Proceedings of the XVIIth International Congress of Roman Frontier Studies]. Zalau.

Guidi, G., Remondino, F., Russo, M. *et al.* (2009) Digitizing the Pompeii Forum, in Crawford, Frischer and Koller (eds), pp. 106–16.

Guitart, J. (2006) *Iesso-Guissona. La descoberta d'una ciutat romana*. Lleida.

Guitart, J. (ed.) (2008) *L'arqueologia i els orígens de les ciutats romanes de Catalunya*. Barcelona.

Guitart, J., Mercado, M., Palet, J. and Rodrigo, E. (2006a) El castellum de Can Tacó/Turó d'en Roïna (Montmeló-Montornès) i la romanització de la Laietània interior. Cap a un estudi arqueològic del jaciment i del territori. *Notes* 21, 241–66.

Guitart, J., Mercado, M., Palet, J. and Rodrigo, E. (2006b) Can Tacó: un recinte fortificat d'època republicana. *Cota Zero* 21, 23–6.

Guitart, J. and Padrós, P. (1990) Baetulo. Cronología y significación de sus monumentos, in Trillmich and Zanker (eds), pp. 165–77.

Guitart, J., Padrós, P. and Fonollá, A. (1994) Aproximació a l'esquema urbanístic fundacional de la ciutat romana de Baetulo (Badalona), in *Actes del XIV Congrés*, pp. 188–91.

Guitart, J. and Pera, J. (1994) La ciutat romana de Iesso (Guissona, La Segarra), in *Actes del XIV Congrés*, pp. 186–7.

Guitart, J. and Pera, J. (eds) (2004) *Iesso I. Miscel.lània Arqueològica*. Barcelona-Guissona.

Guitart, J., Pera, J., Mayer, M. and Velaza, J. (1996) Noticia preliminar y sobre una inscripción ibérica encontrada en Guissona (Lleida), in *La Hispania prerromana* (Acta Salmanticensia 262), pp. 163–70. Salamanca.

Gupta, P.L. (1991) Coins in Rome's Indian trade, in Jha (ed.), pp. 122–37.

Guyon, J., Nin, N., Rivet, L. and Saulnier, S. (1998) *Atlas topographique des villes de Gaule méridionale I : Aix-en-Provence* [*Revue Archéologique de Narbonnaise* Suppl. 30]. Montpellier.

Guzzo, P. and Guidobaldi, M.P. (eds) (2005) *Nuove ricerche archeologiche a Pompei ed Ercolano. Atti del Convegno Nuove ricerche nell'area vesuviana, Roma 28–30 novembre 2002.* Naples.

Guzzo, P. and Guidobaldi, M.P. (eds) (2008) *Nuove ricerche archeologiche nell'area vesuviana (scavi 2003–2006) Atti del Convegno Internazionale, Roma 1–3 febbraio 2007.* Rome.

Guzzo Amadasi, M.G. (1967) *Le iscrizioni fenicie e puniche delle colonie in Occidente* [Studi Semitici 28]. Rome.

Habicht, C. (1976) Royal documents in Maccabees II. *HSCP* 80, 1–18.

Habicht, C. (1990) The Seleucids and their rivals, in *CAH* VIII (2nd edn), pp. 324–87. Cambridge.

Habicht, C. (1997a) *Athens from Alexander to Antony.* Cambridge, MA.

Habicht, C. (1997b) Roman citizens in Athens (228–31 BC), in Hoff and Rotroff (eds), pp. 9–17.

Hackett, J. (ed.) (1989) *Warfare in the Ancient World.* London.

Hales, S. (2003) *Roman Houses & Social Identity.* Cambridge.

Hales, S. and Hodos, T. (eds) (2009) *Material Culture and Social Identities in the Ancient World.* Cambridge.

Hall, J. (1997) *Ethnic Identity in Greek Antiquity.* Cambridge.

Hall, J.M. (2002) *Hellenicity: Between Ethnicity and Culture.* Chicago.

Hall, J.F. (ed.) (1996) *Etruscan Italy: Etruscan Influences on the Civilizations of Italy from Antiquity to the Modern Era.* Provo, UT.

Halpern, C. (ed.) (2004) *Identités: l'individu, le groupe, la société.* Auxerre.

Halstead, P. and O'Shea, J. (eds) (1989) *Bad Year Economics: Cultural Responses to Risk and Uncertainty.* Cambridge.

Hanna, K. (1996) Survey work report. Pompei Forum Project. Institute for Advanced Technology in the Humanities, University of Virginia.

Hansen, I. and Hodges, R. (eds) (2007) *Roman Butrint.* Oxford.

Hansen, M.H. (ed.) (1995) *Sources for the Ancient Greek City-State. Symposium August, 24–27 1994. Acts of the Copenhagen Polis Centre 2* [Historisk-filosofiske Meddelelser 72]. Copenhagen.

Hansen, M.H. (ed.) (2000) *A Comparative Study of Thirty City-State Cultures.* Copenhagen.

Hansen, M. (2006) *The Shotgun Method. The Demography of the Ancient Greek City-State Culture.* London.

Hanson, J.A. (1959) *Roman Theater-Temples.* Princeton.

Harley, J. and Woodward, D. (eds) (1987) *Cartography in Prehistoric, Ancient, and Medieval Europe and the Mediterranean, The History of Cartography*, Chicago.

Harmand, J. (1967) *L'armée et le soldat à Rome de 107 á 50 avant notre ère.* Paris.

Harper, N.K. and Fox, S. (2008) Recent research in Cypriot bioarchaeology. *Bioarchaeology of the Near East* 2, 1–38.

Harris, W.V. (1971) *Rome in Etruria and Umbria.* Oxford.

Harris, W.V. (2001) Steinby's *LTUR* III and IV. *JRA* 14, 539–46.

Harris, W.V. (2007) The Late Republic, in Scheidel, Morris and Saller (eds), pp. 511–39.

Hartswick, K. (2004) *The Gardens of Sallust: A Changing Landscape.* Austin.

Haselberger, L. (2008) *Mapping Augustan Rome* (rev. edn) [*JRA* Suppl. 50]. Portsmouth, RI.

Haselberger, L. and Humphrey, J. (eds) (2006) *Imaging Ancient Rome. Documentation, Visualization, Imagination. Proceedings of the Third Williams Symposium on Classical Architecture Held at the American Academy in Rome, the British School at Rome and the Deutsches Archäologisches Institut, Rome, on May 20–23, 2004.* Portsmouth, RI.

Haselgrove, C. (ed.) (2006) *Celtes et Gaulois. L'archéologie face à l'histoire. Les mutations de la fin de l'âge du fer (Actes de la table ronde de Cambridge 2005).* Glux-en-Glenne.

Haselgrove, C. and Wigg-Wolf, D. (eds) (2005) *Iron Age Coinage and Ritual Practices.* Mainz.

Haskell, F. and Penny, N. (1981) *Taste and the Antique, the Lure of Classical Sculpture, 1500–1900.* New Haven.

Hatzfeld, J. and Roussel, P. (1910) Fouilles de Délos exécutées aux frais de M. le Duc de Loubat. Décrets, dédicaces et inscriptions funéraires (1905–1908) II. *BCH* 34, 355–423.

Häuber, C. (1990) Zur topographie der Horti Maecenatis und der Horti Lamiani auf dem Esquilin in Rom. *KölnJb* 23, 11–108.

Hauschild, T. (1968) Ausgrabungen in Munigua. Neue Kampagne bis 1967. *AA* 358–68.

Hauschild, T. (1988) Excavaciones en la muralla romana de Tarragona. *Butlletí Arqueològic* 5, 6–7, 11–38.

Hauschild, T. (1995/6) Tarraco, Hauptstadt der Provinz Hispania Citerior. *Nürnberger Blätter zur Archäologie* 12, 75–84.

Hauschild, T. (2006) Die römischen Tore des 2. Jhs. v. Chr, in der Stadtmauer von Tarragona, in *Stadttore: Bautyp und Kunstform. Puertas de ciudades. Tipo arquitectónico y forma artística* (*Baetica* 28), pp. 153–71. Madrid.

Häussler, R. (2000) The dynamics of state societies in northwest Italy, in Herring and Lomas (eds), pp. 131–56.

Haverfield, F. (1912) *The Romanization of Roman Britain.* Oxford.

Haverfield, F. (1913) *Ancient Town Planning.* Oxford.

Haynes, S. (2000) *Etruscan Civilization: A Cultural History.* Los Angeles.

Hazlitt, W. (ed.) (1864) *Works of Michael de Montaigne* (2nd edn), 4 vols. New York.

Heiken, G. (ed.) (2006) *Tuffs – Their Properties, Uses, Hydrology, and Resources* [Geological Society of America Special Paper 408]. Boulder, CO.

Heinzelmann, M. (2001) Grabarchitektur, Bestattungsbrauch und Sozialstruktur: Zur Rolle der *familia. Röm Mitteil* 108, 353–81.

Hellenkemper Salies, G. (ed.) (1994) *Das Wrack. Der antike Schiffsfund von Mahdia.* Bonn.

Hemelrijk, J.M. (2007) Review of R. Bonaudo, *La culla di Hermes. BABesch* 82, 277–80.

Hemelrijk, J.M. and Den Boer, E. (2007) Four new Campana dinoi, a new painter, old questions. *BABesch* 82, 365–421.

Henig, M. (ed.) (1990) *Architecture and Architectural Sculpture in the Roman Empire* [Oxford University Committee for Archaeology 29]. Oxford.

Herbert, S. and Ariel, D. (1994) *Tel Anafa: Final Report on Ten Years of Excavation at a Hellenistic and Roman Settlement in Northern Israel* [*JRA* Suppl. 10]. Ann Arbor.

Herbert, S. and Berlin, A. (2003) A new administrative center for Persian and Hellenistic Galilee: preliminary report of the University of Michigan/University of Minnesota excavations at Kedesh. *BASOR* 329, 13–59.

Hermann, W. (1961) *Römische Götteraltäre.* Kallmunz.

Hermanowicz, J.C. and Morgan, H.P. (1999) Ritualizing the routine: collective identity affirmation. *Sociological Forum* 14, 197–214.

Hernández Vera, J.A. (2002) La fundación de Graccurris, in Jiménez and Ribera (eds), pp. 173–82.

Herring, E. (2000) "To see ourselves as others see us!" The construction of native identities in southern Italy, in Herring and Lomas (eds), pp. 45–77.

Herring, E. and Lomas, K. (eds) (2000) *The Emergence of State Identities in Italy in the First Millennium* BC. London.

Herring, E., Whitehouse, R. and Wilkins, J. (eds) (1991) *Papers of the Fourth Conference of Italian Archaeology 2, The Archaeology of Power, Pt. 2.* London.

Hersh, C. and Walker, A. (1984) The Mesagne Hoard. *ANSMN* 29, 103–34.

Herz, N. and Waelkens, M. (eds) (1988) *Classical Marble: Geochemistry, Technology, Trade* [*ASMOSIA 1. Nato ASI Series* 153]. Dordrecht.

Hesberg, H. von (1980) *Konsolengeisa des Hellenismus und der Fruhen Kaiserzeit* [Römischen Mitteilungen Ergänzungsheft 24]. Mainz.

Hesberg, H. von (1992) *Römische Grabbauten*. Darmstadt.

Hesberg, H. von and Panciera, S. (1994) *Das Mausoleum des Augustus. Der Bau und seine Inschriften*. Munich.

Hesberg, H. von and Zanker, P. (eds) (1987) *Römische Gräberstraßen: Selbstdarstellung, Status, Standard. Kolloquium in München 28.–30. 10. 1985*. Munich.

Hibbard, H. (1974) *Michelangelo*. New York.

Higginbotham, J.A. (1997) *Piscinae: Artificial Fishponds in Roman Italy*. Chapel Hill.

Hildebrandt, H. (1979) Die Römerlager von Numantia. Datierung anhand de Münzfunde. *MM* 20, 238–71.

Hill, J.D. (ed.) (1996) *History, Power and Identity: Ethnogenesis in the Americas 1492–1992*. Iowa City.

Hill, M.J., Lanos, P., Chauvin, A. *et al.* (2007) An archaeomagnetic investigation of a Roman amphorae workshop in Albinia (Italy). *Geophysical Journal International* 180, 1–14.

Hin, S. (2008) Counting Romans, in de Ligt and Northwood (eds), pp. 187–238.

Hingley, R. (2005) *Globalizing Roman Culture: Unity, Diversity and Empire*. London.

Hingley, R. (2009) Cultural diversity and unity. Empire and Rome, in Hales and Hodos (eds), pp. 54–75.

Hirschfeld, Y. (1997) *The Roman Baths of Hammat Gader: Final Report*. Jerusalem.

Hitchner, B. (2009) Review article: Roman Republican imperialism in Italy and the west. *AJA* 113, 651–5.

Hobsbawm, E.J. (1990) *Nations and Nationalism since 1780. Programme, Myth, Reality*. Cambridge.

Hobsbawm, E. and Ranger, T. (1983) *The Invention of Tradition*. Cambridge.

Hobson, B. (2009) *Latrinae et Foricae: Toilets in the Roman World*. London.

Hodder, I. (ed.) (2001) *Archaeological Theory Today*. Cambridge.

Hodder, I. and Orton, C. (1981) *Spatial Analysis in Archaeology*. Cambridge.

Hodge, A.T. (2002) *Roman Aqueducts and Water Supply* (2nd edn). London.

Hodges, R. (2000) *Visions of Rome: Thomas Ashby Archaeologist*. London.

Hodos, T. (2009) Local and global perspectives in the study of social and cultural identities, in Hales and Hodos (eds), pp. 3–31.

Hoepfner, W. and Brands, G. (eds) (1996) *Basileia. Die Paläste der hellenistischen Könige, internationales Symposion in Berlin vom 16.12.1992 bis 20.12.1992*. Mainz-am-Rhein.

Hoff, M.C. (1988) The Roman Agora at Athens. PhD thesis. Boston University.

Hoff, M.C. (1989a) Civil disobedience and unrest in Augustan Athens. *Hesperia* 58, 267–76.

Hoff, M.C. (1989b) The early history of the Roman Agora at Athens, in Walker and Cameron (eds), pp. 1–8.

Hoff, M.C. (1992) Augustus, Apollo, and Athens. *Museum Helveticum* 49, 223–32.

Hoff, M.C. (1996) The politics and architecture of the Athenian imperial cult, in Small (ed.), 1996, pp. 185–200.

Hoff, M.C. (1997) *Laceratae Athenae*: Sulla's siege of Athens in 87/6 BC and its aftermath, in Hoff and Rotroff (eds), pp. 33–51.

Hoff, M.C. (2005) Athens honors Pompey the Great, in Blois *et al.* (eds), pp. 327–36.

Hoff, M.C. and Rotroff, S.I. (eds) (1997) *The Romanization of Athens. Proceedings of an International Conference held at Lincoln, Nebraska (April 1996)* [*Oxbow Monograph* 94]. Oxford.

Hoffmann, A. (1979) L'architettura privata, in Zevi (ed.), 1979, pp. 97–118.

Hohlfelder, R. (1983) Sebastos: Herod's harbor at Caesarea. *BiblArch* 46, 133–43.

Hohlfelder, R.L. (ed.) (2008) *The Maritime World of Ancient Rome*. Ann Arbor.

Hölkeskamp, K.-J. (2010) *Reconstructing the Roman Republic: An Ancient Political Culture and Modern Research*. Princeton.

Holland, L.A. (1961) *Janus and the Bridge* [*PAAR* 1]. Rome.

Holleaux, M. (1921) *Rome, la Grèce et les monarchies hellénistiques au IIIe siècle avant J. C. (273–205)*. Paris.

Holliday, P.J. (1980) *Ad triumphum excolendum*: the political significance of Roman historical painting. *Oxford Art Journal* 3–8.

Holliday, P.J. (2002) *The Origins of Roman Historical Commemoration in the Visual Arts*. Cambridge.

Holloway, R.R. (1994) *The Archaeology of Early Rome and Latium*. London.

Holloway, R.R. (2006) The Tomb of the Diver. *AJA* 110, 365–88.

Hölscher, T. (1967) *Victoria Romana. Archäologische Untersuchungen zur Geschichte und Wesensart der römischen Siegesgöttin von den Anfängen bis zum Ende des 3. Jhs. n.Chr.* Mainz.

Hölscher, T. (1987) *Römische Bildsprache als semantisches System*. Heidelberg.

Hölscher, T. (2004) *The Language of Images in Roman Art*. Cambridge.

Hölscher, T. (2006) Greek styles and Greek art in Augustan Rome: issues of the present versus records of the past, in J. Porter (ed.), pp. 237–59.

Holt, F. (1999) *Thundering Zeus: The Making of Hellenistic Bactria*. Berkeley.

Holum, K., Hohlfelder, R., Bull, R. and Raban, A. (1988) *King Herod's Dream: Caesarea on the Sea*. New York.

Holum, K., Raban, A. and Patrich, J. (eds) (1999) *Caesarea Papers II* [*JRA* Suppl. 35]. Portsmouth, RI.

Holz, S. (2009) *Praeda* und Prestige – Kriegsbeute und Beutekunst im (Spät-) Republikanischen Rom, in Coudry and Humm (eds), pp. 187–206.

Hopkins, J.N. (2010) The topographical transformation of Archaic Rome: a new interpretation of architecture and geography in the early city. PhD thesis. University of Texas at Austin.

Hopkins, K. (1966–7) On the probable age structure of the Roman population. *Population Studies* 20, 245–64.

Hopkins, K. (1978) *Conquerors and Slaves. Sociological Studies in Roman History Volume 1*. Cambridge.

Hopkins, K. (1980) Taxes and trade in the Roman Empire (200 BC–AD 400). *JRS* 70, 101–25.

Hopkins, K. (1983) *Death and Renewal*. Cambridge.

Hopkins, K. (1995/6) Rome, taxes, rents and trade. *Kodai* 6/7, 41–75.

Horden, P. and Purcell, N. (2000) *The Corrupting Sea*. Oxford.

Horn, H. and Rüger, C. (eds) (1979) *Die Numider*. Bonn.

Horsnaes, H. (2004) Romanization at Paestum in the 3rd c. BC: a note on the chronology of the ΠΑΙΣΤΑΝΟ coins and the interpretation of the wall-paintings from the Spinazzo Cemetery. *JRA* 17, 305–11.

Horvat, J. (1997) Roman Republican weapons from Šmihel in Slovenia, in Feugère (ed.), pp. 105–20.

Horvat, J. (2002) The hoard of Roman Republican weapons from Grad near Šmihel. *Arheološki vestnik* 53, 117–92.

Hostetter, E. and Brandt, J. (2009) *Palatine East Excavations v.1, Stratigraphy and Architecture*. Rome.

Houghtalin, L. (1985) Coins from the River Liri III. *NC* 45, 67–81.

Howgego, C., Heuchert, V. and Burnett, A. (eds) (2005) *Coinage and Identity in the Roman Provinces*. Oxford.

Hoyos, B. (2003) *Hannibal's Dynasty: Power and Politics in the Western Mediterranean, 247–183 BC*. London.

Hoyos, D. (2007) The age of overseas expansion (264–146 BC), in Erdkamp (ed.), pp. 63–79.

Hughes, A. (1996) Comic stages in Magna Graecia: the evidence of the vases. *Theatre Research International* 21, 95–107.

Humm, M. (1996) Appius Claudius Caecus et la construction de la via Appia. *MÉFRA* 108, 693–746.

Humphrey, J. (1986) *Roman Circuses: Arenas for Chariot Racing*. London.
Humphrey, J. (ed.) (1991) *Literacy in the Roman World* [*JRA* Suppl. 3]. Portsmouth, RI.
Humphrey, J. (ed.) (1995) *The Roman and Byzantine Near East* [*JRA* Suppl. 14]. Portsmouth, RI.
Hunter, F. (2006) Art in Later Iron Age society, in Haselgrove (ed.), pp. 93–115.
Hurst, H. (2006) The Scalae (ex Graecae) above the *Nova Via*. *PBSR* 74, 237–91.
Hurst, H. (2007) The "Murus Romuli" at the northern corner of the Palatine and the Porta Romanula, in Leone, Palombi and Walker (eds), pp. 79–102.
Hurst, H. and Cirone, D. (2003) Excavation of the pre-Neronian *Nova Via*, Rome. *PBSR* 71, 71–84.
Hurst, H. and Owen, S. (eds) (2005) *Ancient Colonizations. Analogy, Similarity and Difference*. London.
Hutchinson, G.E. (ed.) (1970) *Ianula: An Account of the History and Development of the Lago di Monterosi, Latium, Italy*. Philadelphia.
Iacopi, I. (1997) *La decorazione pittorica dell'Aula Isiaca*. Milan.
Iacopi, I. and Tedone, G. (2006) Bibliotheca e porticus ad Apollonis. *RM* 112, 351–78.
IBM Italia; IBM Gallery of Science and Art; Italy. Ministero per i beni culturali e ambientali; Italy. Soprintendenza archeologica di Pompei (1992) *Rediscovering Pompeii*. Rome.
Iglesias, J. and Muñiz, J. (eds) (1999) *Cántabros. La génesis de un pueblo*. Santander.
Ikeguchi, M. (1999/2000) A comparative study of settlement patterns and agricultural structures in ancient Italy: a methodology for interpreting field survey evidence. *Kodai Journal of Ancient History* 10, 1–59.
Ikeguchi, M. (2006) A method for interpreting and comparing field survey data, in Bang, Ikeguchi and Ziche (eds), pp. 137–60.
Insoll, T. (2006) *The Archaeology of Identities. A Reader*. Abingdon.
Iriarte, A., Gil, E., Filloy, I. and Garcia, M.L. (1997) A votive deposit of Republican weapons at Gracurris (Alfaro, La Rioja, Spain), in Feugère (ed.), pp. 233–50.
Isager, J. (1991) *Pliny on Art and Society*. London.
Isayev, E. (2007a) Why Italy?, in Bradley, Isayev, and Riva (eds), pp. 1–20.
Isayev, E. (2007b) *Inside Ancient Lucania: Dialogues in History and Archaeology*. London.
Isayev, E. (2009) Unintentionally being Lucanian: dynamics beyond hybridity, in Hales and Hodos (eds), pp. 201–26.
Isler, H.P. (1981) Contributi per una storia del teatro antico: il teatro greco di Iaitas e il teatro di Segeasta. *NumAntCl* 10, 131–64.
Isler, H.P. (1996) Einflüsse der Makedonischen Palastarchitektur in Sizilien?, in Hoepfner and Brands (eds), pp. 252–7.
Isler, H.P. (2000) *Monte Iato. Guida archeologica*. Palermo.
Istenič, J. (2005) Evidence for a very Late Republican siege at Grad near Reka in western Slovenia, in Jilek (ed.), pp. 77–87.
Izenour, G.C. (1992) *Roofed Theaters of Classical Antiquity*. New Haven.
Izzet, V. (2001) Form and meaning in Etruscan ritual space. *CAJ* 11, 185–200.
Izzet, V. (2007a) *The Archaeology of Etruscan Society*. Cambridge.
Izzet, V. (2007b) Etruria and the Etruscans: recent approaches, in Bradley, Isayev, and Riva (eds), pp. 114–30.
Jackson, M.D., Ciancio Rossetto, P., Kosso, C.K. *et al.* (2011) Building materials of the Theater of Marcellus, Rome. *Archaeometry* 4, 728–42.
Jackson, M.D., Deocampo, D., Marra, F. and Scheetz, B.E. (2010) Mid-Pleistocene volcanic ash in ancient Roman concretes. *Geoarchaeology* 25, 36–74.
Jackson, M.D., Kosso, C., Marra, F. and Hay, R. (2006) Geological basis of Vitruvius' empirical observations of material characteristics of rock utilized in Roman masonry, in Dunkeld *et al.* (eds), vol. 2, pp. 1685–1702.

Jackson, M.D. and Marra, F. (2006) Roman stone masonry: volcanic foundations of the ancient city. *AJA* 110, 403–36.

Jackson, M., Marra, F., Deocampo, D., Scheetz, B.E. and Vella, A. (in press a) Analisi delle componenti geologiche nelle murature del Foro di Cesare. *Scienze dell'Antichità*.

Jackson, M., Marra, F., Deocampo, D., Vella, A., Kosso, C. and Hay, R. (2007) Geological observations of excavated sand (*harenae fossiciae*) used as fine aggregate in ancient Roman pozzolanic mortars. *JRA* 20, 25–53.

Jackson, M.D., Marra, F., Hay, R.L., Cawood, C., and Winkler, E. (2005) The judicious selection and preservation of tuff and travertine building stone in ancient Rome. *Archaeometry* 47, 485–510.

Jackson, M., Scheetz, B. and Marra, F. (2010) Micromorphological textures and pozzolanic cements in Imperial age Roman concretes, in Moreau and Válek (eds), pp. 207–17.

Jackson, M., Vola, G., Všianský, D., Oleson, J.P., Scheetz, B.E., Brandon, C. and Hohlfelder, R.L. (in press b) Cement microstructures and durability in ancient Roman seawater concretes, in Válek, Groot and Hughes (eds).

Jackson, R. (1990) Waters and spas in the Classical world, in R. Porter (ed.), pp. 1–13.

Jackson, R. (1999) Spas, waters and hydrotherapy in the Roman world, in DeLaine and Johnston (eds), pp. 107–16.

Jacobson, D. (2005) *The Hellenistic Paintings of Marisa* [*PEFA* 7]. London.

Jacobson, D. (2007) The Jerusalem Temple of Herod the Great, in Kokkinos (ed.), pp. 145–77.

Jaeger, M. (1997) *Livy's Written Rome*. Ann Arbor.

Jaeger, M. (2003) Livy and the fall of Syracuse, in Eigler *et al.* (eds), pp. 213–34.

Jaeger, M. (2008) *Archimedes and the Roman Imagination*. Ann Arbor.

Jaeger, W. (1938) *Diocles von Karystos*. Berlin.

Japp, S. (2000) *Die Baupolitik Herodes' des Grossen*. Rahden.

Jashemski, W.F. and Meyer. F. (eds) (2002) *The Natural History of Pompeii*. Cambridge.

Jehne, M. (ed.) (in press) *Religiöse Vielfalt und soziale Integration. Die Bedeutung der Religion für die kulturelle Identität und die politische Stabilität im republikanischen Italien*.

Jehne, M. and Pfeilschifter, R. (eds) (2006) *Herrschaft ohne Integration? Rom und Italien in republikanischer Zeit* [Studien zur Alten Geschichte 4]. Frankfurt.

Jenkins, I. (1992) *Archaeologists and Aesthetes*. London.

Jenkins, I. and Sloane, K. (1996) *Vases and Volcanoes. Sir William Hamilton and His Collection*. London.

Jenkins, R. (2008) *Social Identity*. New York.

Jha, A.K. (ed.) (1991) *Coinage, Trade and Economy* [3rd International Colloquium, Indian Institute of Research in Numismatic Studies]. Anjan, India.

Jilek, S. (ed.) (2005) *Archäologie der Schlachtfelder. Militaria aus Zerstörungshorizonten. Proceedings of the Fourteenth Roman Military Equipment Conference* (ROMEC XIV). Vienna.

Jiménez, A. (1989) *La Puerta de Sevilla en Carmona*. Seville.

Jiménez, J. and Ribera, A. (eds) (2002) *Valencia y las primeras ciudades romanas de Hispania*. Valencia.

Jimeno, A. (2002) Numancia: campamentos romanos y cerco de Escipión. *ArchEsp* 75, 159–76.

Jimeno, A. (2005) Numancia y los campamentos romanos: investigación y recuperación del pasado, in Pérez González and Illarregui (eds), pp. 237–49.

Jocelyn, H.D. and Hunt, H. (eds) (1993) *Tria Lustra: Essays and Notes Presented to John Pinsent*. Liverpool.

Johannowsky, B. (1961) Relazione preliminare sugli scavi di Cales. *BdA* 46, 258–68.

Johannowsky, B. (1963) Relazione preliminare sugli scavi di Teano. *BdA* 48, 131–65.

Johannowsky, B. (1976) La situazione in Campania, in Zanker (ed.), pp. 267–99.

Johannowsky, W. and Russo, M. (1991) *Capua antica*. Naples.

Johns, C. (1998) *Antonio Canova and the Politics of Patronage in Revolutionary and Napoleonic Europe*. Berkeley.

Johnson, J. (1935) *Excavations at Minturnae*. Philadelphia.

Johnston, P. (2008) *The Military Consilium in Republican Rome*. Piscataway, NJ.

Jolivet, V. (2009) Céramique á vernis noir de Cosa: Retour sur images. *JRA* 22, 489–94.

Jolivet, V., Pavolini, C., Tomei, M.A. and Volpe, R. (eds) (2009) *Suburbium II: Il suburbio di Roma dalla fine dell'età monarchica alla nascita del sistema delle ville, V–II secolo a.C.* Rome.

Jones, A.H.M. (1931) The urbanization of Palestine. *JRS* 21, 78–85.

Jones, A.H.M. (1948) *Ancient Economic History*. London.

Jones, A.H.M. (1963) The Greeks under the Roman Empire. *DOP* 17, 3–19.

Jones, C. (2009) The inscription from Tel Maresha for Olympiodoros. *ZPE* 171, 100–4.

Jones, H. (ed.) (2004) *Samnium: Settlement and Cultural Change* [Proceedings of the Third E. Togo Salmon Conference on Roman Studies, Hamilton, Ontario, Oct. 1999]. London.

Jones, J.E., Sackett, L.H. and Graham, A.J. (1962) The Dema House in Attica. *BSA* 57, 75–114.

Jones, L. (2007) Eventfulness of architecture: teaching about sacred architecture is teaching about ritual, in Bell (ed.), pp. 251–72.

Jones, R. and Robinson, D. (2005) the structural development of the House of the Vestals (VI.1.6–8, 24–26), in Guzzo and Guidobaldi (eds), 2005, pp. 257–69.

Jones, R. and Robinson, D. (2007) Intensification, heterogeneity and power, in Dobbins and Foss (eds), pp. 389–406.

Jones, S. (1996) Discourses of identity in the interpretation of the past, in Graves-Brown, Jones and Gamble (eds), pp. 62–80.

Jones, S. (1997) *The Archaeology of Ethnicity. Constructing Identities in the Past and Present*. London.

Jordan, D. and Traill, J. (eds) (2003) *Lettered Attica. A Day of Attic Epigraphy* [Proceedings of the Athens Symposium]. Athens.

Judson, S. and Kahane, A. (1963) Underground drainageways in southern Etruria and Latium. *PBSR* 18, 74–99.

Jullian, C. (1908) *Histoire de la Gaule*. Paris.

Jung, F. (1984) Gebaute Bilder. *Antike Kunst* 27, 71–122.

Junkelmann, M. (1986) *Die Legionen des Augustus. Der römische Soldat im archäologischen Experiment*. Mainz.

Juste, M. (2000) Bolskan-Osca, ciudad iberorromana. *Empúries* 52, 87–106.

Kahil, L., Augé, C. and Linant de Bellefonds, P. (eds) (1983) *Iconographie classique et identités régionales* (Actes du Colloque International CNRS 619) [*BCH* Suppl. 14]. Paris.

Kähler, H. (1958) Das Fortunaheiligtum von Palestrina-Praeneste. *Annales Universitatis Saraviensis* 7, 189–240.

Kaimio, J. (2010) *The Cippus Inscriptions of the Museo Nazionale di Tarquinia* [Materiali del museo archeologico nazionale di Tarquinia XVII]. Rome.

Kallet-Marx, R. (1995) *Hegemony to Empire: The Development of the Roman Imperium in the East from 148 to 62* BC. Berkeley.

Kane, S. (2008) Life on the edge: a view from Abruzzo, in Lock and Faustoferri (eds), pp. 93–103.

Karner, D.B., Marra, F. and Renne, P.R. (eds) (2001) The history of the Monti Sabatini and Alban Hills volcanoes: groundwork for assessing volcanic–tectonic hazards for Rome. *Journal of Volcanology and Geothermal Research* 107, 185–219.

Katz, D. and Suchey, J.M. (1986) Age determination of the male os pubis. *American Journal of Physical Anthropology* 69, 427–35.

Keaveney, A. (1987) *Rome and the Unification of Italy*. London.

Keay, S. (1988) *Roman Spain*. London.

Keay, S. (ed.) (1998) *The Archaeology of Early Roman Baetica* [*JRA* Suppl. 29]. Portsmouth, RI.

Keay, S. (2006) The early Roman towns of Tarraconensis: a discussion, in Abad, Keay and Ramallo (eds), pp. 223–37.

Keay, S., Earl, G., Strutt, K. *et al.* (2008) The potential of archaeological geophysics. The work of the British School at Rome in Italy, in Orlando (ed.), pp. 25–45.

Keay, S., Millett, M., Poppy, S. *et al.* (2000) Falerii Novi: a new survey of the walled area. *PBSR* 68, 1–94.

Keay, S. and Terrenato, N. (eds) (2001) *Italy and the West: Comparative Issues in Romanization*. Oxford.

Keller, D. and Rupp, D. (eds) (1983) *Archaeological Survey in the Mediterranean Area* [*BAR-IS* 155]. Oxford.

Kelly, M.G. and Huntley, B. (1991) An 11,000-year record of vegetation and environment from Lago di Martignano, Latium, Italy. *Journal of Quaternary Science* 6, 209–24.

Kemmers, F. (2006) *Coins for a Legion: An Analysis of the Coin Finds from the Augustan Fortress and Flavian "canabae legionis" at Nijmegen* [*Studien zu Fundmünzen der Antike* 21]. Mainz.

Kennedy-Quigley, S. (2001) Visual representations of the birth of Athena/Menrva: a comparative study. *EtrStud* 8, 65–78.

Keppie, L. (1984) *The Making of the Roman Army from Republic to Empire*. London.

Keppie, L. (1997) Military service in the Late Republic: the evidence of inscriptions and sculpture, in Feugère (ed.), pp. 3–11.

Kienast, H.J. (1997) The Tower of the Winds in Athens, in Hoff and Rotroff (eds), pp. 53–65.

Killgrove, K. (2010) Identifying immigrants to Imperial Rome using strontium isotope analysis, in Eckardt (ed.), pp. 157–74.

Kilmer, M.F. (1977) *The Shoulder Bust in Sicily and South and Central Italy*. Göteborg.

King, A. (1990) *Roman Gaul and Germany*. London.

King, A., Rhodes, P.A., Rielley, K. and Thomas, K.D. (1985) I resti animali, in Ricci (ed.), pp. 278–305.

King, C.E. and Wigg, D. (eds) (1996) *Coin Finds and Coin Use in the Roman World, 13th Oxford Symposium*. Berlin.

King, H. (ed.) (2005) *Health in Antiquity*. New York.

Klar, L. (2006) The origins of the Roman scaenae frons and the architecture of triumphal games in the second century BC, in Dillon and Welch (eds), pp. 162–83.

Kleiner, D.E.E. and Matheson, S.B. (eds) (1996) *I Claudia: Women in Ancient Rome*. Austin.

Klingshirn, W.E. (2006) Inventing the *Sortilegus*: lot divination and cultural identity in Italy, Rome, and the provinces, in Schultz and Harvey (eds), pp. 137–61.

Kloner, A. (2003) *Maresha Excavations Final Report I: Subterranean Complexes 21, 44, 70* [IAA Reports]. Jerusalem.

Klynne, A. (2006–7) The Villa Selvasecca revisited. *OpRom* 31–2, 29–58.

Knapp, A. (ed.) (1992) *Archeology, "Annales" and Ethnohistory*. Cambridge.

Knapp, R. (1977a) *Aspects of the Roman Experience in Iberia 206–100 BC*. Valladolid-Vitoria.

Knapp, R. (1977b) The date and purpose of the Iberian denarii. *NC* 137, 1–18.

Knapp, R. (1979) The legions in Spain (200–169 BC). *Revista de la Universidad Complutense* 118 [Homenaje a García y Bellido IV], 87–94.

Knapp, R. (1980) Cato in Spain (195–194 BC), in Deroux (ed.), pp. 21–56.

Knapp, R. (1983) *Roman Corduba*. Berkeley.

Knappett, C. (2005) *Thinking through Material Culture*. Philadelphia.

Knikman, H. (2005) Computer assisted research in Pompeii, in Guzzo and Guidobaldi (eds), 2005, pp. 235–9.

Knight, J. (2001) *Roman France. An Archaeological Field Guide*. Stroud.

Knoepfler, D. (1991) L. Mummius Achaicus et les cités du golfe euboïque. À propos d'une nouvelle inscription d'Érétrie. *MusHelv* 48, 252–80.

Kockel, V. (1983) *Die Grabbauten vor dem Herkulaner Tor in Pompeji.* Mainz.

Kockel, V. (1987) Im Tode gleich? Die sullanischen Kolonisten und ihr kulturelles Gewicht in Pompeji am Beispiel der Nekropolen, in von Hesberg and Zanker (eds), pp. 183–98.

Kokkinos, N. (ed.) (2007) *The World of the Herods: Volume 1 of the International Conference The World of the Herods and the Nabataeans held at the British Museum, 17–19 April 2001* [*Oriens et Occidens* 14]. Stuttgart.

Koon, S. (2010) *Infantry Combat in Livy's Battle Narratives* [*BAR* 2071]. Oxford.

Kopytoff, I. (1986) The cultural biography of things: commoditization as process, in Appadurai (ed.), 1986, pp. 64–91.

Körte, G. (1909) *Das Volumniergrab bei Perugia: Ein Beitrag zur Chronologie der etruskischen Kunst.* Berlin.

Kraft, G., Flege, S., Reiff, F. *et al.* (2006) Analysis of the notches of ancient serrated denars. *Archaeometry* 148 (4), 605–12.

Krause, C. (2004) Das Haus Ciceros auf dem Palatin. *NumAntCl* 33, 293–316.

Kroll, J. (1993) *The Athenian Agora. Results of the Excavations Conducted by the American School of Classical Studies at Athens,* vol. 26, *The Greek Coins.* Princeton.

Kroll, J. (1997) Traditionalism vs. Romanization in bronze coinages of Greece, 42–31 BC. *Topoi* 7, 123–36.

Kromayer, J. and Veith, G. (1928) *Heerwesen und Kriegsführung der Griechen und Römer* [Handbuch der Altertumswissenschaft 4.3.2]. Munich.

Kron, G. (2005) The Augustan census figures and the population of Italy. *Athenaeum* 93, 441–95.

Kron, J.G. (2008a) Animal husbandry, hunting, fishing, and fish production, in Oleson (ed.), pp. 175–224.

Kron, J.G. (2008b) The much-maligned Peasant. Comparative perspectives on the productivity of the small farmer in Classical Antiquity, in de Ligt and Northwood (eds), pp. 71–119.

Kunz, H. (2006) *Sicilia* [Religion der Römischen Provinzen 4]. Tübingen.

Kunze, E. and Schleif, H. (1944) *IV. Bericht über die Ausgrabungen in Olympia.* Berlin.

Künzl, E. and Künzl, S. (1992) Aquae Apollinares/Vicarello (Italien), in Chevallier (ed.), pp. 273–96.

Kuttner, A. (1999) Culture and history at Pompey's Museum. *TAPA* 129, 343–73.

Kuziščin, V.I. (1984) *La grande proprietà agraria nell'Italia romana.* Rome.

La Regina, A. (1975) Centri fortificati pre-Romani nei territori sabellici dell'Italia centrale adriatica. *Posebna Izdanja* 24, 271–82.

La Regina, A. (1976) Il Sannio, in Zanker (ed.), pp. 219–54.

La Regina, A. (1980) Aspetti istituzionali nel mondo sannitico, in *Sannio*, pp. 17–25.

La Regina, A. (1981) Appunti su entità etniche e strutture istituzionali nel Sannio antico. *AION* 3, 120–37.

La Regina, A. (1984) 'Aspetti istituzionali nel mondo sannitico' in *Sannio: Pentri e Frentani dal VI al I sec. a.C. Atti del Convegno*, pp. 17–25. Campobasso.

La Regina, A. (1989) I Sanniti, in Pugliese-Carratelli (ed.), pp. 301–434.

La Regina, A. (1991) C. Papius C.F. Mutilus Imp., in Capini and Di Niro (eds), pp. 149–52.

La Regina, A. (1999) Istituzioni agrarie italiche, in Petrocelli (ed.), pp. 3–18.

La Regina, A. *et al.* (2009–10) *Pietrabbondante: Scavi archeologici dell'annno 2010. Rapporto preliminare.* Rome.

La Rocca, E. (1983/4) Fabio o Fannio. L'affresco medio-repubblicano dell'Esquilino come riflesso dell'arte "rappresentativa" e come espressione di mobilità sociale. *DialArch* 2, 31–53.

La Rocca, E. (1990) Linguaggio artistico e ideologia politica a Roma in età repubblicana, in Ampolo *et al.* (eds), 1990, pp. 289–498.

La Rocca, E. (2006) La tegola angolare di gronda con protome di Acheloo, in Carandini, D'Alessio and Di Giuseppe (eds), pp. 485–500.

La Rocca, E. (2011) La forza della tradizione: l'architettura sacra a Roma tra II e I secolo a.C., in La Rocca and D'Alessio (eds), pp. 1–26.

La Rocca, E. and D'Alessio, A. (eds) (2011) *Tradizione e innovazione: L'elaborazione e del linguaggio ellenistico nell'architettura romana e italic di età tardo-repubblicana* [*StMisc* 35]. Rome.

La Rocca, E., Presicce, C. and Lo Monaco, A. (eds) (2010) *I giorni di Roma. L'età della conquista.* Milan.

La Rocca, E. and Tortorella, S. (eds) (2008) *Trionfi romani.* Rome.

La Romanisation du Samnium aux II° et I° Siècles av. J.-C.: Actes du colloque organisé par le Centre Jean Bérard en collaboration avec la Soprintendenza archeologica e per i BAAAS del Molise et la Soprintendenza archeologica per le Province di Salerno, Avellino e Benevento, Naples, Centre Jean Bérard, 4–5 novembre 1988 (1991) Naples.

La Torre, G. (2011) Reflections on the Lucanian and Bruttians in Calabria between Hannibal and the Principate: *coloniae, civitates foederatae, municipia*, in Colivicchi (ed.), pp. 139–60.

Laffi, U. (2002) La colonización romana desde el final de la guerra de Aníbal a los Gracos, in Jiménez and Ribera (eds), pp. 19–26.

Lafon, X. (2001) *Villa Maritima: recherches sur les villas littorales de l'Italie romaine, IIIe siècle av. J.C.–IIIe siècle ap. J.-C.* Rome.

Lafon, X. and Sauron, G. (eds) (2005) *Théorie et pratique de l'architecture romaine. Études offertes à Pierre Gros.* Aix-en-Provence.

Lagia, A. (2000) Preliminary analysis of the human skeletal remains. *AA* 3, 481–93.

Lagia, A., Eliopoulos, C. and Manolis, S. (2007) Thalassemia: macroscopic and radiological study of a case. *International Journal of Osteoarchaeology* 10, 269–85.

Lagóstena, L. (2001) *La producción de salsas y conservas de pescado en la Hispania romana (II a.C.–VII d.C.)* [Instrumenta 11]. Barcelona.

Lagóstena, L., Bernal, D. and Arévalo, D. (eds) (2007) *Cetariae. Salsas y salazones en Occidente en la Antigüedad* [*BAR-IS* 1686]. Oxford.

Laidlaw, A. (1985) *The First Style in Pompeii: Painting and Architecture.* Rome.

Laidlaw, A. (2007) Mining the early published sources: problems and pitfalls, in Dobbins and Foss (eds), pp. 620–32.

Lamb, H.H. (1995) *Climate, History and the Modern World.* London.

Lamboglia, N. (1950) *Gli scavi di Albintimilium e la cronologia della ceramica romana: Parte prima, Campagne di scavo 1938–1940.* Bordighera.

Lamboglia N. (1952) Per una classificazione preliminare della ceramica campana, in *Atti del I Congresso Internazionale di Studi Liguri*, pp. 139–206. Bordighera.

Lamboglia, N. (1974) Il problema delle mura e delle origini di Tarragona. *Miscelánea Arqueológica* (XXV Aniversario de los cursos de Ampurias (1947–1971)), 397–405.

Lambot, B. (1997) Une monnaie d'Ebusus sur l'oppidum de "Vieux-Laon" à Saint-Thomas (Aisne). Nouvelles données archéologiques sur Bibrax et le camp de Mauchamp (César B.G. II). *Bulletin de la Société archéologique champenoise* 90 (2), 15–29.

Lancaster, L. (2005) *Concrete Vaulted Construction in Imperial Rome, Innovations in Context.* Cambridge.

Lancaster, L., Sottili, G., Marra, F.a d Ventura, G. (2011) Provenancing of lightweight volcanic stones used in ancient Roman concrete vaulting: evidence from Rome. *Archaeometry* 53, 949–61.

Lancel, S. (1995) *Carthage: A History.* Oxford.

Lanciani, R. (1875) Le antichissime sepolture Esquiline. *BullCom* 3, 41–56.

Lanciani, R.A. (1967) *Ancient Rome in the Light of Recent Discoveries* (repr.). New York.

Landau, Y. (1966) A Greek inscription found near Hefzibah. *IEJ* 16, 54–70.

Lapalus, E. (1939) *L'Agora des Italiens* [Exploration archéologique de Délos 19]. Paris.

Lassère, J.-M. (1977) *Ubique populus: peuplement et mouvements de population dans l'Afrique romaine de la chute de Carthage à la fin de la dynastie des Sévères (146 av.C.–235 ap.C.)*. Paris.

Laubenheimer, F. (1990a) *Le temps des amphores en Gaule, vins, huiles et sauces*. Paris.

Laubenheimer, F. (1990b) *Sallèles d'Aude, un complexe de potiers gallo-romain* [Documents d'Archéologie Française 26]. Paris.

Laubenheimer, F. (ed.) (2001) *20 ans de recherches à Sallèles d'Aude*. Besançon.

Laubenheimer, F. (2007) A propos de timbres d'amphores de l'atelier d'Albinia (prov. de Grosseto, Italie). Vin et poisson, in Vitali (ed.), pp. 67–80.

Laubenheimer, F. (2010) La circulation du vin de l'Italie tyrrhénienne en Gaule, in Ouzoulias and Tranoy (eds), pp. 47–57.

Laubenheimer, F. and Marlière, E. (2010) *Echanges et vie économique dans le Nord-Ouest des Gaules. Le témoignage des amphores du IIe s. avant J.-C. au IVe s. après J.-C.* Besançon.

Laubenheimer F., Ménez, Y. and Le Forestier S. (forthcoming) Les amphores de Paule (Côtes d'Armor) et le commerce du vin au second âge du fer dans le nord-ouest de la Gaule, in Olmer (ed.).

Launaro, A. (2011) *Peasants and Slaves: The Rural Population of Roman Italy (200 BC–AD 100)*. Cambridge.

Laurence, R. (1994) *Pompeii: Space and Society*. London.

Laurence, R. (1998) Territory, ethnonyms and geography: the construction of identity in Roman Italy, in Laurence and Berry (eds), pp. 95–110.

Laurence, R. (1999) *The Roads of Roman Italy. Mobility and Cultural Change*. London.

Laurence, R. (2001) Roman Italy's urban revolution, in Lo Cascio and Storchi-Marino (eds), pp. 593–611.

Laurence, R. and Berry, J. (eds) (1998) *Cultural Identity in the Roman Empire*. London.

Laurence, R. and Wallace-Hadrill, A. (eds) (1997) *Domestic Space in the Roman World: Pompeii and Beyond* [*JRA* Suppl. 22]. Portsmouth, RI.

Lauter, H. (1976) Die hellenistischen Theater der Samniten und Latiner in ihrer Beziehung zur Theater-architektur der Griechen, in Zanker (ed.), pp. 413–22.

Lauter, H. (1980–1) Porticus Metelli – Porticus Octaviae. Die baulichen Resten. *BullCom* 87, 37–46.

Lauter, H. (1986) *Die Architektur des Hellenismus*. Darmstadt.

Lavin, S. (1992) *Quatremère de Quincy and the Invention of a Modern Language of Architecture*. Cambridge, MA.

Lazzarini, M.L. and Poccetti, P. (2001) *Il mondo enotrio tra VI e V secolo a.C. L'iscrizione paleo-italica da Tortora* [*Ostraka* Quad. 1/2]. Naples.

Le Bohec, Y. (1992) Notes sur les mines de Sardaigne à l'époque romaine, in Bonello Lai (ed.), pp. 255–64.

Le Bohec, Y. (2005) *Histoire de l'Afrique romaine (146 avant J.-C.–439 après J.-C.)*. Paris.

Le Gall, J. (1953) *Le Tibre fleuve de Rome dans l'antiquité*. Paris.

Le Pera Buranelli, S. and Turchetti, R. (2003) *Sulla Via Appia da Roma a Brindisi. Le fotografie di Thomas Ashby (1891–1925)*. Rome.

Le Roux, P. (2006) L'Vrbs, les provinces et l'Empire de César à la mort de Commode. Autour de la notion capitale, in Belayche (ed.), pp. 231–66.

Le Roux, P. (2010) *La péninsule Ibérique aux époques romaines (fin du IIIe s. av. n.è.–début du VIe s. de n.è.)*. Paris.

Leach, E. (1997) Oecus on Ibycus: investigating the vocabulary of the Roman house, in Bon and Jones (eds), pp. 50–72.

Leach, E. (2004) *The Social Life of Painting in Ancient Rome and on the Bay of Naples*. Cambridge.

Leach, S., Lewis, M., Chenery, C. *et al.* (2009) Migration and diversity in Roman Britain: a multidisciplinary approach to the identification of immigrants in Roman York, England. *American Journal of Physical Anthropology* 140, 546–61.

Leen, A. (1991) Cicero and the Rhetoric of Art. *AJP* 112, 229–45.

Leenhardt, M. (2001) L'atelier de Sallèles d'Aude, fours et bâtiments: mode d'emploi, in Laubenheimer (ed.), pp. 241–56.

Legio VII Gemina (León, 1968) (1970) León.

Lehmann-Hartleben, K. (1923) *Die antiken Hafenanlagen des Mittelmeeres. Beiträge zur Geschichte des Städtebaues im Altertum* [*Klio* Beiheft 14]. Leipzig.

Lehmler, C. (2005) *Syrakus unter Agathokles und Hieron II. Die Verbindung von Kultur und Macht in einer hellenistischen Metropole*. Frankfurt-am-Main.

Leibner, U. (2009) *Settlement and History in Hellenistic, Roman and Byzantine Palestine*. Tübingen.

Lejeune, M. (1990) *Méfitis d'après le dédicaces lucaniennes de Rossano di Vaglio*. Louvain-la-Neuve.

Lemonnier, P. (ed.) (1993) *Technological Choices: Transformation in Material Culture since the Neolithic*. London.

León, M. (ed.) (1996) *Colonia Patricia Corduba. Una reflexión arqueológica (Córdoba, 1993)*. Cordoba.

Leone, A., Palombi, D. and Walker, S. (eds) (2007) *Res bene gestae: ricerche di storia urbana su Roma antica in onore di Eva Margareta Steinby* [*LTUR* Suppl. IV]. Rome.

L'Esculapi: El retorn de déu (2008) Barcelona.

Lethaby, W.R. (1908) *Greek Buildings Represented by Fragments in the British Museum*. London.

Letta, C. (1994) Dall'"oppidum" al "nomen." I diversi livelli dell'aggregazione politica nel mondo osco-umbro, in Aigner Foresti, Barzanò and Bearzot (eds), pp. 387–405.

Leveau, P. (2001) Aqueduct building, financing and costs, in Blackman and Hodge (eds), pp. 84–108.

Leveau, P., Walsh, K., Trément, F. and Barker, G. eds. (1999) *The Archaeology of Mediterranean Landscapes 2, Environmental Reconstruction in Mediterranean Landscape Archaeology*. Oxford.

Lévi-Strauss, C. (2001) *Race et histoire. Race et culture*. Paris.

Levick, B. (1978) Concordia at Rome, in Carson (ed.), pp. 217–33.

Levine, L. (ed.) (1992) *The Galilee in Late Antiquity*. New York.

Lewin, J., Macklin, M.G. and Woodward, J.C. (1995) *Mediterranean Quaternary River Environments*. Rotterdam.

Lewis, M. (2010) Thalassemia: its diagnosis and interpretation in past skeletal populations. *International Journal of Osteoarchaeology* DOI 10.1002/oa.1229.

Lewis, R. (ed. and trans.) (2007) *Asconius: Commentaries on Speeches of Cicero*. Oxford.

Ley, A. and Strauss, R. (1982) Gegenarchitektur. Das Heiligtum der Fortuna Primigenia als Symbol der politischen Selbstbehauptung Praenestes. *Hephaistos* 4, 117–38.

Lézine, A. (1968) *Carthage, Utique, études d'architecture et d'urbanisme*. Paris.

Lézine, A. (1970) *Utique*. Tunis.

Libertini, G. (ed.) (2005) *L'ipogeo di Caivano: Atti del convegno di Caivano del 7 ottobre 2004*. Frattamaggiore.

Lilli, M. (2002) *Ariccia*. Carta archeologica. Rome.

Lilli, M. (2006) I ponti tra il XIX e il XXVII miglio della Via Appia. *Analecta Romana Instituti Danici* 32, 27–41.

Ling, R. (2005) *Pompeii: History, Life and Afterlife*. Stroud.

Lintott, A. (1993) *Imperium Romanum. Politics and Administration*. London.

Lintott, A. (1999) *The Constitution of the Roman Republic*. Oxford.

Lippolis, E. (1987) Organizzazione delle necropoli e struttura sociale nell'Apulia ellenistica: Due esempi, Taranto e Canosa, in von Hesberg and Zanker (eds), pp. 139–54.

Lippolis, E. (1990) *Emergenze e problemi archeologici: Mandria, Taranto, Heraclea*. Mandra.

Lippolis, E. (2008) Il III sec. a.C., in Lippolis and Giammatteo (eds), pp. 323–6.

Lippolis, E. and Giammateo, T. (eds) (2008) *Salpia Vetus: Archeologia di una città lagunare*. Venosa.

Lipps, J. (2011) *Die Basilica Aemilia am Forum Romanum: Der kaiserzeitliche Bau und seine Ornamentik* [*Palili* 24]. Rome.

Little, L. (ed.) (2007) *Plague and the End of Antiquity. The Pandemic of 541–750*. Cambridge.

Liverani, P. (2008) Cesare urbanista, in Gentili (ed.), pp. 43–51.

Lloyd, G. *et al.* (1978) *Hippocratic Writings*. Middlesex.

Lloyd, J. (1991) Farming the highlands. Samnium and Arcadia in the Hellenistic and early Roman Imperial periods, in Barker and Lloyd (eds), pp. 180–93.

Lloyd, J. (1995) Pentri, Frentani and the beginnings of urbanization c.500–80 BC, in Barker (ed.), 1995b, pp. 181–212.

Lloyd, J. and Faustoferri, A. (1998) Monte Pallano: a Samnite fortified centre and its hinterland. *JRA* 9, 5–22.

Lloyd, S. (1967) *Early Highland Peoples of Anatolia*. London.

Lo Cascio, E. (1994) The size of the Roman population: Beloch and the meaning of the Augustan census figures. *JRS* 84, 23–40.

Lo Cascio, E. (2000) The population of Roman Italy in town and country, in Bintliff and Sbonias (eds), pp. 161–71.

Lo Cascio, E. (2001) Il *census* a Roma e la sua evoluzione dall'età "serviana" alla prima età imperiale. *MÉFRA* 113, 565–603.

Lo Cascio, E. (2003) La Spagna degli Annei, in Gualandri and Mazzoli (eds), pp. 9–18.

Lo Cascio, E. (ed.) (2003) *Credito e moneta nel mondo romano, Atti degli incontri capresi de storia dell'economia antica*. Bari.

Lo Cascio, E. (2006) Did the population of Imperial Rome reproduce itself?, in Storey (ed.), pp. 52–68.

Lo Cascio, E. (2009) Introduzione, in Carlsen and Lo Cascio (eds), pp. 7–14.

Lo Cascio, E. (2010) Thinking slave and free in coordinates, in Roth (ed.), pp. 21–30.

Lo Cascio, E. (ed.) (2012) *L'impatto della "peste antonina."* Bari.

Lo Cascio, E. and Malanima, P. (2005) Cycles and stability. Italian population before the demographic transition (225 BC–AD 1900). *Rivista di Storia Economica* XXI (3), 197–232.

Lo Cascio, E. and Malanima, P. (2009) GDP in pre-modern agrarian economies (1–1820 AD). A revision of the estimates. *Rivista di Storia Economica* XXV (3), 391–419.

Lo Cascio, E. and Storchi Marino, A. (eds) (2001) *Modalità insediative e strutture agrarie nell'Italia meridionale in età romana*. Bari.

Lock, G. and Faustoferri, A. (eds) (2008) *Archaeology and Landscape in Central Italy. Papers in Memory of John A. Lloyd*. Oxford.

Lockyear K. (2007) *Patterns and Processes in Late Roman Republican Coin Hoards [BAR-IS 1733]*. Oxford.

Lohman, H. (1992) Agriculture and country life in classical Attica, in Wells (ed.), pp. 29–60.

Lohmann, H. (1979) *Grabmäler auf unteritalischen Vasen*. Berlin.

Lomas, K. (1991) Local identity and cultural imperialism. Epigraphy and the diffusion of Romanisation in Italy, in Herring, Whitehouse and Wilkins (eds), pp. 231–9.

Lomas, K. (1993) *Rome and the Western Greeks. Conquest and Acculturation in Southern Italy*. London.

Lomas, K. (1996) Urban elites and cultural definition: Romanization in southern Italy, in Cornell and Lomas (eds), 1996, pp. 107–20.

Lomas, K. (2003) Public building, urban renewal and euergitism in Early Imperial Italy, in Lomas and Cornell (eds), pp. 28–45.

Lomas, K. (2004) Italy during the Roman Republic, 338–31 BC, in Flower (ed.), pp. 199–224.

Lomas, K. and Cornell, T. (eds) (2003) *Bread and Circuses: Euergetism and Municipal Patronage in Roman Italy*. London.

Lombardo, M. (2006) I paradossi dell'ellenizzazione da Pirro ad Annibale: ideologie e pratiche "ellenizzanti" nell'Italia meridionale di fronte all'espansione romana. *Pallas* 70, 15–26.

Long, L. and Picard, P. (2009) *César: Le Rhône pour mémoire*. Arles.

Longerstay, M. (1995) Les haouanet: état de la question, in Trousset (ed.), pp. 33–53.

López Mullor, A. (2002) Arqueología y restauración: el oppidum del Turó del Montgròs (El Brull, Barcelona), in González and Díez (eds), 2002, pp. 53–65.

Lostal, J. (1992) *Los miliarios de la provincia tarraconense*. Zaragoza.

Loughton, M. (2005) The Republican amphora stamps from the Auvergne, in *SFECAG, Actes du Congrès de Blois*, pp. 683–700. Marseille.

Loughton, M. and Jones, S. (2000) Les amphores républicaines en Auvergne (Puy de Dôme): importations et diffusion avant la conquête. *Revue Archéologique du Centre de la France* 39, 63–81.

Lovejoy, C., Meindl, R., Pryzbeck, T. and Mensforth, R. (1985) Chronological metamorphosis of the auricular surface of the ilium: a new method for the determination of adult skeletal age at death. *American Journal of Physical Anthropology* 68, 15–28.

Lowe, J.J., Davite, C., Moreno, D. and Maggi, R. (1994) Holocene pollen stratigraphy and human interference in the woodlands of the northern Apennines, Italy. *Holocene* 4 (2), 153–64.

Lucy, S. (1992) The significance of mortuary ritual in the political manipulation of landscape. *Archaeological Review from Cambridge* 11 (1), 93–105.

Lucy, S. (2002) *Burial Practice in Early Medieval Eastern England: Constructing Local Identities, Deconstructing Ethnicity*. Oxford.

Lugli, G. (1926) *Forma Italiae, Regio I.1, Latium et Campania, Ager Pomptinus: Anxur-Tarracina*. Rome.

Lugli, G. (1957) *La tecnica edilizia romana*. Rome.

Luik, M. (1997) Die römische Militäranlagen der Iberischen Halbinsel von der Zeit der Republik bis zum Ausgang des Prinzipats. Ein Forschungsüberblick. *JRGZM* 44, 213–75.

Luik, M. (2000) Republikanische Pilumfunde vom "Talamonaccio"/Italien. *ArchKorrBl* 30, 269–77.

Luik, M. (2005) *Der schwierige Weg zur Weltmacht. Roms Eroberung der iberischen Halbinsel 210–19 v. Chr.* Mainz am Rhein.

Luik, M. (2010) The findings of weapons in the Roman camps around Numancia. *Gladius* 30, 61–78.

Lulof, P.S. (2007) L'Amazone dell'Esquilino. Una nuova ricostruzione. *BullCom* 108, 7–31.

Luni, M. (ed.) (2002) *La Via Flaminia nell'Ager Gallicus*. Urbino.

Luzón Nogué, J.M. *et al.* (2011) Casa della Diana Arciazante Pompeya, http://www.dianaarcaizante.com/#/escanometrias (accessed November 17, 2012).

Ma, J. (2002) *Antiochus III and the Cities of Asia Minor*. Oxford.

MacBain, B. (1980) Appius Claudius Caecus and the Via Appia. *CQ* 30, 356–72.

Macchiarelli, R. and I. Salvadei (1986) Topographic distribution of the maxillary carious lesions at Alfedena (VI–V cent. BC, central Italy). *Antropologia Contemporanea* 9, 201–6.

MacDougall, E. (ed.) (1987) *Ancient Roman Villa Gardens*. Washington DC.

MacDougall, E. and Jashemski, W. (eds) (1981) *Ancient Roman Gardens*. Washington DC.

MacIsaac, J. (1995) Coins and the field archaeologist: numismatic finds as artifacts. *ArchNews* 20, 21–5.

MacKinnon, M. (2001) High on the hog: zooarchaeological, literary, and artistic data for pig breeds in Roman Italy. *AJA* 105 (4), 649–74.

MacKinnon, M. (2007) State of the discipline: osteological research in classical archaeology. *AJA* 111, 473–504.

MacMullen, R. (2000) *Romanization in the Time of Augustus*. New Haven.

Macready, S. and Thompson, F. (eds) (1985) *Archaeological Field Survey in Britain and Abroad*. London.

Macready, S. and Thompson, F. (eds) (1987) *Roman Architecture in the Greek World*. London.

Maddison, A. (2007) *Contours of the World Economy, 1–2030 AD. Essays in Macro-economic History*. Oxford.

Maddoli, G. (ed.) (1988) *Strabone e l'Italia antica*. Perugia.

Magallón, M. and Navarro, M. (2010) Las ciudades romanas en la zona central y occidental del Pirineo meridional veinte años después. *Pallas* 82 (*Ab Aquitania in Hispaniam. Mélanges offerts à Pierre Sillières*), 223–53.

Magistrat der Stadt Wien (ed.) (2004) *Enter the Past: The E-way into the Four Dimensions of Cultural Heritage* [*BAR-IS* 1227]. Oxford.

Magness, J. (2002) *The Archaeology of Qumran and the Dead Sea Scrolls*. Grand Rapids.

Maiuri, A. (1930) Studi e ricerche sulla fortificazione di Pompeii. *MonAnt* 33, 113–286.

Maiuri, A. (1934) *I Campi Flegrei; dal Sepolcro di Virgilio all'antro di Cuma*. Rome.

Maiuri, A. (1943) Greci ed Etruschi a Pompei. *Atti della Reale Accademia d'Italia. Memorie* 7 (4), 121–49.

Maiuri, A. (1951) Pompei: Saggi nella cavea del "Teatro grande." *NSc* 5, 126–34.

Maiuri, A. (1973) *Alla ricerca di Pompei preromana*. Naples.

Malagodi, S., Orlando, L., Piro, S. and Rosso, F. (1996) Location of archaeological structures using GPR method: three-dimensional data acquisition and radar signal processing. *Archaeological Prospection* 3, 13–23.

Malfitana, D. (2006) Metodologie, problemi e prospettive di ricerca negli studi di ceramologia ellenistica e romana in Sicilia. Un planning per future ricerche tra archeologia e archeometria, in Malfitana, Poblome and Lund (eds), pp. 399–421.

Malfitana, D. (ed.) (2008) Roman Sicily Project ("RSP"): ceramics and trade. A multidisciplinary approach to the study of material culture assemblages. First overview: the transport amphorae evidence. *Facta. A Journal of Roman Material Culture Studies* 2, 127–92.

Malfitana, D. (2011) The view from the material culture assemblage of Late Republican Sicily, in Colivicchi (ed.), pp. 185–201.

Malfitana, D., Poblome, J. and Lund, J. (eds) (2006) *Old Pottery in a New Century: Innovating Perspectives on Roman Pottery Studies*. Catania.

Mallwitz, A. (1972) *Olympia und seine Bauten*. Munich.

Malone, C. and Stoddart, S. (eds) (1985) *Papers in Italian Archaeology IV.1* [*BAR-IS* 243]. Oxford.

Manacorda, D. (1978) The *Ager Cosanus* and the production of the amphorae of Sestius: new evidence and a reassessment. *JRS* 68, 122–31.

Manacorda, D. (1979) Considerazioni sull'epigrafia della regione di Cosa. *Athenaeum* 57, 73–97.

Manacorda, D. (1981) Produzione agricola, produzione ceramica e proprietari nell'ager Cosanus nel I sec. A.C., in Giardina and Schiavone (eds), vol. II, pp. 55–80.

Mancini, G. (1912) Roma: nuove scoperte nella città e nel suburbio. *NSc* 265–72.

Manderscheid, H. (1988) Römische Thermen. Aspekte von Architektur, Technik und Ausstattung, in Garbrecht (ed.), pp. 99–125.

Manderscheid, H. (2004) *Ancient Baths and Bathing: A Bibliography for the Years 1988–2001* [*JRA* Suppl. 5]. Portsmouth, RI.

Mandy, B. *et al.* (1989) Les fossés du plateau de la Sarra, in Goudineau (ed.), pp. 35–95.

Manning. S. (ed.) (2003) *The Late Roman Church at Maroni Petrera; Survey and Salvage Excavations 1990–1997, and Other Traces of Roman Remains in the Lower Maroini Valley, Cyprus.* London.

Manning, W.H. (1971) The Piercebridge Plough Group. *BMQ* 35 (1/4), 125–36.

Manuwald, G. (2011) *Roman Republican Theatre.* Cambridge.

Mar, R. and Ruiz de Arbulo, J. (1993) *Ampurias romana. Historia, arquitectura y arqueología.* Sabadell.

Marabini Moevs, M.T. (1973) *The Roman Thin Walled Pottery from Cosa (1948–1954)* [*MAAR* 32].

Marabini Moevs, M.T. (1980a) Italo-Megarian ware at Cosa. *MAAR* 34, 161–227.

Marabini Moevs, M.T. (1980b) Aco in northern Etruria: the workshop of Cusonius at Cosa. *MAAR* 231–80.

Marabini Moevs, M.T. (2006) *Cosa: The Italian Sigillata* [*MAAR* Suppl. 3]. Ann Arbor.

Marc, J.-Y. and Moretti, J.-C. (eds) (2001) *Constructions publiques et programmes édilitaires en Grèce entre le IIe siècle av. J.-C. et le Ier ap. J.-C. Acts du colloque organisé par l'École Française d'Athènes 14–17 Mai 1995* [*BCH* Suppl. 39]. Paris.

Marcattili, F. (2009) *Circo Massimo: Architetture, funzioni, culti, ideologia.* Rome.

Marchand, S. (2000) The quarrel of the ancients and moderns in the German museums, in Crane (ed.), pp. 179–99.

Marchegay, S., Le Dinahet, M.-T. and Salles, J.-F. (eds) (1998) *Nécropoles et pouvoir: Idéologies, pratiques et interprétations.* Lyon.

Marchi, M.L. (1999) Il comprensorio venosino: documenti per l'analisi del processo di romanizzazione, in *La Daunia romana: Città e territorio della romanizzazione all'età imperiale, Atti del Convegno Nazionale di Preistoria, Protostoria, Storia della Daunia (S. Severo 6–8 dicembre 1996)*, pp. 111–28. San Severo.

Marchi, M. (2004) Fondi, latifondi e proprietà imperiale nell'ager Venusinus. *Agri Centurati* I, 129–56.

Marchi, M. (2008) Paesaggi rurali della Daunia interna: Il complesso in località "La Santissima" (Spinazzola-Bari), in Gravina (ed.), pp. 425–42.

Marchi, M. and Sabbatini, G. (1996) *Venusta. Forma Italia.* Firenze.

Margaritis, E. and Jones, M. (2008) Greek and Roman agriculture, in Oleson (ed.), pp. 158–74.

Marín Díaz, M.A. (1988) *Emigración, colonización y municipalización en la Hispania republican.* Granada.

Mariotti Lippi, M., Bellini, C. and Mori Secci, M. (2010) Palaeovegetational reconstruction based on pollen and seeds/fruits from a Bronze Age archaeological site in Tuscany (Italy). *Plant Biosystems* 144 (4), 902–8.

Mariotti Lippi, M., Di Tommaso, P.L., Giachi, G. *et al.* (2003) Archaeobotanical investigations into an Etruscan farmhouse at Pian d'Alma (Grosseto, Italy). *Atti della Societa' Toscana di Scienze Naturali (Serie B)* 109, 159–65.

Marquardt, J. (1881) *Römische Staatsverwaltung* (2nd edn). Leipzig.

Marra, F., Karner, D. B., Freda, C. *et al.* (2009) Large mafic eruptions at Alban Hills Volcanic District (Central Italy): chronostratigraphy, petrography and eruptive behavior. *Journal of Volcanology and Geothermal Research* 179, 217–32.

Marra, F. and Rosa, C. (1995) Statigrafia e assetto geologico dell'area romana, in Funiciello (ed.), pp. 50–118.

Marshall, C.W. (2006) *The Stagecraft and Performance of Roman Comedy.* Cambridge.

Martelli, A. (2002) Per una nuova lettura dell'iscrizione Vetter 61 nel contesto del santuario di Apollo a Pompei. *Eutopia* 2, 71–81.

Martí, R. (ed.) (2008) *Fars de l'Islam. Antigues alimares d'Al-andalus (Barcelona-Bellaterra 2006).* Barcelona.

Martin, A.M. (2000) Las termas republicanas de Cabrera de Mar (Maresme, Barcelona), in Fernández Ochoa and García-Entero (eds), pp. 157–62.

Martin, A.M. and García, J. (2002) La romanización del territorio de los layetanos y la fundación de la ciudad romana de Iluro (Hispania Tarraconensis), in Jiménez and Ribera (eds), pp. 195–204.

Martin, Au., Buxó, R., López, J. and Mataró, M. (eds) (1999) *Excavacions arqueològiques a l'Illa d'en Reixac (1987–1992)*. Girona.

Martin, Au. and Plana, R. (eds) (2001) *Territori polític i territori rural durant l'Edat del Ferro a la Mediterrània Occidental [Actes de la Taula Rodona celebrada a Ullastret, 2000]*. Girona.

Martin, H.G. (1987) *Römische Tempelkultbilder. Eine archäologische Untersuchung zur spaten Republik*. Rome.

Martin-Kilcher, S. (1998) Gräber der späten Republik und der frühen Kaiserzeit am Lago Maggiore: Tradition und Romanisierung, in Fasold *et al.* (eds), pp. 191–252.

Martínez Gázquez, J. (1992) *La campaña de Catón en Hispania*. Barcelona.

Martini, K. (1998) Patterns of reconstruction at Pompeii. Institute for Advanced Technology in the Humanities, University of Virginia, http://www2.iath.virginia.edu/struct/pompeii/patterns/sec-00.html#toc (accessed November 21, 2012).

Martins, M. (2006) Bracara Augusta. A Roman town in the Atlantic area, in Abad, Keay and Ramallo (eds), pp. 213–22.

Marzano, A. (2007) *Roman Villas in Central Italy: A Social and Economic History*. Leiden.

Maselli Scotti, F., Mandruzzato, L. and Tiussi, C. (2007) Primo impianto coloniario di Aquileia: l'aerea tra foro e macellum, in Brecciaroli Taborelli (ed.), pp. 35–40.

Massa Pairault, F.-H. (ed.) (1990) *Crise et transformation des sociétés archaiques de l'Italie antique au Ve siècle av. J.-C.* Paris.

Massa Pairault, F.-H. (1992) *Iconologia e politica nell'Italia antica. Roma, Lazio, Etruria dal VII al I secolo a.C.* Milan.

Massó, J. and Sada, P. (eds) (1996) *Actas de la II Reunión sobre escultura romana en Hispania (Tarragona, 1995)*. Tarragona.

Mastelloni, M.A. (2005) Messana e i Mamertini, in Ghedini *et al.* (eds), pp. 275–92.

Masterman, E. (1908) Two Greek inscriptions from Khurbet Harrawi. *PEQ* 41–2, 40–3.

Mastino, A. (2008) Il *dibattito* sull'agorà degli Italici a Delo: un bilancio retrospettivo fra ideologia e urbanistica, in *Le perle e il filo: a Mario Torelli per i suoi settanta anni*, pp. 233–42. Venosa.

Mastino, A. (2009) *Storia della Sardegna antica (La Sardegna e la sua storia*, vol. II) (2nd edn). Nuoro.

Mastrocinque, A. (ed.) (2009) *Grumentum Romana*. Moliterno.

Mateos, P. (ed.) (2006) *El "foro provincial" de Augusta Emerita: un conjunto monumental de culto imperial* [Anejos de *ArchEsp* XLII]. Madrid.

Mateos, P., Celestino, S., Pizzo, A. and Tortosa, T. (eds) (2009) *Santuarios, oppida y ciudades: arquitectura sacra en el origen y desarrollo urbano del Mediterráneo occidental* [Anejos de *ArchEsp* XLV]. Madrid.

Mathers, C., Sadana, R., Salomon, J. *et al.* (2001) Healthy life expectancy in 191 countries, 1999. *The Lancet* 357, 1685–91.

Mattei, M. (2001) Decreto de Gneo Pompeyo Estrabón, in Rodà (ed.), p. 71.

Mattingly, D.J. (1995) *Tripolitania*. London.

Mattingly, D.J. (1996a) First fruit? The olive in the Roman world, in Shipley and Salmon (eds), pp. 213–53.

Mattingly, D.J. (1996b) From one colonialism to another: imperialism and the Maghreb, in Webster and Cooper (eds), pp. 49–69.

Mattingly, D.J. (1997) Africa, a landscape of opportunity?, in Mattingly (ed.), pp. 117–39.

Mattingly, D.J. (ed.) (1997) *Dialogues in Roman Imperialism: Power, Discourse, and Discrepant Experience in the Roman Empire* [*JRA* Suppl. 23]. Portsmouth, RI.

Mattingly, D.J. (2000) Methods of collection, recording and quantification, in Francovich and Patterson (eds), pp. 5–15.

Mattingly, D.J. (2002) Vulgar and weak "Romanization," or time for a paradigm shift? *JRA* 15, 536–40.

Mattingly, D. (2004) Being Roman: expressing identity in a provincial setting. *JRA* 17, 4–25.

Mattingly, D.J. (2009) Cultural crossovers: global and local identities in the Classical world, in Hales and Hodos (eds), pp. 283–95.

Mattingly, D.J. (2010) *Imperialism, Power and Identity: Experiencing the Roman Empire*. Princeton.

Mattingly, D.J. and Witcher R. (2004) Mapping the Roman world: the contribution of field survey data, in Alcock and Cherry (eds), pp. 173–86.

Mattingly, H. and Robinson, E. (1933) The date of the Roman denarius, and other landmarks in early Roman coinage. *ProcBritAc* 18, 211–68.

Mattusch, C. (ed.) (2008) *Pompeii and the Roman Villa: Art and Culture around the Bay of Naples*. London.

Mattusch, C., Donohue, A. and Brauer, A. (eds) (2006) *Proceedings of the XVIth International Congress of Classical Archaeology, Boston, August 23–26, 2003. Common Ground: Archaeology, Art, Science, and Humanities*. Oxford.

Matyszak, P. (2009) *The Roman Conquests: Macedonia and Greece*. Barnsley.

Mau, A. (1879) *Pompejanische Beitrage*. Berlin.

Mau, A. (1882) *Geschichte der decorativen Wandmalerei in Pompeji*. Berlin.

Mau, A. (1906) Das grosse Theater in Pompeji. *RM* 21, 1–56.

Mau, A. (1908) *Pompeji in Leben und Kunst*. Leipzig.

Mayor, A. (2010) *The Poison King. The Life and Legend of Mithradates, Rome's Deadliest Enemy*. Princeton.

Mays, S. and Faerman, M. (2001) sex identification in some putative infanticide victims from Roman Britain using ancient DNA. *JAS* 28, 555–9.

Maxfield, V. and Dobson, M. (eds) (1991) *Roman Frontier Studies 1989*. Exeter.

Mazier, F., Galop, D., Brun, C. and Buttler, A. (2006) Modern pollen assemblages from grazed vegetation in the western Pyrenees, France: a numerical tool for more precise reconstruction of past cultural landscapes. *Holocene* 16, 91–103.

Mazor, G. and Najjar, A. (2007) *Nysa-Scythopolis: The Caesareum and The Odeum*. Jerusalem.

McCall, J. (2001) *The Cavalry of the Roman Republic*. London.

McCann, A.M. (1987) *The Roman Port and Fishery of Cosa*. Princeton.

McCann, A.M. (2002) *The Roman Port and Fishery of Cosa: A Short Guide*. Rome.

McCann, A.M., Bourgeois, J., Gazda, E.K. *et al.* (1987) *The Roman Port and Fishery of Cosa: A Center of Ancient Trade*. Princeton.

McCann, A.M., Bourgeois, J. and Will, E.L. (1977) Underwater excavations at the Etruscan port of Populonia. *JFA* 4 (3), 275–96.

McClellan, A. (1994) *Inventing the Louvre. Art, Politics, and the Origins of the Modern Museum in Eighteenth-Century Paris*. Cambridge.

McDonald, M. and Walton, M. (eds) (2007) *The Cambridge Companion to Greek and Roman Theatre* [Cambridge Companion to Literature]. Cambridge.

McGowan, M. (2000) *The Vision of Rome in Late Renaissance France*. New Haven.

McGuire, W.J., Griffiths, D.R., Hancock, P.L. and Stewart, I.S. (eds) (2000) *The Archaeology of Geological Catastrophes* [Geological Society Special Publication 171]. London.

McKay, A.G. (1975) *Houses, Villas and Palaces in the Roman World*. London.

McKenzie, J. (2007) *The Architecture of Alexandria and Egypt 300 BC–AD 700*. New Haven.

McKern, T. and Stewart, T. (1957) *Skeletal Age Changes in Young American Males, Analyzed from the Standpoint of Identification* [Technical Report EP–45. Headquarters, Quartermaster Research and Development Command]. Natick.

Medri, M., Soricelli G. and Benini, A. (1999) In Baiano sinu: le Piccole Terme di Baia, in DeLaine and Johnston (eds), vol. 2, pp. 207–19.

Meiggs, R. (1973) *Roman Ostia*. Oxford.

Meiggs, R. and Lewis, D. (1969) *A Selection of Greek Historical Inscriptions to the End of the Fifth Century BC*. Oxford.

Meindl, R., Lovejoy, C., Mensforth, R. and Carlos, L. (1985) Accuracy and direction of error in sexing of the skeleton: implications for paleodemography. *American Journal of Physical Anthropology* 68, 29–45.

Meinel, R. (1979) *Das Odeion: Untersuchungen an Überdachten Antiken Theatergebäuden*. Frankfurt am Main.

Meister, F. (2007) *Der Krieg des Sertorius und seine spanischen Wurzeln. Untersuchungem zu Krieg und Akkulturation auf der Iberischen Halbinsel im 2. und 1. Jh. v. Chr.* Hamburg.

Melchor, E., Mellado, J. and Rodríguez Neila, J. (eds) (2005) *Julio César y Corduba: tiempo y espacio en la campaña de Munda (49–45 a.C.) (Córdoba 2003)*. Córdoba.

Mele, A. (1991) I Lucani, in Galasso and Romeo (eds), vol. 1.2, pp. 273–80.

Meloni, P. (1988) La Sardegna romana. I centri abitati e l'organizzazione municipale, in *ANRW* II.11.1, pp. 491–551. Berlin and New York.

Meloni, P. (1990) *La Sardegna romana* (2nd edn). Sassari.

Menichetti, M. (1994) *Archeologia del potere. Re, immagini e miti a Roma e in Etruria in età arcaica*. Milan.

Mercado, M., Rodrigo, E., Flórez, M. *et al.* (2008) El castellum de Can Tacó/Turó d'en Roïna (Montmeló-Montornès del Vallès, Vallès Oriental) i el seu entorn territorial, in *Tribuna d'Arqueologia 2007*, pp. 195–211. Barcelona.

Mercando, L. (ed.) (1998) *Archeologia in Piemonte. L'età romana*, vol. II. Turin.

Mercuri, A.M., Florenzano, A., Massamba N'siala, I. *et al.* (2010) Pollen from archaeological layers and cultural landscape reconstruction: case studies from the Bradano Valley (Basilicata, southern Italy). *Plant Biosystems* 144 (4), 888–901.

Meritt, L.S. and Edlund-Berry, I. (2000) *Etruscan and Republican Roman Mouldings*. Philadelphia.

Merryman, J.H. (2000) *Thinking About the Elgin Marbles. Critical Essays on Cultural Property, Art and Law*. London.

Merryman, J.H. and Elsen, A. (1998) *Law, Ethics, and the Visual Arts* (3rd edn). Philadelphia.

Merryweather, A. and Prag, J. (2003) *Romanization?* [*Digressus* Suppl. 1].

Mertens, D. and De Siena, A. (1982) Metaponto: il Teatro-Ekklesiasterion, parte I. *Bolletino d'Arte del Ministerio della Pubblica Istruzione* 16, 1–57.

Mertens, J. and de Ruyt, F. (1969) *Alba Fucens*, 3 vols. Brussels.

Mertens, J. and Lambrechts, R. (eds) (1991) *Comunità indigene e problemi della romanizzazione nell'Italia centro-meridionale, IV–III secolo a.C.* Brussels.

Meskell, L. (1998) *Archaeology under Fire: Nationalism, Politics and Heritage in the Eastern Mediterranean and Middle East*. London.

Meskell, L. (2001) Archaeologies of identity, in Hodder (ed.), pp. 187–213.

Messenger, P.M. (1989) *The Ethics of Collecting, Cultural Property: Whose Culture? Whose Property?* Albuquerque.

Messineo, G. (1987–8) Località Casale Ghella (circ. XX). *BullCom* 92 (2), 504–9.

Messineo, G., Petracca, L. and Vigna, L. (1985) Via Cassia: Km 11. Località Casale Ghella (circ. XX). *BullCom* 90 (1), 177–84.

Metzler, J., Millett, M., Roymans, N. and Slofstra, J. (eds) (1995) *Integration in the Early Roman West*. Luxembourg.

Metzler, J., Waringo, R., Bis, R. and Metzler-Zens, N. (1991) *Clémency et les tombes de l'artistocratie en Gaule Belgique*. Luxembourg.

Meyboom, P. (1995) *The Nile Mosaic of Palestrina. Early Evidence of Egyptian Religion in Italy.* Leiden.

Meyers, E. (ed.) (1999) *Galilee through the Centuries: Confluence of Cultures.* Winona Lake, IN.

Miccichè, C., Modeo, S. and Santagati, L. (eds) (2007) *La Sicilia romana tra Repubblica e Alto Impero* [Atti del Convegno di Studi, Caltanissetta 20–21 maggio 2006]. Caltanissetta.

Michelini, C. (ed.) (2006) *Guerra e pace in Sicilia e nel Mediterraneo antico (VIII–III sec. a.C.). Arte, prassi e teoria della pace e della guerra, vol. II* [Atti delle quinte giornate internazionali di studi sull'area elima e la Sicilia occidentale nel contesto mediterraneo. Erice, 12–15 ottobre 2003]. Pisa.

Mielsch, H. (1987) *Die römische Villa: Architektur und Lebensform.* Munich.

Miles, M.M. (2002) Cicero's prosecution of Gaius Verres: a Roman view of the ethics of acquisition of art. *International Journal of Cultural Property* 11, 28–49.

Miles, M.M. (2008) *Art as Plunder. The Ancient Origins of Debate about Cultural Property.* Cambridge.

Miles, M.M. (2011) Still in the aftermath of Waterloo: a brief history of decisions about restitution, in Stone (ed.), pp. 29–42.

Miles, M.M. (ed.) (2011) *Cleopatra, A Sphinx Revisited.* Berkeley.

Millán León, J. (1989) *Ilipa Magna.* Alcalá del Río.

Millar, F. (1993) *The Roman Near East: 31 BC–337 AD.* Cambridge, MA.

Miller, A. (2001) *Roman Triumphs and Early Modern English Culture.* New York.

Miller, D. (1985) *Artefacts as Categories. A Study of Ceramic Variability in Central India.* Cambridge.

Miller, D. and Woodward, S. (2007) Manifesto for a study of denim. *Social Anthropology* 15, 335–51.

Miller, M. and DeVoto, J. (1994) *Polybius and Pseudo-Hyginus: The Fortification of the Roman Camp.* Chicago.

Miller, S.G. (1973) The Philippeion and Macedonian Hellenistic architecture. *AM* 88, 189–218.

Miller, S.G. (1995) Architecture as evidence for the identity of the early polis, in Hansen (ed.), pp. 201–42.

Millett, M. (1990a) *The Romanization of Britain.* Cambridge.

Millett, M. (1990b) Romanization: historical issues and archaeological interpretation, in Blagg and Millett (eds), pp. 35–41.

Millett, M. (2000) Dating, quantifying and utilizing pottery assemblages from surface survey, in Francovich and Patterson (eds), pp. 53–9.

Milner, N. (1993) *Vegetius: Epitome of Military Science.* Liverpool.

Minetti, A. (2006) *La Tomba della Quadriga Infernale nella necropoli della Pianacce di Sarteano.* Rome.

Mingazzini, P. (1949) Cagliari: resti di santuario punico e di altri ruderi a monte di Piazza del Carmine. *NSc* 213–74.

Mirabella Roberti, M. (ed.) (1997) *Monumenti sepolcrali romani in Aquileia e nella Cisalpina: Atti della 26 Settimana di studi aquileiesi, 24–28 aprile 1995.* Trieste.

Mitchell, T.N. (1979) *Cicero, the Ascending Years.* New Haven.

Mitens, K. (1988) *Teatri greci e teatri ispirati all'architettura greca in Sicilia e nell'Italia meridionale [AnalRom 13].* Rome.

Moatti, C. (1997) *La raison de Rome. Naissance de l'esprit critique à la fin de la République (IIe–Ier siècle avant J.-C.).* Paris.

Moatti, C. (ed.) (2004) *La mobilité des personnes en Méditerranée, de l'antiquité à l'époque moderne. Procédures de contrôle et documents d'identification. II: La mobilité négociée.* Rome.

Modes de contacts et processus de transformation dans les sociétés anciennes: actes du colloque de Cortone (24–30 mai 1981) organisé par la Scuola normale superiore et l'École française de

Rome avec la collaboration du Centre de recherches d'histoire ancienne de l'Université de Besançon (1983) Pisa.

Molin, D., Castenetto, S., Di Loreto, E. *et al.* (1995) Sismicità di Roma, in Funiciello (ed.), pp. 331–408.

Molina, J. and Valdivieso, A. (2007) Aportaciones sobre la evolución de las murallas de la Córdoba romana a partir de los datos arqueológicos. *Romula* 6, 29–50.

Molist, N. (2000) L'oppidum cossetà d'Olèrdola. L'etapa ibèrica d'un assentament d'ocupació continuada, in Buxó and Pons (eds), pp. 91–106.

Momigliano, A. and Schiavone, A. (eds) (1989) *Storia di Roma*. Turin.

Mommsen, T. (1854–6) *Römische Geschichte*, 3 vols. Leipzig.

Mommsen, T. (1885) *Römische Geschichte: Band 5, Die Provinzen von Caesar bis Diocletian*. Berlin.

Mommsen, T. (1893) *Abriss des römischen Staatsrechts*. Leipzig.

Mommsen, T. (1900) Praetorium. *Hermes* 35, 437–42.

Monteil, M. (1999) *Nîmes antique et sa proche campagne* [Monographies d'Archaéologie Méditerranéenne 3]. Lattes.

Montepaone, C. (ed.) (1996) *L'incidenza del'antico. Studi in memoria di Ettore Lepore*, 3 vols. Naples.

Monterroso, A. (2004) La influencia de los teatros de Roma en la Provincia Baetica: el teatro augusteo de Córdoba, in Burri *et al.* (eds), pp. 279–97.

Monterroso, A. (2007) Forma Urbis y Theatrum Pompei: El fragment 39F de la Planta Marmórea Severiana. *BullCom* 118, 125–44.

Monterroso, A (2009) Via Triumphalis per Theatrum Marcelli, símbolos de arquitectura en la Forma Urbis Marmorea. *RA* 47, 3–51.

Monterroso, A. (2010) *Theatrum Pompei: Forma y arquitectura de la génsis del mondo teatral de Roma*. Madrid.

Moore, A., Taylor, G., Harris, E. *et al.* (eds) (2010) *TRAC 2009. Proceedings of the Nineteenth Annual Theoretical Roman Archaeology Conference. Michigan and Southampton 2009*. Oxford.

Moorman, E. (ed.) (1993) *Functional and Spatial Analysis of Wall Painting* [*BABesch* Suppl. 3]. Leiden.

Mora, G. (2006) Roman military archaeology in Spain: a history of research, in Morillo and Aurrecoechea (eds), pp. 11–24.

Morales, F. (2005) Los campamentos y fuertes romanos del asedio de Numancia, in Pérez González and Illarregui (eds), pp. 251–8.

Moreau, C. and Válek, J. (eds) (2010) *Proceedings of the Second Historic Mortars Conference (HMC 2010)*. Prague.

Morel, J. (1991) Tents or barracks?, in Maxfield and Dobson (eds), pp. 376–86.

Morel, J.-P. (1963) Notes sur la céramique étrusco-campanienne. Vases à vernis noir de Sardaigne et de Arezzo. *MÉFRA* 75, 7–58.

Morel, J.-P. (1965) *Céramique à vernis noir du Forum Romain et du Palatin*. Rome.

Morel, J.-P. (1969) L'atelier des petites estampilles. *MÉFRA* 81, 59–117.

Morel, J.-P. (1981a) *Céramique campanienne: les formes*, 2 vols. Rome.

Morel, J.-P. (1981b) La produzione della ceramica campana: aspetti economici e sociali, in Giardina and Schiavone (eds), vol. II, pp. 81–97.

Morel, J.-P. (1989) The transformation of Italy, 300–133 BC. The evidence of archaeology, in *CAH VIII* (2nd edn), pp. 477–516. Cambridge.

Morel, J.-P. (1998a) L'étude des céramiques à vernis noir, entre archéologie et archéometrie, in Frontini and Grassi (eds), pp. 9–22.

Morel, J.-P. (1998b) Le ceramiche a vernice nera del Piemonte: tipologia, storia, cultura, in Mercando (ed.), pp. 235–52.

Morel, J.-P. (2007) Early Rome and Italy, in Schiedel, Morris and Saller (eds), pp. 487–510.

Morel, J.-P. (2009) Céramique à vernis noir et histoire. *JRA* 22, 477–88.

Morelli, C. (1997) Dalla devozione popolare alla "follia terapeutica": le terrecotte votive, in Campanelli and Faustoferri (eds), pp. 89–98.

Morelli, C., Carbonara, A., Forte, V. *et al.* (2008) The landscape of the Ager Portuensis, Rome: some new discoveries, 2000–2002, in Lock and Faustoferri (eds), pp. 213–31.

Morena, J. (1989) *El santuario ibérico de Torreparedones. Castro del Río-Baena, Córdoba*. Córdoba.

Moret, P. (1996) *Les fortifications ibériques de la fin de l'âge du bronze à la conquête romaine*. Paris.

Moret, P. (2002/3) Los monarcas ibéricos en Polibio y Tito Livio. *Cuadernos de Prehistoria y Arqueología, Universidad Autónoma de Madrid* 28/9, 23–33.

Moret, P. (2003) Fortifications ibériques tardives et défense du territoire en Hispanie Citérieure, in Morillo, Cadiou and Hourcade (eds), pp. 159–84.

Morgan, C. (2009) Ethnic expression on the Early Iron Age and Early Archaic Greek mainland. Where should we be looking?, in Derks and Roymans (eds), pp. 11–36.

Morgan, G. (1971) The Portico of Metellus: a reconsideration. *Hermes* 99, 480–505.

Morichi, R., Paone, R., Rispoli, P. and Sampaolo, F. (2008) Nuova cartografia di Pompei, in Guzzo and Guidobaldi (eds), 2008, pp. 554–5.

Morillo, A. (ed.) (2002) *I Congresso de Arqueología Militar Romana en Hispania, celebrado en Segovia en 1998 [Gladius, Anejo 5]*. Madrid.

Morillo, A. (2006a) Preface, in Morillo and Aurrecoechea (eds), pp. 7–9.

Morillo, A. (2006b) The Roman army in Spain, in Morillo and Aurrecoechea (eds), pp. 85–106.

Morillo, A. (ed.) (2007) *El ejército romano en Hispania. Guía arqueológica*. León.

Morillo, A. and Aurrecoechea, J. (eds) (2006) *The Roman Army in Hispania. An Archaeological Guide*. Léon.

Morillo, Á., Cadiou, F. and Hourcade, D. (eds) (2003) *Defensa y territorio en Hispania de los Escipiones a Augusto*. Madrid.

Morillo, A. and García-Marcos, V. (2002) Twenty years of Roman military archaeology in Spain, in Freeman, Bennett and Fiema (eds), vol. II, pp. 779–87.

Morillo, A., Hanel, N. and Martín, E. (eds) (2009) *Limes XX* (Acts of the XX International Congress of Roman Frontier Studies in León 2006, Anejos of *Gladius* 13), 3 vols. Madrid.

Morley, N. (1996) *Metropolis and Hinterland. The City of Rome and the Italian Economy 200 BC–AD 200*. Cambridge.

Morley, N. (2001) The transformation of Italy, 225–28 BC. *JRS* 91, 50–62.

Morpurgo, L. (1903) Nemus Aricium. *Monumenti Antichi XIII Reale Accademia Nazionale dei Lincei*, 297–368.

Morpurgo, L. (1931) Nemi – Teatro ed altri edifici romani in contrada "La Valle." *NSc* 237–305.

Moscati, S. (1986) *Le stele di Sulcis: caratteri e confronti*. Rome.

Mostalac, A. and Beltrán Lloris, M. (1994) *Colonia Victrix Iulia Lepida-Celsa (Velilla de Ebro. Zaragoza). II. Estratigrafía, pintura y cornisas de la "Casa de los Delfines."* Zaragoza.

Motta, L. (2002) Planting the seed of Rome. *Vegetation History and Archaeobotany* 11 (1–2), 71–7.

Motta, L. (2011) Seeds and the city. Archaeobotany and state formation in early Rome, in Terrenato and Haggis (eds), pp. 245–55.

Motta, L., Camin, L. and Terrenato, N. (1993) Un sito rurale nel territorio di Volterra. *Bollettino di Archeologia* 23–4, 109–16.

Mouritsen, H. (1988) *Elections, Magistrates, and Municipal Elite: Studies in Pompeian Epigraphy*. Rome.

Mouritsen, H. (1998) *Italian Unification. A Study in Ancient and Modern Historiography* [*BICS* Suppl. 7]. London.

Mouritsen, H. (2004) Pits and politics: interpreting colonial fora in Republican Italy. *PBSR* 72, 37–67.

Muensterberger, W. (1994) *Collecting, An Unruly Passion: Psychological Perspectives.* Princeton.

Munzi, M., Cifani, G. and Fusco, U. (1999/2000) Indagini topografiche nel suburbio di Grumentum: le dinamiche insediative. *ArchCl* 51, 439–59.

Munzi, M., Felici, F., Cifani, G. *et al.* (2004) A topographic research sample in the territory of Lepcis Magna: Silin. *Libyan Studies* 35, 11–66.

Munzi, M., Felici, F., Cifani, G. and Lucarini, G. (2004–5) Leptis Magna: città e campagna dall'origine alla scomparsa del sistema sedentario antico. *Scienze dell'Antichità* 12, 433–71.

Munzi, M., Felici, F., Cirelli, E. *et al.* (2010) Il territorio di Leptis Magna: ricognizioni tra Ras el-Mergheb e Ras el-Hammam (2007). *AfrRom* 18, 725–48.

Mura Sommella, A. (2000) "*La grande Roma dei Tarquini.*" Alterne vicende di una felice intuizione. *BullCom* 101, 7–26.

Mura Sommella, A. et al. (2001) Primi risultati dalle indagini archeologiche in Campidoglio nell'area del Giardino Romano e del Palazzo Caffarelli. *BullCom* 102, 261–364.

Murgatroyd, P.S. (2008) Appropriate levels of detail in 3-D visualisation: the House of the Surgeon, Pompeii. *Internet Archaeology* 23.

Murillo, J. (1994) Corduba: de la ciudad indígena a la romana, in *Actes del XIV Congrés*, pp. 303–4.

Murillo, J. and Jiménez, J. (2002) Nuevas evidencias sobre la fundación de Corduba y su primera imagen urbana, in Jiménez and Ribera (eds), pp. 183–93.

Murillo, J. and Vaquerizo, D. (1996) La Corduba prerromana, in León (ed.), pp. 37–47.

Murolo, M. (1959) Il considetto "Odeo" di Pompei ed il problema della sua copertura. *RendNap* 34, 89–101.

Murray, O. (1988) Death and the symposium. *AION* 10, 239–57.

Murray, O. and Tecusan, M. (eds) (1995) *In Vino Veritas.* Oxford.

Musti, D. (1980) La nozione storica di Sanniti nelle fonti greche e romane, in *Sannio*, pp. 71–84.

Musti, D. (2005) *Magna Grecia. Il quadro storico.* Bari.

Muzzioli, M. (1996) La sostruzione della via Appia presso Ariccia: Luigi Canina e Virgino Vespignani. *Journal of Ancient Topography* 6, 169–76.

Ñaco del Hoyo, T. (2003) *Vectigal Incertum. Economía de guerra y fiscalidad republicana en el occidente romano: su impacto histórico en el territorio (218–133 a.C.)* [*BAR-IS* 1158]. Oxford.

Nafissi, M. (2001) I grandi santuari extra-urbani. Riflessioni sull'Heraion del Sele e sul santuario di Apollo Alaios a Punta Alice. *AttiTaranto* 40, 267–316.

Nagy, R. (ed.) (1996) *Sepphoris in Galilee: Crosscurrents of Culture.* Winona Lake, IN.

Nappi, G., De Casa, G. and Volponi, E. (1979) Geologia e caratteristiche tecniche del "Tufo Giallo della Via Tiberina." *Bolletino Societa Geologica Italiano* 98, 431–45.

Nappo, S. (1997) The urban transformation at Pompeii in the late third and early second centuries BC, in Laurence and Wallace-Hadrill (eds), pp. 91–120.

Napoli, M. (1970) *La Tomba del Tuffatore.* Bari.

Naumann, R. (1971) *Architektur Kleinasiens von ihren Anfängen bis zum Ende der hethitischen Zeit* (2nd edn). Tübingen.

Nauta, R.R. and Harder, N. (2005) *Catullus' Poem on Attis. Text and Context.* Leiden.

Naval Intelligence Division (1945) *Italy*, vol. 1. Oxford.

Navarro, F.J. (ed.) (2010) *Pluralidad e integración en el mundo romano* [Actas del II Coloquio Internacional Italica Iberia-Iberica Italica]. Pamplona.

Navarro, M. and Demougin, S. (2001) *Élites Hispaniques.* Bordeaux.

Neer, R. (2005) Connoisseurship and the stakes of style. *Critical Inquiry* 32, 1–26.

Netzer, E. (1975) The Hasmonean and Herodian winter palaces in Jericho. *IEJ* 25, 89–100.

Netzer, E. (1981) *Greater Herodium* [Qedem 13]. Jerusalem.

Netzer, E. (2001) *The Palaces of the Hasmoneans and Herod the Great.* Jerusalem.

Netzer, E. (2006) *The Architecture of Herod, the Great Builder.* Tübingen.

Netzer, E., Kalman, Y., Porath, R. and Laureys-Chachy, R. (2010) Preliminary report on Herod's mausoleum and theatre with a royal box at Herodium. *JRA* 23, 84–108.

Neudecker, R. (2005) Pyramide des Cestius, in Giuliani (ed.), pp. 94–113.

Niccolucci, F. and Hermon, S. (eds) (2010) *Beyond the Artefact – Digital Interpretation of the Past* [Proceedings of CAA2004 – Prato, April 13–17, 2004]. Budapest.

Nicholas, L. (1994) *The Rape of Europa, the Fate of Europe's Treasures in the Third Reich and the Second World War.* New York.

Nicolet, C. (1962) Les *equites campani* et leurs représentations figurées. *MÉFRA* 74, 463–517.

Nicolet, C. (1991) *Space, Geography and Politics in the Early Roman Empire.* Ann Arbor.

Nicolet, C. (1997) *Rome et la conquête du monde méditerranéen (264–27 av. J.-C.)*, vol. II, *Genèse d'un empire* (5th edn). Paris.

Nielsen, I. (1985) Considerazioni sulle prime fasi dell'evoluzione dell'edificio termale romano. *AnalRom* 14, 81–112.

Nielsen, I. (1993) *Thermae et Balnea. The Architecture and Cultural History of Roman Public Baths* (2nd edn), 2 vols. Aarhus.

Nielsen, I. (1994) *Hellenistic Palaces: Tradition and Renewal.* Aarhus.

Nielsen, I. (2002) *Cultic Theatres and Ritual Drama: A Study in Regional Development and Religious Interchange between East and West in Antiquity.* Aarhus.

Nielsen, I. and Poulsen, B. (eds) (1992) *The Temple of Castor and Pollux I* [Lavori e Studi di Archeologia 17]. Rome.

Nielsen, M. (1992) Portrait of a marriage: the old Etruscan couple from Volterra. *Acta Hyperborea* 4, 89–141.

Nijboer, A.J., van der Plicht, J., Bietti Sestieri, A.M. and de Santis, A. (1999–2000) A high chronology for the Early Iron Age in central Italy. *Palaeohistoria* (University of Groningen: Laboratory for Conservation & Material Studies) 41/2, 163–76.

Nilsson, M. (2008) Evidence of Palma Campania settlement at Pompeii, in Guzzo and Guidobaldi (eds), 2008, pp. 81–6.

Nippel, W. (1984) Policing Rome. *JRS* 74, 20–9.

Nissen, H. (1869) *Das Templum.* Berlin.

Nissen, H. (1877) *Pompejanische Studien zur Städtekunde des Altertums.* Leipzig.

Niveau de Villedary, A. (2008) Estado de la cuestión y nuevas perspectivas de la arqueología púnica en la península ibérica: el caso de la bahía de Cádiz, in Vita and Zamora (eds), pp. 81–127.

Noeske, H-C. (1998) Bermerkungen zur Problematik der Siedlungsfunde. *Studien zu Fundmünzen der Antike* 13, 157–65.

Noever, P., Rosenauer, A. and Vasold, G. (eds) (2010) *Alois Riegl Revisited. Beiträge zu Werk und Rezeption.* Vienna.

Noguera, J. (1994) *La escultura romana de la provincia de Albacete (Hispania Citerior-Conventus Carthaginiensis.* Albacete.

Noguera, J. (2005) La escultura hispanorromana en piedra de época republicana, in Abad (ed.), pp. 151–207.

Noguera, J. (2008) Los inicios de la conquista romana de Iberia: los campamentos de la campaña del curso inferior del río Ebro. *ArchEsp* 81, 31–48.

Noguera, J. (2009) Los campamentos romanos en el curso inferior del río Ebro durante la segunda guerra púnica, in Morillo, Hanel and Martín (eds), vol. I, pp. 329–38.

Noguera, J. (2011) *La conquesta romana de Catalunya.* Tarragona.

Noguera, J. and Madrid, M.J. (eds) (2009) *Arx Asdrubalis. La ciudad reencontrada. Arqueología en el Cerro del Molinete (Cartagena).* Murcia.

Noguera, J. and Rodríguez Oliva, P. (2008) Scultura ispànica in epoca repubblicana: note su generi, iconografia e cronologia, in Uroz, Noguera and Coarelli (eds), pp. 379–452.

Noguera, J. and Tarradell-Font, N. (2009) Noticia sobre las monedas del campamento romano de la Segunda Guerra Púnica de La Palma (L'Aldea, Tarragona), in *XII Congreso Nacional de Numismática (Cádiz, 2007)*, pp. 119–42. Madrid and Cádiz.

Nolla, J. (1987) *Girona romana. De la fundació a la fi del món antic.* Girona.

Nolla, J. (2001) El lloc de la ciutat. *Annals de l'Institut d'Estudis Gironins* 42, 151–2.

Nolla, J., Palahí, L. and Vivó, D. (2010) *De l'oppidum a la ciuitas. La romanització inicial de la Indigècia.* Girona.

Nolla, J. and Sanmartí, E. (1984) Algunes consideracions entorn de la romanització al NE de Catalunya, in González and Díez (eds), pp. 13–27.

North, H. (1979) *From Myth to Icon. Reflections of Greek Ethical Doctrines in Literature and Art.* Ithaca.

Oakley, S. (1985) Single combat in the Roman Republic. *CQ* 35, 392–410.

Oakley, S. (1995) *The Hill-forts of the Samnites* [BSR Archaeological Monograph 10]. London.

Oakley, S. (2004) The Early Republic, in Flower (ed.), pp. 15–30.

O'Connor, C. (1993) *Roman Bridges.* Cambridge.

Ogden, J., Keay, S., Earl, G., Strutt, K. and Kay, S. (2010) Geophysical prospection at Portus: an evaluation of an integrated approach to interpreting subsurface archaeological features, in Crawford, Frischer and Koller (eds), pp. 273–84.

Olcese, G. (2009) Produzione e circolazione ceramica in area romana in età repubblicana: linee di ricerca, metodi di indagine e problemi aperti, in Jolivet *et al.* (eds), pp. 143–56.

Olcese, G. and Picon, M. (1998) Ceramiche a vernice nera in Italia e analisi di laboratorio: fondamenti teorici e problemi aperti, in Frontini and Grassi (eds), pp. 31–7.

Olcina, M. (1998) *La ciudad ibero-romana de Lucentum (El Tossal de Manises, Alicante), introducción a la investigación del yacimiento y su recuperación como espacio público.* Alicante.

Olcina, M. (2002) Lucentum, in Jiménez and Ribera (eds), pp. 255–66.

Olcina, M. (2006a) Lucentum. Una ciudad romana de la Hispania Citerior, in Rascón and Sánchez (eds), pp. 77–85.

Olcina, M. (2006b) Lucentum: origin and evolution of a Roman municipium in the Sinus Ilicitanus, in Abad, Keay and Ramallo (eds), pp. 105–18.

Oleson, J.P. (ed.) (2008) *The Oxford Handbook of Engineering and Technology in the Classical World.* Oxford.

Oleson, J.P., Brandon, C., Cramer, S. *et al.* (2004) The ROMACONS Project: a contribution to the historical and engineering analysis of hydraulic concrete in Roman maritime structures. *IJNA* 33, 199–229.

Olesti, O. (1994) Les actuaciones pompeianes a Catalunya: reorganització del territori i fundació de noves ciutats, in *Actes del XIV Congrés*, pp. 316–17.

Olesti, O. (1995) *El territori del Maresme en època republicana (ss. III–I a.C.) [Estudi d'Arqueomorfologia i Història].* Mataró.

Olesti, O. (2000) Integració i transformació de les comunitats ibèriques del Maresme durant el s. II–I a.C: un model de romanització per a la Catalunya litoral y prelitoral. *Empúries* 52, 55–86.

Olinder, B. (1974) *Porticus Octavia in circo Flaminio, Topographical Studies in the Campus Region of Rome.* Lund.

Oliver-Smith, P. and Widrig, W. (1982) Villa rustica romana. Relazione preliminare sulle campagne di scavo 1976 e 1977 nell'agro romano. *NSc* ser. 8, 35, 99–114.

Olmer, F. (2003) *Les amphores de Bibracte–2. Le commerce du vin chez les Eduens d'après les timbres d'amphores* [Collection Bibracte 7]. Glux-en-Glenne.

Olmer, F. (ed.) (forthcoming) *Itinéraires des vins romains en Gaule, IIIe–Ier s. av. J.-C., colloque européen 30 janvier–2 février2007, Lattes, ARALO* [Monographies d'Archéologie Méridionale]. Lattes.

Olmer, F. and Maza, G. (2004) Le marché gaulois, in Brun, Poux and Tchernia (eds), pp. 140–57.

Oransay, A. (2001/2) Roman military equipment at Arycanda. *Journal of Roman Military Equipment Studies* 12–13, 85–7.

Orfila, M. (ed.) (2000) *El forum de Pollentia. Memòria de les campanyes d'excavacións realitzades entre els anys 1996 i 1999.* Alcudia.

Orfila, M. (ed.) (2005) *El món romà a les Illes Balears.* Palma de Mallorca.

Orfila, M., Cau, M. and Chávez, M. (2008) Pollentia y las Baleares en época republicana, in Uroz, Noguera and Coarelli (eds), pp. 43–66.

Orfila, M., Chávez, M. and Cau, M. (2006) Pollentia and the Balearic Islands, in Abad, Keay and Ramallo (eds), pp. 133–45.

Orfila, M., Chávez, M., Merino, J. and Aranegui, C. (2006) *Historia de las Baleares, 4. Las Baleares en época romana y tardoantigua.* Palma de Mallorca.

Orlando, L. (ed.) (2008) *Geofisica per l'archeologia. Possibilita e limiti.* Rome.

Orlin, E.M. (1997) *Temples, Religion and Politics in the Roman Republic.* Leiden.

Ørsted, P., Carlsen, J., Ladjimi Sebaï, L. and Ben Hassen, H. (eds) (2000) *Africa Proconsularis: Regional Studies in the Segermes Valley of Northern Tunisia*, vol. III, *Historical Conclusions.* Aarhus.

Ortalli, J. (1986) *Il teatro romano di Bologna* [Deputazione di Storia Patria per le Province dei Romagna: Documenti e Studi v. 19]. Bologna.

Ortiz de Urbina, E. and Santos, J. (eds) (1996) *Teoría y práctica del ordenamiento municipal en Hispania* [Revisiones de Historia Antigua II, Vitoria-Gasteiz, 1993]. Vitoria.

Ortner, D. (2003) *Identification of Pathological Conditions in Human Skeletal Remains* (2nd edn). San Diego.

Osanna, M. (2006) Architettura pubblica e privata a Kossyra, in Osanna and Torelli (eds), pp. 35–50.

Osanna, M. (ed.) (2009) *Verso la città: Forme insediative in Lucania e nel mondo italico fra IV e III sec. a.C.* Lavello.

Osanna, M. and Serio, B. (2009) Organizzazione dello spazio insediativi e paesaggi agrari in Lucania: il caso di Torre di Satriano, in Osanna (ed.), pp. 89–118.

Osanna, M. and Torelli, M. (eds) (2006) *Sicilia ellenistica, consuetudo Italica.* Rome.

Osborne, R. (1985) Buildings and residence on the land in Classical and Hellenistic Greece: the contribution of epigraphy. *BSA* 80, 119–28.

Osborne, R. (1988) Social and economic implications of the leasing of land and property in Classical and Hellenistic Greece. *Chiron* 18, 279–323.

Osborne, R. (1998) Early Greek colonization? The nature of Greek settlement in the west, in Fisher and van Wees (eds), pp. 251–69.

Osborne, R. and Cunliffe, B. (eds) (2005) *Mediterranean Urbanization 800–600 BC* [Proceedings of the British Academy 126]. Oxford.

Östenberg, I. (2009) *Staging the World: Spoils, Captives, and Representations in the Roman Triumphal Procession.* Oxford.

Ostrow, S.E. (1981) The topography of Puteoli and Baiae. *Puteoli* 3, 77–140.

Otiña, P. and Ruiz de Arbulo, J. (2000) De Cese a Tárraco. Evidencias y reflexiones sobre la Tarragona ibérica y el proceso de romanización. *Empúries* 52, 107–36.

Oulebsir, N. (2004) *Les usages du patrimoine: monuments, musées et politique coloniale en Algérie, 1830–1930.* Paris.

Ouni, K. (1999) Imagerie satellitaire et archéologie. Exemples d'arpentages antiques détectés dans le centre-est tunisien, in Peyras and Tirologos (eds), pp. 13–45.

Ouni, K. and Peyras, J. (2002) Centuriation et cadastres du centre-est tunisien, in Clavel-Léveque and Orejas (eds), Dossier 9.

Ouzoulias, P. and Tranoy, L. (eds) (2010) *Comment les Gaules devinrent romaines.* Paris.

Ovadiah, A. and Turnheim, Y. (2011) *Roman Temples, Shrines and Temene in Israel* [*RdA* Suppl. 30]. Rome.

Overbeck, J. and Mau, A. (1884) *Pompeji in seinen Gebauden, Alterthumen und Kunstwerken.* Leipzig.

Overman, J.A. (2009) Between Rome and Parthia: Galilee and the implications of empire, in Rogers (ed.), pp. 279–301.

Overman, J.A., Olive, J. and Nelson, M. (2007) A newly discovered Herodian temple at Khirbet Omrit in Northern Israel, in Kokkinos (ed.), pp. 177–97.

Overman, J.A. and Schowalter, D. (eds) (2011) *The Roman Temple Complex at Horvat Omrit: An Interim Report* [*BAR-IS* 2205]. Oxford.

Oxé, A. (1909) Die älteste Truppenverteilung im Neusser Legionslager. *BJb* 118, 75–99.

Oxé, A. (1939) Polybianische und vorpolybianische Lagermaße und Lagertypen. *BJb* 143/4, 47–74.

Oxé, A., Comfort, H. and Kenrick, P. (2000) *Corpus Vasorum Arretinorum* (2nd edn). Bonn.

Paci, G. (ed.) (2000) *Epigrafai. Miscellanea Epigrafica in onore di Lidio Gasperini, I.* Macerata.

Packer, J.E., Burge, J. and Gagliardo, M. (2007) Looking again at Pompey's Theater: the 2005 excavation season. *AJA* 111, 505–22.

Padovano, R. (ed.) (2009) *Sorgenti e terme della valle del Sacco.* Padova.

Pagenstecher, G. (1909) *Die calenische Reliefkeramik.* Berlin.

Paine, P.P., Mancinelli, D., Ruggieri, M. and Coppa, A. (2007) Cranial trauma in Iron Age Samnite agriculturalists, Alfedena, Italy: implications for biocultural and economic stress. *American Journal of Physical Anthropology* 132 (1), 48–58.

Paine, R.R., Vargiu, R., Signoretti, C. and Coppa, A. (2009) A health assessment for Imperial Roman burials recovered from the necropolis of San Donato and Bivio CH, Urbino, Italy. *JAS* 87, 193–210.

Painter, K.S. (1980) *Roman Villas in Central Italy: Recent Excavations and Research.* London.

Palagia, O. (1997) Classical encounters: Attic sculpture after Sulla, in Hoff and Rotroff (eds), pp. 81–95.

Palahí, L., Nolla, J. and Vivó, D. (eds) (2007) *De Kerunta a Gerunda. Els orígens de la ciutat.* Girona.

Palermo, L. (1998) Contributo alla conoscenza della ceramics aretina a vernice nera; I materiali dello scavo di Chiusi, Orto del Vescovo, in Frontini and Grassi (eds), pp. 119–30.

Palet, J. and Orengo, H. (2010) Les centuriacions de l'*ager Tarraconensis*: organització i concepcions de l'espai/The centuriations of the Ager Tarraconensis: spatial organisation and conceptualisation, in Prevosti and Guitart (eds), pp. 121–54.

Pallottino, M. (1985) *Civiltà artistica etrusco-italica.* Florence.

Palmer, R.E.A. (1996) Locket gold, lizard green, in Hall (ed.), pp. 12–28.

Palmieri, A. (2009) Le tombe laziali di VI e V sec. a.C. Considerazioni da una prospettiva etrusca, in Drago Troccoli (ed.), pp. 387–96.

Pamment Salvatore, J. (1993) Roman tents "replicated" in stone-built barracks of the 2nd century BC in Spain. *Journal of Roman Military Equipment Studies* 4, 23–31.

Pamment Salvatore, J. (1996) *Roman Republican Castrametation. A Reappraisal of Historical and Archaeological Sources* [*BAR-IS* 630]. Oxford.

Pamment Salvatore, J. (1999) The emergence of the legionary cohort as an army unit in Roman Spain in the Late Republican period: evidence from archaeology and elsewhere, in Gudea (ed.), pp. 261–6.

Panciera, S. (1966) Il sepolcro ostiense di C. Cartilius Poplicola ed una scheda epigrafica di Gaetano Marini. *ArchCl* 18, 54–63.

Panei, L. (2011) The Tuffs of the "Servian Wall" in Rome. *ArchéoSciences* 34, http://archeosciences.revues.org/2599 (accessed November 18, 2012).

Panella, C. and Zeggio, S. (2004) Indagini tra Palatino e valle del Colosseo: Nuovi dati. *Workshop di Archeologia Classica* I, 65–87.

Panosa, M. (2009) *De Kese a Tarraco*. Tarragona.

Panzram, S. (2002) *Stadtbild und Elite: Tarraco, Corduba und Augusta Emerita zwischen Republik und Spätantike*. Stuttgart.

Panzram, S. (ed.) (2007) *Städte im Wandel: bauliche Inszenierung und literarische Stilisierung lokaler Eliten auf der Iberischen Halbinsel*. Hamburg.

Papi, E. (1998) "Domus est quae nulli villarum mearum cedat" (Cicero *Fam.* 6.18.5). Osservazione sulle residenze del Palatino alla metà del I sec. a.C., in Cima and La Rocca (eds), pp. 45–70.

Papi, R. (1990) *Dischi-corazza abruzzesi a decorazione geometrica nei musei italiani*. Rome.

Parcak, S.H. (2009) *Satellite Remote Sensing for Archaeology*. London.

Pareschi, M.T., Stefani, G., Varone, A. *et al.* (2000) A Geographical Information System for the archaeological area of Pompeii, in McGuire *et al.*, eds., pp. 143–59.

Parise Badoni, F. and Ruggeri Giove, M. (1980) *Alfedena: la necropoli di Campo Consolino*. Chieti.

Parisi Presicce, C. (2000) *La lupa capitolina*. Rome.

Parker, A.J. (1992) *Ancient Shipwrecks of the Mediterranean and the Roman Provinces [BAR-IS 580]*. Oxford.

Parker, H.T. (1965) *The Cult of Antiquity and the French Revolutionaries, A Study in the Development of the Revolutionary Spirit*. New York.

Parkin, T. (1992) *Demography and Roman Society*. Baltimore.

Parslow, C. (2007) Entertainment at Pompeii, in Dobins and Foss (eds), pp. 212–23.

Pasqualini, A. (ed.) (1996) *Alba Longa, mito storia, archeologia* [Atti dell'incontro di studio, Roma-Albano Laziale 27–29 gennaio 1994]. Rome.

Pasquinucci, M. (1972) *La ceramica a vernice nera del Museo Guarnacci di Volterra*. Rome.

Pasquinucci, M. (ed.) (1988) *Terme romane e vita quotidiana*. Modena.

Pasquinucci, M., Alessi, D., Bianchini, S. *et al.* (1998) Ceramica a vernice nera dall'Etruria settentrionale costiera. Primo contributo alla caratterizzazione delle produzioni locali e delle importazioni, in Frontini and Grassi (eds), pp. 101–18.

Pastor, M. (2006) *Viriato. O héroi lusitano que lutou pela liberdade do seu povo*. Lisbon.

Patané, R. (2002) Centuripe in età ellenistica: i rapporti con Roma, in Rizza (ed.), pp. 127–67.

Patané, R. (2006) Ceramiche e reti commerciali nella Sicilia repubblicana: il caso di Centuripe, in Malfitana, Poblome and Lund (eds), pp. 477–86.

Patrich, J. (2001) The *carceres* of the Herodian hippodrome/stadium at Caesarea Maritima. *JRA* 14, 269–83.

Patterson, H. (2004) Introduction, in Patterson (ed.), pp. 1–8.

Patterson, H. (ed.) (2004) *Bridging the Tiber: Approaches to Regional Archaeology in the Middle Tiber Valley* [*BSR* Archaeological Monograph 13]. London.

Patterson, H. and Di Giuseppe, H. (2009) Il Dibattito storiografico intorno alla South Etruria Survey e nuovi resultati del progetto Valle del Tevere, in Jolivet *et al.* (eds), pp. 7–26.

Patterson, H., Di Giuseppe, H. and Witcher, R. (2004) Three south Etrurian "crises": first results of the Tiber Valley Project. *PBSR* 72, 1–36.

Patterson, H. and Millett, M. (1998) The Tiber Valley Project. *PBSR* 66, 1–20.

Patterson, J.R. (1992) The city of Rome: from Republic to Empire. *JRS* 82, 186–215.

Patterson, J.R. (2006) Rome and Italy, in Rosenstein and Morstein-Marx (eds), pp. 606–24.

Patterson, J.R. (2010) The city of Rome revisited: from Mid Republic to Mid Empire. *JRS* 100, 210–32.

Payá, X., Puig, F. and Reyes, T. (1994) Primeres datacions dels nivells fundacionals d'Aeso. *Revista d'Arqueologia de Ponent* 4, 151–72.

Peacock, D. and Williams, D. (1991) *Amphorae and the Roman Economy*. London.

Pearce, M. and Tosi, M. (1998) *Papers from the EAA Third Annual Meeting at Ravenna 1997 [BAR-IS 719]*. Oxford.

Pearce, S.M. (1992) *Museums, Objects and Collections: A Cultural Study*. Leicester and London.

Pearce, S.M. (1995) *On Collecting, An Investigation into Collecting in the European Tradition*. London.

Peddie, J. (1994) *The Roman War Machine*. Stroud.

Pedley, J.G. (1990) *Paestum: Greeks and Romans in Southern Italy*. London.

Pedley, J.G. (1994) Pre-Roman Lucania: indigenous populations in and around Roccagloriosa. *JRA* 7, 320–5.

Pedley, J.G. and Torelli, M. (eds) (1993) *The Sanctuary of Santa Venera at Paestum*, vol. 1. Rome.

Pékary, T. (1968) *Untersuchungen zu den römischen Reichsstrassen*. Bonn.

Pelagatti, P., Di Stefano, G. and De Lachenal, L. (eds) (2006) *Camarina. 2600 anni dopo la fondazione. Nuovi studi sulla città e sul territorio. Atti del Convegno Internazionale, Ragusa, 7 dicembre 2002/7–9 aprile 2003*. Ragusa and Rome.

Pelgrom, J. (2008) Settlement organization and land distribution in Latin colonies before the Second Punic War, in de Ligt and Northwood (eds), pp. 333–72.

Pelikan Pittenger, M.R. (2008) *Contested Triumphs: Politics, Pageantry, and Performance in Livy's Republican Rome*. Berkeley.

Peña, J.T. and McCallum, M. (2009a) The production and distribution of pottery at Pompeii: a review of the evidence; part 1, production. *AJA* 113, 57–80.

Peña, J.T. and McCallum, M. (2009b) The production and distribution of pottery at Pompeii: a review of the evidence; part 2, the material basis for production and distribution. *AJA* 113, 165–201.

Pena, M. (1988) Nota sobre Livio XLIII, 3. La fundación de la colonia de Carteia. *Espacio,Tiempo y Forma* 1, 267–76.

Pena, M. (1994) Conquête et colonisation dans la péninsule ibérique pendant le IIIe siècle av n.è., in Clavel-Lévêque, Jouffroy and Vignot (eds), pp. 247–58.

Penney, J. (2009) The Etruscan language in its Italic context, in Swaddling and Perkins (eds), pp. 88–93.

Pensabene, P. (1975) Cippi funerari di Taranto. *RM* 92, 263–97.

Pensabene, P. (1988) Scavi nell'area del tempio della Vittoria e del santuario della Magna Mater sul Palatino. *Bollettino di archeologia* 11–12, 11–51.

Pensabene, P. and D'Alessio, A. (2006) L'immaginario urbano. Spazio sacro sul Palatino tardo repubblicano, in Haselberger and Humphrey (eds), pp. 30–50.

Pera, J. (2005) Pervivencia de la lengua ibérica en el siglo i a.C. El ejemplo de la ciudad romana de Iesso (Guissona, Lleida). *Paleohispanica* 5, 315–31.

Pera, J. (2008) Les torres romanes a Catalunya. Entre la tradició erudita i l'evidència arqueològica, in Martí (ed.), pp. 17–38.

Percival, J. (1976) *The Roman Villa: An Historical Introduction*. London.

Pérez Almoguera, A. (1994) Sobre las fundaciones republicanas en Hispania. El caso de Ilerda, in *Actes del XIV Congrés*, pp. 325–7.

Pérez González, C. and Illarregui, E. (eds) (2005) *Arqueologia militar romana en Europa/Roman Military Archaeology in Europe (Segovia 2001)*. Segovia.

Pergola, P., Volpe, R. and Santangeli Valenzani, R. (eds) (2003) *Suburbium: il suburbio di Roma dalla crisi del sistema delle ville a Gregorio Magno*. Rome.

Perkins, P. (1999) *Etruscan Settlement, Society, and Material Culture in Central Coastal Etruria*. Oxford.

Perkins, P. (2007) *Aliud in Sicilia?* Cultural development in Rome's first province, in van Dommelen and Terrenato (eds), pp. 33–53.

Perkins, P. (2009) DNA and Etruscan identity, in Swaddling and Perkins (eds), pp. 94–111.
Perkins, P. and Attolini, I. (1992) An Etruscan farm at Podere Tartuchino. *PBSR* 60, 71–134.
Perkins, P. and Walker, L. (1990) Survey of an Etruscan city at Doganella in the Albegna Valley. *PBSR* 45, 1–144.
Pesando, F. (1990) Fasi di Urbanizzazione lungo la Via Consolare, in Carocci *et al.* (eds), pp. 193–205.
Pesando, F. (1997) *Domus. Edilizia private e società pompeiana fra III e I secolo a.C.* Rome.
Pesando, F. (2008) Casa di età medio-sannitica nella *Regio* VI: tipologia edilizia e apparati decorativi, in Guzzo and Guidobaldi (eds), 2008, pp. 159–71.
Pesando, F. and Guidobaldi, M.P. (2006) *Pompei, Oplontis, Ercolano, Stabiae* [Guide archeologiche Laterza 14]. Rome.
Pesic, P. (2003) *Seeing Double. Shared Identities in Physics, Philosophy and Literature.* Cambridge, MA.
Peters, W.J.Th. (ed.) (1993) *La Casa di Marcus Lucretius Fronto a Pompei e le sue pitture.* Amsterdam.
Peterse, K. (1999) *Steinfachwerk in Pompeji: Bautechnik und Architektur.* Amsterdam.
Petrocelli, E. (ed.) (1999) *La civiltà della transumanza: storia, cultura e valorizzazione dei tratturi e del mondo pastorale in Abruzzo, Molise, Puglia, Campania e Basilicata.* Isernia.
Peyras, J. (1994) Remarques sur les centuriations et les cadastres de l'Afrique proconsulaire, in Clavel-Lévêque, Jouffroy and Vignot (eds), pp. 223–45.
Peyras, J. and Tirologos, G. (eds) (1999) *L'Afrique du Nord antique: cultures et paysages: Colloque de Nantes – Mai 1996.* Paris.
Pfeilschifter, R. (2007) The Allies in the Republican Army and the Romanisation of Italy, in Roth and Keller (eds), pp. 27–42.
Phang, S.E. (2008) *Roman Military Service: Ideologies of Discipline in the Late Republic and Early Principate.* Cambridge.
Philip, H. and Koenigs, W. (1979) Zu den basen des L. Mummius in Olympia. *AM* 94, 193–216.
Pietilä-Castrén, L. (1987) *Magnificentia publica, The Victory Monuments of the Roman Generals inthe Era of the Punic Wars.* Helsinki.
Pietri, C. (ed.) (1987) *L'Urbs: espace urbain et histoire* [CÉFR 98]. Rome.
Piganiol, A. (1962) *Les documents cadastraux de la colonie romaine d'Orange* [*Gallia* Suppl. 16]. Paris.
Pina, F. (1997) Las comisiones senatoriales para la reorganización de Hispania (App. *Iber.* 99–100). *Dialogues d'Histoire Ancienne* 23 (2), 83–104.
Pina, F. (2000) La conquista, in Beltrán Lloris, Martín-Bueno and Pina (eds), pp. 31–7.
Pina, F. and Pérez Casas, J. (1998) El oppidum de Castra Aelia y las campañas de Sertorius en los años 77–76 a.C. *JRA* 11, 245–64.
Pina Polo, F. (2003) Minerva, custos urbis de Roma y de Tarraco. *ArchEsp* 76, 111–19.
Pina Polo, F. (2006) Deportation, Kolonisation, Migration. Bevölkerungsverschiebungen im republikanischen Italien und Formen der Identitätsbildung, in Jehne and Pfeilschifter (eds), pp. 171–206.
Pina Polo, F. (2007) Kontinuität und Innovation im Städtenetzwerk der nordöstlichen Iberischen Halbinsel zur Zeit der Republik, in Panzram (ed.), pp. 25–35.
Pina Polo, F. (2008) Hispania of Caesar and Pompey. A conflict of clientelae?, in García-Bellido, Mostalac and Jiménez (eds), pp. 41–8.
Pina Polo, F. (2009) Hispania y su conquista en los avatares de la República tardía, in Andreu, Cabrero and Rodà (eds), pp. 223–36.
Pinzone, A. (1999a) *Provincia Sicilia. Ricerche di storia della Sicilia romana da Gaio Flaminio a Gregorio Magno* [Testi e studi di storia antica 7]. Catania.
Pinzone, A. (1999b) L'immigrazione e i suoi riflessi nella storia economica e speiale della Sicilia del II sec. a.C., in Barra Bagnasco, De Miro and Pinzone (eds), pp. 381–402.

Pisani Sartorio, G. (1982) L'area sacra dei templi della Fortuna e della Mater Matuta nel Foro Boario, in *Roma Repubblicana fra il 509 e il 270 BC*, pp. 51–6. Rome.

Pisani Sartorio, G., Colini, A.M. and Buzzetti, C. (1986) Portus Tiberinus, in Quilici Gigli (ed.), 1986, pp. 157–97.

Pisani Sartorio, G. and Quilici, L. (eds) (1983) *L'archeologia in Roma capitale tra sterro e scavo. Catalogo Mostra Roma Capitale 1870–1911*. Venice.

Pisani Sartorio, G. and Quilici Gigli, S. (1987/8) A proposito della Tomba dei Cornelii. *BullCom* 92, 247–64.

Piso, I. (ed.) (2008) *Die römischen Provinzen. Begriff und Gründung (Cluj-Napoca, 2006)*. Cluj-Napoca.

Pittenger, M. (2008) *Contested Triumphs. Politics, Pagentry and Performance in Livy's Republican Rome*. Berkeley.

Pitts, M. (2007) The emperor's new clothes? The utility of identity in Roman archaeology. *AJA* 111, 693–713.

Pitts, M. and Versluys, M.J. (eds) (in press) *Globalisation and the Roman World. Perspectives and Opportunities.*

Pobjoy, M. (2000a) The first Italia, in Herring and Lomas (eds), pp. 187–211.

Pobjoy, M. (2000b) Building inscriptions in Republican Italy: euergitism, responsibility and civic virtue, in Cooley (ed.), pp. 77–92.

Poblome, J. and Waelkens, M. (2003) Sagalassos and Alexandria. Exchange in the eastern Mediterranean, in Abadie-Reynal (ed.), pp. 179–92.

Poccetti, P. (1982) Mefitis. *Annali del Seminario di Studi del Mondo Classico: Sezione Linguistica* 4, 237–60.

Poccetti, P. (1990) Il testo della laminetta, in Gualtieri and Fracchia (eds), 1990, pp. 141–50.

Poehler, E.E. (2006) The circulation of traffic in Pompeii's Regio VI. *JRA* 19, 53–74.

Poggesi, G. (2000) Prato-Gonfienti: la ricerca archeologica nell'area dell'Interporto. Il contesto, in Bettini and Poggesi (eds), pp. 61–9.

Poggesi, G. (2004) *Cinta muraria di Cosa. L'intervento di restauro 2003–2004*. Roccastrada.

Polito, E. (1996) Sepolture di guerrieri nell'Italia meridionale fra V e IV secolo a.C. Alcune riflessioni. *RendNap* 66, 177–87.

Pollitt, J.J. (1974) *The Ancient View of Greek Art, Criticism, History and Terminology*. New Haven.

Pollitt, J.J. (1978) The impact of Greek art on Rome. *TAPA* 108, 155–74.

Pollitt, J.J. (1986) *Art in the Hellenistic Age*. Cambridge.

Pomey, P. (1997) *La navigation antique*. Aix-en-Provence.

Pommier, E. (1999) Réflexions sur le problème des restitutions d'oeuvres d'art en 1814–1815, in *Dominique-Vivant Denon, l'oeil de Napoleon* (Exhibition catalog), pp. 254–7. Paris.

"Pompeii Forum Project under way" (1994) *CSA Newsletter* VII (2), http://csanet.org/newsletter/aug94/nl089406.html (accessed November 21, 2012).

Pons, E. (2002) *Mas Castellar de Pontós (Alt Empordà). Un complex arqueològic d'època ibèrica (excavacions 1990–1998)*. Girona.

Pons Pujol, Ll. (ed.) (2010) *Hispania et Gallia: dos provincias del Occidente romano* [*Instrumenta* 38]. Barcelona.

Pontrandolfo, A. (1983) Per una puntualizzazione della cronologia delle monete a leggenda ΠΑΙΣΤΑΝΟ. *AIIN* 30, 63–81.

Pontrandolfo, A. (1987a) Le necropoli nella città greca alla colonia latina. *Atti Taranto* 27, 225–65.

Pontrandolfo, A. (1987b) Un'iscrizione poseidoniate in una tomba di Fratte di Salerno. *AION* 9, 55–63.

Pontrandolfo, A. (1995) Simposio e élites sociali nel mondo Etrusco e Italico, in Murray and Tecusan (eds), pp. 167–95.

Pontrandolfo, A. and D'Agostino, B. (1990) Greci, Etruschi e Italici nella Campania e nella Lucania tirrenica, in Massa Pairault (ed.), pp. 101–16.

Pontrandolfo, A. and Rouveret, A. (1992) *Le tombe dipinte di Paestum*. Modena.

Pontrandolfo Greco, A. (1994) Le pitture di Paestum. *Archeo* 9, 48–97.

Pontrandolfo Greco, A. (1998) Uso dello spazio, gerarchie sociali, distinzioni di sesso e di età nelle necropoli dell'Italia meridionale, in Marchegay, Le Dinahet and Salles (eds), pp. 125–39.

Porath, Y. (1995) Herod's amphitheater at Caesarea, in Humphrey (ed.), 1995, pp. 15–27.

Porath, Y. (2003) Herod's circus at Caesarea: a response to J. Patrich (*JRA* 14). *JRA* 16, 451–5.

Portale, E., Angiolillo, S. and Vismara, C. (2005) *Le grande isole del Mediterraneo occidentale: Sicilia Sardinia Corsica*. Rome.

Porter, J. (ed.) (2006) *Classical Pasts. The Classical Traditions of Greece and Rome*. Princeton.

Porter, R. (ed.) (1990) *The Medical History of Waters and Spas* [*Medical History* Suppl. 10]. London.

Posluschny, A., Lambers, K. and Herzog, I. (eds) (2008) *Layers of Perception. Proceedings of the 35th International Conference on Computer Applications and Quantitative Methods in Archaeology (CAA)*. Berlin (CD-ROM).

Potter, D. (2004) The Roman army and navy, in Flower (ed.), pp. 66–88.

Potter, D.S. and Mattingly, D.J. (eds) (1999) *Life, Death and Entertainment in the Roman Empire*. Ann Arbor.

Potter, T.W. (1979) *The Changing Landscape of South Etruria*. London.

Potter, T.W. (1987) *Roman Italy*. Berkeley.

Potter, T.W. (1989) *Una Stipe Votive da Ponte di Nona*. Rome.

Potter, T.W. and Wells, C. (1985) A Republican healing sanctuary at Ponte di Nona near Rome and the Classical tradition of votive medicine. *Journal of the British Archaeological Association* 138, 23–47.

Potts, A. (1994) *Flesh and the Ideal, Winckelman and the Origins of Art* History. New Haven.

Poucet, J. (2008) Quand l'archéologie, se basant sur la tradition littéraire, fabrique de la "fausse histoire": le cas des origines de Rome. *Folia Electronica Classica* 16 (1), http://bcs.fltr.ucl.ac.be/fe/16/FausseHistoire.htm (accessed November 18, 2012).

Poulle, B. (2000) Les vases acoustiques du théâtre de Mummius Achaicus. *RA*, 37–50.

Poulsen, B. (2008) Trench V, in Cullhed and Slej (eds), pp. 361–71.

Poulsen, F. (1922) *Etruscan Tomb Paintings: Their Subjects and Significance*. Oxford.

Poux, M. (2004a) *L'Âge du vin*. Montagnac.

Poux, M. (2004b) De Midas à Luern, le vin des banquets, in Brun, Poux and Tchernia (eds), pp. 68–95.

Prados, F. (2008) *Arquitectura púnica. Los monumentos funerarios*, Madrid.

Prag, J. (2006a) But who were the "Punickes"? *PBSR* 74, 1–37.

Prag, J.R.W. (2006b) Il miliario di Aurelius Cotta (*ILLRP* n. 1277): una lapide in contest, in Michelini (ed.), pp. 733–44.

Prag, J. (2007a) Ciceronian Sicily: the epigraphic dimension, in Dubouloz and Pittia (eds), pp. 245–71.

Prag, J. (2007b) *Auxilia* and *gymnasia*: a Sicilian model of Roman imperialism. *JRS* 97, 68–100.

Prag, J. (2009a) Republican Sicily at the start of the 21st century: the rise of the optimists? *Pallas* 79, 131–44.

Prag, J. (2009b) Identità siciliana in età romano-repubblicana, in Ampolo (ed.), 2009, pp. 87–99.

Prag, J. (2011) Kinship diplomacy between Sicily and Rome, in Bonnet *et al.* (eds), pp. 179–206.

Prag, J. and Quinn, J. (eds) (forthcoming) *The Hellenistic West*. Cambridge.

Prentice, W. (1901) A mosaic pavement and inscription from the bath at Serdjilla, central Syria. *RA* 39, 62–76.

Prevosti, M. and Guitart, J. (eds) (2010) *Ager Tarraconensis I. Aspectes històrics i marc natural* [Documenta 16]. Tarragona.

Principal, J. (2006) Late Hellenistic black gloss wares in the northeastern Iberian Peninsula: production, traditions and social practices, in Malfitana, Poblome and Lund (eds), pp. 41–56.

Prontera, F. (1988) L'Italia meridionale di Strabone, in Maddoli (ed.), pp. 93–109.

Prosdocimi, A.L. (2000) Il Sannita, in La Regina (ed.), pp. 202–13.

Pucci, G. and Mascione, C. (eds) (2003) *Manifatture ceramica etrusco-romana a Chiusi: il complesso produttivo di Marcianella.* Bari.

Pugliese Carratelli, G. (1972) Sanniti, Lucani, Brutti e Italioti nel IV secolo a.C. *AttiTaranto* 11, 37–54.

Pugliese-Carratelli, G. (ed.) (1989) *Italia omnium terrarum parens.* Milan.

Puig, A. and Martin, A. (2006) *La colònia grega de Rhode (Roses, Alt Empordà).* Girona.

Purcell, N. (1985) Wealth and wine in ancient Italy. *JRS* 75, 1–19.

Purcell, N. (1987) Town in country and country in town, in MacDougall (ed.), pp. 187–203.

Purcell, N. (1990) The creation of provincial landscape. The Roman impact on Cisalpine Gaul, in Blagg and Millett (eds), pp. 7–29.

Purcell, N. (1994) South Italy in the fourth century BC. *CAH* VI, 381–403.

Purcell, N. (1996) Rome and the management of water: environment, culture and power, in Shipley and Salmon (eds), pp. 180–212.

Purcell, N. (2003) The way we used to eat: diet, community, and history at Rome. *AJP* 124, 329–58.

Purcell, N. (2007a) Urban spaces and central places. The Roman world, in Alcock and Osborne (eds), pp. 182–202.

Purcell, N. (2007b) The *horti* of Rome and the landscape of property, in Leone, Palombi and Walker (eds), pp. 361–77.

Purcell, N. (forthcoming) Coloniae and the Capitol, in Stek and Pelgrom (eds).

Quatremère de Quincy (1989) *Lettres à Miranda sur le Déplacement des Monuments de l'Art de l'Italie.* Paris.

Quesada Sanz, F. (1997a) *El Armamento ibérico. Estudio tipológico, geográfico, funcional, social y simbólico de las armas en la Cultura Ibérica (siglos VI–I a.C.)*, 2 vols. Montagnac.

Quesada Sanz, F. (1997b) Montefortino-type and related helmets in the Iberian Peninsula: a study in archaeological context, in Feugère (ed.), pp. 151–66.

Quesada Sanz, F. (1997c) *Gladius hispaniensis*: an archaeological view from Iberia, in Feugère (ed.), pp. 251–70.

Quesada Sanz, F. and de Prado, E.K. (2006) Roman Republican weapons, camps and battlefields in Spain: an overview of recent and on-going research, in Morillo and Aurrecoechea (eds), pp. 65–84.

Quilici, L. (1974) *Collatia* [Forma da Italiae I, X]. Rome.

Quilici, L. (2002) Da Roma alle foci del Garigliano. Per uno parco regionale della Via Appia Antica. *AnalRom* Suppl. 30, 77–86.

Quilici, L. and Quilici Gigli, S. (eds) (2001) *Fortificazioni antiche in Italia. Età repubblicana* [Atlante Tematico di Topografia Antica 9]. Rome.

Quilici, L. and Quilici Gigli, S. (eds) (2007) *Architettura pubblia e privata nell'Italia antica.* Rome.

Quilici Gigli, S. (1976) *Blera: Topografia antica della città e del territorio.* Mainz.

Quilici Gigli, S. (ed.) (1986) *Il Tevere e le altre vie d'acqua del Lazio* [*Archeologia Laziale* 7.2]. Rome.

Quilici Gigli, S. (ed.) (1990) *La Via Appia. Decimo incontro di studio del Comitato per l'Archeologia Laziale.* Rome.

Quilici Gigli, S. (ed.) (1997) *Uomo, Acqua e Paesaggio. Atti dell'incontro di studio sul tema Irreggimentazione delle acque e trasformazione del paesaggio antico.* Rome.

Quinn, J. (2003) Roman Africa?, in Merryweather and Prag (eds), pp. 7–34.

Quinn, J. (2004) The role of the 146 settlement in the provincialization of Africa. *AfrRom* 15, 1593–601.

Quinn, J. (2009) North Africa, in Erskine (ed.), 2009, pp. 260–72.

Quinn, J. (forthcoming) Monumental power: "Numidian royal architecture" in context, in Quinn and Prag (eds).

Quinn, J. and Prag, J. (eds) (forthcoming) *Identifying the Punic Mediterranean.*

Raban, A. and Holum, K. (eds) (1996) *Caesarea Maritima – A Retrospective after Two Millennia.* Leiden.

Rainini, I. (1985) *Il Santuario di Mefite in Valle d'Ansanto.* Rome.

Rakob, F. (1976) Hellenismus in Mittelitalien. Bautypen und Bautechnik, in Zanker (ed.), pp. 366–86.

Rakob, F. (1983) Architecture royale numide, in *Architecture et société de l'archaisme*, pp. 325–48.

Ramallo, S. (1992) Un santuario de época tardo-republicana en La Encarnación, Caravaca, Murcia. *Cuadernos de Arquitectura Romana* 1, 39–66.

Ramallo, S. (1993) Terracotas arquitectónicas del santuario de La Encarnación (Caravaca de la Cruz, Murcia). *ArchEsp* 66, 71–98.

Ramallo, S. (1997) Templi e santuari nella Hispania repubblicana, in Arce, Ensoli and La Rocca (eds), pp. 253–66.

Ramallo, S. (2006a) Carthago Nova: urbs opulentissima omnium in Hispania, in Abad, Keay and Ramallo (eds), pp. 91–104.

Ramallo, S. (2006b) Carthago de Hispania. Puerto privilegiado de la costa mediterránea, in Rascón and Sánchez (eds), pp. 97–121.

Ramallo, S., Fernández, A., Madrid, M. and Ruiz, E. (2008) Carthago Nova en los dos últimos siglos de la República: una aproximación desde el registro arqueológico, in Uroz, Noguera and Coarelli (eds), pp. 573–602.

Ramallo, S. and Ruiz, E. (1994) Transformaciones urbanísticas en la ciudad de Carthago Nova, in *Actes del XIV Congrés*, pp. 342–3.

Ramallo, S. and Ruiz, E. (1998) *El teatro romano de Cartagena.* Murcia.

Ramallo, S.and Ruiz, E. (2002) Carthago Nova. Capital de Hispania Citerior, in Jiménez and Ribera (eds), pp. 122–33.

Ramallo, S., Ruiz, E., Moneo, R. and Murcia, A. (2009) *Museo Teatro romano de Cartagena.* Murcia.

Ramos, R. and Uroz, J. (1992) Ilici. *DialArch* 3rd ser., 10, 95–104.

Ranieri, S. (1998) Contribution to the study of stratigraphic sequence under AD 79 layers of Pompeii. *Opuscula Pompeiana* 8, 135–41.

Ranini, I. (1985) *Il santuario di Mefite in Valle D'Ansanto.* Rome.

Rascón, S. and Sánchez, A. (eds) (2006) *Civilización.* Madrid.

Rasmussen, T. (1997) The Tarquins and "Etruscan Rome," in Cornell and Lomas (eds), 1997, pp. 23–30.

Rast, W. (1992) *Through the Ages in Palestinian Archaeology.* Philadelphia.

Rathbone, D. (1981) The development of agriculture in the "Ager Cosanus" during the Roman Republic: problems of evidence and interpretation. *JRS* 71, 10–23.

Rathbone, D. (1983) Review: the slave mode of production in Italy. *JRS* 73, 160–8.

Rathbone, D. (2008) Poor peasants and silent sherds, in de Ligt and Northwood (eds), pp. 305–32.

Rathje, A. and van Kampen, I. (2001) The distribution of space and materials in domestic architecture in early Rome, in Brandt and Karlsson (eds), pp. 383–8.

Raubitschek, A.E. (1954) Epigraphical notes on Julius Caesar. *JRS* 44, 65–75.

Raubitschek, A.E. (1959) The Brutus statue in Athens, in *Atti del terzo congresso internazionale di epigrafia greca e latina*, pp. 15–21. Rome.
Rauflaub, K. (ed.) (2005) *Social Struggles in Archaic Rome* (2nd rev. edn). Oxford.
Rauh, N.K. (1992) Was the agora of the Italians an établissement de sport? *BCH* 116, 293–333.
Rauh, N.K. (1993) *The Sacred Bonds of Commerce: Religion, Economy, and Trade Society at Hellenistic Roman Delos, 166–87 BC*. Amsterdam.
Rawlings, L. (2007) Army and battle during the conquest of Italy (350–264 BC), in Erdkamp (ed.), pp. 45–62.
Rawson, E. (1971) The literary sources for the pre-Marian army. *PBSR* 39, 13–31.
Rawson, E. (1975) Architecture and sculpture: the activities of the Cossutii. *PBSR* 43, 36–47.
Rawson, E. (1981) Chariot-racing in the Roman Republic. *PBSR* 49, 1–16.
Rawson, E. (1985a) Theatrical life in Republican Rome and Italy. *PBSR* 53, 97–113.
Rawson, E. (1985b) *Intellectual Life in the Late Roman Republic*. London.
Reale, O. and Dirmeyer, P. (2000) Modelling the effects of vegetation on Mediterranean climate during the Roman Classical Period. Part 1, climate history and model sensitivity. *Global and Planetary Change* 25, 163–84.
Reay, B. (2005) Agriculture, writing, and Cato's aristocratic self-fashioning. *ClAnt* 24 (2), 331–61.
Reddé, M. (2006) Chaussée-Tirancourt, in Reddé *et al.* (eds), pp. 249–51.
Reddé, M. (2008) Les camps militaires républicains et augustéens: paradigmes et réalités archéologiques, in Cadiou *et al.* (eds), pp. 61–71.
Reddé, M., Brulet, R, Fellmann, R. *et al.* (eds) (2006) *Les fortifications militaires. L'architecture de la Gaule romaine*. Bordeaux.
Reddé, M. and Schnurbein, S. von (eds) (2001) *Alésia. Fouilles et recherches franco-allemandes sur les travaux militaires romains autour du Mont-Auxois (1991–1997)* [Mémoires de l'académie des inscriptions et belles-lettres. Tome XXII]. Paris.
Redfern, R. and DeWitte, S. (2011) A new approach to the study of Romanization in Britain: a regional perspective of cultural change in Late Iron Age and Roman Dorset using the Siler and Gompertz-Makeham models of mortality. *American Journal of Physical Anthropology* 144, 269–85.
Reece, R. (1996) The interpretation of site finds – a review, in King and Wigg (eds), pp. 341–55.
Regev, E. (2004) Family burial, family structure and the urbanization of Herodian Jerusalem. *PEQ* 136, 109–32.
Reifferscheid, A. (1860) *C. Suetoni tranquilli praeter Caesarum libros reliquiae*. Leipzig.
Remesal, J. (2011) *La Bética en el concierto del Imperio romano*. Madrid.
Remolà, J. and Pociña, C. (2005) La font dels lleons, in Sada (ed.), pp. 53–66.
Renfrew, C. (1994) The archaeology of religion, in Renfrew and Zubrow (eds), pp. 47–54.
Renfrew, C. (2000) *Loot, Legitimacy and Ownership. The Ethical Crisis in Archaeology*. London.
Renfrew, C. and Bahn, P. (1991) *Archaeology, Theories, Methods and Practice*. London.
Renfrew, C. and Cherry, J.F. (eds) (1986) *Peer Polity Interaction and Social Change*. Cambridge.
Renfrew, C. and Zubrow, E. (eds) (1994) *The Ancient Mind: Elements of Cognitive Archaeology*. Cambridge.
Reusser, G. (2002) *Vasen für Etrurien*. Zurich.
Revell, L. (2009) *Roman Imperialism and Local Identities*. Cambridge.
Reyes, T., González Villaescusa, R. and García, J. (1998) Estudi de l'ager Aesonensis (Isona i Conca Dellà, Pallars Jussà). *Revista d'Arqueologia de Ponent* 8, 39–59.
Ribera, A. (1998) *La fundació de València*. Valencia.
Ribera, A. (2006) The Roman foundation of Valencia and the town in the 2nd–1st c. BC, in Abad, Keay and Ramallo (eds), pp. 75–89.

Ribera, A. (2008) Valentia (Hispania Citerior) una fundación itálica de mediados del siglo II a.C. Novedades y complementos, in Uroz, Noguera and Coarelli (eds), pp. 169–97.

Ribera, A., Olcina, M. and Balester, C. (2007) *Pompeya bajo Pompeya: las excavaciones en la Casa de Ariadna*. Valencia.

Ricci, A. (ed.) (1985) *Settefinestre. Una villa schiavistica nell'Etruria romana. La villa e suoi reperti*, vol. 3. Modena.

Ricciardi, L. (1987) Blera, in Barbieri (ed.), pp. 83–7.

Rich, J. (2007) Warfare and the army in early Rome, in Erdkamp (ed.), pp. 7–23.

Richards, M., Hedges, R., Molleson, T. and Vogel, J. (1992) Stable isotope analysis reveals variations in human diet at the Poundbury Camp cemetery site. *JAS* 25, 1247–52.

Richardson, A. (2003) Space and manpower in Roman camps. *OJA* 22 (3), 303–13.

Richardson, A. (2004) *Theoretical Aspects of Roman Camp and Fort Design* [*BAR-IS* 1321]. Oxford.

Richardson, J. (1998) *The Romans in Spain*. Oxford.

Richardson, L., jr. (1988) *Pompeii: An Architectural History*. Baltimore.

Richardson, P. (1996) *Herod: King of the Jews and Friend of the Romans*. Columbia, SC.

Richmond, I. (1982) *Trajan's Army on Trajan's Column*. London.

Rickman, G. (1971) *Roman Granaries and Store Buildings*. Cambridge.

Ridgway, B.S. (1984) *Roman Copies of Greek Sculpture: The Problem of the Originals*. Ann Arbor.

Ridgway, D. (2010) Greece, Etruria and Rome: relationships and receptions. *Ancient West & East* 9, 43–61.

Ridgway, D. and Ridgway, F. (eds) (1979) *Italy before the Romans*. London.

Ridgway, D., Ridgway, F.S., Pearce, M. *et al.* (eds) (2000) *Ancient Italy in Its Mediterranean Setting. Studies in Honour of Ellen Macnamara*. London.

Ridgway, F.S. (2004–6) Revisiting the Etruscan underworld. *Accordia Research Papers* 10, 127–41.

Ridley, R.T. (2005) Unbridgeable gaps: the Capitoline Temple at Rome. *BullCom* 106, 83–104.

Riera, M. and López Mullor, A. (2004) Intervencions recents (1997–2001) a l'oppidum del Turó del Montgròs (El Brull, Osona). *Tribuna d'Arqueologia* 2000/1, 135–86.

Rife, J., Morison, M., Barbet, A. *et al.* (2007) Life and death at a port in Roman Greece: the Kenchreai Cemetery Project 2002–2006. *Hesperia* 76, 143–81.

Riggs, C. (2005) *The Beautiful Burial in Roman Egypt. Art, Identity and Funerary Religion*. Oxford.

Ripollès, P.P. (2002a) La ordenación y cronología de las emisiones, in Ripollès and Llorens (eds), pp. 276–91.

Ripollès, P.P. (2002b) El panorama moneta monetario, in Ripollès and Llorens (eds), pp. 331–9.

Ripollès, P. (2005) Coinage and identity in the Roman provinces: Spain, in Howgego, Heuchert and Burnett (eds), pp. 79–94.

Ripollès, P. (2007) *Las acuñaciones de la ciudad ibérica de Saitabi*. Valencia.

Ripollès, P. and Llorens, M. (2002) *Arse-Saguntum. Historia monetaria de la ciudad y su territorio*. Sagunto.

Rispoli, P., Candela, M., Di Maio, G. and Jacobelli, L. (2008) L'Insula Occidentalis VII, 16 di Pompei: indagini preliminari al progetto di restauro, in Guzzo and Guidobaldi (eds), 2008, pp. 536–38.

Riva, C. (2009a) *The Urbanisation of Etruria: Funerary Practices and Social Change, 700–600 BC*. Cambridge.

Riva, C. (2009b) Ingenious inventions: welding ethnicities east and west, in Hales and Hodos (eds), pp. 79–113.

Riva, C. (2010) Trading settlements and the materiality of wine consumption in the north Tyrhennian Sea region, in van Dommelen and Knapp (eds), pp. 210–32.

Riva, C. (forthcoming) Discussion, in Gleba and Horsnaes (eds).

Rives, J. (1995) *Religion and Authority in Roman Carthage from Augustus to Constantine*. Oxford.

Rix, H. (1997) The pre-Lucanian inscriptions of southern Italy, in Adams (ed.), pp. 144–54.

Rix, H. (2002) *Sabellische Texte. Die Texte des Oskischen, Umbrischen und Südpikenischen*. Heidelberg.

Rizakis, A. and Lepenioti, C. (eds) (2010) *Roman Peloponnese III, Society, Economy and Culture under the Roman Empire: Continuity and Innovation*. Athens.

Rizza, G. (ed.) (2002) *Scavi e ricerche a Centuripe* [Studi e Materiali di Archeologia Mediterranea 1]. Catania.

Roberto, C., Plambeck, J. and Small, A. (1985) The chronology of sites of the Roman period around San Giovanni: methods of analysis and conclusions, in Macready and Thompson 1985, (eds), pp. 136–45.

Roberto, C. and Small, A. (1994) The field survey, in Small and Buck (eds), pp. 19–22.

Robinson, H.R. (1975) *The Armour of Imperial Rome*. New York.

Robinson, M. (2008) La stratigraia nello studio dell'archeologia preistorica e protostorica a Pompei, in Guzzo and Guidobaldi (eds), 2008, pp. 125–38.

Robinson, O.F. (1980) The water supply of Rome. *Studia et Documenta Historiae et Iuris* 46, 44–86.

Rodà, I. (1986–9) Els lloctinents de Juli Cèsar, primers patrons d'Empúries. *Empúries* 48–50, 246–9.

Rodà, I. (1993) Escultura republicana en la Tarraconense: el monumento funerario de Malla, in *Actas de la I Reunión sobre Escultura Romana en Hispania (Mérida, 1992)*, pp. 207–19. Mérida.

Rodà, I. (1998) La difícil frontera entre escultura ibérica y escultura romana, in *Actas del Congreso Internacional "Los Iberos. Príncipes de Occidente." Estructuras de poder en la sociedad ibérica (Barcelona, 1998)*, pp. 265–73. Barcelona.

Rodà, I. (2001) La cronología de las fortificaciones de las ciudades romanas de Hispania Citerior (al norte del Ebro y del Duero). *Histria Antiqua* 7, 27–31.

Rodà, I. (ed.) (2001) *Tarraco, puerta de Roma*. Barcelona.

Rodà, I. (2002) Indígenes i romans a l'ibèric final. *Limes* 8, 153–8.

Rodà, I. (2004) Datación e iconografía del "Asclepio" de Emporiae, in *IV Reunión sobre escultura romana en Hispania (Lisboa-Cascais 2002)*, pp. 307–20. Madrid.

Rodà, I. (2005) La figura de Agripa en Hispania, in Pérez González and Illarregui (eds), pp. 319–31.

Roddaz, J.-M. (1994) *Marcus Agrippa*. Paris.

Roddaz, J.-M. (1998) Les Scipions et l'Hispanie. *RÉA* 100, 341–58.

Roddaz, J.-M. (2002) Hispania pacata: l'empereur et les Espagnes aux deux premiers siècles de l'Empire, in Urso (ed.), pp. 201–23.

Roddaz, J.-M. (2006) D'une péninsule à l'autre: l'episode sertorien, in Sartori and Valvo (eds), pp. 99–115.

Roddaz, J.-M. (2009) Des Nîmes à Lesbos: Agrippa et le patronage des cités, in Christol and Darde (eds), pp. 53–7.

Rodrigo, E. (2004) L'estructuració del territory de Iesso en època romana, in Guitart and Pera (eds), pp. 171–86.

Rodríguez Almeida, E. (1984) *Il monte Testaccio: ambiente, storia, materiali*. Rome.

Rodríguez Colmenero, A. (ed.) (1998) *Los orígenes de la ciudad en el noroeste hispánico. Actas del Congreso Internacional (Lugo, 1996)*. Lugo.

Rodríguez Colmenero, A. and Ferrer, S. (ed.) (2006) *Excavaciones arqueológicas en Aquis Querquennis. Actuaciones en el campamento romano (1975–2005)* [Anejos de *Larouco* 4]. Lugo.

Rodríguez Colmenero, A. and Rodà, I. (eds) (2007) *Murallas de ciudades romanas en el Occidente del Imperio: Lucus Augusti como paradigma (Lugo 2005)*. Lugo.

Rodríguez Gutiérrez, O. (2011) *Hispania arqueológica. Panorama de la cultura material de las provincias romanas*. Seville.

Rodríguez Neila, J. (1992) *Confidentes de César. Los Balbos de Cádiz*. Madrid.

Rodríguez Neila, J. (2006) *Cornelii Balbi Gaditani*: de las élites locales de *Hispania*, in Sartori and Valvo (eds), pp. 117–31.

Rodríguez Neila, J. and Melchor, E. (2006) *Poder central y autonomía municipal. La proyección pública de las elites romanas de Occidente*. Córdoba.

Rodríguez Neila, J. and Navarro, F. (1999) *Elites y promoción social en la Hispania romana*. Pamplona.

Rodríguez Oliva, P. (1996) Primeras manifestaciones de la escultura romana en la Hispania meridional, in Massó and Sada (eds), pp. 13–30.

Rogers, Z. (ed.) (2009) *A Wandering Galilean: Essays in Honour of Seán Freyne*. Leiden.

Roldán, L., Bendala, M., Blánquez, J. and Martínez, S. (1998) *Carteia*. Madrid.

Roldán, L., Bendala, M., Blánquez, J., Martínez, S. and Bernal, D. (2003) *Carteia II*. Madrid.

Roldán Hervás, J. (1996) Conquista y colonización en la Bética en época republicana, in Ortiz de Urbina and Santos (eds), pp. 27–39.

Roldán Hervás, J. and Wulff Alonso, F. (2001) *Citerior y Ulterior. Las provincias romanas de Hispania en la era republica*. Madrid.

Rolland, H. (1969) *Le mausolée de Glanum* [*Gallia* Suppl. 21]. Paris.

Roller, D. (1998) *The Building Program of Herod the Great*. Berkeley.

Roma medio repubblicana. Aspetti culturali di Roma e del Lazio nei secoli IV e III a.C. (1973). Rome.

Roman, D. and Roman, Y. (2007) *Aux miroirs de la Ville: images et discours identitaires romains (IIIe s. avant J.-C.–IIIe s. après J.-C.)*. Brussels.

Romano, D.G. (2003) City planning, centuriation, and land division in Roman Corinth, in Williams and Bookidis (eds), pp. 279–301.

Romano, D.G. (2005) Urban and rural planning in Roman Corinth, in Schowalter and Friesen (eds), pp. 25–59.

Romano, D.G. (2010) Romanization in the Corinthia: urban and rural developments, in Rizakis and Lepenioti (eds), pp. 155–72.

Romano, I.B. (1994) A Hellenistic deposit from Corinth: evidence for interim period activity (146–44 BC). *Hesperia* 63, 57–104.

Romiopoulou, C. (1974) Un nouveau milliaire de la Via Egnatia. *BCH* 98, 814–16.

Romizzi, L. (2001) *Ville d'otium dell'Italia antica (II sec. a.C.–I sec. d.C.)*. Naples.

Rook, T. (1978) The development and operation of Roman hypocausts. *JAS* 5, 269–82.

Root, M.C. (1973) An Etruscan horse race from Poggio Civitate. *AJA* 77, 121–37.

Rosafio, P. (1994) Slaves and coloni in the villa system, in Carlsen, Ørsted and Skydsgaard (eds), pp. 145–58.

Roselaar, S. (2008) Regional variations in the use of the "ager publicus," in de Ligt and Northwood (eds), pp. 573–604.

Roselaar, S. (2009) Competition for land and privatization of "ager publicus," in Carlsen and Lo Cascio (eds), pp. 281–301.

Rosenberg, S. (2008) *Hasmonean and Herodian Palaces at Jericho*, vol. 4, *The Decoration of Herod's Third Palace at Jericho*. Jerusalem.

Rosenberger, M. (1975) *The City Coins of Palestine*. Jerusalem.

Rosenstein, N. (2004) *Rome at War. Farms, Families, and Death in the Middle Republic*. London.

Rosenstein, N. (2008) Aristocrats and agriculture in the Middle and Late Republic. *JRS* 98, 1–26.

Rosenstein, N. (2009) Aristocrats and agriculture in the Late Republic: the "high count," in Carlsen and Lo Cascio (eds), pp. 243–58.

Rosenstein, N. and Morstein-Marx, R. (eds) (2006) *A Companion to the Roman Republic*. Oxford.

Rosenstein, N. and Morstein-Marx, R. (eds) (2009) *A Companion to the Roman Republic*. Oxford.

Rossiter, J. (1981) Wine and oil processing at Roman farms in Italy. *Phoenix* 35 (4), 345–61.

Rostovtzeff, M. (1957) *The Social and Economic History of the Roman Empire* (2nd edn). Oxford.

Roth, R. (2006) Black gloss wares from the Acropolis of Capena (La Civitucola, Provincia di Roma). *PBSR* 74, 119–63.

Roth, R. (2007a) Ceramic integration? Typologies and the perception of identities in Republican Italy, in Roth and Keller (eds), pp. 59–70.

Roth, R. (2007b) *Styling Romanisation: Pottery and Society in Central Italy*. Cambridge.

Roth, R.E. (2008) Review of Scott 2008. *BMCR* 17.06.2008.

Roth, R. and Keller, J. (eds) (2007) *Roman by Integration: Dimensions of Group Identity in Material Culture and Text* [*JRA* Suppl. 66]. Portsmouth, RI.

Roth, U. (ed.) (2010) *By the Sweat of Your Brow. Roman Slavery in Its Socio-economic Setting*, London.

Rotroff, S. (1997) Coins and stratigraphy, in Sheedy and Papageorgiadou-Banis (eds), pp. 8–16.

Rous, B. (2009) No place for cult: the sacred landscape of Latium in the Late Republic. *BABesch* 84, 53–84.

Rousseau, D. and van der Veen, A.M. (2005) The emergence of a shared identity. An agent-based computer simulation of idea diffusion. *Journal of Conflict Resolution* 49 (5), 686–712.

Rouveret, A. (1992) Les langages figuratifs de la peinture funéraire paestane. *Atti Taranto* 27, 179–282.

Rouveret, A. and Greco Pontrandolfo, A. (1983) Pittura funeraria in Lucania e Campania. Puntualizzazioni cronologiche e proposte di lettura. *DialArch* 1, 91–130.

Rowland, I. (1998) *The Culture of the High Renaissance. Ancients and Moderns in Sixteenth-Century Rome*. Cambridge.

Rowland, I. (2008) Marriage and mortality in the Tetnies Sarcophagi. *EtrStud* 11, 151–64.

Rowland, R.J. (2001) *The Periphery in the Center. Sardinia in the Ancient and Medieval Worlds* [*BAR-IS* 970]. Oxford.

Royal, J. (2012) Illyrian Coastal Exploration Program (2007–2009): the Roman and Late Roman finds and their contexts. *AJA* 116, 405–60.

Royo, M. (1999) *Domus Imperatoriae. Topographie, formation et imaginaire des palais impériaux du Palatin*. Rome.

Ruby, P. (2006) Peuples, fictions? Ethnicité, Identité ethnique et sociétés anciennes. *RÉA* 108, 25–60.

Ruddy, F.S. (1975) *International Law in the Enlightenment. The Background of Emmerich de Vattel's Le Droit des Gens*. New York.

Ruegg, S.D. (1995) *Underwater Investigations at Roman Minturnae: Liris-Garagliano River*. Gothenburg.

Ruggeri, M. (2003) Dall'età del Bronzo finale alla prima età del Ferro, in *Atti della XXXVI Riunione Scientifica Preistoria e Protostoria dell'Abruzzo*, pp. 451–64. Chieti-Celano.

Ruggeri, M. and Tavano, G. (eds) (2007) *Guerrieri e re dell'Abruzzo Antico*. Pescara.

Ruggiero, I. (1991/2) Ricerche sul tempio di Portuno nel Foro Boario: per una rilettura del monumento. *BullCom* 94, 253–86.

Ruiz, E. (2009) De Qart Hadast a Carthago Nova: apuntes para una síntesis, in Noguera and Madrid (eds), pp. 50–8.

Ruiz de Arbulo, J. (1992) Tarraco, Carthago Nova y el problema de la capitalidad en la Hispania Citerior republicana, in *Miscellània Arqueològica a Joseph M. Recasens*, pp. 115–30. Tarragona.

Ruiz de Arbulo, J. (1995) El santuario de Asklepios y las divinidades alejandrinas en la Neapolis de Ampurias (s. II –I a.C.). Nuevas hipótesis. *Verdolay* 7, 327–38.

Ruiz de Arbulo, J. (2002) La fundación de la colonia de Tarraco y los estandartes de César, in Jiménez and Ribera (eds), pp. 137–56.

Ruiz de Arbulo, J. (2002/3) Santuarios y fortalezas. Cuestiones de indigenismo, helenización y romanización en torno a Emporion y Rhode (s. VI–I a.C). *Cuadernos de Prehistoria y Arqueología, Universidad Autónoma de Madrid* 28/9, 161–202.

Ruiz de Arbulo, J. (2006) Scipionum opus and something more: an Iberian reading of the provincial capital (2nd–1st c. BC), in Abad, Keay and Ramallo (eds), pp. 33–44.

Ruiz de Arbulo, J. (2007) Las murallas de Tárraco. De la fortaleza romano-republicana a la ciudad tardo-antigua, in Rodríguez Colmenero and Rodà (eds), pp. 569–92.

Ruiz de Arbulo, J. (2009a) La legio Martia i la fundació de la colonia Tarraco, in *Tarraco pedra a pedra*, pp. 36–55. Tarragona.

Ruiz de Arbulo, J. (2009b) Arquitectura sacra y fundaciones urbanas en las Hispanias tardo-republicanas. Corrientes culturales, modelos edilicios y balance de novedades durante el siglo II a.C., in Mateos *et al.* (eds), pp. 253–97.

Rüpke, J. (ed.) (2007) *A Companion to Roman Religion*. Oxford.

Russo, A. (1992) Moltone di Tolve – Complesso residenziale, in De Lachenal (ed.), pp. 39–42.

Russo, A. and Di Giuseppe, H. (eds) (2008) *Felicitas Temporum. Dalla terra alle genti: La Basilicata settentrionale tra archeologia e storia*. Potenza.

Russo, F. (2003) Il sistema insediativo sannitico nelle fonti letterarie. *Rivista di Cultura Classica e Medioevale* 45, 277–304.

Russo, L. (2004) *The Forgotten Revolution: How Science Was Born in 300 BC and Why It Had to Be Reborn*. Berlin.

Rutter, N.K. (ed.) (2001) *Historia Numorum: Italy*. London.

Ryberg, I.S. (1955) *Rites of the State Religion in Roman Art* [*MAAR* 22].

Rykwert, J. (1988) *The Idea of a Town*. Cambridge.

Sabbatini, G. (2001) *Ager Venusinus I (Forma Italiae 40)*. Firenze.

Sabin, P., van Wees, H. and Whitby, M. (eds) (2007) *The Cambridge History of Greek and Roman Warfare*, vol. I, *Greece, the Hellenistic World and the Rise of Rome*. Cambridge.

Sada, P. (ed.) (2005) *Tàrraco i l'aigua*. Tarragona.

Sadori, L., Mercuri, A.M. and Mariotti Lippi, M. (2010) Reconstructing past cultural landscape and human impact using pollen and plant macroremains. *Plant Biosystems* 144 (4), 940–51.

Sage, M. (2008) *The Republican Roman Army: A Sourcebook*. London.

Sakai, S. (2000–1) La storia sotto il suolo del 79 d.C. Considerazioni sui dati provenienti dalle attività archeologiche svolte sulle fortificazioni di Pompei. *Opuscula Pompeiana* 8, 87–100.

Salinas de Frías, M. (1995) *El gobierno de las provincias hispanas durante la República romana (218–27 a.C.)*. Salamanca.

Sallares, R. (2002) *Malaria and Rome: A History of Malaria in Ancient Italy*. Oxford.

Salmon, E.T. (1967) *Samnium and the Samnites*. Cambridge.

Salmon, E.T. (1970) *Roman Colonization under the Republic*. Ithaca.

Salom, C. (2006) El *Auguraculum* de la Colonia Tárraco: *Sedes inaugurationis Coloniae Tarraco*. *ArchEsp* 79, 69–87.

Salviat, F. (1977) Orientation, extension et chronologie des plans cadastraux d'Orange. *Revue Archaologique de Narbonnaise* 10, 107–18.

Sammartano, R. (2003) Riflessioni sulla "Troianità" degli Elimi, in Quarte Giornate *Internazionali di Studi sull'Area Elima (Erice, 1–4 dicembre 2000). Atti*, pp. 1115–48. Pisa.

Sampson, G. (2010) *The Crisis of Rome: The Jugurthine and Northern Wars and the Rise of Marius*. Barnsley.

Sánchez-Moreno, E. (ed.) (2008) *Protohistoria y antigüedad de la península ibérica, II. La Iberia prerromana y la Romanidad*. Madrid.

Sanciu, A. (1997) *Una fattoria d'età romana nell'agro di Olbia*. Sassari.

Sanciu, A. (1998) Insediamenti rustici d'età tardo-repubblicana nell'agro di Olbia. *AfrRom* 12, 777–99.

Sanders, G.D.R. (1984) Reassessing ancient populations. *BSA* 79, 251–62.

Sannio. Pentri e Frentani dal VI al I sec. a.C. (1980) [Exhibition catalog]. Rome.

Santillo Frizell, B. (2009) Changing pastures, in Bjur and Santillo Frizell (eds), pp. 39–59.

Sartori, A. and Valvo, A. (eds) (2006) *Hiberia-Italia. Italia-Hiberia* [Convegno Internazionale di Epigrafia e Storia Antica, Brescia, 2005]. Milan.

Sassatelli, G. (2004) The Etruscans on the Po Plain, in Camporeale (ed.), pp. 168–91.

Sassatelli, G. and Govi, E. (eds) (2005) *Culti, forma urbana e l'artigianato a Marzabotto. Nuove prospettive di ricerca*. Bologna.

Sauron, G. (1987) Le complexe pompéien du champ de Mars, in *L'Urbs. Espace urbain et histoire*, pp. 457–73. Rome.

Sauron, G. (2001) Les Propylées d'Appius Claudius Pulcher à Éleusis: L'art Neo-Attique dans les contradictions idéologiques de la noblesse romaine à la fin de la République, in Marc and Moretti (eds), pp. 267–83.

Sax, J.L. (1999) *Playing Darts with a Rembrandt, Public and Private Rights in Cultural Treasures*. Ann Arbor.

Scaparro, M. *et al.* (eds) (1994) *Teatri greci e romani alle origini del linguaggio rappresentato: censimento analitico*. Rome.

Scardigli, B. (2002) Trent'anni di studi sertoriani, in Urso (ed.), pp. 143–61.

Schachter, A. (1992) Policy, cult and the placing of Greek sanctuaries, in Schachter (ed.), pp. 1–58.

Schachter, A. (ed.) (1992) *Le santuaire grec [Entretiens sur l'Antiquité Classique 37]*. Geneva.

Schäfer, T. (2006) Decorazione architettonica e stucchi di Cossyra, in Osanna and Torelli (eds), pp. 57–67.

Schapp, A. (1996) *The Discovery of the Past*. London.

Schattner, T. (2005) La Puerta de Sevilla en Carmona y otras puertas romanas en la península ibérica. *Romula* 4, 67–98.

Scheibler, I. (1995) *Griechische Töpferkunst. Herstellung, Handel und Gebrauch der antiken Tongefäße* (2nd edn). Munich.

Scheid, J. (1991) Sanctuaires et thermes sous l'Empire. *Les thermes romains: Actes de la table ronde organisée par l'École française de Rome* (*CÉFR* 142), 205–16.

Scheidel, W. (2001) Progress and problems in Roman demography, in Scheidel (ed.), pp. 1–81.

Scheidel, W. (ed.) (2001) *Debating Roman Demography*. Leiden.

Scheidel, W. (2004) Human mobility in Roman Italy, I: the free population. *JRS* 94, 1–26.

Scheidel, W. (2005) Human mobility in Roman Italy. II: the slave population. *JRS* 95, 64–79.

Scheidel, W. (2007) Demography, in Scheidel, Morris and Saller (eds), pp. 38–86.

Scheidel, W. (2008) Roman population size: the logic of the debate, in de Ligt and Northwood (eds), pp. 17–70.

Scheidel, W. and Friesen, S.J. (2009) The size of the economy and the distribution of income in the Roman Empire. *JRS* 99, 61–91.

Scheidel, W., Morris, I. and Saller, R. eds. (2007) *The Cambridge Economic History of the Greco-Roman World*. Cambridge.

Schepartz, L., Fox, S. and Bourbou, C. (eds) (2009) *New Directions in the Skeletal Biology of Greece* [*Hesperia* Suppl. 43]. Princeton.

Scheuer, L. and Black, S. (2004) *The Juvenile Skeleton*. London.

Schiavione, A. and Giardina, A. (eds) (1990) *Storia di Roma*. Turin.

Schiffman, L. (1975) *The Halakah at Qumran*. Leiden.

Schindler, M. (1967) *Archäologische Forschungen zu den Grabungen des Magdalensberges 1. Die "Schwarze Sigillata" des Magdalensberges*. Klagenfurt.

Schlereth, T. (ed.) (1988) *Material Culture : A Research Guide*. Lawrence, KS.

Schlereth, T. (1999) *Material Culture Studies in America: An Anthology*. Walnut Creek, CA.

Schnapp, A. (1993) *The Discovery of the Past: The Origins of Archaeology*, trans. I. Kinnes and G. Varndell. New York.

Schneider, G. (1990) Chemische Analysen römischer Keramik aus dem mittleren Rheingebiet. *Rei Cretariae Romanae Fautorum Acta* 27/8, 135–46.

Schnurbein, S. von (2001) Camps et *castella*, in Reddé and Schnurbein (eds), pp. 507–13.

Schnurbein, S. von (2006) Formes, taille, terminologie, configuration générale, in Reddé *et al.* (eds), pp. 68–71.

Schoonhoven, A.V. (2006) *Metrology and Meaning in Pompeii: The Urban Arrangement of Regio VI*. Rome.

Schowalter, D. and Friesen, S. (eds) (2005) *Urban Religion in Roman Corinth: Interdisciplinary Approaches*. Cambridge, MA.

Schröter, M.-G. (2008) Das Theaterkomplex des Cn. Pompeius Magnus in Kontext seiner Politik, in Albers (ed.), pp. 29–44.

Schulten, A. (1905) *Numantia. Eine topographische-historische Untersuchung. Abhandlungen der königlichen Gesellschaft der Wissenschaften zu Göttingen, philologische-historische Klasse*. Berlin.

Schulten, A. (1914) *Numantia. Die Ergebnisse der Ausgrabungen 1905–1912. Band I. Die Keltiberer und ihre Kriege mit Rom*. Munich.

Schulten, A. (1926) *Sertorius*. Leipzig.

Schulten, A. (1927) *Numantia. Die Ergebnisse der Ausgrabungen 1905–1912. Band III. Die Lager des Scipio*. Munich.

Schulten, A. (1928) Römische Lager in Spanien. *Forschungen und Fortschritte* 4, 41–3.

Schulten, A. (1929) *Numantia. Die Ergebnisse der Ausgrabungen 1905–1912. Band IV. Die Lager bei Renieblas*. Munich.

Schulten, A. (1931) *Numantia. Die Ergebnisse der Ausgrabungen 1905–1912. Band II. Die Stadt Numantia*. Munich.

Schulten, A. (1933) *Geschichte von Numantia*. Munich.

Schulten, A. (1937) *Fontes Hispaniae Antiquae, IV*. Barcelona.

Schultz, C.E. and Harvey, P.B. (eds) (2006) *Religion in Republican Italy*. Cambridge.

Schulze-Oben, H. (1989) *Freigelassene in den Städten des römischen Hispanien*. Bonn.

Schwinzer, E. and Steingräber, S. (eds) (1990) *Kunst und Kultur in der Magna Graecia: Ihr Verhältnis zum griechischen Mutterland und zum italischen Umfeld. Referate vom Symposium des Deutschen Archäologen-Verbandes: Städtisches Museum Schloß Reydt, Mönchengladbach, 8.–10.1.1988*. Tübingen.

Scibona, G. and Tigano, G. (eds) (2009) *Alaisa-Halaesa. Scavi e ricerche (1970–2007)*. Messina.

Scopacasa, R. (2009) Identity as social practice in the funerary sites of central Apennine Italy ("Samnium") 6th–3rd century BC. PhD thesis. University of Exeter.

Scopacasa, R. (2010) Beyond the warlike Samnites. Rethinking grave goods, gender relations and social practice in ancient Samnium (Italy), in Moore *et al.* (eds), pp. 120–31.

Scoton, P.D. (2005) A new fragment of an inscription from the Julian Basilica at Roman Corinth. *Hesperia* 74, 95–100.

Scott, A.R. (2008) *Cosa: The Black-glaze Pottery 2*. Ann Arbor.

Scott, R.T. (1988) The Latin Colony of Cosa. *DialArch ser.* 3, 6/2, 73–7.

Scott, R.T. (2005) The contribution of archaeology to early Roman history, in Rauflaub (ed.), pp. 98–106.

Scott, R.T. (ed.) (2009) *Excavations in the Area Sacra of Vesta (1987–1996)* [*MAAR* Suppl. 8]. Ann Arbor.

Scrinari, V.S. (1968 [1972]) Tombe a camera sotto via S. Stefano Rotondo presso l'ospedale di S. Giovanni in Laterano. *BullCom*, 17–23.

Scullard, H.H. (1982) *From the Gracchi to Nero: A History of Rome from 133 BC to AC 68*. London.

Sear, F. (1990) Vitruvius and Roman theater design. *AJA* 94, 249–58.

Sear, F. (2006) *Roman Theatres: An Architectural Study*. Oxford.

Seel, O. (1964) *Römertum und Latinität*. Stuttgart.

Segal, A., Mlynarczyk, J., Burdajewicz, M. *et al.* (eds) (2004) *Hippos-Sussita: Fifth Season of Excavations and Summary of All Five Seasons (2000–2004)*. Haifa.

Séguier, J.-M. (forthcoming) La consommation des vins italiens chez les Sénons, les Meldes et les *Parisii*, in Olmer (ed.).

Sekunda, N. (2007) Military forces. A. Land forces, in Sabin, van Wees and Whitby (eds), pp. 325–57.

Senatore, F. (ed.) (1999) *Pompei, il Vesuvio e la Penisola Sorrentina, Atti del secondo ciclo di conferenze di geologia, storia e archeologia, Pompei, ott. 1997–feb. 1998*. Rome.

Senatore, F. (2006) *La lega sannitica*. Capri.

Settis, S. and Catoni, M.L. (eds) (2008) *La forza del bello: l'arte greca conquista l'Italia*. Mantova.

Sewell, J. (2010) *The Formation of Roman Urbanism, 338–200 BC: Between Contemporary Foreign Influence and Roman Tradition* [*JRA* Suppl. 79]. Portsmouth, RI.

Shatzman, I. (1972) The Roman general's authority over booty. *Historia* 21, 177–205.

Shear, jr., T.L. (1981) Athens: from city-state to provincial town. *Hesperia* 50, 356–77.

Sheedy, K. and Papageorgiadou-Banis, C. (eds) (1997) *Numismatic Archaeology/Archaeological Numismatics: Proceedings of an International Conference held to Honour Dr. Mando Oeconomides in Athens 1995*. Oxford.

Shennan, S. (ed.) (1989) *Archaeological Approaches to Cultural Identity*. London.

Shipley, G. and Salmon, J. (eds) (1996) *Human Landscapes in Classical Antiquity. Environment and Culture*. London.

Shirley, E. (2001) *Building a Roman Legionary Fortress*. London.

Siebert, G. (1987) Quartier de Skardhana: la fouille. *BCH* 111, 629–42.

Sievers, J. (1990) *The Hasmoneans and their Supporters*. Atlanta.

Sievers, S. (1997) Alesia und Osuna: Bemerkungen zur Normierung der spätrepublikanischen Bewaffnung und Ausrüstung, in Feugère (ed.), pp. 271–6.

Sillières, P. (1997) *Baelo Claudia. Una ciudad romana de la Bética*. Madrid.

Simpson, E. (ed.) (1997) *The Spoils of War, World War II and Its Aftermath: The Loss, Reappearance, and Recovery of Cultural Property*. New York.

Sinn, F. (1987) *Stadtrömische Marmorurnen*. Mainz.

Sirena, G. (2011) *Via Pompeia. L'antico tracciato tra Messina e Siracusa*. Acireale and Rome.

Sjöquist, E. (1962) Excavations at Morgantina (Serra Orlando) 1961, preliminary report VI. *AJA* 66, 135–43.

Small, A.M. (ed.) (1992) *Gravina: An Iron Age and Republican Settlement in Apulia*, 2 vols. London.

Small, A.M. (ed.) (1996) *Subject and Ruler: The Cult of the Ruling Power in Classical Antiquity. Papers Presented at a Conference Held in the University of Alberta on April 13–15, 1994* [*JRA* Suppl. 17]. Ann Arbor.

Small, A.M. (2000) The use of javelins in central and south Italy in the 4th century BC, in Ridgway *et al.* (eds), pp. 221–34.

Small, A.M. (2001) Changes in the pattern of settlement and land use around Gravina and Monte Irsi, in Lo Cascio and Storchi Marino (eds), pp. 36–53.

Small, A.M. and Buck, R. (eds) (1994) *The Excavations of San Giovanni di Ruoti*, vol. 1, *The Villas and Their Environment*. Toronto.

Small, A.M. and Small, C. (2002) The Basentello Valley Survey (Apulia, Basilicata), in Attema *et al.* (eds), pp. 83–6.

Small, A.M., Small, C., Campbell, I. *et al.* (1998) Field survey in the Basentello Valley on the Basilicata-Puglia border. *Echos du Monde Classique/Classical Views* 42, 337–71.

Small, D.B. (1983) Studies in Roman theater design. *AJA* 87, 55–68.

Small, J.P. (1983) Choice of subject on late Etruscan funerary urns, in Kahil, Augé, and Linant de Bellefonds (eds), pp. 87–92.

Smith, A. (1991) *National Identity*. London.

Smith, C.J. (1996) *Early Rome and Latium. Economy and Society c. 1000 to 500 BC*. Oxford.

Smith, C.J. (2005) The beginnings of urbanization in Rome, in Osborne and Cunliffe (eds), pp. 91–111.

Smith, C.J. (2000) Early and Archaic Rome, in Coulston and Dodge (eds), pp. 16–41.

Smith, C.J. (2006) *The Roman Clan: The Gens from Ancient Ideology to Modern Anthropology*. Cambridge.

Smith, C.J. (2008) Ethnic identity in Roman politics. *JRA* 21, 387–8.

Smith, C.J. (ed.) (forthcoming) *The Cambridge Ancient History, New Edition: Plates to Volumes VIII.2 to IX (500–133 BC)*.Cambridge.

Smith, C.J. and Serrati, J. (eds) (2000) *Sicily from Aeneas to Augustus: New Approaches in Archaeology and History*. Edinburgh.

Smith, R.R.R. (1988) *Hellenistic Royal Portraits*. Oxford.

Smith, S.K. (2010) Differential diagnosis and discussion of a large nasal neoplasm from a Late Bronze Age Athenian male. *International Journal of Osteoarchaeology* 20, 731–6.

Söderlind, M. (2002) *Late Etruscan Votive Heads from Tessennano. Production, Distribution, Sociohistorical Context*. Rome.

Sogliano, A. (1937) *Pompeii nel suo sviluppo storico. Pompei preromana (dalle origine all'anno 80 a.C.)*. Rome.

Solé, J. *et al.* (eds) (2007) *Planimetria arqueològica de Tarraco 2007* [*Documenta* 5, ICAC]. Tarragona.

Sommella, P. (1978) *Forma e urbanistica di Pozzuoli romana* [Puteoli. Studi di Storia Antica 2]. Naples.

Soraci, C. (2011) *Sicilia frumentaria. Il grano siciliano e l'annona di Roma, V a.C.–V d.C.* Rome.

Sordi, M. (ed.) (1982) *Politica e religione nel primo scontro tra Roma e l'oriente*. Milan.

Sordi, M. (2006) La più antica iscrizione latina in Hispania, in Sartori and Valvo (eds), pp. 1–4.

Sotgiu, G. (ed.) (1985) *Studi in onore di Giovanni Lilliu per il suo settantesimo compleanno*. Cagliari.

Soto, P. de (2010) Análisis de la xarxa de comunicacions i del transport a la Catalunya romana: estudis de distribució i mobilitat. PhD thesis. Universitat Autónoma de Barcelona.

Sourlas, D. (2008) ηεότερα στοιχεία για τη ρωμαικη αορά της αθήνας, in Vlizos (ed.), pp. 99–114.

Spanò Giammellaro, A. (ed.) (2005) *Atti del V Congresso Internazionale di studi fenici e punici: Marsala–Palermo, 2–8 ottobre 2000*. Palermo.

Spanò Giamellaro, A., Spatafora, F. and van Dommelen, P. (2008) Sicily and Malta: between sea and countryside, in van Dommelen and Gómez Bellard (eds), pp. 129–58.

Spatafora, F. and Montali, G. (2006) Palermo: nuovi scavi nell'area di Piazza della Vittoria, in Osanna and Torelli (eds), pp. 133–51.

Spawforth, A.J. (1997) The early reception of the imperial cult in Athens, in Hoff and Rotroff (eds), pp. 183–201.

Spina, L., Arciero, G. and Sampaolo, V. (1997) *L'Anfiteatro campano di Capua*. Naples.

Spinazzola, V. (1915) Pozzuoli. Rovine di un second anfiteatro. *NSc* 12, 409–15.

Spurr, M.S. (1986) *Arable Cultivation in Roman Italy c.200 BC–c.AD 100*. London.

Squarciapino Floriani, M. (1958) *Le necropoli. Parte I: Le tombe di età repubblicana e Augustea*. Rome.

Staccioli, R. (1961) Terme minori e balnea nella documentazione della "Forma Urbis." *ArchCl* 13, 92–102.

Staccioli, R.A. (2003) *The Roads of the Romans*. Los Angeles.

Stambaugh, J.E. (1978) The functions of Roman temples, in *ANRW* II.16.1, pp. 554–608.

Stamper, J.W. (2005) *The Architecture of Roman Temples*. Cambridge.

Stanco, E.A. (1999) La ceramica a vernice nera dello scavo di Lungotevere Testaccio. *BullCom* 100, 7–36.

Stanco, E.A. (2009) La seriazione cronologica della ceramica a vernice nera etrusco-laziale nell'ambito del III secolo a.C., in Jolivet *et al.* (eds), pp. 157–93.

Stanco, F., Battiato, S. and Gallo, G. (eds) (2011) *Digital Imaging for Cultural Heritage Preservation: Analysis, Restoration, and Reconstruction of Ancient Artworks*. Boca Raton.

Stark, M.T. (ed.) (1998) *The Archaeology of Social Boundaries*. London.

Staveley, E.S. (1959) The political aims of Appius Claudius Caecus. *Historia* 8, 410–33.

Stazio, A. (ed.) (2002) *Taranto e il Mediterraneo: Atti del Quarantunesimo Convegno di studi sulla Magna Grecia. Taranto, 12–16 ottobre 2001*. Taranto.

Steel, C.E.W. (2002) *Cicero, Rhetoric and Empire*. Oxford.

Stein, G. (ed.) (2005) *The Archaeology of Colonial Encounters. Comparative Perspectives*. Oxford.

Steingräber, S. (1981) *Etrurien. Städte – Heiligtümer – Nekropolen*. Munich.

Steingräber, S. (2006a) *Abundance of Life: Etruscan Wall Painting*. Los Angeles.

Steingräber, S. (2006b) *Affreschi etruschi: Dal periodo geometrico all'Ellenismo*. San Giovanni Lupatoto.

Stek, T.D. (2009) *Cult Places and Cultural Change in Republican Italy*. Amsterdam.

Stek, T.D. (in press) Questions of Cult and continuity in Late Republican Roman Italy: "Italic" or "Roman" sanctuaries and the so-called pagus-vicus system, in Jehne (ed.).

Stek, T.D. and Burgers, G.-J. (eds) (in press) *The Impact of Rome on Cult Places and Religion in Italy: New Approaches to Change and Continuity*. Amsterdam.

Stek, T.D. and Pelgrom, J. (eds) (forthcoming) *Roman Republican Colonisation: New Archaeological and Hstorical Perspectives.*

Stella, M.S. and Laidlaw, A. (2008) Nuove indagini nella Casa di Sallustio (VI 2,4), in Guzzo and Guidobaldi (eds), 2008, pp. 147–57.

Stephanidou-Tiveriou, T. (2008) Tradition and Romanization in the monumental landscape of Athens, in Vlizos (ed.), pp. 11–40.

Stern, E. (ed.) (2008) *The New Encyclopedia of Archaeological Excavations in the Holy Land Supplementary Volume 5*. Jerusalem.

Stevens, J. and Stevens, A. (2007) Via dell'Abbondanza Project, www.pompeiiperspectives.org (accessed November 18, 2012).

Stevens, S.T. (1991) Charon's obol and other coins in ancient funerary practice. *Phoenix* 45, 215–29.

Stewart, A.J.S. (2004) *Attalos, Athens, and the Akropolis. The Pergamene "Little Barbarians" and Their Roman and Renaissance Legacy*. Cambridge.

Stewart, T.D. (ed.) (1970) *Personal Identification in Mass Disasters*. Washington DC.

Stillwell, R. (1964) The Theater of Morgantina. *Kokalos* 10–11, 579–88.

Stillwell, R. (1967) Excavations at Morgantina (Serra Orlando) 1966, preliminary report IX. *AJA* 71, 245–50.

Stilp, F. (2001) *Mariage et suovetaurilia: étude sur le soi-disant "Autel Ahenobarbus"* [*RdA* Suppl. 26]. Rome.

Stirling, L., Mattingly, D.J. and Ben Lazreg, N. (eds) (2001) *Leptiminus (Lamta). Report no. 2, The East Baths, Cemeteries, Kilns, Venus Mosaic, Site Museum and Other Studies* [*JRA* Suppl. 41]. Portsmouth, RI.

Stolle, F. (1912) *Das Lager und Heer der Römer*. Strasbourg.

Stone, D.L. (2004) Problems and possibilities in comparative survey: a North African perspective, in Alcock and Cherry (eds), pp. 132–43.

Stone, D.L. (2007a) Monuments on the margins: interpreting the late 1st millennium BCE rock-cut tombs (*haouanet*) of North Africa, in Stone and Stirling (eds), pp. 43–74.

Stone, D.L. (2007b) Burial, identity, and local culture in North Africa, in van Dommelen and Terrenato (eds), pp. 126–44.

Stone, D.L. and Stirling, L. M. (eds) (2007) *Mortuary Landscapes of North Africa* [*Phoenix* Suppl. 43]. Toronto.

Stone, P. (ed.) (2011) *Cultural Heritage, Ethics, and the Military*. Woodbridge.

Stopponi, S. (ed.) (1985) *Case e palazzi d'Etruria*. Milan.

Storchi Marino, A. (ed.) (1995) *L'incidenza dell'antico. Studi in memoria di Ettore Lepore vol. 1*. Naples.

Storchi Marino, A. (ed.) (1996) *L'incidenza dell'antico. Studi in memoria di Ettore Lepore vol. 2*. Naples.

Storey, G. (ed.) (2006) *Urbanism in the Preindustrial World*. Tuscaloosa, AL.

Strange, J. (1992) Six campaigns at Sepporis, in Levine (ed.), pp. 339–56.

Strazzulla, M.J. (1973) *Il santuario sannitico di Pietrabbondante* [Documenti di antichità italiche e romane 1]. Molise.

Strazzulla, M.J. (1997) La testa in bronzo di personaggio virile da S.Giovanni Lipioni, in Campanelli and Faustoferri (eds), pp. 8–13.

Strazzulla, M.J. (2006) Ancora sul frontone di via S. Gregorio: alcune considerazioni. *Scienze dell'Antichità* 13, 249–64.

Strocka, V. (1991) *Casa del Labirinto* [Häuser in Pompeji Band 4]. Munich.

Studi sull'Italia dei Sanniti (2000) Rome.

Stylow, A. (1996) De Corduba a Colonia Patricia. La fundación de la Corduba romana, in León (ed.), pp. 37–47.

Suano, M. (1991a) Alcune osservazioni sui cinturoni di bronzo di tipo sannitico, in Capini and Di Niro (eds), pp. 135–9.

Suano, M. (1991b) A princeless society in a princely neighbourhood, in Herring, Whitehouse and Wilkins (eds), pt. 2, pp. 65–72.

Suano, M. (2000) Il cinturone sabellico-sammnita come abbigliamento sociale, in La Regina (ed.), pp. 183–91.

Suano, M. (in press) *I vivi fra i loro morti: frequentazione rituale nella necropoli di Tornareccio*. Chieti.

Suchey, J.M. (1979) Problems in the aging of females using the os pubis. *American Journal of Physical Anthropology*, 51, 467–70.

Sumner, G. (1970) The legion and the centuriate organization. *JRS* 60, 67–78.

Swaddling, J. and Perkins, P. (eds) (2009) *Etruscan by Definition: Papers in Honour of Sybille Haynes*. London.

Swift, E. (2009) *Style and Function in Roman Decoration*. London.

Symposium de Arqueología. Pollentia y la romanización de las Baleares. XXI Centenario de la fundación de Pollentia (Alcudia-Baleares) (Alcúdia, 1997) (1983) Palma de Mallorca.

Tagliamonte, G. (1994) *I figli di Marte. Mobilita, mercenari e mercenariato italici in Magna Grecia e Sicilia*. Roma.

Tagliamonte, G. (1996) *I Sanniti. Caudini, Irpini, Pentri, Carricini, Frentani*. Milan.

Tagliamonte, G. (2004) Il mercenariato itálico nel mondo italiota del IV secolo a.C., in *Alessandoro il Molosso e i "condottieri" in Magna Grecia* (Atti del quarantatreesimo Convegno di studi sulla Magna Graecia, Taranto), pp. 135–64. Taranto.

Tagliamonte, G. (2007) Considerazioni sull'architettura santuariale di età tardo-reppubblicana tra Campania e Sannio, in Quilici and Quilici Gigli (eds), pp. 53–68.

Talamo, E. (1998) Gli Horti di Sallustio a Porta Collina, in Cima and La Rocca (eds), pp. 113–69.

Talamo, E. (2008a) La scenografia del trionfo nella pittura funeraria, in La Rocca and Tortorella (eds), pp. 62–71.

Talamo, E. (2008b) La fronte depinta del sepolcro degli Scipioni, in La Rocca and Tortorella (eds), p. 119.

Talamo, E. (2008c) Frammento di affresco con scena storica, in La Rocca and Tortorella (eds), pp. 168–9.

Talbert, R. (ed.) (2000) *Barrington Atlas of the Greek and Roman World*. Princeton.

Taliercio Mensitieri, M. (1996) La monetazione di Poseidonia-Paestum dall'occupazione lucana alla colonia latina, in Cipriani, Longo and Viscione (eds), pp. 210–14.

Taloni, M. (1973) La necropolis dell'Esquilino, in *Roma medio repubblicana*, pp. 188–200.

Tanner, J. (2000) Portraits, power and patronage in the late Roman Republic. *JRS* 90, 18–50.

Tanner, J. and Osborne, R. (eds) (2007) *Art's Agency and Art History*. London.

Taplin, O. (1993) *Comic Angels and Other Approaches to Greek Drama through Vase-Paintings*. Oxford.

Tarpin, M. (2002) *Vici et pagi dans l'Occident romain*. Rome.

Tatum, W. (1992) Review of N. Rosenstein, *Imperatores Victi. Military Defeat and Aristocratic Competition in the Middle and Late Republic*. *AJP* 113, 637–41.

Taylor, D.M. (1957) Cosa: the black glaze pottery. *MAAR* 25, 65–193.

Taylor, L.R. (1960) *The Voting Districts of the Roman Republic: The Thirty-five Urban and Rural Tribes* [PAAR 20]. Rome.

Taylor, R. (2002) Temples and terracottas at Cosa. *AJA* 106, 59–83.

Taylor, R. (2003) *Roman Builders: A Study in Architectural Process*. Cambridge.

Taylor, R. (2008) Review of: *Roman Theatres: an Architectural Study* by F. Sear; *The Roman Amphitheatre: from Its Origins to the Colosseum* by K.E. Welch. *JSAH* 67, 443–5.

Tchernia, A. (1986) *Le vin de L'Italie romaine*. Rome.

Tchernia, A. (2009) L'exportaion du vin : interprétations actuelles de l'exception gauloise, in Carlsen and Lo Cascio (eds), pp. 91–113.

Tchernia, A. and Brun, J-P. (1999) *Le vin romain*. Grenoble.

Tchernia, A., Pomey, P. and Hesnard, A. (1978) *L'épave romaine de la Madrague de Giens*. Paris.

Temelini, M. (2006) Pompey's politics and the presentation of his theatre-temple complex, 61–52 BCE. *Studia Humaniora Tartuensia* 7, 1–14.

Terrenato, N. (1992) Velia and Carinae: some observations on an area of Archaic Rome. *Papers of the Fourth Conference of Italian Archaeology* 4 (2), 31–47.

Terrenato, N. (1998a) The Romanization of Italy. Global acculturation or cultural bricolage?, in Forcey, Hawthorne and Witcher (eds), pp. 20–7.

Terrenato, N. (1998b) Tam firmum municipium: the Romanization of Volaterrae and its cultural implications. *JRS* 88, 94–114.

Terrenato, N. (2000) The visibility of sites and the interpretation of field survey results: towards an analysis of incomplete distributions, in Francovich and Patterson (eds), pp. 60–71.

Terrenato, N. (2001a) The auditorium site and the origins of the roman villa. *JRA* 14, 5–32.

Terrenato, N. (2001b) Introduction, in Keay and Terrenato (eds), pp. 1–6.

Terrenato, N. (2001c) A tale of three cities: the Romanization of northern coastal Etruria, in Keay and Terrenato (eds), pp. 54–67.

Terrenato, N. (2004) Sample size matters! The paradox of global trends and local surveys, in Alcock and Cherry (eds), pp. 36–48.

Terrenato, N. (2007) The essential countryside: the Roman world, in Alcock and Osborne (eds), pp. 139–61.

Terrenato, N. (2008) The cultural implications of the Roman conquest, in Bispham (ed.), pp. 234–64.

Terrenato, N. (2011) The versatile clans: Archaic Rome and the nature of early city-states in central Italy, in Terrenato and Haggis (eds), pp. 231–44.

Terrenato, N. (2012) The enigma of "Catonian" villas: the *De agricultura* in the context of second century BC Italian architecture, in Becker and Terrenato (eds), pp. 69–93.

Terrenato, N. and Haggis, D. (eds) (2011) *State Formation in Italy and Greece: Questioning the Neoevolutionist Paradigm*. Oxford.

Teutsch, L. (1962) *Das Städtewesen in Nordafrika in der Zeit von C. Gracchus bis zum Tode des Kaisers Augustus*. Berlin.

Thesaurus Cultus et Rituum Antiquorum, I–VI (2004–6). Los Angeles.

Thomas, M.L. (2000) The technology of daily life in a Hellenistic Etruscan settlement. *EtrStud* 7, 107–8.

Thomas, M. and Meyers, G. (eds) (forthcoming) *Monumentality in Etruscan and Early Roman Architecture: Ideology and Innovation*. Austin.

Thomas, R. and Wilson, A. (1994) Water supply for Roman farms in Latium and south Etruria. *PBSR* 62, 139–96.

Thompson, H.A. (1987) The impact of Roman architects and architecture on Athens, in Macready and Thompson (eds), 1987, pp. 1–17.

Thompson, H.A. and Wycherley, R.E. (1972) *The Athenian Agora. Results of the Excavations Conducted by the American School of Classical Studies at Athens* [*The Agora at Athens* 14]. Princeton.

Thomsen, R. (1957–61) *Early Roman Coinage: A Study of Chronology*, 3 vols. Copenhagen.

Thuillier, J.-P. (1975) Denys d'Halicarnasse et les jeux romains (*Antiquités Romains* VII, 72–73). *MÉFRA* 87, 563–81.

Thuillier, J.-P. (1982) Le programme "athlétique" des ludi circenses dans la Rome républicaine. *RÉL* 60, 105–22.

Thuillier, J.-P. (1985) *Les jeux athlétiques dans la civilisation étrusque* [BÉFAR 256]. Rome.

Thünen, J.H. von (1826) *Der isolierte Staat in Beziehung auf Landwirtschaft und Nationalökonomie*. Jena.

Tinner, W. and Theurillat, J.-P. (2003) Uppermost limit, extent, and fluctuations of the timberline and treeline ecocline in the Swiss Central Alps during the past 11,500 years. *Arctic, Antarctic, and Alpine Research* 35 (2), 158–69.

Tirelli, M. (1993) *Il Museo archeologico nazionale e le aree archeologiche di Altino*. Padova.

Tirelli, M. (1997) Horti cum aedificiis sepulturis adiuncti: I monumenti funerari delle necropoli di Altinum, in Mirabella Roberti (ed.), pp. 175–210.

Tirelli, M. (2003) Nuovi dati da Altino preromana, in Braccesi and Antonelli (eds), pp. 223–34.

Todd, T.W. (1920) Age changes in the pubic bone. I. The male white pubis. *American Journal of Physical Anthropology* 3, 285–334.

Tolle-Kastenbein, R. (1994) *Das Olympieion in Athen*. Cologne.

Tomassetti, A. (2000) Un edificio antico lungo il clivus Argentarius: Il monumentum di Gaio Publicio Bibulo. *BullCom* 101, 39–80.

Tomei, D. (2008) *Gli edifici sacri della Sardegna romana: problemi di lettura e di interpretazione*. Ortacesus.

Tomei, M. (ed.) (2006) *Roma: Memorie del sottosuolo: Ritrovamenti archeologici 1980/2006*. Milan.

Torelli, M. (1975) *Elogia Tarquiniensa*. Florence.

Torelli, M. (1980) Per il Sannio tra IV e I sec. a.C.: note di archeologia, in *Sannio*, pp. 27–34.

Torelli, M. (1985) I duodecim populi Etruriae. *Annali della fondazione per il Museo Claudio Faina* 2, 37–57.

Torelli, M. (1990) La formazione della villa, in Schiavone and Giardina (eds), pp. 123–32.

Torelli, M. (1992) Paestum Romana, in *Atti Convegni Magna Grecia 27*, pp. 33–115. Naples.

Torelli, M. (1993) Le iscrizioni, in Pedley and Torelli (eds), pp. 195–221.

Torelli, M. (1995) *Studies in the Romanization of Italy* (ed. and trans. H. Fracchia and M. Gualtieri). Edmonton.

Torelli, M. (ed.) (1995) *Ostraka* 4 (1), 9–190.

Torelli, M. (1996) La romanizzazione del Sannio, in del Tutto Palma (ed.), pp. 27–44.

Torelli, M. (1999) *Tota Italia. Essays in the Cultural Formation of Roman Italy*. Oxford.

Torelli, M. (2000) The Etruscan city-state, in Hansen (ed.), 2000, pp. 189–203.

Torelli, M. (2003) Un avo della *domi nobilis*. Mineia M.F. in una nuova iscrizionde lucana di Paestum. *Ostraka* 12, 103–6.

Torelli, M. (2006) The topography and archaeology of Republican Rome, in Rosenstein and Morstein-Marx (eds), pp. 81–101.

Torelli, M. (2007) L'urbanistica di Roma regia e repubblicana. La città medio-reppublicano, in Gros and Torelli (eds), pp. 81–157.

Torelli, M. (2010) The topography and archaeology of Republican Rome, in Rosenstein and Morstein-Marx (eds), pp. 81–101.

Torelli, M. (2011) Concluding remarks, in Colivicchi (ed.), pp. 203–15.

Torelli, M. (2012) The early villa: Roman Contributions to the development of a Greek protoype, in Becker and Terrenato (eds).

Torelli, M., Menichetti, M. and Grassigli, G. (2008) *Arte e archeologia del mondo romano*. Milan.

Torelli, M. and Sgubini, A.M. (eds) (2008) *Etruschi. Le antiche necropoli del Lazio*. Verona.

Tosi, G. and Lucia, B. (2003) *Gli edifici per spettacoli nell'Italia romana*. Rome.

Toynbee A. (1965) *Hannibal's Legacy: The Hannibalic War's Effects on Roman Life*, 2 vols. London.

Tranoy, A. (1981) *La Galice Romaine*. Paris.

Treloar, A. (1971) The Roman Shield: Polybius vi.23.2. *The Classical Review* 21 (1), 3–5.

Trendall, A.D. (1967) *Phlyax Vases* (2nd rev. and enlarged edn). London.

Trendall, A.D. (1976) *South Italian Vase Painting* (2nd edn). London.

Trigger, B.G. (1989) *A History of Archaeological Thought*. Cambridge.

Trillmich, W. and Zanker, P. (eds) (1990) *Stadtbild und Ideologie. Die Monumentalisierung hispanischer Städte zwischen Republik und Kaiserzeit (Madrid, 1987)*. Munich.

Tronchetti, C. (1990) *Cagliari fenicia e punica*. Sassari.

Trotter, M. (1970) Estimation of stature from intact long bones, in Stewart (ed.), pp. 71–83.

Trotter, M. and Glesser, G. (1952) Estimation of stature from long bones of American whites and negroes. *American Journal of Physical Anthropology* 10, 463–514.

Trotter, M. and Glesser, G. (1958) A reevaluation of estimation of stature based on measurements of stature taken during life and of long bones after death. *American Journal of Physical Anthropology* 16, 79–123.

Trousset, P. (1977) Nouvelles observations sur la centuriation romaine a l'est d'el Jem. *AntAfr* 11, 175–207.

Trousset, P. (1978) Les bornes de Bled Segui. Nouveaux aperçus sur la centuriation romaine du sud Tunisie. *AntAfr* 12, 125–78.

Trousset, P. (ed.) (1995) *Monuments funéraires, institutions autochtones en Afrique du Nord antique et médiévale* [Actes du colloque sur l'histoire et l'archéologie de l'Afrique du Nord 6.2]. Paris.

Trümper, M. (1998) *Wohnen in Delos: eine baugeschichtliche Untersuchung zum Wandel der Wohnkultur in hellenistischer Zeit*. Rahden.

Trümper, M. (2003) The material and social environment of Graeco-Roman households of the east: the case of Hellenistic Delos, in Balch and Osiek (eds), pp. 19–44.

Trümper, M. (2008) *Die "Agora des Italiens" in Delos: Baugeschichte, Architektur, Ausstattung und Funktion einer späthellenistischen Porticus-Anlage, I–II.* Rahden.

Tsafrir, Y. (1975) The location of the Seleucid Akra in Jerusalem. *RBibl* 82, 501–21.

Tsakirgis, B. (1995) Morgantina: a Greek town in central Sicily, in Fischer-Hansen (ed.), pp. 123–47.

Tsiolis, V. (2001) Las Termas de *Fregellae*. Arquitectura, tecnología y cultura balnear en el Lacio durante los siglos III y II a.C. *Cuadernos de Prehistoria y Arqueología Universidad Autónoma de Madrid* 27, 85–114.

Tsiolis, V. (2006) Fregellae: Il complesso termale e le origini degli edifici balneari urbani nel mondo romano, in Osanna and Torelli (eds), pp. 243–55.

Tsiolis, V. (2008a) Modelli di convivenza urbana. Fregellae e la questione dell'introduzione delle pratiche termali nel Lazio meridionale, in Corsi and Polito (eds), pp. 133–43.

Tsiolis, V. (2008b) El modelo balnear republicano entre Italia e Hispania, in Uroz, Noguera and Coarelli (eds), pp. 285–306.

Tucci, P. (2005) "Where high Moneta leads her steps sublime", the "tabularium" and the Temple of Juno Moneta. *JRA* 18, 7–33.

Tuck, A.S. (1994) The Etruscan seated banquet: Villanovan ritual and Etruscan iconography. *AJA* 98, 617–28.

Tuck, S.L. (2000) A new identification for the "Porticus Aemilia." *JRA* 13, 175–82.

Turfa, J.M. (2006) Votive offerings in Etruscan religion, in De Grummond and Simon (eds), pp. 90–115.

Tusa, S. (2010) *Archeologia e storia nei mari di Sicilia.* Udine.

Tybout, R.A. (1989) *Aedificiorum Figurae. Untersuchungen zu den Architekturdarstellungen des frühen zweiten Stills.* Amsterdam.

Tybout, R.A. (2001) Roman wall painting and social significance. *JRA* 14, 33–56.

Tzifopoulos,Y. (1993) Mummius' dedications at Olympia and Pausanias' attitude to the Romans. *GRBS* 34, 93–100.

Ubelaker, D.H. and Rife, J. (2007) The practice of cremation in the Roman-era cemetery of Kenchreai, Greece. *Bioarchaeology of the Near East* 1, 35–57.

Ubelaker, D.H. and Rife, J. (2011) Skeletal analysis and mortuary practice in an early Roman chamber tomb at Kenchreai, Greece. *International Journal of Osteoarchaeology* 21, 1–18.

Uggeri, G. (1990) La Via Appia nella politica espansionistica di Roma, in Quilici Gigli (ed.), 1990, pp. 21–8.

Uggeri, G. (ed.) (1992) *L'età annibalica e la Puglia.* Mesagne.

Uggeri, G. (2004) *La viabilità della Sicilia in età romana.* Galatina.

Uggeri, G. (2007) La formazione del sistema stradale romano in Sicilia, in Micciché, Modeo and Santagati (eds), pp. 228–43.

Ulbert, G. (1984) *Cáceres el Viejo. Ein spätrepublikanisches Legionslager in Spanisch-Extremadura* [Madrider Beiträge 11]. Mainz.

Ulrich, R. and Quenomoen, C. (eds) (in press) *Blackwell Companion to Roman Architecture.* Oxford.

Untermann, J. (ed.) (2000) *Monumenta Linguarum Hispanicarum V.1.* Wiesbaden.

Uroz, J., Noguera, J. and Coarelli, F. (eds) (2008) *Iberia e Italia: modelos romanos de integración territorial (Murcia, 2006).* Murcia.

Urso, P. (ed.) (2002) *Hispania terris omnibus felicior. Atti del Convegno Internazionale (Cividale del Friuli, 2001).* Pisa.

Valdiserri Paoletti, I. (1980) Cippi funerari cilindrici dal territorio di Marruvium. *MemLinc* 8, 193–216.

Válek, J., Groot, C. and Hughes, J. (eds) (in press) *Historic Mortars: Characteristics and Tests.*

Valenti, M. (ed.) (2010) *Monumenta: I mausolei romani, tra commemorazione funebre e propaganda celebrativa. Atti del Convegno di Studi Monte Porzio Catone, 25 ottobre 2008.* Rome.

Valenza Mele, N. (1996) Una nuova tomba dipinta a Cuma e la "legio linteata," in Storchi Marino (ed.), 1996, pp. 325–60.

Valeri, C. (2010) Il paesaggio funerario a Roma tra il III e il I secolo a.C., in La Rocca, Presicce and Lo Monaco (eds), pp. 137–47.

Vallance, J.T. (1996) Dietetics. *Oxford Classical Dictionary* (3rd edn). Oxford.

Van Deman, E.B. (1912) Methods of determining the date of Roman concrete monuments. *AJA* 16, 230–51.

Van Deman, E.B. (1934) *The Building of the Roman Aqueducts.* Washington DC.

van der Leeuw, S. (1993) Giving the potter a choice: conceptual aspects of pottery techniques, in Lemonnier, ed., pp. 238–88.

Vandermersch, C. (1994) *Vins et amphores de Grand Grèce et de Sicile, IVe–IIIe s. avant J.-C. (Études I).* Naples.

Van der Poel, H. (1986) *Corpus Topographicum Pompeianum.* Rome.

van Dommelen, P. (1998a) *On Colonial Grounds. A Comparative Study of Colonialism and Rural Settlement in First Millennium* BC *West Central Sardinia.* Leiden.

van Dommelen, P. (1998b) Punic persistence. Colonialism and cultural identities in Roman Sardinia, in Laurence and Berry (eds), pp. 25–48.

van Dommelen, P. (2000) Post depositional effects and ceramic analysis: comment on Jeremy Taylor, "Depositional and post-depositional effects on ploughsoil ceramic assemblages," in Francovich and Patterson (eds), pp. 27–8.

van Dommelen, P. (2001) Cultural imaginings. Punic tradition and local identity in Roman Republican Sardinia, in Keay and Terrenato (eds), pp. 68–84.

van Dommelen, P. (2006) Punic farms and Carthaginian Colonists: Surveying Punic Rural Settlement in the Central Mediterranean. *JRA* 19, 7–28.

van Dommelen, P. (2007) Beyond resistance: Roman power and local traditions in Punic Sardinia, in van Dommelen and Terrenato (eds), pp. 55–70.

van Dommelen, P. and Finocchi, S. (2008) Sardinia: diverging landscapes, in van Dommelen and Gómez Bellard (eds), pp. 159–202.

van Dommelen, P. and Gómez Bellard, C. (eds) (2008) *Rural Landscapes of the Punic World* [Monographs in Mediterranean Archaeology 11]. London.

van Dommelen, P. and Knapp, A.B. (eds) (2010) *Material Connections in the Ancient Mediterranean. Mobility, Materiality and Identity.* London.

van Dommelen, P. and Terrenato, N. (2007) Introduction: local cultures and the expanding Roman Republic, in van Dommelen and Terrenato (eds), pp. 7–12.

van Dommelen P. and Terrenato, N. (eds) (2007) *Articulating Local Cultures: Power and Identity under the Expanding Roman Republic* [*JRA* Suppl. 63]. Portsmouth, RI.

van Kampen, I., Rathje, A. *et al.* (2005) The Pre-Republican habitation layers of the Via Sacra in Giacomo Boni's excavations of the Sepolcreto Arcaico, in *Papers of the Sixth Conference of Italian Archaeology* 6, pp. 745–53. Oxford.

Van Tilburg, C. (2007) *Traffic and Congestion in the Roman Empire.* London.

Van Wonterghem, F. (1999) Il culto di Ercole e la pastorizia nell'Italia centrale, in Petrocelli (ed.), pp. 413–28.

Vann, R. (ed.) (1992) *Caesarea Papers* [*JRA* Suppl. 5]. Portsmouth, RI.

Vannini, G. (ed.) (1985) *L'antico palazzo dei vescovi a Pistoia I. Le indagini archeologiche.* Florence.

Vannini, G. (ed.) (1987) *L'antico palazzo dei vescovi a Pistoia II. I documenti archeologici.* Florence.

Vaquerizo, D. (2006) Corduba, una ciudad puente, in Rascón and Sánchez (eds), pp. 123–41.

Vaquerizo, D. (ed.) (2010) *Las áreas suburbanas en la ciudad histórica. Topografía, usos, function* [Monografías de Arqueología Cordobesa 18]. Córdoba.

Varone, A. (1988) La struttura insediatira di Pompei: l'avvio di una indagine computerizzata per la lonoscenza della realtà economica e social di una città della prima età imperiale, in Franchi dell'Orto (ed.), pp. 25–48.

Varone, A. (2005) Il progetto di scavo e pubblica fruizione dell'insula pompeian dei Casti Amanti (Insula IX 12), in Guzzo and Guidobaldi (eds), 2005, pp. 191–9.

Varone, A. (2008) Per la storia recente, antica e antichissima del sito di Pompei, in Guzzo and Guidobaldi (eds), 2008, pp. 349–60.

Vasaly, A. (1993) *Representation, Images of the World in Ciceronian Oratory*. Berkeley.

Vassallo, S. (2000) Abitati indigeni ellenizzati della Sicilia centro-occidentale dalla vitalità tardo-arcaica alla crisi del V sec. a.C, in *Terze giornate internazionali di studi sull'area elima. Atti II*, pp. 983–1008. Pisa and Gibellina.

Vassallo, S. (2009) La villa rustica di Contrada San Luca (Castronovo di Sicilia). *Ampolo* 2009, 671–7.

Veal, R. (2008) Fuel supplies for Pompeii. Pre-Roman and Roman charcoals of the Casa delle Vestali, in Fiorentino and Magri (eds), pp. 287–98.

Venit, M.S. (2002) *Monumental Tombs of Ancient Alexandria. The Theater of the Dead*. Cambridge.

Ventura, A., León, M. and Márquez, C. (1998) Roman Córdoba in the light of recent archaeological research, in Keay (ed.), pp. 87–107.

Ventura, A., Márquez, C., Monterroso, A. and Carmona, M. (eds) (2002) *El teatro romano de Córdoba*. Córdoba.

Versluys, M.J. (2004) *Isis Capitolina* and the Egyptian cults in Late Republican Rome, in Bricault (ed.), pp. 421–28.

Versluys, M.J. (2008) Exploring identities in the Phoenician, Hellenistic and Roman east. A review article. *BibO* 65, 342–56.

Versluys, M.J. (2010) Understanding Egypt in Egypt and beyond, in Bricault and Versluys (eds), pp. 6–37.

Versluys, M.J. (in press) *Nemrud Da and Commagene under Antiochos I. Material Culture, Identity and Style in the Late Hellenistic World*. Cambridge.

Versnel, H.S. (1990/4) *Inconsistencies in Greek and Roman Religion*, 2 vols. Leiden.

Verzár, M. (1976) Archäologische Zeugnisse aus Umbrien, in Zanker (ed.), pp. 116–42.

Verzár-Bass, M. (1998) Grab und Grabsitte in Aquileia, in Fasold *et al.* (eds), pp.143–77.

Vetter, E. (1953) *Handbuch der italischen Dialekte*. Heidelberg.

Veyne, P. (1979) L'hellénisation de Rome et la problématique des acculturations. *Diogène* 196, 1–29.

Veyne, P. (1990) *Bread and Circuses: Historical Sociology and Political Pluralism*. London.

Vicente, J.D., Punter, M.P. and Ezquerra, B. (1997) La catapulta tardo-republicana y otro equipamiento militar de "La Caridad" (Camineal, Teruel), in Feugère (ed.), pp. 167–99.

Villaronga, L. (1982) Les seques ibèriques catalanes: una síntesi. *Fonaments* 3, 172–3.

Ville, G. (1969) La guerre et le munus, in Brisson (ed.), pp. 185–95.

Ville, G. (1981) *La gladiature en occident des origins à la mort de Domitien*. Rome.

Vincenti, V. (2008) Pavimenti dalla prima fase delle terme di *Fregellae* (FR). Cenni preliminari, in Angelelli and Rinaldi (eds), pp. 407–18.

Vincenti, V. (2009) *La tomba Bruschi di Tarquinia*. Rome.

Virtual Theatre District (2007) http://artscool.cfa.cmu.edu/~hemef/pompeii/index.html (accessed November 22, 2012).

Visconti, A. (1999) *Aristosseno di Taranto. Biografia e formazione spirituale*. Naples.

Vismara, C., Pergola, P., Martorelli, R. and Istria, D. (2011) *Sardinien und Korsika in römischer Zeit*. Mainz-am-Rhein.

Vita, J. and Zamora, J. (eds) (2008) *Nuevas perspectivas II: La Arqueología fenicia y púnica en la península ibérica* [*Cuadernos de Arqueología Mediterránea* 18]. Madrid.

Vitali, D. (1990) *Monterenzio e la valle dell'Idice. Archeologia e storia di un territorio.* Bologna.

Vitali, D. (ed.) (2007) *Le fornaci e le anfore di Albinia.* Bologna.

Vitelli, K.D. (ed.) (1996) *Archaeological Ethics.* Walnut Creek.

Vitti, M. (2011) Il tempio dei Castori "in circo": La stato della questione, in La Rocca and D'Alessio (eds), pp. 109–34.

Vivó, D. (2007) El monument de Malla, in Palahí, Nolla and Vivó (eds), pp. 21–3.

Vlizos, S. (ed.) (2008) *Η Αθήνα κατα τη Ρωμαικη Εποχή:* Πρόσφατες ανακαλύψεις, νέες έρευνες [Mouseio Benaki 4th Suppl.]. Athens.

Volpe, G. (1990) *La Daunia nell'età della romanizzazione: paesaggio agrario, produzione, scambi.* Bari.

Volpe, R. (2007) *Centocelle II. Roma S.D.O. Le indagini archeologiche.* Soveria Mannelli.

Volpe, R. (2009) Vino, vigneti ed anfore in Roma repubblicana, in Jolivet *et al.* (eds), pp. 369–81.

Volpe, R. (2012) Republican Villas in the *suburbium* of Rome, in Becker and Terrenato (eds), pp. 94–110.

Waelkens, M., de Paepe, H. and Moens, L. (1988) Quarries and the marble trade in Antiquity, in Herz and Waelkens (eds), pp. 11–28.

Walbank, F. (1992) *The Hellenistic World.* Cambridge.

Walbank, M. (2005) Unquiet graves: burial practices of the Roman Corinthians, in Schowalter and Friesen (eds), pp. 249–80.

Walker, A. (1997) Excavation coins: the use and misuse of numismatic evidence in archaeology, in Sheedy and Papageorgiadou-Banis (eds), pp. 17–26.

Walker, D. (1988) The Roman coins, in Cunliffe (ed.), pp. 281–356.

Walker, S. and Cameron, A. (eds) (1989) *The Greek Renaissance in the Roman Empire. Papers from the Xth British Museum Classical Colloquium* [*Bulletin of the Institute of Classical Studies* Suppl. 55]. London.

Wallace-Hadrill, A. (ed.) (1989) *Patronage in Ancient Society.* New York.

Wallace-Hadrill, A. (1994) *Houses and Society in Pompeii and Herculaneum.* Princeton.

Wallace-Hadrill, A. (1996) Engendering the Roman house, in Kleiner and Matheson (eds), pp. 104–15.

Wallace-Hadrill, A. (1998) Horti and Hellenization, in Cima and La Rocca (eds), 1998, pp. 1–12.

Wallace-Hadrill, A. (2001) *The British School at Rome: One Hundred Years.* London.

Wallace-Hadrill, A. (2007) The creation and expression of identity: the Roman world, in Alcock and Osborne (eds), pp. 355–80.

Wallace-Hadrill, A. (2008) *Rome's Cultural Revolution.* Cambridge.

Wallrodt, J. (2011) Image handling at PAPP:PS. *Paperless Archaeology.* http://paperlessarchaeology.com/2011/02/22/image-handling-at-parpps (accessed November 18, 2012).

Walt, S. (1998) Review of A. Grandazzi, *The Foundation of Rome. Myth and History. BMCR* 1998.04.02.

Walthall, D.A. (2011) Magistrate stamps on grain measures in Early Hellenistic Sicily. *ZPE* 179, 159–69.

Wang, W. (ed.) (1994) *Ten Major Museums of Shaanxi.* Hong Kong.

Ward, R. (1992) Women in Roman baths. *HTR* 85, 125–47.

Ward-Perkins, J.B. (1974) *Cities of Ancient Greece and Italy: Planning in Classical Antiquity.* New York.

Ward-Perkins, J.B. (1977) *Roman Architecture.* New York.

Ward-Perkins, J.B. and Kahane, A. (1972) The Via Gabina. *PBSR* 40, 91–126.

Warden, P.G. (1982) Bullae: Roman custom and Italic tradition. *OpRom* 14, 69–75.

Warden, P.G. (2008a) The blood of animals: predation and transformation in Etruscan funerary representation, in Bell and Nagy (eds), pp. 198–219.
Warden, P.G. (2008b) Ritual and representation on a Campana dinos in Boston. *EtrStud* 11, 121–133.
Warden, P.G. (ed.) (2008) *From the Temple and the Tomb.* Dallas.
Warden, P.G. (2009) Remains of the ritual at the sanctuary of Poggio Colla, in Gleba and Becker (eds), pp. 107–21.
Warden, P.G. (2011a) The temple is a living thing: fragmentation, enchainment, and the reversal of ritual at the Acropolis Sanctuary of Poggio Colla, in de Grummond and Edlund-Berry (eds), pp. 55–67.
Warden, P.G. (2011b) The Chimaera of Arezzo. Made in Etruria? *AJA Online Forum* 115 (1).
Warren, K.J. (2004) A philosophical perspective on the ethics and resolution of cultural properties issues, in G. Anderson (ed.), pp. 303–24.
Watson, P. (2006) *The Medici Conspiracy. The Illicit Journey of Looted Antiquities.* New York.
Waurick, G. (1975) Kunstraub der Römer: Untersuchungen zu seinen anfängen anhand der Inschriften. *JRGZM* 22, 1–46.
Webster, J. (2001) Creolising the Roman provinces. *AJA* 105, 209–25.
Webster, J. and Cooper, N. (eds) (1996) *Roman Imperialism: Post-colonial Perspectives.* Leicester.
Weinberg, S. (1971) Tel Anafa: the Hellenistic town. *IEJ* 21, 86–109.
Weiss, C. (2009) Determining function of Pompeian sidewalk features through GIS analysis, in Crawford, Frischer and Koller (eds), pp. 363–72.
Weiss, Z. and Netzer, E. (1996) Hellenistic and Roman Sepporis: the archaeological evidence, in Nagy (ed.), pp. 29–37.
Welch, K.E. (1991) Roman amphitheatres revived. *JRA* 4, 272–81.
Welch, K.E. (1994) The Roman arena in Late Republican Italy: a new interpretation. *JRA* 7, 59–80.
Welch, K.E. (2003) A new view of the origins of the Basilica: the Atrium Regium, Graecostasis, and Roman diplomacy. *JRA* 16, 5–34.
Welch, K.E. (2006a) Art and architecture in the Roman Republic, in Rosenstein and Morstein-Marx (eds), pp. 496–542.
Welch, K.E. (2006b) Domi militiaeque: Roman domestic aesthetics and war booty in the Republic, in Dillon and Welch (eds), pp. 91–161.
Welch, K.E. (2007) *The Roman Amphitheatre: From Its Origins to the Colosseum.* New York.
Wells, B. (ed.) (1992) *Agriculture in Ancient Greece. Proceedings of the Seventh International Symposium at the Swedish Institute at Athens 1990.* Stockholm.
Wells, P.S. (1999) *The Barbarians Speak. How the Conquered Peoples Shaped Roman Europe.* Princeton.
Wesolowsky, A.B. (1973) The skeletons of Lerna Hollow. *Hesperia* 42, 340–51.
White, K.D. (1970) *Roman Farming.* London.
White, K.D. (1975) *Farm Equipment of the Roman World.* Cambridge.
White, T., Black, M. and Folkens, P. (2011) *Human Osteology* (3rd edn). Waltham, MA.
White, T.D. and Folkens, P. (2000) *Human Osteology* (2nd edn). San Diego.
Widrig, W. (1986) Excavations on the ancient Via Gabina: second preliminary report. *NSc* ser. 8.37, 141–82.
Widrig, W. (1987) Land use at the Via Gabina villas, in MacDougall (ed.), pp. 223–60.
Wigg-Wolf, D. (2005) Coins and ritual in Late Iron Age and Early Roman sanctuaries in the territory of the Treveri, in Haselgrove and Wigg-Wolf (eds), pp. 361–79.
Wikander, O. (2000) *Handbook of Ancient Water Technology.* Leiden.
Wilke, N. and Coulson, W. (eds) (1985) *Contributions to Aegean Archaeology.* Minneapolis.
Wilkins, J.B. (1990) Nation and language in ancient Italy: problems of the linguistic evidence. *Accordia Research Papers* 1, 53–72.

Wilkins, J., Harvey, D. and Dobson, M. (eds) (1995) *Food in Antiquity*. Exeter.
Will, E.L. (1987) The Roman amphoras, in McCann *et al.*, 1987, pp. 171–220.
Williams, II, C.K. (1993) Roman Corinth as a Commercial Center, in Gregory (ed.), pp. 31–46.
Williams, II, C.K. and Bookidis, N. (eds) (2003) *Corinth. Results of Excavations Conducted by the American School of Classical Studies at Athens*, vol. 20, *The Centenary*. Princeton.
Williams, J.H.C. (2001) *Beyond the Rubicon. Romans and Gauls in Republican Italy*. Oxford.
Wilson, A. (2004) Tuscan landscapes: surveying the Albegna Valley. *JRA* 17, 569–76.
Wilson, J. (2004) *Caesarea Philippi: Banias, the Lost City of Pan*. London.
Wilson, R.J.A. (1980–1) Sardinia and Sicily during the Roman Empire: aspects of the archaeological evidence. *Kokalos* 26–7, 219–42.
Wilson, R.J.A. (1985) Changes in the pattern of urban settlement in Roman, Byzantine and Arab Sicily, in Malone and Stoddart (eds), pp. 313–44.
Wilson, R.J.A. (1990a) *Sicily under the Roman Empire: The Archaeology of a Roman Province, 36 BC–AD 535*. Warminster.
Wilson, R.J.A. (1990b) Roman architecture in a Greek world, in Henig (ed.), pp. 67–90.
Wilson, R.J.A. (1996) Sicily, Sardinia and Corsica, in Bowman, Champlin and Lintott (eds), pp. 434–48.
Wilson, R.J.A. (2000a) Rural settlement in Hellenistic and Roman Sicily: excavations at Campanaio (AG) 1994–1998. *PBSR* 68, 337–69.
Wilson, R.J.A. (2000b) Ciceronian Sicily: an archaeological perspective, in Smith and Serrati (eds), pp. 134–60.
Wilson, R.J.A. (2005) La sopravivenza dell'influenza punica in Sicilia durante il dominio romano, in Spanò Giammellaro (ed.), pp. 907–17.
Wilson, R.J.A. (forthcoming a) Hellenistic Sicily, c. 270–100 BC, in Prag and Quinn (eds).
Wilson, R.J.A. (forthcoming b) Sicily, c. 300 BC–133 BC, in Smith (ed.).
Wilson, R.J.A. (forthcoming c) Agorai in Hellenistic and Roman Sicily: an overview, in Ampolo (ed.).
Wilson Jones, M. (1993) Designing amphitheatres. *RM* 100, 391–442.
Winter, F.E. (2006) *Studies in Hellenistic Architecture*. Toronto.
Winter, N.A. (2009) *Symbols of Wealth and Power. Architectural Terracotta Decoration in Etruria & Central Italy, 640–510 BC*. Ann Arbor.
Winter, N.A., Iliopoulos, I. and Ammerman, A.J. (2009) New light on the production of decorated roofs of the 6th c. BC *at sites in and around Rome*. *JRA* 22, 6–28.
Wirth, G. (ed.) (1982) *Romanitas-Christianitas. Untersuchungen zur Geschichte und Literatur der römischen Kaiserzeit*. Berlin.
Wiseman, J. (1979) Corinth and Rome I: 228 BC–AD 267, in *ANRW* II.7.1, pp. 438–548.
Wiseman, T.P. (1970) Roman Republican road building. *PBSR* 38, 122–52.
Wiseman, T.P. (1974) The Circus Flaminius. *PBSR* 42, 3–26.
Wiseman, T.P. (1976) Two questions on the Circus Flaminius. *PBSR* 44, 44–7.
Wiseman, T.P. (1981) The Temple of Victoria on the Palatine. *AntJ* 61, 35–52.
Wiseman, T.P. (1987) Conspicui postes tectaque digna deo: the public image of aristocratic and imperial houses in the Late Republic and Early Empire, in Pietri (ed.), pp. 393–413.
Wiseman, T.P. (1990) The central area of the Roman Forum. *JRA* 3, 245–7.
Wiseman, T.P. (1993) Rome and the resplendent Aemilii, in Jocelyn and Hunt (eds), pp. 181–92.
Wiseman, T.P. (1995) *Remus: A Roman Myth*. Cambridge.
Wiseman, T.P. (2000) Review of A. Carandini, *La Nascita di Roma: Dei, Lari e Uomini all'Alba di una Civiltà*, and M. Bettelli, *Roma. La città prima della città: i tempi di una nascita*. *JRS* 90, 210–2.
Wiseman, T.P. (2001) Review of A. Carandini and R. Cappelli, *Roma. Romolo, Remo e la fondazione della città*, and A. Carandini, *Giornale di scavo: pensieri sparsi di un archeologo*. *JRS* 91,182–93.

Wiseman, T.P. (2004) Where was the *Nova Via*? *PBSR* 72, 167–83.
Wiseman, T.P. (2007a) Andrea Carandini and Roma Quadrata. *Accordia Research Papers* 10, 103–25.
Wiseman, T.P. (2007b) Three notes on the triumphal route, in Leone, Palombi and Walker (eds), pp. 445–9.
Wiseman, T.P. (2008) *Unwritten Rome*. Exeter.
Wiseman, T.P. (2009) Review of A. Carandini and D. Bruno, *La casa di Augusto dai "Lupercalia" al Natale*. *JRA* 22, 527–45.
Wiseman, T.P. (2012) Roma Quadrata, archaic huts, the House of Augustus, and the orientation of Palatine Apollo. *JRA* 25, 371–87.
Witcher, R. (2000) Globalisation and Roman imperialism, in Herring and Lomas (eds), pp. 213–25.
Witcher, R. (2005) The extended metropolis: *urbs, suburbium* and population. *JRA* 18, 120–38.
Witcher, R. (2006a) Agrarian spaces in Roman Italy: society, economy and Mediterranean agriculture. *Arqueología espacial (Paisajes agrarios)* 26, 341–59.
Witcher, R. (2006b) Broken pots and meaningless dots? Surveying the rural landscapes of Roman Italy. *PBSR* 74, 39–72.
Witcher, R. (2006c) Settlement and society in Early Imperial Etruria. *JRS* 96, 88–123.
Witcher, R. (2008) Regional field survey and the demography of Roman Italy, in de Ligt and Northwood (eds), pp. 273–304.
Witcher, R. (2009) The Middle Tiber Valley in the Imperial period, in Coarelli and Patterson (eds), pp. 467–86.
Witcher, R. (2011) Missing persons? Models of Mediterranean regional surveys and ancient populations, in Bowman and Wilson (eds), 2011, pp. 36–75.
Witten, A. (2006) *Handbook of Geophysics and Archaeology*. London.
Wobst, H.M. (1999) Style in archaeology, or archaeologists in style, in Chilton (ed.), pp. 118–32.
Wodko, S. (2000) *Wörterbuch der keltiberischen Inschriften*. Wiesbaden.
Wonder, J. (2002) What happened to the Greeks in Lucanian-occupied Paestum? Multiculturalism in southern Italy. *Phoenix* 56, 40–55.
Wood, J., Milner, G., Harpending, H. and Weiss, K. (1992) The osteological paradox: problems of inferring prehistoric health from skeletal samples. *CurrAnthr* 33, 343–58.
Wood, S. (2010) *Horti* in the city of Rome: emulation and transcendence in the Late Republic and Early Empire, in Moore *et al.* (eds), pp. 75–90.
Woolf, G. (1995) The formation of Roman provincial cultures, in Metzler *et al.* (eds), pp. 9–18.
Woolf, G. (1996–7) Beyond Romans and natives. *WorldArch* 28, 339–50.
Woolf, G. (1998) *Becoming Roman: The Origins of Provincial Civilization in Gaul*. Cambridge.
Woolf, G. (2000) Urbanization and its discontents in early Roman Gaul, in Fentress (ed.), pp. 115–31.
Woolf, G. (2001) The Roman cultural revolution in Gaul, in Keay and Terrenato (eds), pp. 173–86.
Woolf, G. (2004) The present state and future scope of Roman archaeology: a comment. *AJA* 108, 417–28.
Wootton, W. (2002) Another Alexander mosaic. Reconstructing the Hunt mosaic from Palermo. *JRA* 15, 264–74.
Wulff, F. (ed.) (2004) *Historia de Numancia. Adolf Schulten*. Pamplona.
Wynia, S.L. (1982) The excavations in and around the House of M. Lucretius Fronto, in De Franciscis (ed.), pp. 329–40.
Yarrow, L. (2006) Lucius Mummius and the spoils of Corinth. *Scripta Classica Israelica* 25, 57–70.

Yavetz, Z. (1962) The policy of C. Flaminius and the Plebiscitum Claudianum. *Athenaeum* 40, 325–44.

Yegül, F. (1979) The small city bath in Classical Antiquity and a reconstruction study of Lucian's "Baths of Hippias." *ArchCl* 31, 108–31.

Yegül, F. (1992) *Baths and Bathing in Classical Antiquity.* Cambridge, MA.

Yegül, F. (1996) The thermo-mineral complex at Baiae and *De Balneis Puteolanis. ArtB* 78 (1), 137–61.

Yegül, F.K. (2010) *Bathing in the Roman World.* Cambridge and New York.

Yegül, F.K. (forthcoming) Roman Baths in Isthmia and sanctuary baths in Greece, in Gregory and Gebhard (eds).

Yegül, F.K. and Couch, T. (2003) Building a Roman bath for the cameras. *JRA* 16, 153–77.

Yntema, D. (1974) Messapian painted pottery. Analysis and provisory classification. *BABesch* 49, 3–84.

Yntema, D. (1993a) *In Search of an Ancient Countryside.* Amsterdam.

Yntema, D. (1993b) The settlement of Valesio, southern Italy: final report on the field survey. *BABesch* 68, 49–70.

Yntema, D. (2002) Introduction: advances in regional archaeological research in the Mediterranean, in Attema *et al.* (eds), pp. 1–5.

Yntema, D. (2005) *Conspectus Formarum of Apulian Grey Gloss Wares* (2nd edn). Amsterdam.

Yntema, D. (2006) The birth of a Roman southern Italy: a case study: ancient written sources and archaeological evidence on the early Roman phase in the Salento District, southern Italy (3rd–1st c. BC). *BABesch* 81, 91–133.

Yntema, D. (2008) Polybius and the field survey evidence from Apulia, in de Ligt and Northwood (eds), pp. 373–86.

Zaccaria Ruggiu, A. and Maratini, C. (2008) Saggi e ricerche nell'insula 7 della *Regio* VI, informatizzazione dei dati, GIS, in Guzzo and Guidobaldi (eds), 2008, pp. 177–87.

Zambon, E. (2008) *Tradition and Innovation. Sicily between Hellenism and Rome.* Stuttgart.

Zamora, D. (2006/7) *L'oppidum de Burriac. Centre del poder polític de la Laietània ibèrica* [*Laietania* 17]. Barcelona.

Zancani Montuoro, P. (1973) Qualche documento dell'arte e dell'artigianato. *La monetazione di bronzo di Poseidonia-Paestum* [*AIIN* Suppl. 18–19], 11–23.

Zancani Montuoro, P. (1979) Il faro di Cosa in un ex-voto a Vulci ?. *RivIstArch* s.3, II, 5–29.

Zangenberg, J. (2010) Archaeological news from the Galilee: Tiberias, Magdala and rural Galilee. *Early Christianity* 1, 471–84.

Zanker, P. (ed.) (1976) *Hellenismus in Mittelitalien [Kolloquium in Gottingen vom 5. bis 9. Juni 1974].* Göttingen.

Zanker, P. (1988) *The Power of Images in the Age of Augustus.* Ann Arbor.

Zanker, P. (1999) *Pompeii. Public and Private Life.* Cambridge, MA.

Zanker, P. (2007) *Die römische Kunst.* Munich.

Zanker, P. (2008) L'arte greca a Roma, in Settis and Catoni (eds), pp. 165–73.

Zecchini, G. (2002) Scipione in Spagna: un approccio critico alla tradizione polibiano-liviana, in Urso (ed.), pp. 87–103.

Zevi, F. (1976) L'identificazione del tempio di Marte "in circo" e altre osservazioni, in *L'Italie préromaine et la Rome républicaine. Mélanges offerts à Jacques Heurgon* (CÉFR 27) 2, pp. 1047–64. Rome.

Zevi, F. (ed.) (1979) *Pompeii 79, Raccolta di Studi per il Decimonono Centenario dell'Eruzione Vesuviana.* Naples.

Zevi, F. (1991) La città sannitica. L'edilizia privata e la Casa del Fauno, in Zevi (ed.), pp. 47–74.

Zevi, F. (ed.) (1991) *Pompei.* Naples.

Zevi, F. (1996) Pompei dalla città sannitica alla colonia sillana: per un'interpretazione dei dati archeologici, in Cébeillac-Gervasoni (ed.), pp. 125–38.

Zevi, F. (2003) L'ellenismo a Roma nel tempo della colonizzazione in Italia. *RendLinc* 19, 53–104.

Zevi, F. (2005) Considerazioni finali. *AttiTaranto* 44, 801–26.

Zhmodikov, A. (2000) Roman Republican heavy infantrymen in battle (IV–II centuries BC). *Historia* 49, 67–78.

Zifferero, A. (2002) The geography of the ritual landscape in complex societies, in Attema *et al.* (eds), pp. 246–65.

Ziolkowski, A. (1988) Mummius' Temple of Hercules Victor and the Round Temple on the Tiber. *Phoenix* 42, 309–33.

Ziolkowski, A. (1992) *The Temples of Mid-Republican Rome and Their Historical and Topographical Context*. Rome.

Zink, S. (2012) Augustus' Palatine sanctuary and the so-called House of Augustus: new archaeological documentation. Unpubl. paper in "Changing Historical Perspectives," Columbia University, 2012.

Zucca, R. (1996) Inscriptiones latinae liberae rei publicae Africae, Sardiniae et Corsicae. *AfrRom* 11, 1425–89.

Zucca, R. (2003) *Insulae Sardiniae et Corsicae. Le isole minori della Sardegna e della Corsica nell'antichità*. Rome.

Zucca, R. (2004) *Sufetes Africae et Sardiniae. Studi storici e geografici sul Mediterraneo antico*. Rome.

Index

Note that Roman names are listed under their cognomina, if these are better known (e.g. Cicero and not Tullius); if they are so listed, the gens entry will provide a cross-reference to the cognomina (e.g. Tullius, see Cicero). Where ancient authors are mentioned, their views are discussed in the text; the index does not include each citation. Ancient cities are normally listed under their ancient names, with modern names, if known, provided in parentheses. Page numbers referring to figures are in italics.

Printed and bound by CPI Group (UK) Ltd, Croydon, CR0 4YY

15/09/2026

14962394-0001